WATCHMAKERS AND CLOCKMAKERS
OF THE WORLD

VOLUME 2

WATCHMAKERS
AND CLOCKMAKERS
OF THE WORLD

VOLUME 2

SECOND EDITION

BRIAN LOOMES, BA, FSG

N.A.G. PRESS, COLCHESTER, ESSEX

First edition 1976
Reprinted 1980, 1984
Second edition 1989

Author's additional works
The Early Clockmakers of Great Britain
Yorkshire Clockmakers
Lancashire Clocks and Clockmakers
Westmorland Clocks and Clockmakers
The White Dial Clock
County Clocks and their London Origins

British Library Cataloguing in Publication Data

Loomes, Brian
  Watchmakers and Clockmakers of the World
  Volume 2. — 2nd ed.
  1. Horology — International
  I. Title
  681'.11' 0922

ISBN 0-7198-0250-4

Printed and bound in Great Britain by
Courier International Ltd, Tiptree, Essex

PREFACE

This book is written as a supplement to the late G. H. Baillie's *Watchmakers and Clockmakers of the World.* Its function is to complement it, not replace it. The reader searching for facts on a maker should first of all turn to Baillie, since I have not repeated in this present book any maker already detailed in *Volume One,* unless new facts have since come to light.

This book contains entries of three distinct types: Firstly, makers from about 1820 to about 1875 (later in some instances) — Baillie did not attempt to extend his list beyond about 1825; Secondly, I have included any makers of any period who were men not known to Baillie; Thirdly, I have included some makers listed by Baillie where further information has since come to light that either extends the working period of the maker or occasionally corrects an error.

In instances where Baillie's dates differ from my own — usually by virtue of our having covered different periods — then his dates are also included in brackets. Thus: Saunders, Daniel. *London* (B. 1802-24) 1832-44 means that this maker was recorded by Baillie as working in the 1802-24 period, by me as working 1832-44, and therefore worked 1802-44. Dates I have shown are only those which I can confirm personally and I have not attempted any comprehensive searching before 1820.

Generally speaking, if I have shown dates both of my own finding and of Baillie's, this indicates that these two are complementary, not contradictory. In those few instances where I have found errors in Baillie's dates, I have replaced his dates entirely by my own. In the case of the important London makers of the seventeenth century I have re-entered makers where I have found even minor variations in the span of their working lives, and have also included for completeness the names of masters of those men who completed their apprenticeships, so that it may be seen what 'school' they were taught in.

It is important for the reader to understand fully the system adopted before using the book, otherwise there is a danger of misinterpreting the facts presented. As far as possible I have tried to follow Baillie's system of abbreviations. The reader is advised to read carefully details of the Conventions before attempting to use the work. Failure to understand these fully is bound to cause confusion.

As far as possible dubious entries have been omitted, as I wanted facts, not vague estimations of periods. The book contains about 35,000 entries, bringing the total content of both volumes to over 70,000 makers. I cannot claim this to be comprehensive. However, we are now 35,000 makers closer to comprehensiveness, if nothing else.

The task of checking a work of this nature is one which requires extreme dedication, involving most of the work and none of the glory. It was done by my dear wife, Joy, without whose help the book might never have been done.

In compiling this present volume I have gained a greater appreciation of the magnitude of the task the late Mr Baillie performed. Only someone who sets to and produces such a volume knows the extent of the work involved. If therefore the reader fails to locate a maker he seeks, I hope he will show tolerance of our shortcomings.

BRIAN LOOMES

Nidderdale, 1976

PREFACE TO SECOND EDITION

Since publication of the First Edition in 1976 many thousands of new facts and figures about clockmakers have come to light, both in the form of new details about known makers and information about makers who were previously unrecorded. Revised entries remain within the main body of the text. New names appear in the Addenda section starting on page 264.

I have found it very difficult to maintain an on-going index of this size. At one time an index card system of over 30,000 cards filled an entire wall of my study and ultimately became so unwieldy they had to be disposed of. Computerisation might seem to have been a solution looking back, but as the original entries were not computerised, it is now impossible to provide the time or expense of doing this.

Experience has shown that those who use this book often misunderstand the information recorded there. It is essential that the user should read carefully the Preface and Conventions sections before attempting to interpret the details shown for each maker.

A considerable number of regional books on clockmakers and watchmakers has appeared in print since 1976, and these are set out on page viii.

I apologise in advance for any errors or omissions.

BRIAN LOOMES

Nidderdale 1989

ACKNOWLEDGEMENTS

A great many authors, historians and researchers have graciously allowed me to use facts from their works. A list of those published works follows, to the authors and publishers of which I acknowledge a deep debt. Those seeking more extensive information on makers concerned are strongly advised to consult the appropriate book.

Amongst correspondents who have made available facts from their private researches (as yet unpublished), I am especially grateful to the following:

Peter C. Nutt, who again helped admirably with the research. John F. Marbrook for help with Kent makers. E. Legg for help with Buckinghamshire makers. W. A. Seaby, MA, FSA, FMA, for help with facts from his forthcoming book on the makers of Northern Ireland. William G. Stuart for dates from his proposed book on the makers of Ireland. Dr. J. E. S. Walker for help with the makers of Holderness. A. E. Truckell, MBE, MA, FSA(Scot), FMA, Curator of Dumfries Burgh Museum, for lists of local makers. John Sapwell, MA, MB, for help with East Anglian makers. J. K. Antill, FLA, of States of Jersey Library Service for local makers. David Halton with Essex and Irish makers. Andrew Nicholls, author of 'Clocks in Colour'. Mrs. M. J. Lodey, Alan Redstone, E. J. Tyler, Paul Lavoie for help with Canadian makers; C. J. Glazebrook, John Houghton, David Barron, D. J. Eyre, David Barker, Brian Morrison, Dr. K. W. Shanks, Hugh Thornton, E. J. Collins, Beresford Hutchinson. To the many others, just too numerous to mention individually, my sincere thanks.

Allix, Charles. 'Carriage Clocks'. Antique Collectors Club.

Beeson, Dr. C. F. C. 'Clockmaking in Oxfordshire'. Antiquarian Horological Society.

Bellchambers, J. K. 'Devonshire Clockmakers'. Devonshire Press Ltd, Torquay.

Bellchambers, J. K. 'Somerset Clocks and Clockmakers'. Antiquarian Horological Society.

Bruton, Eric. 'The Longcase Clock'. Hart-Davis, McGibbon, Ltd.

Burrows, G. Edmond, CA. 'Canadian Clocks and Clockmakers'. Kalabi Enterprises Ltd, Ontario.

Daniell, J. A. 'Leicestershire Clockmakers'. Leicestershire Museums & Art Galleries & Records Service, and Leicestershire County Council.

Edwardes, E. L. 'The Grandfather Clock'. John Sherratt & Son Ltd.

Loomes, Brian. 'Yorkshire Clockmakers'. Dalesman Publishing Co. Ltd, Clapham, Yorks.

Loomes, Brian. 'The White Dial Clock'. David & Charles Ltd.

Loomes, Brian. 'Westmorland Clocks and Clockmakers'. David & Charles Ltd.

Loomes, Brian. 'Lancashire Clocks and Clockmakers'. David & Charles Ltd.

Loomes, Brian. 'Country Clocks and their London Origins'. David & Charles Ltd.

Mason, Bernard, OBE. 'Clock and Watch Making in Colchester'. Country Life Books.

Miles-Brown, H. 'Cornish Clocks and Clockmakers'. David & Charles Ltd.

Palmer, Brooks. 'The Book of American Clocks'. MacMillan Publishing Co., New York.

Peate, Dr. Iorwerth C. 'Clock and Watch Makers in Wales'. National Museum of Wales.

Royer-Collard, F. B. 'Skeleton Clocks'. NAG Press Ltd.

Roberts, Kenneth. 'The Contributions of Joseph Ives to Connecticut Clock Technology'. American Clock and Watch Museum Inc., Bristol, Conn.

Smith, John. 'Old Scottish Clockmakers'. Oliver and Boyd Ltd and E. P. Publishing Ltd, East Ardsley, Yorks.

Tebbutt, Laurence. 'Stamford Clocks and Watches'. Dolby Bros. Ltd, Stamford Lincs.

Tyler, E. J. 'European Clocks'. Ward Lock & Co. Ltd.

Additional Books:

Bruton, Eric. The Wetherfield Collection of Clocks, N.A.G. Press

Bates, Keith. Clockmakers of Northumberland and Durham. Pendulum Publications, 1980.

Dowler, Graham. Gloustershire Clock & Watch Makers. Phillimore, 1984.

Elliott, D. J. Shropshire Clocks & Clockmakers. Phillimore, 1977

Hagger, A. L., & Miller, L. F. Suffolk Clocks & Clockmakers. Antiquarian Horological Society, 1974.

Hughes, R. G. Derbyshire Clock & Watch Makers. Derby Museum, 1976.

Lee, R. A., The Knibb Family Clockmakers. Manor House Publications, 1963

Legg, Edward. Clock & Watch Makers of Buckinghamshire. Bradwell Abbey Field Centre for the Study of Archaeology, 1976.

Loomes, Brian. The Early Clockmakers of Great Britain. N.A.G. Press Ltd, 1982.

McKenna, Joseph. Watch & Clock Makers of Warwickshire. Pendulum Press, 1988.

McKenna, Joseph. Watch & Clock Makers of Birmingham. Pendulum Press, 1988.

Continued **X**

CONVENTIONS

DATES — As in Volume One dates do not imply any finality unless b. or d. is used. Dates are working dates unless otherwise indicated and are positive dates unless preceded by ca. (= circa). Dates in brackets indicate known dates but not necessarily at that same place. Dates within brackets with initial B. refer to Baillie's datings in Volume One for comparison.

WORKS — No examples of works, e.g. in museums, are mentioned. The terms C., W., T.C., etc. are used only if examples are known to exist or if the man is specifically known as a maker of them. A man known as a maker of clocks may well have made watches and vice versa. No distinction is made between a maker and a retailer of horological works. Terms such as 'fine' and 'eminent' are not used, as these are meaningless if, as is usual, they come from obituaries.

NAMES — These are in strict alphabetical order. Alternative spellings have not been cross-referenced so that if, for example, what is sought is not found under Whitaker, one should try Whittaker.

PERIOD — The work attempts to include all known makers up to about 1875-80, occasionally a little later. With the newer countries — USA and Canada — they are included as late as the 1920s.

ABBREVIATIONS

a.	apprentice or apprenticed
B.	Baillie (Watchmakers and Clockmakers of the World, Volume 1)
b.	born
BC.	Blacksmiths' Company
C.	clockmaker
c.	when after 18, etc., century
ca.	when preceding a date, *circa* (approximately)
CC.	Clockmakers' Company. CC. 1665, free of CC. in that year
chrons.	maker of chronometers
d.	died
mar.	married
prob.	probably
T.C.	Turret Clockmaker
W.	watchmaker

Books Continued

Mather H. Clock & Watch Makers of Nottinghamshire. Friends of Nottingham Museum, 1979.

Moore, Nicholas. Chester Clocks & Clockmakers. Grosvenor Museum, Chester, no date.

Norgate, J. & Hudson F. Dunfermline Clockmakers. pub. F. Hudson, 1982.

Penfold, John. The Clockmakers of Cumberland. Brant Wright Associates, 1977.

Ponsford, C. N. & Authers, W. P. Clocks & Clockmakers of Tiverton. pub. W. P. Authers, 1977.

Ponsford, Clive N. Time in Exeter. Headwell Vale Books, 1978.

Ponsford, Clive N. Devon Clocks & Clockmakers. David & Charles, 1985.

Pryce, W. T. R. & Davies, T. Alun. Samuel Roberts - Clockmaker. National Museum of Wales, 1985.

Seaby, W. A. Clockmakers of Warwick & Leamington. Warwicks Museum, 1981.

Snell, Michael. Clocks & Clockmakers of Salisbury. Hobnob Press, 1986

Tribe, T. & Whatmoor, P. Dorset Clocks & Clockmakers. Tanat Books, 1981.

Tyler, E. J. The Clockmakers of Sussex. Watch & Clock Book Society Ltd, no date.

Wallace, William. Marking Time in Hamilton (Lanarkshire). pub. by author, 1981.

GAZETTEER OF LESSER-KNOWN PLACES

Adstock. Buckinghamshire
Aldwincle. Northamptonshire
Altarnun. Cornwall
Alverstoke. Hampshire
Alyth. Scotland
Ambleside. Westmorland
Ards. co. Down
Ash. Kent
Ashen. Essex
Aspley Guise. Bedfordshire
Auchinblae. Scotland
Audlem. Cheshire
Augher. co. Tyrone

Bacup. Lancashire
Bale. Norfolk
Balfron. Scotland
Ballingdon. Essex
Ballybofey. co. Donegal
Barking. Essex
Barrie. Scotland
Barton-in-the-Beans. Leicestershire
Basford. Nottinghamshire
Batley. West Yorkshire
Bedworth. Warwickshire
Belper. Derbyshire
Bentham. Yorkshire
Berkeley. Gloucestershire
Bervie. Scotland
Bildeston. Suffolk
Bingley. West Yorkshire
Binham. Norfolk
Birkenshaw. West Yorkshire
Blakeney. Norfolk
Bloxham. Oxfordshire
Boreham. Sussex
Boughton. Kent
Bourton. Gloucestershire
Bowland Bridge. Westmorland
Bowness. Westmorland
Boxford. Suffolk
Bredbury. Cheshire
Brighouse. West Yorkshire
Brightlingsea. Essex
Brill. Buckinghamshire
Broadclyst. Devon
Broughton. Lancashire
Bruton. Somerset
Burnham. Gloucestershire
Burtonwood. Lancashire
Burwell. Cambridgeshire

Cambourn. Cornwall
Castleford. Yorkshire
Cavendish. Suffolk
Cawston. Norfolk

Chacewater. Cornwall
Charlbury. Oxfordshire
Childrey. Berkshire
Church Stretton. Shropshire
Clay Cross. Derbyshire
Cleator. Cumberland
Clevedon. Somerset
Cley. Norfolk
Clun. Wales
Coatbridge. Scotland
Cold Brayfield. Buckinghamshire
Coldingham. Scotland
Coleford. Somerset
Colmonell. Scotland
Cradley Heath. Staffordshire
Crawcrook. Durham
Crowle. Lincolnshire
Cumrie. Scotland
Cupar, Fife. Scotland
Curry Rivel. Somerset

Daddry Shiels. Durham
Dalton. Lancashire
Dawley. Shropshire
Denny. Scotland
Dingleton. Scotland
Disley. Cheshire
Docking. Norfolk
Dockum. Holland
Dore. Derbyshire
Drayton. Shropshire
Dronfield. Derbyshire
Dufftown. Scotland
Dunchurch. Warwickshire
Dungannon. co. Tyrone
Dunscore. Scotland
Dunster. Somerset

Earls Barton. Northamptonshire
Earl Shilton. Leicestershire
Earlstown. Lancashire
Ebchester. Durham
Egremont. Cumberland
Escrick. East Yorkshire
Everdon. Northamptonshire
Eydon. Northamptonshire

Falkingham. Lincolnshire
Fazeley. Staffordshire
Ferrybridge. West Yorkshire
Ferry Hill. co. Durham
Filey. East Yorkshire
Foleshill. Warwickshire
Foulsham. Norfolk
Fowey. Cornwall

Galby. Leicestershire
Gargrave. Yorkshire
Garragh. co. Londonderry
Grampound. Cornwall
Grasmere. Westmorland
Grays Thurrock. Essex
Great Appleby. Leicestershire
Great Haseley. Oxfordshire
Great Wakering. Essex
Groton. Suffolk
Gunnerside. North Yorkshire

Hailsham. Sussex
Hawkhurst. Kent
Haworth. Yorkshire
Hayle. Cornwall
Heaton Norris. Cheshire
Hebden Bridge. Yorkshire
Heckmondwike. Yorkshire
Hednesford. Staffordshire
Hedon. Yorkshire
Hempnall. Norfolk
Henfield. Sussex
Heywood. Lancashire
Hillsborough. co. Down
Holmfirth. West Yorkshire
Hook Norton. Oxfordshire
Horley. Oxfordshire
Hugglescote. Leicestershire
Huish Episcopi. Somerset
Huntly. Scotland
Hurstpierpoint. Sussex
Husbands Bosworth. Leicestershire

Ibstock. Somerset
Idle. West Yorkshire
Ilkley. West Yorkshire
Ironbridge. Staffordshire
Iver. Buckinghamshire
Iwerne. Dorset

Kenninghall. Norfolk
Kibworth Beauchamp. Leicestershire
Kidsgrove. Staffordshire
Kilkeel. co. Down
Kilwinning. Scotland
Kincardine O'Neil. Scotland
Kingsland. Hampshire
Kingstown. Ireland
Kingussie. Scotland
Kirkby Malzeard. North Yorkshire
Kirkcabbin. co. Down

Landport. Hampshire
Leyburn. North Yorkshire
Leyland. Lancashire
Leytonstone. Essex
Lisnaskea. co. Fermanagh
Liswatty. co. Londonderry
Litcham. Norfolk
Little Weldon. Northamptonshire
Long Buckby. Northamptonshire

Longnor. Staffordshire
Long Stratton. Norfolk
Lostwithiel. Cornwall
Lydney. Gloucestershire
Lyndhurst. Hampshire

Madron. Cornwall
Maiden Newton. Dorset
Malvern Wells. Worcestershire
Marazion. Cornwall
Marham Church. Cornwall
Markinch. Scotland
Marple. Cheshire
Martham. Norfolk
Masham. Yorkshire
Matching Green. Essex
Melcombe Regis. Dorset
Melbourn. Derbyshire
Methwold. Norfolk
Micheldean. Gloucestershire
Middleham. Yorkshire
Mildenhall. Suffolk
Millom. Cumberland
Milton. Hampshire
Mitchel Dean. Gloucestershire
Monksilver. Somerset
Moulsham. Essex

Nether Stowey. Somerset
New Brompton. Kent
New Buckenham. Norfolk
Newburgh. Scotland
Newton Longville. Buckinghamshire
Ninfield. Sussex
North Curry. Somerset
Norwood. Surrey

Oldbury. Worcestershire
Ollerton. Nottinghamshire
Orsett. Essex
Osmotherley. North Yorkshire
Ossett. West Yorkshire
Over. Cheshire

Padiham. Lancashire
Padside. West Yorkshire
Patrington. East Yorkshire
Penge. Surrey
Peniston. Yorkshire
Penkridge. Staffordshire
Penryn. Cornwall
Pewsey. Wiltshire
Pilton. Somerset
Plumstead. Kent
Pocklington. East Yorkshire
Port Dinorwic. Wales
Puckeridge. Hertfordshire
Pulham. Norfolk

Radford. Nottinghamshire
Radstock. Somerset
Ramsbury. Wiltshire

Rayleigh. Essex
Reckleford. Somerset
Redhill. Surrey
Reeth. North Yorkshire
Ribchester. Lancashire
Romaldkirk. North Yorkshire
Ronne. Denmark
Rotherfield. Sussex
Ryde. Isle of Wight

Saffron Walden. Essex
St. Day. Cornwall
Saintfield. co. Down
St. Germans. Cornwall
St. Just-in-Penwith. Cornwall
St. Leonards. Sussex
St. Mary Cray. Kent
Saltash. Cornwall
Scole. Suffolk
Sculcoates. East Yorkshire
Shanklin. Isle of Wight
Shefford. Bedfordshire
Shelton. Staffordshire
Shepshead. Leicestershire
Shipdham. Norfolk
Shirley. Hampshire
Sibford. Oxfordshire
Soham. Cambridgeshire
South Heigham. Norfolk
South Weald. Essex
Springfield. Essex
Staindrop. co. Durham
Stansfield. Suffolk
Stapleford. Nottinghamshire
Stockbridge. Hampshire
Stourport. Worcestershire
Stradbroke. Suffolk
Sturminster. Dorset
Swanage. Dorset
Swineshead. Leicestershire

Tattershall. Lincolnshire
Thornton-in-Lonsdale. Yorkshire
Thorpe-le-Soken. Essex
Ticknall. Derbyshire
Tingewick. Buckinghamshire
Tipton. Staffordshire
Titchfield. Hampshire
Tonystick. co. Fermanagh
Totley. Derbyshire
Totton. Hampshire
Troon. Scotland

Upwell. Cambridgeshire

Ventnor. Isle of Wight

Wadebridge. Cornwall
Wanstead. Essex
Warrenpoint. co. Down
Warton. Lancashire
Welwyn. Hertfordshire
Westerleigh. Gloucestershire
West Rudham. Norfolk
Wheathampstead. Hertfordshire
Whitstable. Kent
Whittlesey. Cambridgeshire
Whitwick. Leicestershire
Wick. Scotland
Wickhambrook. Suffolk
Willingham. Cambridgeshire
Winchcomb. Gloucestershire
Winster. Westmorland
Wirksworth. Derbyshire
Wotton-under-Edge. Gloucestershire
Worsborough. West Yorkshire
Wragby. Lincolnshire
Wrotham. Kent.

Yardley. Worcestershire
Yeardsley-cum-Whalley. Cheshire
Yeldham. Essex

A

A, H. Monogram of H. Acier, q.v.
A, T., *Hawkshead.* Probably monogram of Thomas
  Armstrong, late 18c.
AARON—
  Benjamin Wolfe. *London.* 1844-57.
  Esther. *Birmingham.* 1835.
  Levi. *Birmingham.* 1816.
AARONS, Morris. *Toronto.* 1874-77.
ABBATY, John. *London.* Early 17c. W.
ABBEY—
  David. *Pittsburgh, USA.* ca.1830.
  William. *Hull.* 1858. W.
ABBOTT—
  ——. place unknown. ca.1810-ca.20. Clock dial maker.
  (England).
  Charles. *Bildeston.* 1830-39.
  Charles. *Lavenham.* 1839-75.
  Francis. *London.* 1844.
  and Garnet. *Farnworth.* 1834-51. W.
  George. *Philadelphia.* 1822.
  George. *London.* 1869.
  H. C. and Bros. *Birmingham, USA.* ca.1901.
  Henry. *New York.* 1881-92. W.
  J. H. *Bildeston.* 1865.
  James. *Farnworth.* 1806. W.
  James. *London.* 1857.
  John. *Bold.* a.1757-1800. W.
  John. *Prescot.* 1818-23. W.
  John. *Liverpool.* 1822.
  John. *Cheltenham.* 1850.
  Joseph Hines. *Lavenham.* 1875-79.
  Moses. *Sutton, NH, USA.* ca.1820. C. Later to
    Pomfret, Vermont.
  Nathan. *Farnworth.* a.1747-mar.1757 W.
  Nathan. *Farnworth.* 1833. W.
  Samuel. *Boston, USA,* ca.1810-32. C.
  Thomas. *Farnworth.* 1788. W.
  W. *Coventry.* 1880.
  William. *Burtonwood.* 1776. W.
  William. *Prescot.* 1818-23. W.
ABDELL, ——. *Richmond (prob. Yorks).* ca. 1750-65. C.
ABEL—
  Bros. *Farringdon.* 1877.
  Charles. Leeds Co., *Canada.* pre-1867.
  1867.
  E. *Farringdon.* 1864.
  Robert K. *Philadelphia.* 1840.
  T. L. Leeds Co., *Canada.* pre-1867.
  William. *Falkingham.* 1850.
  William T. *Billingborough.* 1850.
ABELING, William. *London.* (B. 1817) 1828-44.
ABERNETHY, Scott. *Leith.* 1836-50.
ABLITT, John. *Ipswich.* 1803-39. believed to be from
  London. Partner with William Kirk qv. at
  Stowmarket. 1810-14.
ABRAHALL, James. *Mitchell-Dean.* 1830.
ABRAHAM—
  Alfred. *London.* 1851-63.
  B. & S. S. *Norwich.* 1858.
  Barnett. *London.* 1857. Successor to Alfred.

ABRAHAM—continued.
  Ebenezer. *Olney.* b.1735-d.1815.
  Godfrey. *London.* 1832-44.
  H. *London.* 21, Bevis Marks. 1811. W.
  Henry. *Southampton.* 1840-78. W.
  Jacob. *Canterbury.* 1826.
  James. *Olney.* bro. of John. mar.1802-15.
  John. *Olney.* bro. of James. 1798-1814.
  Joseph. *Bridgewater.* ca. 1810. C.
  Josiah. *Liskeard.* 1796-1879.
  Levi. *Wellington, Salop.* 1842.
  Miles. *Olney* and *Newport Pagnell.* mar.1768-1824.
  Phineas. *Hull.* 1822, then *Leeds.* 1826-37.
  Samuel, Simon and James. *London.* 1839-44.
ABRAHAMS—
  Aaron. *Lostwithiel.* pre-1821, then *Plymouth.* W.
  Abraham. *London.* 1839.
  Abraham. *Canterbury.* 1838-74.
  Abraham. *Plymouth.* a.1764 to Richard Temple.
  Abraham. *Southampton.* 1839.
  Alexander. *Portsmouth.* 1878.
  Benjamin. *Norwich.* 1865-75.
  Bros. *London.* 1832.
  and Co. *London.* 1832.
  Henry. *London.* (B. 1802-)1828.
  Isaac. *Sheerness.* 1802 W.
  J. *Amesbury.* 1859.
  Levy. *London.* a.1767 to Thomas Sheafe.
  Lewis. *London.* 1863-81.
  Michael. *Birmingham.* 1868-80.
  Mrs. Ann. *London.* 1844-69. ?wid. of Abraham.
  Mrs. Elizabeth. *London.* 1839-51.
  S. & M. *London.* 1863
  Samuel. *Sandwich.* (B. 1783-)1803.
  V. *Grimsby.* 1868.
ABRAHAMSON—
  William. *Liverpool.* 1851.
  Wolff. *London.* 1863-81.
ABRAM—
  Pere & Fils. *Montecheroux.* ca.1800-60. Toolmakers.
  George. *Poulton.* 1834-58.
  Thomas. *Prescot.* 1823. W
ACASTLE, Robert. *Taunton.* (B. ca.1660-) d.1677. W.
ACEY, Peter. *York.* d.1639. C.
ACHESON—
  John. *Londonderry.* 1784.
  William. *Dublin.* 1858.
ACIER—
  Emile A. *Paris.* 1889. Carriage clocks.
  H. *Paris.* 1867. Carriage clocks.
ACKERS—
  Edward. *St. Helens.* 1767-83. W.
  John. *Prescot.* 1788-96 W.
ACKLAM, John Philip. *London.* (B. 1820-25)-39.
ACLAND, Henry. *Fowey.* 1844.
ACKRILL, Samuel. *Worcester.* 1850.
ACKROYD, Thomas. *Leeds.* 1777-1807. W.
ADAM—
  Andrew. *Irvine.* 1860.

ADAM—*continued.*
John. *Alloa.* 1837.
John. *Lanark.* 1835-60.
John. *Paisley.* 1820-38.
Joseph. *Glasgow.* 1837.
ADAMS—
Bros. *Coventry.* 1868.
C. K. *Montreal.* 1866.
Charles. *Erie, Pa. USA.* ca.1860. C.
E. W. *Seneca Falls, NY, USA.* ca.1830-50. C.
(William) and EATON (Samuel A.). *Boston, USA.* 1816-21.
Edward, *Dublin.* Free 1727.
Francis Bryant (& Sons). *London.* (B. a.1821-) 1828-32 (-75).
Freuerick Ward. *Doncaster.* 1834-51.
G. *Coventry.* 1880.
George Burke. *Bury St. Edmunds.* 1839.
H. B. *Elmira, NY, USA.* 1842.
(William) & HARLAND (Henry). *Boston, USA.* 1813.
Henry. *Birmingham.* 1880.
Henry. *Coventry.* 1880.
Hubert. *Yeovil.* 1883.
J. *Coventry.* 1880.
J. *Northampton.* 1864-69.
J. *Ringwood.* 1859.
James. *Coventry.* 1868-80.
James. *Croyden.* 1878.
Jeremiah. *Tipton.* 1850.
John. *Crediton.* d.1639.
John. *Dublin* a. 1763 to Alexander Gordon.
John. *Middlewich.* 1828-48.
John C. *Chicago.* 1860.
John P. *Newburyport, Mass, USA.* 1858-60.
John Q. *Toronto.* 1861.
Jonas. *Rochester, NY, USA.* 1834.
Joseph. *Southampton.* 1878.
Joseph. *Stourbridge.* 1850-60.
Joseph Warry. *Bristol.* 1859.
Joshua. *Magherafelt, co. Londonderry.* 1824-68.
Nathaniel. *Danvers, Mass, USA.* 1783-ca.90, later Boston.
Nathaniel. *Liverpool.* 1851.
Nathaniel. *London.* 1839.
and PERRY Watch Mfg. Co. *Lancaster, Pa, USA.* 1874, later became Hamilton Watch Co (1892).
Robert. *Deal.* 1874.
Robert. *Sherborne.* 1867-75.
Robert Folkard. *Stowmarket.* 1823-39. Also tobacco pipe maker.
Samuel. *Bromley, Kent.* (B. pre-1775-)1792.
Thomas. *Bingley.* 1853.
Thomas. *Coventry.* 1880.
Thomas. *Dublin.* 1868-80. C.
Thomas. *London.* 1851-75.
Thomas. *Oxford.* 1621-d.1664. T.C.
Thomas. *Plymouth.* ca.1685. Lantern clock.
Thomas. *Shelburne, Nova Scotia.* 1786-d.1837.
Thomas F. *Baltimore.* 1804-07; *Edenton, NC,* 1809; *Boston, USA,* 1810.
(William) and TROTT (Andrew C.). *Boston, USA.* 1810.
Walter. *Bristol, USA.* 1810-80.
William. *Exeter.* s. of John. Free 1780-95. W.
William. *Milton, Hants.* 1878.
William Lee. *Selby.* 1846-58 C & W.
ADAMSON—
——. *Kilmarnock.* 1850.
Chárles. *Montrose.* 1820-37.
Henry. *Liverpool.* 1834.
John. *Cupar, Fife.* 1837.
S. *Tadcaster.* ca.1780-ca.1820. C.
Stephen. *Wakefield.* a.1743. C.
William. *Exeter.* 1803. C.
William. *Forfar.* 1860.
ADCOCK—
and ENGLISH. *Montreal.* 1871-76.
Frederick. *Watton.* 1875.
George. *Watton.* 1875.
J. *East Dereham.* 1858-65.
John W. *East Dereham.* 1875.
Mrs. E. *Watton.* 1858-65.

ADCOCK—*continued.*
Richard. *Buckingham.* 1869-77.
Samuel. *Shipdham.* 1836.
Samuel. *Watton.* 1836-46.
Theophilus. *Methwold.* 1875.
W. *Methwold.* 1865.
William. *Aldwincle.* 1747 (?a. in London).
ADDERLEY, William. b. 1824 *Birmingham.* At *Leamington* 1851-68. W.
ADDINAL, George. *Selby.* 1822.
ADDISON—
G. *Thirsk.* ca.1750-60. C.
J. *Kirkby Malzeard.* ca.1800 C (see Iddison).
James. *Dublin.* 1868.
John. *York.* a.1769. Free 1789. C.
John. *Bridgenorth.* 1842-50.
Matthew. *Renfrew, Ontario.* 1871.
Thomas. *Ulverstone.* 1744-d.1766. C.
ADDOR, Augustus and Julian. *London.* 1828. Also musical snuffboxes.
ADIE, Thomas. *Matlock.* 1876.
ADKINS—
H. B. *Wimborne.* 1867.
Henry B. *Lewes.* 1878.
Henry Bolton. *London.* 1863.
Henry Boulton and Edward. *Lewes.* 1870.
J. *Burslem.* 1868.
J. (& Sons). *Coventry.* 1860(-80).
Jesse. *Alcester.* 1860-80.
John. *Longton.* 1842-50.
John Baker. *Deal.* 1832.
ADKINSON, Edward. *Newcastle-on-Tyne.* see Atkinson.
ADLARD, James. *Louth.* 1849-76.
ADNUM, George. *London.* 1851-57.
ADOLPH, Duff. *Philadelphia.* 1837.
ADSON, Thomas. *Oundle.* 1777. W.
AEITS, J. G. *Tongres* (?*France*) ca.1810. C.
AGAR—
Charles, s. of John I. *York.* a.1770-84. Later Pontefract.
Francis. *York.* In 1808 succeeded uncle, Thomas, q.v.
John I. *York.* b.1730-d.1815. C.
John II. s. of John I. b.ca.1750, a.1766. At Malton by 1784, d.1814.
John III. *Malton.* s. of John II, b.ca.1782-d.1871.
Mary (& Sons). *Bury.* 1848-58. Successor to Thomas.
Seth. *York.* b.1717, free 1743.
Thomas. *York.* s. of John I. b.ca.1757, a.1774. Free 1779, d.1807.
Thomas. *Bury.* 1822-34.
W. & Co. *Bolton.* 1858.
AGNEW—
James. *St. John, Nova Scotia.* 1834-45.
M. F. *Liverpool, Nova Scotia.* 1866.
Samuel A. *Quebec.* 1864-71.
Thomas & Son. *Manchester.* 1874. C.
W. H. *Ballymacarrett, Belfast.* 1858.
William. *Ballymacarrett, Belfast.* 1854.
William and Robert. *Belfast.* 1839-50.
AGUILLON & ROCHET. ?*Geneva.* ca.1790.
AHERNS, Adolph. *Philadelphia.* 1837.
AHRONSBERG—
I. *Landport, Hants.* 1859.
Isaac. *London.* 1857.
AIANO, C. *Canterbury.* ca.1840.
AICHER, Hillary. *Bristol.* 1863-79.
AICKEN—
George. *Cork.* Free 1770-d.1813.
Gr(e)aves. *Lisburn, Co. Antrim.* ca.1770-78; to Lurgan. ca.1778-90; to Newry, ca.1790-1820. C.
James. *Cork.* b.ca.1711, d.1795.
Robert. *Dublin.* a.1709.
AIKEN, Andrew. *Greenock.* 1860.
AINSWORTH—
Henry. *Romford.* 1855-74.
and HILL. *Birmingham.* 1880.
J. B. *Croydon.* 1851-78.
James. *Prescot.* 1834.
John. *St. Helens.* 1774-84.
W. *Leatherhead.* 1855-62.
AIRD—
David. *Middletown, Conn. USA.* 1785. W.
James. *Glasgow.* 1860.

AIRD—continued.
and THOMPSON, 19c. br. clock.
AIREY, George. Sunderland. 1827-34.
James. Grasmere. b.1804-d.1864. C.
L. Monkwearmouth. 1844.
Mrs. Sunderland. 1847.
Richard. Bowness. 1869. C. & W.
Robert. Sunderland. 1834-56.
Smith. Sunderland. 1827-56.
Thomas. Liverpool. 1848-51.
AITCHISON—
Hector. Edinburgh. 1860.
James. Edinburgh. 1860.
John. Edinburgh. 1850.
Lawrence. Glasgow. 1860.
AITKEN—
David, jun. Carnwath. 1840-75.
James. Glasgow. 1841.
James. Kelso. 1860.
James. London. 1828. C.
James. Markinch. 1837.
John. Dalry. 1850-60.
John. Philadelphia. 1785-1813. C.
Peter, jun. Glasgow. 1841(-60).
Robert. Galashiels. 1836-60.
William. Auchinblae. 1860.
William. Haddington. 1805-60.
AKED, Joseph. Halifax. 1728.
AKEHURST, William Henry. Brighton. 1839.
AKERS—
Edward. Baltimore, USA. 1843-82.
Thomas. London. 1844.
AKRILL, J. C. Llanfyllin. l.c. clock — ?mid-19c.
ALABONE, A. Newport, Isle of Wight. 1848.
ALAIN, Pierre. Quebec. 1890.
ALAIS, C. F. Peckham. 1878.
ALBEE, Willard W. Newburyport, USA. 1860.
ALBERT, John. Huntington, Pa. USA. ca.1816. C.
'ALBERTA'. 1914 and later. Product of Pequegnat Clock Co.
ALBERTI(S), Jo. Baptiste. Italy. 1685. C.
ALBINO—
Donello. Cheltenham. 1840-56.
J. Cheltenham. 1879 (see ALBINS).
Vitore. Bourton, Glos. 1850.
W. Stratford-on-Avon. 1868.
William. Stroud. 1870.
ALBINS, Joseph. Cheltenham. 1870 (probably Albino, q.v.).
ALBRIGHT—
R. E. Flushing, N.Y. USA. ca.1860.
Thomas F. Philadelphia. 1837-45.
ALCE, Henry, jun. Lewes. s. of Henry, sen. a.1743 to Thomas Best.
ALCOCK—
——. Kilkenny. ca.1790. C.
Charles and Albert. Cheltenham. 1870.
Henry. Sheffield. 1871.
John. Prescot. 1781. W.
John. St. Helens. 1782. W.
Joseph. Prescot. mar.1757. W.
Thomas. Farnworth. mar.1785. W.
ALDER—
C. G. Blyth. 1858. (not 18c.)
W. D. New York. 1868. chron.
ALDERSLADE, J. T. London. 1875-81.
ALDERSLEY—
George. Whitstable. 1838.
John. Chatham. 1832.
ALDERSON—
——. Richmond (?Yorks). Late 18c. C.
John. Romaldkirk. d.1821. C.
Walton. Leyburn. 1823-31. Sundials.
William. Daddry Shiels. 1827.
William. St. John Weardale. 1834.
ALDOUS, W. Abingdon. 1864.
ALDRED—
E. R. Yarmouth. 1846.
S. H. & Son. Yarmouth. 1830-75.
ALDRET, Samuel. Yarmouth. 1836.
ALDRIDGE—
Alfred. London. 1857.

ALDRIDGE—continued.
Joseph. London. 1863-75.
Thomas. Oldham. a.1752. C.
William. London. 1828. C.
ALDWINCKLE, George. London. 1869.
ALDWORTH, Samuel. Formerly a. of John Knibb.
Oxford 1689-97; then London till ca.1720; then Childrey, Berks till ca.1725. C.
ALDY, Edward. Lincoln. ca.1750. C.
ALECOCK, Edward. Beccles. 1875-79.
ALEXANDER—
A. Kirkland (Kendal). 18c. C.
Alexander. Elgin. 1845-60. Also silver, jewelry & photographer & Co. London. (B. 1820-)1839.
David C. Greenock. 1860.
David Crichton. Kilmarnock. 1837.
George. Leith. 1813-25. W.
George. Turriff. s. of James. 1838-60.
George. Hull. 1823-58. W.
J. Thornaby. 1898.
James. Cork. 1847. W.
James. Kincardine O'Neil. 1846.
James. Elgin. 1820-45. W.
James. Aberdeen. 1860.
James. Balfron. 1860.
James. Stewartstown, Co. Tyrone. 1846-80.
James. Turriff. b.1796-d.1838.
John. Carrickfergus. ca.1825-40. C.
Levi. Colchester. b.ca.1750, d.ca.1806. W.
Mary. Turriff. 1837.
Morris. London. 1839-51.
Mrs. F. London. 1857 (?wid. of Morris).
Nehemiah. Slough. 1877.
Peter. London. 1857.
Richard. Chippenham. (B. 1795-) 1830-48.
Robert. Leith. 1751-1826. More than one. (One b.1746-d.1830. W.).
Samuel. Philadelphia. 1787-1808.
Samuel. Coventry. 1880.
W. A. Glasgow. 1836.
William. Glasgow. 1836.
William. Hexham. 1848-58.
William. London. 1828-44.
William. Southampton. 1878.
William. Toronto. 1874-77.
William Paterson. Balfron. 1836.
William (& Son). Glasgow. 1841 (-60).
ALEXANDRE, ——. Paris. ca.1700.
ALEY, Thomas. London. 1839-69.
ALFORD—
James. Brighton. 1870-78.
William H. Ventnor. 1867-78.
ALGOOD, Alfred. Ledbury. ca.1830. C.
ALIETTI, Christopher. Highworth. 1842. Also jeweller.
ALKER—
James. Chorley. 1851-58.
James. Wigan. 1832. C.
John I. Wigan. 1797-d.?1832. C.
John II. Wigan. 1851-58. W.
John & Nicholas. Leyland. 1851. W.
Nicholas. Wigan. 1830-50; then Chorley. 1851-58. W.
Thomas. Wigan. 1822-48.
ALLAMOND—
Brothers. London. 1869.
Louis. London. 1875-81.
ALLAN—
C. & J. Toronto. 1877.
James. Aberdeen. 1836-46.
James. Kilmarnock. (B. 1807-) 1820-37.
Thomas & Co. Montreal. 1875.
William. Kilwinning. 1837-50.
William Archibald. Dumfries. 1887.
ALLANSON—
William. Buckingham. 1842-54.
William. Canterbury. 1795. W.
William. London. 1795. W.
William. Winchester. 1830.
ALLATSON—
Henry William. Thetford. 1875.
Mrs. Margaret. London. 1875.
W. Dorking. 1862.
William. Brightlingsea. 1839.

ALLATSON—*continued.*
William. *Colchester.* b.ca.1795-1842. W. & C.
William. *London.* 1851-69.
William. *Witham.* 1828-39.
ALLAWAY, Joshua. *Taunton.* Early 18c. C.
ALLBUT, Richard. *Birmingham.* 1850.
ALLCOCK—
E. *Congleton.* 1878.
Thomas. *London.* 1851.
Thomas. *Sandbach.* 1878.
ALLCOTT, W. *Emsworth.* 1867-78.
ALLDIS, Alfred Augustus. *London.* 1869-81.
ALLDRIDGE, Edwin. *Birmingham.* 1850-60. Also dial
maker.
ALLDRITT, Joseph. *Burton-on-Trent.* 1842.
ALLEBACH—
Henry. *Reading, Pa. USA.* 1829.
M. B. *Philadelphia.* ca.1800.
ALLEN—
Alexander. *Rochester, USA.* 1860.
Augustine. *Barnsley.* 1822.
Augustus. *Thrapstone.* 1841-77.
Charles. *Bewdley.* 1835-60.
Charles. *Fillonglen.* 1814. W.
Charles. *Higham Ferrers.* 1824-41. Also gunsmith and
bellhanger.
Charles R. *Rye.* 1870-78.
Edward. *Dungarvan.* 1858. W.
Francis. *Canterbury.* 1768. W.
Francis. *Maidstone.* 1768. W.
Frederick. *Leeds.* 1850-53. W.
G. J. *Blandford.* 1867.
George. *Bath.* 1866-75.
George. *London.* 1828-44.
George. *Sheffield.* 1871. Clock cleaner.
George James. *Bryn-mawr.* 1887.
H. *Pangbourne.* 1847-54.
Henry. *Pangbourne.* 1837-54.
Henry. *Reading.* 1830.
J. T. *Rochester, USA.* 1844-46.
James. *Boston, USA.* 1684.
James. *Bristol.* a.1757-74.
James. *Londonderry.* 1858 C. Also hardware.
James. *Watford.* 1874.
Jared T. *Batavia, NY, USA.* 1832-36.
Job. *Bath.* 1866-83.
John. *Dublin.* 1880.
John. *Macclesfield.* ca.1780. C.
John. *New York.* 1798.
John. *Twerton, nr Bath.* 1883.
John B. *Norwich.* 1858-75.
Joseph. *Rye.* 1855-78.
Owen. *Bath.* 1875-83.
& SKINNER. *Dundas, Canada.* 1861-66.
Thomas. *Barnsley.* 1846-71. W.
Thomas. *Bewdley.* 1828.
Thomas. *Buckingham.* ca.1790-1805.
Thomas. *Rye.* 1839-51.
Thomas. *Tenterden.* 1823.
Thomas. *Wantage.* 1800-37.
Thomas. *Wolverhampton.* 1828-35. W.
W. *Alford.* 1861.
William. *London.* 1832. Watch cases.
William. *Longton.* 1860-76.
William. *Port Tobacco, Md. USA.* 1773-80s.
William. *Market Harboro.* 1849-76.
William H. *Trenton, Canada.* 1857-63.
ALLERDING, Frederick. *London.* 1857.
ALLETT, George. *London.* (B. a.1683-CC. 1691)-97.
ALLEY, Jerome. *Kilkenny.* 1792. W.
ALLGOOD—
Alfred. *Ledbury.* 1830-70.
Miss E. *Ledbury.* 1879.
ALLIEZ & BERGER. *Geneva.* ca.1850. W.
ALLIN, Thomas. *Manchester.* 1787-1824. C.
ALLING—
James. *Great Dunmow.* 1828.
James. *London.* 1839-63.
ALLINSON, ——. *Malton.* ca.1790. C. See Atkinson
William.
ALLIS—
Henry Jacob. *Bristol.* 1850-79.
John Hagger. *Bristol.* 1850-70.

ALLISON—
A. B. *Sunderland.* ?late 18c. C.
E. *Alnwick.* 1858.
Francis & Co. *Richmond, Yorks.* 1866. W.
Thomas W. *Wakefield.* 1822-37.
William. *Liverpool.* 1795. W.
ALLKINS—
Sampson. *Horncastle.* 1828-35.
T. *Birmingham.* 1854.
Thomas. *Fazeley.* 1828.
Thomas. *West Bromwich.* 1842-50.
ALLMAN, William. *Chipping Barnet.* 1828.
ALLMARK, C. *Chester.* 1865.
ALLOTT—
John. *Bradford.* 1822-37. Later Allott & Wilson.
& WILSON. *Bradford.* 1850. Later Wilson & Fairbank.
ALLOWAY, William. *Ithaca, NY, USA.* 1823.
ALLOWAY, Joshua. *Taunton.* 1755. W.
ALLPORT—
——. *London. Knightsbridge.* ca.1820. br. clock.
Samuel. *Birmingham.* (B. 1790-) 1828-60.
W. B. *Uttoxeter.* 1868-76.
ALLSON, William. *Wheathampstead.* 1874.
ALLSOPP—
& BURTON. *Nottingham.* 1855.
Elijah. *Nottingham.* 1864-76.
John. *London.* ca.1715. C.
John. *London.* 1828.
T. S. *Middlesboro.* 1898.
ALLYN—
John. *Hartford, Conn. USA.* 1657. (May be Van
Allen.)
Nathan. *Hartford, Conn. USA.* 1808.
ALMAN, J. M. *Bristol.* 1801.
ALMEY, Thomas. *Sheepshead* (?*Shepshed*). ca.1810. W.
ALMOND—
Francis. *Peterboro, Canada.* 1851-76.
J. H. *Sarnia, Canada.* 1861-63.
John. *London.* s. of Ralph. CC. 1671-80.
R. P. *Newark.* 1855.
Ralph. *London.* (B. a.1637-CC. 1646)-d.1680. C.
William. *Oxford.* a.1750. C.
ALMY, James. *New Bedford, Mass. USA.* 1836.
ALOQUIER & Fils. *Geneva.* ca.1830. W.
ALRICHS—
Henry S. *Wilmington, Del. USA.* s. of Jacob whose
business he continued 1857.
Wessell. *New Castle, Del. USA.* ca.1670-1734, later
Salem, NJ.
ALSOP, Thomas. *Philadelphia.* 1842-50.
ALSTON—
Robert. *Dunfermline.* 1860.
William. *Dundee.* 1860.
ALTMORE, Marshall. *Philadelphia.* 1819-32.
ALTSON, Isaac. *Middlesboro.* 1866. W.
ALVES, James. *Coatbridge.* 1860.
ALVEY—
George. *London.* 1869-81.
I. *Nottingham.* 1876.
ALVORD, Philo. *Utica, NY, USA.* 1812-78.
ALWIN, Thomas. *Deptford.* 1874.
ALYEWARD, John. *Guildford.* 19c. Skeleton clocks.
ALYN, John. *London.* mid-18c. C.
ALZINGRE, ——. *Montbeliard.* 1860. Toolmaker.
AMAS—
Thomas. *Cleobury.* ca.1765. C.
Thomas. *Coleorton.* ca.1795.
AMATT, George & Son. *Portsea.* 1878.
AMBERMAN, John S. *Brooklyn, NY, USA.* 1856.
AMBRIDGE, James. *London.* 1828-32.
AMBROSE—
George H. *West Ham.* 1866-74.
James Christopher. *Sudbury.* 1830-58.
Martin. *Wisbech.* 1830-40.
Robert. *Leeds.* 1866. W.
Unice. *Dedham.* 1839.
William. *Coggeshall.* 1839.
AMBROSINI, Francis. *London.* 1875-81.
& Son. *Brighton.* 1851.
AMERICAN—
Clock Co. *New York.* 1849-ca.1880. Also opened at
Chicago ca.1868 and Philadelphia ca.1880.

AMERICAN—*continued.*
Horologe Co. *Boston, USA.* 1850-53.
Repeating Watch Factory. *Elizabeth, NJ, USA.* 1885-1905.
Watch Co. *Birkenhead.* 1878 (James Roberts, manufacturer).
Watch Co. *London.* 1875-81.
AMES
Ernest. *Norwich.* 1875.
Horace. *New York.* ca.1860.
Thomas K. *Norwich.* 1875.
'AMHERST'. 1914 and later, product of Pequegnat Clock Co.
AMIDON, L. *Bellows Falls, Vt, USA.* ca.1860.
AMOORE, Thomas. *Worthing.* 1851-78.
AMYOTT, Thomas. *Norwich.* 1698-1767.
ANDERSON—
——. *Gravesend.* ca.1820. C.
——. *Huddersfield.* ?error for John Anderton.
Alexander. *Aberdeen.* 1860.
Alexander. *Liverpool.* ca.1760-87. C. & W.
Alexander. *London.* 1869.
Andrew. *Comrie.* 1837.
Arthur. *London.* 1863-81.
& Co. *Hamilton, Ontario.* 1880.
David M. *Waynesburg, Pa, USA.* ca.1820, later *Honey Brook.* ca.1862.
E. C. *Gravesend.* 1851.
Frederick B. *Gravesend.* 1823-51.
George. *Aberdeen.* 1837.
George. *Liverpool.* ca.1800. W.
George. *St. Andrews.* (b.1815) 1860-d.1892.
Hercules. *Bervie.* 1837.
Hugh. *Gravesend.* 1839-66.
J. *Belfast.* 1809.
J. *Jersey.* 1832. C.
J. F. *Newburgh.* 1860.
James. *Downpatrick.* 1868.
James. *London.* 1875-81.
John. *London.* 1857-63.
John. *Newton, nr. Barrie.* 1860.
John. *Quebec.* 1850-80, later *New Glasgow, Nova Scotia.*
Joseph. *Montreal.* 1850-77. Later *Toronto.*
Michael. *Dublin.* 1874-80. C.
Richard. *Preston.* 1792-d. pre-1817. W.
Robert. *Rochdale.* 1773-74. W.
Robert. *Windsor, Nova Scotia.* 1816.
Thomas. *Belfast.* 1785.
Thomas W. *Toronto.* 1833-51.
William. *Bervie.* 1860.
William. *Bristol.* 1840-79.
William. *Dundee.* 1860.
William. *Lancaster.* a.1750, post-1758 partner with Robinson, d.1801.
William. *Liverpool.* 1848.
William. *London.* 1844.
William. *St. Andrews.* 1860.
ANDERTON—
J. *Liverpool.* ca.1800. W.
J. H. *Croydon.* 1878. Clock cases.
John. *Gargrave.* 1838. W.
John. *Huddersfield.* 1822-37.
Miles. *Backbarrow.* d.1806. C.
William. *Shrewsbury.* 1879.
ANDOUIN, Peter. *Dublin.* a.1719.
ANDREW—
Alexander. *Portsoy.* 1830.
Benjamin. *St. Austell.* 1755-d.1797. C. & W.
Charles. *Grantham.* 1861-76.
Charles John. *Maldon.* 1866-74.
David. *Stewarton.* 1860.
Edward G. *London.* 1857-81.
John. *Salem, Mass. USA.* 1769.
John. *Somerton.* 1750-ca.1820. C.
William. *Huntly.* 1837-60.
ANDREWS—
——. *London.* ca.1840. C.
Alexander. *London.* 1828-32.
& Co. *Londonderry.* 1880-90. W.
Ezekiel. *London.* (B. a.1674)-d.1684, aboard ship on voyage to East Indies. C.

ANDREWS—*continued*
Franklin C. *New York* and *Bristol, USA.* 1848.
George. *London.* 1875-81.
Haydock. *Dublin.* 1736-96.
J. *Clapham.* 1878.
James. *Coleraine.* ca.1765-ca.85. C.
James. *London.* 1869-75.
James. *New York.* ca.1815.
John. *Wincanton.* 1741. C.
John Edward. *Whitstable.* 1866-74.
(Lucius M.) & FRANKLIN. *Bristol, USA.* 1837-43.
N. & T. *Meriden, Conn, USA.* 1832.
Nathan. *Yatebank, Lancs.* a.1724. Prob. later at *Sheffield* where d.1782.
Potter. *Saxmundham.* 1839-75.
R. *Little Weldon,* 1847-54.
Richard. *Dover.* Free 1797.
Richard. *Kettering.* 1864-77.
Robert Somers. *Pilton.* 1819-93. C.
Thomas. *Dover.* 1773-1802. W.
Thomas. *New Buckenham.* 1836-75.
W. *Diss.* 1858-75.
W. *Middlewich.* 1857-65.
W. Theodric. *Kings Lynn.* 1830-46.
William. *Londonderry.* 1839-68. C. & W.
William. *Coleraine.* ca.1820-ca.35. C.
William. *Westerham* and *Edenbridge.* 1847.
William. *Middlewich.* 1828-48.
William. *Royston.* 1840.
William Henry, sen. *Royston.* (B. ca.1790-) 1828-30.
ANEAR—
C. *Chacewater.* 1856.
Charles Henry. *Truro.* 1873. W.
Frederick. *Truro.* 1834-73. W.
Henry. *Chacewater.* 1852.
Henry. *Camborn.* 1873.
ANGEL(L)—
C. *Portsea.* 1848.
John. *Wootton Bassett.* 1842. Clock dealer.
Otis N. *Johnston, RI, USA.* 1834.
ANGELUS, Clock Co. *Philadelphia.* 1874.
ANGIER, Robinson. *Dublin.* a.1714.
ANGUENOT, ——. ?*Montbeliard.* 1890. Escapement makers.
ANGUS—
Charles. *Castletown, Thurso.* 1860.
William. *Liverpool.* ca.1870. C.
ANGWORTHY & Co. *Quebec.* 1811.
ANIS, A. *London.* 1878-86 (repairer).
ANNISTON, Isaac. *Philadelphia.* 1785.
ANSDELL, James. *St. Helens.* 1755-58.
ANSDLE, John. *St. Helens.* 1761-74. W.
ANSELL—
David. *Leeds.* 1871.
Edward. *Abingdon.* 1810-30
Henry. *Chatham.* 1839-55.
Henry Samuel (or Simon). *Birmingham.* 1868-80.
Hyam. *London.* 1828-32.
John. *Ramsgate.* 1845.
ANSLEY, Richard. *Witney.* 1823.
ANSLOW, Samuel. *Shrewsbury.* 1879.
ANSONIA—
Brass & Battery Co. *Ansonia, Conn, USA.* ca.1870.
Brass & Clock Co. *Ansonia, USA*—same as Ansonia Clock Co. q.v.
Clock Co. *Ansonia, USA.* 1851-78, then *New York.* 1878-ca.1930.
Clock Co. *London.* 1881.
ANSPETH, J. *Rollin, Buffalo, NY, USA.* ca.1880.
ANSTEE, Jonah. *Bridgend.* 1887.
ANTAME, Joseph Ameel. *Philadelphia.* ca.1784.
ANTES, John. *Fulneck, nr. Leeds.* b.1740-d.1811. Inventor.
ANTHONY—
Isaac. *Newport, RI, USA.* ca.1750.
Jacob (sen. & jun.). *Philadelphia.* 1802.
James. *Truro.* 1698-99. Kept church clock.
John Bray. *St. Ives, Cornwall.* 1844-73.
L. D. *Providence, RI, USA.* 1848.
William. *Truro.* 1688-d.1768. C.
ANTONEY, Walter. *Ashburton.* 1499-1500. Made church clock.

ANTRIM, Charles. *Philadelphia.* 1837-47.
ANTROBUS—
John. *Manchester.* mar.1772. C.
Philip, sen. *Manchester.* ca.1740. C.
Philip, jun. b.1741. d.1820. C.
APDOVEL, William. *Thirsk.* see MacDougal.
APP, Samuel. *Philadelphia.* 1835-ca.1850.
APPEL, John Ellis. *Dover, Pa, USA.* 1778-80.
APPERLAY, Edward. *London.* ca.1680. Prob. APPLEY, q.v.
APPLEBY, G. *Tipton.* 1868.
APPLEGARTH, Thomas. *London.* (B. a.1664, CC. 1674)-1680.
APPLETON—
George B. *Salem, Mass, USA.* 1859-64.
John. *Prescot.* 1783-96. W.
Richard. *Cronton.* a.1756. W.
Thomas. *Prescot.* 1795-98. W.
APPLEY, Edmund. *London.* From *Westmorland.* a.1670, CC. 1677-d.1688, whilst on visit to Edinburgh.
APPLEWHITE, William. *Columbia and Camden, SC, USA.* 1830s.
APPS—
Robert. *Battle.* b.ca.1752-d.1821. C. & W.
William. *Maidstone.* 1840.
APRILE, Joseph & Napoleon. *Sudbury.* 1839-64. (Brothers).
APSEY, William. *Monksilver.* 1699. Made church clock.
APTHORP, John. *Rumsey.* ca.1780.
ARBOE—
Otthe. *Bornholm.* ca.1775. C.
Ottho Poulson. *Ronne.* b.1719-1773. C.
ARBUCKLE—
Joseph. *Paisley.* 1836.
Joseph. *Philadelphia.* 1847.
ARCHARD—
Alfred. *London.* 1881. Successor to Henry.
George & Co. *Yarmouth.* 1875 (also at *London*).
Henry. *London.* 1851-75.
Leonard E. *Yarmouth.* 1875.
ARCHBOLD—
George. *Berwick-on-Tweed.* 1848-60.
Henry. *Yarmouth.* 1830-58.
Thomas. *Greenock.* 1837.
ARCHDEACON, Michael. *Dublin.* 1764-1800. W.
ARCHER—
George. *Cork.* 1770. W.
George. *Retford.* 1828.
George. *Rochdale.* 1824.
George. *Wetherby.* 1822.
Henry. *London.* As early as 1622 (B.BC. 1628-d.1649).
J. *Burton-on-Trent.* 1860.
J. *Chipping Ongar.* 1657.
J. G. *Montreal.* 1849.
James. *Cork.* 1769. W.
John. *Henley-on-Thames.* 1532-41. Rep'd. ch. clocks.
John. *Liverpool.* 1767-84. W.
John. *London.* d.1603.
John. *London.* (B. a.1650, CC. 1660)-1662. Engraver.
John. *London.* 1863.
Percy. *Liverpool.* a.1742. W.
Samuel. *London.* 1803-11.
Samuel. *London.* 1828.
Samuel. *London.* 1857-81.
Samuel William. *London.* 1851-57.
Thomas. *Colchester.* 1836.
William. *Hastings.* 1862-78.
William. *London.* 1844-57.
William. *London.* 1881.
William. *New York.* ca.1830.
William. *Stratton.* 1590. Kept ch. clock.
William, jun. *Salem, Mass, USA.* 1846-50.
ARCHETTO, Louis. *London.* 1869-75.
ARDEN—
Henry, sen. *Malton.* 1834-51.
Henry, jun. *Malton.* 1851-66.
John. *Liverpool.* ca.1775-d.1780. W.
ARDOUIN—
C. J. *Quebec.* 1854-64.
C. J. R. *Quebec.* 1822-64.
M. & Son. *Quebec.* 1844-52.

ARGENT—
William. *Colchester.* ca.1846-d.1852. C. & W.
William. *Rhyl.* 1869-90.
ARDGRAVE, George P. *Canterbury.* 1846-74.
ARIEL, James. *London.* 1828. Watch case maker.
ARIS—
Philip. *Wakefield.* ca.1790-1800. C.
Thomas. *Uppingham.* 1835-76.
William. *Uppingham.* 1828-35.
ARKELL, James. *Canajoharie, NY, USA.* 1880-86.
ARKINSTALL
Francis. *Drayton.* 1828-42.
James. *Newport, Salop.* 1879.
ARLANDY, John. *London.* (B. CC. 1682)-97.
ARLAUD—
Jaques. *Geneva.* b.1663.
Jaques Antoine. *Geneva.* 1668-1743.
ARLE—
Alfred. *London.* 1881. Successor to William Henry.
William Henry. *London.* 1875.
ARLOT, William. *Sunderland.* (B. 1801-) 1820-27.
ARLOW, George. *Dromore.* 1865-68. C.
ARMAND—
J. W. *Copenhagen.* 1777. chron.
Theophile. *Quebec.* 1890.
ARMITAGE—
Benjamin. *London.* 1851.
J. *Bradford.* 1866.
J. H. *Warrington.* 1858.
ARMOUR—
& Co. *Glasgow.* 1860. Wholesalers.
Morris. *Toronto.* 1859.
ARMS, Edwin Herbert & QUIGLEY, Robert J. *Toronto.* 1879-83. Watch cases.
ARMSON, John Edward. *Alfreton.* 1876.
ARMSTRONG—
——. *Warton.* ?early 19c. C. (see Armstrong, T).
A. *Woodford Green.* 1866.
Asnath. *Manchester.* 1851. Imported clocks.
Barzillia. *Colchester.* ca.1846-ca.55. W. & C.
Edward. *Manchester.* 1851. Imported clocks.
George. *Hull.* 1834-38.
George Booth. *Manchester.* s. of Joseph. b.1836-87.
Henry. *Louth.* 1850.
Hugh. *Alston.* 1879.
John. *Hetton le Hole.* 1834.
John. *York.* 1756.
John. *York.* 1841-66. W.
Joseph. *Manchester.* 1829-51. W.
Joseph B. *Douglas.* 1860. Also optician.
Josepn Boyd. *Manchester.* s. of Joseph. b.1831-59. W.
Robert. *Leeds.* 1826.
Robert. *Manchester.* mar.1811-34.
Robert. *Stokesley.* 1823.
T. *Hawkshead.* late 18c. C.
T. *Kendal.* late 18c. C.
T. *Warton.* 1858 C.
Thomas. *Lancaster.* 1851.
Thomas. *Leeds.* 1834.
Thomas. *Manchester.* s. of Joseph. b.1829-73. W.
Thomas. *Manchester.* b.1755-d.1835. C.
Thomas. *Milnthorpe.* 1836. C.
Thomas. *New York.* ca.1830.
William. *Hawkshead.* 19c. C.
William. *Saffron Walden.* 1874.
ARNABOLDI, A. *Buckingham.* 1869-77.
ARNELL—
John. *London.* 1844.
John Christopher. 1851.
ARNOLD—
——. *Stony Stratford.* ca.1820. C.
Ebenezer N. *London.* 1881.
Edward. *Leicester.* 1783-95. C.
E. G. *Caernarvon.* ICC. – date unknown.
& Frodsham. *London.* 1844-51.
Frodsham Charles. *London.* 1857.
George. *Merthyr Tydfil.* 1887.
Henry. *Farnworth.* mar.1803. W.
J. *Burton-on-Trent.* 1876.
Jacob. *Philadelphia.* 1848.
Jared. *Amber, NY, USA.* ca.1830.

ARNOLD—*continued.*
John I. *Bodmin.* s. of Richard II. 1754. C.
John II. *Bodmin.* 1736-99, later *Chigwell.*
John. *Launceston.* 1754. C. May be the *Bodmin* man.
Joseph. *Tadcaster.* 1834-37.
Joshua. *London.* 1769-76. W. Aldgate.
Mrs. & Son. *Tadcaster.* 1844.
Richard. *Bodmin.* 1692-d.1724. C.
Richard II. *Bodmin.* s. of Richard I. 1737-d.1755.
& Son. *Fareham.* 1878.
W. D. *Fareham.* 1859-67.
William S. *London.* 1857-63.
William S. *Ryde.* 1867-78.
William Simpkin. *Boston.* 1850.
William Simpkin. *Stony Stratford.* 1842-47. Successor to George Berrill.
ARNOLDI & COMENS. *Montreal.* 1807.
ARNOT(T)—
James. *Ards.* 1622. C.
John. *Chesham.* 1842.
Richard. *London.* (B. CC. 1808-) 1828-32.
Thomas. *Edinburgh.* 1723.
William. *London.* 1881.
ARON, Adolphe. *London* and *Paris.* 1881. Luminous C. & W.
ARONSON—
Harris. *Caernarvon.* 1835.
John. *Bangor.* 1841-74.
Saul & John. *Bangor.* 1835.
ARRISON, John. *Philadelphia.* 1837.
ARROWSMITH—
John. *Farnworth.* mar.1791. W.
John. *St. Helens.* 1825. W.
ARSBORN, Thomas. *Bloxham.* 1830-75 (see also Osborne).
ARTER—
Martha. *Bristol.* 1842-50 (?wid. of William).
William (?senior). *Bristol.* 1840.
William (?junior). *Bristol.* 1863.
William. *Pensford, Bristol.* 1875-83.
ARTHUR—
H. G. *Boston, USA.* 1830.
James. *Hull.* 1834-58.
James. *New York.* 1842-1930.
Arthur P. *Madron.* 1856.
Peter. *Penzance.* 1844-d.1857. C.
ARTHURS, Joseph. *Dorking.* 1866-78.
ARUNDEL, John Cundill. *York.* b.1820-61.
ARWEN, William. *Huddersfield.* 1771-95. C.
ARWIN, William. *Albany, NY, USA.* 1837.
AS(H)BURY, Joseph Fletcher. *Hurstpierpoint.* 1862-70.
ASCHER—
& Co. *Montreal.* 1862-67.
G. I. *Montreal.* 1849-54.
G. I. *Toronto.* 1857.
ASCOLI, A. *Manningtree.* 1866.
ASCROFT, James. *Wigan.* 1718. Sundial.
ASH(E)—
David. *London.* 1881.
H. *Portsea.* 1867.
J. *Birmingham.* 1860.
Lawrence. *Baltimore, USA.* 1773. From Philadelphia, perhaps later at Boston.
Ralph. *London.* (B. CC. 1646-) 1656.
ASHALL—
Charles. *Bolton.* 1848-58.
John. *Bolton.* 1834.
William. *Bolton.* 1848-51.
William. *Toronto.* 1856-77.
ASHBURNER, Robert. *Preston.* ca.1750. C.
ASHBURTON, ——. *Liverpool.* ca.1785. W.
ASHBY—
James. *Boston, USA.* From *Cumberland, England.* (B. 1769) -73. W.
Joseph. *London.* (B. a.1664, CC. 1674)-97.
ASHCRAFT, O. *New York.* 1840.
ASHCROFT, John. *Liverpool.* a.1752. Late 18c. (B. 1816-29). W.
ASHDOWN—
Charles. *London.* 1844.
Charles. *Straiford, Essex.* 1839-51.
Frank. *Tonbridge.* 1874.

ASHDOWN—*continued.*
G. *Sevenoaks.* 1866.
W. *London.* 1857-69.
ASHE, Edward. *Rotherham.* 1833-37.
ASHER, G. *Sunderland.* 1851.
ASHFORD—
Robert. *Crawley.* 1828-39.
William. *Ipswich.* 1839-53.
William Henry. *Birmingham.* 1880.
ASHLEY—
David. *Whittlesey.* 1830.
E. F. *London.* 1878. W.
Edward. *London.* (B. a.1812-21) 1828-44.
Edward & Son. *London.* 1881.
Robert. *Sheffield.* 1871. Clock dealer.
Thomas. *Liverpool.* 1851.
ASHMAN, G. *Mildenhall.* 1858.
ASHMORE—
George. ?*Derby.* 1733. C.
William. *Bawtry.* 1822.
ASHTON—
C. *Philadelphia.* 1762-97.
Edward. *Dublin.* a. 1671 to G. Southwarck.-1680. W. & C.
G. *Leek.* 1860-68.
Isaac. *Philadelphia.* 1790.
James. *Manchester.* 1717-38. W.
John. *Leek.* (B. 1795-) 1828-50.
John. *Prescot.* 1825. W.
Martin. *Ashbourne.* s. of Samuel. b.1762-1829.
Nicholas. *Prescot.* 1796. W.
Samuel. *Macclesfield.* b.1728-62, then *Ashbourne* till d.1804. C.
Samuel. *Manchester.* 1793. C.
Samuel. *Philadelphia.* 1790s.
Samuel & Son. *Ashbourne.* 1792-1804 (Samuel & Martin).
Thomas. *London, Ontario.* 1875.
Thomas Heard. *Bridgend.* 1871-75.
Thomas Heard. *Milford Haven.* 1868.
W. *Philadelphia.* 1762-97.
William. *Barton-on-Humber.* 1828-35.
William. *Wakefield.* 1862-71.
ASHWELL, Nicholas. *London.* (B. a.1642, CC. 1649)-1657.
Josiah. *Leicester.* b.ca.1680-a.1702-ca.1750. C.
ASHWORTH—
J. *Rochdale.* 1858.
James. *Sheffield.* 1871. Clock cleaner.
John. *Bacup.* 1848-51.
ASK—
William. *Barnsley.* 1834-37.
William. *Sheffield.* 1862-71.
William. *Wakefield.* 1851-53.
William, junior. *Wakefield.* 1871.
ASKEW—
James. *Charleston, SC, USA.* 1770-90.
James. *Birmingham.* 1880.
John. *Dublin.* 1858.
ASKEY—
J. T. *Coventry.* 1854.
T. *Coventry.* 1860.
Thomas. *Coventry.* 1850.
ASKWITH, John. *York.* Free 1740-58. W.
ASPDIN, James. *Leeds.* b.1742-d.1788. W.
ASPIN(W)ALL—
Charles. *Goderich, Canada.* 1861.
(Charles), BEEMER & CO. *Hamilton, Canada.* 1856.
H. *Kirkby Lonsdale.* Late 18c. C.
James. *Liverpool.* 1800. W.
Peter. *Ashton in Makerfield.* Free 1664-d.1677. C. Also guns and spurs.
William. *Bolton.* 1848-58.
Zalmon. *Boston, USA.* 1809-13.
ASPREY, Ebenezer. *Olney.* 1798.
ASSELIN(NE), Francis. *London.* (B. CC. 1687)97.
ASSELTINE, George N. *Gananoque, Canada.* 1871.
ASSHE, Rainey. *Dublin.* 1800-08.
ASTLE, Simon. *Derby.* 1864-76.
ASTLEY, Edward. *Liverpool.* 1834.
ASTON—
Joseph. *Cirencester.* 1870-79.
Samuel. *Birmingham.* 1880.

ASTON—*continued.*
Thomas. *Newburyport, Mass, USA.* 1860.
Thomas. *Toronto.* 1861.
William Matthew. *London.* 1851.
ASWORTH, Richard. *Manchester.* 1848.
ATHA—
Job. *Pontefract.* 1871.
Joseph. *Pontefract.* 1866.
ATHERLY, ——. *Wolverhampton.* 1724. W.
ATHERTON—
Frederick. Mark of U.S. Watch Co. 1867.
John. *Prescot.* mar.1767. W.
John. *Prescot.* 1823. W.
Matthew. *Philadelphia.* 1837-40.
Otis. *New York.* 1798.
S. *Liverpool.* ca.1785. W.
Thomas. *Liverpool.* ca.1770-ca.80. W.
Thomas. *Prescot.* mar.1820. W.
ATKIN—
Alfred. *Sheffield.* 1871.
Charles Wheelock. *Tullamore.* 1858.
Francis. *Liverpool.* 1800-29. W.
John. *Sheffield.* 1862.
John. *Yeardsley-cum-Whalley.* 1878.
Moses. *Stapleford.* 1876.
Paul. *Sheffield.* 1862.
S. *Alford.* 1861.
W. *Ticknall.* 1849-64.
ATKINS—
Alden A. *Bristol, USA.* 1835-46.
Alvin. *Rochester, NY, USA.* 1849.
Clock Co. *Bristol, USA.* 1859-79.
Clock Manufacturing Co. *Bristol, USA.* 1855-58.
(Alden A.) & Co. *Bristol, USA.* 1837-46.
& COLE. *London.* 1869-75.
(Rollin) & DOWNS (Anson). *Bristol, USA.* 1831-32.
Eldridge G. *Bristol, USA.* 1835-42.
Francis. *London.* 1811. W.
George. *Arundel.* 1878.
George. *London.* 1828.
George & Son. *London.* 1839-57.
Henry. *London.* 1828.
Irenus. *Bristol, USA.* 1792-1882.
Irenus & Co. *Bristol, USA.* 1847-57.
J. *Coventry.* 1860. Clock cleaner.
J. *Bristol.* 1863-79.
Jearum. *Joliet, Ill., USA.* 1848.
Joel. *Middletown, Conn, USA.* 1777. C.
John. *Bristol.* 1842-56.
Joshua. *Chipping Norton.* ca.1800-23. C. & W.
Merritt W. & Co. *Bristol, USA.* 1846-56.
Michael (Thomas). *London.* 1828(-32).
(Merritt W.) & PORTER (Henry H.). *Bristol, USA.* 1840-46.
R. (Rollin) & I. (Brothers). *Bristol, USA.* 1833-37.
Robert. *London.* 1863-69.
Rollin. *Bristol, USA.* bro. of Irenus. 1790-1844.
S. *London.* 1811. Watch case maker.
& Son. *Bristol, USA.* ca.1870.
& WELTON. *Bristol, USA.* 1835-36.
Whiting & Co. *Bristol, USA.* 1850-54.
William. *Chipping Norton.* mar.1764-78. C. & W. (?a. in London).
William (& Co.). *London.* 1832-51 (57-63).
ATKINSON—
Abraham. *Sheffield.* 1871. Clock cleaner.
Anna Maria Leroy. *Baltimore, USA.* Wife of Wilmer. ca.1819-23.
Charles. *Barnsley.* 1834.
Edward. *Newcastle-on-Tyne.* a.1757-77. C.
Elizabeth. *Barrow-in-Furness.* 1869. W.
George. *Louth.* 1828-35.
George. *Spilsby.* 1876.
J. *Douglas.* ca.1790. C.
James. *Boston, USA.* from *London.* 1744-d.1756. W.
James. *Bowness.* 1879. W.
James. *Dunse.* a. of Nean Davidson there in 1808.
John. *Barnsley.* 1837.
John. *London.* 1844.
John William. *Dublin.* 1796. W.
Jonathan. *Manchester.* mar.1833. C.
Leroy. *Baltimore, USA.* 1824-30.

ATKINSON—*continued.*
Matthew & William. *Baltimore, USA.* 1787.
Peabody. *Concord, NH, USA.* 1790.
Robert. *Liverpool.* 1800 (-B. 1825). W.
Samuel, sen. *Bridlington.* ca.1730-ca.1760. C.
Samuel, jun. *Bridlington.* b.1772. C.
Thomas. *Carlisle.* ca.1830. C.
Thomas. *Dublin.* a.1753. wkg. 1765-1806. W.
Thomas. *Ulverston.* 1862; *Dalton.* 1866.
Thomas. *Newcastle-on-Tyne.* 1848.
Thomas. *Ormskirk.* c.1780-86. C. & W. ?from Lancaster.
Thomas. *South Shields.* 1851-56.
W. *Coventry.* 1880.
W. *Sheffield.* ca.1740-50. C.
William. *Keighley.* 1822. C.
Wilmer. *Lancaster, Pa, USA.* mar.1748.
ATMAR, Ralph. *Charleston, SC, USA.* ca.1800.
ATMORE, ——. *London.* 1725 (B.-1735). W.
ATTELEE, John. *Henley-on-Thames.* Rep'd. ch. cl. 1412-48.
ATTENBOROUGH—
George (& Son). *London.* 1869(-75).
James. *London.* 1857.
ATTENVILLE, James. *London.* Early 19c. C.
ATTERBURY, John. *London.* 1881.
ATTLEY, Edwin Joseph. *London.* 1869.
ATTMORE, Marshall. *Philadelphia.* 1821-37.
ATTWELL—
John Stewart. *Maidenhead.* 1830.
W. *Kettering.* 1864-69.
William. *Brockville, Canada.* 1851.
William. *Bryn-mawr.* 1875.
William Henry. *Romford.* 1828-74.
ATTWOOD—
——. *Lewes.* c.1780. C.
Edward. *London.* 1828.
George. *London.* (B. 1805-21) 1828-32.
Mathias W. *Hamilton, Ontario.* 1865-95.
William. *Bingley.* 1853.
William. *Lewes.* 1828.
ATWELL, Robert. *London.* 1811. W.
ATWOOD—
Anson L. *Bristol, USA.* 1816-1907.
B. W. *Plymouth, USA.* ca.1860.
& BRACKETT, *Littleton, NH, USA.* ca.1850.
William. *Hurstpierrepoint.* 1839.
AUBER, J. *Burton-on-Trent.* 1868-76.
AUBERT—
Charles A. & Co. *London.* 1863-69.
& KLAFTENBERGER. *London.* 1839-57.
& KLAFTENBERGER. *London.* 1881.
& LINTON. *London.* 1863.
Moesa. *London.* 1832.
& WATSON. *London.* 1869.
AUBREY, Richard Stephen. *London.* 1875.
AUDEMARS, Louis. *London.* mid-19c. Also *Paris, Switzerland, Vienna, New York, St. Petersburg.* (W. H. Garnish, agent).
AUDOUIN, William. *London.* ca.1760. br. clock.
AUDUS, Francis. *Halifax.* 1871.
AUERBACH, Z. *Montreal.* 1875.
AUGHAVAN, Patrick. *Louth.* 1876.
AUGUSTE, ——. *Paris.* ca.1840-80. Carriage clocks.
AUGUSTINE, George (place unknown). ca.1810. C.
AUKER, John. *Kings Lynn,* 1875.
AULD, William. *Belfast.* ca.1772-1800. W. (Perhaps then at *Edinburgh.* 1800-23).
AULKIN, Richard. *Oxford.* a.1770.
AULT—
James. *Belper.* 1822-29.
Joseph. *Derby.* 1849-52.
John. *Montreal.* 1850.
John. *Toronto.* 1856 from *Montreal.*
Thomas. *London.* (B. 1822-25) 1828.
AURORA Watch Co. *Aurora, Ill, USA.* 1885-92.
AUSTEN—
Ambrose J. *London.* 1881.
John. *Dundee.* 1836.
John. *London.* 1857-81.
Richard. *Liskeard.* Grandson of Thomas. 1823-d.1845. W.

AUSTEN—*continued.*
Thomas. *Cork.* 1795-1809. W.
Thomas. *Liskeard.* ca.1760. C.
AUSTIN—
Aaron. *Bristol.* mar. pre-1770. (B. 1775-97).
Benjamin. *Kalamazoo, Mich, USA.* ca.1850.
Ezekiel. *Earls Barton.* 1877.
Frederick William. *Bath.* 1883.
Isaac. *Philadelphia.* 1785-d.1801.
Isaac. *Upper Delaware Ward, Pa, USA.* 1783.
James. *Aldershot.* 1878.
John. *Philadelphia.* ca.1830-40.
John. *Uxbridge.* ca.1710. C.
John. *Watford.* ca.1710. C.
Joseph. *London.* ca.1740. C.
Josiah. *Salem, Mass, USA.* 1853-57, later *Boston.*
Matthew L. *Halifax, Nova Scotia.* 1864-67.
Orrin. *Waterbury, Conn, USA.* ca.1820.
Seymour. *Hartford, Conn, USA.* ca.1800.
T. *Bristol.* 1856.
Thomas. *Bristol.* 1830.
William. *Portsmouth.* 1878.
AUTEL, J. d'. *Quebec.* 1855.
AVENALL—
——. *Winton.* ca.1760. C.
George. *Farnham.* Early 18c. C.
AVENELL—
Charles. *London.* 1869-81.
Philip, sen. *Farnham.* d.1783. C.
Philip, jun. *Farnham.* ca.1765-d.1807. C.
William. *Gravesend.* ca.1770. C.
AVERILL, John. *London,* see d'AVERILL.
AVERY, James. *Birmingham.* 1880.
AVES, N. E. *West Hartlepool.* 1898.
AVISE, M. *Reading, Pa, USA.* 1827.

AVISSE, Charles. *Baltimore, USA.* 1812.
AXFORD, ——. *Bath.* 18c.
AXMANN, Frederick Otto. *London.* 1869.
AYERS, Edward. *Beccles.* 1839-58.
Mrs. *Beccles.* 1865.
Robert. *Beccles.* 1875.
Thomas. *East Dereham.* 1836.
AYLING, Allen. *Petworth.* 1878.
AYL(E)WARD—
John. *Aylesbury.* 1691. TC.
John. *Guildford.* Late 17c-ca.1720. W. & lant. clock.
AYNSWORTH—
——, *London.* ca.1685. C.
——. *Westminster.* ca.1710. C.
& Son. *Blackburn.* Late 19c. C.
AYRE—
Henry. *Gateshead.* (B. ca.1820). d.1827. C.
Mary. *Gateshead.* 1827-33.
Thomas. *Dublin.* (B.1824-   ) 1827-39. W.
AYRES—
Alexander. *Essex Co., Va, USA.* pre-1776; then *Lexington, Ky.* 1790-1823; then *Danville, Ky.* 1823-24.
(E) & BEARD (Evans C.) *Louisville, Ky, USA.* 1816-31.
E. *Beccles.* 1846.
E. *Louisville, Ky, USA.* ca.1816-31.
Hamilton. *New Holland, Pa, USA.* ca.1820.
Samuel. *Lexington, Ky, USA,* then *Danville, Ky.* 1776-1824.
T. *Yarmouth.* 1846-58.
AZANS, Charles. *London.* 1881.
AZULAY, Raphael. *London.* 1869-81.
AZUR, Hippolite. *London.* 1851.

B

B, D. L. Mark of Delepine-Barrois, q.v.
B, P. Monogram of Paulus Braun of *Augsburg*, q.v.
B, P. *Cleobury.* Early 18c. C.
BAAB, John Martin, *London.* 1844-81.
BABBAGE, John. *Bristol.* 1870.
BABBITT, H. W. *Providence, RI, USA.* 1849.
BABCOCK—
   Alvin. *Boston, USA.* 1810-13.
   & Co. *Philadelphia.* 1831-33.
BABINGTON, Richard. *Dublin.* a.1756.
BACH—
   Albert. *London.* 1881.
   Anton. *London.* 1857-81.
BACHAN, Henry. *London.* mid-18c. (B. d.1768). W. &
   C.
BACHMAN—
   Jacob. *Lampeter, Pa, USA.* 1766-98.
   John, *Bachmanville, Pa, USA.* S. of Jacob, b.1798,
   mar.1822.
   Joseph. *New York.* 1855.
BACK—
   George. *London.* 1832. Importer of Dutch clocks.
   Robert, *Peterboro.* 1854-77.
BACKES, J. P. *Charleston, SC, USA.* ca.1850.
BACKHOUSE—
   Benjamin. *Masham.* 1823-34.
   Frederick. *London.* 1863.
   James I. *Lancaster.* (B. 1726) mar.1744, d.1747. W.
   James II. *Lancaster.* a.1749.
   John, *London.* 1869.
BACKWELL, Thomas, *London.* 1662.
BACON—
   A. T. *Scole.* 1865.
   Charles, *London.* 1869-81.
   James. *Chertsey.* 1862-78.
   John. *Bristol, USA.* 1833-45.
   John. *Sleaford.* 1850s. Worked for John Hyde,
   whose daughter he married. To *Dover.* ca. 1853-74.
   Succ. by dtr. until 1937. C.
   John. *Dublin.* 1820-23. W.
   William, *Colchester.* ca.1642-ca.1680. Lantern cl.
BADCOCK, ——. *Kibworth.* Early 19c. C.
BADDER, Isaac. *Dayton, Ohio, USA.*, ca.1830.
BADELY, Thomas, *Boston, USA.* 1712-20.
BADER—
   H. *Holyhead.* ICC. Date unknown, see next entry
   though.
   Joseph. *Bangor.* 1844, then *Holyhead.* 1848-74.
BADGER, James. *New York.* ca.1840.
BADLAM, Stephen. *Boston, Lower Mills & Donchester,
   Mass, USA.* 1751-1851. Clock casemaker.
BADMAN, Joseph. *Colebrookdale, Pa, USA.* ca.1800. C.
BADOLLET—
   J. J. & Co. *Geneva.* 1880.
   John M. (& Co.). *London.* 1844-51 (57-75). Succ. in
   1881 by Huguenin, Son & Hall.
   ·Paul, *New York.* 1798.
BAER(R)—
   L. *Swansea.* 1899.
   William, *Weaverdale, Cal, USA.* ca.1860.
BAETENS, Joseph. *London.* 1832. Bronze and ormolu
   clocks.

BAGGS, Samuel, *London.* (B. 1817-30)-32.
BAGLEY—
   Nicholas. *Cork.* 1802. W.
   Richard. *Cork.* 1787-1824. W.
   William. *Cork.* 1802. W. (B. 1820).
   William. *Kinsale.* 1858. W.
BAGNALL—
   Benjamin. *Boston, USA.* mar.1712-ca.1730. (B.
   -ca.1740).
   Benjamin, jun. s. of Benjamin, sen. *Boston,* later
   Philadelphia, Providence and Newport, RI, USA.
   1715-42.
   Henry. *London.* 1832.
   J. *Birmingham.* 1854.
   James Eustace. *Birmingham.* 1850.
   Matthew. *Cork.* 1770- d.1795. W.
   R. *Talk*? ca.1710. C.
   Robert. *Penkridge.* 1828-50.
   Thomas. *Belfast.* ca.1790-1820. C.
   William Henry. *London.* 1839-57.
BAGOT, John. *Lancaster.* b.1808-69. W.
BAGSHAW, David. *London.* 1863-81.
BAILE—
   John. *Carmarthen.* 1796- early 19c.
   William. *Carmarthen.* mar.1765-91. C.
BAILLAN, F. B. Paris? ca.1750. W.
BAIL(L)E—
   James I. *Downpatrick.* ca.1825. C.
   James II. *Downpatrick.* ca.1843-68. C.
   John, *Belfast.* ca.1770-85. C.
   John, *Dromore.* 1766, then *Downpatrick.* 1768. C.
   John. *Ballynahinch.* 1865-68.
   R. B. *Ballynahinch.* d.1901.
   Samuel, *Dromore.* ca.1765-85. C.
   William. *Belfast.* ca.1787-1824, then *Downpatrick.*
   1824-d.1827.
   William. *Kirkcabbin.* ca.1800-1810, *Downpatrick.*
   1843.
BAILEY—
   BANKS & BIDDLE & CO. *Philadelphia, USA.* Post
   1878 to present day.
   Brothers. *Utica, NY, USA.* 1846-52.
   Calvin, *Hanover, Mass, USA.* Bro. of John. 1761-1828.
   Capt. John S. *Philadelphia, USA,* ca.1855-90.
   & Co. *Philadelphia, USA.* 1846-78.
   Edward. *Brighton.* 1878.
   Edward R. *London.* 1869-81.
   F. *Ringwood.* 1867.
   G. S. & Co. *Danbury, Conn, USA.* ca.1860.
   Gamaliel. *Mount Holly, NJ, USA.* 1807, then *Phila-
   delphia* till 1833.
   George. *Luton.* 1877.
   H. *Bristol.* ca.1840. C.
   I. G. *Conn, USA.* ca.1848.
   J. *Brixton.* 1862.
   J. H. *Louth.* 1861-68.
   J. T. *Philadelphia.* 1828-33.
   James. *Henfield.* 1828-39.
   John. *Hanover & Lynn, Mass, USA.* ca.1770.
   mar.1780-1808.

BAILEY—*continued.*
John. *Brixton.* 1878.
John. *Egham.* 1878.
John. *Louth.* 1835.
John. *London.* 1857.
John Horncastle. (B. pre-1807)-1828-35.
John Sampson. *Horncastle.* 1849-68.
Joseph. *Higham & New Bedford & Lynn, Mass, USA.* ca.1800-1840.
Joseph. *Liverpool.* a.1756. W.
Joseph. *Yoxford.* 1830-39. Also jeweller.
Junius. *London.* 1857-63.
and KITCHEN, *Philadelphia.* 1832-46.
Lebbeus. Bro. of John. *Hanover, USA.* 1763-1827, later *No. Yarmouth, Me, USA.*
and OWEN. *Abbeville, SC, USA.* 1848.
Parker, *Rutland, Vt. USA.* ca.1860.
Putnam. *N. Goshen, Conn, USA.* 1830-40. C.
Samuel. *Newcastle-on-Tyne.* 1852-58.
Thomas. *Birmingham.* 1828.
Thomas. *Liverpool.* ca.1830-48.
Thomas. *London.* Early 18c.
W. *Bromley.* ca.1830.
W. *Leicester.* Early 19c. C.
William. *Aldershot.* 1867-78.
William. *Glasgow.* 1860.
William. *Portaferry.* ca.1790. C.
BAILHACHE, P. *Jersey.* ca.1865.
BAILLOD, August. *London.* 1832. Watch cases.
BAILLY, Comte et Fils. *Morez du Jura.* 1851. Carr. Clocks.
BAILY—
J. *Leicester.* 1876.
Joel. *West Bradford, Pa, USA.* 1732-97. C. Also gunsmith, etc.
and WARD. *New York.* 1832.
William. *London.* 1828-39.
William. *Philadelphia.* 1816-22.
BAIN, George. *Brechin.* 1837-60.
BAINBRIDGE, Charles. *London.* 1875.
Elizabeth, *Dublin.* 1802. W.
George. *Dublin.* 1761-1801. W.
and STERLING. *Dublin.* 1803-05.
BAIN(E)S—
Benjamin. *Chelmsford.* Pre-1849.
John. *Snaith.* 1790-1803. C.
John. *Tadcaster.* ?ca.1800. C.
T. *Middlesbrough.* 1898.
BAINTON, James. *London.* 1839.
BAIRD—
Clock Co. *Plattsburg, NY, USA.* c.1892.
George. *Carlisle.* (B. 1811-29) 1828-34. Also jeweller.
Walter. *Glasgow.* 1848.
William. *London.* 1801. W.
William and John. *London.* 1811-28. W.
BAIRSTOW, John. *Wakefield.* 1862.
BAITSON, Thomas. *Beverley.* 1807-23.
BAKER—
——. Billingshurst. ?ca.1780. C. (Probably David B., q.v.)
A. *Sittingbourne.* 1874.
Alexander. *Newry.* 1846.
Alexander. *New York.* 1854.
Alfred. *Boughton (Kent).* 1855-66.
Ann. *Tamworth.* 1828-42.
Booth. *Hull.* ca.1760. C.
Charles. *Dublin.* 1798-1808.
Charles. *London.* 1881.
Charles Frederick. *London.* 1869-81.
David. *Billingshurst.* 1839-70.
Edward. *Birmingham.* 1842-68.
Edward. *London.* 1844-57.
Edward (or Edmond?). *Gravesend.* 1858-74.
Eleazar. *Ashford, Conn, USA.* 1764-mar.1787-1849. Later at *Hampton & Middletown, Conn.*
Elias. *New Brunswick, NJ, USA.* ca.1840.
Frederick. *Farnborough (Kent).* 1874.
Frederick G. *Shanklin.* 1878.
G. *Birmingham.* 1860.
George. *Lenham.* 1867. W.
George. *London.* 1869-81.
George. *Salem, Mass. USA.* 1790.

BAKER—*continued.*
George. *New York.* 1868.
H. *Ashby-de-la-Zouche.* 1849.
Henry. *Lewes.* a.1726 to Henry Barrett.
Henry. *Boughton (Kent).* 1826-51.
Henry. *Great Appleby.* 1835-49. Also bellhanger.
Henry. *Town Malling.* ca.1705. C. (B. pre-1768-1784).
J. *Crowle.* 1868.
J. *Malton.* Early 18c. C.
J. *Woodbridge.* 1853-65.
James M. *Philadelphia.* 1842.
John. *Hull.* Believed late 18c. C.
John. *Plymouth Dock.* (B. 1795). mar.1798.
John. *Sevenoaks.* Early 18c.
John. *Stafford.* 1876.
John Charles. *Highworth.* 1859-75.
Jonas. *St. Austell.* ca.1770. C. & W.
Joseph. *Great Appleby.* ca.1750. C. (B. pre-1784. W.)
Joseph. *Birmingham.* 1842-50.
Joseph. *Philadelphia.* 1772.
Mrs. Thomas Waghorn. *London.* 1875. (Presumably widow of Thomas Waghorne.)
Richard. *Birmingham.* 1842.
Robert. *Boughton (Kent).* 1827. C. & W.
Samuel. *Birmingham.* 1839-50. Dial maker.
Samuel. *New Brunswick, NJ, USA.* 1855.
Samuel & Son. *Birmingham.* 1858. Dial maker.
Stephen, *Salem, Mass, USA.* 1791-1827.
Thomas. *Birmingham.* 1839-54. Also dial maker.
Thomas. *Devizes.* 1830-59.
Thomas. *Falmouth.* ca.1810. C.
Thomas. *West Malling.* 1768. W.
Thomas junior. *Haverhill, Mass, USA.* 1793-1820.
Thomas Waghorne. *London.* 1857-69.
Thomas William. *London.* 1869.
William. *Birmingham.* 1822. Dial maker.
William. *Birmingham.* 1854-68. Anglo-American clocks.
William. *Birmingham.* 1854-68. Clock case maker.
William. *Liverpool.* 1851.
William. *London.* 1832-75.
William David. *Horsham.* 1855-78.
BAKKER, Carol Willem. *Goor (Holland).* ca.1770-81. C.
BALAQUIER, James. *Dublin.* a.1707-24.
BALCH—
B. & Son. *Salem and Boston, Mass, USA.* 1832-42.
Benjamin. *Salem, Mass, USA.* 1796-1818 (B. 1837).
Charles Hodge. *Newburyport, Mass, USA.* 1787-1817.
(Daniel) & CHOATE (John). *Boston, USA.* 1816.
Daniel, sen. *Newburyport, Mass, USA.* 1734-90. C.
Daniel, jun. *Newburyport, Mass, USA.* Son of Danl., sen. 1761-1835.
Ebenezer. *Hartford, Conn, USA.* 1744-1808.
James. *Salem, Mass, USA.* 1806.
Joseph, son of Ebenezer. *Wethersfield, Conn, USA.* 1760-94, then *Williamstown, Mass,* then in 1810 to *Johnstown, NY, USA.*
(James) & LAMSON (Charles). *Salem, Mass, USA.* 1842-64.
Moses Phippen. *Lowell, Mass, USA.* 1810-32, then *Lynn, Mass.* 1844.
(Benjamin) & SMITH (Jesse, jun.). *Salem, Mass, USA.* 1818.
Thomas Hutchinson. *Newburyport, Mass, USA.* Bro. of Daniel. 1771-1817.
BALCHIN, Josiah Charles. *London.* 1881.
BALCON, James. *London.* 1844.
BALDOCK—
Mrs. Sarah. *London.* 1857.
William. *London.* 1869.
BALDR(E)Y,
James. *Southwold.* 1823.
James. *Wrentham nr. Wangford.* 1844-65
BALDWIN—
Benjamin. *Loughborough.* 1864-76.
David. *Dublin.* 1868-80.
Ebenezer. *Nashua, NH, USA.* ca.1810-30.
Edgar. *Troy, NY, USA.* 1848-50.
George W. *Sadburyville, Pa, USA.* 1777-1844.
Harland. *Sadburyville, Pa, USA.* bro. of George. c.1809. (Cases only?)
Henry. *London.* 1863.
Henry Holmes. *Castleford.* 1862.

BALDWIN—*continued.*
J. *Sittingbourne.* 1770.
Jabez, *Salem, Mass, USA.* ca.1777-1815. Then *Boston* till 1829.
James Mears. *Huddersfield.* 1871.
Jedidiah. *Northampton, Mass, USA.* 1791, then *Hanover, NH,* then *Morrisville, NY.* 1818-20, then *Fairfield, NY,* and *Rochester, NY.* 1834-44.
John. *Faversham.* 1739-60. W.
Joseph. *Burton-on-Trent.* 1828-42.
Matthais. *Philadelphia.* Pre-1830.
Oliver P. s. of George. *Coatesville, Pa, USA.* ca.1860-1890.
Robert. s. of George W., *Coatesville, Pa, USA.* ca.1860.
S. S. & Son. *New York.* ca.1830.
(Jedidiah) & STORRS (Nathan). *Northampton, Mass. USA.* 1792-94.
Thomas. *Dublin.* 1792-94. C. & W.
Thomas. *London.* 1839-44.
Thomas F. H. *Downington* and *Coatesville, Pa, USA.* 1807-58.
BALE—
Thomas. *Bristol.* 1850-70.
Thomas. *Cheadle.* b.1830-51. C.
BALERNA—
Lewis. *Halifax.* 1834-53.
Richard. *Halifax.* 1837. bro. of Lewis. C. Also weatherglass maker.
BALES, J. *Worksop.* 1855.
BALFE, T. *Potton.* 1847-64.
BALFOUR, James. *Brechin.* ca.1850. C.
BALL—
Albert. *Poughkeepsie, NY, USA.* 1832-35.
and BAKER. *Gravesend.* 1851.
Charles. *Poughkeepsie, NY, USA.* 1840-42.
and EDWARDS. *Birmingham.* 1854-60. Chrons.
Isaac. *Prescott.* 1819. W.
John. *Liverpool.* 1699-d.1716. C.
John. *Newport Pagnell.* 1731-60. W.
& MACAIRE. *London.* 1811-39. Watch case maker.
Nathaniel. *Leicester.* 1876.
Philip. *Truro.* 1769-1837. W.
S. *Black Rock, NY, USA.* 1826.
Samuel. *High Wycombe.* 1786.
Thomas. *Wigan.* a.1752. mar.1767 at *Warrington.* W.
Thomas. B. R. *Coventry.* 1854-80.
Webb Clock Co. *Cleveland, Ohio, USA.* 1879-94.
William. No place. marq. lc. cl. Believed late 17c. Probable the Bicester man.
William I. *Bicester.* mar.1735-ca.1740. C. & W.
William II. *Bicester.* s. of William I. b.1738-d.1823. C. & W.
William. *Farnworth.* 1807. W.
William. *High Wycombe.* 1798-1839. Also innkeeper.
William. *Liverpool.* 1822.
William. *Philadelphia.* 1729-1810.
William. *Prescot.* a.1760. W.
William. *Truro.* 1844-47.
BALLAM, Thomas. *Malton.* 1844.
BALLAN—
William. *Leamington.* 1842.
William. *York.* 1834-38.
BALLANTINE, James. *Lisburn.* 1804.
BALLANTYNE—
——. *Paisley.* 1846.
William. *London.* (B. 1815-) 1828-51.
BALLARD—
& Co. *Cranbrook.* 1845.
E. *Bradford, Wilts.* 1875.
Frederick & William (F. & W.). *Cranbrook.* 1847-66.
George. *Frome.* 1861-83.
H. *Edmonton.* 1851.
Henry. *Cranbrook.* 1847-58.
Henry. *London.* 1857-63.
Isaacs. *Lamberhurst.* 1838. C.
J. *Hertford.* 1859.
James. *Lamberhurst.* 1826-55. C.
John. *Lamberhurst.* 1885.
John. *Leytonstone.* 1874.
John Thomas. *Staplehurst.* 1844-48.
Joseph. *Lamberhurst.* 1839-74.

BALLARD—*continued.*
William. *Brompton* (*Kent*). 1857-74.
William. *Leominster.* 1879.
William (or Frederick William). *Cranbrook.* 1826-66.
BALLERAY, Joseph. *Longueil, Quebec.* 1830-40.
BALLISTON, Thomas. *London.* 1839-44.
BALLO, E. *Titchfield.* 1848.
BALLY, Francis. London. ca.1760. C. (?cf. Bailey).
BALMER—
Thomas. *Liverpool.* 1814-34.
William. *Prescot.* 1795-97. W.
BALMFORTH, A. Place unknown. *Northern England.* ca.1820-30. C.
BALSILLIE, Andrew. *Cupar, Fife.* 1835.
BALSTER, John C. *Sarnia, Canada.* 1861-63.
BAMBER—
Abraham. *Preston.* 1851-58. W.
John. *Aughton, Lancs.* a.1701. W.
S. *Blackpool.* 1858. W.
Samuel. *Bolton.* 1848.
BAMBRICK, George. *Dublin.* 1786. W.
Richard. *Dublin.* 1831. W.
BAMBRIDGE, H. *Yarmouth.* 1865.
BAMFORD, James William. *London.* 1863-75.
Edward, *Dublin,* ca.1800. W.
BANCROFT—
——. *Stockport.* ca.1700. C.
E. J. *Darlington.* 1898.
G. P. *Granville, O. USA.* 1831. Clock cases.
Isaac. *Derby.* 1835-58.
John. *Scarborough.* ca.1780-1807. Supposedly from *London.* C.
Titus. *Sowerby Bridge* and *Halifax.* 1809-38. C. and TC.
William. *Scarborough.* 1823. C.
William Sheldon. *Derby.* 1864-76.
BAND—
Henry & Son. *Coventry.* 1880.
James. *Nottingham.* 1864-76.
BANGER, Edward. *London.* b.ca.1668. a.1687. CC. 1695. d.1720. Nephew of Tompion with whom he worked till ca.1708.
BANISTER—
A. *Hawkhurst.* 1882.
Henry. *London.* 1851-69.
Joseph. *Colchester.* 1803-53. C.
Thomas. *Colchester.* 1835. C. & W.
BANK—
John. *Llangefni.* 1886-90. C.
John. *Port Dinowric.* 1886. W.
BANKES—
James B. *Runcorn.* 1865-78.
John. *Oldham.* Late 18c.
BANKS—
Austin(e). *London.* 1660. Believed clockmaker.
Edward P. *Portland, ME, USA.* 1834.
Ellen. *Preston.* 1848.
George. *Darlaston.* 1835-50.
George. *London.* 18c. C.
George. *Wednesbury.* 1828.
Henry. *London.* Early 18c. C.
Henry. *Woolwich.* 1866-74.
James. *Ormskirk.* 1848-54.
John. *Dawley (Shropshire).* b. ca. 1817-1879. W.
John. *Dublin.* Free 1735-43. W. and Goldsmith.
John. *London.* 1828-32.
John. *York.* ca.1750. C.
Ralph. *Plymouth Dock.* mid-18c.
Solomon. *Leicester.* 1828-35.
Thomas. *Preston.* ca.1835. C.
BANNATYNE Watch Co. *Waterbury, Conn, USA.* 1905-11. Succ. by Ingraham Co. of *Bristol.*
BANNEKER, Benjamin. *Baltimore, MD, USA.* ca.1754. C.
BANNERMAN—
Alexander. *London.* 1811. W.
Gilbert. *Banff.* 1821.
BANNISTER—
——. *London.* 1815. W.
Anthony. *Liverpool.* Early 18c. C. May be from *London* where one such working 1715-36.
Henry. *Coleshill.* ca.1750. W.

BANNISTER—*continued.*
Henry. *London.* 1869-81.
James. *Derby.* 1858-71.
James. *London.* (B. CC. 1818-35). 1828-63.
James. *Wrexham.* 1765-d.1780. C.
John Charles. *Middlesbrough.* 1866.
Richard. *Coleshill.* 1783-1828.
Thomas. *Brecon.* 1712-d.1737. C.
William. *Liverpool.* a.1719-1754. W.
BANQUET, ——? *London.* ca.1630-40. W. (Probably Bouquet, qv.)
BANTEL, Philip. *New York.* 1858.
BANTING & GARDINER. *London.* 1881.
BANYARD, Charles. *Upwell.* 1840. W.
BARBECK, C. G. *Philadelphia.* 1835.
BARBER—
——. *Melton Mowbray.* ca.1830.
Abraham. *London.* 1839-57.
Albert W. *Birmingham.* 1860-68.
Aquila. *Bristol.* (B. 1797-1830). 1810-42. W. & C.
CATTLE & NORTH. *York.* 1834. Later BARBER & NORTH, q.v.
Charles. *Nottingham.* 1828.
Edwin. b. 1816. *Farnworth, Staffs.* To *Bridgnorth* 1856.
E. *Bury.* 1858. W.
James. *Antrim.* Late 18c-early 19c. C.
James. *Philadelphia.* 1842.
James. *York.* Free 1814-d. 1857. Senior partner of Barber, Cattle & North.
John. *Bridgnorth.* 1861-68. Bro. of Edwin.
John. *Clun.* 1879.
John. *Macclesfield.* a.1762 to Charles Cock.
John. *Newark.* a.1768 to William Barnard, whom he succ. 1786-95. C.
John & Co. *Coventry.* 1868-80.
Jonas. *London.* (b.1652 *Otley*). CC. 1682-d.1698. C.
Jonas, sen. Nephew of Jonas of *London.* b.1688 *Skipton,* then at *Bowland Bridge.* c.1717-27, then Winster 1727-d.1764.
Jonas, jun. s. of Jonas, sen. *Winster, Westmorland.* b.c.1718-d.1802. C.
Jonathan. *Skipton.* b.ca.1770, mar.1790-1837. C. & W.
Mrs. Mary. *Bridgnorth.* 1870.
& NORTH. *York.* 1840-46. Later James Barber alone, q.v.
Robert. *Cawston.* 1875.
Samuel. *Glossop.* 1846-84.
Stephen Price. *Bristol.* 1830-65.
Thomas. *Dublin.* 1783-88. W.
Thomas. *Yarmouth.* 1830-36.
(James) & WHITWELL (William). *York.* 1815-22. Later BARBER, CATTLE & NORTH, q.v.
William. *Aylsham.* 1830-36.
William. *Eye.* 1830-39.
William. *Philadelphia.* c. 1846.
William junior. *Coventry.* 1868.
BARBORA, Charles. *Cardiff.* l.c. clock, date unknown.
BARBOULT, Adam. *Dublin.* 1707-d.1751. W.
Soloman. *Dublin.* ca.1720-d.1758. W.
BARCHAM, Asber. *Tonbridge.* 1826-45.
BARCLAY—
David. *Edinburgh.* 1860.
David. *Montrose.* 1830.
James. *London.* 1819-57. W.
John & Thomas. *London.* 1863-69.
Robert. *Paris, Ontario.* 1857-76.
BARD, Emil. *London.* 1881.
BARDEN—
J. *Boreham.* 1851. C.
W. *Birkenhead.* 1865. Chrons.
BARFOOT—
——. *Brighton?* Late 18c. l.c. clock (Prob. William, q.v.)
C. H. *Bournemouth.* 1859.
Cornelius Henry. *Huntingdon.* 1877.
John. *Atherstone.* 1860.
R. E. *Dorchester.* 1875.
Richard Edwin. *Castle Cary.* 1875-83.
Robert Thomas. *Brighton.* 1862-78.
William. *Brighton.* 1839-62.
William. *London.* 1875-81.
BARGEMAN, Morris. *London.* 1875.

BARGER, George. *Philadelphia.* 1844.
BARHAM—
George. *Hawkhurst.* 1835-74.
George. *Sevenoaks.* 1874.
BARK, William. *Kidderminster.* 1835-68.
BARKER—
B. B. *Croydon.* 1878.
Charles. *Dublin.* 1792-1806.
Daye. *Wigan.* b.1747-ca.1800. C.
George. *Colchester.* 1827-45. C. & W.
George. *Dedham.* 1839.
George. *London.* CC. 1653.
George Osborne. *St. Mary Cray.* 1874.
H. *Pewsey.* 1867.
Henry & Joseph. *Easingwold.* 1857-66.
J. *Chichester.* 1851-55.
James. *London.* 1839-44.
James. *Pewsey.* 1875.
James F. *Palmyra, NY, USA.* 1826.
Jonathan. *Worcester. Mass, USA.* d.1807 (of BARKER & TAYLOR).
Joseph. *Easingwold.* 1807-34. W.
Robert. *Chichester.* 1812-39. W.
Samuel Keer. *Framlingham.* b. 1801 s. of Thomas, qv. Working 1823-65.
(Jonathan) & TAYLOR (Samuel). *Worcester, Mass, USA.* pre-1807.
Thomas. *Barnsley.* 1754-ca.1810. C.
Thomas. *Wigan.* 1737. C.
W. *Pewsey.* 1848-59.
William. *Barnsley?* Late 18c. C.
William. *Birmingham.* 1880.
William. *Boston, USA.* 1800-25.
William. *Framlingham.* 1879.
William. *Wigan.* 1748-d.ca.1786. C. Also gunsmith.
William Keer. *Beccles.* 1830.
BARKLAY—
J. *Baltimore, USA.* 1817-24.
J. & S. *Baltimore, USA.* 1812-16.
BARKSHIRE, James. *Reading.* 1877.
BARLE, A. *London* (*Stepney*). Early 19c. C.
BARLING—
Benjamin. *London.* 1839.
Joseph. *Maidstone.* 1847-74.
BARLOW—
B(enj?). *Oldham.* ca.1750-(B. ca.1780). C.
Edmund. *Chester.* 1848. C.
Edward. *Oldham.* ca.1750-70. C.
James. *Eccles.* 1788. C.
Richard. *Manchester.* 1851.
Robert. *Liverpool.* 1848.
S. *Stockport.* 1857.
Thomas. *Worksop.* 1842-55.
Thomas. *Sale.* 1878.
BARNACLE, David. *Coventry.* 1880.
BARNARD—
Benjamin. *Canterbury.* 1832.
& CRABTREE. *London.* 1875.
Francis (Franz). *London.* 1839-51.
Franz John. *London.* 1857.
Frederick. *Sheerness.* 1826-55. W.
J. *Maidstone.* 1865.
James. *Erith.* 1851-66.
James. *Malmesbury.* 1875.
James. *Sittingbourne.* (B. late 18c-) 1823-47.
James. T. *London.* 1857.
James Thompson. *London.* 1881.
John. *Eton.* 1837.
R. J. *Middlesbrough.* 1898.
Samuel. *Cheltenham.* 1830.
Samuel. *Utica, NY, USA.* 1844.
Thomas. *London.* 1851.
Thomas (& Son). *Cirencester.* 1850-70 (-79).
W. *Brighton.* 1878.
BARNBY—
Bishop. *Hull.* 1834. W.
& RUST. *Hull.* 1858 & later. Succ. to Barnby & Son.
& Son. *Hull.* 1851. Succ. to Bishop Barnby.
BARNES—
Alphonso. *Bristol. USA.* 1804-77.
(Edward) & BACON (John). *Bristol, USA.* 1833-48. C.

BARNES—*continued.*
& BAILEY. *Middletown & Berlin, Conn, USA.* 1831.
Brothers. (Wallace, Carlyle F. & Harry). *Bristol, USA.* 1880-84.
Carlyle F. *Bristol, USA.* 1880-84 (s. of Wallace).
Charles. *Grays Thurrock.* 1839. W.
David. *Ipswich.* mar. 1820 -39 W.
E. *Malden.* ca.1790-1800. C.
Edward. *Alford.* 1828.
Frederick. *Framlingham.* 1839.
Frederick. *Peckham.* 1845.
G. F. *Tamworth.* 1876.
George. *Birmingham.* 1828.
George. *Lincoln.* 1835.
George. *Manchester.* 1848-51.
George & Co. *Gainsborough.* 1861-76.
Harry. *Bristol, USA.* s. of Wallace. 1880-84.
J. G. *London.* 1857-63.
James. *Gainsborough.* 1850.
(Alphonso) & JEROME (Andrew J.). *Bristol, USA.* 1833-37.
John. *London.* 1857-63.
John. *Philadelphia.* ca.1759.
(Thomas, jun.) & JOHNSON (William). *Bristol, USA.* 1819-23.
Joseph. *Alford.* 1835.
Matthew. *Northampton.* 1663-85.
Robert. *Liverpool.* 1751-61. W. (B. later at Birmingham).
Samuel. *Oundle.* 1777-ca.1800. W. & C.
and Sharp. *Northampton.* ca.1820.
Stephen. *Bristol, USA.* 1771-1810.
Thomas. *Gainsborough.* 1828-35.
Thomas. *London.* 1869-81.
Thomas. *Taunton.* 1846.
Thomas, jun. *Bristol, USA.* 1773-1855.
Thomas, jun. & Co. *Bristol, USA.* 1819-23.
Thomas Randall. *Wareham.* 1830.
W. *Chesterfield.* 1855.
Wallace. *Bristol, USA.* s. of Alphonso. 1827-93.
(Thomas, jun.) & WATERMAN (Samuel). *Bristol, USA.* 1811.
(Thomas, jun.) & WELCH (Elisha N.). *Bristol, USA.* 1832-34. C.
& WHITFORD. *Armagh.* 1843. C. (prob. error for BURNS & WHITEFORD).
William. *Sheffield.* 1862. W.
BARNET(T)—
——. *London.* 1839. C.
Abraham. *Hull.* 1858-67.
Barnard. *London.* 1828.
Barnett & Son. *London.* 1844.
D. & Son. *London.* 1851.
George. *London.* 1857-81.
Isaac. *London.* 1811. W.
Isaac. *London.* 1857-75.
Israel. *Ramsgate.* 1826-55.
J. *Prescott, Canada.* 1862.
J. *Spalding.* 1861.
John. *Edinburgh.* 1846-60.
John. *Lewes.* 1839.
Joseph. *Hull.* 1834-40. W. (may be Barret).
Josiah. *London.* 1828-32. Watch case-maker.
Montague. *London.* 1844.
S. H. & Co. *Coburg, Canada.* 1862.
Samuel T. (or Samuel G.). *Spalding.* 1868-76.
Thomas H. *Brigg.* 1868-76.
William. *Llangefni.* 1868-90.
BARNHART, Simon. *Kingston, O. USA.* ca.1850.
BARNISH—
Charles. *Rochdale.* b.1775-d.1812. C.
John. *Rochdale.* b.1760-d.1829. C.
William. *Rochdale.* b.1734-d.1776. C.
BARNITZ, A. E. *York, Pa, USA.* ca.1850.
BARNS, John. *West River, Md, USA.* 1756. C.
BARNSBERY, Thomas. *Battersea.* 1862-78. (or. Barnsby).
BARNSBY—
J. *Reckleford.* 1861.
James. *Maidstone.* 1865-74.
John. *Yeovil.* 1875.
Thomas. *Battersea.* 1862-78 (or Barnsbery).

BARNSDALE—
——. *Burnham.* Late 19c.
Alfred John. s. of William. *London,* later *Whitstable.* b.1861-d.1932. W.
John. *Burnham.* Later 18c. C.
John. *Norwich.* 1830-36.
John. *London.* 1863.
Thomas, sen. *Bale.* 1742-d.ca.1823. C.
Thomas, jun. Nephew of Thomas, sen. *Bale.* b.1779-mar.1803, then to *London.*
Stanley. s. of William James. *London.* b.1894-d.1973.
William. s. of John of *London.* b.1834. Succeeded 1869, d.1921. Chrons.
William James. s. of William. *London.* b.1860-d.1935.
BARON—
Louis. *Montreal.* 1801-33.
William Alfred. *London.* 1881.
BARR—
——. *Bolton.* ca.1815-20. C. (B.1790 seems incorrect.)
David. *Hamilton.* 1860. Clock cleaner.
David. *Woodstock, Canada.* 1851-62.
Fidele. *Newcastle-on-Tyne.* 1834-36. C.
Fidele & Thomas. *Edinburgh.* 1850.
George and William. *Dublin.* ca.1805. W.
Henry. *Prescot.* a.1751. W.
J. *Alford.* 1849-61.
John. *Port Glasgow, NY, USA.* ca.1840.
Joseph. *Alford.* 1850.
Mark. *Lanark.* 1836.
Mrs. Sophia. *Alford.* 1868-76.
Thomas. *Lanark.* b.1819-d.1890.
W. *Stony Stratford.* 1869.
Walter. *Londonderry.* ca.1773-85. C.
William. *Glasgow.* 1860.
William. *Hamilton.* 1808-37. W.
William. *Port Glasgow.* ca.1790.
William Brookes. *Cannock.* 1876.
BARRACLOUGH—
James. *Thornton, nr. Bradford.* 1871.
John I. *Haworth.* b.ca.1774-d.1835. C.
John II. s. of John I. *Haworth.* b.1802. To *Thornton, nr. Bradford.* 1837. d.1880. C.
Thomas. s. of John I. *Haworth.* b.1818-1866.
Z. *Paris* (imported clocks sold by Z. Barraclough of *Haworth,* q.v.).
Zerrubbabel I. *Haworth.* s. of John I. 1838-d.1878.
Zerrubbabel II. *Haworth.* s. of John II. b.1824. To *Leeds* 1887.
BARRAGANT, Peter. *Philadelphia.* ca.1820
BARRATT—
George. *Norwich.* 1836.
James. *Bradford.* 1853.
Peter. *St. Day.* 1844-d.1845. C. & W.
BARRAUD—
Hilton P. *London.* 1851-69.
and LUND (& Sons). *London.* 1839-63 (69-81).
BARRELL—
Colborn. *Boston, USA.* 1772.
George. *Woolwich.* 1838-74.
Louis Howe. *London.* 1881.
BARRETT—
——. *London.* Regency wall clock.
Alexander. *Whitchurch.* 1879.
Daniel. *Chatham.* 1792-1800.
E. *Blandford.* 1848-55.
Edward. *Blandford.* (B. late 18c) 1824.
Edward. *Lewes.* 1656. Lantern cl.
Harry. *London.* 1875-81.
Henry. *Lewes.* Took Henry Baker as a. in 1726. C.
Henry. *London.* (B. a.1684, CC. 1692)-97.
Henry. *Pembroke Dock.* 1850-68.
Henry William. *London.* (B. CC. 1802-40)-1844.
James. *Manchester.* 1678. W.
James. *Norwich.* 1820. W.
John. *Canterbury.* s. of Thomas, sen. Free 1705. Clocksmith.
John. *London.* 1869-81.
John. *Pembroke Dock.* 1830-40.
John. *Portsmouth.* 1830-48.
John. *Shipley.* 1871.
John & Son. *Pembroke Dock.* 1844.
& LADD. *Amhurst, Nova Scotia.* 1835-40. From

BARRETT—*continued.*
Yarmouth.
Mary. *Blandford.* 1830.
Moses. *Yarmouth, Nova Scotia.* 1830-70, later at *Amhurst.*
Mrs. Emma Alice. *London.* 1875 (?widow of William).
Richard. *Brantford, Canada.* 1853-62.
& SHERWOOD. *San Francisco, Cal, USA.* ca.1850-88.
T. *Little Waltham, Chelmsford.* ca.1859.
Thomas. *Billericay.* 1804-39. W.
Thomas. *Colchester.* b.1772. a.1786.
Thomas. *Eastby nr. Skipton.* 1729-32. C.
Thomas. *Lewes.* 1713-36. C.
Thomas. *Lewes.* Third quarter 18c. C.
Thomas senior. *Canterbury.* 1660-92. C.
Thomas junior. *Canterbury.* Free 1726. C.
William. *Ashford.* 1638. T.C.
William. *Halifax.* 1720.
William Henry. *London.* 1832. (Prob. same as Henry William, QV).
William Henry. *Pembroke Dock.* 1875.
William (& Son). *London.* 1844-63 (-69).
BARRIE—
——. 1914 and later, product of Pequegnat Clock Co.
Andrew. *Edinburgh.* 1840-d.1907.
John Lyall. *Coatbridge.* 1860.
BARRIGER, John. *Hull.* d.1801. C.
BARRINGTON—
Isaac. *Dublin.* 1805-30. W.
Isaac. *Liverpool.* 1834.
Joseph. 1792 *Dumfries, Va, USA,* then 1826 *Salisbury, NC, USA,* then 1832-39 *Tarboro, NC, USA.*
BARRISCALE, Richard. *Blackheath.* 1874.
BARRITT—
David T. *Macclesfield.* 1878.
George. *Witney.* a.1800.
BARRON—
Charles. *Queenstown.* 1858.
& GREY. *Aberdeen.* 1846.
James. *Prescot.* 1823-25. W.
John. *Leeds.* 1850-71.
John. *London.* Mid 18c.
John. *Queenstown.* 1858.
John & Son. *Aberdeen.* 1836.
Thomas. *Liverpool.* a.1710-24. W.
William. *London.* ca.1780. C.
BARROW(S)—
Benjamin Francis. *London.* 1851-81.
David Ratcliffe. *Prescot.* 1819-26. W.
Edward. *Warrington.* a.1728. Watch tool maker.
G. *Cheltenham.* 1879.
James M. *Tolland, Conn, USA.* 1832.
John. *Bradford.* 1866-71.
John. *Prescot.* 1823-26. W.
John. *Ross.* 1830-63.
Robert. *Farnworth.* 1848-51.
Thomas. Place unknown. Early 19c. W.
Thomas. *Farnworth.* 1818. W.
William. *Bradford.* 1834-53.
William. *Dublin.* ca.1805. W.
William. *Halewood, Lancs.* 1713. Watch spring maker.
BARRS, Samuel. *Windsor.* 1830-54.
BARRY—
Edward. *Micheldean.* 1856-79.
J. *Chippenham.* 1848.
James. *London.* 1863-75.
James Trotter. *Cardiff.* 1844-75. W.
John. *Bolton.* 1824.
John Charles. *Neath.* 1871-75.
M. *Geneva.* ca.1880. W.
& Sons. *Cardiff.* 1877.
Thomas. *Bolton.* 1822-29.
Thomas. *Coventry.* 1880.
Thomas. *Ormskirk.* 1786. W. & C.
BARTENS & RICE. *New York.* 1878.
BARTER—
George. *London.* 1869.
George Gannaway. *London.* 1875-81.
Joseph. *Dorchester.* 1855-75.
Thomas. *Fordingbridge.* 1830.
Thomas W. *Fordingbridge.* 1878.
BARTHELEMY, Lewis Charles. *London.* 1857.

BARTHOLOMEW—
——. *Maidstone.* 1582. Clock keeper.
(William G.), BROWN (Jonathan C.) & Co. *Bristol, USA.* 1833.
E(Eli) & G(George W.). *Bristol, USA.* 1829-33. C.
Eli & Co. *Bristol, USA.* pre-1828-33. C.
George W. *Bristol, USA.* 1805-97.
Harry Shelton. s. of George W. *Bristol, USA.* 1832-1902.
Josiah. *London.* (B. 1799-1842) 1844.
Josiah. *Sherborne.* (s. of Edward). b. 1729. d. 1835. W.
BARTLE—
G. *Wellingborough.* Early 19c. C.
George, sen. *Brigg.* 1828-76.
George, jun. *Brigg.* 1876.
George. *Caistor.* 1849-50.
J. *Caistor.* 1861.
J. *Grimsby.* 1868.
James. *Leicester.* 1876.
John. *Keighley.* 1871.
William. *Kirton-in-Lindsey.* 1849-61.
William. *Market Rasen.* 1868-76.
BARTLETT—
Alfred. *London.* 1844.
B. B. *Augusta, Canada.* pre-1860 Clock dealer.
F. J. *Maidstone.* 1865. W.
Horatio (& George). *London.* 1828 (-32). Watch case makers.
James. *London.* 1839.
Samuel. *Maidstone.* 1821-55. W.
& Son. *Maidstone.* 1851.
Walter Charles. *Cirencester.* 1879.
BARTLEY—
Andrew. *Bristol.* (B. 1825) 1830-56.
Mark. *Bristol.* (B. 1816) 1830-50.
Mrs. S. N. *Bristol.* 1856.
Thomas. *Denbigh.* 1781-1811. C.
Thomas. *Pwllheli.* ca.1800-pre-1846. C.
BARTLIFF—
Charles. *York.* 1841, then *Malton.* 1851-58. W.
George. *York.* b.1777. Successor to great uncle, Thomas Kidd, in 1823. d.1846 at *Malton.*
Robert. *York.* b.1784, free 1807. Moved to *Malton* by 1823-d.1855.
BARTMANN, George. *London.* 1851-57.
BARTON—
Benjamin. *Aberford.* b.1771-1799.
Benjamin. *Alexandria, Va, USA.* ca.1830.
D. *Leicester.* 1864-76.
& ESPLIN. *Wigan.* 1831-38.
Fanny. *Wicklow.* 1858.
George. *London.* 1863.
J. *Walsall.* 1868.
J. *Brampton, Canada.* 1853.
J. *Newark.* 1849-55.
James. *Birmingham.* (B. 1804)-1816. W.
James. *Dewsbury.* 1837-53. C. & W.
James I. *Chester.* 1696, then *Ormskirk.* mar. 1698 -d. 1718. C.
James II. *Ormskirk.* Grandson of James I.1718. C.
James. *Selby.* 1834.
John. *Dublin.* 1715.
John. *Henley on Thames.* mar.1685. C. & W.
John. *Salem, Mass, USA.* 1846-48. *Newburyport.* 1849-50.
Joseph. *Liverpool.* 1834.
Joseph. *Stockbridge, Mass, USA.* 1764-1804. *Utica, USA.* 1804-32.
O. G. *Fort Edward, NY, USA.* ca.1830 W.
R. H. *Weybridge.* 1878.
Richard. *Coventry.* 1860.
Richard. *Walton, Lancs.* a.1722. C.
Rix. *Brinton.* 1836.
Robert. *Omagh.* 1840.
Robert H. *Toronto.* 1843-52.
Robert Henry. *Walton, Surrey.* 1878.
Stephen. *London.* 1875.
T. *North Collingham, Notts.* 1864.
William. *Dungannon.* 1843-59. C.
William. *Wigan.* 1834.
William Cleveland. *Salem, Mass, USA.* 1813.
BARTOW, John. *Haverhill, Mass, USA.* 1853.

BARTRAM—
Austin. *London.* 1811. W.
Simon. *London.* 1630-believed d.1667.
BARTSH, August J. H. *St. John, New Brunswick, Canada.* 1857.
BARWICK—
Bros. *Keighley.* 1871.
Bros. *Skipton.* 1871.
Samuel. *Boston.* 1876.
BARWISE—
——. *London.* ca.1840. C.
John. *London.* 1780. W. (B. 1790-d.1842).
John. *London.* 1832-81.
Weston & John. *London.* 1828.
BASCHET, ——. *Paris.* 1855. Carr. clock.
BASCOM (Asa) & NORTH (Noah). *Torrington, Conn, USA.* 1813.
BASFORD, Daniel. *Wem.* (B. 1795) 1828-35.
BASHAM, David. *Boston.* 1876.
BASIL, ——. From *Baltimore* to *Albany, NY, USA.* ca.1773.
BASKELL, Martin & Co. *Cheltenham.* mid-19c.
BASKERVILLE, George Lee. *Leek.* 1868-76.
BASNET(T)—
James. *Liverpool.* (B. 1825) 1834-51.
Thomas. *Prescot.* 1816-34. W.
BASS—
G. & Son. *Northampton.* 1877.
George. *Northampton.* 1841-69.
BASSETT—
Benjamin. *St. Germans.* 1847.
J. & W. H. *Cortland, NY, USA.* ca.1815, then *Albany.* 1820.
Jean Jacques Louis. *York.* mar.1767-1772. W.
Joseph. *Chipping Norton.* ca.1780. C.
Joseph. *Hereford.* 1870.
Nehemiah. *Albany, NY, USA.* 1795-1819.
BASSNETT—
& Co. *London.* 1881.
Thomas. *Coventry.* 1880.
BASSOLD, Edward. *London.* 1881.
BASTIEN, ——. *Paris.* Early 18c.
BASTUBAK, Johann. *Kaarlela.* ca.1740. C.
BATCHELDER—
Andrew. *Danvers, Mass, USA.* 1772, mar.1802-45.
Ezra. bro. of Andrew. *Danvers, Mass, USA.* 1789-1840.
BATCHELOR—
& BENSEL. *New York.* 1848-54.
Charles. *New York.* 1854.
Henry F. *Eye.* 1839-58.
N. *New York.* 1846.
Samuel. *Harrogate.* 1822-37.
William. *Battle.* a.1731 to Obadiah Body. C.
BATCOCK, Francis. *Penrice, Gower.* Late 18c.
BATE—
Anthony, *Dublin.* 1786-88.
& BIRD. *Dublin.* 1794. W. & jeweller.
Isaac. *Whitby.* 1851-66.
James Nurthall. *Birmingham.* 1880.
William. *London.* 1881.
BATEMAN—
Bros. *London.* 1863-75.
H. *Dublin.* 1802-05.
John. *Kendal.* b.1829-51. W.
John. *Mold.* 1856.
John (& Sons). *London.* 1844-51(-57). Clock case maker (dial clocks).
Sarah. *London.* 1832. Clock case maker.
Teresa. *London.* (B. 1820-25) 1828-32.
BATES—
Alexander. *Dublin.* a.1753. W.
Alfred. *Coventry.* 1880.
Daniel. *Coventry.* 1880.
E. *Lyndhurst.* 1859.
Edmund. *Downton.* 1867-75.
Edward. *Cuckfield.* (B. 1791-) 1828-39. W.
Edward. *Kingsland* (?*USA*). 1858. C.
Edward (& Son). *Kettering.* pre-1811 (-24).
F. *Market Harborough.* ca.1820. C. (may be Thomas B.).
G. J. *Stalham.* 1858-65.

BATES—*continued.*
George. *Kettering.* 1830.
Henry. *Coventry.* 1880.
Henry. *Ramsgate.* 1874.
Isaac. *Halifax.* 1840.
J. *Brighton.* 1862.
J. *Cuckfield.* 1851-62.
James. *Coventry.* 1880.
James C. *Haverhill, Mass, USA.* 1879-89.
John. *Hythe.* 1847.
John. *Kettering.* mar.1756-1800.
John. *Kettering.* 1830-54.
John. *Ramsgate.* 1849.
John. *Uppingham.* 1846.
Joseph. *Coventry.* 1850-54. Watch cases.
Joseph. *Dundas, Canada.* 1853.
Joshua. *Huddersfield.* 1814-37. W.
Richard. *London.* 1806. W.
Robert George. *London.* 1881.
Thomas. *Exeter.* 1785.
Thomas. *Kettering.* b.1733-1777. W.
Thomas. *London.* 1881.
Thomas. *Market Harborough.* (B. 1783-ca.1800) 1828-35.
William. *Dublin.* 1728-44.
William Henry. *Cuckfield.* 1870-78.
BATH—
Barten. *New York.* 1850.
J. Long. *Bath.* 1866-83.
BATKIN, William & Son. *Birmingham.* 1803. Dial-maker.
BATT—
Frank James. *Bath.* 1883.
John. *Petersfield.* 19c.
BATTELL—
George E. *Newburyport, Mass, USA.* 1860.
BATTEN, John. *London.* (B. CC. 1668)-d.1686.
BATTERBEE, John. *Llanfair Caereinion.* 1868-90.
BATTERS, Richard. *Middlesbrough.* 1866. W.
BATTERSON—
James (B. a.1696. *London*). From *London* via *Philadelphia* to *Boston* (*USA*) 1707-27. Also *New York* and *Charleston, SC.*
John. *Annapolis, Md, USA.* 1723. W.
BATTIE, James. *Sheffield.* 1787 (B. 1797).
BATTING—
Joseph. *Philadelphia.* 1850.
William. *London.* 1839. Watch cases.
BATTINSON, John. *Colne.* 1807-12, then *Burnley.* 1822-25. C.
BATTLES, A. T. *Utica, NY, USA.* 1795-1846.
BATTS, George Thomas. *London.* 1828. Watch cases.
BATTY—
Anthony. *Worsborough.* mar.1735. C. (may have worked at *Wakefield*).
Benjamin. *Hamilton, Ontario.* 1861-78.
James. *Halifax.* 1770. C.
John. *Halifax.* 1770. C.
John. *Hyde.* 1857-78.
John. *Stockport.* 1848.
Joseph. *Halifax.* a.1752-d.1801. C.
W. & Sons. *Manchester.* Late 19c.
William. *Manchester.* 1851.
BAUER, John R. (or N.). *New York.* 1832.
BAULLER & Fils. Place unknown. 1824. C.
BAUM—
& Co. *Coventry.* 1868.
W. *Leamington.* 1854.
BAUMAN—
Eugene. *Pontypridd* and *Tonypandy.* 1875-87. C.
George. *Sugar Creek, O, USA.* 1815.
BAUMANN, C. H. *Nottingham.* 1876.
BAUME—
Bros. *London.* 1851.
Celestin. *London.* 1875. Also Switzerland (successor to Baume & Lezard).
& Co. *London.* 1881. (Successor to Celestin B.). "estd. 1834".
& LEZARD. *London.* 1857-69 (successor to Baume Bros.).

BAUMEYER—
& DE MEYER. *London.* 1881. (Successor to Matthew Baumeyer).
Matthew. *London.* 1869-75.
& MAYER. *London.* 1857-63.
BAUMGART, Charles. *London.* 1839-81.
BAUR, James. *London.* 1851-81.
BAURLE—
Alexander. *London.* 1863-81.
L. *Rochester.* 1866.
L. & F. *Chatham.* 1874.
BAURLEY, Philip. *Scarborough.* 1858. C.
BAUSCH, F. *Newcastle on Tyne.* 1852.
BAVEUX, Alfred & Louis. *St. Nicolas d'Aliermont.* ca.1857-89. Carr. clocks.
BAWD & DOTTER. *New York.* Post-1800.
BAWDEN—
A. *Bridgwater.* 1866.
Charles. *Briton Ferry.* 1868.
Edward. *London.* 1881.
W. *Weston super Mare.* 1861-83.
William. *Bridgwater.* 1861-75.
BAWDYSON, Alan. *London.* 1540-68. T.C.
BAX—
Edward. *Chichester.* 18c. W.
John. *Brighton.* 1870.
Mrs. Emily. *Brighton.* 1878 (?wid. of John).
BAXANDALE, Squire. Born 1792. *Hull* till 1820, then *Sculcoates.* 1822.
BAXTER—
Elizabeth. No town. l.c. clock dated 1789.
George. *Nottingham.* 1876.
Isaac. *Sheffield.* 1862. W.
John. *Dunkeld.* 1836.
John. *Edinburgh.* 1797.
John. *Ibstock.* 1864-76.
John. *London.* 1828-32. Watch case maker.
John. *Lurgan.* 1824. C.
John. *Monaghan.* 1824-68.
John. *Perth.* 1841-43.
John. (b. 1781) *Shrewsbury* 1812 -d. 1835. W.
John Vincent. *Bristol.* Bankrupt 1763 then to Paris. W.
Manister. *St. Neots.* Early to mid-18c. C.
Robert. *Redditch.* 1872-76.
T. *Bury St. Edmunds.* 1846-79.
William. *London.* 1662.
William. *Quebec.* 1844-67.
BAY(C)LIFFE, John. *Halifax.* 1765-70.
BAYER, Matthias. *London.* 1881.
BAYFIELD, Joseph. *Norwich.* 1858-75.
BAYFORD, George. *London.* b.c.1667, a.1682, mar.1692. W.
BAYLAND & UPJOHN. *London.* Late 18c. W.
BAYLE (?), Richard. *London.* 1660.
BAYLES, Charles. *London.* 1775. C.
BAYLEY—
Bros. *Peckham.* 1878.
Charles. *London.* 1851-63.
F. *Uttoxeter.* ca.1760. C.
Frederick George. *London.* 1869.
J. *Bristol.* ca.1800. C.
John. *Boston, USA.* 1803-16.
John. *Bridgwater.* ca.1730. C.
John. *Harrow on the Hill.* 1725. C. in *Warwick Lane, Hatch End.*
John. *Romney.* 1795. W.
John, jun. *Hingham, Mass, USA.* s. of John. 1787-1883.
Joseph. *Newcastle-under-Lyme.* 1828-35.
& STREET. *Bridgwater.* ca.1760. C. Also bellfounders.
T. *Newcastle-on-Tyne.* 1850.
Thomas. *Bridgwater.* ca.1750-68. C.
Thomas. *Bristol.* a.1724-31.
Thomas. *Burton (Westmorland).* pre-1764. W. (B. ca.1790).
W. & A. *Peckham.* 1878.
William. *Maidstone.* 1784.
William. *Newcastle-under-Lyme.* 1850-68.
BAYLIFF, John. *Manchester.* 1816. C.
BAYLIS(S)—
H. J. *Shepton Mallet.* 1883.
J. *Birmingham.* 1854-60.

BAYLIS(S)—continued.
J. *Tewkesbury.* ca.1760-80. C.
John Potter. *Stockport.* 1878.
William. *Bristol.* 1840-79.
BAYLY—
John Phillip. *London.* 1857-63.
Richard senior. *Ashford.* 1752-74. W.
Robert. *Wells.* 1683 rep'd ch.cl.
T. *Weybridge.* 1878.
BAYNE, William. *Alston.* 1828-48.
BAYNES—
B. B. *Lowell, Mass, USA.* 1835.
R. *Bolton.* 1858. W.
Richard. *Knaresborough.* 1866. W.
Robert. *London.* 1832.
BAYNHAM, Benjamin. *Limerick.* 1820-24.
BAYNTON, F. *Reading.* 1837.
BAYNTUN—
Andrew. *Gosport.* 1859-78.
F. *Southsea.* 1848-67.
Harry. *Landport.* 1867-78.
BEACH—
& BYINGTON. *Plymouth, Conn, USA.* ca.1840.
Charles. *Bristol, USA.* 1876-94.
& HUBBELL. *Bristol, USA.* 1859-63.
HUBBELL & HENDRICK. *Bristol, USA.* 1853.
J. *Birmingham.* 1850-80. Dialmaker.
Miles. *Litchfield, Conn, USA.* 1743-1828 (of BEACH & SANFORD).
(Miles) & SANFORD (Isaac). *Litchfield, Conn, USA.* 1785. Then moved to *Hartford* 1785-88, when dissolved partnership.
(Miles) & Son (John). *Hartford, USA.* 1813-28.
(Miles) & WARD (James). *Hartford, USA.* 1790-97.
William. *?Hartford, USA.* 1834.
BEAL(E)—
Charles Wilson. *Salisbury.* 1842.
Henry. *Fairford.* 1863-79.
John. *Fairford.* 1830-50.
John. *Oundle.* 1830-54.
Michael. *Sheffield.* 1871.
Michael & Son. *Sheffield.* 1862.
Richard. *Sheerness.* 1847-55.
Samuel. *Sheffield.* 1834-37. C.
Thomas. *Oundle.* 1864-77.
Walter. *Sheffield.* 1862. C. & W.
BEALL—
Joseph. *St. Ives (Hunts).* 1830.
Mary. *St. Ives (Hunts).* 1839 (wid. of Joseph?).
Thomas. *Lindsay, Ontario.* 1828-1912.
BEALS—
J. J. *Boston, USA.* 1847-74.
J. J. & Co. *Boston, USA.* 1848-53.
J. J. & Son. *Boston, USA.* 1854.
J. J. & W. *Boston, USA.* 1848.
William. *Boston, USA.* 1838-54.
William & J. J. *Boston, USA.* 1846.
BEAMONT—
Charles John. *Greenwich.* 1847-58.
John. *Harringworth.* mar.1755. C.
BEAN—
Edward. *Kingston (Kent).* (B. ante-1792. W.). 1800. W.
John. *Quebec.* 1819-26.
BEAR—
Robert. *Hertford.* 1770-1855.
Robert. *London.* 1828. W.
BEARD—
Evan C. *Louisville, KY, USA.* With E. AYER & CO. 1816-31, then alone till 1875.
J. *Tonbridge.* ca.1800. C.(?).
Philip. *Devonport.* ca.1820. C.
Philip. *Needham Market.* 1830-58.
William Henry. *London.* 1869-81.
BEARDMORE—
J. *Southampton.* 1848.
Joseph. *Bristol.* 1842.
BEARDSLEY—
Charles. *Nottingham.* 1864-76.
N. *Nottingham.* 1855.
H. P. *Curonna, Mich, Canada.* 1870.
Joseph. H. *Nottingham.* 1864-77.

BEARN, William (& Son). *Wellingborough.* 1824-30 (41-54).
B(E)ASELEY, Thomas. *London.* b.c.1662. mar.1684-1711. Watch case maker.
BEATH—
    & ELLERY. *Boston, USA.* ca.1810.
    John. *Boston, USA.* 1805.
BEATTY—
    Albert L. *Philadelphia.* 1833.
    Benjamin. *Hamilton.* See BATTY, Benjamin.
    Charles A. *Georgetown, USA.* 1812. C.
    D. *Quebec.* 1897.
    George. *Harrisburg, Pa, USA.* 1808-50. C.
BEATY—
    T. K. *Guelph, Canada.* 1857.
    Thomas. *Toronto.* 1856.
BEAUCHAMP—
    Alderic. *Montreal.* 1866-77.
    F. X. *Montreal.* 1872.
    John. *Croydon.* a.1735 to Thomas Bugden. W.
BEAUDRY—
    Armand. *Quebec.* 1881.
    & DUFRESNE. *Montreal.* 1871-77.
    E. E. *Montreal.* 1862-67.
    Leonard. *Quebec.* 1871.
    Narcisse. *Montreal.* 1862-80.
BEAUMONT—
    Henry. *Ilkeston.* 1876-95
    Joseph. *Flanshaw nr. Wakefield.* ca.1710-20. C.
BEAUMONT—
    Joseph. *Howden.* mar.1737. d.1750.
    Joseph. *Huddersfield.* 1871.
BEAUTY, 1914 and later product of Pequegnat Clock Co.
BEAUVAIS, Paul. *London.* (B. ante-1704-ca.1730). 1730. W. at St. Martins in the Fields.
BEAVEN—
    John Richard. *Woolwich.* 1838-51.
    William. *Cradley Heath.* 1868-76.
BEAVER—
    1914 and later product of Pequegnat Clock Co.
    ——. *Wakefield.* pre-1828. C.
    Louis. *Manchester.* 1848-51.
    William. *Wigan.* 1822.
BEAVERS, T. *Halifax.* 1824.
BEAVINGTON, Charles. *Stourbridge.* 1828-42.
BEAVIS, George. *London.* (B. CC. 1687)-97.
BEAZLEY, Frederick William. *London.* 1881.
BEB(B)INGTON—
    Alfred. *Swansea.* 1887-99.
    Creasey. *Burslem.* 1835.
BECHEL, Charles. *Bethlehem, Pa, USA.* c.1850. T.C.
BECHLER, Hans. *Magdeburg.* 1590. TC.
BECHTEL, Henry. *Philadelphia.* 1817.
BECHTLER—
    C. & Son. *Spartanburg, SC, USA.* ca.1857.
    Christopher, sen. *New York.* 1829. *Philadelphia.* 1830. *Rutherford, NC.* 1831. d.1842.
    Christopher, jun. from *Europe. New York.* 1829. *Rutherford, NC.* 1831. *Spartanburg, SC.* 1842-ca.1857.
BECK—
    Charles Edwin. *Brighton.* 1878.
    Constantine. *London.* 1863-75.
    Henry. *Philadelphia.* 1837-39.
    I. *Henley-on-Thames.* Early 19c. W. & C.
    J. *Andover.* 1859.
    J. *Stockbridge.* 1859.
    Jacob. *Hanover, Pa, USA.* ca.1820. C.
    John & Edward. *Andover.* 1839-48.
    Richard. *Bradford.* 1866-71.
    Thomas. From *Philadelphia.* ca.1784. To *Trenton, NJ, USA.*
    W. *Battersea.* 1866. Clock cleaner.
    W. H. *Andover.* 1867.
BECKER—
    Charles. *Cleveland, O, USA.* ca.1830.
    Edward. *London.* 1881.
    John. *London.* 1863.
BECKERLEGGE, John. *Penzance.* 1864. W.
BECKET(T)—
    Gilbert. *London.* (B. 1678)-98. May have left trade then.
    J. F. *Lowestoft.* 1879.

BECKET(T)—*continued.*
    James. *Dover.* Early 18c.
    John M. *Blairgowrie.* 1860.
    Robert Anderson. *Montreal.* 1898.
BECKMAN, Edwin. *London.* 1851.
BECKNER, Ignaz. *Halifax, Nova Scotia.* 1877.
BECKWITH—
    Dana. *Bristol, USA.* 1818-37.
    John. *Richmond, Yorks.* b.1766-d.1834.
    Lewis Myers. *Hull.* 1838-40. W.
    William. *Barking.* 1828.
    William. *Dartford.* 1832-39 (or BECKWORTH).
BEDARD—
    Joseph. *Quebec.* 1860-65.
    Simon. *Quebec.* 1855-63.
BEDDOWS, James Henry. *Hednesford.* 1876.
BEDELL—
    Benjamin. *Hull.* 1858.
    Peter. *Hull.* 1813-51. C. & W.
BEDFORD—
    1914 and later, product of Pequegnat Clock Co.
    E. *Batavia, NY, USA.* 1816.
    Henry E. *Halifax.* 1871.
    Isaac. *Dublin.* 1788-1826.
    William. *Hingham.* 1830-58.
    William. *Hook Norton.* 1842. C. & W.
    William. *Norwich.* 1840. W.
    William Thomas. *Halifax.* 1866-71.
BEDWARD—
    T. W. *Richmond, Surrey.* 1851-55. (B. one such a.1809).
    T. W. *East Sheen.* 1862-66.
    Thomas. *Richmond, Surrey.* 1839.
BEEBE, William. *New York.* 1833.
BEECH—
    John. *Newcastle-under-Lyme.* 1734-ca.1740. C.
    Walter & Co. *Coventry.* 1880.
BEECROFT, Edward. *Leeds.* 1837-71.
BEEDLE, James Francis. *Woolwich.* 1874.
BEEFORTH—
    John. *York.* b.1656. Free 1680. W.
    Robert. *York.* 1658. Clocksmith and locksmith.
BEEGELAAR, Jan. *Amsterdam.* ca.1750. C.
BEELS, Robert. *Kings Lynn.* (Error for BEETS, q.v.)
BEEMER, & Co. *Hamilton, Canada.* 1852-65.
BEER—
    Alfred. *Versailles, Ind, USA.* ca.1870.
    Robert. *Olean, Versailles, Ind, USA.* 1825.
BEESLEY—
    George & Richard. *Liverpool.* (B. 1825) 1834.
    James. *Farnworth.* 1799-1812. W.
    James. *Lancaster.* 1869.
    James. *Prescot.* 1819-25. W.
    John. *Farnworth.* 1809. W.
    John. *Prescot.* 1778. W.
    John. *Prescot.* 1823. W.
    Joseph. *Farnworth.* 1782. W.
    Joseph. *Farnworth.* 1806. W.
    Joseph. *Farnworth.* 1810. W.
    Richard & George. *Liverpool.* (B. 1825) 1828-48. W.
    Thomas. *Farnworth.* 1796. W.
    William. *Farnworth.* 1810. W.
    William. *Prescot.* 1817-26. W.
BEESTON—
    Bros. *Southampton.* 1859.
    J. *Southampton.* 1867.
    James. *Lymington.* 1878.
    John. *Burslem.* 1876.
BEETH, Robert. *Dublin.* 1798.
BEETLES, John & William. *London.* 1839.
BEETS, Robert. *Kings Lynn.* d.1761.
BEGG—
    J. G. *Ingersoll, Canada.* 1861.
    J. G. *London, Canada.* 1862.
    John. *Belfast.* 1793. W. (One such Dublin 1786.)
    John. *Portsea.* 1830-39. Also chrons.
BEGGS, James. *Rongford nr. Templepatrick.* ca.1825-d.1845. C.
BEGLEY, Alexander. *Coleraine.* 1785.
BEGUIN, ——. *Paris.* ca.1870. Carriage clocks.
BEHA—
    & Co. *Norwich.* 1858.

BEHA—*continued.*
Isidor. *Chester.* 1848. German clocks.
J. *Nottingham.* 1864.
Leander. *Bridgend.* 1868-99. W.
LICKER & SCHWERER. *Norwich.* 1875.
& SCHWERER. *Norwich.* 1865.
Thadeus. *Nottingham.* 1828.
BEHENNA, Richard. *Penryn.* ca.1810. C.
BEHMBER, Augustus. *London.* 1875.
BEHN, M. H. *New York.* ca.1860. Self-illuminating
alarm clock.
BEHRENS, Jacob. *Nottingham.* 1835-49.
BEIDT, Julius. *Philadelphia.* 1848.
BEIGEL, Henry. *Philadelphia.* 1816.
BEILBY & HAWTHORN. *Newcastle-on-Tyne.* (B.
1790-1802. W). ca.1801-11. Clock dial makers.
BEILBY, Ralph. *Newcastle-on-Tyne.* (B. 1767) 1782.
BELAND—
Francis. *London.* 1863.
George. *London.* 1839.
BELCHER—
Albert. *Wantage.* 1837-54.
Benjamin. *London.* 1863.
J. *Farringdon.* 1854.
J. *Surbiton.* 1862.
John. *Oxford.* a.1723. W (may be J. Belchier below).
BELCHIER, John. *Abingdon.* 1731. W (may be J.
Belcher above).
BELFIELD, Arthur John. *Bristol.* 1879.
BELK, William. *Philadelphia.* (? from London where one
such 1790) 1797-1800.
BELKNAP, Ebenezer. *Boston, USA.* 1809-30.
BELL—
———. *Mevagissey.* ca.1770. W.
A. *Whalley.* 1858. C.
Archibald. *Dingleton.* 1860.
Benjamin. *London.* (B. a.1649-98.) a.1649. CC. 1657-
d.1691. bro. of John. W.
Benjamin. *Uttoxeter.* 1842-50.
Bros. *Doncaster.* 1871.
& DAMS. *Uttoxeter.* 1868-76.
David. *Glasgow.* 1860.
Dawson. *Strabane* and *Omagh, Co. Tyrone.* 1820-59.
Edward. *Ashbourne.* 1828-35.
Edward (& Son). *Uttoxeter.* (B. 1795-)1828(-35).
Francis. *Macclesfield.* 1828-34.
Frank Godfrey. *Rochester.* 1839.
Frederick & Thomas. *Rochester.* 1838.
G. *Spalding.* 1861.
J. *Ashbourne.* 1849.
J. *Bath.* 1856-61.
J. *Burnley.* 1851-58. W.
J. *Chatham.* 1838.
J. *Garstang.* ca.1750. C.
James. *Blyth, Notts.* 1828.
James. *Doncaster.* 1834-62.
James. *Edinburgh.* 1850.
John. *Bath.* 1866-83.
John. *Cheshunt.* 1828.
John. *Cupar.* 1860.
John. *Hexham.* (B. ca.1790)-1827-34. W.
John. *Jedburgh.* 1837.
John. *Kirkby Lonsdale.* ca.1828. C.
John. *Kirkby Stephen.* 1816.
John. *Lancaster.* 1825. C.
John. *Leyburn.* 1823.
John. *London.* CC. 1667. math. inst. maker.
John. *London.* (B. a.1671, CC. 1685)-91. bro. of
Benjamin.
John. *London.* 1828-75. Musical clocks.
John. *Lurgan.* 1820-24. C.
John. *Moy Charlemont (Co. Armagh).* 1854. C.
John. *Richmond, Yorks.* 1866.
John. *South Shields.* 1851-56.
John. *Strabane.* 1843-46.
John. *Stretford (Manchester).* 1848.
John. *York.* a.1819. Free 1830-1866.
Joseph. *Doncaster.* b.1757. a. 1771 to John Jullion &
Son until 1778. Worked for Mudge
then back to *Doncaster.* d. 1834.
Joseph. *Norwich.* 1830-46.
Joseph Widdowson. *Leeds.* 1866-71. W.
L. *Bridgend.* ca.1850. C.
Mrs. Susan. *Toronto.* 1859-65.

BELL—*continued.*
Peter. *Garstang.* (B. ca.1770-95) 1824.
R. (R.). *Gateshead.* ca.1825-30. C.
Robert. *Douglas, IoM.* 1860.
Robert. *Norwich.* 1846-75.
Thomas. *Hexham.* ca.1830-54 C. & W.
Thomas. *London.* 1844-63.
Thomas Finch. *Leeds.* 1834-71.
W. *Lurgan.* 1824 (?error for John).
William. (No place). br. clock ca.1800.
William. *Coleraine.* ca.1780. W.
William. *Lancaster.* b.1836-69. W.
William. *Toronto.* 1846-67.
William. *Wick.* 1837-60.
BELLAIRS, Elias. *Spalding.* 1858-76.
BELLAS—
———. *Kilrea.* Early 19c. C.
Hugh & Thomas. *Coleraine.* 1839-44.
James. *Liswatty.* (b.1752) ca.1780-d.1842. At *Ballyar-
ton, Co. Derry.*
BELLAMY, Joseph. *Grimsby.* 1828-50.
BELLATTI—
Louis. *Grantham.* 1828.
Louis Lawrence. *Lincoln.* 1849-68.
BELLEFONTAINE, Augustus (& Son). *London.* 1832-
51 (69-75).
BELLER, John. *Reading.* 1830.
BELLEROSE, G. S. H. *Three Rivers, Quebec.* 1790-
1843.
BELLES, Joseph. *Fixby* nr. *Huddersfield.* ca.1810-20. C.
BELLEVILLE. 1914 and later, product of Pequegnat
Clock Co.
BELLEW, Matthew. *Belfast.* 1792.
BELLING—
Bennett M. *Hamilton, Ontario.* 1867.
James E. *Hamilton, Ontario.* 1862-69.
John. *Toronto.* 1851-53.
John I. *Bodmin.* b.1685-ca.1761. C.
John II. *Bodmin.* b.1738. s. of John I. d.1807.
John III. *Bodmin.* s. of John II. 1791-1814. W.
John IV. *Bodmin.* 1823-56. W.
John. *Lostwithiel.* 1823. W.
Mrs. Elizabeth. *Bodmin.* 1873.
BELLION—
C. *Chester.* 1878.
Charles. *Kington.* 1870-79.
Edward. *Liverpool.* 1848-51. Imported clocks.
William. *Liverpool.* 1877. W.
BELLMAN—
———. *Ambleside.* 1845. W.
———. *Kendal.* ca.1820-30. C.
Daniel. *Broughton.* b.1799. s. of William. d.1865. C.
Thomas. *Ulverston.* s. of William. b.1806-51. W.
William. *Broughton.* 1790-ca.1812. C.
BELLON (? = BELLOME), Captain Peter. *London.*
mar. pre-1659 to dtr. of Josias Cuper. 1659-62.
Watch case maker.
BELLONI, Frederick. *Shaftesbury.* 1830-55.
BELLRINGER, Francis. *London.* 1857.
BELOZZI, John B. *Weymouth.* 1830.
BELSEY—
John. *London.* 1844-75.
Mrs. Mary Ann. *London.* 1881. (?wid. of John.)
BELSHAW, James. *Lurgan.* 1846. C.
BELT, Thomas. *Guisborough.* 1823. W.
BELTON, W. P. *Stafford.* 1868.
BELLWOOD, John. (No town) prob. *Kepwick, N.
Riding, Yorks.* ca.1760.
BELZONI, Frederick. *Sheftesbury.* 1830 (prob. BEL-
LONI, q.v.).
BEMIS—
Augustus. *Paris, Me, USA.* ca.1810.
Samuel. *Boston, USA.* 1789-1881.
BENBOW, Thomas. *Wellington.* (b. 1773 at *Prees)*
d. 1833. W.
BENDON, George & Co. *London.* 1881.
BENEDICT—
A. *Syracuse, NY, USA.* 1835.
Andrew. *New York.* ca.1810-30.
Bros. *New York.* (B. founded 1819) 1836.
(Deacon Aaron) & BURNHAM (Co.). *Waterbury, Conn,
USA.* 1850-c.1880. Later Waterbury Watch Co.
Martin. *New York.* ca.1830.

BENEDICT—*continued.*
Samuel W. *New York.* 1845.
& SCUDDER. *New York.* ca.1830.
BENFORD, John & George. *London.* 1875-81.
BENHAM—
Augustus. *New Haven, Conn, USA.* ca.1840.
John. *Collumpton.* ca.1830. C. (B. 18c).
John H. *New Haven, Conn, USA.* 1840-56.
BENITZ, A. & Co. *Kettering.* 1841.
BENJAMIN—
Abraham. *London.* 1863.
Asher. *London.* 1832.
E. *Jersey.* 1832. C.
E(verard) & Co. *New Haven, Conn, USA.* 1846.
Edward & Abraham. *London.* 1857.
(Everard) & FORD (George H.). *New Haven, Conn, USA.* 1848.
Joel (& Co.). *London.* (B.1822-)1828-32(-39).
John. *Stratford, Conn, USA.* 1730. mar.1753-96.
& LAZENBY. *London.* 1828.
Lewis. *London.* 1839.
Michael. *London.* 1832-51.
BENKEL, Louis. *London.* 1869-81.
BENNAM, John. *London.* 1839.
BENNER, Johannes. *Augsburg.* 17c. C. (B. ca.1650).
BENNETT—
——. *Coggeshall.* 1781. W. (see also Thomas B.).
Alfred. *Philadelphia.* 1837-47.
Anthony. *Kettering.* 1841. C.
Bros. *Halifax, Nova Scotia.* 1875.
& CALDWELL. *Philadelphia.* 1843-48.
& CLENCH. *Salisbury.* 1859-75.
Ebenezer. *Romford.* 1828.
Elizabeth. *Lee (Kent).* 1839-55.
Ephraim. *Lower Gornall, Staffs.* 1850-76.
G. *Bexley Heath.* 1855.
G. *Ipswich.* 1846.
G. W. *Blackheath.* 1851.
G. W. *Woolwich.* 1851-55.
J. *Fowey.* 1856.
J. *Stockport.* 1857-78.
J. *Truro.* ca.1830. C.
J. & H. *London.* 1839.
James. *New York.* 1768-73.
James. *Romford.* 1839.
James. *Salisbury.* 1842.
John. *Cork.* 1760. W.
John. *Grays Thurrock.* 1866-74.
John. *Helston.* ca.1775-79. C. & W.
John. *London.* (B. a.1670, CC. 1677-)1713.
John. *London.* 1839. Watch cases.
John. *Lostwithiel.* 1873. C. & W.
John. *Norwich.* ca.1790. C.
John. *St. Austell.* 1841-retired 1844. W.
John Edmund. *London.* 1839-69.
John, FRAS. *London.* 1857-63 (later Sir John 1869-81).
John R. *Halifax, Nova Scotia.* 1873.
Joseph. *Dublin.* a.1759.
Joseph. *London.* 1828-32.
Mrs. E. S. *Greenwich.* 1851.
Robert. *Millbrook (Cornwall).* 1856-73. C. & W.
Samuel. *Helston.* 1763-68. C.
T. *Tipton.* 1868.
T. N. *Canandaigua, NY, USA.* ca.1860.
T. P. *Cradley Heath.* 1876.
& THOMAS. *Petersburg, Va, USA.* ca.1820.
Thomas. *Coggeshall.* 1828.
Thomas. *Dublin.* 1858.
Thomas. *Horndean, Swinton, Scotland.* 1860.
W. *Yeardsley-cum-Whaley, Cheshire.* 1865. C.
William. *Liverpool.* 1824.
William. *Truro.* 1847.
William Cox. *Blackheath.* 1866.
William Cox. *London.* 1866.
Wing & Co. *London.* 1839.
BENNING, John. *New Windsor, Berks.* (B. pre-1758) 1775. W. & C.
BENNIT, George. *Manchester.* 1766. W.
BENNOCK, James. *Thornhill, Scotland.* 1860.
BENNY, Jonathan. *Easton, Md, USA.* 1798.
BENOIT, Joseph E. *Pembroke, Ontario.* 1871.

BENRUS—
Manufacturing Co. *Waterbury, Conn, USA.* 1936.
Peter C. *New York.* 1874.
BENSEL, Leonard. *New York.* 1880s.
BENSLON, John. *Martham.* ?mid-18c.
BENSON—
Duncan. *Glasgow.* 1843.
James William. *London.* 1857-87. C.
John. *Belfast.* 1753 (from *Whitehaven*).
John W. *Quebec.* 1871.
Joseph. *Stourbridge.* 1842.
Samuel S. *London.* 1857.
Thomas. *Kirkham.* 1848.
BENT—
George. *Bishop Auckland.* a.1724. C.
John. *Garboldisham, Norfolk.* 1836-46.
William. *London.* 1869-81.
BENTHAM, Thomas. *Newent.* 1870.
BENTLEY—
——. *Darlington.* mid-late 18c.
Eli. *West Whiteland, Pa, USA.* 1752-mar.1772. Later to *Taneytown, Md.* after 1778.
John. *Llanelly.* 1865-87.
John. *Long Buckby.* ca.1780. C.
John. *Thirsk.* 1791-1807. C. & W.
Thomas. *Gloucester* and *Boston, Mass, USA.* 1764-1804.
Thomas. *Neath.* 1835.
BENTON—
& Son. *London, Wigmore Street.* ca.1790. C.
William. *Liverpool.* 1848.
BENZIE—
James. *Basingstoke.* 1878.
Mrs. M. A. E. *Maidstone.* 1866.
Simpson. *West Cowes.* 1867-78.
BEQUILLARD, EMIN & Co. *St. Johns, Quebec.* 1871.
BERAGEZ, ——. *Paris.* ca.1740. C.
BERCHUNE, L. *Ottawa.* 1862.
BERENDT, Theodore. *London.* 1857.
BERESFORD—
Charles & James. *Leeds.* 1871. Clock case makers.
Henry. *Carmarthen.* 1847-52.
John Hewson. *Pontypridd.* 1871-75, then *Swansea.* 1887. W.
BERGENSTEIN, S. *London.* 1844-51.
BERGER—
L. *London.* 1839.
Michael. *London.* 1844-69.
BERGHAUSER, A. A. *Eichstaett.* ca.1775.
BERGMAN, David. *London.* 1875.
BERHULT, John. *New York.* ca.1860.
BERI & DELARA. *Leek.* 1860.
BERINGER—
Bros. (? Peter & Joseph). *Belfast.* 1865-1900. C. & W.
Fidelis. *Penzance.* 1864. W.
John. *Helston.* 1847-50. W. & C.
John. *Penzance.* 1844-47.
Joseph. *Helston.* 1844.
Joseph & Sons. *Falmouth.* 1873.
Joseph & Sons. *Helston.* 1873. W.
& SCHWERER. *Falmouth.* 1847. Also outfitters.
& SCHWERER. *Penzance.* 1856-73. C. & W.
SCHWERER & Co. *Redruth.* 1844-73.
& SCHWERER. *Truro.* 1849.
BERK, S. *London.* 1776. W.
BERK(E)L(E)Y—
& Co. *Worcester.* 1860.
J. *Lewisburg, Pa, USA.* ca.1800-20. C.
Joseph. *Gloucester.* 1879.
Lucas. *Ledbury.* 1856-79.
Lucas. *Ross.* 1850.
BERLIN—
1914 and later, product of Pequegnat Clock Co.
B. *Windsor.* 1877.
& HECKSCHER. *Windsor.* 1854.
J. *Windsor.* 1847.
BERMAN—
Charles. *London.* 1857-63.
Jacob (& Co.). *London.* (B. 1820) 1828(-32). C.
John. *Bristol.* 1842.
John. *London.* 1832.
Joseph & Co. *London.* 1863.

BERNARD, Elias. *Southampton.* ca.1700-ca.1720. W. & C.
BERNASCONE, Innocent. *High Wycombe.* 1830.
BERNEY, William. *Dublin.* 1731. W.
BERNHAR(D)T—
Richard. *London.* 1863-75.
Ullerich & Co. *London.* 1851-63.
Wolf. *Graz.* 1567. rep'd. town clock.
BEROLLA, ——. *Paris.* ca.1830-80. Carriage clocks.
BERQUEZ, Francis. *London.* 1828-44. (B. has BERQUIZ 1815-30).
BERRES, T. *London.* 1819. W.
BERRESFORD, Samuel. *Chesterfield.* 1876.
BERGANT, Peter. *Philadelphia.* 1829-33.
BERRICK, Bernard. *Liverpool.* 1848-51.
BERRIDGE, Robert. *Ramsey, Hunts.* 1830-47.
BERRILL, George. *Stony Stratford.* b.1792-1842. C.
BERRINGER—
A. J. *Albany, NY, USA.* 1834.
Jacob. *Albany, NY, USA.* 1835-43.
BERRINGTON—
James. *Bolton.* 1834.
James. *St. Helens.* 1822-34.
John Johnson. *Bolton.* 1822-51.
Nathaniel. *Castle Donnington.* ca.1795. C.
BERROLLAS—
J. A. *London.* 1844-51.
Joseph Anthony & Augustus. *London.* (B. 1800-30). 1832.
Joseph A. *London.* 1844.
BERROW, E. J. *Solihull.* 1880.
BERRY—
Arthur. *Liverpool.* (B. 1815-24). 1824-29. chrons.
Charles. *High Wycombe.* 1885-1908. Successor to Charles Strange.
D. *Huntingdon.* 1847.
Francis. *Northampton.* 1711.
Frederick. *London.* 1844-81.
G. F. *West Hartlepool.* 1898.
George. *Gloucester.* 1879.
George. *Hull.* 1826-50.
George. *Whitby.* 1851-66. W.
George Allan. *London.* 1863.
James. *New York.* 1790s. C.
James. *Pontefract.* ca.1770. C. & W.
James. *Prescot.* 1769. W.
James. *Stonehaven.* 1840.
James (& Son). *Aberdeen.* 1836-46 (-60).
John. *Manchester.* a.1717-31. (B. 1738-62).
Joseph. *Manchester.* 1848.
Mrs. Jane M. *London.* 1869-81.
R. *Mitcham, Surrey.* 1862.
R. E. *Sutton, Surrey.* 1862.
Richard. *Pontefract.* 1826-37.
Richard Edward. *London.* 1869-81.
Thomas. *London.* 1832.
Thomas. *Stowmarket.* 1875. Clock cleaner.
William. *Edinburgh.* 1750. Seal cutter.
BERTELE, Jacob. *Salzburg.* ca.1775. C.
BERTHIAUME, Louis. *St. Hyacinthe, Canada.* 1851.
BERTISH, M. *Swindon.* 1875.
BERTRAND, William. *London.* 1863-81.
BERWICK, John Thomas. *Shipley.* 1871.
BESSE, Jeremy. *London.* 1839-44.
BESSONET, John P. *New York.* 1793.
BESSONETT, John Staynor. *Halifax, Nova Scotia.* 1827-31.
BEST—
George Christopher Henry. *Cardiff.* 1887.
H. *Shipston.* 1860. Prob. same as next entry.
Henry. *Shipston.* 1828-50.
John. *Padstow.* 1772-ca.1780. C.
Richard. *London.* Late 18c. br. clock. (May be same as next man).
Richard. *London.* 1839-44.
Robert. *London.* (B. CC. 1783-1820). 1828. W.
Thomas. *Lebanon, O, USA.* 1808-26.
Thomas. *Lewes.* Took Henry Alce. a.1743. C. & W.
Walter. *Pontarddulais.* 1887. *Cardiff.* 1899.
William. *Ballymena.* (b.1788) ca.1810-d.1830. C.
BESTWICK, Henry. *London.* s. of Richard. (B. a.1678, CC. 1686)-97.

BESWICK—
Edward J. *Shirley (Southampton).* 1878.
W. *Southampton.* 1859-67.
William. *Farnworth.* 1828. W.
BETAGH—
Thomas. *Londonderry.* From *London.* ca.1774.
William. *Dublin.* a.1743-d.1770. C.
BETCH, John. *Hull.* 1846. C. & W.
BETERTON, ——. *Dublin.* ca.1805. W.
BETHEL(L)—
A. *Ringwood.* 1848.
Francis. *Toronto.* 1861-67.
John. *Warrington.* (B. ca.1820). 1822-34.
& M'QUILLAN. *Montreal.* 1842-45.
BETJEMANN, George & Sons. *London.* 1875.
BETTELBROCK, Johannes. *Augsburg.* ca.1730. W.
BETT(E)RIDGE—
J. P. *Coventry.* 1860-80.
Mrs. M. *Birmingham.* 1854.
Richard Ezekiel. *Birmingham.* 1842-80.
BETTLE, J(ames). *London.* 1863-75.
BETTS—
Fairfax & Co. *Birmingham.* 1854-60.
Thomas. *Birmingham.* 1850.
BEVAN—
Edward. *Birkenhead.* 1865-78.
G. A. *Rugeley.* 1868.
J. *Woolwich.* 1838.
Jabez. *Cinderford.* 1879.
James. *Stratton.* 1847.
John. *Brecon.* 1840-49.
John. *Builth.* 1868-71.
John. *Hay.* 1868-87.
John. *Swansea.* 1887-99.
John. *Ystradgynlais.* 1887.
Thomas. *Stratton.* ca.1820-45.
& WEARE. *Birkenhead.* 1857.
William. *Monmouth.* 1868.
William John. *Monmouth.* 1858-80.
BEVANS—
James. *Haverfordwest.* 1822-50.
Mrs. Catherine. *Haverfordwest.* 1871-75.
S. & C. (?error for J. & C.). *Haverfordwest.* 1868.
BEVENS, William. *Philadelphia.* 1810-13. Then Norris-town till 1816.
BEVERIDGE—
Robert. *Kirkcaldy.* 1837-43. C.
Robert. *Newburgh (Fife).* 1781-1835. W.
BEVES, Samuel. *Dorking.* 1839-78.
BEVIN, Edward. Imprint of Newark Watch Co. (*USA*). 1867-69.
BEWES, Daniel. *Quebec.* 1844-51.
BEWLAY, John. *York.* 1800-01.
BEYER, G. C. *Newton Abbot.* ca.1810.
BEYNON—
James (?error for Jane, q.v.). *Merthyr Tydfil.* 1875.
Jane. *Merthyr Tydfil.* 1868-87.
Lewis. *Merthyr Tydfil.* 1844-52.
BEZANT—
A. W. *Abingdon.* 1854.
Aaron Walter. *Hereford.* 1863-79.
BIANCHI—
B. *Tunbridge Wells.* 1855-77. W.
J. *Windsor.* 1847-64.
BIBBY—
George. *Prescot.* 1795. W.
H. *Bolton.* 1858.
John. *Newport, Mon.* 1875.
Ralph. *Prescot.* 1819. W.
Samuel Jordan. *Caernarvon.* 1874-90.
Thomas. *Liverpool.* (B. 1825). 1828.
BIBLIE, Edward. No place (*Somerset*). Same as Bilbie, q.v. lant. clock ca.1720.
BICHAULT, James. *Boston, USA.* From *London, England.* 1729.
BICKERTON—
George. *Kidderminster.* 1828-50.
T. *Carlisle.* 1858.
Thomas Osten. *Bridgnorth* 1870-1900.
W. *Coventry.* 1860-80.
William. *Stourport.* 1835-76.

BICKLEY—
Henry. *London.* 1875-81.
John. *Coventry.* 1880.
John. *London.* 1832.
John. *London.* 1863-69.
W. *Coventry.* clock.
BICKNALL, John. *Tintern.* ca.1690. lant. clock.
BICKNELL—
& Co. *London.* 1811. New Bond St.
John. *Chepstow.* ca.1750. C.
BICKNILL, John. *Cirencester.* Early 18c. clock.
BIDDELL, G. A. *London.* ca.1860. C.
BIDDLE—
George. *Birmingham.* 1828-60.
George. *London.* 18c. C (no dates known).
Horatio. *Birmingham.* 1842.
Owen. *North Ward, Philadelphia, USA.* 1737-d.1799.
P. Edwin. *Merthyr Tydfil.* 1871-87.
Philip. *Mountain Ash (Wales).* 1875-87.
BIDDLECOMBE, Joseph. *Clinton, Canada.* 1857-62.
BIDDULPH, William. *Newcastle-under-Lyme.* 1868-76.
BIDLAKE—
James Hodgson. *London.* (B. a.1801, CC. 1809-18). 1839-44.
James & James Hodgson. *London.* 1828-32.
William S. *London.* 1844-51.
BIDMEAD, Jonathan William. *Coventry.* 1880.
BIDWELL, J. W. *Montreal.* 1843.
BIEGEL, Henry William. *Philadelphia.* 1810-13.
BIERSHING, Henry. *Hagerstown, Md, USA.* 1815-43.
BIGELOW—
& Bros. *Boston, USA.* 1842.
John B. *Boston, USA.* ca.1840.
BIGGER, Gilbert. *Dublin* 1776-83, then *Baltimore USA.* 1784-1816. W.
BIGGERSTAFF—
James. *Stratford-on-Avon.* 1828-35.
Robert. *Stratford-on-Avon.* 1842.
BIGGIN, Christopher. *Thorpe-le-Soken.* 1765. also at *Beccles* and *Harwich* 1766.
BIGGS—
Bryant. *Cardiff.* 1875-87.
Thomas. *Philadelphia.* Early 19c.
BIGNELL, Thomas. *Oxford.* a.1764-1812. C. & W.
BIGRAVE, John. *London.* 1828-32.
BIJOU, 1914 and later, product of Pequegnat Clock Co.
BILBIE—
Abraham. *Chewstoke.* 1728. C.
John. *Axbridge.* 1734-67.
John. *Chewstoke.* Late 18c. C.
Thomas. *Chewstoke.* 1742. C. (B. 1774 C.).
Thomas. *Cullompton.* l.c. clock (no date known).
Thomas Webb. *Chewstoke.* d.1829.
BILCLIFF see **BILTCLIFF**
BILES, George. *Bury St. Edmunds.* 1865-79.
BILL—
Joseph. *Middletown, Conn, USA.* 1841.
Joseph R. *Middletown, Conn, USA.* 1831. C.
Thomas. *Derby.* 1876.
BILLETER, Frederick Arnold. *London.* 1875.
BILLIARD—
& GAY. *London.* 1857.
Lewis. No place. 1540-68. T.C.
BILLIET, Victor. *London.* 1851-81.
BILLINGE—
Chalwell. *Wadebridge.* 1802-04. W.
George. *Birmingham.* 1850-60.
James. *Liverpool.* 1795-1814. W.
John. *Wigan.* 1671. W.
Topping. *Liverpool.* 1790. W. (B. has Topham).
BILLINGES, John. *London.* CC. 1637.
BILLINGS—
Andrew. *Poughkeepsie, NY, USA.* ca.1810.
Charles. *Bristol.* 1879.
Edmund. *Bristol.* 1879.
Frederick. Used by New York Watch Co. 1871-75.
G. *Birmingham.* 1854.
John. *Shanklin.* 1878.
John. *Ventnor.* 1878.
Jonathan. *Reading, Pa, USA.* ca.1730.
L. *Northampton, Mass, USA.* 1837.
BILLINGTON—
Edward. *Doncaster.* b.ca.1728-d.1797. C.
Edward. *Leeds.* ca.1770. C.

BILLINGTON—*continued.*
E(verard). *Market Harborough.* 1740-70.
J. *Leicester.* 1864.
John. *Aspley Guise.* 1864-77.
BILLON, Aime. *Chaux de Fonds.* ca.1830.
BILLS—
Elijah. *Plymouth, Conn, USA.* 1830-38. C.
J. *Dover.* 1855.
BILLYEALD, Alfred. *Nottingham.* 1876.
BILSBROUGH—
William, sen. *Gargrave.* 1707-09. C.
William, jun. *Gargrave.* mar.1727 d.1750. C.
BILSBUROW, William. *Gargrave.* Same as preceding.
BIL(T)CLIFF—
Charles. *Penistone.* 1871.
George. *Penistone.* 1834-62.
John. *York.* Free 1617-39. C.
Robert. *York.* Free 1627. d.1641. W.
BINCH, James. *Liverpool.* 1767-77. C. (B. 1777-1781.)
BINDER—
George. *Montreal.* 1862.
L. *Vienna.* ca.1820. C.
BINDLEY—
Joseph. *London.* 1875-81.
William. *London.* 1832-69.
BING, Mrs. *Canterbury.* 1865.
BINGHAM—
B. D. *Nashua, NH, USA.* 1830-40.
Charles. *Birmingham.* 1816. Dial maker.
Henry. *London.* 1844.
T. *Birmingham.* 1860. Dial maker.
Thomas. *Birmingham.* 1828-50. T.C.
Walter. *Dublin.* Free 1680-d.1727. C.
BINKS—
Anthony. *Chester-le-Street.* 1847-50.
Anthony. *Darlington.* 1851-66.
Joseph. *Worksop.* 1828-35.
Thomas. *Barningham* nr. *Barnard Castle.* (Not *Birmingham*). b.1737-d.1807. C.
BINLEY, John. *Market Harborough.* ca.1790. C.
BINNEY, Horace. *Philadelphia.* ca.1800-40. C.
BINNS—
George. *London.* 1828-32.
James. *Halifax.* 1720-63. C.
John. *Halifax.* 1801. C.
Joseph. *Halifax.* 1769-1802.
Robert. *Halifax.* 1710-d. 1729. C.
William. *Halifax.* mar.1822. C.
BIOLETTI, ——. *Wincanton.* 1777-1869. C.
BIRCH—
& GAYDON. *London.* 1881. (Successor to Wm. Birch.)
James. *Liverpool.* (B. 1811-29). 1834.
James. *Pembrokeshire.* ?1723-95. W. Later *London.*
Joseph. *Farnworth.* 1793. W.
& MASTERS. *Tenterden.* 1836-47. W. Partnership of John Masters & William Birch.
Richard. *Heywood.* 1828. W.
Thomas. *Hoddesdon.* 1828.
W. *Derby.* 1849-64.
William. *Bradford.* 1866.
William. *Leicester.* 1864-76.
William. *London.* 1844-75.
William. *Odiham.* 1867-78.
William. *Tenterden.* 1823-39. W.
William & Son. *Bradford.* 1871.
BIRCHALL—
George. *Shrewsbury.* mar. 1722, again in 1725, d.1738. W.
George. *Prescot.* mar.1766. W. (B. ca.1820).
James. *Prescot.* 1823-25. W.
John. *Liverpool.* d.1686. W.
John. *Prescott.* 1805-20. W.
Joshua. *Liverpool.* 1781-d.1789. W.
Mary. *Liverpool.* 1790. W. wid. of Joshua.
Peter. *London.* 1857-63.
William. *London.* (B. 1809-24 W.). 1828-44.
William. *Prescot.* 1797. W.
William. *Prescot.* 1818. W.
BIRCHBY, Thomas. *Wragby.* 1835.
BIRCHLEY—
Matthew. *Halifax.* 1838.
Mrs. Jane. *Halifax.* 1850.

BIRCKLEY—
Frederick. *Manchester.* 1834.
Frederick. *Rochdale.* 1848.
BIRD—
Adam. *Barnard Castle.* ca.1800-34. W.
Charles. *Stroud.* 1840.
Edward. *Bristol.* a.1777 to Joseph Wood, whom he succeeded in 1801-1818. W.
George. *London.* 1832.
J. *Stroud.* 1856-70.
J. D. *Sarnia, Canada.* 1857.
John. *Belfast.* 1654-67. C.
John. *London.* 1839-44.
John Thomas. *Birmingham.* 1880.
Joseph. *Stroud.* Early 19c. C. (See J. Bird).
Michael I. *Oxford.* From *London.* ca.1654-d.1689. C. & W.
Michael II. *Oxford.* s. of Michael I, to whom a.1672. Prob. worked *London.*
Nathaniel. *Oxford.* s. of Michael I, to whom a.1678. Prob. worked *London.*
Robert. *Yeldham.* c.1760. C. (See also BRID).
Thomas H. *Halifax, Nova Scotia.* 1873.
William. *Seagrave.* ca.1750. C. (B. pre-1778 W.).
Wright. *Oxford.* s. of Michael I, to whom a.1682-cancelled 1686.
BIRDLAKE, James. *Ewell.* 1878.
BIRDSLEY, E. C. & Co. *Meriden, Conn, USA.* 1831.
BIRGE—
(John), CASE (Hervey, Erastus) & Co. *Bristol, USA.* 1833-34.
(John) & FULLER (Thomas). *Bristol, USA.* 1844-48.
(John) & GILBERT (William C.). *Bristol, USA.* 1837.
(John) & GILBERT & CO. *Bristol, USA.* 1834-37.
& HALE. *Bristol, USA.* 1823.
HAYDEN & CO. *St. Louis, Mo, USA.* ca.1850.
& IVES. *Bristol, USA.* ca.1832-33.
John. *Bristol, USA.* b.1785-d.1862. See other company titles.
John & Co. *Bristol, USA.* 1848.
John, jun. *Bristol, USA.* 1815-18.
MALLORY (Ransom) & Co. *Bristol, USA.* 1838-43.
PECK & CO. *Bristol, USA.* 1849-59.
& TUTLE. *Bristol, USA.* 1823.
BIRKET, Henry. *Pickering.* mar.1793. W.
BIRKETT—
John. *Grasmere.* 1877. W.
John. *Keswick.* 1869-79.
Robinson. *Grasmere.* 1879. W.
BIRKLE—
Bros. *Croydon.* 1851-55.
David. Bros. & Co. *London.* 1857-81.
D. & Co. *Ipswich.* 1853-65.
BIRKLEY—
Bros. *Birmingham.* 1868.
George. *Birmingham.* 1868.
John. *Malton.* 1866.
Joseph. *York.* 1866.
M. & Z. *Birmingham.* 1860.
Mrs. Maria. *Birmingham.* 1880.
Primus. *Newport, Mon.* 1868-99.
BIRKS, Henry & Sons Ltd. *Canada.* 1905.
BIRLEY—
George & William. *Birmingham.* 1868-80.
J. *Birmingham.* 1854.
J. G. *Birmingham.* 1860.
Samuel. *Birmingham.* 1816-42.
& Son. *Birmingham.* 1850.
BIRNIE—
John. *Templepatrick.* ca.1730-d.1763.
John II. *Templepatrick.* s. of John I. ca.1750-ca.1785. C.
Laurence. *Ballymoney.* s. of William. a.1767-74. Later to *Philadelphia.* C.
William. *Templepatrick* and *Parkgate.* ca.1745-78. C. & W.
BIRNSTEIN, Richard. *London.* 1881.
BIRTLE, Thomas. *Husbands Bosworth.* 1835.
BIRTLES—
Edward. *Liverpool.* 1777-d.1797. C.
J. *Kibworth Beauchamp.* 1849-76.
John. *Market Harborough.* 1795-1849.

BIRTS, Thomas Pacey. *Woolwich.* 1874.
BISBROWN, Thomas. *Halifax, Nova Scotia.* 1790-1799.
BISH—
Edward. *Hemel Hempstead.* 1839-51.
William. *Hemel Hempstead.* 1866-74.
BISHOP(P)—
(James) & BRADLEY (Lucius B.). *Watertown, Conn, USA.* 1827-32. C.
Charles. *Brighton.* 1878.
David. *Bourton.* 1863-79.
Edwin James. *London.* 1869.
George. *Lincoln.* 1835-76.
Homer. *Bristol, USA.* Post-1837.
J. *Allentown, Pa, USA.* ca.1810. C.
James. *London.* 1875-81.
James. *Watertown, Conn, USA.* ca.1825-30. C.
James Griffin. *London.* (B. 1815-24). 1828-32.
John. *Ledbury.* 1870-79.
John. *Maidstone.* 1733.
John. *Redmile.* Early 19c. C.
John. *Sherborne.* 1830.
Joseph. *Philadelphia.* 1829-33.
Joseph. *Wilmington, NC, USA.* 1817-22.
Leman. *London.* 1875.
Martha. *Maidstone.* a.1720.
Moritz. *Easton, Pa, USA.* 1786-88.
(Daniel E.) & NORTON (Daniel F.). *Bristol, USA.* 1853-55.
Richard. *London.* 1839.
Rufus. *Mt. Joy, Pa, USA.* ca.1850.
S. R. *Woolwich.* 1866.
Samuel. *London.* 1851.
Samuel Robert. *London.* 1857.
(Thomas) & TREAT (Sherman). *Bristol, USA.* 1831-33. C.
William. *Trowbridge.* 1842-59.
BISPHAM, Samuel. *Philadelphia.* 1696.
BISS—
———. *Bath.* 1856.
E. H. *Bath.* 1861-66.
BISSELL, David E. *Windsor, Conn, USA.* ca.1830. Also dentist.
BISSET, David. *Sligo.* 1776. C.
BISSET, W. *Sunderland.* ca.1790(?). C.
BISWANGER—
Bernard, sen. *Prague.* s. of Wolfgang. ca.1750.
Bernard, jun. *Prague.* s. of Bernard, sen. 1778.
Johann. *Prague.* ca.1808-15.
Leopold. *Prague.* 1775-1808.
Wolfgang. *Folback.* Early 18c.
BITTNER, William. *London.* 1839-44.
BIVEN, Edward. *London.* 1857.
BIXFIELD, James Henry. *New Brompton.* 1874.
BIXLER—
Christian, sen. *Reading, Pa USA.* 1760-1800. Clock dealer.
Christian, jun. *Reading* and *Easton, Pa, USA.* 1763-1840. C.
Daniel. *Philadelphia.* ca.1810.
BLACK—
Alexander. *Kirkcaldy.* 1860.
Andrew. *Alloa.* 1830-72.
Andrew. *Colinsburgh, Scotland.* 1837.
Andrew. *Leslie, Scotland.* 1837-60.
Daniel. *Colinsburgh, Scotland.* 1830.
George. *Arbroath.* 1860.
H. *Stewartstown, Canada.* 1851.
Hugo. *Toronto.* 1875-77.
James. *Dublin.* 1756. W.
James. *London.* 1844.
James, jun. *Berwick-on-Tweed.* 1834-36.
John. *Aberdeen.* 1846.
John. *Banff.* 1860.
John N. *Nuneaton.* 1860-80.
John. *Philadelphia.* 1839-50.
John & Thomas. *Kirkcaldy.* 1860.
Jonathan. *Lurgan.* ca.1830-46. C.
Jonathan. *Milton, Canada.* 1851.
Thomas. *Dumfries.* 1837.
Thomas. *Lurgan.* ?Early 19c. C.
William. *North Shields.* 1827-58. C.

BLACKBURN—
——. *Oakham.* ca.1720-ca.1750. C.
David. *Horncastle.* 1861-76.
George. *Whitehaven.* 1858-79.
James. *Kirkby Lonsdale.* 1849-69.
James. *Prescot.* a.1742. W. (B. d.1774).
M. *Lincoln.* 1861.
Stephen. *Oakham.* ca.1770. C.
BLACKFORD, Edward. *New York.* ca.1830.
BLACKHAM—
Charles. *Warrenpoint.* 1846.
George. *Newry.* 1821-57. W.
Maxwell. *Newry.* 1860-68. W.
William. *Newry.* 1820-58.
BLACKHURST—
George. *Warrington.* 1851-58.
J. *Weaverham.* 1857-65.
James. *Over.* 1848-78.
John. *Crewe.* 1878.
T. *Lymm.* 1865.
T. *Middlewich.* 1857.
BLACKIE—
J. R. *Leith.* 1850.
John R. *Edinburgh.* 1848-60. Electric clocks.
BLACKLEY—
J. W. *Canterbury.* 1805.
Thomas. *London.* Early 19c. W.
BLACKLOCK, Joseph. *Lockerbie.* 1860. Dealer.
BLACKMORE—
John. *Exeter.* Free 1699. W. (One such a. *London* 1688
— may be him).
John Lane. *London.* 1863-81.
Thomas. *Liverpool.* 1824.
BLACKNER, John L. *Cleveland, O, USA.* 1837.
BLACKSLEY, Richard. *Dayton, O, USA.* 1829.
BLACKWELL—
W. B. *Birkenhead.* 1857.
William. *London.* 1857.
William B. *Holywell.* 1868.
BLACKWOOD—
John Thomas. *North Shields.* 1856.
William. *North Shields.* (B. 1820-27). 1856.
William & John Thomas. *North Shields.* 1848-51.
BLADES—
James. *Bradford.* 1866-71.
John. *Kirkby Moorside.* 1823-40.
William. *Brackley.* 1824.
William. *Horncastle.* 1828-35.
BLADON, Thomas. *Uttoxeter.* 1842.
BLAGBURN—
J. *Gateshead.* 1851-56. (Believed d.1878). C.
John. *Sunderland.* 1834.
BLAGBOROUGH, Benjamin. *Thorne.* 1822-37.
BLAIK, James. *Peterhead.* ca.1790-ca.1800. C.
BLAIR—
Alexander. *Port Glasgow.* 1860.
D. W. *Montreal.* 1857.
Elisha. *Brooklyn, NY, USA.* ca.1840.
James. *Kilwinning.* 1836.
BLAISDELL—
Abner. s. of Isaac. *Chester, NH, USA.* 1771-1812.
Charles C. *Ossipee, NH, USA.* 1838-1924.
David I. s. of Jonathan. *Amesbury, Mass, USA.*
1712-56. C.
David II. s. of David I. *Amesbury,, Mass, USA.*
1736-94.
David III. s. of Isaac. *Chester, NH and Peacham, NH,
USA.* 1767-1807.
Ebenezer. *Chester,,NH, USA.* s. of Isaac. 1778-1813.
Isaac I. s. of David. *Chester, NH, USA.* 1738-1791. C.
Isaac II. s. of Isaac I. *Chester, NH, USA.* 1760-97.
Jonathan. *Amesbury, Mass, USA.* 1678-1748.
Nicholas. s. of David. *Amesbury, Mass and Portland,
Me, USA.* 1743-1800. C.
Richard. s. of Isaac. *Chester, NH, USA.* 1762-90. C.
BLAKE—
——. *Coleshill.* ca.1790. C.
E. G. *Farmington, Me, USA.* ca.1850.
Joseph. *Bristol.* 1818. (B. 1825).
Paul. *Sheffield.* 1871.
Richard. *London.* ca.1790. W.
Thomas. *Ballinamore.* 1858.

BLAKE—*continued.*
Thomas. *London.* 1839. Watch cases.
BLAKEBOROUGH—
Charles. *Todmorden.* 1848.
Henry. *Burnley.* 1822-51. C.
Henry. *Leyburn.* 1813-18. W.
Henry. *Ripon.* 1837.
John. *Bacup.* 1848-51.
John Mangle (or Mangie?). *Burnley.* 1858. W.
Jonathan S. *Keighley.* 1834-71.
Richard. *Pateley Bridge.* b.1778-1817. Then *Otley.*
1817-37. C.
Richard. *Ripon.* 1851-71. C. & W.
William R. *Heywood.* 1851.
BLAKEBURN, W. *Padiham* (nr *Burnley*). 1858.
BLAKELY, John. *Liverpool.* 1784. C.
BLAKEMAN, John R. *Audlem.* 1857-78.
BLAKENEY, Thomas. Place unknown. 18c. C.
BLAKESLEE—
Edward K. *Cincinnati, O, USA.* ca.1840-50.
Jeremiah. *Plymouth, Conn, USA.* 1841-49.
M(arvin) & E(dward). *Plymouth, Conn, USA.* 1832-37.
C.
Milo. *Plymouth, Conn, USA.* ca.1824.
R., jun. *New York.* 1848. C.
William. *Newtown, Conn, USA.* s. of Ziba. 1820.
Ziba. *Newtown, Conn, USA.* b.1768 mar.1792-d.1834.
BLAKESLY, Harper. *Cincinnati, O, USA.* ca.1830.
BLAKEWAY, Thomas. *Broseley.* 1835-50.
BLAKEY, George. *Gainsborough.* 1828-35.
BLANCHARD—
Asa. *Lexington, Ky, USA.* 1808-38.
E. *Camberwell.* 1866.
William. *Hull.* 1813-23. (B. 1836).
William & Son. *Hull.* 1838-51.
BLANCKENSEE—
Joel & Co. *Birmingham.* 1880.
Solomon. *Birmingham.* 1850-54.
& Son. *Birmingham.* 1860. Wholesalers.
BLAND—
George. *London.* 1844-51.
Patrick. *Dublin.* ca.1810. W.
Samuel. *Philadelphia.* 1837-50. W.
Thomas. *Leeds.* 1807. C.
BLAN(D)FORD, George. *London.* (B. a.1783).
1839-63.
BLANKFORD, William. *Chicago, USA.* b.*London.*
1838. To *USA.* 1899-d.1920.
BLANSHARD, Joseph. *York.* a.1756.
BLASE, James. *Colchester.* b.ca.1810-35.
BLATT, John. *Philadelphia, USA.* ca.1841.
BLAXLAND, John. *Sittingbourne.* 1866.
BLAYLOCK—
& DUDSON. *Carlisle.* 1869-73.
John. *Carlisle.* 1834-58.
John (II?). *Carlisle.* 1879.
William. *Carlisle.* 1828.
BLEAR, James. *Cookstown.* Early 19c. C.
BLEIBEL, John. *Cardiff.* 1875-99.
BLENGINI, ——. *Turin.* ca.1800. W.
BLENKARNE, George. *Stowbridge.* 1868-76.
BLETHYN—
George. *Tenby.* 1813-50.
Richard. *Pembroke.* 1840-50. C.
William. *Carmarthen.* l.c. clock. Date unknown. (Prob.
same man as below.)
William. *Pembroke.* 1791. (B. 1795).
BLEW—
B. *London.* 1845. W.
William. *London.* 1766-1812. W.
BLIGHT—
Edward. *Callington.* 1873. W.
Walter. *London.* 1881.
BLISS(E)—
Ambrose. *Canterbury.* Free 1647. W. Must be same
man as at London where CC. 1653-62. Working in
Cornhill.
& Co. (John). *New York.* ca.1891-98.
G. F. *Stony Stratford.* 1869.
J. *Newport Pagnell.* ca.1790. C.
John. *Brill.* 1823-42.
John. *Weston Underwood.* Late 18c. C.

BLISS(E)—*continued.*
John. *Zanesville, O, USA.* 1815. C.
Joshua. *London.* Early mid-18c. C.
William. *Cleveland, O, USA.* ca.1810.
BLOCH, George. *Kings Lynn.* 1875.
BLOMFIELD—
Edwin. *Bexhill.* 1870.
Isaac Samuel. *Lowestoft.* 1853-58.
BLOMGREN, Bernard. *London.* 1857-75.
BLONDEAU, ——. *Paris.* (B. 1815) 1827-37. C.
BLONDEL, Nicholas. *Guernsey.* ca.1740. C. (One such in Paris 1743-89. B.)
BLONDER, Robin. *London.* 1881.
BLOOD, Alfred. *Peterborough.* 1876.
BLOODWORTH, Mark. *Swansea.* 1887.
BLOOMER & SPERRY. *New York.* ca.1840.
BLOOMFIELD—
Samuel. *London.* 1857.
William Valentine, *Maidenhead.* 1877.
BLOOR, John. *Newcastle-under-Lyme.* 1828-42.
BLOORE, Edward. *Birmingham.* 1842.
BLORE, William. *Liverpool.* 1752-56. W.
BLOUNT—
Bridget. *Monmouth.* 1868. (?wid. of James Henry)
George. *Dumfries.* 1792.
J. *Belper.* 1855-64.
James Henry. *Monmouth.* 1858.
BLOW, J. *Tamworth.* 1860-68.
BLOWE, George. *Philadelphia.* 1837-50.
BLUETT, Frederick. *Swansea.* 1868.
BLUMANTHALL, J. *Lowestoft.* 1858.
BLUMBERG & Co. Ltd. *London* and *Paris.* ca.1880. Carriage clocks.
BLUMENTAL, Adolphe. *London.* 1863-81.
BLUMER—
(Jacob) & GRAFF (Joseph). *Allentown, Pa, USA.* 1799. C.
Jacob. *Allentown, Pa, USA.* 1798-1820.
Joseph. *Allentown, Pa, USA.* 1798-ca.1820.
BLUNDELL—
Henry. *London.* 1839-51.
John. *Horley.* ?From *London* (where one such a.1678). 1700. C.
John. *London.* 1839.
Joseph. *Dublin.* Free 1704-d.1732. C.
Joseph. *Port Glasgow.* 1860.
Robert. *Dublin.* 1830. W.
Thomas. *Dublin.* Free 1733 -d. 1775. W. & C.
Thomas. *Limerick.* 1858.
Thomas. *Liverpool.* 1822-83. W.
Thomas H. *Toronto.* 1875-77.
William. *Limerick.* 1858.
BLUNDY, Aaron. *London.* 1758 (-83?). W.
BLUNT—
& Co. *New York.* 1869. Chrons.
D. & Son. *Northampton.* 1847-54.
Deleney. *Northampton.* 1824-41.
BLURTON—
Edward. *Stourbridge.* 1860-76.
John. *Stourbridge.* 1828-50.
John & William. *Dudley.* 1828.
BLYNDENBURG, Samuel. *New York.* 1833.
BLYTH, R. H. *Montreal.* 1844.
BOANAS, John. *Osmotherley.* mar.1825. W. (cf. BOWNESS).
BOARDMAN—
Chauncey. *Bristol, USA.* 1789-1857.
Chester. *Plymouth, USA.* 1842-45.
Samuel. *Manchester.* a.1742, d.pre-1760. C.
(Chauncey) & SMITH (Samuel B.). *Bristol, USA.* 1832.
(Chauncey) & WELLS (Joseph A.). *Bristol USA.* (B. 1815) 1832-43. C.
William. *Birmingham.* 1860-68.
BOARDS, William. *Hull.* 1838-58. C. & W.
BOAST, Henry. *London.* 1869.
BOATE, Benjamin, jun. *Ironbridge.* 1879.
BOATMAN, James. *Orsett.* 1874.
BOCHEMSDE, Frederick. *Boston, USA.* 1809-21.
BODDINGTON—
W. *Altrincham.* 1857.
William. *Birmingham.* 1828-35.

BODDINGTON—*continued.*
William. *Rhyl.* 1868-90.
William & Sons. *Coventry.* 1842.
BODE, John. *Chelmsford.* 1665-86.
BODELEY, Thomas. *Boston, USA.* ca.1720.
BODKIN—
John. *Belfast.* ca.1825-68. Prob. same as below.
John. *Donaghadee.* 1800-25. Prob. same as above.
BODLEY—
Frederick. *London.* 1828.
Sydney. *Ramsgate.* 1874.
BODY—
Abraham. *Cranbrook.* s. of Obadiah. Succeeded ca.1774 to George THATCHER, d.ca.1779.
Henry. *Battle.* b.ca.1748. s. of Obadiah. d.1818 C.
Henry. *Cranbrook.* 1858.
Obadiah. *Cranbrook.* b.ca.1702. From *Westfield, Sussex* in 1716 as apprentice to George Thatcher of *Cranbrook, Kent.* At *Cranbrook* till 1730, then *Battle* where d.1767. C.
BOEHME, Charles. *Baltimore, USA.* 1774-1868.
BOESE—
B. *Kidderminster.* 1868-76. prob. same as next man.
B. *London.* 1863. prob. same as previous man.
BOFENSCHEN, Charles. *Camden, SC, USA.* 1854-57.
BOFFI, Leopold. *Battle.* ca.1835-39.
BOGARDUS—
Everardus. *New York.* 1675. Free 1698, mar.1704.
James. b.1800-74. C. *New York* and *Henrietta, NY, USA.*
BOGER—
John E. *Salisbury, NC, USA.* 1845-53.
(John R.) & WILSON (William R.). *Salisbury, NC, USA.* 1846-53.
BOHAN, Stephen. *Montreal.* 1819.
Stephen. *Waterford.* 1809. W.
BOHL, H. F. *Ottawa.* 1875.
BOHLE—
(Peter) & HENDRY (Robert). *Montreal.* 1850-55.
Peter. *Montreal.* 1786-1865.
BOHME, Friedrich. *London.* 1869-75.
BOIVIN—
& BEAUDRY. *Montreal.* 1857.
L. P. *Montreal.* 1833-54.
BOLD—
John. *Liverpool.* 1851.
John. *Prescot.* 1794. W.
Matthew. *Bold* (nr. *Farnworth*). ca.1780. C.
William. *Liverpool.* ca.1820-34. C.
BOLDING, Joseph. *Burton.* Early 19c.
BOLEY, Richard. *Leicester.* ca.1750. C.
BOLITHO—
Henry W. *Portsea.* 1859-78.
S. *Constantine* (*Cornwall*). Early 19c. C.
BOLLAND, John. *Halifax.* 1691.
BOLLMAN, William. *Atlanta, Ga, USA.* ca.1870.
BOLLOTTE, ——. *Paris.* ca.1860. C.
BOLOQUET, Marcel. *New Bern, NC, USA.* ca.1760. C.
BOLT—
——. *Teignmouth.* ca.1790.
Frederick T. *Bristol.* 1879.
T. & Co. *Birkenhead* and *Liverpool.* 1857.
BOLTE, H. N. *Atlantic City, NJ, USA.* (estd. 1864.)
BOLTON—
Edward. *Birkenhead.* 1878.
George Alfred. *Lossiemouth.* 1860.
H. *Birkenhead.* 1857.
Henry. *Liverpool.* 1848-51.
Lancelot. *Chester-le-Street.* 1827-56.
Robert. *Wigan.* 1791-1848. C.
T. *Liverpool.* 1857. W.
Thomas. *Halifax, Nova Scotia.* 1809-38.
William James. *Workington.* 1879.
BOLUS, Alfred. *London.* 1881.
BOLVILLER, ——. *Paris.* 1830-70. Carriage clocks.
BONCEUR, ——. *Copenhagen.* 1720. C.
BOND—
Charles. *Manchester.* pre-1798 -ca. 1810. W. & C.
Charles. *Boston, USA.* 1830-42.
George. *Gloster.* 1879.
H. *Shrewsbury.* 1856.
Henry Charles. b. 1833 *Kew. Bishopcastle (Shropshire)* 1851-56. *Church Stretton* by 1868-91. W.

BOND—*continued.*
Henry. *Cirencester.* 1850-70.
James Charles. *London.* 1857.
John Turner. *Blackpool.* 1851.
John Turner. *Leeds.* 1866. W.
John Turner. *Preston.* 1848.
Joseph. *Bentham.* ca.1845-50. C.
Samuel. *Liverpool.* 1848.
William. *Liverpool.* 1848-51.
William. *Portland, Me, USA.* ca.1785.
William & Son. *Boston, USA.* 1813-42 and later.
BONE—
B. & Son. *Rougham, Norfolk.* 1865.
Richard. *Litcham, Norfolk.* 1875.
Samuel. *London.* 1881.
William. *Elmham, Norfolk.* 1875.
BONES, Thomas. *Colchester,* a.1760-67, then *Maldon*
1768-72. C. & W.
BONETTO, Peter A. *London.* 1857.
BONEY—
Caleb I. *Padstow.* 1747-d.1827. C.
Caleb. *Penzance.* prob. same as above man. ca.1820. C.
Caleb II. *Padstow.* s. of Caleb I. 1828. C.
Caleb III. *Saltash.* Grandson of Caleb I. 1847-56. W.
John. *Padstow.* s. of Caleb I. 1844-56.
BONFIGLIO, B. *Cardiff.* 1875.
BONHAM & PERKINS. *Coventry.* 1880.
BONINGTON & THORPE. *London. Red Lion St.,*
Clerkenwell, London. 1811. Clock case makers.
BONNARD, M. *Philadelphia.* 1799.
BONNER—
George. *Colchester.* Partner with Henry Clow
TURNER. 1893-1913.
T. *Brighton.* 1851-62.
BONNET, Jean A. & Co. *London.* 1851.
BONNIE. 1914 and later prod. of Pequegnat Clock Co.
BONSALL—
John. *London.* 1869. prob. same man as below.
John. *Lowestoft.* 1879. prob same man as above.
BONTHOUX, Alphonse. *Duck Lake, Saskatchewan.*
1924.
BONYA(?) (or **BONYER**), James. No place stated.
ca.1710. Lantern clocks. English.
BONYER (?or **BONVER**), Peter. *Redlench.* ca.1770-
1790. C.
BOOKER—
Godfrey. *London.* 1828-32. Dialmaker.
Henry. *Walsall.* 1876.
BOON & CLIFTON. *London.* 1881.
BOORD, John. *Wells.* 1675. Rep'd ch. clock.
BOORE, John. *Liverpool.* 1782-retired 1784. W.
BOORMAN, John Luke. *Gravesend.* 1839-66.
BOORSMA, G. *Dockum.* 1856. C.
BOOSEY, Nathaniel. *Yeldham.* 1851-74.
BOOTH—
Benjamin. *Pontefract.* b.1746-d.1806. W.
Charles. *Manchester.* mar.1792. W.
& Co. *Rochester.* 1794.
Dudley. *Bytown, Canada.* 1851.
Edward. *Birmingham.* 1880.
George. *Manchester.* 1760-d.1788. W.
George. *Manchester.* 1822-25.
George. B. s. of Benjamin. *Selby.* 1807-20, then
Pontefract. 1822-26. W.
George F. *London.* 1844.
Hiram N. *New York.* 1849.
James. *Auchinblae.* 1837-60.
James. *Dublin.* 1823-68.
James & Son. *Dublin.* 1874-1950s.
James. *Kirkburton.* d. 1749. W.
James. *Pontefract.* mar.1810-22. W.
James Bawker. *Manchester.* (B. 1772-1814). s. of
George, q.v. 1822-24.
John. *Huddersfield.* 1759-61. W.
John. *Kirkburton* 1755 - d. 1793. C. & W.
John. *Stalybridge.* 1824.
John. *St. John, New Brunswick.* 1795.
John. *Wakefield.* 1760. W. May be same man as at
Huddersfield.
John Richard. *Ashton-under-Lyne.* 1834.
Joseph. *Bridport.* 1824-48.
Joseph. *Dublin.* 1716-24.

BOOTH—*continued.*
Joseph. *Sandwich.* 1711.
P. *Dukinfield.* 1857.
R. *Woolwich.* 1812-17.
William. *Leeds.* 1817-26. W. & G.
William. *Skipton.* 1834. C.
William. *Wexford.* 1824-58.
BOOTHROYD—
James. *Reeth.* 1840.
Richard. *Bedale.* 1866. W.
BOOTLE(S), Thomas. *Wigan.* 1851-58. W.
BORDEN, Gail. Used by National Watch Co., *Chicago*
1871-75.
BORDER, George. *Sleaford.* 1828-68.
BOREHAM, Samuel. *London.* 1857-69.
BOREL, Henri Justin. *Chaux-de-Fonds.* 1851. C.
BORELLI—
Charles. *Farnham, Surrey.* 1851-78.
Gaeteno. *Reading.* 1877.
BORHEK, Edward. *Philadelphia.* 1829-40.
BORLEY, John Henry. *London.* 1857-75.
BORLINA, Edward. *London.* b.1806-51. W.
BORORET, Nicholas. *France.* ca.1690 W.
BORRELL—
George Henry. *London.* 1832.
John Henry. *London.* 1839.
Louis Howe. *London.* 1875.
Maximillian. *London.* 1847. W.
Maximilian John. *London.* 1832-63.
BORRETT, ——, *Newcastle-on-Tyne.* mid-18c. C.
BOSCO, ——. *Comber.* ca.1790-1830. Itinerant Italian.
Moondial painter.
BOSKELL, John. *Liverpool.* 1816. W.
BOSKET, R. *Liverpool.* Early 19c. W.
BOSQUET, F. T. *London.* 1869.
BOSS—
James. *Philadelphia.* 1846-59.
& PETERMAN. *Rochester, NY, USA.* 1841.
BOSSANGE, MOREL & CO. *Quebec.* 1852.
BOS(S)TOCK—
Joseph. *Capstone.* ca.1780. C.
Thomas. *Hyde.* 1878.
Thomas. *Sandbach.* 1834-57.
BOSTON CLOCK CO. *Chelsea, Mass, USA.* 1888-97.
succ. to Eastman Clock Co.
BOSTWICK & BURGESS MANUFACTURING CO.
Norwalk, O, USA. 'Columbus Clocks'. 1892.
BOSWELL, John. *Belfast.* ca.1786-1809. W.
BOSWORTH—
George. *Leicester.* 1864-76.
Reuben. *Nottingham.* Succ. W. Hall 1832-77 C.
BOTHAMLEY—
Benjamin. *Boston.* 1828-50.
H. W. *Boston.* 1861-76.
BOTHROYD, Richard. *Ferry Hill.* 1851. W.
BOTLY, George. *Reading.* 1837-54.
BOTSFORD—
J. S. *Troy, NY, USA.* ca.1840.
L. F. *Albany, NY, USA.* 1830.
Patrick. *New Haven, Conn, USA.* ca.1840.
S. N. *Hampden, Whitneyville, Conn, USA.* ca.1850.
BOTT—
Alfred Thomas. *Birkenhead.* 1878.
Edward H. *Leicester.* 1876.
Edwin. *Redditch.* 1850-76.
Jarvis. *Blackfordby.* Early 19c. W.
John. *Redditch.* 1828-35.
Samuel. *Loughborough.* 1835.
Samuel. *Melbourne.* 1828.
Thomas & Co. *Liverpool.* 1848-53.
T(homas & Co.). *Birkenhead.* 1865 (-78).
BOTTEN—
J. H. & J. *London.* 1851.
Jabez Henry. *London.* 1857.
Josiah. *London.* 1857-69.
BOTTERELL, James Hosking. *Liskeard.* 1873.
W.
BOTTGER, Gustavus Adolphe. *London.* 1857.
BOTTING, Ebenezer. *London.* 1881.
BOTTOM, John. *Sheffield.* 1862.
BOTTOMLEY—
James. *Long Melford.* 1858-64.

BOTTOMLEY—*continued.*
John. *Bradford.* 1743-49. C.
BOTTRALL, Thomas. *St. Just-in-Penwith.* ca.1820. C.
BOUCHET, ——. (*Paris?*). ca.1830. W.
BOUCKER, John. *Drayton.* 1850.
BOUCLY, Edward. *London.* 1869-81.
BOUDEWINS, Jan. *Antwerp.* ca.1580-ca.1600. Table clocks.
BOUDRY—
Gustavus. *London.* 1839-44.
John. *London.* 1832.
BOUGHELL, Joseph. *New York.* 1786-97.
BOUL, J. *Spalding.* 1849.
BOULT, Thomas. *Oakingham (Berks).* 1837.
BOULTER, Samuel. *London.* 1839-44.
BOULTON—
Henry. *London.* 1863-81.
T. *Coventry.* 1854.
BOULTWOOD—
——. *Matching Green.* ca.1760. C.
Thomas. *Cheshunt.* 1839.
BOURA—
——. *St. Mary Cray.* 1839.
W., sen. *Sevenoaks.* mar.1789-1802.
William, jun. *Sevenoaks.* 1823-51.
BOURDELOT, Edward L. *London.* 1851-81.
BOURDIN, A. E. *Paris.* 1844-67. C.
BOURE, Narcisse. *Quebec.* 1854-65.
BOURN(E)—
Aaron C. *London.* 1863-69.
Henry. *Rye.* 1828.
Loder. *Canterbury.* 1750-88.
BOURNER, T. *Ninfield.* 1862.
BOURQUIN—
Francis Henry. *London.* 1869. prob. same as man below.
Frederick Henry. *London.* 1844. prob. same as man above.
BOUSSET, V. *Morez (Jura).* ca.1880. C.
BOUSTON, John. *Philadelphia.* 1790.
BOUTCHER, ——. *Broadclyst.* mid-18c. C.
BOUTE, Lewis Charles. *Philadelphia.* 1839.
BOVE, Francis. *Dublin.* a.1687.
BOVET, Freres & Co. *London.* 1857. Fabian Sebastian Notermann, agent.
BOVEY, Mrs. Jane. *London.* 1863.
BOUVRIER, Daniel. *Zanesville, O, USA.* From France 1816.
BOUWER, William. *London.* 1869-81. (B. has one such Amsterdam ca.1800).
BOWDEN—
George. *Rainhill, nr Liverpool.* a.1747-81 (B. 1784). W.
Leopold Quinton. *Bristol.* 1877.
BOWELL, Alfred. *Croydon.* 1878.
BOWEN—
David. *Alfreton.* 1849-76.
David. *Swansea.* d.1791. C.
Francis. *London.* (B. a.1647, CC. 1655)-56.
George A. *Boston, USA.* 1842.
Henry. *Monmouth.* 1868-99.
James. *Matlock.* 1835.
John. *Bristol.* 1718-34. (May be from *London,* where one CC. 1709.)
John. *London. Long Acre.* 1811.
John. *London.* 1828-81. May be same man as above.
Owen. *Merthyr Tydfil.* 1822.
Thomas. *Bridgend.* 1791. (B. 1795).
Thomas. *Swansea.* First half 19c.
Thomas & Co. *Quebec.* 1890.
BOWER—
George. *Buxton.* 1828-35.
Henry. *Philadelphia.* ca.1820. C.
John H. *Landport.* 1859-78.
S. *Chesterfield.* 1855-64.
William. *Charleston, SC, USA.* 1772-83.
William. *Chesterfield.* 1800-29
BOWERS—
Andrew. *Cupar, Fife.* 1842.
George. *Philadelphia.* 1850.
John. *Leeds.* 1837.
BOWES—
James. *Bradford.* 1837-1871. C.

BOWES—*continued.*
John. *Helmsley.* 1851-66. W.
BOWKER, G. *Market Drayton.* b. ca. 1814 -75. W.
BOWLER—
John. *Brill.* 1831. W.
Joseph. *Manchester.* 1851. Foreign clocks.
Richard. *Manchester.* mar.1802. C.
BOWLES—
C. *Norwich.* 1865-75.
Edward. *Ringwood.* 1830.
Edward. *Thame.* mar.1783. C. & W.
John. *Dunstable.* 1847-77.
John. *Poole.* 1793 (B. 1795). W. & silversmith.
Samuel. *Wimborne.* 1824-30.
BOWLEY—
Devereux. *London.* b. 1696, a.1710, CC. 1718, d.1773.
William. *Melton Mowbray.* 1876.
BOWLING, William. *Leeds.* 1798-1807. C. & W.
BOWMAN—
Arthur Hart. *London.* 1881.
Edward. *London.* 1832.
Edward. *London.* 1857.
Edward Charles. *London.* 1875.
Ezra F. *Lancaster, Pa, USA.* 1847-1901.
George. *Columbus, O, USA.* 1840s. C.
John. *Bishop Auckland.* 1834.
Johnston. *Carrickfergus.* 1854-92. W. & C.
Richard. *Manchester.* 1851.
BOWNE (or **BROWNE**), Samuel. *New York.* mar.1778-99. C. (B. 1751?).
BOWNESS—
George. *Lancaster.* b.1836-69. W.
John. *Osmotherly.* 1840. C. Cf. BOANAS.
BOWRA—
George. *Sevenoaks.* 1835-51.
James. *Sevenoaks.* ca.1760. C.
John. *London.* (B. 1820)-28.
William (E.). *Sevenoaks.* 1823-55 (-66).
BOWRING—
B(enjamin). *Exeter.* Free 1734-ca.1790. C.
Benjamin. *St. Johns, Newfoundland.* 1815.
BOWTEY, John. *London.* 1759. W.
BOX—
E. & J. *Dursley.* 1856.
Edward. *Dursley.* 1863-79.
John. *Berkeley.* 1850.
John. *Dursley.* 1830-50.
John. *Launceston.* 1823-47.
John. *Wotton-under-Edge.* ca.1800. C.
Joseph. *Cheltenham.* 1863-70.
William. *Marhamchurch.* ca.1830. C.
William. B. *London.* 1851-81.
BOXALL—
T. *Brighton.* 1851, but see BOXELL.
T. *Hurstpierpoint.* 1855.
BOXELL—
F. *Cuckfield.* 1855.
T. *Brighton.* (1851)-55-78.
BOXER—
John, sen. s. of Michael. *London.* 1779-pre-1804, then *Folkestone* where free 1804. W.
John, jun. s. of John, sen. *Folkestone.* b.1779-d.1852.
Michael. *Folkestone.* 1780-?d.1810. W.
BOYALL, Joseph. *Spilsby.* 1861-76.
BOYCE—
Abraham. *London.* Not sure whether a clockmaker, but certainly trained some. a.via B. Co. 1607-1641.
Barrington. *Norwich.* 1875.
Benjamin M. *Boston, USA.* 1880-95.
Joseph Alfred. *London.* 1875-81.
Robert. *Ballymacarrett.* 1868. Watch glasses.
S. H. *Dereham.* ca.1780-ca.1800. C. (prob. too early estimate for next man.)
Samuel H. *East Dereham.* 1830-65.
Thomas. *Gravesend.* 1839.
BOYD—
Alexander. *Belfast.* 1814. W.
George. *Belfast.* ca.1825-d.1827.
George. *Quebec.* s. of George of Belfast. d.1832. W.
H. *Blairsville, Pa, USA.* ca.1830. C.
Hugh. *Newry.* 1785.
John. *Lowell, Mass, USA.* 1861.

BOYD—*continued.*
John. *Sadsburyville, Pa, USA.* 1805-60. W. & C.
Lawrence. *Puddleton.* Early 18c. C.
Michael. *Galway.* 1858.
Robert. *Kilkeel.* 1858-90. C.
Samuel. *Quebec.* 1855-58.
William. *Johnstone (Scotland).* 1860.
BOYER, Jacob. *Boyertown, Pa, USA.* 1754-90.
    C.
BOYLAN—
Patrick. *Dublin.* 1770-1828.
Robert. *Dublin.* 1831-38.
BOYLE—
Philip. *Ballinsloe.* 1858. W.
William. *London.* (B. 1800-21) 1828-51.
BOYNTON—
Calvin. *Buffalo, NY, USA.* 1859.
Frederick. *Southsea.* 1878.
John E. *Manchester, Iowa, USA.* ca.1870.
BOYTER, Daniel. *Boston, USA.* 1800. Also *Pough-keepsie, NY, USA.* 1803.
BRABINER, W. A. *Halifax, Nova Scotia.* 1816.
BRABROOK, Arthur. *Bury St. Edmunds.* 1865-79.
BRABY, J. *Tunbridge Wells.* 1866.
BRACE—
Joshua. *Chepstow.* (B. 1763-) ca.1780. C.
(Rodney) & PACKARD (Isaac). *Brockton, Mass, USA.* ca.1825.
Thomas. *Leicester.* ca.1790-ca.1820. C. & W.
William H. *St. Albans* and *Greenville, O, USA.* 1826-27.
BRACEBRIDGE—
Edward Charles & Co. *London.* 1851-81.
James. *London.* 1863.
James & Edward Charles & Co. *London.* 1828-44.
BRACEWELL—
Hartley. *Scarborough.* 1823.
Henry. *Scarborough.* 1866.
William. *Scarborough.* 1834-58.
BRACHER—
R. *Reading.* 1864.
& SYDENHAM. *Reading.* 1877 - 99.
BRACKETT, J. R. *Boston, USA.* 1842.
BRADBERRY—
Matthew. *Bolton.* 1824.
Robert. *Richmond (Yorks).* 1830. C.
BRADBORNE, John. *Westerham.* (B. 1744-) 1764. C. & W.
BRADBROOK, George Richard. *London.* 1875.
BRADBURY—
Jacob. *New York.* 1847.
Matthew. *Leyburn.* 1806-18. W.
Thomas. *Nottingham.* 1876.
W. *Melbourne.* 1864.
William. *Castle Donnington.* 1835-64.
William. *Melbourne.* 1835-56.
BRADCHURCH, George. *Whitchurch.* ca.1800. C.
BRADDOCK—
George. *Martham.* 1836-75. W.
James. *Cheadle.* 1860-76.
James. *Colchester.* b.1777-d.1854. W.
John. *Manchester.* a.1744. C. From Derbyshire.
Samuel. *Macclesfield.* 1878.
BRADFORD—
——. *Newton Abbot..* l.c. clock. Date not known.
E. *Kenilworth.* 1860.
Henry. *London.* (B. 1820) 1839.
J. *Tiverton.* 1795.
James. *Cardiff.* 1875.
John. *Castle Donnington.* 1828-55.
John. *East Ashby.* ?early 19c. l.c. clock.
John. *Kenilworth.* 1880.
Nathaniel. *Enniskillen.* 1824-54. C.
O. C. *Binghampton, NY, USA.* 1841.
Phillip. *London.* 1875-81.
Richard & William. *Cork.* 1858.
Robert. *Rathfriland.* 1814-d. 1831. W.
Robert. *Enniskillen.* 1858.
S. *Melbourne.* 1849-64.
Thomas. *Buckingham.* 1738-75. C.
BRADLEY—
——. *Halifax.* 1755.

BRADLEY—*continued.*
——. *Huddersfield.* 1814.
& AUSTIN (Orrin). *Straitsville, Conn, USA.* ca.1830. C.
B. & Co. *Boston, USA.* 1862. Successor to Jerome Mfg. Co.
& BARNES. *Boston, USA.* 1856.
D. W. *New York.* ca.1850.
DIMON, *Cairo, NY, USA.* 1814.
G. C. *Binghampton, NY, USA.* 1841.
George. *Sheffield.* 1871.
H. *Marietta, O, USA.* ca.1810.
Horace P. *Rochester, NY, USA.* 1832-40.
& HUBBARD. *Meriden, Conn, USA.* 1854-56.
& HUBBARD MFG. CO. *Meriden, USA.* post-1854-ca.1915. C.
James. *Tadcaster.* 1837-51.
James Gibbon. *Liverpool.* (B. 1825) 1828-34.
Jeremiah. *Long Sutton (Lincs).* 1828-68.
John. *Blackburn.* 1824.
John. *Louth.* 1876.
John. *Todmorden.* 1822.
Joseph. *Hanley.* 1860-76.
Joseph. *Shelton.* 1842.
Joseph. *Stoke-on-Trent.* 1850.
Lucius B. *Watertown, Conn, USA.* 1800-70. C.
Nelson. *Plymouth, Conn, USA.* 1840.
Richard. *Hartford, Conn, USA.* 1825-39.
Samuel. *Birmingham.* 1880. Importer.
Samuel. *Worcester.* 1728. W. (-d.1783).
Thomas. *Newent.* 1830-42.
W. *Chesterfield.* 1864.
William. *Lancaster.* From *Liverpool.* pre-1831-51. C.
William. *Liverpool.* b.1785. to *Lancaster* pre-1831.
William. *Newent.* 1850.
William. *Petersfield.* 1878.
Z & Son. *New Haven, Conn, USA.* 1840.
BRADNOR, William. *Birmingham.* 1842.
BRADSHAW—
George. *Ellesmere.* 1835.
George. *Whitchurch.* ca.1790. C. May be too early estimate of next man.
George. *Whitchurch.* (b. ca. 1796) 1828-56.
Henry Charles. *Wye (Kent).* 1832. W.
James. *Blackburn.* 1834-58.
James. *Prescot.* 1818-25. W.
John. *Liverpool.* (B. 1825). 1834-51. W.
John. *Manchester.* (B. 1765-1814). 1822-34. W. See next entry also.
John. *York.* a.1754-70. Believed later at *Manchester.*
Joseph. *Whitchurch.* 1863-79.
Peter. *St. Helens.* 1767-80. W.
& RYLEY. *Coventry.* 1816. W.
S. *Twyford.* 1854.
Samuel. *Tring.* 1874.
T. & W. *Bolton.* 1858.
Thomas. *Canterbury.* ca.1790. W. (B. 1804).
Thomas. *Prescot.* mar.1793-97. W.
Thomas. *St. Albans.* 1828-51.
W. *Bolton.* 1858.
William. *Blackburn.* 1851.
William. *Liverpool.* (B. 1810-)1814-34.
BRADY—
Daniel Henry. *Brompton (Kent).* 1847-58.
Edward. *Tuam.* 1824-58. W.
John. *Philadelphia, USA.* 1835.
Sampson. *Antrim.* ca.1790-1825. C. & W.
BRADYCAMP, Lewis. *Lancaster, Pa, USA.* 1836-60. C.
BRAENDL, ——. *Vienna.* ca.1815. C.
BRAGG—
Edward. *Windsor.* 1877.- 99.
Robert. *Stroud.* 1863-70.
BRAHAM—
Joseph. *London.* 1839-51.
Joseph. *London.* 1875.
Joseph John. *London.* 1857.
BRAITCH, A. *Midhurst.* 1855.
BRAITHWAITE—
Charles. *Bedale.* 1758. C.
John. *Hawkshead.* Believed early 18c. C.
Lemuel. *Belfast.* 1850. W.
Samuel. *Belfast.* 1864-68.

BRAITHWAITE—continued.
William. Hawkshead. b.1782 d.1829.
William. Reeth. b.1740-1800. C.
BRAKE, Frederick. Albany, NY, USA. 1840-44.
BRALY, Jo. London. ca.1720. W.
BRAMBLE—
Abraham. Andover. 1830.
E. London. 1863-81.
Edward. London. 1844-51.
G. A. USA. Place Unknown. ca.1858.
Mary. London. 1839.
Joseph. London. (B. 1805-20). 1828-32. C.
BRAMBLES, Robert. Leeds. 1866.
BRAME, Philip & Co. London. 1881.
BRAMER, Paulus. Amsterdam. mid-18c. C.
BRAMLEY—
Thomas. Andover. 1867-78.
W. Andover. ca.1790. C.
William. Newbury (and Basingstoke). 1826-43.
BRAMPTON. 1914 and later, prod. of Pequegnat Clock Co.
BRAMWELL—
A. Stockton. 1898.
& HARBRON. Darlington. 1898.
Thomas. Alston. 1869-79.
BRANCH—
C. Ixworth. 1853.
R., jun. Norwich. 1858-75.
BRAND—
Edward. London. 1851.
George. Howden. 1792-97. W.
James. Boston, USA. 1711. From London.
John. Boston, USA. 1711-14. W. From London.
Robert. Dumfries. 1788-1816.
William. Aberdeen. 1860.
BRANDEGEE, Elishana. Berlin, Conn, USA. ca.1830.
BRANDON—
1914 and later, prod. of Pequegnat Clock Co.
Aron Jozef. London. 1875.
BRANDRETH & WALKER. Birmingham. 1880.
BRAN(D)T—
A. & Co. Philadelphia, USA. ca.1803-04.
Adam. New Hanover, Pa, USA. ca.1750.
Laurent. London. 1881.
BRANFORD—
Edward. Ashford. 1874.
BRANSCOMBE, Joseph. Exeter. a. of Will Hoppin, locksmith. Free 1648-73. Attended ch. clock
BRANSON, John. Hull. 1768. C.
BRANTFORD. 1914 and later, prod. of Pequegnat Clock Co.
BRANWELL, Robert Matthews. Lostwithiel. ca.1798. Then Penzance. 1798-1807. W.
BRASEA, N. Montreal. 1862.
BRASHER, Abraham. New York. 1786-1805.
BRASTOW, Adison & Co. Lowell, Mass, USA. 1832-1837.
BRAULT (DIT. POMINVILLE), Joseph. Montreal. b.1770-d.1834.
BRAUN, Paulus. Augsburg. ca.1600. C.
BRAUND—
Thomas. Launceston. 1844. W.
William. Dartford. 1823-55.
BRAY—
H. T. Shirley. 1867.
William Thomas. London. 1839-63.
BRAYLEY—
Joseph. London. 1881.
Joseph W. London. 1863-69.
William. London. 1857-69.
BRAZEAU—
Edmond. Montreal. 1862.
Guillaume. Montreal. 1866.
L. E. L. Montreal. 1862.
Napoleon & Co. Montreal. 1862.
BRECKENRIDGE (also BREAKENRIDGE)—
Alexander (& Son). Kilmarnock. 1799-d.1848. (cont'd. by dtr. till 1898.) (B. has Edinburgh?.)
J. M. New Haven and Meriden, Conn, USA. 1809-96.
James. Edinburgh. ca.1800-ca.1825. C.
William. Kilmarnock. 1850.
BRECKNALL, Charles. Walsall. 1876.

BRECKNELL—
J. Birmingham. 1860.
J. Wolverhampton. 1868.
Job. London. 1881.
Job. Peckham. 1878.
Noah junior. Birmingham. 1860-68.
& Son. Wolverhampton. 1860.
T. Birmingham. 1854.
BRECKWELL, John. Chester Springs, Pa, USA. 1831-35.
BREESE—
James. London. 1839-51.
James. Wisbeach. 1830.
Lyman. Wellsburg, Va, USA. ca.1830.
BREMER, Paulus. Amsterdam. 18c. C.
BREMNER—
Alexander. Portsoy. 1860.
William. Kirkwall. 1835.
BRENCHLEY, George. London. 1869-81.
BRENEISER, Samuel I. Adamstown and Reading, Pa, USA. 1799. C.
Samuel II. Believed s. of Samuel I. Reamstown and Reading, Pa, USA. ca.1830-60. C.
William. bro. of Samuel I. Womelsdorf, Pa, USA. ca.1800. C.
BRENFTER, Walter. Canterbury, Conn, USA. ca.1800. C.
BRENKELEAR, Jan. New York. pre-1750. May be from Amsterdam where one such recorded early 18c.?.
BRENNAN, Barnabas. Philadelphia, USA. 1843.
BRENTNALL—
Thomas. Sutton Coldfield. (B. 1795)-1828.
William. Mold. 1868.
William. Sutton Coldfield. 1835-60.
BRERETON, J. W. Worcester. 1868-76.
BRESLAR, Raphael. London. 1863.
BRETAR, Bayley. Cork. 1775-77. W.
BRETON, Thomas. Cork. 1858.
BRETT—
Robert. London. 1639. CC.
Thomas. London. ca.1760. C.
BRETTELL, Walter. London. 1881.
BREW, Philip. Dublin. d.1773. W.
BREWER—
——. Middletown, Conn, USA. 1800-03.
——. Mevagissey. 1712. rep'd. clock.
John. Rochdale. b.1693-d.1720. C.
Jonathan. London. 1811.
Richard. London. Oxford Street 1783. Then to Lancaster. C.
Richard. Norwich. ca.1720.
Richard. Prescot. a.1760. W.
Thomas. Clitheroe. 1822.
Thomas. Preston. 1817-58. W.
Thomas. Selby. ca.1760-70. C.
Thomas A. Philadelphia, USA. 1830-ca.1850.
William. Blackburn. (B. 1814-) 1824-28.
BREWERTON—
E. Longnor. 1860.
S. Northfleet nr. Gravesend. 1851-74.
BREWIN—
George. March. 1858-75.
George. Wisbeach. 1840-65.
Thomas D. Leicester. 1864-76.
BREWSTER—
Abel. Canterbury, Conn, USA. 1796-1800. Then Norwich Landing. 1800-07.
& BROWN. Bristol. 1839-40.
Charles E. Portsmouth, NH, USA. ca.1840.
& Co. (Elisha C.). Bristol, USA. 1840-43.
& Co. N. L. Bristol, USA. 1861.
E. C. & Son (Noah). Bristol, USA. 1855.
Elisha C. Bristol, USA. 1791-1800. C.
George G. Portsmouth, NH, USA. ca.1850.
(Elisha) & INGRAHAMS (Elias & Andrew). Bristol, USA. 1844-52. Also New York. 1848.
(Elisha) & IVES. Bristol, USA. pre-1844.
MFG Co. Bristol, USA. 1852-54.
Noah Lewis & Co. London. 1863. American clocks.
& WILLARD. Portsmouth, NH, USA. 1830s.
BREWSTON, Samuel. Northfleet (Kent). 1847-51.
BRIANT, Robert. London. 1857-81.

BRICE—
——. *St. Anns.* ca.1860. C.
Francis Henry. *Chippenham.* 1875.
John. *Sandwich.* a.1723. Free 1740. d.1755. C.
W. *Sandwich.* s. of John. a.1752-1803.
BRICKLAND, Humphry. *Oxford.* 1723-d.1750. C. &
W.
BRID, Robert. *Yeldham.* ca.1760. C. See also BIRD.
BRIDEKIRK, John. *Scarborough.* 1851.
BRIDGE—
A. *Midhurst.* 1851.
E. *London.* 1777. W.
Henry. *Farnworth.* mar.1794. W.
John. *East Dulwich.* 1878.
Samuel. *Colchester.* 1768-96. Watch case maker. prob.
worked mostly in *London* where one such a.1762.
Thomas. *Bolton.* d.1717. C.
Thomas. *Manchester.* mar.1739. C.
Thomas. *St. Helens.* mar.1771-1807. W.
William. *London.* 1674. (Clockmaker or blacksmith).
BRIDGES—
John. b. ca. 1775 *Ipswich.* a. *London* then *Ipswich*
ca. 1798 -d. 1831. Also silversmith. Bro. of William
William. *Colchester.* b.ca.1774-1814.
BRIDGMAN, Jeremiah. *Dublin.* 1784-1815. Watch case
maker and goldsmith.
BRIDGMAN, Richard. *London.* 1881.
BRIERLEY—
John. *Manchester.* 1871.
John. *Saddleworth.* 1866.
Joseph. *Ashton-under-Lyne.* 1848-51.
BRIGDON, George S. *Norwich, Conn, USA.* ca.1810-
1830.
BRIGGS—
E. H. J. *Norwich.* 1875.
G. *Wisbech.* 1865.
George. *Grantham.* 1849-50.
J. *Ollerton.* 1849.
J. *Tuxford.* 1855-64.
J. T. *Spalding.* 1850-76. W.
John. *Leeds.* 1866.
John. *Skipton.* 1822-53. C.
John C. *Concord, NH, USA.* 1855-70s.
Luke. *Holbeach.* 1828.
Luke. *Wisbech.* 1830-58.
Nathaniel. *Boston, USA.* 1820.
Robert. *Doncaster.* d.1631. Clock keeper.
Robert. *Doncaster.* 1663.
Stephen. *London.* 1881.
William. *Ilkeston.* 1864-76.
William. *Nuneaton.* 1850.
BRIGHT—
——. *Paris.* ca.1800. C.
Edward. *Brighton.* 1862-78.
H. *Leamington.* 1860-68.
Henry. *Birmingham.* 1842.
Henry & Edward. *Leamington.* 1850-54.
Horace. *South Weald.* 1866-74.
Isaac & Son(s). *Sheffield* ca. 1797-1820, then
Buxton ca. 1817-88, also *Leamington* 1835-42.
Jerome Denny. *Saxmundham* b. 1792, s. of Jerome
senior whom he succ. pre-1829. Mar. 1818 and again
1827. Ret'd. ca. 1840, d. 1871. Succ. by nephew Jerome
Wooltorton.
John & Son. *Sheffield.* 1804-37.
Philip. *Doncaster.* b.1784 T.C. & W. d.1841.
Samuel. *Caterham.* 1878.
Selim. *Buxton.* 1876.
& Sons. *Scarborough.* 1858.
& WOODMANSEY. *Doncaster.* 1851. Successor to
Philip Bright. See also WOODMANSEY, Henry.
BRIGHTWELL, Thomas. *Colchester.* b.1791-d.1823.
W.
BRILLMAN & CO. *London.* mid-late-19c.
BRIMBLE & ROUCKLIFFE. *Bridgwater.* ca.1770.
BRIMILOW, Peter. *Prescot.* b.1766 d.1808.
BRINCKERHOF, Dirk. *New York.* 1756.
BRINDLE, Ralph. *Liverpool.* (B. 1825). 1834.
BRINDLEY—
A. *Newcastle-on-Tyne.* ca.1790. C.
James. *Newcas'le-under-Lyme.* 1835-50. W.
John Beavis. *Newcastle-under-Lyme.* 1850.
Joseph. *Leeds.* 1871.

BRINE—
F. *Iwerne Minster.* 1875.
J. *Iwerne Minster.* 1848-67.
BRINKLEY, William. *Canterbury.* From *London.* 1768 W.
BRINKMAN—
& GOLLIN. *London.* 1844.
Henry. *London.* 1832-57.
BRINKWORTH, William. *Dundas, Canada.* 1862.
BRINTON, Ralph. *Brighton.* 1878.
BRISBROWN, Thomas. From *Edinburgh* to *Quebec.*
pre-1785.
BRISCALL—
James. *Birmingham.* 1842-68.
Robert. *Birmingham.* 1854-80.
BRISTER, James. *Wanstead.* 1874.
BRISTOL—
Brass & Clock Co. *Bristol, USA.* 1850-1903.
Brass Clock Manufactory. *Bristol, USA.* 1818-19.
Clock Case Co. *Bristol, USA.* 1854-57.
Clock Co. *Bristol, USA* and *New York.* 1843-54.
Manufacturing Co. *Bristol, USA.* 1837.
BRISTOW—
Richard. *Maidstone.* 1874.
William. *London.* 1828-32.
BRITISH—
& COLONIAL & GOOD TEMPLAR (The Watch Co.).
London. 1875. (R. Squire, manager).
INDUSTRIAL WATCH & CLOCK ASSOCIATION. *Lon-
don.* 1869-75. (Lewis Hasluck, managing director
1869; Paul Nooncree Hasluck, managing director
1875).
WATCH & CLOCK MAKING CO. *London.* 1844.
BRITT—
W. *Bristol.* 1856.
W. *Melcombe Regis.* 1867.
BRITTAIN—
Henry. *Norwich.* (B. 1795-)d.1822. C.
J. *Docking.* 1836. (See also BRITTON).
Joseph. *Bakerstown, Pa, USA.* ca.1830.
Robert. *London.* 1839.
BRITTAN, John. *London.* 1857-69.
BRITTON—
J. *Docking.* 1858. C. (See also (BRITTAIN).
J. J. *Stockton.* 1898.
John. *Nottingham.* 1876.
Sandys. *London.* 1832.
BROAD—
John. *Bodmin.* s. of Richard. 1809-56. W.
John Butler. *Wadebridge.* 1856-73.
Joseph. *Bodmin.* s. of Richard. 1805-13. C.
Richard. *Bodmin.* ca.1785-1825. W.
Robert. *London.* 1828.
William. *Liskeard.* 1873. C. & W.
William. *London.* (B. CC. 1792-1820). 1828-44. W.
William. *Lostwithiel.* 1844-56. W. & C.
William Henry. *Bodmin.* 1873. W.
BROADBELT—
George. *Newcastle-on-Tyne.* 1834.
Michael. *Bishop Auckland.* 1760-d.1796. C.
BROADBENT—
John. *Ashton-under-Lyne.* 1834-48.
T. *Ashton-under-Lyne.* 1858.
BROADBRIDGE, James. *Newburgh, NY, USA.*
1806-32.
BROADHEAD—
E. *Belper.* 1849-64.
E. *Derby.* 1876.
BROADWATER, Hugh. *Oxford.* Free 1688-89. Then
moved to *London* where CC. 1692-97. W.
BROADWAY, Edward. *Upton-on-Severn.* 1828-50.
BROADY, J. *West Hartlepool.* 1898.
BROCK—
——. *Axbridge.* ca.1785. C.
——. *Lewisham.* ca.1830. C.
——. *Woolwich.* 19c. C.
George. *Axbridge.* Early 18c. C.
James. *Axbridge.* Late 18c. C.
James. *London.* 1857-81.
John. *Lewisham.* 1839. chrons.
John. *London.* 1844-51.
& SCOTT. *London.* 1851. (Successor to John Brock).
Thomas. *Bristol.* 1792-1807. C. & W.

BROCK—*continued.*
William. *Axbridge.* ca.1720. C.
BROCKBANK—
& ATKINS. *London.* (B. 1815-35)-39.
& ATKINS. *London.* 1863-81. (?perhaps son of first ATKINS).
ATKINS & SON. *London.* (B. 1840). 1844-57.
William. *London.* 1851-63.
BROCKEDON—
F. *Totnes.* 1787-ca.1805. C.
William. *Totnes.* b.1787-d.1877. W.
BROCKHURST, Benjamin. *Coventry.* ca.1720-40. C.
BROCKHUYSEN, A. V. *?Rotterdam.* 18c. W.
BROCKLEBANK—
H. *Coventry.* 1854-60.
William. *London.* 1844.
BROCKWELL, John. *London.* 1839-44. Clock case maker.
BROCOT, Achille (& Son). *London.* (B. b.1817-d.1878). 1863-69. (1875-81).
BRODERICK—
Edward. *Newbottle.* 1834.
Francis. *Whittlesey.* 1830.
Sarah. *Peterborough.* 1841.
T. *Peterborough.* 1854-64.
T. & Son. *Holbeach.* 1828.
Thomas. *Holbeach.* 1835.
Thomas. *Kirton* (*Lincs*). mid-18c.-ca.1770. C.
Thomas, jun. *Spalding.* 1828(-35).
William. *Boston.* 1835-76.
William. *Dublin.* 1825-58. C. & W.
William, *Swineshead.* 1835.
BRODERSEN—
Christian. *Russellville, Ky, USA.* 1834-87.
Emil. *Cincinnati, O, USA.* bro. of Christian. From *Denmark.* pre-1860.
BRODIE—
George. *Hertford.* 1839-59.
James. *St. Albans.* 1866-74.
John. *Wooler.* 1858. (B. early 19c.).
BRODRICK, Thomas. *Preston.* 1834.
BRODY, Solomon. *London.* 1863.
BROGAN—
Daniel. *Dublin.* a.1718. W.
John. *New York.* ca.1810.
BROGDEN—
George. *Knaresborough.* 1769-d.1779. C.
Joseph. *York.* 1774-1807. W.
Robert. *York.* 1713-39. (B. 1763).
BROKAW—
Aaron. *Elizabethtown, NJ, USA.* s. of Isaac. 1768-1850s.
Isaac. *Hillsborough, NJ, USA.* 1746-1826.
John. *Woodbridge, NJ, USA.* bro. of Aaron.
BROKESMIELD, Joseph. *Cincinnati, O, USA.* ca.1830
BROMFIELD, J. *Coventry.* 1860.
BROMILOW—
George. *Prescot.* 1794. W.
Thomas. *St. Helens.* 1820-37. W.
BROMLEY—
Charles. *Halifax.* 1820-34.
E. H. *Bexley Heath.* 1851-55.
Edward. *Halifax.* 1834-66.
Henry. *Bexley Heath.* 1847-74.
J. & Son. *Horsham.* 1855.
John. *Horsham.* (B. 1804-) 1828-39.
John. *London.* 1828.
L. *Halifax.* mid-18c.
M. *Horsham.* 1862-78.
T. *Woolwich.* 1851-55.
Thomas P. *Erith.* 1866-74.
William. *Bexley Heath.* 1832-39.
William. *Halifax.* 1801-66.
BROMWICH—
J. *Coventry.* 1860-80. Clock cleaner.
J. *Dudley.* 1860.
Joseph. *Nottingham.* 1876.
Mrs. H. *Dudley.* 1868.
T. *Coventry.* 1860.
BRON, Templeton. *Montreal.* 1842.
BRONNEN, Thomas. *Buckingham.* ca.1775. C.

BRONSON—
Bennet. *Waterbury, Conn, USA.* ca.1800.
I. W. *Buffalo, NY, USA.* 1825-30.
John. *Stowmarket.* pre-1766.
John. *Walsham.* ca.1760-ca.1780. C.
Pharris. *Waterbury, Conn and Cairo, NY, USA.* 1814.
BROOK(E)—
A. *Coventry.* 1880.
Alfred. *Longton.* 1876.
Charles. *Normanton.* 1871.
Frederick Herbert. *Builth.* 1887.
H. *Coventry.* 1868.
Henry. *Tamworth.* 1876.
John. *Battle.* a.1726 to Samuel Hammond. mar.1737. C.
John. *London.* 1832.
John. *New York.* c.1830.
John. *Wakefield.* b.1732 a.1751.
John Walter. *London.* 1828.
Richard. *Rotherham.* 1862. W.
Sarah. *Coventry.* 1835.
Thomas. *Coventry.* 1828.
Thomas. *Coventry.* 1860-80.
Thomas. *Maidstone.* 1604. Clocksmith.
Thomas Harvey. *London.* 1863.
BROOKER, R. *East Grinstead.* 1851-66.
BROOK(E)S—
Abel. *Stalybridge.* 1848.
B. F. *Utica, NY, USA.* 1828-37.
B. F. & Co. *Utica, NY, USA.* 1831.
Bernard. *Salem Crossroads, Pa, USA.* ca.1830. C.
Charles. *Stamford.* 1849-50.
Charles V. *Utica, NY, USA.* 1834-40.
Eli. *Sevenoaks.* 1847-55.
F. O. *Madison, Ind, USA.* ca.1860.
& GRISWOLD. *Utica, NY, USA.* 1832.
Henry. *Cork.* 1858.
Henry. *Coventry.* 1842-54.
Hervey. *Goshen, Conn, USA.* ?1830s.
J. *Boston.* 1849.
J. E. *Birmingham.* 1860.
James. *Middleham.* 1866.
John. *Birmingham.* 1868.
John. *Boston.* 1850-61.
John. *London* (*4 Bridgewater St.*). 1811.
John. *Stratford-on-Avon.* 1842-50.
L. S. *Mass, USA.* pre-1848.
Mrs. Sarah. *Stow.* 1850-56 (?wid. of William).
Richard. *Oxford.* a.1765.
Samuel. *London.* 1832. Watch case maker.
Thomas. *Cork.* 1794-1810. W.
W. *Stow.* ca.1790-ca.1800. C. (Perhaps too early estimate for William, q.v.).
Watts. *Goshen, Conn, USA.* s. of Hervey. 1830s.
William. *Dursley.* 1842.
William. *Kenilworth.* 1842-54.
William. *London.* 1828.
William. *Stonehouse.* 1850-79.
William. *Stow-on-Wold.* (B. ca.1800). 1830-42.
William Penn Beal. *Boston, USA.* 1856. W.
BROOKHOUSE—
Joseph. *Derby.* 1835.
R. *Barton-on-Humber.* 1861.
Richard Burton. *Newcastle-under-Lyme.* 1850.
Robert. *Easingwold.* 1866. W. (B. has ca.1800-?too early).
Sarah. *Sheffield.* 1862. W.
BROOKSBANK, William. *Bradford.* 1837-71.
BROOM—
Charles. *Prescot.* mar.1823. W.
H. *Ipswich.* 1865.
BROOMALL, Lewis R. *Philadelphia.* 1845-50.
BROOMFIELD, Louis. *Glasgow.* 1860.
BROS, John. *London.* 1832-44.
BROSELO, Pietro. (Place unknown—*Europe*). ca.1800.
BROTHERS—
James. *Coventry.* 1880.
John. *Coventry.* 1868-90.
BROTHERSON & MACKAY. *Dalkeith.* 1830-44. C. (Also BROTHERSTONE).
BROTHERTON, George. *London.* 1851-75.

BROUGHTON—
Henry. *Swansea.* 1868-99.
John. *Ash.* ca.1770. C.
BROUWER, A. J. *Tiel.* ca.1851.
BROWER, S. Douglas. *Troy, NY, USA.* 1841-42.
BROWN(E)—
——. *Elgin.* post-1820.
——. *Liverpool.* ca.1735. C. prob. Joshua BROWNE, q.v.
——. *Maidstone.* 19c. Regulator.
——. *Newtownards.* ca.1790-1805. C.
——. *Sevenoaks.* 1793. W. 'Next door to John Osbourn'.
A. *Rayleigh.* 1855.
Aaron. *Erith.* pre-1769-76. W.
Abraham. *Wirksworth.* 1864-76.
Adam. *Comber.* 1824-46.
Albert. *Columbia, Pa, USA.* ca.1850.
Alexander. *Coatbridge.* 1847.
Alexander. *Dudley.* 1868-76.
Alexander (& Co.). *Glasgow.* 1836. (37-41).
Alfred. *Halifax, Nova Scotia.* 1866-72.
Alfred. *London.* 1857.
Alfred. *London.* 1875.
Alfred. *Winchester.* 1867-78.
(Thomas W.) & ANDERSON. *Wilmington, NC USA.* 1850-71.
Benjamin. *Newcastle-on-Tyne.* 1731. Journeyman of Abraham Fromanteel. (B. pre-1746. W.).
Brothers. *Chester.* 1865-78.
Brothers. *Coventry.* 1850-80.
& BUCK. *Columbus, O, USA.* 1850.
& BUTLER. *Bristol, USA.* 1853.
C. *Bridport.* 1848.
C. F. *Coventry.* 1854.
& CHALMERS. *Leith.* 1842.
Charles. *Eccleshall.* 1850. (*Staffordshire*).
Charles. *Edinburgh.* 1850-60.
Charles. *London.* 1857.
Charles. *Selby.* 1834-46. T.C.
Charles. *Stranraer.* d.1844.
& Co. *London.* ·1881.
Daniel. *Glasgow.* 1783.
David. *Edinburgh.* a.1772-81.
David. *Providence, RI, USA.* 1834-50.
E. *Coventry.* 1860.
Edward. *Manchester.* 1848-51.
Edward. *Norwich.* 1726. W. (B. pre-1756-ca.1775. C.).
Edward L. *Newburyport, Mass, USA.* 1860.
Eli. *Norwich.* 18c. C. (?perhaps Edward).
Elisha. *Northwich.* 1828.
G. *Chester.* 1865.
G. *Kings Lynn.* 1858.
Gawen. *Boston, USA.* b.1719. *England.* There by 1749-1801. C.
George. *Airdrie.* 1817-37.
George. *Arbroath.* 1834-60.
George. *Bingley.* 1871.
George. *Dunscore.* 1855. W.
George. *Edinburgh.* a.1772-1825.
George. *Farnworth.* mar.1831. W.
George. *Glasgow.* 1830.
George. *Lewes.* a.1720 to Richard Turner. C.
George. *London.* 1844-57.
George. *Macclesfield.* 1848.
George. *Minehead.* 1875-83.
George. *Newark.* 1855-76.
George. *Selby.* 1837.
George. *Wrexham.* 1874.
George B. *Leith.* 1836.
George Henry. *Bradford.* 1871.
H. *Bridgwater.* 1890-97.
H. *Winchester.* 1867.
H. & Sons. *Halifax, Nova Scotia.* 1871.
Harris Leon. *Sheffield.* 1871.
Henry. *Bristol.* 1801.
Henry. *Halifax, Nova Scotia.* 1838-63.
Henry. *Halifax, Nova Scotia.* 1864-72.
Henry. *Liverpool.* a.1752-d.1773. (B. 1761-96 must be wrong.)
Henry. *St. Helens.* 1786. W.
Henry. *St. Helens.* 1809-13. W.

BROWN(E)—*continued.*
Henry John. *Doncaster.* 1837-51.
Henry, jun. *Halifax, Nova Scotia.* 1866-72.
Henry Thomas. *London.* 1857-75.
Ilisha. *Chester.* 1848.
Isaac. *London.* 1832-39.
J. *Gloucester.* 1863.
J. *Kingsbridge.* ca.1800. C.
J. *Prescot.* 1858. Chrons.
J. F. *Napanee, Canada.* 1853.
J. H. *Des Moines, Ia, USA.* ca.1850.
J. & T. *Aberford.* 1844. W.
J. W. *Chichester.* 1855.
Jacob & Co. *Birmingham.* 1860-80.
James. *Bristol.* 1679.
James. *Colchester.* 1849-d.1863.
James. *Edinburgh.* 1860.
James. *Epsom.* 1828-39.
James. *Liverpool.* 1814-34. W.
James. *Liverpool.* 1888. W.
James. *London.* (B. 1799-1840). 1851. Portman Square.
James. *London.* 1844. Goswell Rd.
James. *Melton Mowbray.* ca.1820. C.
James. *Nuneaton.* 1828.
James. *Oxford.* a.1663-?96. W.
James. *Oxford.* 1852. C. & W.
James. *Winchester.* 1830.
James. *Wootton-under-Edge.* 1870.
Jeremy. *Liverpool.* 1744. W.
John. *Aberford.* 1826-34.
John. *Amherstburgh, Canada.* 1861-67.
John. *Brighton.·*a.1728 to George Browne. C.
John. *Cookstown.* 1854.
John. *Elgin.* 1837.
John. *Glasgow.* 1785.
John. *Harleston.* 1808-36. C.
John. *Helmsley.* 1810. C.
John. *Irvine.* 1829.
John. *Kirkaldie.* l.c. clock, date unknown.
John. *Lancaster, Pa, USA.* Late 18c. C.
John. *Llancarfan.* a.1778. C.
John. *London (56 Charing Cross).* 1811.
John. *Monaghan.* 1786.
John. *Plymouth.* mar.1764-83.
John. *Southampton.* 1830.
John. *St. Jacobs, Canada.* 1857.
John. *Wem.* 1828.
John James. *Andover, Mass, USA.* ca.1840.
John James. *?London.* 1857-69. Clock case makers.
John Joseph. *London.* 1857-81.
Jonathan. *Harwich.* 1813-28. W.
Jonathan Clark. *Bristol, USA.* 1807-72.
Joseph. *Birmingham.* 1868-80.
Joseph. *Kidderminster.* 1828.
Joseph. *Ledbury.* 1830.
Joseph. *London.* 1875.
Joseph. *Manchester.* 1834.
Joseph. *Providence, RI, USA.* ca.1850.
Joseph. *Whitby.* 1840.
Joseph C. *Coventry.* 1860-80.
Joseph Edward. *London.* 1857-69. Clock cases.
Joseph Thomas. *Coventry.* 1850-60.
Joshua. *Liverpool.* mar.1711-73. (?two men). C.
& KIRBY. *New Haven, Conn, USA.* 1840.
Lawrence. *Ipswich.* 1853-64.
Lamont. *Utica, NY, USA.* 1838-41.
& LEWIS. *Bristol.* 1839.
Macmanus. *Kinsale.* 1787.
Matthew. *Dublin.* 1651. C.
Matthew. *Edinburgh.* 1860.
& MELETZ. *London.* 1839.
Michael. *Pontefract.* 1815-37. W.
Michael Septimus & Co. *Halifax, Nova Scotia.* b.1818-d.1886.
Morris. *Leeds.* 1853-66. W.
Mrs. J. *Edinburgh.* 1837.
Nathaniel. *Manchester.* ca.1760-70. C.
Peter. *Ayr.* d.1847.
Peter. *Manchester.* ca.1850 or later. C.
R. Balfour. *Yarmouth, Nova Scotia.* 1864-67.
Richard. *Farnworth.* mar.1828. W.

BROWN(E)—*continued.*
R(ichard). *Newport, IoW.* 1848-67(-78).
Richard. *Prescot.* 1819-24. W.
Richard. *Swanage.* 1875.
Robert. *Barrowford* nr *Colne.* 1851-58.
Robert. *Cheltenham.* 1842-79.
Robert. *Gloucester.* 1830-56.
Robert. *Newtown* (*Mon*). 1835-44.
Robert (& Son). *Baltimore, USA.* 1829-31 (-33).
Roger. *London.* 1839-51.
S. *Norristown, Pa, USA.* ca.1850.
Samuel. *Bingham, Notts.* 1828-49.
Samuel. *Cookstown.* 1824-46.
Samuel. *Disley.* 1834.
Samuel. *Dover.* 1815. W.
Samuel. *Leicester.* 1864. French clocks.
Samuel. *New Mills, Derbyshire.* 1835.
Samuel. *New York.* ca.1820.
Samuel A. *Lowell, Mass, USA,* 1835-37.
(J. R.) & SHARPE. *Providence, RI, USA.* 1856.
& SKELTON. *Edinburgh.* 1784-87. W.
Solomon. *Swansea.* 1868-75.
& Son (?James). *Winchester.* 1839-59.
T. *Maidstone.* 1838-65.
T. M. *Brighton.* 1878.
Templeton. *Peterboro, Canada.* 1851-76.
Thomas. *Belfast.* 1854-80. W.
Thomas. *Berwick-on-Tweed.* 1849.
Thomas. *Birmingham.* 1800.
Thomas. *Birmingham.* 1880.
Thomas. *Bristol.* 1643.
Thomas. *Coventry.* 1850.
Thomas. *Douglas, IoM.* 1860.
Thomas. *Halifax, Nova Scotia.* 1851-70.
Thomas. *Leith.* ca.1800. Regulator.
Thomas. *Liverpool.* (B. 1807-24). 1848-51.
Thomas. *Manchester.* 1824-51. W.
Thomas. *Zanesville, O, USA.* Clock case maker.
Thomas William. *New York, USA.* 1803-mar.1822-72.
Titus F. *Napanee, Canada.* 1857.
W. *Bedworth.* 1868.
W. *Chester.* 1857-65.
W. *Foleshill.* 1860-80.
W. *Totton.* 1859.
W. F. *Gloucester.* 1863.
W. J. *Toronto.* 1875-77.
Walter. *Birmingham.* 1880.
William. *Bangor* (*Wales*). l.c. clock, date unknown (prob. mid-19c).
William. *Barnsley.* 1862-71. W.
William. *Bradford.* 1866-71.
William. *Bungay.* 1865-79.
William. *Dumfries.* 1745.
William. *Dumfries.* 1787-d.1795.
William. *Enniskillen.* 1810.
William. *Farnworth.* mar.1815. W.
William. *London.* 1844.
William. *Magherafelt.* 1846-54.
William. *Manchester.* mar.1791-96. C.
William. *North Shields.* ca.1810-58. C.
William. *Prescot.* 1794. W.
William. *Prescot.* mar.1818-24. W.
William. *Preston.* 1851-58. C. & W.
William. *Rothesay.* 1860.
William. *Sheffield.* 1862-71. W.
William. *St. Helens.* 1766-86. W.
William. *St. Helens.* 1836. W.
William. *Stroud.* 1830.
William. *York, Pa, USA.* ca.1840.
William Henry. *Birmingham.* 1880.
BROWNBILL—
Edmund. *Prescot.* 1796. W.
Henry R. *Leeds.* 1767-90. W.
James. *Liverpool.* Several such 1795-1851.
James. *Poulton.* 1851-58. W.
John. *Leeds.* 1871.
John. *Liverpool.* Several such 1769-1834.
John. *Prescot.* 1818-23. W.
R. S. *Preston.* 1858. W.
Robert. *Liverpool.* 1851.
Thomas. *Eccleston.* mar.1816-19. W.
Thomas. *Leeds.* s. of Henry. b.1773-1866. W.

BROWNING—
E. *Nether Stowey.* 1875-83.
Frederick. *London.* 1869-81.
J. *North Curry.* 1866.
Joseph. *Street* nr *Bath.* 1861-83.
Samuel. *Boston, USA.* 1815. Math. inst. maker.
William. *Markham, Canada.* 1857.
BROWNLESS—
John, sen. *Staindrop.* ca.1760-d.1800. C.
John, jun. *Staindrop.* s. of John, sen. 1787-1827. C.
BROWNLIE—
——. *Hamilton.* 1784. W. (B. William 1800.)
Archibald. *Strathaven.* d.1844. C.
James. *Glasgow.* 1836.
BROWNBRIDGE, Henry Joseph. *London.* 1881.
BROWNSCOMBE. *Exeter.* See Branscombe.
BROWNSWORD—
John. *Nottingham.* 1828-49.
Peter. *Derby.* 1814-57.
Robert. *Nottingham.* 1840-52.
Robert. *Whitwick.* 1855-76.
BRUCE—
John. *Altofts* (*Normanton*). a.1723. C.
William. *Workington.* 1879. Also jewellers.
BRUDNO, R. *Whitehaven.* 1873.
BRUFF—
Abraham. *Dublin.* a.1702.
Charles Oliver. *New York.* s. of James. 1769.
James. *Elizabeth, NJ, USA* and *New York.* 1748-d.1780.
BRUGELL, Giovanni Adamo. *Venice.* ca.1700.
BRUGGER—
A. (not Andrew). *London.* 1863.
Andrew (& Co.). *London.* 1851 (63-81).
Barnard. *London.* 1857.
BECK & CO. *London.* 1844-51.
Conrad. *Kings Lynn.* 1830.
J. *Derby.* 1849.
John (& Co.). *London.* 1832(-39).
John & Co. *Sheffield.* 1862. C.
John Paul. *London.* 1869.
Joseph & Bernhard. *London.* 1844-51.
L. *Bishopwearmouth.* 1844.
Martha Elizabeth. *Kings Lynn.* 1836. (?wid. of Conrad.)
Matthew & Lawrence. *Newcastle-on-Tyne.* 1848-52.
& STRAUB. *London.* 1844-75.
BRUGHNER, Jacob. *Brooklyn, New York, USA.* ca.1840-50.
BRUL(E)FER, Louis. *London.* mid-17c. C.
BRUMFIT, Joseph. *Bradford.* 1853.
BRUMHEAD, John. *Stamford.* 1828-35.
BRUNDAGE, Lawrence Foster. *St. John, New Brunswick.* 1834.
BRUNDELL, William. *Blofield, Norfolk.* 1836-46.
BRUNEAU, ——. *Paris.* 1839. C.
BRUNELOT, Jules. *Paris.* 1878-89. C.
BRUNER, J. *Brentwood.* 1866.
BRUNET, Philemon. *Quebec.* 1867-71.
BRUNKER, Thomas. *Dublin.* 1858-80.
BRUNNER—
Engelbert. *Hull.* 1838-58. C. & W.
Ignatius. *Birmingham.* 1835-54. Also musical boxes.
Leopold. *Birmingham.* 1860-80.
BRUNSDON, William. *Marlborough.* 1830.
BRUNSKILL, John. *Kendal.* b.1836-75. W.
BRUNTON, John. *London.* 1863-69.
BRUSH—
James (& Son). *Dublin.* 1772-1804 (1804-09). W. Also jeweller.
James. *Londonderry.* 1785.
Thomas. *London.* 1662.
BRUTON, Augustus. *London.* 1869.
BRYAN—
Benjamin. *Belfast.* 1835.
James. *Stockport.* 1848-57.
John. *Belfast.* 1822-40. W. (Successor to James Carruthers 1828.)
R. H. *Boston.* 1861.
Richard. *London.* (B. a.1683, CC. 1696)-97.
Robert Hoff. *Lincoln.* 1849-50.
Stephen. *London.* 1828-32. Watch case maker.

BRYANT—
E. D. *New York.* 1850.
Frederick Benjamin. *London.* 1863-69.
J. M. *Pembroke Dock.* ca.1840. C.
John. *St. Agnes* (*Cornwall*). Dates unknown. C.
Thomas. *Rochester, NY, USA.* ca.1800.
Thomas. *Old Buckenham* (*Norfolk*). 1830.
William. *Dublin.* 1868-80.

BRYARS—
Arthur. *Chester.* 1828. (B. has *Liverpool* 1800-03, then *Chester.*)
Samuel. *Chester.* 1834.

BRYDEN—
T. *Halifax, Nova Scotia.* 1819.
Thomas. *Johnshaven* (*Scotland*). 1837.
William. *Lockerbie.* 1860. Dealer.

BRYER—
John. *London.* 1844-81.
W. *Dorchester.* 1848.

BRYMER, Alexander. *Halifax, Nova Scotia.* 1866.
BRYOM, William. *Cork.* 1787-1810.

BRYSON—
Charles. *Glasgow.* 1837.
J. *Dartford.* 1866.
John. *Dalkeith.* 1836-82.
& Son. *Dalkeith.* ca.1785-93. W.

BUARD, Charles W. *Philadelphia, USA.* 1849.
BUBB, John. *Bradford* (*Wilts*). 1830.

BUCHAN—
Alexander. *Blairgowrie.* 1860.
Alexander. *Perth.* 1833.
H. *Bristol.* 1863.
H. T. *Bristol.* 1863.
Henry. *London.* 1839.

BUCHAN(N)AN—
A. D. *Tattershall.* 1835.
Alfred Daniel. *Boston.* 1828-35.
Andrew. *Denny.* 1860.
Andrew. *Greenock.* 1836.
Archibald. *Dublin.* 1781-1815. C.
Daniel. *Glasgow.* 1860. Chrons.
George. *Strabane.* ca.1785-95. C.
John. *Ashton-under-Lyne.* 1822-28.
Robert, jun. *Glasgow.* 1835.
Thomas. *Dublin.* 1816-28. W.

BUCHER, Jesse. *Dayton, O, USA.* ca.1830.

BUCK—
Solomon. *Glens Falls, NY, USA.* 1827-28.
T. *Dorking.* 1851.
William. *Hawes.* 1823-34.

BUCKENHAM—
Charles. *Norwich.* 1830-36.
J. C. *Tombland, Norfolk.* 1858.

BUCKINGHAM—
James. *London.* ca.1700. W.
Joseph. *London* (in ye Minories). ca.1685. C.
Mrs. E. *Ipswich.* 1853.

BUCKINHILL, John. *Worcester.* 1719. W. (One such in *London* a.1664-85 may be same man.)

BUCKLAND—
Friend William. *Maidstone.* 1839. C.
George. *London.* 1881.
Henry. *Birmingham.* 1868-80.
Henry. *Handsworth.* 1868-76.
William. *Thame.* 1786-1821. C. & W.

BUCKLE—
Jacob. *Pittsburgh, Pa, USA.* Ca.1830.
Robert. *Crosby Garrett.* 1858.
William. *Bradford.* 1837.

BUCKLER, Aaron. *Bowmanville, Canada.* 1851-63.

BUCKLEY—
Charles. b. ca. 1830 *Long Itchington* (*Warwicks*). *Warwick* 1851-80.
George. *Canterbury.* 1823.
James. *Manchester.* 1871-78.
Thomas. *Canterbury.* 1784. W.

BUCKMASTER—
George. *Royston.* 1828-40.
Mrs. S. *Royston.* 1859-66.
S. *Royston.* 1851.

BUCKNALL—
W. *Burslem.* ca.1800. C.
William. *Stoke-on-Trent.* 1828-35. (Also Bucknell.)

BUCKNELL—
——. *Burslem.* ca.1790. C.
A. H. *Midhurst.* 1878.
Thomas. *Berkhampstead.* Early 18c. lantern clock.
William. *London.* (B. 1816-25). 1828. W.

BUCKNEY, Daniel. *London.* 1875-81.
BUCKSHER, John. *London.* (B. a.1775-1820). 1828.
BUCKTROUT, Richard. *Wakefield.* From *London.* 1770. W.
BUCKWELL, Edward, jun. *Brighton.* (B. 1819-)1839. W.

BUDD—
James H. *Alverstoke.* 1867-78.
Joseph. *New Mills, NJ, USA.* ca.1806-20.

BUDDS, Charles Albert. *Huntingdon.* 1877.

BUDGE—
Mrs. Elizabeth. *Callington.* 1873. W. (?wid. of W.)
W. *Callington.* 1856. W. & C.

BUERGER, Laurenz. *Neustadt* (*Prague*). 1589.

BUFFET—
George Francis. *Philadelphia.* ca.1796.
John. *Colchester.* b.1692 d.1758. C. & W.

BUFFHAM, Reuben. *Spalding.* 1876.

BUGDEN—
David. *London.* (B. a.1804). 1828-32.
David (II?). *London.* 1857-69.
John. *Croydon.* 1828-51.
Thomas. *Croydon.* 1735. W.

BUGLAS, James. *Coldingham.* 1860.

BUHRER—
Abraham. *London.* 1839-57.
Joseph. *Brighton.* 1839.

BULFORD, John. *London.* 1857-69.
BULKLEY, Joseph. *Fairfield, Conn, USA.* 1755. mar.1778-1815. C.

BULL—
Brothers. *Bedford.* 1877.
Edmund. *London.* Free of B. Co. 1607. Petitioner for CC. in 1622-1630. W.
Emanuel. *London.* s. of Randolph. 1617. W.
George. *London.* B. Co. 1617. Petitioner for CC. 1622.
Isaac, jun. *Dublin.* 1769-1802. W.
James P. *Newark, NJ, USA.* 1800-57.
John. *Ashford.* 1845-55.
John. *Bedford.* Est. 1817-92 W. & C.
John. *London.* (B. 1630)-62. *Temple Bar.*
John. *York, Pa, USA.* Early 19c.
Joseph. *Coventry.* 1880.
Lionel. *Oxford.* a.1761. C.
Randolph. *London.* ca.1582. Clockmaker to the Queen 1591-1613. d.1617. C., W. & goldsmith.
& Son. *Bedford.* 1869.
Thomas. *Bristol.* 1840-42.
Thomas. *Oxford.* 1592. T.C.
William. *Leicester.* 1835.
William. *Loughborough.* 1828.

BULLARD, Charles. *Boston, USA.* 1816. mar.1822-71. Dialpainter.

BULLEN—
Benjamin. *Hempnall.* 1836-46.
John. *Hempnall.* 1858-75.
Joseph. *Pulham.* 1836.

BULLER, Robert. *Banbury.* b.1705 a.1719. Later *Abingdon* where one such worked pre-1759.

BULLINGHAM, C. *Bromsgrove.* 1860-72.

BUL(L)MAN—
——. *Liverpool.* ca.1800. C.
John. *Liverpool.* mar.1707-08. C.
Thomas. *Liverpool.* 1701-04. C.

BULLOCK—
A. *Bradford* (*Wilts*). 1859-75.
E. *Nether Knutsford.* 1865.
E. *Wrentham.* 1846.
Ezekiel. *Lurgan.* ca.1700-40. C.
Gilbert. *Bishopscastle.* 1724 -d. 1773. C.
Grace. *Bath.* 1848.
J. *Chippenham.* 1859-67.
J. *Westbury.* 1848.
J. N. *Ipswich.* 1879.
James. *Bath.* 1845.
John. *Bishop's Waltham.* 1775. (B. 1785.) W.
John. *Melksham.* 1830-48.

BULLOCK—*continued.*
John W. *London.* 1857-69.
Joshua. *Merkethill, Armagh.* 1858.
R. *Beccles.* 1846-65.
Richard. *Ellesmere.* ca.1740. C.
Robert Henry. *Bungay.* 1830-53.
Thomas. *Corsham.* 1842-75.
William. *Bradford (Wilts).* 1830-48.
William (?I). *Dublin.* 1815-31.
William (?II). *Dublin.* 1858-68.
William. *Trowbridge.* 1842-59.
BULOVA Watch Co. *New York.* 1875.
BUMEL, Michael. *Nurnberg.* (B. 1601). ca.1630-d.1660. W.
BUMFORD—
Mrs. N. *Knighton.* 1887.
Richard. *Knighton.* 1875.
BUMPUS, Francis W. *Loughborough.* 1876.
BUMSEL, M. *Toronto.* 1871.
BUNCE—
John. *Birmingham.* 1842-60.
John. *Farringdon.* 1830.
Thomas. *Wantage.* (B. 1795)-1830.
BUNDY (W. L.). Time Recording Co. *Binghampton, NY, USA.* 1893.
BUNN, R. *Norwich.* 1858. C.
BUNNELL, Edwin. *Bristol, USA,* 1850.
BUNSTER, T. *Truro.* ca.1750-70.
BUNSTON, Joseph. *Chard.* 1861-83.
BUNTING—
Daniel. *Philadelphia.* 1844.
John. *Long Buckby.* ca.1700-ca.1780. C.
William. *London.* (B. a.1637, CC. 1645.) Prob. died 1650.
BUNYAN & GARDNER. *Salford.* 1848-51. C.
BURBAGE, Henry, *Sturminster.* 1824-67.
BURBERY, Abner. *Reigate.* 1828.
BURB(R)IDGE—
Adam. *London.* 1832.
Elizabeth. *Manchester.* 1851.
Joseph. *Manchester.* 1848.
BURCH—
Robert. *Maidstone.* 1845-65.
William. *Chelmsford.* d.1811. W.
William. *Hamilton, Canada.* 1856.
William, jun. *Maidstone.* a.1815-66. W.
William. *Tenterden.* 1823.
BURCHELL, William Silvans. *Northampton.* 1767.
BURCKBY, Joseph. *Halifax.* 1837.
BURKHAM, ——. *London.* 1732. W.
BURCKHARDT, Ferdinandt. *Fridberg.* ca.1720. W.
BURDALL, John. *Market Deeping.* 1835-49.
BURDEN—
A. *Coventry.* 1881.
A. R. *Havant.* 1867.
William. *Birmingham.* 1880.
BURDESS, Adam. *Coventry.* 1868-80.
BURDGE—
Louisa. *Galway.* 1858. W.
Nicholas. *Galway.* 1817-24. W. (And prob. later too.)
BURDICK—
M. H. *Bristol, USA.* ca.1849.
S. P. Place unknown, *USA.* ca.1880.
BURDITT—
J. *Woodbridge.* 1858.
J. W. *Banbury.* 1868. C.
Mrs. E. *Woodbridge.* 1865.
BURDON, Edward. *London.* 1357-69.
BUREGAR, S. *Greenwich.* 1851.
BURGAR—
John W. *London.* 1844.
John W. *Birmingham.* 1850.
BURGER, Adolphua. *London.* 1875-81.
BURGES(S)—
Bezaliel. *Liverpool.* Pre-1751-69. W.
Charles Smith. *Ipswich.* b. ca. 1807, working there 1843 -d. 1886.
E. *Wandsworth.* 1878.
Edward. *London.* ca.1665-ca.1690. C.
Edward. *London.* 1832-57. Clock case maker.
H. *Melksham.* 1867.
Henry. Place unknown. Late 17c. to early 18c. Lantern clock.

BURGES(S)—*continued.*
Henry. *Norwood.* 1878.
I. *Stockport.* 1865.
J. *Hatfield.* 1866.
John. Place unknown. ca.1720. C. (Prob. next man.)
John. *Gosport.* 1726. C. (?B. has Thomas B. pre-1740-ca.-50).
John. *Manchester.* 1851.
Joseph. *Coventry.* 1850.
& LANGTON. *Liverpool.* 1767. W.
William. *Hatfield.* 1874.
William. *London.* 1881.
William. *Quebec.* 1822-43.
BURGHART, Augustin. *Bolton.* 1834. *Salford.* 1848-51. Imported clocks.
BURGI, Frederick. *Trenton, USA.* 1774-75.
BURGMANN, Jan. Place unknown. 1442. C.
BURITT, P. *Ithaca, NY, USA.* 1821.
BURK(E)—
——. *London.* 1633. CC.
Charles. *Philadelphia.* 1848.
James. *Dublin.* 1820-23. W. & jeweller.
W. H. & Co. *London.* 1869.
William. *Galway.* d.1806. W.
BURKELOW, Samuel. *Philadelphia.* 1790-1813.
BURKMAR, Thomas. *?Boston, USA.* 1776.
BURLEY, George C. *London.* 1857-63.
BURLINGHAM—
D. *Peterborough* and *King's Lynn.* 1864.
Daniel Catlin. *King's Lynn.* 1858-75.
Elizabeth. *King's Lynn.* 1830-46.
BURLINSON—
Henry. *Ripon.* b.1800-d.1879. C.
John. *Durham.* 1834-56.
BURMAN, Samuel. *Bristol.* 1870-79.
BURNAP—
Daniel. *East Windsor, Conn, USA.* 1759-1800. Then to *Coventry (Andover), Conn.* 1800-d.1838.
Ela. *Boston, USA.* 1810.
(Ela) & JAMES (Joshua). *Boston, USA.* 1809.
BURNET(T)—
Charles. *Dundee.* 1860. C.
Charles. *London.* 1773. W.
J. *Portsmouth.* 1839.
James Robert. *Newport, Salop.* 1879.
John. *Tarves (Scotland).* (B. 1810-46). 1860.
Smith. *Newark, NJ, USA.* 1770-1830.
William. *Durham.* 1851-56. C.
BURNHAM—
——. *Danvers, Mass, USA.* 1831.
E. B. *Salisbury, NC, USA.* 1821.
Enoch. *Paris Hill* and *Rumford, Mass, USA.* Also *Portland.* 1806-35. C.
J. W. *Salisbury, Conn, USA.* ca.1825-35. C.
BURNS—
Alfred. *Dowlais.* 1887.
Hugh. *Philadelphia.* 1809-12.
James I. *Armagh.* 1830-46.
James II. *Armagh.* 1858-90.
James. *London.* 1839-51.
James. *St. John, New Brunswick.* 1810-13.
John. *Brighton.* 1839.
John. *St. John, New Brunswick.* 1815.
Richard. *Manchester.* 1778-d.1806. W.
Samuel. *Armagh.* 1854-58.
BURNSTEIN—
Nathan. *Quebec.* 1853-63.
Samuel J. *Quebec.* 1862-66.
BURPUN, John. *?London.* ca.1700-ca.1715. C. Place unknown but believed *London.*
BURPUTT, James. *London.* ca.1700-ca.1720. C.
BURQUHART, Augustin. *Manchester.* 1848.
BURR—
C. A. *Rochester, NY, USA.* 1841.
& CHITTENDEN. *Lexington, Mass, USA.* 1831-37. Later BURR went to Chicago.
Ezekiel & William. *Providence, RI, USA.* 1792.
Horace. *Dundas, Canada.* 1830-35.
James. *Bristol.* a.1741-75. (B. 1783.)
Jonathan. *Lexington, Mass, USA.* 1831-37.
Richard. *Rye.* 1828.
BURRAGE, John. *Baltimore, USA.* 1769. C.

BURRELL—
Benjamin. *London.* 1832.
George & J. C. *Sheffield.* 1861-71. W.
John. *Beverley.* 1858.
M. & Louis. *London.* 1869.
Mary. *Lewes,* from *Wiston.* a.1718 to Isaac Guepin. W.
T. *North Collingham, Notts.* (B. has 1800). 1849-64.
BURRINGTON, Thomas. *Exeter.* mar.1764-67.
BURRITT, Joseph. *Ithaca, NY, USA.* 1795-1889. C.
BURROUGHS—
Edward. *Fordham.* ca.1770-ca.1780. C.
James. (b. 1795 *Tilstock) Ironbridge* 1826-48.
Dawley 1849-68 C.
James. *Rochdale.* mar.1703-d.1721. C.
John. *Brampton.* b.ca.1752, mar.1775-78.
John. *Ironbridge.* ca.1800-30. C.
John. *Newtown* (*Mon.*). 1868.
Joseph. *Montgomery.* 1886-90.
BURROW(S)—
A. *Stradbroke.* 1858-79.
A. Millar. *Wellington, Somerset.* 1883.
Edward. *London.* 1832-69.
George. *Milton, Canada.* 1871.
Henry. *Norwich.* 1875.
James. *Ironbridge.* 1828-56.
James. *London.* 1828.
James. *Wellington* (*Som.*). 1861-75.
John. *Farnworth.* mar.1782. W.
John. *Strasbourg, Pa, USA.* ca.1808. C.
& MILLS. *Birmingham.* 1880.
R. & Co. *London.* 1881.
Richard. *Liverpool.* 1848.
Richard. *Newport Pagnell.* ca.1770. C.
Robert. *London.* 1875-81.
Rowland Kitson. *Ironbridge.* b. 1833 s. of James.
-1875. also at *Dawley* 1870-75.
Thomas. *Liverpool.* 1848-51.
Thomas. *Lowell, Mass, USA.* 1834.
BURT—
& CADY. *Kansas City, Mo, USA.* ca.1870.
David. *Portsea.* 1839-48.
BURTCHNELL, G. *Brighton.* 1862.
BURTON—
——. *Eastry.* ca.1820. C.
Alfred. *London.* 1857.
& AVENELL. *Battersea Park.* 1866.
Emanuel I. *Kendal.* ca.1720-30. C.
Emanuel II. *Kendal.* b.1732-d.1809. C.
Emanuel III. *Kendal.* s. of Emanuel II. b.1765-d.1830.
C.
Emanuel. *Thornton-in-Lonsdale.* 1810. C. May be the
third *Kendal* man.
Frederick. *London.* 1851.
Frederick. *Lewisham.* 1855-74.
George. *Tebay.* ca.1750. C. (?from *Skipton.*)
George. *Knottingley.* 1866-71. W.
George. *Uttoxeter.* 1828-50.
H. *Peckham.* 1862.
Henry. *Penge.* 1866.
Isaac, sen. *Backbarrow.* b.1797. To *Ulverston* pre-1821-
65. C.
Isaac, jun. *Ulverston.* b.1828. s. of Isaac, sen. -1858. C.
James. *London.* Lincolns Inn Gate 1811. W.
James. *Norwich.* Late 18c.
John. *Bradford.* 1853.
John. *Dublin.* Free 1710-17. W.
John. *Halifax.* 1782-d.1783. C.
John. *Huddersfield.* 1834-37. C.
John. *Old Sampford, Essex.* 1866-74.
Jonathan, sen. *Backbarrow.* 1786-1819, then *Ulverston*
till 1825. C.
Jonathan, jun. *Ulverston.* s. of Isaac, sen. b.1826-69.
Lancaster. *Birstall.* 1850. C.
& PATTISON. *Halifax.* 1800. C.
S. *Peckham.* 1855.
S. & Son. *Lewisham.* 1851.
Samuel. *London.* 1857.
Samuel. *Salisbury.* 1842-48.
Thomas. *Hanley.* 1835.
Thomas. *Hawkshead.* 1745. C.
Thomas. *Leeds.* 1866-71. C.
Thomas. *Skipton.* mar.1739. C.
Thomas. *Ulverston.* 1808-11. Clock repairer.

BURTON—*continued.*
W. *Rawtenstall.* 1858.
William. *Kendal.* 1732-ca.45. C.
Wolsey. *Philadelphia.* Early 18c.
BURTONWOOD, William. *Farnworth.* mar.1801. W.
BURTT, Thomas F. *Stoneham, Mass, USA.* 1855, then
Danvers 1872, then *Stoneham* again 1890.
BURUT, Andrew. *Baltimore.* 1819-37.
BURWELL (Elias) & CARTER (W. W. & L. E.). *Bristol,*
USA. 1859-61. C.
Elias. *Bristol, USA.* 1851-67.
BUSBKY, Isaac. *London.* 1839.
BUSBY—
& BAXTER. *Guildford.* 1851-66.
Samuel. *Dublin.* 1763-85. W.
William. *London.* (B. 1805-08). 1839.
William. *Oakville, Canada.* ca.1870-ca.1880.
BUSFIELD, Thomas. *London.* 1697.
BUSH—
Edward. *Norwich.* 1865-75.
George James. b. ca. 1836 *Devizes,* worked with
William Enock at *Warwick* 1851. *Devizes* again
1867-75.
& GUY. *Devizes.* 1842.
Harry. *Wincanton.* 1815. C.
Henry. *Cincinnati, O, USA.* 1841.
James. *London.* (B. 1820.) 1828-44.
Robert. *Wincanton.* 1815-21. C.
Thomas. *London.* ca.1800. W.
Thomas Samuel. *Norwich.* 1876.
William. *Bristol.* a.1786-1830.
BUSHE, James. *Londonderry.* ca.1780-ca.85. C.
BUSHELL—
George. *Ramsbury.* 1830-59.
Matthew. *Bold.* a.1761. W.
Robert. *Liverpool.* mar.1709-34. W.
BUSHMAN, William. *Stratford, Essex.* (B. 1805-08).
1828.
BUSSELL, Thomas. *Collumpton.* ca.1790-1800. C.
BUSSEY—
Alfred. *Plumstead.* 1866-74.
James, jun. *Bristol.* (B. 1825-)1840.
BUSWELL, Edwin. *Brecon.* 1875-99.
BUTCHER—
Benjamin Thomas. *Bedford.* 1864-77.
& FRASER. *Luton.* 1864.
H. S. *Wellingborough.* 1847-77.
J. A. *Dunstable.* 1864.
John William. *Leominster.* 1863-70.
John William. *Salisbury.* 1875.
BUTLAND, William. *London.* 1839.
BUTLE, James. *Halifax, Nova Scotia.* 1750.
BUTLER—
(Henry W.) & BARTLETT (Edward M.). *West Chester,*
Pa, USA. 1834.
Bernard. *London.* 1869.
Caleb. *Christchurch.* (B. 1791-)1830.
Edward. *Tutbury.* (B. 1795-)1828.
Franklin. *Philadelphia, USA.* 1846-ca.50.
George. *Bournemouth.* 1878.
George. *Christchurch.* 1839-67.
(William) & HENDERSON (James). *New Hartford,*
Conn, USA. 1828-33. C.
& HENDERSON. *Annapolis, Nova Scotia.* 1824.
HENDERSON & CO. *Clement, Nova Scotia.* 1830.
Henry. (b. ca. 1816 *Whitchurch). Wem.* 1840-75.
I. R. *Cornwall, Canada.* 1862.
J. *Coventry.* 1868.
J. *Kingston, Surrey.* 1862.
J. *Reading.* 1854.
James. *Boston, USA.* 1713-76.
James & John. *London.* 1875-81.
Joseph. *London.* (B. a.1799)-1828.
Mrs. *Bolton.* 1822.
Nathaniel. *Utica, NY, USA.* 1760-1829.
& RACE. *London.* 1881.
Robert. *Greenock.* 1836.
Samuel James. *London.* 1839. Watch cases.
W. *Chatham.* 1851-55.
W. *St. Helens.* 1851-58. Also jeweller.
BUTT—
F. *Chester.* 1865-78.
John Henry. *Weymouth.* 1875.
BUTTALL, John. *Plymouth.* Late 18c. C.

BUTTERFIELD—
——. *Paris.* ca.1700. Sundial. (B. d.1724.)
John. *Todmorden.* 1770-1824.
John. *Wath-on-Dearne.* 1862-71. W. & C.
BUTTERLEY, St. *Dartford.* ca.1720-ca.1740. C. (B. 1791.)
BUTTERWORTH—
H. *Bacup.* 1858. W.
H. *Rawtenstall.* 1858. W.
H. *Rochdale.* ca.1800. C.
J. & W. *Leeds.* 1798. Engravers.
John. *Leeds.* 1759. 'Engraver of clock faces'.
BUTTERY—
John. *Monmouth.* 1822-52.
John Benjamin. *London.* 1828.
BUTTON—
George. *Wakefield.* 1851.
James. *London.* 1857.
John & Son. *Wakefield.* 1834.
Thomas. *Hitchin.* 1828-39.
Thomas (& Son). *Wakefield.* 1814-22(-37). C.
BUTTRESS, William. *Crediton.* ca.1785.
BUXTON—
Frederick William. *London.* 1881.
George Samuel. *Colchester.* b.1780-d.1851. C. & W.
J. *Wickhambrook.* 1858-65.
Samuel. *Diss.* ca.1760-70. C. & W.

BUXTON—*continued.*
William. *Bishop Auckland.* 1827-56. C.
William. *Lavenham.* 1875-79.
BYAM, C. C. *Amesbury, Mass, USA.* ca.1850.
BYINGTON—
(Loring) & Co. *Bristol, USA.* 1843-49.
(Loring) & GRAHAM. *Bristol, USA.* 1852.
Loring (Lawler). *Bristol, USA.* 1797-1889. Also at *Newark, NJ.* ca.1830.
BYRAM, Ephraim. *Sag Harbor, NY, USA.* ca.1840-60. C.
BYRNE—
——. *Liverpool.* Late 19c. regulator.
Daniel. *Bryn-mawr.* 1868-75.
F. *Birmingham,* ca.1780-ca.1810. Dial maker.
James. *Philadelphia.* 1784. *New York.* 1789-97. *Elizabeth, NJ.* 1799. C. & W.
Samuel. *Wexford.* 1858.
BYROM, James. *Rochdale.* 1695-d.1730. C.
BYRON—
John. *Prescott.* mar.1822-26. W.
Thomas. *Prescott.* mar.1820-25. W.
(?William). *Cork.* 1787-1824. C.
BYWATER, Bon. *Stamford.* Late 18c. C.
BYWORTH—
Mrs. Mary. *London.* 1839-51. (?wid. of Thomas I).
Thomas (?II). *London.* 1857. Successor to Mrs. Mary.

C

C, D. DC is mark of Drocourt of *Paris*, QV.
C, FC. *Neuchatel*. ca.1810. W.
C, M. D. Place unknown. ca.1820. W.
C, R. RC is mark of C. A. Richard & Cie.
CABLE, Stephen. *New York*. 1848-54.
CABRIER, Charles. *London*. 1730 at the Dial, Token-house Yard, in the parish of *St. Margaret, Lothbury*. W.
CACHOT, Felix. *Bardstown, Ky, USA*. 1839.
CADDICK, Richard. *Liverpool*. 1834.
CADE—
George, sen. *Market Weighton*. 1823-51. W.
George, jun. *Market Weighton*. 1851-58.
Robert. b. 1820. mar. *London* 1842. To *Ipswich* 1843-59 after which became a photographer. Thompson William. *Northallerton*. 1834-66. W. & C.
CADELACE, Richard. *London*. ca.1780-1800. C.
CADMAN—
Jesse. *Mansfield*. 1842-64 C.
Joseph. *Birmingham*. 1880.
CADOLA, ——. *Wincanton*. ca. 1820.
CADOT, ——. *Paris*. 1740. Sundial.
CADWELL—
Charles. *Maidstone*. a.1800.
Edwin. *New York*. 1840.
CAHAN, Lewis. *London*. 1863.
CAHOON, John. *Belfast*. 1868-92. C. & W. Later CAHOON Brothers.
CAIN—
Henry. *Luton*. 1805-30. W.
Henry Charles. *Halifax*. 1853-66.
Michael. *Albany, NY, USA*. 1831-32.
CAINEN—
Michael. *Dublin*. 1798-1830.
Patrick. *Dublin*. 1794-1806. Also jeweller.
CAINES, Alfred. *Bristol*. 1879.
CAIRD, D. *Auchinblae*. 19c. C.
CAIRNS—
Brothers. *Brampton*. 1873-79.
David. *Belford*. 1827. C.
David. *Bilshopwearmouth*. 1850-56.
David. *Morpeth*. 1834-48.
James. *Deal*. 1866-74.
John. *Haltwhistle*. 1848.
John. *Providence, RI, USA*. 1784. W.
John. II. *Providence, RI, USA*. 1840-53.
Ralph. *Alston*. 1828.
Ralph. *Brampton*. 1834-69. Also auctioneer.
Ralph. *Newcastle-on-Tyne*. ca.1810. C.
CAIRNZ, John. *Liverpool*. 1848.
CAITHNESS, David. *Dundee*. 1787.
CAKEBREAD, Richard. *Oxford*. 1566-92. Rep'd. ch. clock.
CALAME—
Alcide & Co. *London*. 1869.
Olivier. *Frederick, Md, USA*. 1819.
CALBROOK, James. *Manchester*. 1848. Clock dealer.
CALCOT(T)—
Arthur. *Malpas*. mid-19c.
Peter. *Roscommon*. 1824-58.
Thomas. *Malpas*. 1828-48.

CALCULATOR, B. *Portsmouth*. 1867.
CALCUTT, Peter. *Athlone*. 1792-94.
CALDER, John. *Edinburgh*. 1819-25. C.
CALDERBANCK, Richard. *Farnworth*. mar.1783. W.
CALDERWOOD—
Andrew. *London*. 1869-81.
Andrew. *Stewartstown*. ca.1785-1802. Later *Philadelphia*. (1802-20).
CALDWELL—
James. *Coleraine*. ca.1780-1824. C.
(James E.) & Co. *Philadelphia*. 1850.
James E. *Philadelphia*. 1840-50.
CALENDAR—
& AUGER. *Meriden, Conn, USA*. ca.1860.
Clock Co. *Glastonbury, Conn, USA*. 1856.
'CALIFORNIA', used by Otay Watch Co. 1889.
CALLAGHAN—
Daniel. *Cork*. 1858.
John. *Cork*. 1781. C.
CALLAM, Alexander. *London*. (B. 1790-1808 CC.). 1811. *Little Hermitage St.* W.
CALLCOTT—
Arthur. *Malpas*. 1878.
John. *Cotton (nr. Wem)*. b. 1753 *Edstaston*, neph. of Richard C. Mar. *Wem* 1776, d. 1830 C. & W. S. *Malpas*. 1857-65.
CALLDWELL, John. Place unknown, but may be *Ireland*. ca.1740. C.
CALLENDER, James. *Edinburgh*. a.1731-d.1762.
CALLISHER, David L. & Co. *Toronto*. 1859-63.
CALLON, T. *South Shields*. ca.1820. C.
CALLOW(E)—
Harry. *London*. Late 17c. C. (Worked for Tompion as CALLOT 1695.)
James. *Ramsey, IoM*. 1860.
CALLWOOD—
John. *Liverpool*. 1790-d.1801. C.
Susannah. *Liverpool*. 1800. C. (B. 1803-05.)
CALOR, C. H. *Plainville, NJ, USA*. 1888-89. Clock cases.
CALVER, Mrs. Susan. *Eye*. 1830-53.
CALVERT—
——. *Stokesley*. Early 18c. C.
(T. G.) & BURNETT (B. L.). *Lexington, Ky, USA*. 1855.
George. *Tadcaster*. 1871.
John P. *Leeds*. 1853.
CAMB, T. *Manchester*. 1858. C.
CAMBRIDGE, Watch Co. *New York*. 1900.
CAMERDEN & FORSTER. *New York*. Late 1800s.
CAM(M)ERER, Andrew & Co. *London*. 1828-32. C.
Fuller & Co. *London*. 1839.
Kuss & Co. *London*. 1844-63.
Kuss & Co. *London*. 1881-present day.
Kuss, Tritschler & Co. *London*. 1869-75.
M. *London*. 1844.
P. *London*. 1844.
CAM(M)ERON—
Alexander. *Dundee*. 1823-37.
Alexander. *Liverpool*. 1848. Chrons.
Alexander. *London*. 1857.

38

CAM(M)ERON—*continued.*
James. *Dundee.* 1828-60.
James. *Selkirk.* 1832.
CAMERON—
James. *Stewarton.* 1837.
John. *Aberfeldie.* 1836.
John. *Barrhead.* 1836.
John. *Kilmarnock.* 1850-60.
John. *London.* 1875. Dutch clocks.
John R. *Liverpool.* 1851. Chrons.
Matthew. *London.* 1839.
CAMFIELD, Thomas. *Tunbridge Wells.* 1823-47.
CAMM—
Martin. *Stroud.* 1840.
Richard. *Stroud.* 1830-50.
CAMMACK, Robert. *Ormskirk.* 1848-58.
CAMOZZI—
Charles. *Bicester.* 1832-50.
Eleanor. *Bicester.* 1852. (?wid. of Charles.)
CAMP—
Ephraim. *Salem Bridge, Conn, USA.* ca.1830.
Hiram. *Plymouth* and *Bristol, USA.* b.1811-d.1893.
Reuben. *Derby.* 1864-76.
CAMPANI, Giuseppe & Pier Tomasso. *Rome.* 1660.
CAMPANUS, Petrus Thomas. *Rome.* 1666-1683. C.
CAMPBELL—
——. *Paris.* 1839. Carr. clock.
Alexander. *Brooklyn, USA.* ca.1840-50.
Alexander. *Hamilton, Ontario.* 1875-81.
Andrew. *London.* 1863-81.
Archibald. *Newry.* 1784.
& BEATON. *London.* 1839.
Benjamin. *Hagerstown, Md, USA.* 1775-92. Then
   Uniontown, Pa. 1792-1830.
Charles. *Inverness.* 1860.
Charles. *London.* 1875.
Charles. *New York.* 1798.
Charles. *Philadelphia.* 1794-1803.
& Co. *Belfast.* 1850-94. W.
Colin. *Glasgow.* 1860.
Francis. (b. 1768) *Oswestry.* 1822 -d. 1841.
H. J. *Aurora, Canada.* 1862.
J. R. *Gateshead.* Early 19c. W.
James. *Johnstone.* 1836.
John. *Bangor, N. Ireland.* 1858.
John. *Belfast.* 1804.
John. *Dungannon.* 1789.
John. *Hagerstown, Md, USA.* ca.1773. C.
John. *Kintore, Inverury.* 1860.
John. *Moy, Charlemont.* 1846-58.
John. *New York.* ca.1840.
Joseph M'Gregor. *London.* 1869-81.
(John) & McCARTNEY. *Bangor, N. Ireland.* 1835-65.
   C. Then Campbell alone.
Marshall. *Colmonell.* 1850.
Mrs. C. *Leeds.* 1871.
Neal. *Aylesbury.* 1718-48. W.
Oliver. *Moira.* (b.1805) ca.1830-d.1845.
Richard. *Portaferry.* 1813.
Robert. *Baltimore, USA.* 1799-1872.
Robert G. *Ravenna, O, USA.* 1829.
Thomas. *Birkenhead.* 1878.
Thomas. *Broughton-in-Furness.* 1858-66. C.
Thomas. *Egremont.* 1860.
Thomas. *New York.* ca.1810.
Thomas. *Strabane.* ca.1780-1820. C.
William. *Carlisle, Pa, USA.* 1763.
William. *London.* 1881.
William L. *Glasgow.* 1860.
CAMPER—
& BUTLAND. *London.* 1828.
James. *London.* 1839.
CAMPLIN, James. *Bristol.* a.1812.
CANADA—
1914 and later, product of Pequegnat Clock Co.
Clock Co. *Whitby, Canada.* 1872-75.
Clock Co. *Hamilton, Ontario.* 1884. New name of
   Hamilton Clock Co.
CANADIAN TIME, 1914 and later, product of Peque-
   gnat Clock Co.
CANADINE, Joseph. *Stourbridge.* 1876.
CANAWAY, T. *Landport.* 1859-67.

CANBY—
George. *Selby.* mar.1665. d.1705. Lantern clock.
& NIELSON (Alex). *West Chester, Pa, USA.* 1819.
CANDEE & McEWAN *Edgefield, SC, USA.* 1858.
CANDWELL, Zerah. *London.* 1875.
CANEY—
William. *Market Rasen.* 1849-76.
William. *Tunbridge Wells.* 1839.
CANFIELD—
(Samuel) & FOOTE (William). *Middletown, Conn
   USA.* 1795-96.
Samuel. *Middletown, Conn* and *Lansingburg, NY,
   USA.* 1780-ca.1800. C.
CANHAM—
E. *Watford.* 1851-66.
Etham. *Croydon.* 1839.
William. *Ipswich.* 1876.
CANN, G. F. *Taunton.* 1866.
CANNALL, George. *Birkenhead.* 1878.
CANNIFF, Jacob. *St. Thomas, Canada.* 1851.
CANNIN, Joseph & Co. *Liverpool.* 1848.
CANNINGE, John. *London.* 1622.
CANNON—
John. *Greenwich.* 1874.
Robert. *London.* 1881.
S. *London.* 1760. W.
William. *Canterbury.* 1676.
William. *London.* 1863-81.
CANOVA, Peter Manzia. *Halesworth.* (Italian),
   pre-1846 -d. 1882.
CANSDALE, Solomon. *Ipswich.* 1830-58.
CANT—
Edward Bromley. *London.* 1863-75.
Edward & Son. *London.* 1881.
T. *Falkingham.* 1849.
CANTER—
Joseph. *Dumfries.* 1849.
Nicholas. *Sedgefield.* 1851. W.
CANTIN—
Isidore. *Quebec.* 1890.
Victor. *Quebec.* 1890.
CANTON, Richard. *West Hartlepool.* 1856.
CANTONI, Brazio Carfax. *Horsham.* 1839.
CANTOR, Jacob. *London.* 1857-63.
CANTY, G. W. *Barton-on-Humber.* 1876.
CANUCK, 1914 and later, product of Pequegnat Clock
   Co.
CAPELL—
David. *Birmingham.* 1850-80.
James. *Birmingham.* 1850.
CAPELLA, J. *Newport, Mon.* 1868-75.
CAPITAIN(E)—
Henry. *London.* 1863.
& WEHRLE. *London.* 1869-81.
CAPITOL. 1914 and later, product of Pequegnat Clock
   Co.
CAPLOSS, George. *London.* 1839.
CAP(EH)ORN, W. P. *Coventry.* 1860-68.
CAPELLA, J. *Newport, Mon.* 1868.
CAPPER—
——. *Nantwich.* ?18c. C.
Elizabeth. *Nantwich.* 1828.
CAPPO—
Anthony. *Belfast.* 1819-54.
Joseph. *Belfast.* 1835-80. Weatherglasses.
CAPSTICK, Thomas. *Skipton.* mar.1741-66. C.
CAPT—
Henry (of *Geneva*). *London.* 1881-ca.1890. C.
Henri Daniel. *Paris.* 1850-80.
CARBE(?), John. *London.* 1697.
CARCY, Jacques Page. *Quebec.* 1713.
CARD, Joseph. *London.* 1851-81.
CAREY—
Edward. *Pentre Rhondda.* 1887.
Thomas. *Dublin.* 1783-85. W.
CARIOLI, A. *Whitby.* 1839. W.
CARKEEK, George. *Truro.* 1803-23. (Also KARKEEK.)
CARLETON—
Dudley. *Bradford, Mass* and *Newbury, Vt, USA.*
   1748-76.
James H. *Haverhill, Mass, USA.* 1853.

CARLETON—*continued.*
John C. *Bradford, Vt, USA.* ca.1800. C.
Michael. *Bradford* and *Haverhill, USA.* 1757. mar.1795-1836. C.
Thomas. *Dublin.* 1713. W.
CARLEY—
Enock. *Bungay.* 1830-39.
George. *Diss.* 1836.
George (& Co.). *London.* 1844-51(57-81).
Jonathan. *Thetford.* 1830-36.
Mrs. S. *Bungay.* 1846.
Richard. *Bungay.* 1853-79.
CARLHIAN & CORBIERE. *London.* 1857-63.
CARLILL, James Bellamy. *York.* (B. 1799-). 1801-04. W.
CARLISLE, Dinsdale. *Kendal.* b.1847-79. W.
CARLOSS, ——. *London.* 1825. Chron.
CARLTON, William. *Cork.* 1844. W.
CARLYON—
John. *Falmouth.* 1729-53. C. & sundials.
John. *Penzance.* 1771. W.
CARMALT, John. *Ringwood.* 1830-48.
CARMAN—
Samuel. *Brooklyn, USA.* ca.1850.
Samuel. *Harleston.* 1830-75.
CARMICHAEL—
J. *Gateshead.* ca.1800. C.
Robert. *Newcastle-on-Tyne.* 1848-58.
CARNALL, M. *Napanee, Canada.* 1862.
CARNE—
Geoffrey. *Plymouth.* 19c. W.
John. *Penzance.* 1752-75.
Robert. *Pont-y-cymer.* 1887.
CARNEGIE—
——. *Arbroath.* 1850.
Brothers. *Toronto.* 1861-63.
Charles. *Toronto.* 1867-77.
D. G. *Toronto.* 1857-60.
James. *Kemptville, Canada.* 1857-63.
Robert. *Auchinblae.* 1837.
Robert. *Stonehaven.* 19c. C.
Robert (& Brothers). *Kineft* and *Drumlithie (Scotland).* 1838.
CARNES, John. *Liverpool.* 1851. W.
CARNILL, Joseph. *Basford.* 1876.
CARNOVA, P. *Halesworth.* 1846.
CARO, Michael. *London.* a.1765 to William Storck. *St. Clements, Middlesex.*
CAROLAN, Edward. *Carrickmacross.* 1784.
CAROLI, Christian. *Koenigsberg.* ca.1675. C.
CARPENDALE, William. *Hull.* 1846.
CARPENTER—
A. W. *New Holland, Pa, USA.* s. of Anthony. 1814-69.
Brothers. *Carshalton.* 1866.
C. H. *Middleboro, Mass, USA.* ca.1870.
David. *London.* 1857-63.
Edward. *London.* a.1639 through CC.
James. *Dover.* 1866-74.
James. *London.* 1875.
Joseph. *Norwich, Conn, USA.* 1747. mar.1775-1804.
Lumen. *Oswego, NY, USA.* 1845-47.
Thomas. *London.* 1869-75.
Thomas & Richard. *London.* 1811. Watch case makers.
William. *Banbury.* 1834-53. C. & W.
William. *Birmingham.* 1880.
William. *London.* 1828-44.
William. *Yardley.* 1876.
CARR—
Andrew. *Belfast.* 1865.
Frank B. *Fall River, Mass, USA.* ca.1870.
George William. *Thetford.* 1858-75.
Hamilton. *?Dublin.* 1858-68.
Henry. *Penge.* 1866-78.
J. P. *New York.* ca.1820.
James. *Chipping nr Clithero.* 1828.
James. *Garstang.* 1848-58. W.
James. *Halifax, Nova Scotia.* 1863-97.
James. *Lowell, Mass, USA.* 1834.
James. *Malton.* 1823-40.
James. *Ohio, USA.* ca.1830. Travelling clock-seller.
John. *London.* 1863-81.
John E. & Son. *Swaffham.* 1830-65. C.

CARR—*continued.*
John Elliner. *Swaffham.* 1875.
Joseph. *Hexham.* 1775-d.1776. C.
Joseph. *London.* 1869.
Lyman. *Manchester, NH, USA.* 1855.
Patrick. *Drogheda.* 1858.
Samuel. *Birmingham.* 1828-35.
Thomas. *Runcorn.* 1878.
CARRELL—
Daniel. *Philadelphia.* pre-1790, *Charleston.* ca.1790-ca.1801. *Philadelphia* again 1801-06.
Edward F. *Jersey.* ?19c. C.
John. *Philadelphia.* 1789-94. C.
CARRICK—
Andrew. *Troon.* 1860.
Robert. *Belfast.* 1790-ca.1810. C.
Samuel. *Portadown.* 1865-90. C.
CARRIER, YOUNG & MARSHALL. *Toronto.* 1877.
CARRINGTON, William. *Charleston, SC, USA.* ca.1830, mar.1845-d.1901. W.
CARROL(L)—
B. *High Wycombe.* 1839 (prob. error for CARRORI, q.v.).
George. *Mansfield.* mid-19c. C.
John. *Cork.* 1810. W.
Jonathan. *London.* 1839.
T. J. & Co. *Hamilton, Ontario.* 1884.
W. J. *London* (20 *London Street*). Early 19c. C.
William. *Cork.* 1820-28.
CARRON, Samuel. *London.* (B. 1689, CC.)-1697.
CARRORI—
B. *Buckingham.* 1854.
Benjamin. *High Wycombe.* 1830-54.
CARRUTHERS—
Charles. *Langholm.* 1860. Also cutler.
David. *Ecclefechan, Scotland.* 1840.
David. *Lockerbie.* 1840, same man as at *Ecclefechan.*
George. *Langholm.* (b.1790) 1836-d.1866.
James. *Belfast.* 1808-28. C. & W.
James. *Carlisle.* 1834-48.
James. *Longtown.* 1828. C.
John. b.1813 at *Carlisle. Lancaster* 1851. W.
CARRY, O. *Paris.* 1912. C.
CARRYER (Rupert Samuel), YATES (John) & WADDINGTON (Thomas). *Newcastle-under-Lyme,* also *Burslem* and *Longton.* 1876.
CARRYL, Patrick. *New York.* 1758. C.
CARSE, Seth. *Ardmillan.* ca.1750. Clock case maker.
CARSON—
Arthur. *Dumfries.* 1798.
George. *York, Pa, USA.* ca.1819.
Henry. *Halifax, Nova Scotia.* 1866.
Thomas. *Albany, NY, USA.* ca.1830.
Thomas. *Belfast.* d.1849. W.
William. *Halifax, Nova Scotia.* 1866.
CARSWELL—
George. *London.* 1863-81.
John. *London.* (B. a.1804, CC. 1819-)1828.
John. *Toronto.* 1837-57.
CARTEN, Andrew. *Liverpool, Nova Scotia.* 1864 (prob CARTER, q.v.).
CARTER—
Albert George. *Haverhill.* 1875-79.
Andrew. *Liverpool, Nova Scotia.* 1866.
Andrew W. *Picton, Nova Scotia.* 1879.
Charles. *Rochford.* 1839.
Christopher. *Galby.* b.1676-ca.1700. Lantern clock.
G. *Baldock.* 1859-66.
G. *Coventry.* 1868.
George. *Cavendish.* 1846-79.
George. *Hitchin.* 1851-74.
George. *Linton, Cambridgeshire.* 1858-75.
George. *London.* 1863-81.
George. *Seaham Harbour.* 1851.
George. *Sittingbourne.* 1874.
H. *Stansfield.* 1846.
Henry. *Long Melford.* 1865-79.
Henry. *Oxford.* 1852. C. & W.
Henry. *Ripon.* b.1795. s. of William Edward. d.1876.
Henry. *Southminster.* 1866-74.
Henry & Son. *Ripon.* 1834-51.
J. *Birmingham.* 1854. Clock case maker.

CARTER—*continued.*
J. *Brackley, Northants.* 1864-69.
J. F. *Hamilton, Ontario.* 1852-53.
Jacob. *Concord, NH, USA.* ca.1840.
James, sen. *Warrington.* 1822-48. Also silversmith, china dealer and swordsmith.
James, jun. *Warrington.* 1834.
James Ellis. *Bristol.* a.1809-18.
John. *Shoreham.* 1878.
John. *Whitby.* 1781-89. C. & W.
John Ashwell. *Baldock.* 1839.
Jonathan. *Faversham.* 1874.
Jonathan. *Herne Bay.* 1874.
Joseph. *London.* 1839.
Joseph. *Warrington.* With James in 1848; alone 1851.
Robert. *Warrington.* 1822.
Samuel. *Stansfield.* 1853-79.
& Son, *London. Cornhill.* ca.1810. C.
Thomas. *Bishop Auckland.* 1724-45. C.
Thomas. *Coventry.* 1880.
Thomas. *London.* 1869-81.
Thomas. *Mitchell Dean.* 1850.
Thomas Sayer. *Bristol.* 1850.
W. W. & L. F. *Bristol, USA.* 1863-68.
William. *Haverhill.* 1844-64.
William. *Cambridge.* (B. pre-1714-)1721. W. & C.
William. *Chelmsford.* ca.1801-39.
William. *Hampstead.* ca.1750.
William. *London.* 1839. Watch case maker.
William. *Moulsham nr Chelmsford.* 1828-39.
William. *Oxford.* a.1779. C.
William. *Salisbury.* 1830-75.
William Edward. *London.* 1881.
William Edward. b.1760. *Ripon* till 1829, then *London* where d.1842. C. & W.
William Henry. *Crawley.* 1878.
CARTIER—
A. *Ottawa.* 1875.
& AMEZ-DROZ. *London.* 1875-81.
CARTLEDGE, John. *London.* 1828.
CARTMELL, John. *Kirkham.* 1851-58. W.
CARTO, Joseph. *Putney.* 1828-39.
CARTWRIGHT—
——. *Nantwich.* ca.1760. C.
Charles. *Birmingham.* 1854-60. American clocks.
Charles & Henry. *Birmingham.* 1880.
Charles & Sons. *Birmingham.* 1868.
George. *London.* 1811.
H. W. *Chertsey.* 1851-78.
J. *Worksop.* 1864.
S. G. & Co. *London.* 1869.
Thomas. *Halifax.* 1834-37.
William. *London.* a.1705-59. C. & W. Later *Oxford.*
CARVALKS, D. N. *Philadelphia.* 1846.
CARVALLO, N. *Charleston, SC, USA.* ca.1780.
CARVER—
Alfred. *Falmouth.* 1873.
G. *Thetford.* 1846.
CARWITHIN, William. *Charleston, SC, USA.* ca.1733. Clock cases.
CARY, William. *Trowbridge.* 1842-59.
CARYER, Jesse. *Sittingbourne.* 1874.
CARYL—
Christopher. *Crewe.* 1878.
George. *Crewe.* 1878.
John. *Crewe.* 1857-78.
CASE—
(Erastus, Harvey) & BIRGE (John). *Bristol, USA.* 1830-37.
Charles James. *London.* 1844-63.
DYER, WADSWORTH & Co. *Augusta, Ga, USA.* ca.1835.
Erastus. *Bristol, USA.* 1830-37.
Harvey. bro. of Erastus. *Bristol, USA.* 1830-37.
Henry. *Prescot.* 1826. W.
James. *Prescot.* mar.1810-26. W.
(Everett G.) & ROBINSON (Abel T.). *Bristol, USA.* 1850-51. Clock cases.
(Erastus, Harvey), WILLARD (Sylvester) & COX. *Bristol, USA.* 1835.
CASEY, John & Thomas. *London.* 1869.
CASHMORE, John. *London.* 1857-81.

CASHORE, Andrew. *Strabane.* 1787.
CASKEY, Joseph W. *Coleraine.* 1839. C.
CASPAR, S. *Toronto.* 1859.
CASPER—
E. *London (Grace's Alley).* 1811. W.
Elias. *London.* 1828.
Levy. *London.* 1828.
Lewis. *Chatham.* 1832-51.
Nathan. *London.* (B. 1825-40)-51.
Philip N. *London.* 1857-75.
Samuel. *Toronto.* 1859-65.
CASS, John. *Woolwich.* 1823.
CASSAL, Abraham. *Philadelphia.* 1835. W.
CASSEL(S), James. *Lanark.* b.1833-d.1906.
CASSIDY, Hugh. *Maghera.* 1846.
CASTELBERG—
Antoine. *London.* 1875.
& Co. *London.* 1880.
Edward. *Birmingham.* 1880.
CASTENS, J. M. *Charleston, SC, USA.* ca.1770.
CASTLE, John. *Ballaugh, IoM.* 1860.
CASTLEBERB, M. *London.* 1857.
CASTLEBARG, Motel. *London.* 1863.
CASTON, T. *Norwich.* 1858.
CASTRO, Peter. *London.* 1839. Watch cases.
CASWELL, A. *New York.* ca.1860.
CASY, Haas. *Sheerness.* 1874.
CATALANO, Bouchet Antonio. *Italy.* 1481. C.
CATCHPOLE—
George A. *Kendal.* 1879.
J. *Feltwell (Norfolk).* 1865.
J. *Thetford.* 1865.
R. *Harleston.* 1858.
R. *Kenninghall.* 1865-75.
CATCHPOOL—
George. *Atherstone.* 1880.
William. *Chipping Barnet.* 1839.
William. *Dover.* Free 1801. W.
William. *London.* (B. 1826) 1828-32.
CATES, Joseph & Co. *Leamington.* (b. ca. 1821) 1842.
CATFORD—
H. *Clevedon.* 1861-66.
William. *Wellington, Somerset.* 1883.
CATHCART, A. H. *Marshall, Mich, USA.* ca.1870..
CATHERWOOD—
——. *Tanderagee.* Early 19c. (probably SITHERWOOD).
John. *Charleston, SC, USA.* ca.1760.
Joseph. *London.* (B. 1790-1815) 1832-39.
William. *London.* 1832-57.
CATHRO—
George & Robert. *London.* 1828-32.
J. & R. *London.* 1811. W.
Thomas George. *Quebec.* 1822-44. W.
CATLIN, Joel. *Augusta, Ga, USA.* ca.1820.
CATON, Richard. *St. John, New Brunswick.* 1818.
CATTANEO/CATTANIO—
Anthony. *Malton.* 1840-58.
& Co. *Leeds.* 1866.
Henry. *York.* 1837-58.
Joseph. *Folkestone.* 1858-74.
Joseph. *York.* 1851. W.
Natal. *Kidderminster.* 1850.
Pasqual. *Croydon.* 1839-55.
Peter. *Croydon.* 1851.
Peter & Co. *Reigate.* 1839.
S. *Stockton.* ca.1831. W.
Vincent. *Stockton.* 1851-56.
William. *Malton.* 1866. W.
CATELL, G. D. *Bath.* 1856.
CATTERALL—
James. *Prescot.* 1797. W.
Joseph. *Bolton.* 1828.
CATTERMOLE, George. *Toronto.* 1837.
CATTLE—
& BARBER. *York.* 1809. W.
George. *York.* Free 1785-d.1807. W.
John. *London.* 1632-33.
Robert. *York.* b.1767-d.1842. W.
CATTLIN—
Henry. *London.* 1857-81.
James. *London.* 1839-44.

CATTON, Richard. *Quebec.* From *London, England.* 1819-33.

CAUDLE, ——. *Torquay.* ca.1830. C.

CAULSFIELD, William. *Londonderry.* 1854-59. C.

CAUSER, R. *Coventry.* 1860. Watch case maker.

CAUSTON, Robert Henry. *St. Albans.* 1828.

CAUTHERS (or CARRUTHERS), William. *Glenarm.* 1846.

CAVE—
Arthur. *London.* 1881.
Douglas. *London.* 1875.
John R. *Deal.* 1847-51.
Richard. *Dover.* Free 1826-51. W.
Samuel. *London.* 1832-39.
W. J. *Dover.* 1855.
William. *Eastry.* 1832-51.
William John & Son. *Dover.* 1855-89.
William Richard. *Deal.* 1874.

CAVIT, John. *Bedford.* 1830.

CAWCUT, Obadiah. *London.* 1699. Believed clockmaker.

CAWDLE—
——. *Torquay.* ca.1830. C.
William. *Torquay.* c.1760. C. (Perhaps too early estimate of above.)

CAWDREY, Walter. *London.* a.1692.

CAWDRON—
Edward. *Wells.* 1865-75.
R. C. *Wells.* 1865.
Robert Charles. *Cley.* 1875.

CAWLEY, John. *Manchester.* 1744. W.

CAWSE, R. J. C. *Biggleswade.* 1877.

CAWSON—
Edward. *Lancaster.* b.1759. d.post-1811 and pre-1817.
Ellen. *Liverpool.* 1834.
James. b.1757. *Lancaster. Greta Bridge.* 1779-ca.1817.
James. *Liverpool.* 1822.
Mary. *Liverpool.* 1828.
William. s. of James of *Greta Bridge. Liverpool.*1817.W.

CAWTHORNE—
William. *Grantham.* 1835-61.
William B. *Port Hope, Canada.* 1857-63.

CAWTON, Thomas. *Sheffield.* 1724. C.

CAYEA, O. *Ottawa.* 1875.

CAYGILL—
Christopher. *Askrigg.* b.1747-d.1803. C.
George. *Ripon.* 1866. W.
Matthew. *Askrigg.* b.1775-d.1819.C. s. of Christopher.

CEALEY—
W. *Wainfleet.* 1861.
William. *Bourn.* 1868-76.

CEAREY, William. *Dublin.* a.1718.

CEARSWELL, George. *Aberystwyth.* 1816-30.

CECILL, Septimus. *Dublin.* 1736-d.1758. W.

CELLAR, John. *Bath.* 1798.

CELLERS, John. *Chillicothe, O, USA.* 1814.

CETTI—
John. *Buckingham.* 1842-54.
Paul. *Wellington (Shropshire).* (b. Italy). 1841-79. Sometime partner with Goetuno del Vecchio, Dominic & Samuel Frankel.
William. *Buckingham.* 1843.¯W.

CHABONER, Henry. *Dublin.* a.1678. W. & C.

CHADWICK—
——. *London (Holborn).* ca.1820. C.
Benjamin. *Liverpool.* 1841-51. Chron.
J. *Derby.* 1849.
James. *Congleton.* 1868-76.
James. *St. Helens.* 1848-51. Also optician.
John. *Manchester.* mar.1797. C. (B. 1804-13.)
Joseph. *Boscawen, NH, USA.* 1787-1831.
Joseph. *London.* (B. CC. 1815.) 1839-57.
Joseph. *St. Helens.* 1834. W.
Lewis. *London.* 1832.
William. *Stockport.* ca.1820. C.

CHALFONT—
John William & Son. *London.* 1881.
Mrs. M. *London.* 1851.

CHALK, Frederick. *Ipswich.* 1879.

CHALKER-
John. *Maiden Newton (Dorset).* 1848-75.

CHALKER—*continued*
Thomas. *London.* 1881.

CHALKLEN, Mrs. *Canterbury.* Widow of John. post-1766 and pre-1795.

CHALLEN, William. *Brighton.* 1855-78.

CHALLONER, William. *London.* 18c. br. clock.

CHALMERS—
Alexander Thomson. *Aberdeen.* 1836.
& Co. *Ottawa.* 1871.
George. *Dublin.* 1748-80, at the Half Moon in Capel Street.
James C. *Morrisburgh, Canada.* 1871.

CHALTON, Robert. *Woolwich.* 1826-32.

CHAMBERLAIN—
Benjamin Moses. *Salem, Mass, USA.* 1842-70.
Charles. *Philadelphia.* 1833-39.
John. *Bury St. Edmunds.* b. ca. 1643. Free 1687 d. 1727.
Joseph. *Chelmsford.* d.1692.
Lewis. *Philadelphia.* 1829-42.
Nathaniel. *London.* (B. a.1651, CC. 1659-94)-97.
Richard T. *London.* 1869.
T. *Hugglescote.* Early 18c. C.
Thomas. *Chelmsford.* d.1633.
W. J. *Leicester.* 1876.
William. *Towanda, Pa, USA.* ca.1830-50.
Webb. *Bristol.* From *Salisbury.* 1781.
Webb. *Salisbury.* pre-1781, later to *Bristol.*

CHAMBERLIN—
Cyrus. *New Haven, Conn, USA.* ca.1850.
Lewis. *Elkton, Md, USA.* ca.1824, then *Philadelphia.* 1831-40.

CHAMBERS—
Denis. *Pucklechurch* in *Bristol.* Early 18c.
Edward. *Canterbury.* Free 1678 (may be from London, where one such a.1670 to Evan Jones).
George. *Warminster.* 1859-75.
J. *Hoddesdon.* 1866.
James. *Liverpool.* 1828-34.
James. *Ore (nr. Hastings).* 1870.
N. J. *Dunville, Canada.* 1861.
R. *Bolton.* 1858. W.
W. J. *Ruabon.* ?19c. C.
William. *Boston.* 1828.
William. *St. Neots.* 1877.
William A. *Faversham.* 1847.
William (& Sons). *London.* 1851-63 (69-75).

CHAMBL(E)Y—
E. *Wolverhampton.* 1869.
William. *Tamworth.* 1835-50.

CHAMP, William. *London.* mid-18c. C. (See also CHAMPS.)

CHAMPAGNE, ——. *Montreal.* 1760.

CHAMPION—
Daniel Thomas. *London.* 1832. Tortoiseshell clock case maker.
Emile. *Paris.* 1889.
George. *London.* 1857-81.
John. *Wells.* 1641.
Richard. *Dublin.* ca.1795.
Richard. *Liverpool.* ca.1790-1800. W.
Thomas. *Egloshayle (Cornwall).* 1831. W.

CHAMPLIN, John. *New London, Conn, USA.* 1773.

CHAMPNEY—
Charles. *York.* a.1764.
& FELTON. *Troy, NY, USA.* ca.1860.
L. C. *Troy, NY, USA.* 1830. W.
Lewis C. *Philadelphia.* 1845.

CHAMPS, William. *London.* ca.1720. br. clock (see also CHAMP).

CHANCE—
John. *Chepstow.* ca.1760-91. C. & W.
John. *Worcester.* 1850.

CHANCELLOR—
John. *Dublin.* 1811-1900.
William. *London.* Late 18c. W.

CHANDLAR, Thomas. *Odiham.* ca.1740. C.

CHANDLEE—
Benjamin I. b. *Ireland* 1685, to *Philadelphia* 1702, to *Nottingham, Md.* 1712-41, then *Wilmington* 1741-45.
Benjamin II. s. of Benjamin I, whom he succeeded 1723-91.

CHANDLEE—*continued.*
Benjamin III. *Winchester, Va* and *Baltimore, USA.* s. of Goldsmith.
Ellis. *Nottingham, Md, USA.* s. of Benjamin III. 1755-1816.
Goldsmith. *Stephensburg, Va.* and later *Winchester, Va, USA.* s. of Benjamin II. To 1821.
(Benajmin III) & HOLLOWAY (Robert). *Baltimore, USA.* 1818-23.
Isaac. *USA.* 1760-1813.
John. *Nottingham, Md, USA.* 1757-1813.
Veazey. s. of Ellis. *USA.* 1804-ca.1846.
CHANDLER—
Abiel. *Concord, NH, USA.* 1807-81.
Anthony. *Drayton Parslow.* 1673.
E. *Leamington.* 1868-80.
Henry. *Birmingham.* 1868.
John. *Suffolk, Md, USA.* 1774.
Major Timothy. *Concord, NH, USA.* 1760-1846. C. (Ret'd 1829.)
Perley. *Barre, Vt, USA.* ca.1870-90. W.
CHANNING—
George. Used by U.S. Watch Co. 1868.
John. *Brighton.* 1878.
CHANT, James Cook. *Brigg.* 1861-76.
CHANTRILL, James. *Coventry.* 1880.
CHAPIN—
Aaron & Elliphalet. *Windsor* and *Hartford, Conn, USA.* 1780-1800. C.
Aaron & Son. *Hartford, Conn, USA.* 1825-38.
Edwin G. *Buffalo, NY, USA.* 1836.
S. & A. *Northampton, Mass, USA.* 1834.
CHAPLIN—
——. *Ashby.* ca.1750. C.
R. J. & Sons. *London.* 1869-81.
Robert, jun. *London.* 1857.
Robert John H. *London.* 1863.
CHAPMAN—
——. *Oxford.* 1852-85. C., W. & T.C.
Ann. *Liverpool.* 1834.
Benjamin. *Toronto.* 1875-77.
C. H. *Easthampton, Mass, USA.* ca.1860.
Charles. *Woodbridge.* 1838-75. W.
Daniel. *Chatham.* 1826-45. C.
Edwin. *Chatham.* 1832-55.
Francis. *Glasgow.* 1840.
Francis. *Pollokshaws.* 1836.
G. *Wolverhampton.* 1876.
George. *Lincoln.* 1828-35.
Isaac. *Colchester.* 1874.
J. *Royston.* 1859-66.
J. G. *Bristol.* 1840.
James. *Belfast.* ca.1843-64, when believed went to USA.
James. *Lincoln.* 1828-35.
James. *London.* 1869-75.
James. *Melbourn, Cambridgeshire.* 1875.
John. *Clerkenwell.* 1828-32. Clock case maker.
John. *Dublin.* 1858.
John. *Lewes.* 1870-78.
John. *Loughborough.* b.1760s-d.1840s. C.
John. *Watford.* (B. 1795-)1828-51.
Joseph. *Liverpool.* 1822.
Josiah. *London.* 1875-81.
Julia Ann. *Belfast.* ca.1836-58 (wid. of William) d.1876 at Hollywood.
Moses. *Liverpool.* 1848. Chrons.
Mrs. S. *Portsea.* 1878.
R. George. *Barton-on-Humber.* 1849-76.
Thomas. *Gravesend.* 1823-32.
Thomas. *Rodley* (nr. *Leeds*). 1743. C.
W. *Portsea.* 1859-67.
William. *Belfast.* (b.1790) from *Dublin* where worked 14 years, ca.1830-d.1836. W. & C.
William. *Bristol.* 1850-79.
William. *Lincoln.* 1849-76.
William. *Sleaford.* 1868-76.
CHAPPELL, (Samuel) & SARTWELL (James). *Jamestown, NY, USA.* ca.1840.
CHARGOIS, A. *Dublin.* 1868-80.
CHARLES—
Andrew. *Cookstown.* 1880. C.

CHARLES—*continued.*
Charles. *Haverfordwest.* 1840. (Successor to W. Gibbs, q.v.)
Charles. *Swansea.* 1865-71.
George. *London.* 1785. W.
J. *Paris.* ca.1810. C. (B. 1817-25.)
John. *London.* 1869-81.
John. *Portsea.* 1830-48.
John H. *Cookstown.* 1858.
Lewis. *Philadelphia.* 1837.
Samuel. *Belfast.* 1880-87. C.
William. *Cookstown.* 1843-68. C.
William. *Trelegg (Wales).* ?18c.
CHARLESTON, John. *St. Helens.* 1836. W.
CHARLSON, G. *London.* 1759. W.
CHARLTON—
C. *London.* 1869.
Edward. *West Auckland.* 1827-34.
Eugene. *Hetton-le-Hole.* 1851-56.
J. G. *Stockton.* 1898.
James. *Leicester.* 1876.
James. *London.* 1839-69.
Jesse. *Staindrop.* 1834-56.
John. *Woolwich.* 1838-47.
Robert. *Newcastle-on-Tyne.* 1834.
Robert. *Woolwich.* 1839.
William. *Epsom.* 1851-78.
CHARPENTIER, ——. *Besançon.* 1867-70.
CHARROT, Bros. *Paris.* ca.1820.
CHARTER—
——. *Philadelphia.* 1785. Watch case maker.
Philip. *London.* 1655.
CHARTERIS, William. *Dumfries.* 1837-49.
CHARTRAIN—
Jaques. *Quebec.* 1854-65.
Louis. *Quebec.* 1852-65.
CHARTRES—
Charles. *London.* 1832-75.
James William. *London.* 1839.
CHASE—
George. *New York.* 1854.
John F. *Newark, NJ, USA.* 1851-54.
Samuel. *Newburyport, Mass, USA.* 1794-1851.
Timothy. *Belfast, Me, USA.* 1826-40.
William H. *Salem, Mass, USA.* ca.1840.
CHASTON—
James. *Yarmouth.* 1865-75.
Mrs. E. *Yarmouth.* 1858.
CHATAIGNER, ——. *La Rochelle.* ca.1700. W.
CHATELAIN, Edmond. *London.* 1881.
CHATER—
Henry. *Ringwood.* b.1726-d.1767. C.
Jonathan. *Watford.* 1874.
William. *London.* (B. 1805-24) 1828-44.
CHATFIELD—
Amos. *Brighton.* 1870.
E. *Longton.* 1860-68.
Henry. *Uckfield.* 1878.
CHATHAM. 1914, and later product of Pequegnat Clock Co.
CHATTAWAY, D. *Cheltenham.* 1879.
CHATTLE, William. *London.* 1828.
CHATWIN—
Charles. *Huddersfield.* 1866. W.
Charles. *Leeds.* 1871.
CHAUDRON & RASCH. *Philadelphia.* ca.1830. Clock importers.
CHAULK, Stephen. *London.* 1869-75.
CHAUNTER, Thomas. *Exeter.* mar.1749.
CHAVAE, Zacharia. *?London.* a.1765 to Charles Mervile.
CHEAD, J. *Bath.* 1856.
CHEASEBROUGH, Aaron. *Penrith.* (B. 1689) ca.1710. C.
CHEEDAL, Joseph. *Dublin.* 1775. W.
CHEES(E)MAN—
James L. *New York.* ca.1830.
John. *Battle.* a.1725-44.
John. *Rye.* From *Battle,* a.1725 to William Reeve. W.
CHEETHAM—
J. *Leigh.* 1858. W.
Joseph. *Leeds.* 1794-1826, then *Rodley* 1834-38.

CHEETHAM—*continued.*
Richard. *Leeds.* 1826. W.
CHEEVERS, Bettson. *Dublin.* 1825.
CHELSEA CLOCK CO. *Chelsea, Mass, USA.* 1897-1904.
CHELSHAM, Thomas. *Horsham.* From *Rudgwick,* a.1732 to John Inkpen. C.
CHENERY, Frederick. *Bungay.* 1846-79.
CHENEY—
Asahel (Ashel). *East Hartford, Conn, USA.* s. of Benjamin. 1759, then *Northfield, Mass, USA.* ca.1790. C.
Benjamin, sen. *E. Hartford, Conn, USA.* ca.1745-1815.
Benjamin, jun. *E. Hartford* or *Manchester, Conn, USA.* s. of Benjamin, sen.
& DWIGHT. *Montreal.* 1809-19.
Elisha. *Middletown* and *Berlin, Conn, USA.* s. of Benjamin, sen. 1793-1833.
Martin. *Montreal.* 1800-20.
Martin. *Windsor, Vt, USA.* ca.1790. s. of Benjamin, sen.
Olcott. *Middletown* and *Berlin, Conn, USA.* s. of Elisha. 1835.
Russell. *East Hartford, Conn,* and *Putney* and *Thetford, Vt, USA.* s. of Benjamin, sen.
CHERBONNEAU, Rene. *London.* 1832.
CHERRY—
——. *Dromore.* Early 19c. C.
Andrew. *Banbridge.* 1868.
Andrew. *Lurgan.* ca.1835-68.
James. *Banbridge.* 1843-46. W.
James. *Philadelphia.* 1849.
M. *Birmingham.* 1860.
CHESHIRE—
Clock Co. *Cheshire, Conn, USA.* ca.1880.
James. *London.* 1832.
Watch Co. *Cheshire, Conn, USA.* 1883-ca.1893.
CHESNEY, John. *Clonmel.* d.1778. W.
CHESTER—
G., jun. *Leamington.* 1880.
Richard. *Hanover, Pa, USA.* 1795-1816.
William. *London.* (B. 1802-24) 1828-39. Also silversmith.
CHESTERTON, Thomas. *Leicester.* 1876.
CHESWORTH—
Thomas. *Prescot.* 1819-24. W.
William. *Prescot.* 1825. W.
CHETHWORD, John. *St. Helens.* 1774. W.
CHETTLE—
Thomas. *Battle.* 1870-78.
Thomas Robert. *Farnham.* b.1820 (at *Lambeth*) 1846. Then at *Battle.* ca.1848-ca.65. C.
William. *London.* 1832-57. ·
CHEVALIER, James. *London.* 1832-51. Clock and chronometer case makers.
CHEVIS, James. *Andover.* 1878.
CHEVOB, Ernest. *London.* 1875.
CHEVRIER, ——. *London.* mid-18c. br. clock.
CHEW—
A. E. *Nailsworth.* 1863.
John. *Prescot.* 1794-97. W.
Thomas. *Prescot.* 1819-26. W.
Walter. *Mansfield.* 1780-1832.
Walter. *Pateley Bridge.* 1822-25. Prob. man above.
CHICHESTER, Charles, *Dungannon.* 1820. C.
CHICK—
Peter. *Steyne.* 1862-70.
Peter. *Worthing.* 1878.
R. H. *South Wimbledon.* 1878.
Revett Hart. *London.* 1851-75.
CHIDWICK, William. *Dover.* 1832.
CHIERY, ——. *Paris.* ca.1780. C.
CHIFFELLE, Lina. ?*France.* ca.1874. W.
CHIGNALL—
John F. *London.* 1869.
Robert. *Hoxton.* 1768. W.
CHIL(L)COTT—
Thomas. *Bristol.* 1772.
William & Alfred. *Bristol.* 1850.
CHILD—
Benjamin. *Great Missenden.* 1830-69.
Henry. *Chelmsford.* ca.1840-48.

CHILD—*continued.*
Henry & Lawrence. *Chelmsford.* ca.1839 or before.
Henry T. *Philadelphia.* 1840-42.
John. *Philadelphia.* 1813-47.
Mrs. E. *Landport.* 1878.
Percy. *London.* 1881.
Robert William. *London.* 1857.
S. J. *Lyons, NY, USA.* ca.1830. W.
S. T. & T. T. *Philadelphia.* 1848.
Samuel. *Birmingham.* 1860-80. Importer.
Samuel T. *Philadelphia.* 1843-48.
Thomas. *Birmingham.* 1880.
Thomas T. *Philadelphia.* 1845.
Walter. *London.* 1875.
CHILDS—
Ezekiel. *Philadelphia.* 1830-35.
W. *Melbourne.* 1849.
William. *Southwell.* 1855-76.
CHILVER, S. *Halesworth.* 1846.
CHINN—
Edward William. *Huddersfield.* 1853-76. C.
James. *Birmingham.* 1868. Clock cases.
CHIPP, Robert. *London.* (B. a.1679) to Robert Seignior, and still with him in 1687.
CHIPPENDALE, Gilbert. *Halifax.* (B. from *London* 1776) 1780-87. C.
CHIPPERFIELD—
James. *Chelmsford.* d.1839.
N. W. & Co. *New York.* 1854.
CHIPPING, C. J. *Norwood (Surrey).* 1866-78.
CHISHOLM & CO. *London.* 1828.
CHITCHENER, Thomas. *London.* 1828. Cf. Thitchener.
CHITTENDEN—
Austin. *Lexington, Mass, USA.* 1831-37. C.
Isaac. *Yalding.* 1806. W.
CHITTEY, William. *Croydon.* 1728. W.
CHITTOCK, J. *Norwich.* ca.1820.
CHITTY, Caleb. *London.* 1844-51.
CHIVERS—
Joseph. *Marlborough.* 1859-75.
R. *Cricklade.* 1859-67.
Samuel. *Radstock.* 1861-83.
CHOLLOT (or **CHOLLIET**), John B. *Philadelphia.* 1816-19.
CHORLEY—
Matthew. *Prescot.* mar.1817-25. W.
William. *Prescot.* mar.1823-25.
CHOUINARD, Joseph Jacob. *Quebec.* 1847.
CHRISTENSON—
F. J. *Ventnor.* 1848.
Frederick J. *West Cowes.* 1859-78.
P. *West Cowes.* 1839-48. Also chrons.
Peter. *Newport, Isle of Wight.* 1830. W.
CHRISTIAN—
J. *Pilkington* nr *Manchester.* 1858. W.
John. *Calthorpe.* ca.1770. C.
John. *Liverpool.* 1834.
CHRISTIE—
Alexander. *Dublin.* 1758-d.1801. W.
& BARRIE. *Arbroath.* 1860. C.
George. *Wallingford.* 1864-83.
George. *Aylesbury.* 1851.
Henry. *London.* 1839-44.
James. *Brechin.* 1860.
James. *Cumberclaudy.* ca.1830. C.
James. *Perth.* (B. 1820-)1860. Two such.
Robert L. *Leith.* 1860.
William. *Laurencekirk.* 1860.
William. *London.* (B. 1825-) 1839-81.
William. *Stirling.* 1860.
CHRISTMAS—
D. S. *Quebec.* 1847-53.
James. *Yarmouth.* 1830-36.
CHRISTOPHER, William. *Oundle.* 1841.
CHRISTY, Alexander. *Campbeltown.* 1860.
CHRYSTLER, William. *Philadelphia.* 1828-35.
CHUMBLEY—
Joseph. *Sheffield.* 1862. W.
William. *Retford.* 1835. C. (Successor to UNDY, q.v.).
William. *Sheffield.* 1837. W.
CHUNN, William. *Leicester.* 1876.

CHURCH—
Edward. *New Haven, Conn, USA.* ca.1850.
James. *London.* (B. a.1793?) 1844.
John. *London.* 1851-69.
John Thomas. *London.* 1844.
Joseph. *Hartford, Conn, USA.* 1825-38.
Lorenzo. *Hartford, Conn, USA.* 1846.
Thomas. *Gainsborough.* 1876.
Thomas. *Norwich.* 1712-1801.
Thomas Robert. *London.* 1839.
William F. *Skowhegan, Me, USA.* ca.1830.
CHURCHER, John. *Southampton.* 1878.
CHURCHILL—
C. *London.* 1869.
George. *Curry Rivel.* 1866-75.
George. *Downton.* 1867-75.
CHURTON—
I. *Shrewsbury.* 1856-63.
Joseph. *Whitchurch.* 1799-1836.
CHUTER, William. *Farnham, Surrey.* 1851-78.
CIAPPESSONI, F. *Eastbourne.* 1862.
CINCINNATI TIME RECORDER CO. *Cincinnati, O, USA.* 1936.
CIPPRIAN, Michael. ?*London.* pre-1660.
CITADEL. 1914 and later, product of Pequegnat Clock Co.
CITO, J. C. *Boston, USA.* ca.1820. C.
CITY CLOCK CO. *London.* 1881.
CLACHER, Alexander. *Glasgow.* 1860.
CLACKNER, John. *Troy, NY, USA.* 1833-37.
CLAGGETT, William. b.1696. *Wales.* mar. *Boston, USA.* 1714. To *Newport (RI)* 1716-d.1749. C.
CLAMP, John. *Bournemouth.* 1878.
CLAPHAM, George. *Brigg.* (B. 1767-)ca.1790. C.
CLAPMAN, Robert. *Barton-on-Humber.* 1850. Cf. CHAPMAN.
CLAPP—
Benjamin. *Chard.* ?early 19c. C.
John. *Roxbury, Mass, USA.* ca.1800.
Preserved. *Amherst, Mass, USA.* 1731-76.
CLAPPISON, John Ward. *Hull.* 1851-58. C. & W.
CLAPSON—
John. *Sheffield.* 1871. T.C.
L. *Sheffield.* 1871.
CLARE—
Henry James. *London.* 1832-51.
J. *Sandy.* 1869.
J. A. *Coventry.* 1854.
J. & Son. *Coventry.* 1860.
Joseph. *Abergavenny.* 1880. Also optician.
Joseph. *Kimbolton.* 1847-77.
Thomas. *Bedford.* 1828. C.
Thomas P. *Bedford.* 1830-69. W.
CLARENCE—
Frederick William. *Birmingham.* 1860-80.
Frederick William. *Smethwick.* 1868-76.
CLARENDON—
George. *Kingstown* (Ireland). 1868-80.
George H. *Drogheda.* 1858. W.
John N. *Kingstown.* 1858-80.
CLARIDGE—
Charles. *Chepstow.* 1852.
George. *Chepstow.* 1844-80.
George. *Lydney.* 1879.
William. *Bristol.* 1840-42.
CLARINGBOWL, F. *Hamilton, Ontario.* 1880-95.
CLARK(E)—
——. *Liskeard.* 1831. W.
——. *Redruth.* ca.1795. C.
A. *Lowestoft.* 1865.
A. N. *Plainville, Conn, USA.* ca.1888-89. Watch cases
Alexander. *London.* mid-18c. W.
Alfred J. *London.* 1869.
Ambrose. *Baltimore, USA.* 1784-87.
Ambrose. *Dublin.* 1781-86.
Amos. *Lewisberry, Pa, USA.* ca.1780-1800.
B. H. *Newcomerstown, O, USA.* 1846.
Benjamin S. *Wilmington, Del, USA.* ca.1850.
Benjamin W. *Philadelphia.* 1831.
Brothers. *Rutland, Vt, USA.* ca.1830. W.
C. *St. Ives, Hunts.* 1847-64.
Charles. *Epping.* 1839.

CLARK(E)—*continued.*
Charles. *Epworth.* 1861-76.
Charles. *London.* 1857-63.
Charles Thomas. *London.* 1869-75.
Christopher. *Dublin.* 1736-56. W.
Christopher, jun. *Dublin.* 1770-92. W.
Christopher. b.ca.1668 *England.* mar. *Amsterdam* 1694. Dtr of Ahasuerus Fromanteel II, with whom he became partner. Later partner with Roger Dunster. d.1735 in London. W. & C.
& Co. *Augusta, Ga, USA.* ca.1820.
(Daniel), COOK (Zenas) & CO. *Waterbury, Conn, USA.* ca.1815.
(A.N.) & COWLES. *Plainville, Conn, USA.* ca.1870-89.
David. *Charleston, SC, USA.* ca.1770.
& DUNSTER. *Amsterdam.* Partnership between Christopher Clarke & Roger Dunster, ca.1720-ca.30. C. & W.
E. *Westerham.* 1866.
E. J. *Melcombe Regis.* 1855.
Edward. *Swaffham Prior.* 1858-75.
Edwin. *London.* 1857.
Edwin. *Sturminster.* 1824. Also stationer.
Edwin. *Wakefield.* 1834.
Edwin. *Weymouth.* 1830.
Ephraim & Charles. *Philadelphia, USA.* 1806-11.
Fleetwood. *Wellingborough.* 1777. W.
Francis. *St. John, New Brunswick.* 1838.
G. *Hinckley.* 1855.
Gabriel. *Baltimore, USA.* 1831-39.
George, *Aberdeen.* (b.1816) d.1852.
George. *Hinckley.* 1835.
George. *London.* ca.1710-1774. W. & C.
George. *London.* 1844-75. Wholesaler.
George. *Long Buckby.* 1830-47.
George. *Tring.* 1859-74.
George D. *Coosada, Ala, USA.* ca.1878-1935. C.
George G. *Providence, RI, USA.* ca.1824.
George R. *Utica, NY, USA.* ca.1820.
GILBERT, & BROWN. *New York.* ca.1840.
GILBERT & CO. New York. 1848.
(Lucius), GILBERT (William) & CO, *Winsted, Conn, USA.* 1841 and later. C.
H. *Bury St. Edmunds.* 1865.
H. B. *London.* 1837.
H. G. *Lindsay, Canada.* 1862.
Harry. *Bristol.* 1879.
Heman. *Plymouth, Conn, USA.* 1783-d.1838. C.
Henry. *Reading.* 1864-77.
Henry J. *Bishop's Stortford.* 1874.
Horace G. *Weston, Vt, USA.* ca.1830.
Hugh G. *Bowmanville, Canada.* 1851.
J. Earith, *Hunts.* 1854-64.
J. *Farnham, Surrey.* 1866.
J. *St. Neots.* 1847.
J. M. *Horncastle.* 1861.
J. M. *Sleaford.* 1868.
J. W. *Coventry.* 1868.
James. *Bath.* 1861-83.
James. *Edinburgh.* mar.1791-1860.
James. *Finmere, Oxon.* 1858. T.C.
James. *Frome.* 1705-20. C. & W.
James. *Kirkcaldy.* 1843.
James. *Morpeth.* ca.1750.
James Henry Edward. *London.* 1881.
James William. *London.* 1881.
James Z. *London.* 1844-57.
Jesse W. & Co. *Philadelphia.* 1809-14.
John. *Bristol.* a.1643-79. Lant. clock (B. ca.1635 seems too early).
John. *Bury.* ca.1770. C.
John. *London.* 1828.
John. *London.* 1857-63.
John. *New York.* 1770-90.
John. *Peterboro, Canada.* 1857-76.
John. *Philadelphia.* 1797-1845. May be two such.
John. *Stamford, Lincs.* 1696.
John. *Toronto.* 1851.
John (& Son). *Greenock.* 1820-36 (-1860).
John T. *London.* 1857-63.
John V. *Longtown.* 1879.
John Thomas. *London.* 1832-44.

CLARK(E)—*continued.*
Joseph. *Exeter.* mar.1724.
Joseph. *Kirkcaldy.* 1843.
Joseph. *New York.* 1768-77. *Dadbury, Conn,* 1777-1811, later *Alabama.* (mar.1787).
Josiah. *London.* ca.1710. C.
& LATHAM. *Charleston, SC, USA.* 1780.
Lucius. *Winsted, Conn, USA.* from *Bristol (USA)* 1841.
Matthew. *Burwell.* 1840.
Matthew. *Swaffham Prior.* 1830.
Matthew. *Swaffham Prior.* 1858-75.
Matthew Henry. *Birmingham.* 1842-50.
Michael. *Hull.* 1814-16. W.
Michael. *Morpeth.* 1827-34.
Michael. *Stetworth.* ca.1770. C.
& MORRIS. *Liverpool.* 1851.
& MORSE. *Plymouth, Conn, USA.* ca.1820-30. C.
Mrs. H. R. *Birmingham.* 1854.
Mrs. Rosina. *London.* 1881. (?wid. of Charles Thomas?).
Nathan. *Crewkerne.* 1883.
Peter. *Newent.* 1879.
Philip. *Albany, NY, USA.* 1837-43.
Philip & Sons. *Bailieborough.* 1846-90.
R. *Liverpool.* 1824. W.
Richard. *Charleston, SC, USA.* 1767-72.
Richard. *Dublin.* 1782. W.
Richard. *Newport, Isle of Wight.* 1830.
Richard, jun. *Worcester.* 1850.
Richard & Son. *Bradford.* 1822. C.
Robert. *Cerne Abbas.* 1830-55.
Robert. *Downpatrick.* ca.1800.
Robert. *Kilmarnock.* 1850.
Robert. *Newburgh (Fife).* 1836.
Robert. *Norwich.* 1830-36.
Robert. *York* 1807, later *Cottingham* nr *Hull.* d.pre-1826.
Russell. *Woodstock, Vt, USA.* 1830(-69?).
Russell & Randall. *Woodstock, Vt, USA.* 1869.
& Son. *Norwich.* 1846.
& Son. *St. Ives, Hunts.* 1877.
Stafforth. *Oundle.* 1864-77.
Stephen William. *London.* 1869.
Sylvester. *Salem Bridge, Conn, USA.* 1830. C.
Theodore. *Kirkby Lonsdale.* ca.1820. C.
Thomas. *Ashton-under-Lyne.* d.1712. C.
Thomas. *Boston, USA.* 1764.
Thomas. *Burton-on-Trent.* 1850.
Thomas. *Crawcrook.* 1856 (d.1884?). C.
Thomas. *Epworth.* 1828-76.
Thomas. *London.* 1875.
Thomas. *Stella (Co. Durham).* 1834.
Theodore Cuthbert. *Ulverstone.* 1825-28. C.
Thomas. *Ulverstone.* 1770-1801.
Thomas. *Windle.* d.1726. Clocksmith.
Thomas W. *Philadelphia.* 1830-50.
& TURNER. *Fayetteville* and *Wadesborough, NC, USA.* 1820-23.
W. *Clapham.* 1851.
W. *Forest Hill.* 1866.
W. *Napanee, Canada.* 1862.
W. *Oundle.* 1847-54.
W. *Willingham.* 1858.
W. B. *Coventry.* 1868-80.
William. *Belford.* ?18c. C.
William. *Cartmel.* 1767. W.
William. *Cerne Abbas.* 1824.
William. *Dublin.* Free 1798-d.1833.
William. *Grasmere.* May be same man as at Kendal.
William. *Kendal.* b. 1716 d. 1763. C. Quaker.
William, jun. *King's Lynn.* ca.1760-ca.1790. C.
William. *Liverpool.* 1824-28. C.
William. *London. Rotherhithe.* ca.1790. C.
William. *London.* 1832-51.
William. *Long Buckby.* 1854-77. C.
William. *Picton, Canada.* 1851-57.
William. *Rotherhithe.* ca.1790. C.
William. *St. Ives, Hunts.* 1839.
William. *York.* a.1748. d.1796. C.
William H. *Palmer, Mass, USA.* ca.1860.
William R. *Norwood, Surrey.* 1878.

CLARK(E)—*continued.*
William & Sons. *London.* 1839-81.
William Staffort. *Chatteris.* 1875
William Thomas. *London.* 1863-69.
CLARKSON—
Edward. *Stockport.* ca.1790. C.
George Finlay. *Northallerton.* 1866-72 (trained by Cade).
John. *London.* 1869.
Michael. *Halifax.* 1840.
Thomas. *Beverley.* 1834-40.
CLASSIC. 1914 and later, product of Pequegnat Clock Co
CLATON, Elias B. *Philadelphia.* 1848.
CLATWORTHY, Thomas. *Ystalyfera, Pontardawe, Allt Wen* and *Ynyscedwyn.* 1872-87. C.
CLAUDE—
Abraham. *Annapolis, Md, USA.* 1773-ca.1779. From London.
(Abraham) & FRENCH. *Baltimore, USA.* 1771, then *Annapolis.* 1773-83.
CLAUDON, John-George. *Charleston, SC, USA.* From *London.* 1743. W.
CLAUGHTON, Joseph. *Prescot.* mar.1759. W.
CLAUS, Henry. *London.* 1869-81.
CLAVEL, P. *Geneva.* (B. ca.1770)-ca.1790.
CLAY—
Alfred. *Kirton-in-Lindsey.* 1868.
C. *Kirton (Lincs.).* 1849.
Charles. *Gainsborough.* (B. 1762-)1828.
Charles. *London.* (B. ca.1730-)1761. W.
Charles & Son. *Lincoln.* 1876.
Edwin. *Birmingham.* 1868-80.
H. *Gainsborough.* 1868.
James. *Birmingham.* 1850-60.
James. *Manchester.* 1851. Foreign clocks.
John. *Hunmanby (Filey).* 1851-67.
S. *Lincoln.* 1849.
Samuel William. *Gainborough.* 1849-76.
Thomas. *Chelmsford.* ca.1650?
Thomas. *Liverpool.* 1851.
CLAYFIELD, W. *Wootton-under-Edge.* 1856.
CLAYMORE, Robert. *Germantown, Pa, USA.* 1765-80. Formerly a slave.
CLAYTON—
E. *Castleford.* 1871.
George. *Marple.* ca. 1680 -d. 1716. C.
Henry. *Puckeridge.* 1874.
J. *London.* 1844.
J. H. *Darlington.* 1898.
John B. *Bradford.* 1866-71.
Joseph. *London.* 1832.
Martin. *Manchester.* mar.1789-1824. W.
Richard. *Cincinnati, O, USA.* From England. 1834-56.
CLEAK(E)—
Ezekiel. *Exeter.* Free 1710. C. (s. of Ezekiel C., locksmith).
John. b. 1763. s. of Adam Cleak. of *Luppitt (Devon).* To *Bridport* where he mar. 1790. d. 1833. C.
Samuel. b. 1824 *Taunton.* to *London* ca. 1860. d. there 1896. C.
CLEEVE, John. *Belfast.* May be a clockmaker — known to have made weights for a clock. 1665.
CLEGG—
James. *Manchester.* 1848.
John. *Sunderland.* 1789. C.
Joseph. *Newton-Stewart.* 1860.
Samuel James. *London.* 1875-81.
William. *Hyde.* 1848.
William. *Manchester.* 1848-51. (Three shops.)
William Frederick. *Manchester.* 1851.
CLEGHORN—
John & Son. *London.* 1828. Dialmakers.
Samuel. *London.* (B. 1790-1808) 1828-32.
CLEIN, John. *Philadelphia.* 1830-33.
CLELAND, Mrs. Margaret. *Edinburgh.* Wid. of John. 1790-96.
CLEMENCE, Brothers. *London.* 1869-81.
CLEMENS, Moses. *New York.* 1749.
CLEMENT—
——. *Bodmin.* 1695. Rep'd ch. clock (may be the Exeter man, q.v.).
& BOURJOIS. *Morez (Jura).* ca.1830-51. C.
Edward. *Exeter.* 1671-1710. C. (Prob. the one a. in

CLEMENT—*continued.*
London 1662, who went to Exeter ca.1671).
Elizabeth. *Tring.* 1828.
F. *Paris.* 1690.
& Sons. *Tring.* 1866-74.
T. & Co. *Tring.* 1859.
William. *Totnes.* 1695-d. 1736. W. Also haberdasher of hats.

CLEMENTS—
Allen. *London.* 1641. W.
Handley & Co. *London.* 1881.
James Thomas. *Walsall.* 1842.
John. *Oxford.* 1808-53. C. & W.
(Henry) & SMITH (John). *Dublin.* Partners. 1783-90. W. & C.
William. *London.* bca.1638, CC. 1677-d.1704. C. Sometimes credited with invention of anchor escapement.
William. *London.* 1863-75.
William. *Portsmouth.* 1830.

CLEMENTSON, Edward. *West Ham.* 1874.
CLEMETSON, ——. *Dunstable.* ca.1785. C.
CLEMINSON, George. *Ulverstone.* b.ca.1700-d.1776. C.
CLEMENS, Isaac. *Shelburne, Nova Scotia.* 1786.
CLEMMOW, C. *Grampound.* 1856.
CLEMSON, Samuel. *Birmingham.* 1828.
CLENCH, Robert. *Dublin.* ca.1800. W.
CLENDINNEN, John. *Montreal.* 1848-54.

CLERC—
——. *Fleurier (Switzerland).* ca.1800. W.
Henry. *New York.* ca.1820.

CLERKE—
Frederick William. *London.* 1863-81.
G. J. *London.* 1839-44.
George. *London.* (B. a.1792. CC. 1802-25) 1828-32.
S. *Dorchester.* 1867.
William. *London.* 1828.

CLEVELAND—
Benjamin Horton. *Newark, NJ, USA.* 1767, mar. 1789-1837.
Francis. *Zanesville, O, USA.* 1815. C.
William. *Norwich, Conn, USA.* 1770-ca.-1800, then *Salem, Mass,* ca.1800.

CLEVERDON—
J. R. *Halifax, Nova Scotia.* 1840.
William H. *Halifax, Nova Scotia,* 1878-97.

CLEVERL(E)Y—
Henry. *Coventry.* 1860-80.
Henry John. *Melksham.* 1875.
Mrs. H. *Melksham.* 1867.
CLEWER, William Henry. *Todmorden.* 1853-71.

CLEWL(E)Y—
Alfred. *London.* 1851-75.
James. *London.* 1832.

CLIFF—
George. *Tutbury.* 1828. C.
John. *St. Helens.* 1769. W.
Thomas. *Hull.* ca.1740-d.1781. C.
CLIFFORD, Ebenezer. *Exeter, NH, USA.* 1746-88.

CLIFTON—
John. *London.* 1646. Lothbury, failed to observe CC. rules.
John. *Liverpool.* 1777-d.1794. C.
Joseph. *London.* 1663. *Bull Head Yard, Cheapside.* Prob. maker of clock *cases* for the Fromanteels.
CLINCH, George. *London.* ca.1720. C.
CLINKUNBROOMER, Charles. *Toronto.* 1833-67.
CLINTON, Richard. *Eccleshall, Staffs.* 1828-42. W.

CLITHEROE/CLITHEROW—
John. *Farnworth.* mar.1831. W.
John. *London.* 1832.
John. *Rainhill.* 1819. W.
Thomas. *Eccleston.* 1819. W.
Thomas. *Liverpool.* (B. 1825) 1828-34.
Thomas. *St. Helens.* mar.1770. W.
Walter. *Beverley.* 1719. W.
William. *Prescot.* 1796. W.
William. *Prescot.* 1823-25. W.
CLOAKE, Thomas. *London.* 1851-69.
CLODE, William. *Camelford.* ca.1760.
CLOGGETT, William. Place unknown. ca.1745. C.

CLOSE, Samuel. *Belfast.* 1700. W.
CLOTHIER, James. *London.* 1851-57.
CLOUGH—
John. *Manchester.* 1744-mar.1756. C.
Mrs. Caroline. *Leeds.* 1871.
Thomas. *Colchester.* b.ca.1570-d.1626. Kept ch. clock.
William. *Settle.* 1853.
CLOW, Andrew. *Glasgow.* 1860.
CLOWES—
D. *Liverpool.* 1806. W.
John Joseph. *Liverpool.* 1828-51. Imported clocks.
CLUCAS—
Joseph. *Peel, IoM.* 1860.
William. *Douglas, IoM.* 1844-60. W.
Cluley, William. *Manchester.* mar.1802. C.
CLYMER—
Elizabeth. *Bristol.* 1840. (?wid. of Marmaduke.)
John. *Bristol.* 1850.
John Marmaduke. *Bristol.* 1863-79.
Marmaduke. *Bristol.* (B. 1785-1830) 1830-42.
COAD—
James Thomas. *Hayle.* 1873.
R. *Otley.* Late 19c. W.
Rev. Frank A. M. *Hillsboro, NH, USA.* ca.1800.
COAKE, George. *London.* 1869.
COALES—
C. J. *Luton.* 1847-69.
Dennis. *Newport Pagnell.* 1847-77. Partner with William Peck COALES.
Mrs. S. *Newport Pagnell.* 1869.
& Sons. *Newport Pagnell.* 1877.
W. P. & D. *Newport Pagnell.* 1847-54.
William Peck. *Newport Pagnell.* mar.1820-d.1855. Partner with Dennis.
COATES—
Andrew. *Romford.* 1874.
Archibald II. *Wigan.* s. of Archibald I. ca.1795-ca.1810.
Benjamin Ellis. *Wakefield.* 1807-22. C. & W.
George. *Coventry.* 1880.
Giles. *London.* 1844.
H. G. *Whitby.* Late 19c. C.
Isaac. *Philadelphia.* 1835-39.
James. *London.* 1839.
James & Robert. *Wigan.* Sons of Archibald I. Partners from ca.1780. Robert d.1800. Thereafter James alone till ca.1811. C. & W.
John. *Bristol.* 1870.
John Henry. *Wakefield.* 1837.
Richard. *Dudley.* 1863-76.
Thomas. *Hamilton.* ca.1780. C.
W. *Ottawa.* 1862.
William. *Brockville, Canada.* 1862-71.
William. *Buffalo, NY, USA.* 1835-39.
William. *Prescott, Canada.* 1853-58.
COATSWORTH, Joseph. *Middleton-in-Teesdale.* 1827-34.
COBB—
George. *Ollerton.* 1835-55.
George. *Retford.* 1828-44.
James. *Liverpool.* 1851. Imported clocks.
William. *York.* Free 1659-73. W.
Z. B. *Cincinnati, O, USA.* 1850.
COBHAM, Joshua. *Liverpool.* 1674-81. W. & Watch cases.
COCHRAN(E)—
——. *Ballymoney.* Early 19c. C.
Edward. *Belfast.* ca.1815-39. W.
George. *Portadown.* 1868.
George. *West Chester, Pa, USA.* 1799-1807. W.
Hugh Hall. *Lurgan.* ca. 1825 C. Later at *Portadown,* then *Tanderagee.* d. 1879 aged 74.
& SHAW. *Belfast.* ca.1815-25. W.
Thomas. *London.* 1832-39.
Thomas. *New York.* ca.1810.
W. *Brighton.* 1855.
William. *Paisley.* 1836.
William. *London.* 1832.
William. *London.* 1857.
COCK—
Charles. Place not stated. Third quarter 18c. C. (prob. the next maker, q.v.)

COCK—*continued.*
Charles. *Macclesfield.* 1762 (-B. d.1782). C.
John. *London.* Late 17c. Marq. l.c. clock.
Josiah. *Bristol.* 1824.
Patrick. *Nottingham.* b. 1713, mar. 1739-55. C.
Richard. *London.* 1857-81.
Richard. *Penryn.* ca.1720-29. C.
Richard. *Penzance.* ca.1720. W.
William. *Penryn.* mar.1745. C.
COCKAYNE, Thomas. *London.* 1863.
COCKBURN—
Andrew Thompson. *Berwick.* 1843.
Henry. *Richmond.* 1862.
J. R. *Richmond, Surrey.* 1851-55.
James. *Richmond, Surrey.* 1828-39.
John. *Richmond, Surrey.* 1862-78.
John. *Sherbrooke, Canada.* 1851-63.
Thomas. *Huntingdon, Canada.* 1851.
COCKERAM, Isaac. Place unknown. mid-18c. C.
COCKEY—
William. *Warminster.* ca.1680-1700. C.
William. *Wincanton.* 1692-1721. C.
William. *Yeovil.* ca.1730.
COCKINGS—
Benjamin, *Bridgwater.* 1795-1860. C.
Edward. *Andover.* 1839.
John W. *Andover.* 1878.
Richard. *Andover.* 1830-67.
Richard. *Deal.* 1874.
Spencer Benjamin. *Bridgwater.* 1875-83.
W. *Bridgwater.* 1861-66.
COCKLE, Thomas. *Deptford.* 1866-74.
COCKLING, William. *Sandwich.* b.1725, mar.1752-55.
COCKRELL, James. *Philadelphia.* 1843.
COCKS—
George. *Bristol.* ca.1830-40.
William. *Wells.* 1858-75.
COCQUE, George. *Ath.* 1604. W.
COCKSHO(O)T(T)—
John. *Liverpool.* 1796. W.
William. *St. Helens.* 1848.
COCKSON, Thomas. *Bradford.* 1853.
COE—
& Co. *New York.* 1854.
Russell. *Meriden, Conn, USA.* 1856.
COGAN—
Edmond. *Dublin,* a.1733.
Richard. 1751-84. W.
COGDON, Thomas Samuel & Son. *London.* 1881.
COGGER—
Richard. *Newport, Isle of Wight.* 1830-67.
Thomas. *Hastings.* 1828-55.
William. *Maidstone.* 1838.
COGGLESHALL, G. *Bristol, USA.* Early 19c. C.
COGILL, H. W. *Chatham.* 1855-66.
COGSWELL—
Henry. *Salem, Mass, USA.* s. of John C. 1845-47.
John Cleaveland. *Salem, Mass, USA.* 1793-1853.
Joseph. *Norwich, Conn, USA.* Late 18c.
Robert. *Halifax, Nova Scotia.* 1869-97.
Robert. *Salem, Mass, USA.* b.1791-mar.1815-1837.
COHAN—
Asher & Son. *Liverpool.* 1848-51.
John. *Liverpool.* 1834. C., W. & Chrons.
COHEN—
Abraham. *Leeds.* 1866. W.
Abraham. *Spitalfields.* 1828. Clock case maker.
Alexander. *London.* 1839. Clock case maker.
Benjamin. *Sheffield.* 1862. W.
Benjamin L. *Quebec.* 1862.
Daniel. *Shearness.* 1823-32.
David. *London.* 1832-39.
E. *Abergavenny.* ca.1800.
Emanuel. *Redruth.* 1766-d.1849.
Henry. *London.* 1857.
Isaac. *Birmingham.* 1868.
Isaac. *Hastings.* 1828-39.
Isaac. *London.* 1863.
J. *Birmingham.* 1860.
J. *Coventry.* 1860.
J. *Louth.* 1861.
Jacob & Co. *Birmingham.* 1869-75.

COHEN—*continued.*
Jacob Moses. *London.* 1828-32
James. *Birmingham.* 1869-75.
Joseph. *Coventry.* 1850.
Lewis. *Long Sutton.* 1876.
M. *Dover.* 1851.
Maxmillian. *Manchester.* ca.1850. W.
Morris. *Dublin.* 1858.
Morris. *Hull.* 1858. C. & W.
Morris. *Leeds.* 1866.
Moses. *Sheffield.* 1862-71.
Mrs. H. *Louth.* 1868.
Philip. *Coventry.* 1854-68.
S. P. *Neath.* 1822.
& SAMUEL. *Coventry.* 1880.
& SAMUEL. *London.* 1881.
Simeon. *Liverpool.* 1834.
Solomon. *Edinburgh.* 1860.
Solomon. *Lincoln.* 1850.
Thomas. *Chillicothe, O. USA.* 1814, then *St. Louis.*
ca.1845.
W. O. *Dublin.* 1880.
COHN, S. J. *Halifax, Nova Scotia.* 1873-81.
COIFFIER, Alexander. *London.* 1869. Wholesale clocks.
COIGLEY, James. *Liverpool.* 1822-34. C. W. & chrons.
COILEY, James Joseph. *London.* 1851-57.
COLBURN, John. *York.* 1866-81. W.
COLE—
Charles. *London.* 1832.
Charles. *Whitchurch, Hants.* 1830-59.
Charles J. *West Bromwich.* 1876.
David. *New York.* ca.1840.
Henry. *Birmingham.* 1808.
J. *Sherborne, Dorset.* 1848-67.
J. *Stratford-on-Avon.* 1860.
James. *Norwich.* (B. ante-1758. W.) 1830.
James C. *Rochester, New Hants, USA.* 1812. W.
James, Ferguson & Thomas. *London.* 1828.
John. *Barum.* mid-18c. C.
John. *Newport.* mid-18c.
John. *Shaftesbury, Dorset.* 1824-30.
John. *Stratford-on-Avon.* 1850.
Joseph. *Coventry.* 1854.
R. P. *Ludlow, Vt, USA.* ca.1830. W.
R. S. *Ipswich.* 1858.
Richard Alfred. *London.* 1863.
Richard Stinton. *Ipswich* b. 1809 s. of Richard Cole
senior. Ret'd. 1865.
Robert William. *Cambridge.* 1875.
Shubael. *Rochester, NH, USA.* s. of James, ca.1830-50.
& Son. *Melcombe Regis.* 1875.
& Son. *Sherborne, Dorset.* 1875.
Thomas. *London.* 1851-69.
Thomas. *London.* b.1800-d.1864. C.
W. *Devizes.* 1848-75.
William. *London.* 1863.
COLEMAN—
Andrew. *Drogheda.* 1858.
Andrew. *Lisnaskea.* 1846. C.
Benjamin & Co. *Liverpool.* 1834.
Charles W. *Toronto and Lindsay, Ontario.* 1861-77.
& CHAPMAN. *Liverpool.* 1824-29. C. & W.
Henry. *London.* 1863.
Henry S. *Sandwich.* b.1810-67.
James. *Belfast.* ca.1815-35.
James. *London.* 1844.
James. *Philadelphia.* 1833.
John. *Cork.* 1844. W.
John. *Dublin.* 1794-98. Watch cases.
John. *Ipswich.* ?ca.1740-ca.1770?
John. *Philadelphia.* 1848.
Nathaniel. *Burlington, NJ, USA.* bro. of Samuel.
1765-1842.
Obadiah. *Bristol.* 1760-d.1800.
Samuel. *Antrim.* ca.1795-ca.1810. C.
Samuel. *Trenton, NJ, USA.* 1761-1842.
Thomas. *Kington.* 1879.
Thomas. *Leith.* 1811.
COLES—
Angelo. *Scarborough.* Late 18c. C. (but prob. error
for Michael qv.
Edward. *Dublin.* a.1755.
John. *Barnstaple.* rep'd ch. clock 1736, d.pre-1760.
John. *Cold Brayfield.* 1798.

COLES—*continued.*
Richard, sen. *Buckingham.* (b.1757) mar.1786-d.1819.
Richard, jun. Buckingham. s. of Richard, sen. b.1792-
d.1853.
Richard. *Chipping Norton.* a.1771. C. May be later at
Buckingham (1786).
T. E. *Stratford-on-Avon.* 1880.
COLHOUN—
James & Son. *Londonderry.* (b.1753), ca.1795-d.1833.
C. & W.
James, jun. Londonderry. s. of James C., sen. 1820-24.
Margaret. *Londonderry.* wid. of Will. 1837-54.
William. *Londonderry.* (b.1788) ca.1824-35 (d.1837).
COLLARD, John (Brothers). *Bowmanville, Canada.*
1857-63.
COL(L)BRAN, Thomas. *Hailsham.* 1851-78.
COLLES—
Charles I. *Kilkenny.* d.1768. C.
Charles II. *Kilkenny.* 1778.
Christopher. ?*London.* 1685. C.
Thomas. *Clonmel.* 1799. W.
COLLET(T)(E)—
G. or J. *Fakenham.* 1858-65. (prob. Jonas.)
Jonas. *Fakenham.* 1830-46.
Joshua. *Lebanon, O, USA.* 1825. W.
Julius. *Little Ilford.* 1874.
Mrs. Susannah. *Fakenham.* 1875.
COLLIER—
Archibald. *London.* (B. ċa.1790-1825) 1828.
David. No town - the *Gatley/Etchells* man.
David. *Kingussie.* 1860.
G. *Cheadle.* 1860-68.
James. *Rochdale.* 1834. Clock dealer.
John. b. 1762. s. of David Collier of *Gatley.*
To *Cheadle* ca. 1780.
Robert. *Blackburn.* 1828. May be same as next man.
Robert. *Salford* nr *Manchester.* 1848. May be from
Blackburn?
Robert. *Richmond, Surrey.* 1828.
Samuel. b. 1750. s. of David Collier senior.
Worked at *Eccles.* d. 1806 C.
Samuel. *Cheadle.* b. 1785 s. of Samuel Collier
of *Eccles* qv. Worked at *Cheadle.* d. 1865. C.
Thomas. *Glossop.* 1828-35.
Thomas. *Manchester.* 1834.
COLLIN—
B. *London.* 1839.
John. b.1829 *IoM.* 1851 at *Kendal.* W.
COLLINGE, W. *Burnley.* 1858. W.
COLLINGRIDGE—
John. *London.* 1851-75.
Thomas. *London.* 1839-69.
COLLINGS—
Alfred. *Thornbury.* 1842-79.
Alfred, jun. *Thornbury.* 1879.
Harris. *Coleford.* 1842.
Harris. *Newport, Mon.* 1848-52.
Harry. *Llantrisant.* 1887.
James. *Barnard Castle.* Early 19c. C.
James. *Southsea.* 1830.
John. *Sodbury.* Early 19c. C.
John. *Stroud.* 1850.
Joseph. *Cardiff.* 1858-75. W.
Joseph. *Thornbury, Glos.* 1840-42.
Joseph. *Usk.* 1852.
Mrs. Margaret Robertson. *Cardiff.* 1887.
Samuel. *Cardiff.* 1858-75.
Samuel. *Downend (Bristol).* ca.1770. C.
Samuel. *Sodbury.* 1842.
Samuel. *Thornbury, Glos.* 1830-70.
Thomas. *Sodbury.* 1830.
Thomas George. *Dunstable.* 1830-64.
COLLINGWOOD—
C. *Streatham.* 1851-62.
& Co. *West Hartlepool.* 1898.
Henry. *Rochdale.* 1834-58.
Mathew. *Alnwick.* 1827-58. C.
Matthew George. *Middlesborough.* 1855-66. W.
Matthew George. *Redcar.* 1866.
& RAINTON. *Rochdale.* 1828.
Robert. *Rochdale.* 1824.
& Son. *Middlesborough.* 1898.
COLLINS—
Charles Mason. *Worcester.* 1876.
& Co. *Cranston, RI. USA.* ca.1840. C.

COLLINS—*continued.*
Elijah. *Boston, USA.* 1727. W.
James. *Barnard Castle.* ca.1800. C.
James. *Goffstown, NH, USA.* s. of Stephen. 1802-44.
James. *Stratford, Essex.* 1874.
John. *St. John, New Brunswick.* 1847.
John Bull. *Northampton.* (B. 1808-)1830.
John F. *Whitby, Canada.* 1872-75, also *Hamilton,*
Canada. 1876.
Joseph. *Havant.* 1830.
Metal Watch Factory. *New York.* 1873. W.
Richard. *Margate.* 1780-1827. W.
Richard. *Truro.* 1873. W.
Robert. *Hartlepool.* 1834.
S. *East Harling.* 1858.
S. *Wattisfield.* 1846-65.
Samuel. *Swindon.* 1867-75.
Simon. *Chartsey.* ca.1730. C.
Solomon. *Chertsey.* ca.1730. C.
Stephen. *Goffstown, NH, USA.* 1847.
T. *Botesdale.* 1846-65, prob. same man as below.
Thomas. *Botesdale.* 1830-39, see also previous entry.
Thomas. *Doncaster.* mar.1737. C. (B. pre-1766.)
Vincent Bull. *Northampton.* 1830.
W. A. *Troy, NY, USA.* 1840.
William Bull. *Northampton.* 1806-24.
William. *Herne Hill, Surrey.* 1878.
William. *London.* 1881.
William. *New York.* 1854.
COLLINSON, Benjamin. *Kendal.* 1816-57. C.
COLLIS—
R. *Mildenhall.* 1858.
Richard. *Romford.* (B. a.1768-1824) 1828.
William. *Brentwood.* 1828-51.
William Humphrey. *Bury St. Edmunds.* 1858-79.
COLLISON, Alexander. *Stonehaven.* 1834-60. Also in-
spector of weights & measures.
COLLOM, David W. *Philadelphia.* 1846.
COLLYER—
Brothers. *Glastonbury.* 1875.
& Son. *Glastonbury* and *Bridgewater.* 1883.
COLMAN—
Francis. *Ipswich.* mar. 1668. C. & W.
Robert. *Ashwellthorpe.* 1858-75.
COLONIAL—
——. 1914 and later, product of Pequegnat Clock Co.
Manufacturing Co. *Zeeland, Mich, USA.* 1929-42.
COLQUHOUN—
John. *Glasgow.* 1860.
John Duncan. *Wales (Canada).* 1886.
COLSTON, James. *Cardiff.* 1887.
COLTMAN, William. *Dewsbury.* 1871.
COLTON, ——. *Great Yarmouth.* 18c. C.
COLUMBUS Watch Co. *Columbus, O, USA.* 1882-1902.
W.
COLYER—
Allen. *Dover.* 1845-47.
Allen. *Ramsgate.* 1874.
Allen. *Witham.* 1866-74.
COMBERBACH—
Edward Stephen. *Blackburn.* 1858. W.
E. S. & Son. *London.* 1881.
COMBS, Thomas. *London.* 1869-81.
COMENS, W. Benjamin. *Montreal.* From *Windsor,*
Vermont. 1806-11.
COMMANDER, Samuel. *London.* (B. a.1796-)1851-57.
COMMON—
James. *Coldstream.* ca.1770. C.
James. *Coldstream.* d.1849.
James. *Coldstream.* s. of James, sen. 1849-60.
COMPIGNE, David. From *Caen, France.* ca. 1680 to
Winchester. C. At *Spital Street, Shoreditch,*
London. 1705. W.
COMPTON—
——. *London.* 1695. Goldsmith & watch seller.
John. *Calne.* (B. 1795-)1830. C.
Samuel. *Frome.* 1875.
William. *Rochester, NY, USA.* 1844-46. W.
COMTESSE—
Louis. *London.* 1832-39. Watch case maker.
Son. DUBOIS & PETTERMAND. *London.* 1828. Watch
case makers.
CONANT—
H. *New York.* 1889.

CONANT—*continued.*
Nathaniel Peabody. *Danvers, Mass, USA.* 1819-ca.50. W.
& SPERRY. *New York, USA.* ca.1840.
W. S. *New York.* ca.1820.
CONCONI, Louis. *Margate.* 1832-66.
CONDART, Solomon. *Dublin.* Free 1707. W.
CONDE, Edward Jacques. *Bristol.* 1840-56.
CONDER, J(abez?). *Croydon.* ca.1750. C.
CONDLIFF, James. *Liverpool.* 1822-51.
CONEN—
George. *London.* 1832. Watch cases.
George. *London.* 1881.
James. *London.* 1839. Watch cases.
CONEY, George. *Birmingham.* 1868. Clock case maker.
CONIBERE, Samuel. *Ashburton.* 1790-1800.
CONKEY, William. *Comber.* 1858.
CONN, Robert. *Pershore.* 1868-76.
CONNECTICUT CLOCK CO. *New York.* post-1850-72.
CONNELL—
Hugh. *Glasgow.* 1860.
William. *London.* (B. a.1817-) 1839-69.
William George. *London.* 1881.
CONNOLD—
Thomas. *Aylsham.* 1830-65.
Thomas. *Hastings.* 1862-70.
William. K. *York.* 1858. W.
CONNOLLY—
John. *London.* 1851-81.
John. *Dublin.* 1823-31.
CONNOR—
Benjamin. *Eniscorthy.* 1858. Also dentist.
J. W. *Treorchy.* 1875.
John. *Dublin.* 1820-31. W.
O. *London.* 1819. W.
CONOVER—
George Samuel. *Georgetown, Canada.* 1886.
William Wallace. *Toronto.* 1886.
CONRAD—
Osborne. *Philadelphia.* 1841. C.
Osborne & Co. *Philadelphia.* ca.1850.
CONROY, John. *St. Johns, Canada.* 1857.
CONSCIENCE, A. & Co. *London.* 1851.
CONSTABLE, Alexander. *Dundee.* 1838.
CONSTANT, Francis. *Kingston, NY, USA.* 1860.
CONVERSE—
& CRANE. *Boston, USA.* ca.1870.
Pascal. *New Haven, Conn, USA.* ca.1850.
CONWAY—
John. *Sturminster.* 1824-48. W.
Thomas A. *Baltimore, USA.* 1819-24.
William. *Poole, Dorset.* 1824-30. Also gunsmith.
COOCHEY, Charles. *Kingston, Surrey.* 1878.
COOK(E)—
Alexander. *Cannonsburg, USA.* 1841. C.
Benjamin. *Northampton, Mass, USA.* ca.1846.
Brothers. *London.* 1863-81.
C. *Earl Shilton.* 1849.
Charles. Place unknown (*England*). ca.1750-60. C.
Edward. *Bognor.* 1878.
Erastus. *Rochester, NY, USA.* 1815-45.
Frederick. *Columbia, Pa, USA.* 1825-32, then *York, Pa.* 1832-42.
Frederick D. *London.* 1863-81.
G. *Eastbourne.* 1862. Clocksmith.
George. *London.* 1875.
H. S. *Shiffnal.* 1856.
H. W. *Southwell.* 1849.
H. W. *Walsall.* 1860.
Harry. *Poughkeepsie, NY, USA.* 1815-17.
Henry & Co. *Nottingham.* 1828.
Henry Winter. *Nottingham.* 1864-76.
J. *Nottingham.* 1855.
James. *Dumfries.* (b.1812)-d.1874.
James. *Leeds.* ca.1850-71. C.
James. *London.* 1828.
James. *Strichen, Scotland.* 1846.
James George. *Birmingham.* 1858-80. Dialmaker.
Job. *Pershore.* 1872-76.
John. *Aylesbury.* 1752-59. W.
John. *Hexham.* 1848-58.

COOK(E)—*continued.*
John. *London.* Alien working in 1622 as apprentice of Cornelius Mellin in Blackfriars.
John. *London.* 1869-81.
John. *Oakham.* 1864-76. s. of Thomas.
John. *Pershore.* 1850-68.
John. *Strabane.* 1820-65.
John. *Trafford.* ca.1750-ca.60. W.
John. *Whitchurch, Hants.* (B. 1791-) 1830-59.
Joseph. *Aylesbury.* 1737-d.1759. W. & C. (B. ante-1777. W.)
Joseph. *Sheffield.* 1871.
Joshua. *London.* 1857-81.
Owen & Sydney. *London.* 1863.
Richard. *Colchester.* (from *Ipswich*) b.1691-ca.1720, then *Ipswich* again ca.1720-d.1746. C.
Robert. *London.* 1811. Chrons.
Samuel. *Waterbury, Conn, USA.* 1830. C.
Samuel John. *London.* 1875.
& STILLWELL. *Rochester, NY, USA.* ca.1850.
Thomas. *Derby.* 1843-60.
Thomas. *London.* (B. a.1649-)1662.
Thomas. *London.* 1851-81.
Thomas. *Loughborough.* (B. ante-1822-)1828.
Thomas. *Oakman.* 1828-49.
Thomas. *York.* b.1807-d.1868. T.C.
William. *Aberdeen.* 1651.
William. *Liverpool.* 1834.
William G. *Baltimore, USA.* 1817-24.
COOKE'S SONS, B. G. *Philadelphia, USA.* 1853.
COOKSON, Thomas. *Ulverston.* 1822.
COOLEY, Henry P. *Cooperstown* and *Troy, NY, USA.* 1834-43.
COOLIDGE Henry J. *New Haven, Conn, USA.* 1787. C. & W.
COOMBE—
Joseph. *Bristol.* a.1773-84.
William. *Ballynahinch.* 1805.
William. *Exeter.* b.1752, free 1790. W. (s. of William C., tinplate worker.)
COOMB(E)S—
C. *Stratford-on-Avon.* 1860.
Joseph. *Melksham.* (B. 1791-95) 1830.
William. *Saintfield.* ca.1840-68. C.
COOPER—
A. *Odiham.* 1859-67.
Charles. Place unknown. ca.1780. C.
Charles. *Lebanon, Pa, USA.* Early 19c.
Charles. *Maidenhead.* 1877.
D. A. *Worcester.* 1868-76.
Duncan Anderson. *Huddersfield.* 1850-66. W.
G. *Hastings.* 1851.
G. *Thornaby.* 1898.
George. *Scarborough.* 1858-66. W.
H. W. *Stalybridge.* 1878.
& HEDGE. *Colchester.* Partnership of William Cooper and Nathaniel Hedge during years 1733-39 only.
I. & Son. *Norwich.* 1865-75.
Isaac. *Nantwich.* 1813. W.
Isaac Daniel. *Reigate.* 1878.
J. *Derby.* 1855.
J. *Eccles.* 1848-58. W.
James. *Croydon.* 1878.
James. *London.* 1811. W.
James. *London.* 1832.
James. *Wolverhampton.* 1835-50.
John. *Colchester.* b.1790-1835, later *Norwich.*
John. *Great Haseley.* 1818. Church dial painter.
John. *London.* 1705.
John. *New York.* 1846.
John. *Norwich.* From *Colchester.* 1836-ca.1840.
John. *Sevenoaks.* 1845.
Joseph. *Columbia, SC, USA.* 1843-60.
Joseph. *Greenville, SC, USA.* ca.1855.
Joseph B. *Philadelphia, USA.* 1842-46.
Mark. *London.* 1881.
Matthew. *Hull.* 1834-40. W.
Mrs. F. *Odiham.* 1878 (?successor to A.)
& NEWTON. *Norwich.* Partnership of John Cooper and F. Newton, q.v. 1835-36.
R. *Coventry.* 1860. Cleaner.
R. *Worverhampton.* 1860.

COOPER—*continued.*
Reuben. *Brierley Hill.* 1876.
Robert. *Great Wakering.* 1866-74.
Robert. *Ringwood.* 1839.
Robert H. *Philadelphia.* 1850 and later.
Samuel B. *Philadelphia.* 1840.
Stephen. *Wolverhampton.* 1828.
T. *Earlstown.* mid-19c. W.
T. *West Rudham.* 1865.
Thomas. *Derby.* 1828.
Thomas. *Everdon.* 1787-1800.
Thomas. *Hamilton.* 1836.
Thomas. *Leicester.* ca.1820. C.
Thomas. *London.* 1857-81.
Thomas. *Sandbach.* 1878.
Thomas. *Tamworth.* 1835-50.
Thomas & Dirintea. *Tamworth.* 1828.
Thomas Earle. *Rye.* a.1720 to William Reeve. W.
Thomas Frederick. *London.* (B. 1820-40) 1839-75.
Thomas William. *London.* 1863-81.
W. H. *Litcham.* 1858-65.
William. *Birmingham.* 1880.
William. *Colchester.* From *East Bergholt, Suffolk.* b.1706-57. C. & W.
William. *London.* 1857. Also clock cases.
William. *Sheffield.* 1871.
COPE—
Benjamin. *Bewdley.* Late 18c. C.
Brothers (G. & W.). *Nottingham.* 1864-76. T.C.
G. & W. *Radford, Notts.* ca.1855-ca.1876. C.
Henry. *Dukinfield.* 1878.
Jacob. *Watsontown, Pa, USA.* ca.1800. C.
John. *Lancaster, Pa, USA.* ca.1800.
& MOLYNEUX. *London.* 1859. W.
Thomas. *Dublin.* Free 1715-d.1739. W.
Thomas. *Nottingham.* 1835.
W. H. *Birmingham.* 1860.
William. *Radford.* 1876. T.C.
COPELAND, Robert. *Baltimore, USA.* 1796.
COPPELL—
John. *Liverpool.* 1767. W. (B. 1816-18.)
Zallel. *Liverpool.* 1848.
COPELSTONE, William. *Plymouth Dock.* ca.1775. C.
COPLEY, ——. *London.* d.ca.1635.
COPPERT, T. *Olneyville, RI, USA.* ca.1840.
COPPING(E)—
George. *London.* (B. a.1654-)1667,
Richard. *Bury St. Edmunds. ca. 1660 -ca. 1700.*
Richard. *London.* ca.1654-ca.1661.
COPPINGER, Francis. *Cork.* 1846. W.
COPPLESTONE, William. *Plymouth.* ca.1790. C. (see also COPPELSTONE).
COPPOCK, Thomas. *Prescot.* mar.1769. W.
COPPUCK, G. W. *Mount Holly, NJ, USA.* 1804-82. C.
COPSEY, Samuel. *Gravesend.* 1839.
COQUELLE, ——. *Paris.* 1889. C.
CORBET(T)—
Hugh. *Oxford.* 1589-1616. rep'd. ·ch. clock.
John. *Comber.* 1846-80.
John L. *Belfast.* 1868. W.
Robert. *Glasgow.* (B. 1822-41) 1860.
Samuel. *Coventry.* 1860-80.
William. *Adstock.* 1798. W.
CORBIERE—
HAINES & BRINDLE Ltd. *London.* 1881.
& Son. *London.* 1869-75.
CORBIT, Jonathan. *London.* 1828-32.
CORBOLD, John. *Wallingford.* 1864-77.
CORBY, Charles. *London.* 1875-81.
CORDELL, Mrs. M. A. *Brighton.* 1855-62.
CORDEUX, Charles. *Bristol.* 1856-79.
CORDING—
Charles. *London.* 1811. W.
John. *London.* 1819-39. W.
CORDINGLEY—
Thomas. *Leeds.* b.1780-d.1836. C. & W.
Thomas. *Wick.* 1836.
William. *Leeds.* 1866. W.
CORDNER—
John. *Newry.* 1840. C.
William Henry. *Newry.* 1864-90. W.
William Henry & Co. *Newry.* 1846.

CORDWELL, Robert. *London.* B. mar.1625-37. CC.
COREY, P. *Providence, RI, USA.* ca.1850.
CORFIELD, W. *Lichfield.* 1860-68.
CORKE, William. *Wolverhampton.* 1850-68.
CORKEN, Archibald. *b. ca. 1815 Scotland. Oswestry* by 1842-52.
CORKER—
D. S. *Lower Tooting.* 1851-78.
Daniel. *London.* (B. 1817-40) -1863.
E. *London.* 1863-69.
John. *Newcastle-under-Lyme.* 1828-35.
John. *Stafford.* 1850-68.
Nathaniel Shipston. *London.* 1844-75.
Thomas. *Eccleshall, Staffordshire.* 1835.
Thomas. *Stafford.* 1828-42.
CORL, Abraham. *East Nantmeal, Pa, USA.* 1779-1842. C.
CORLETT—
Edward. *Castletown, IoM.* 1860.
James. *Liverpool.* 1848.
T. H. *Ironbridge.* 1879.
CORLISS, James. *Weare, NH, USA.* ca.1800.
CORMACK, Francis. *Upper Pulteney. nr. Wick.* 1860.
CORNAH, James. *Lancaster.* Free 1777, d.1795 at *Manchester.* W.
CORNE—
Frederick. *Farnborough.* 1878.
Frederick. *London.* 1863-69.
CORNELIUS—
Jacob. *London.* 1635-46.
Julius. b.1825. *Prussia, via France and England to New York* 1835, then *Halifax, Nova Scotia* where d.1916.
CORNELL—
Charles M. *Royston.* 1840-74.
George. *Maidstone.* 1874.
Walter. *Newport, RI, USA.* ca.1800. C.
Watch Co. *Chicago, USA.* 1870-74.
William E. *Toronto.* 1867-74.
CORNER, Benjamin. *Richmond (Surrey).* 1878.
CORNFORTH—
Elizabeth. *Macclesfield.* 1848.
Joseph William. *Macclesfield.* 1828.
William. *Macclesfield.* 1834.
CORNU, Leon & Co. *London.* 1881.
CORNWALL—
James. *Liverpool.* 1824.
M. A. *Aurora, Canada.* 1864.
Thomas. *Dublin.* 1745-77. W.
CORNWELL—
Daniel. *Billericay.* 1775 (-B. 1791) C. & W. (one such in *London.* ca.1750. May be same man.)
Nathan. *Darien, Ga, USA.* ca.1820.
CORONA WATCH CO. *New York.* 1900.
CORRALL—
Edwin. *Hanley.* 1876.
Francis. *Lutterworth.* b. 1809 s. of William - 1850. C.
George. *Mansfield.* 1849-76.
Mrs. M. A. *Lutterworth.* 1855.
William. *Lutterworth.* 1868-76.
CORRAN, Henry. *St. Johns, Canada.* 1851-57.
CORRIE, James. *London.* 1832-57.
CORRIGAN, Samuel *Belfast.* 1843-54.
CORRINGHAM—
R. *Canterbury.* 1858-66.
Richard. *Strood.* 1874.
CORRIVEAU, Eugene. *Quebec.* 1890.
CORRY, James. *Colchester.* b.ca.1691-d.ca.1740. C. & W.
CORSAN, Thomas. *London.* 1832-51.
CORSON—
James. *Maryport.* 1848-58.
Joseph. *Whitehavem.* 1834-48.
CORTELYOU, Jacques. Place unknown, *USA.* 1781-1822.
CORTHORN, John. *Lincoln.* 1850-61.
CORTLAND, ——. *Cortland, NY, USA.* 1840s.
CORVASIER, Edward. *Philadelphia.* 1846.
CORWIN, Ebenezer M. Place unknown, *USA.* 1867.
CORY, Lewis. *Rayway, NJ, USA.* ca.1830.
CORYNDON—
William. *Launceston.* 1752-ca.1790. C.
William. *Plymouth.* Pre-1765, later at *St. Keyne.* W.

COSENS—
James. *Osmotherley.* 1829. W.
Nathaniel. *Bristol.* s.1773-81 (B. 1791).
Nicholas. *York.* 1638 d.1654. Hour glass maker.
COSGROVE, Henry. *Leek.* 1876.
COSHAM, Caleb. *Brighton.* 1862-78.
COSINE, John. *London.* (B. 1622 alien). 1622. With two
apprentices in St. Bride's parish by the Conduit in
Fleet Street. prob. brother of Alexander.
COSSON, Samuel. *London.* 1828-44.
COSSOR, William C. *London.* 1851-57.
COSTELL, Stacey. *Philadelphia, USA.* 1830. W.
COSTELLOW, John. *Chichester.* b.1677-d.1744. C.
COSTEN—
Adam. *Kirkham.* ca.1750-ca.1780. C.
John. *Kirkham.* bro. of William. d.1803. W.
William. *Kirkham.* With bro., John till 1803, then
alone. W.
COSTER—
Charles. *Henley-on-Thames.* 1853. C. & W.
J. *Stoney Stratford.* 1854.
James. *Henley & Great Marlow.* 1798-1830. W. & C.
James. *Maidenhead.* 1837.
John. *Great Marlow.* 1823.
Robert. a. 1695 to Luke Wise of *Reading*
(believed b. there 1677). Worked at *Reading* until
ca. 1707, then *Newbury.* d. 1749 at *Avington nr.
Newbury.* C.
COSTIGAN, Robert. *Dublin.* 1827-68. C. & W.
COTTAM, Joseph. *Maidstone.* Free 1694-99.
COTTEBRUNE, Charles (Son). *London.* 1863-75 (-81).
COTTEE—
J. *Sutton, Surrey.* 1862.
William P. *Springfield.* 1866-74.
William Pettica. *Chelmsford.* ca.1870.
COTTELL—
——. *Ilminster.* ca.1800. C.
& COUSSENS. *Crewkerne.* ca.1810. C.
Joseph. *Crewkerne.* 1795. C.
Simon. *Crewkerne.* ca.1791. W.
COTTEN, William. *Yarmouth.* ca.1800. C.
COTTERELL/COTTERILL—
Harry. *Brighton.* 1878.
Henry. *Shrewsbury.* 1879.
John. *Wirksworth.* 1828-49.
Joshua. *St. Helens.* 1785. C.
COTTEY, Abel. *Philadelphia, USA.* b.1655 *Exeter.* To
Philadelphia 1682, d.1711. C. & W.
COTTISWOLD, Thomas. *Thame.* 1458-1464. Rep'd.
ch. clock.
COTTON—
Benjamin. *Great Marlow.* d.1763. C.
William Brown. *Yarmouth.* 1830-65.
COTTRELL—
Henry. *Stratford, Essex.* 1874.
John. *Wrexham.* ?18c. C.
COUCH—
George, *Gravesend.* 1847.
Robert. *London.* ca.1750. W.
W. J. *St. Columb.* ca.1820. C.
COUCHMAN, T. *St. Leonards.* 1878.
COUET, Charles. *Paris.* 1862-67. C.
COUGHIN, James. *Manchester.* 1828-34.
COUILLER, James. *Market Rasen.* 1828.
COULES, George G. *Windsor.* 1830-54.
COULON & MOLITOR. *Herimoncourt.* 1889.
COULSON—
Henry. *Penzance.* 1772-1834. C. & W.
John. *Crewe.* 1878.
John. *Ebchester.* 1856.
William. *Helston, Cornwall.* ca.1820. C.
William. *North Shields.* (B. 1811-)1827-34. C.
William Vigurs. *Newark.* 1855-76.
COULTER, William. *Saltcoats.* 1837-60.
COULTON—
Francis. *London.* (B. CC. 1690-)1700. 'In *St. Anns'*.
Francis. *York.* Free 1757. C. s. of John.
John. b.1793 at *Kendal.* By 1822 in *Kendal* till 1851. W.
John. *York.* Free 1701-d.1739. C.
Thomas. *Ulverston.* 1826.
William. *York.* Free 1739-84. C.
COUNSELL—
Edwin. *Faringdon.* 1854-99.

COUNSELL—*continued.*
Edwin J. *London.* b.1829-51. W. From *Buckingham-
shire.*
John Webb. *Ross.* 1835-50.
COUNT, J. C. *Sleaford.* 1861-68.
COUPAR—
John. *Philadelphia.* ca.1774. W.
Robert. *Philadelphia.* 1774. W.
COUPE, William. *Birmingham.* 1850.
COUPER, Andrew. *Edinburgh.* 1836. Clock case
maker.
COURLANDER, N. *Richmond, Surrey.* 1878.
COURT—
Cornelius. b. 1817 *Redditch.* s. of Thomas Court qv.
At. *Leamington* ca. 1835, then to *Warwick*
by 1843. d. 1870.
Isaac. b. ca. 1823 s. of Thomas Court. Working
1851-68 *Leamington.*
James. *Southwold.* 1858-79.
Joseph. *Bromsgrove.* 1835.
Mary. *Bromsgrove.* 1842.
Thomas. *Bromsgrove.* 1828.
Thomas. b. ca. 1790 *Henley in Arden.* To *Leamington*
by 1825-54. At Redditch ca. 1790-1825.
COURTER, William. *Ruthin.* ca.1780. C.
COURVOISIER, Freres. *London.* 1875-81.
COUSENS—
Basil R. *Swansea.* 1887.
Frederick. *London.* 1857-63.
Henry John. *London.* 1881.
John. *Langport.* ca.1800. C.
Mrs. C. *London.* 1869. (wid. of Frederick?).
Richard William. *Swansea.* 1868-75.
Robert William. *London.* 1839-44.
& STOESSIGER. *London.* 1875.
& WHITESIDE. *London.* 1844-51.
COUSINS—
George. *Chichester.* 1828.
George. *Whitby.* 1851-66.
& Sons. *Bath.* 1883.
& WHITESIDE. No place. Early 19c. (See COUSENS &
WHITESIDE).
COUSSENS—
J. *Bristol.* ca.1780. C.
Roger. *Crewkerne.* ca.1820. C.
COUTTS—
John. *Aberdeen.* 1860.
John Francis. *Woolwich.* 1874.
COUTURE—
David. *Quebec.* 1848-53.
Pierre. *Quebec.* 1843-52. Succeeded by widow.
COUZENS, J. *Langport.* 1861.
COVELL, GRAY & CO. *New York.* 1872. C. Imported
French clocks.
COVELLER, Ann. *Alford.* 1828.
COVENEY, Zebulon. *Dover.* a.1815.
COVENTEN, Joseph. *Clerkenwell.* 1828-81. Clock case
maker.
COVENTRY—
Carr. *London.* a.1649 to Samuel Davis, CC. 1657-68.
Henry. *London.* 1857.
James. *London.* 1832.
COVERDALE, William. *Darlington.* 1851-56. C.
COVINGTON—
J. *Wimborne.* 1848.
James. *Salisbury.* 1830.
John. *Stony Stratford.* 1798-1823. C.
William. *Stony Stratford.* (B. 1785) mar.1788-1802. C.
COWAN—
Hugh. *Thurso.* 1837.
William. *Glasgow.* 1806-22.
William. *Lennoxtown.* 1836.
William. *Richmond, Va, USA.* ca.1819.
COWARD—
J. *Warrington.* 1858. C.
James. *Bognor.* 1870-78.
COWBURN—
Henry. *Kirkham.* 1848-58.
J. *Preston.* 1858.
COWDEROY—
Richard F. *London.* 1857-63.
William. *London.* 1863-81.

COWELL—
David. *Manchester.* 1851.
Henry. *Liverpool.* 1848. W.
COWELLS & PENNINGTON. *?London.* ca.1795. br. clock.
COWEN—
Benjamin. *Toronto.* 1861.
David. *Manchester.* 1848. Dealer.
George. *Dublin.* 1830. W. & jeweller.
William. *Dundalk.* 1858.
COWHAM, Thomas. *Hull.* 1846-51.
COWLAND, William. *Moulsham.* 1828.
COWLES—
Alfred. *London.* 1875-81.
Irving C. Place unknown, *USA.* 1876.
Nathan. *London.* 1869.
COWLEY—
C. *Brighton.* 1862.
E. *Stockton.* 1898.
G. *Coventry.* 1868.
L. D. *La Grange, Ind, USA.* ca.1870.
Thomas. *Whitby.* ca.1790. C.
COWLING—
C. *Altarnun.* 1856. C.
Joseph. *Chelmsford.* ca.1826-39.
Joseph. *Moulsham.* 1839.
COWPE, James. *London.* CC. 1654-56.
COWTON—
Samuel & Sons. *Bridlington.* 1840.
Thomas. *Bridlington Quay.* (b. 1795) 1846 -d. 1873 W.
COX—
Adam. *Reading.* 1854-77.
Alfred. *London.* 1857.
Andrew. *Southsea.* 1839.
& BARROW. *Christchurch.* 1830. Fuzee, chain and hook makers.
Benjamin. *Bristol.* From *London.* 1761.
Benjamin. *Earlsdon* nr *Coventry.* 1868-80.
Benjamin. *Philadelphia.* 1809-13.
C. & Co. *Christchurch.* 1839-67. Fuzee, chain and hook manufacturers.
C. H. (& Son). *Northampton.* 1864(-69).
Charles Humphrey. b. ca. 1816 *Wallingford.* To *Leamington 1840-51.* To *Coventry* by 1860
& CLARK. *New York.* 1832.
Edward. *Lindfield.* 1828.
Edward. *Nottingham.* 1835.
F. *Cuckfield.* 1851.
Frederick. *Bath.* 1875-83.
Frederick. *Langport.* 1883.
Frederick. *London.* 1857-81.
H. *Ramsey, Hunts.* 1877.
J. M. *Bath.* 1856-66.
James. *New York.* ca.1810.
Jason. *London.* 1723 W. (B. 1747-51).
John. *Whitehaven.* 1848.
Martha. *Kidderminster.* 1835.
Nathaniel. *London.* 1832-57.
Richard. *Bristol.* a.1809 (B. 1825-30).
Richard. *London.* 1863.
Richard Workman. *Bristol.* 1830-50.
Robert Nathaniel. *Pickering.* 1823-51. C.
S. W. *Market Harborough.* 1849-55.
Stephen. *Goderich, Canada.* 1857.
Stephen. *Kidderminster.* 1828.
Thomas. *Cromhall* nr *Thornbury.* ca.1750-ca.1760. C. prob. same man as d.1784 at *Thornbury.*
W. *Bath.* 1866.
William. *Bath.* 1883.
William. *London.* 1869-81. (Successor to Richard).
COXALL, Samuel. *Royston.* b.1734 d.1815. W.
COXHEAD, William (& Son). *Reading.* 1830-77.
COXON—
Thomas. *Nottingham.* 1849-76.
W. *Birmingham.* 1854.
COZENS—
J. B. (or Joseph). *Philadelphia.* 1823-29.
MATTHEWS & THORPE. *London.* 1851-81.
William & Son. *London.* 1828-44. Wholesale watches.
CRABFORD(?), Matthew. *London.* 1662.
CRACK, Walter. *Bury St. Edmunds.* 1858-79.
CRACKLES, Samuel. *Hull.* b.1797-d.1866. C. & W.

CRADOCK—
Edward. *London.* 1832.
Edward. *London.* 1851-63.
William. *London. Shadwell Dock.* 1789. W. (B. 1802-08).
CRAFT, Jacob. *Shepperdstown, W. Va, USA.* 1775-1815.
CRAFTER—
Edward John. *London.* 1869-81.
J. B. *Landport.* 1848.
John. *London.* 1863-69.
CRAGG—
Edwin Newton. *Colchester.* Succeeded Thomas Mason 1853-75. See also CRAGG & NEWTON.
James. *Lancaster.* Free 1779, then *Milnthorpe* till ca.1818, though may have worked at *Manchester* ca.1802. C.
James. *Manchester.* ca.1802. May be same as previous entry.
John. *Liverpool.* 1822. W.
John. *London.* 1828-63. Wholesale watches.
(Edwin Newton) — LAMBERT. *Colchester.* ca.1856-70.
Smith Isaac. *London.* 1828. Wholesale watches.
W. *Haltwhistle.* 1858.
CRAGGS, Richard. *London.* (B. a.1652, CC. 1660)-1662.
CRAIG—
Charles. *Dublin.* 1761-88. W.
James. *Ballymoney.* 1865.
James. *Belfast.* 1858.
James. *Williamsburg, Va, USA.* ca.1772.
John. *Morpeth.* 1834-58.
John. *Newcastle-on-Tyne.* ca.1790-1827.
John. *York.* 1851.
John B. *Pittsburgh, Pa, USA.* ca.1845.
Joseph. *Maryport.* 1828.
Peter. *Glasgow.* 1837-. Clock dial maker.
Richard. *Dublin.* 1818-30. W.
William. *Haltwhistle.* 1834-54. C.
CRAIGHEAD—
Andrew. *Iverury.* 1860.
& WEBB. *London.* 1851-63.
CRAIGMYLE, John. *London.* 1839-75.
CRAIK(E)—
Robert. *London.* ca.1770. C.
William. *Dalkeith.* 1860.
CRAKE—
A. *Lowestoft.* 1858.
E. *Lowestoft.* 1846-58.
Edmund. *Loddon.* 1836-46.
Edmund. *Lowestoft.* 1830-39.
Edmund Fisher. *Lowestoft.* 1858-79.
Frederick. *Ipswich.* 1864-68.
F. *Lowestoft.* 1858.
J. *Loddon.* 1858-75.
& Son. *Lowestoft.* 1853.
CRAMB, J. *Birkenhead.* 1865.
CRAMP—
Jury. *Horsham.* 1878.
Richard. *Canterbury.* 1768-75. C.
CRAMPTON—
John. *Dublin.* Free 1704-28. C.
Thomas. *Dublin.* Free 1718-d.1751. W. & goldsmith.
CRANAGE—
John. *Liverpool.* 1824-34. C., W. & chrons.
Joseph. *Liverpool.* 1824.
CRANBROOK, William. *Deal.* 1823-28.
CRANCH, Richard. *Salem, Mass, USA.* ca.1754-67, then *Braintree, Mass.* 1773, then *Boston.* 1775.
CRANE—
Aaron D. *Caldwell, NJ, USA.* 1829, then *Newark, NJ.* 1841.
Edwin Francis. *St. Leonard.* 1878.
George. *Worcester.* (B. 1819) 1828-35.
John. *E. Lowell, Mass, USA.* ca.1850.
John & George. *Bromsgrove.* Early 19c. C. Perhaps same as following.
John & Son. *Bromsgrove.* 1820-22.
Jonas. *Newark, NJ, USA.* 1850-55.
Joshua. *Bromsgrove.* 1828.
Simeon. *Canton, Mass, USA.* ca.1810. C.
William. *Canton, Mass, USA.* ca.1780, then *Stoughton, Mass, USA.* ca.1800. C. & W.

CRANEFIELD, (Corba?). *Cley.* Late 18c.
CRANEY, G. *London.* 1863-69.
CRANMER, J. *Ipswich.* 1853-58.
CRAPP, William. *St. Austell.* d.pre-1803. C.
CRAPPER, Charles William. *Halifax.* 1809. C.
CRATE George. *New Barnet.* 1874.
CRATHORN, Jonathan. *Beverley.* ca.1700. W.
CRAUFORD, Alexander. *Whitby.* mar.1779. C. & W.
CRAVEN—
    Alfred. *Philadelphia.* 1843.
    Joseph. *Cork.* 1765-94. W. and jeweller.
CRAW, James. *Forfar.* 1837.
CRAWFORD—
    Alexander, sen. *Scarborough.* 1790-1823.
    Alexander, jun. *Scarborough.* 1823-34.
    Archibald. *Largs.* 1836-50.
    Daniel. *Lochwinnoch.* 1860.
    George. *Dublin.* 1858.
    George. *Falkirk.* 1836.
    J. *Glasgow.* ca.1850. C.
    James. *Johnstone.* 1836.
    James. *St. Catherines, Canada.* 1861-78.
    Robert. *Halifax, Nova Scotia.* 1822.
    Thomas William. *Toronto.* 1887.
    W. *Belfast.* 1868.
    Walton. *Scarborough.* 1834-66. W.
    William. *Ballymena.* ca.1800-46. C.
    William. *Guelph, Canada.* 1870.
    William. *Halifax, Nova Scotia.* 1816-67.
    William. *Markinch.* 1837.
    William. *Oakham, Mass, USA.* 1745. C.
    William. *Whitby.* mar.1809. W.
CRAWLEY—
    C. *Philadelphia.* 1829-33.
    Thomas. *London.* a.1655, CC. 1660-62. *Westminster.*
CRAWSHAW—
    Andrew. *Rotherham.* b.1780-d.1844. C.
    Edward. *Worsborough.* Attended ch. clock. d.1724.
    Thomas. *Barnsley.* Pre-1790. C.
    William. Town not known. ca.1750. C. prob. same man as below.
    William. *Worsborough.* s. of Edward. b.1694-1749.
CREAK & SMITH. *London.* ca.1740. br. clock.
CREASER, Thomas. *York.* 1815-41. W.
CREASY, Joseph. *Barking.* 1839-51. German clocks.
CREE, John. *Glasgow.* 1829-41 (-60?).
CREED—
    R. (& Son). *Peterborough.* 1864 (-69).
    Thomas. *Oxford.* a.1657. W. Later *London.* Where one such CC. 1668-99.
CREEDY, Joseph Henry. *London.* 1869.
CREEK—
    David. *Whittlesford.* 1858-75.
    Edward. *Newry.* 1784.
CREES, Thomas S. *Lyme Regis.* 1855-75.
CREHORE, Charles Crane. *Milton, Mass, USA.* 1793-1828, then *Boston* 1828-d.1879. Clock cases.
CREIGHTON—
    David. *Kingstown.* 1858.
    John. *Broughshane.* ca.1765. To *Ballymena.* ca.1770-85. C.
CREMER, John Thomas. *Chelmsford.* ca.1801-32. W.
CRESCENT Watch Case Co. *Newark, NJ, USA.* 1923.
CREUSE, Benjamin. *Philadelphia.* 1772.
CREW(E)—
    ——. *Ledbury.* Early 19c.
    Cyrus. *Tetbury.* 1830-56. Also jeweller.
    Frederick. *Tetbury.* 1863-70.
    Henry. *Ledbury.* 1835-79.
    John. *Leeds.* 1817. C.
CREYK—
    James. *Montreal.* 1843-50.
    William. *Montreal.* 1842-54.
CRIBB, William E. *London.* 1851-75.
CRICHLOW, Thomas. *Liverpool.* mar.1704-12. C.
CRIGHTON—
    Archibald. *Dublin.* 1868-74.
    George. *Mid-Calder.* 1829. W.
CRICK—
    ——. *London.* ca.1810. C.
    Francis. *Newmarket.* 1875.
    Francis Richard. *Bishop's Stortford.* 1874.

CRIGHTON—
    James. *Manchester.* mar.1799. C.
    Walter. *Haddington.* 1850.
CRILLY, Thomas. *Ballymacarrett,* nr *Belfast.* 1843. Watch glasses.
CRINGAN, Robert. *Carluke.* 1860. W.
CRISP(E)—
    Brothers. *London.* 1881.
    Stedman Shrubsole. *London.* 1881.

    William Baker. *London.* 1851-81.
CRISSWELL, Isaac. *Philadelphia.* ca.1840.
CRITCHETT, James. *Candia, NH, USA.* Early 19c.
CRITCHFIELD, James. *London.* 1828. Watch case maker.
CRITCHLEY—
    Henry. *St. Helens.* 1836. W.
    Joseph. *Hyde.* 1848.
    Joseph. *Manchester.* 1851.
    Lawrence. *Prescot.* 1824. W.
CRITTALL, Francis B. *Braintree.* 1866-74.
CRITTENDEN—
    Charles. *London.* 1857.
    Charles. *Tallmage. O. USA.* ca.1830.
    George. *Chatham.* 1838-74.
    John. *Chatham.* 1838-47.
    Simeon. *Guildford, Conn, USA.* 1796-1867.
CROAL, Robert. *Alyth.* 1860.
CROASDALE, Thomas Roscoe. *Bury.* 1848-58. W.
CROCKER—
    & BRADBROOK. *London.* 1869.
    Brothers. *Kingston, Surrey.* 1878.
    J. R. *Valley Falls, RI, USA.* 1860s.
    Orasmus. *East Meriden, Conn, USA.* ca.1831-37. C.
    William. *Philadelphia.* 1834, then *New York.* ca.1840.
    William Henry. *London.* 1839-69.
CROCKETT—
    Edward. *Pontypridd.* 1887.
    John. *London.* ca. 1695. Lantern clock.
    John. *Pontypridd.* 1860-75. W.
CROFT—
    G. & Co. *Coventry.* 1880.
    Thomas. *London.* 1832.
    William & Co. *Toronto.* 1871.
CROFTS—
    James. *Bristol.* 1870.
    James. *London.* 1844.
    James John. *London.* 1857.
    John. *Daventry.* 1738.
    John. *London.* 1828-39. Watch case maker.
    Thomas. *Daventry.* 1716-38.
    Thomas. *Halton,* nr *Leeds.* 1752-63. C.
    Thomas, jun. *Newbury.* 1793 (*North Brook Street*) -1830.
CROLL, William. *Dundee.* 1837.
CROMAR, Henry. *Southampton.* 1878.
CROMBIE—
    David. *Auchtermuchty.* 1860.
    John. *Peterhead.* 1860. Also nautical instruments.
CROMEY, William. *Bristol.* 1863-70.
CROMWELL, William. *New York.* 1820.
CRONAGE, Thomas. *Liverpool.* 1814. See also CRANAGE.
CRONE, William. *Aberdeen.* 1846.
CROOK—
    Albert. *London.* 1875-81.
    James. *London.* 1828.
CROOKELL, Richard. *Farnworth.* mar.1822. W.
CROOKES—
    Thomas John. *Sheffield.* 1871.
    & TWIST. *Sheffield.* 1871.
CROPPER, Samuel. *Nottingham.* 1849-76.
CROSBY—
    C. *Coventry.* 1854.
    C. A. W. *Boston, USA.* Late 19c. C.
    Charles. *Angelica, NY, USA.* 1835.
    Charles. *Coleshill.* 1828-50.
    Charles. *Daventry.* 1824.
    D. S. *New York.* ca.1850.
    Joseph. *Bridlington.* 1840-79.
    & WASHBURGH. *New York.* ca.1860.
CROSIER, J. M. *Hartlepool.* 1898.

CROSLAND, William. *Wakefield.* (B. 1802-)1822-37.
CROSS—
Adam. *Saintfield.* ?early 19c.
Alexander. *Dublin.* ca.1800. W.
C. *Newcastle.* 1855.
& CARRUTHERS. *Edinburgh.* 1837.
Edward. *Wincanton.* 1809-11.
J. *Kingsland.* 1867.
James. *Prescot.* mar.1788-98. W.
Jean. *Montmagny, Quebec.* 1792-1837.
John. *Trowbridge.* ca.1830. C.
John Berryhill. *London.* 1839-63.
John Henry. *Brigg.* 1876.
Joseph. *Bradford, Wilts.* 1830.
Joseph. *Bratton, Wilts.* 1859-75.
Mark. *Bristol.* 1879.
Martin. *Wincanton.* 1811.
Osmond. *Wincanton.* 1746-60. C.
Richard. *Eccleston.* mar.1801-18. W.
Richard. *London.* 1632. CC.
Richard. *Wallingford.* (B. pre-1821-)1830-37.
Robert. *Prescot.* mar.1790-96. W.
S. *Landport.* 1848.
& Sons. *Kettering.* ca.1780.
Theodore. *Boston, USA.* pre-1775.
William. b. ca. 1791 *Baschurch (Shropshire).*
Ellesmere 1841-63.
William. *Landport.* 1839.
William. *London.* 1851-75.
William. *Portsea.* 1830.
William. *Prescot.* mar.1819-24. W.
CROSSKILL, George. F. *Halifax, Nova Scotia.*
1878-97.
CROSSLAND—
J. *Stalybridge.* 1865.
James. *London.* 1828-39.
James. *London.* 1881.
Nathaniel. *London.* 1828. Watch case maker.
CROSSLEY—
Henry. *Manchester.* 1824-28.
Humphrey. *Manchester.* 1822. prob. error for previous
man.
James. *Morpeth.* 1771. C.
CROSTHWAITE—
& HODGES. *Dublin.* 1804-09. W. & C.
John (& Son). *Dublin.* 1775-96 (1796-1803). W.
CROSSWELL, John. *London.* 1814. W.
CROSTI, Joseph. *Wednesbury.* 1876.
CROUCH—
Alfred Daniel. *Aylesbury.* 1877.
Arthur. *London.* 1875.
B. *Cambridge.* 1858.
Charles. *St. Albans.* 1874.
Daniel. *Dover.* 1847-55.
Edmund. *London.* 1857.
Edmund. *Clapham, Surrey.* 1862-78.
Henry. *Glastonbury.* 1875-83.
Henry Byron. *Cardiff.* 1887-99.
J. *St. Albans.* 1851.
John. *Aylesbury.* s. of William. b.1832-69.
T. *Glastonbury.* 1861-66.
Thomas. *Aylesbury.* 1850.
Thomas Henry. *Worthing.* 1870.
W. *Aylesbury.* 1854.
William. Place unknown (*U.K.*) ca.1750.
William. *St. Albans.* b.1790-ca.1850, then *Aylesbury*
till 1864.
William. *St. Albans.* 1828-66. May be same as above
man.
William (Son). *Edinburgh.* 1850(-60).
CROUCHLEY—
Lawrence. *Prescot.* 1796. W.
Thomas. *New York.* ca.1840.
Thomas. *Prescot.* d.1773. Another man of same name
d.1782. W.
CROUDACE—
John. *Durham.* 1827.
John & Son. *Durham.* 1847 (Also CRUDDAS).
CROW(E)—
——. *Faversham.* 1802.
Edward. *Birmingham.* 1839. Dial maker.
Edward. *Faversham.* 1823-45.
Edward. *Hillington.* 1865-75.

CROW(E)—*continued.*
Francis. *Margate.* 1823.
George, jun. *Wilmington, Del, USA.* 1802.
John. *Liverpool.* 1848-51.
John. *Wilmington, Del, USA.* 1770-98. C.
T. *Birmingham.* 1849. Dial maker.
CROWDER, W. *Grimsby.* 1861-68.
CROWFOOT, Henry James. *London.* 1863-69.
CROWHURST, Thomas. *Brighton.* 1878.
CROWLE, John. *St. Austell.* 1679. rep'd. ch. clock.
CROWLEY—
John. *Liverpool.* 1834.
John. *Manchester.* mar.1821. W.
CROWN, Marcus. *Birmingham.* 1880.
CROWTHER, William. *Halifax.* 1834-71.
CROYDON, Charles Henry. *Ipswich.* 1875-79.
CROZIER—
——. *Newry.* Early 19c. C.
James. *Belfast.* 1858-68. W.
James. *Dromore.* 1854. prob. later to *Belfast.*
& MARTIN. *Birmingham.* 1880.
Thomas. *Hillsborough.* 1854-68.
CRUDDAS, John (& Son). *Durham.* 1834 (1851-60).
CRUIKSHANK, Alexander. *London.* 1869.
CRUM & BARBER. *Unionville, Conn, USA.* 1830-40. C.
CRUMP—
Francis. *London.* 1832.
Henry. *London.* 1662. Watch case maker (B. CC.
1667-88).
W. *Brighton.* 1851.
W. *Ryde.* 1859.
CRUMPSTY, Thomas & Co. *Liverpool.* 1822-25.
CRUNDEL, W. J. *Leominster.* 1879.
CRUNDWELL—
Benjamin. *Edenbridge.* 1874.
John, *Edenbridge.* 1829-55.
Joseph. *Hadlow.* 1855-74.
S. *Yalding.* 1866.
Samuel. *Tunbridge Wells.* 1847-74.
Stephen. *Tonbridge.* 1823-66.
William. *Edenbridge.* 1838.
CRUNN, William. *Haverfordwest.* 1791-1805. W.
CRUTTENDEN, Thomas. *York.* b.1657. *London*
trained. *York* from 1679-d.1698. C.
CUDDON, H. R. *Colchester.* 1866.
CUENDET, Samuel, *London.* (B. 1815-24) 1832. Watch
cases.
CUFF—
——. *Wells.* 1725. rep'd. ch. clock.
James. *Axbridge.* ca.1750. C.
William Frederick. *Dulverton.* 1883.
William Frederick. *Ferndale (Wales).* 1887-99.
William Treasure. *Bristol.* 1856-79.
CUFFLIN, Robert Stephens. *London.* 1863-81.
CULLEN, John. *Armagh.* 1820-24. C.
CULLEY, Frederick. *London.* 1857-63. Clock cases.
CULLIDON, Andrew. *Dublin.* 1780-83. W. & C.
CULLIFORD, John. *Bristol.* 1680-1718. C.
CULLWICK, B. *Hastings.* 1878.
CULVERWELL, Richard Major. *Liverpool.* 1834.
CULYER—
G. *Aldborough (Suffolk).* 1853.
George. *Halstead.* 1866-74.
CUMINE, James. *Belfast.* 1858.
CUMMENS—
William, sen. *Boston* or *Roxbury, Mass, USA.* 1768.
mar.1793-1834. C.
William, jun. *Boston, USA.* s. of William, sen. 1793-
1816.
CUMMIN & DOUGLAS, *Quebec.* 1781.
CUMMING—
Alexander. *Inverary.* 1775.
James. *Craigellachie nr Dufftown.* 1860.
John. *Albany, NY, USA.* bro. of Alexander. 1774-79,
possibly to *Europe.* ca.1830.
John (& Son). *London.* 1828-51 (1857).
Richard. *Alford.* 1828.
CUMMINGS—
J. *Battersea.* 1866.
Job. *Sandwich.* 1858.
Robert. *London.* 1869-75.
Robert Daniel. *London.* 1881.

CUMMINGS—*continued.*
S. *Concord, NH, USA.* Late 18c.
W. H. *Middlesborough.* 1898.
CUMMINS—
Charles. *London.* 1837-51. W.
Mrs. Elizabeth. *London.* 1857. (?wid. of Charles.)
Thomas. *London.* (B. CC. 1806-20)-1832.
CUNDALL, J. T. *London.* 1851-69.
CUNDITT—
Robert & Frederick. *Brighton.* 1851-70.
William R. *Brighton.* 1878.
CUNDY, George Albert. *London.* 1869-81.
CUNNINGHAM—
A. J. *Charleston, SC, USA.* 1830-50. W.
Alexander. *Forres.* 1860.
Hugh. *Dublin.* 1754-d.1776. W.
J. *Long Stratton.* 1846.
John. *Dublin.* 1754. W.
John. *Dublin.* 1800-1812. W.
Joseph. *Dublin.* a.1757.
R. *Kingston, Canada.* 1851-53.
Robert. *Long Stratton.* 1858-75.
Thomas. *Annan.* 1860.
William. *Sanquhar.* 1837-60.
CUNO, Jacob. *Frankfurt.* 1583.
CUPER, Josias. *London* (*Chelsea*). From *France.* 1622.
BC. 1628, CC. 1632, d.1660.
CURE—
John. *Bradford.* 1804-ca.1830. C.
Jule F. *Philadelphia.* 1839-40.
Louis (Lewis). *Philadelphia.* 1811-19, then *Brooklyn* 1830s.
CURGENVEN, Henry. *Helston.* 1763-ca.1785. C. May be Curgensen from *London?*
CURNOW, Jacob. *Gillingham.* 1867-75.
CURRAGH, ——. *Newtownards.* d.1840.
CURRAN—
Henry. *St. Johns, Canada.* See CORRAN.
Thomas. *Liverpool.* 1824.
CURRER—
John. *Hillsborough.* 1846. W.
John. *Peebles.* 1840.
Robert. *Peebles.* 1836.
CURRIE—
Archibald John. *Maidstone.* 1826-40. W.
John. *Lennoxtown.* 1860.
Robert. *Falkirk.* 1860.
CURRIER—
Edmund. *Salem, Mass, USA.* 1828.
Eliphalet. *Haverhill, Mass, USA.* 1774-1831. W.
(Edmund) & FOSTER. *Salem, Mass, USA.* 1831-37. C.
John. *Salem, Mass, USA.* 1831.
& TROTT. *Boston, USA.* 1833-39.
CURRY, Thompson, *Hexham.* ca.1800. C.
CURRYER—
Barrett. *London.* 1875.
Thomas. *Brixton.* 1866-78.
William Barratt. *London.* 1881.
CURSON—
A., jun. *King's Lynn.* 1846.
Alexander, jun *Hartlepool.* 1851.
Charles. *Coventry.* 1880.
Edward. *Binham.* 1858-75.
M. *Yarmouth.* 1858-65.
Matthew. *Fakenham.* 1830.
Matthew. *Wells.* 1836-46.

CURTAIN—
Simon. *Cork.* 1717-d.1766. W.
William. *Cork.* 1748. W.
CURTIS (& CURTICE)—
& CLARK. *Plymouth, Conn, USA.* ca.1825. C.
(Lemuel) & DUNNING (J. L.). *Concord, Mass, USA.* ca.1813-18, later, *Burlington, Vt, USA.*
F. R. *Dorchester.* 1867-75.
Greenaway. *Oxford.* a.1688-d.1702. C. & W.
H. & Co. *Meriden, Conn, USA.* ca.1830. C.
J. & Co. *Cain, NY, USA.* ca.1830. C.
Joel. *Waterbury, Conn, USA* and *Cairo, NY, USA.* 1814.
John. *Laugharne (Wales).* 1835.
John William. *Neath.* 1848-50.
Joseph. *Chatteris.* 1875.
Lewis. *Farmington, Conn, USA.* 1774-1820, then *St. Charles, Mo, USA.* ca.1820, then *Hazel Green, Wis, USA,* where d.1845.
Samuel. *Boston, USA.* bro. of Lemuel. 1788-1879.
(Samuel) Manufacturing Co. 1853. Forerunner of Waltham Watch Co.
Thomas. *Knighton (Wales).* 1875.
Thomas. *Presteign.* 1871-75.
W. *Newburyport, Mass, USA.* Early 19c. C.
William. *Axminster.* 1790-1800.
William. *Battersea.* 1878.
William. *Exeter.* (B. 1795-) 1803-16. W.
William. *Neath.* 1844.
CURZON, Joseph. *Coventry.* 1880.
CUSHING, Theodore *Hingham, Mass, USA.* ca.1808. Clock cases.
CUSTER—
Daniel. *Reading, Pa, USA.* ca.1810-20.
Isaac. *Norristown, Pa, USA.* bro. of Jacob. ca.1830-50. Later in *St. Louis.*
CUTBUSH—
Charles. *Maidstone.* s. of John. a.1734 to father.
George. *Maidstone.* a.1725 to father, John.
John. *Maidstone.* a.1722-32(-44?).
John. *Maidstone.* a.1765 to father, Richard (or father, Robert?).
Richard. *Maidstone.* a.1761 to father, Robert -1779.
Robert. *Maidstone.* 1761-65. C.
Thomas. *Maidstone.* a.1702. Locksmith.
William. *Maidstone.* a.1727 to father, John.
CUTHBERT—
& ALEXANDER. *Guelph, Canada.* 1862.
William. *Whitchurch, Hants.* 1878.
CUTHBERTSON, Alexander. *Newlands* nr *Linton, Peeblesshire.* 1860.
CUTLER—
John N., sen. *Albany, NY, USA.* 1822-50.
John N., jun. *Albany, NY, USA.* 1842 and later.
Richard. *Carrickfergus.* 1782. C.
Richard. *Cork.* 1787.
CUTLEY, Able. *Philadelphia.* 1710 (?error for Abel Cottey).
CUTTLIFFE, Robert. *London.* 1857-63.
CUTLOVE, John. *Beccles.* ca.1740-ca.60. C.
CUTMORE, John. *London.* 1851-63.
CUZNER—
Henry Lionel. *Alcester.* 1880.
Jeremiah. *Shepton Mallet.* 1861-83.
CYCLO CLOCK CO. *New York.* Sold up 1892.

D

D, A. AD is mark of A. Dumas of *Paris*. q.v.
D, J. JD is mark of J. Dejardin, q.v.
D, M. MD is mark of Drugeon of *Paris*, q.v.
DACLIN, J. H. *Lyon*. 1872. C.
DADIN, Louis. *Charleston. SC.* ca.1840.
DADSWELL—
  T. *Rotherfield.* ca. 1710. C.
  Thomas I. *Burwash.* 1732 -d. 1752. C.
  Thomas II. *Burwash.* a.1735 to Thomas Dadswell
  (sen.?) C. (B. 1790).
DADSWORTH, Thomas. *Burwash.* ca.1750. C. prob.
  Dadswell, q.v.
DAFFRON, Joseph. *Dublin.* 1784-1823. W. & jeweller.
DAFT—
  Thomas. *Nottingham.* b.1703. mar.1746. C.
  Thomas. From *Philadelphia* to *New York.* 1786-90.
DAGGETT—
  James. *St. Louis, Mo, USA.* ca.1820.
  T. *Providence, RI, USA.* ca.1840.
DAGLISH, Joseph. *Alnwick.* 1790-1834. C. & W.
DAGNALL, Henry. *Liverpool.* mar.1834. W.
DAILLE, ——. *Paris.* 1722-60. C.
DAISY. 1914 and later, product of Pequegnat Clock Co.
DALBY—
  & NICHOLSON. *Chelmsford.* ca.1823-28.
  Robert. *Chelmsford.* 1828.
DALDORPH, Adolph. *London.* 1881.
DALE—
  Alfred Samuel. *Llanelly.* 1868-75.
  Charles. *Andover.* 1830.
  James. *Eccleshall.* 1876.
  Joseph. *Reading.* 1864-77.
  W. F. *Lee, Kent.* 1866-74.
  W. S. *Hartlepool.* 1898.
  William. 1874.
DALGET(T)Y—
  Alexander. *Deptford.* 1847.
  Alexander. *London.* 1851.
  Alexander. *Peckham.* 1862-78.
DALGRESS, Samuel. *London.* a.1737 to Thomas But-
  terfield. W.
DALLAS, Alexander. *Inverness.* 1850.
DALLAWAY—
  Thomas. *Bexhill.* 1828-62.
  William & Son. *Edinburgh.* 1785-1812. Japanner of
  clock dials.
DALRYMPLE—
  Hannah. *Dublin.* (wid. of John, sen.) 1780-83.
  James. *Dublin.* 1792-98. W.
  James. b.1765. *Ireland,* to *USA* (*Salem, Mass*) by
  1795. mar.1806-39.
  John, sen. *Dublin.* 1754-d.1779. W.
  John, jun. *Dublin.* a. of John, sen. to whom a.1768, free
  1789, d.1823. W. and silversmith.
DALTON—
  John. *Hartlepool.* 1850-56.
  John. *York.* 1830. W.
  W. H. *Dudley.* 1868.
  William Howard. *Birmingham.* 1880.
DALZIEL, John. *New York.* 1798.

DAMER—
  John. *Jersey.* 1834-53.
  W. H. *Jersey.* 1875.
DAMIENS—
  ——, DUVILLIER. *Paris.* 1855-67.
  Emile. *Paris.* 1880. C.
DAMMANT, Barnaby. *Colchester.* b.1683-d.1738. C. &
  W.
DAMPER—
  Mrs. J. *Tunbridge Wells.* (?wid. of William.). 1866.
  William. *Biggleswade.* 1864.
  William. *Tunbridge Wells.* 1838-55.
DANA—
  Daniel. *Providence, RI, USA.* Early 19c.
  Payton & Nathaniel. *Providence, RI, USA.* ca.1830.
  (George) & WHITTAKER (Thomas). *Providence, RI,
  USA.* 1805-ca.1825.
DANAGHY, John. *Cork.* 1828.
DANBY—
  George, *Norwich.* 1836.
  George. *London.* 1857-81.
DANCE, Joseph. *Ipswich.* 1857-85.
DANDY, 1914 and later, product of Pequegnat Clock Co.
DANEYGER, L. *Newcastle-on-Tyne.* 1856.
DANIEL(L)—
  C. J. *Great Malvern.* 1860-76.
  Charles James. *London.* 1839.
  F. H. *Marazion.* 1845.
  G. L. *Rome, NY, USA.* 1840-43.
  George. C. *Elizabeth City, NC, USA.* 1829, to *Halifax*
  1831.
  Henry. *Bristol.* (B. 1825-)1830-42.
  Henry. *London.* 1857-63.
  James. *London.* 1832-44.
  Matthew. *St. Ives., Cornwall.* 1873. W.
  Michael. *Penzance.* 1847.
  Nicholas. *Penzance.* 1844. C. & W.
  Nicholas Charles. *St. Just* in *Penwith, Cornwall.*
  1847-73. W.
  R. *Bristol.* mar.1782. W.
  Richard Thomas. *Marazion.* 1823-44.
  Robert. *Plymouth Dock.* mar.1771-91. W.
  Thomas. *London.* a.1646 to Daniel Fletcher, CC.
  1656-62.
  Thomas. *York.* a.1761. C.
  William. *Cork.* 1768.
  William. *Maldon.* a.1783-1820.
DANIEL(L)S—
  Alfred. *Minchinhampton.* 1879.
  Charles James. *London.* 1857-63.
  David. *Carmarthen.* 1868.
  Henry & John. *Liverpool.* 1824-51.
  Isaac. *Hull.* 1840-58.
  J. *New Romney.* ?Early 18c. Lantern clock (B. 1779-
  d.1788).
  James. *Kilrea.* Early 19c. C.
  James. *St. Helens.* 1823-26. W.
  John Thomas. *London.* 1881.
  W. *Richmond Hill. Ontario.* 1862.
DANKS, ——. *Edinburgh.* 1819-35. Watch case maker.

DANN—
James. *Wisbech.* Successor to P. Vurley. 1865-75.
R. B. *Greenwich.* 1874.
W. *Maidstone.* 1826-66.
DANZIGER, A. & CO. *Coventry* and *Birmingham.* 1860.
DARBY—
Charles. *Handsworth.* 1876.
John. *London.* (B. 1802-40) 1844.
William. *Wishaw.* 1856. W.
DARBYSHIRE, Roger. *Wigan* 1662-d.1690. W. First
Wigan maker.
DARE, J. *Chertsey.* 1862.
DARLEY, F. H. *Folkestone.* 1866.
DARLING—
George Lacey. *Simcoe, Canada.* 1851-71, then *Hamilton.* 1874-d.1899.
William. *York.* a.1820-58. C.
& WOOD. *York.* 1866. W. Successor to William
Darling.
WOOD & ANFIELD. *York,* post-1866, successor to
DARLING & WOOD.
DARLINGTON—
Benedict. *Westtown, Pa, USA.* 1810-12. C.
J. *Burslem.* 1860-76.
DARNTON, Cuthbert. *Chester-le-Street.* 1765-76. C.
DARRAH, E. Coventry. 1880.
DARROW—
Elijah. *Bristol, USA.* 1824-56.
Franklin C. *Bristol, USA.* 1857-65.
& MATTHEWS. *Bristol, USA.* 1824-26.
William & Co. *Bristol, USA.* 1834-36.
DART—
Henry. *London.* 1881.
Lewis. *Jersey City, NJ, USA.* ca.1850.
DARVILL, Henry. *London.* 1857.
DASER, John George. *London.* (B. 1817-24) 1839.
DAVENPORT—
——. *Liverpool.* pre-1752. W.
George. *Ashbourne.* 1864-76. C.
James. *Macclesfield.* 1848-65.
James. *Salem, Mass, USA.* 1784. C. & W.
Joseph. *Newton Upper Falls, Mass, USA.* 1773-1849.
C.
Thomas. *Chesterfield.* ca.1770. C.
William. *Ashbourne.* 1849-76.
William. *Philadelphia.* Early 1800s.
DAVERILL (also (D'Averill), John. *London.* Alien 1622.
CC. 1636-41. Worked for Lewis Cook at first.
DAVEY—
Ebenezer. *Truro.* 1823. Also DAVY.
G. M. & T. *Lewes.* 1862.
John. *Liskeard.* 1847.
John. *Needham Market.* 1823-55. C.
Samuel. *Norwich.* 1772. W. (B. 1784.)
DAVID—
Thomas. *Bridgend.* 1835-52.
W. *Morristown (Wales).* ca.1850-55.
DAVIDSON—
Adam. *London.* 1832-44.
Andrew. *Stranraer.* 1820.
Barzillai. *Norwich, Conn, USA.* 1775.
Barzillai, *New Haven, Conn, USA.* 1825. T.C. May be
s. of Barzillai I?
C. *New York.* 1830. C.
& Co. *Toronto.* 1871.
Edward. *Hamilton, Ontario.* 1895.
& FELTUS. *Halifax, Nova Scotia.* 1890-97.
James. *Dunbar.* 1813-60.
James. *Old Deer, Aberdeenshire.* 1836-60.
John. *Falmouth.* 1847.
John. *Glasgow.* 1836.
John. *Wick.* 1892.
Robert. *Halifax, Nova Scotia.* 1830.
Robert. *Lerwick.* 1836.
Samuel. *Maryland, USA.* 1774.
W. *Streetsville, Canada.* 1857.
DAVIE—
Joseph. *London.* (B. 1820-40) 1828-57.
Samuel E. *Campbeltown.* 1860.
DAVIES—
——. *Llandeilo.* 1799.
Alfred A. *Brandon.* 1875-79.

DAVIES—continued.
Arthur Price. *Newcastle Emlyn.* 1887.
B. F. & T. M. *Ithaca, NY, USA.* ca.1870. C.
Benjamin. *Carmarthen.* 1820-44.
Benjamin. *Hay.* 1830-45.
Charles. *Toronto.* 1871-77.
Charles William. *London.* 1851-81.
D. *Builth.* 1791.
D. *Stratford, Essex.* 1874.
Daniel. *Cheltenham.* 1830-42.
David. *London.* Second half 18c. C.
David. *Machynlleth.* 1840-68.
David Roger. *Newcastle Emlyn.* 1835-44.
E. *Shrewsbury.* 1879.
Edward. *Barrow.* 1866-69, when listed as DAVIS
Brothers.
Edward. *Dalton, Lancs.* 1866, same man as at Barrow.
Edward. *Ellesmere.* 1835.
Edward. *Minwere (Pembroke).* 1726. C.
Emanuel. *Swansea.* 1875.
Evan. *Kidwelly (Wales).* 1830-75. W. & town crier.
Evan. *Lampeter.* 1841-76.
Evan. *Llansawel.* 1871. W.
Evan. *Newbridge* (i.e. *Pontypridd*). 1799-1888.
Evan. *Runcorn.* 1857-78.
Evan. *Tregaron.* 1840. C.
Francis. *Exeter.* mar.1752.
G. *Hereford.* 1856-63.
George. *Todmorden.* 1837.
George Alfred. *Holywell.* 1835-56. W.
Griffith. *Dolgellau.* 1800-84. C. & W.
H. *Weobley.* 1863.
Harris. *Sheffield.* 1862-71. W.
Henry. *Prescot.* mar.1769. W.
Henry Langdon. *Birmingham.* 1880.
Herbert. *Carmarthen.* 1847-52.
& HODGENS. *New York.* 1873.
Howell. *Dolgellau.* 1844-d.1875. C.
J. *Chester.* 1865.
J. A. *Ebbw Vale.* 1891.
James. *Hastings.* 1878.
James. *Haverfordwest.* ca.1790 C.
James. *London.* 1881.
James J. *Blackheath.* 1874.
James Swithen. *Bristol.* a.1810-18.
Job. *Dowlais.* 1871-75 from *Pen-y-darren.*
Job. *Pen-y-darren.* 1852-68, later *Dowlais.*
John. *Brecon.* 1844-71. W. Successor to John Lloyd.
John. *Cardigan.* 1830-44.
John. *Carmarthen.* 1818-22.
John. *Chepstow.* 1807. W.
John. *Chester.* 1848-57.
John. *Llanbryn Mair.* 1783-1855.
John. *Pembroke Dock.* 1887.
John. *Pontypool.* 1822.
John. *St. Harmons (Wales).* Early 19c. C.
John. *Swansea.* 1800-1830. C.
Joseph. *Bristol.* 1850-70.
Langdon & Co. *Birmingham.* 1880. Importer.
Michael. *Gwter Fawr (= Brynamau).* mid-19c.
Michael. *Llandovery.* Late 18c. C.
Morgan. *Carmarthen.* 1830.
Owen. *Llanidloes.* 1835-57. C. & W.
Owen Woolaston. *Porth (Wales).* s. of Howell. 1860-
1936.
Pierce. *Abergele.* 1868-87. W.
R. & Co. *Gloucester.* 1879. Chrons.
R. G. *Carmarthen.* 1847-52.
Rev. Llewellyn. *St. Harmons (Wales).* 1745-74. C. and
vicar.
Robert. *Ffestiniog (Wales).* 1874-87. W. & C.
Robert & Co. *Gloucester.* 1879.
Samuel. *Abergavenny.* 1880.
Samuel. *Brecon.* 1830.
Samuel. *Merthyr Tydfil.* 1875.
Samuel. *Treherbert (Wales).* 1875.
Sarah (Mrs.) *Dowlais.* 1887.
Theophilus. *Llanrwst.* 1856-76.
Thomas. *Carmarthen.* mar.1782-1835. W.
Thomas. *Llandovery.* 1868-75.
Thomas. *Llanelly.* 1830-35. Also auctioneer.
Thomas. *Monmouth.* (B. 1777-)1791. C. & W.

DAVIES—*continued.*
Thomas. *Morriston (Wales)*. 1868.
Thomas. *Narberth (Wales)*. 1887.
Thomas. *New York*. 1846.
Thomas. *Utica, NY, USA*. ca.1820.
Tobias. *London*. a.1646 to William Selwood, free 1653-70.
Uvedale. *Weobley*. 1835-56.
William. *Dudley*. 1872-76.
William. *Farnworth*. mar.1818. W.
William. *Hay*. 1868-87.
William. *Llandingad*. 1799.
William. *Llandovery*. ca.1800. W.
William. *London*. 1844. Clock case maker.
William. *Merthyr Tydfil*. 1825. W.
William. *Swansea*. 1810.
William A. *Carmarthen*. 1840-62. C. & W.
William A. & Sons. *Carmarthen*. 1868-87. C. & W.
William A. & Sons. *Neath*. 1868-87. C. & W.
William Ayeleway. *Cardigan*. mar.1844.
William Christmas. *Mountain Ash (Wales)*. 1887-99.
William J. *Harlech*. 1833-74. C. & W.

DAVIS—
A. *Clapham*. 1878.
Aaron David & Co. *London*. 1839-63.
Alexander. *Chichester*. 1839.
Alexander. *Havant*. 1830.
Alfred & Co. *London*. 1832.
Alphonso. *Nailsworth*. 1879.
Ann. *Leominster*. 1830-44.
Ari. *Boston, USA*. ca.1840.
& BABBITT. *Providence, RI, USA*. ca.1810.
Caleb. *Woodstock, Va, USA*. ca.1816. Later *Clarksburg*.
Charles. *Belfast*. 1827.
Charles Orlando. *Bath*. 1875.
Clock Co. *Columbus, Mass, USA*. ca.1880.
D. P. *Boston, USA*. 1842-49.
D. & Son. *Glasgow*. 1860.
Daniel. *Shrewsbury*. b. 1804-64. W.
David. *London*. 1828-32.
David. *London*. 1869-81 (may be same as above).
Edward. *North Shields*. 1793-d.1805. W.
Edwin. *Bradford*. 1866-71.
Emanuel. *Nailsworth*. 1850-70.
Gabriel. *Mannheim, Pa, USA*. ca.1780.
H. *Ingersoll, Canada*. 1861.
H. *Reckleford*. 1861.
H. & Co. *Montreal*. 1862.
H. & Son. *London, Ontario*. 1831-1900.
Henry. *Bath*. 1875.
Henry. *London, Ontario*. 1851-71.
J. *Cambridge*. 1865.
J. S. *Wiveliscombe*. 1866.
James. *Enniskillen*. 1858.
James. *London*. 1844. Clock case maker.
James. *London*. 1881.
James. *Wakefield*. 1815.
John. *Coleraine*. ca.1770.
John. *Fairfield, Conn, USA*. 1750. C.
John. *New Holland, Pa*. ca.1802-18. C.
John. *Shiffnal*. b. 1799 s. of William Davis of *Shifnal* - 1875. C.
John. *Tonystick*. 1846-59.
John. *Windsor*. (B. 1739-) 1745-79 (?-84). T.C.
John. *Wolverhampton*. 1835. Also 'dialist'.
John Frederick. *London*. 1881.
John Wheelwright. *Salem, Mass, USA*. 1800-24.
Joseph. *London*. 1828.
Joseph. *London*. 1881.
Joseph W. *Toronto*. 1875-77.
Josiah. *Heaton Norris*. 1857.
Josiah S. *Providence, RI, USA*. 1827.
Matthew. *Coventry*. 1854-68.
& McCULLOUGH. *Hamilton, Ontario*. 1884-95.
Nicholas. *Ballybofey*. 1787.
PALMER & Co. *Boston, USA*. 1842.
Phineas. *York, Pa, USA*. 1809.
Richard. *Philadelphia, USA*. 1837.
Riley A. *New Bern, NC, USA*. 1850.
Robert. *Burnley*. 1723. C. Believed first maker there.
Robert. *London*. 1851-57. Clock case maker.
Samuel. *Boston. USA*. 1830.

DAVIS—*continued.*
Samuel I. *London*. a.1640 to William Selwood, CC. 1647-82. Lothbury. C.
Samuel. *Pittsburgh, Pa, USA*. 1815-50.
Thomas. *London*. 1832.
Thomas. *London*. 1857.
Thomas. *Worcester*. 1828-50.
TOKLAS & Co. *San Francisco, Cal, USA*. 1880.
W. *Shifnal*. Early 19c. C.
W. *Shrewsbury*. 1863.
W. M. *Morrisville, NY, USA*. 1825.
William. *Belfast*. 1858 (-68?).
William. *Bristol*. 1781.
William. *Croft, Leicestershire*. 1720. T.C.
William. *London*. (B. a.1763?) 1768. W.
William. *London*. 1857-81.
William. *Shiffnall*. (b. ca. 1764) 1798 -d. 1842. C.
William Henry. *London*. 1863-81.
William Henry. *Wellington, Salop*. 1835.
William & Sons. *Birmingham*. 1854-80.
William & Sons. *London*. 1863.
Wolf. *Brighton*. 1862-78.

DEVISE, Robert. *Granville, Nova Scotia*. 1838.

DAVISON—
——. *Ecclesham*. ca.1770-ca.1775. C.
C. *New York*. 1830.
Elizabeth. *Beaumaris*. 1874.
Evan. *Beaumaris*. 1887-90.
George. *Wooler*. 1827-34. C.
J. *Lutterworth*. Early 19c. C.
John. *Darlington*. 1638. C.
Jonathan. *Olney*. mar.1806. W.
R. *Liverpool*. ca.1790. W.
T. *Newport Pagnell*. 1847.
Thomas. *Newton Lingville*. 1847.
Thomas. *North Shields*. 1834.
William. *Beaumaris*. 1835-68. C.
William. *Eccleshall*. mar.1774. W.
William. *Kettering*. 1841-77.
William. *Toronto*. 1861-63.

DAVY—
——. *Kenninghall*. ca.1830-58. C. prob. Robert q.v.
E. (jun.?). *Yarmouth*. 1846-65.
Ebenezer Lovell. *Lewes*. 1828-39.
Edward. *Norwich*. 1830-36.
Elijah, sen. *Yarmouth*. 1830-65.
John M. *Fakenham*. 1830-65.
R. *East Harling*. 1858.
R. *Yarmouth*. 1846-65.
Richard. *London*. 1869-75.
Robert. *Kenninghall*. 1830-58.
William. *Burnham, Norfolk*. 1836.
William. *Wells*. 1830.

DAWBIN, Joseph. *London*. 1844.

DAWES—
James. *Attleborough*. (B. 1799-)1830-36.
James. *Whitehaven*. 1848-69.
John. *Ulverston*. mar.1792.
John. *Whitehaven*. (B. 1820-)1828-34.
Miss A. *Kenninghall*. 1846.
Richard. *Southampton*. 1830-48. C.
Robert. *Boston, USA*. 1842.
Thomas. *Chester*. ca.1760 (B. d.1779). C.
William. *Attleborough*. 1846-75.
William M. *Gravesend*. 1823-27.
William Matthias. *London*. 1832-81.

DAWKES, Robert. *Whiston*. ca.1700. Lantern clock.

DAWKINS, T. *Llanstephan*. 19c. C.

DAWREY, Charles D. *Brooklyn, USA*. 1850. Clock dial painter.

DAWSON—
Charles. *Boxford*. 1855-74. W. Also at *Bures (Suffolk)*.
Charles. *Groton*. 1875-79.
David. *Johnstone*. 1860.
David. *Tarbolton, Scotland*. 1837.
E. *Ballingdon*. 1851-65.
Edward Charles. *London*. 1881.
George. *Sudbury*. (B. 18c.) 1810-55, succ. to John Clark.
James. *Bradford*. 1871.
John. *London*. b.ca.1666, a.1681 to Daniel Beckman, CC. 1688, mar.1689-1718. C.
John. *Sheffield*. 1871.

DAWSON—*continued.*
M. *London.* 1851.
Mrs. E. *East Dereham.* 1875.
Mrs. Susan. *Ballingdon.* 1866-74.
Stewart & Co. *Liverpool.* 1878-88. W.
Thomas. *London.* CC. 1635-39.
W. *Ballingdon.* 1855.
W. *East Dereham.* 1865.
William. *Foulsham.* 1830-58.
William. *Holbeach.* 1835.
William. *Wellington.* (b. ca. 1806). 1840-46
William Henry. *Putney.* 1878.
DAWTREY, James. *London.* From *Petworth.* a.1736 to
John Memes. C.
DAY—
——. *Keninghall.* ca.1815. C.
D. N. & R. *Westfield, Mass, USA.* ca.1830.
E. *Olney.* 1813. W.
Henry. *London.* 1881.
Henry James. *London.* 1863-81.
James. *Swansea.* 1887.
James. *Tredegar.* 1850-68.
John. *Deal.* 1832-45.
John D. *Brooklyn, NY, USA.* ca.1840-50.
Mrs. S. *Peckham.* 1855.
Nathan Henry. *Birmingham.* 1880.
Robert. *Stony Stratford.* 1830.
Thomas. *London.* 1881.
William. *Dewsbury.* 1871.
William Henry. *Newport, Isle of Wight.* 1859-78.
DA(Y)KIN—
J. *Grassington.* ca.1860. C. prob. same man as at
Skipton.
Jonathan. *Skipton.* 1866-71. W.
R. *Gunnerside.* ?Late 19c. C.
DAYNES—
J. *Norwich.* 1858.
W. *Norwich.* 1858.
DEACE, John. *Dublin.* ca.1780. C.
DEACON—
Arthur William Scripps. *Sandwich.* 1862-1909.
C. A. *Norwood.* 1878.
Edward. *Luton.* 1869-77.
G. *Swindon.* 1859-67.
H. *Lewes.* 1862.
Hubert John. *Swindon.* 1875.
J. P. *Bradford, Wilts.* 1867.
John. *Leicester.* b.ca.1770 at *Barton,* s. of Samuel. To
Leicester by 1795, d.ca.1830.
John Charles. *London.* 1828-57.
S. *Barton in the Beans.* 1849.
Samuel. *Barton in the Beans.* b.*Ratby.* 1746, to *Barton*
1771. d.1816. C. & W.
Samuel. *Market Bosworth.* 1835.
T., sen. *Barton in the Beans.* 1864.
T., jun. *Barton in the Beans.* 1864.
Thomas & Son. *Barton in the Beans.* 1876.
William. *Leicester.* ca.1790. W.
DEADMAN, Arthur. *Aylesbury.* 1847.
DEAKIN, W. *Dawley, Salop.* 1856.
DEAL—
John. *Richmond, Surrey.* 1878.
William. *London.* 1851-57.
DEAN—
Charles. *Bradford.* 1871.
(William) & Co. (George). *Salem, Mass, USA.*
1800-37.
G. *Macclesfield.* 1857-65. C.
George. *Salem, Mass, USA.* ca.1778-1831.
J. *Leigh.* 1858. W.
J. *Stroud.* 1863.
James. *West Malling.* 1874.
John. Place not stated. 18c. C. May be next man.
John. *Dublin.* 1752-71. W.
John. *Pudsey.* 1822. C.
Joseph. *Birkenshaw.* b.1740-d.1814. C.
Joseph. *Clitheroe.* 1824.
Richard. *Leigh.* 1828-51.
Robert. *Dublin.* 1780-86. W.
Samuel. *Birmingham.* 1842-50.
Thomas. *Eccles.* 1828-58.
Thomas. *Glasgow.* 1841.

DEAN—*continued.*
Thomas. *Leigh.* 1824.
Thomas. *London.* (B. CC. 1819-)1839-44.
William. *Halifax, Nova Scotia.* 1791-97.
William. *Pleasant Valley, NY, USA.* 1821.
William. *Salem. Mass, USA.* ca.1775-1800. then DEAN
& Co to 1837.
DEAR, James G. *Baldock.* 1851-74.
DEARLING, G. W. *Croydon.* 1878.
DEASE, John. *Youghal (Ireland).* 1820-24. W.
DEBENHAM—
J. *Lavenham.* 1846.
William. *Sudbury.* (B. 18c?) 1830-46.
DE BERARD, ——. *Utica, NY, USA.* ca.1820.
DEBNAM—
Charles W. *Newport, Essex.* 1839-74.
Isaac. *Warminster.* ca.1760. C.
DEBNEY—
J. *Birkenhead.* 1857.
James. *Wednesbury.* 1850.
DE BOO—
Philip. *London.* 1863-81.
William. *London.* 1844-69.
DE BOOS, Augustus. *London.* 1869.
DEBRUHL, Michael Samuel. *Charleston, SC, USA.*
1798-1806.
DECAUX, Lucas. *Norwich.* 1718-31. W.
DECELLE, Martha. *St. Hyacinthe, Canada.* 1857.
DE CHARMES, Simon. *Paris.* ca.1720. prob. from
London, where one such 1688-ca.1730.
DECOOL, J. F. *Namur.* ca.1800. C.
DEDLOW, L. *Little Dockray, Penrith.* 1858.
DEEBLE, Richard. *Cork.* b.1675-d.1719. C.
DEEKS—
J. *Norwich.* 1846-58.
John. *South Heigham.* 1875.
DEEME, Henry. *Honiton.* ca.1740-ca.1760. C.
DEEMS, John. *Baltimore, USA.* 1842.
DEERING, John. *Belfast.* 1858. C.
DE FONTAINE—
Lewis (Louis). *London.* 1828-44.
Louis W. *London.* 1851-69.
DE FOREST & CO. *Salem Bridge, Conn, USA.* 1832. C.
& W.
DEFRANC, ——. *St. Nicolas d'Aliermont.* 1889. C.
DE FRECE, Jacob. *London.* 1881.
DEFRIES, Jonas & Sons. *London.* 1875-81.
DEGAN, Joseph. *Balham.* 1878.
DE GIY, Lewis. *Boston, USA.* ca.1800.
DE GRUCHY, Thomas. *Jersey.* 1832.
DE HAYNE, ——. *Woodbridge.* ca.1770. C.
DEHOLME, Henry. *London.* 1832.
DEHORTER, ——. *Paris.* ca.1860.
DEHUFF, A. *York, Pa, USA.* ca.1850.
DEIMAND, A. *Eye.* 1846.
DEJARDIN, J. *Paris.* ca.1880. C.
DELACOUR—
George. *Chatham.* 1823-55.
H. W. *Bath.* 1856.
Henry W. *London.* 1851.
Mrs. J. L. *Bath.* 1866.
W. *Bath.* 1861.
DELACOURT, Bernard. *Beverley.* ca.1750. C. & W.
DE LA FEUILLE, Henry. *Jersey.* ca.1690-d.1736. C.
DELAFONS, Henry. *London.* 1811. W.
DE LA FOSSE, Samuel. *London.* (B. 1689-94) 1697.
DELAGARDE—
Albans. *London.* 1851.
Gustavus. *London.* 1839-44.
DE LAGRAVE, Francois. *Quebec.* 1819-31. Partner with
J. G. Hanna.
DE LAHAYE, B. *Quebec.* 1862-64.
DE LA HOYDE, Barnaby (or Bernard). *Dublin.* 1771-
1806. W.
DELANY—
Bernard. *Birmingham.* 1860-80.
John. *P. Dublin.* 1874.
DELAPLAINE, James R. *New York.* 1786-99.
DELARUE, Jean Anthoine. *London.* b.ca.1655,
mar.1682. W.
DE LA SALLE—
James. *Fareham.* 1839-48.

DE LA SALLE—*continued.*
James Thomas. *London* (B. CC. 1816-) 1828-44.
John. *London.* (B. 1809-11) 1828.
DÉLAUNCE, James. *Frome.* ca.1690-ca.1735. C. (One such London CC. 1678).
DELAVRESPEIRE (also **DE LAVRESPIERRE** and other spellings), William. *London.* Free CC. 1650-61.
DEL CAMPO, B. *Amsterdam.* Mid-18c. C.
DELEPINE—
Antoine. *St. Nicolas D'Aliermont.* 1849-d.1867.
Antoine. *Paris.* 1849-d.1867.
-BARROIS. *St. Nicolas d'Aliermont.* 1867-1913. C.
& CAUCHY. *St. Nicolas d'Aliermont.* ca.1847-59.
DELETTREZ, Jean B. *London.* 1863-75. Wholesaler.
DELIGHT, 1914 and later. Product of Pequegnat Clock Co.
DELILE, ——. *Dublin.* d.1716.
DELL—
Brothers & Co. *Bristol.* 1850-79.
Isaac. *Bristol.* 1840-42. Also chrons.
Marcus. *Wiveliscombe.* 1883.
DELLESSER, Ellis. *Liverpool.* 1824.
DELMAS, ——. *Paris.* 1815.
DELOCHE—
Constantine. *London.* 1851-57.
Henry. *London.* 1844-81.
DE LONG, Charles. *Glen Falls, NY, USA.* 1886. *Troy* 1887, *Albany* 1880-89.
DELONGUEMAIRE, ——. *Charleston, SC, USA.* ca.1700.
DELORTHE, Esias. *Dublin.* a.1719.
DE LOSTE, Francis. *Baltimore, USA.* 1817.
DELUNESY, Nicolas Pierre. *Paris.* 1764-83. C.
DELUXE Clock and Manufacturing Co. *New York.* ca.1929. C.
DEL VECHIO & DOTTI. *Wellington, Salop.* 1850.
DELVES, Alfred. *Burslem.* 1868-76. Repairer.
DELVIN, Mark H. *Salem, Mass, USA.* 1864.
DEMAINE, Anthony. *York.* b.1746, a.1762.
DEMEZA, George Ellis. *London.* (B. 1825) 1828. W.
DE MILT—
Benjamin. *New York.* 1802-18.
Samuel. *New York.* 1820-40.
Thomas. *New York.* 1798-1818.
Thomas & Benjamin. *New York.* 1805.
DE MOUCHIE, John. *London.* 1622. Alien journeyman with Cornelius Mellin in *Blackfriars.*
DEMOUY, ——. *Paris.* ca.1850. C.
DEMPSEY—
Denis. *Dublin.* 1858.
James. *Halifax.* 1871.
DEMPSTER—
Anthony. *York.* 1777-1830. W.
Mark Anthony. *York.* b.1800-29, then *Richmond* 1834-1841. d. at *Shedforth, Co. Durham.* 1875.
DENETT, Abraham. *London.* 1662.
DENHAM—
Charles. *Durham.* (B. 1820) 1827-44.
John. *Philadelphia.* 1848.
John. *Barden.* 1881.
W. *Huish Episcopi.* 1861-66.
DENIERE, J. *Paris.* ca.1850-60. C.
DENIS, Freres & Cie. *St. Nicolas d'Aliermont.* 1874.
DENMAN—
Edward George. *London.* 1875-81.
John Flaxman. *London.* 1839-44.
DENNE—
J(ohn). *London.* 1811 (B. 1817-24).
John. *Cheltenham.* 1830-40.
DENNETT—
Charles William. *Luton.* 1877.
James. *St. Helens.* 1822-48. C.
John. *Wigan.* 1851-58.
Thomas. *St. Helens.* 1851.
DENNEY, George. *London.* 1844-51. Clock case maker.
DENNING—
& HENDERSON. *London.* 1857.
Isaac. *London.* 1863.
DENNIS, Thomas. *Canterbury.* Free 1684-85. W. (One such a. *London* 1672 to James Wightman. Prob. same man).

DEN(N)ISON—
Benjamin. *Bradford.* 1853-71.
Robert. *Shipley.* Late 18c. C.
Thomas. *Halifax.* 1814.
DENNY, David, *Quebec.* 1822-26.
DENT—
E. & Co. *London.* 1863-81. Successor to Frederick.
Edward John. *London.* 1844-51.
Frederick. *London.* 1857. Successor to Edward John.
George. *Kirkby Stephen.* ca.1740. C.
James. *Kendal.* b.1857-71. W.
Joseph. *Hull.* 1790-95 (?Error for Denton, q.v.).
M. F. (& Co.). *London.* 1863 (69-81). Successor to Richard Edward). C. W. & chrons.
Richard Edward & Co. *London.* 1857.
Robert. *London.* 1863.
DENTON—
& FOX. *Hull.* 1802. C.
Frederick. *Rotherham.* 1837.
George. *Prescot.* 1795. W.
J. *Stoke-on-Trent.* 1860.
John. *Doncaster.* 1834-37.
John. *Wath-on-Dearne.* 1822.
Joseph. *Doncaster.* b.1741-d.1805. C.
Joseph. *Hull.* 1779-1814. C. & W.
Joseph. *Reading.* 1877.
Joseph. *Rotherfield.* 1870.
Robert. *Oxford.* 1730-d.1769. C. & W.
Samuel. I. *Oxford.* 1756-d.1795. C. & W.
Samuel. II. *Oxford.* s. of Samuel I. From *London.* 1802-27. C. & W.
William. *Oxford.* s. of Robert. 1756-74. C. & W.
William. *Warrington.* a.1728. W.
William & Samuel. *Oxford.* bros. of Robert. Partners 1756-74. C. & W.
DE PAEP, Baltasar. *Belgium.* ca.1600. W.
DEPLANCHES-VASSY. *St. Nicolas d'Aliermont.* 1889.
DE RAAY—
Franz. *London.* 1875-81.
Joseph. *Malvern Wells.* 1872.
DE RABOURS, ——. *Paris.* 18c. C.
DERBY—
Charles. *Salem, Mass, USA.* 1846-57.
Charles, jun. *Salem, Mass, USA.* 1859. W.
John. *New York.* 1816.
William. *Wishaw.* 1860.
DE REIEMER & MEAD. *Ithaca, NY, USA.* 1830-31.
DERIEMER, George. *Portsmouth.* 1808. W.
DERMOTT, Richard William. *London.* 1869-81.
DE ROCHES, Francois. *France.* 1750.
DERQUET, Cyrille. *Quebec.* 1841-1922. (Also written DUQUET).
DERRICK, J. *Blagdon, Bristol.* 1861.
DE ST. PAUL—
Johan. *Oxford.* 1576-96. T.C.
Triumph. *Oxford.* s. of Johan. 1601-1651. T.C.
DE SAULES & Co. *New York.* ca.1830.
DESBOIS, Daniel (& Sons). *London.* 1844-75 (1881-).
DESCHAMPS, Francis. *Philadelphia.* 1846-49.
DESCOMBES, Edward J. *Baltimore, USA.* 1840.
DESFONTAINES, Maison Leroy & Son. *Paris.* 1851-67.
DESFORGES, Samuel. *Spilsby.* 1850.
DESH, Leonard. *Glasgow.* 1860.
DESHAYS, ——. *Paris.* 1834-39.
DESJARDIN, P. H. *Montreal.* 1875.
DESLANDES, ——. *Amiens.* ca.1700.
DE SOLLA, Jacob & Son. *London.* 1857-81. Swiss clocks.
DESPLANCHES-VASSY, same as DEPLANCHES, q.v.
DESPREY & GREEN, *Bristol.* 1850.
DESPREZ, Charles. *Bristol.* 1863-70. C.
DESSIEUX. *St. Nicolas d'Aliermont.* 1889.
DESTACHES (sometimes **DUTACHE**), John. *London.* 1646. CC. 1661-62. Engraver.
DE ST. LEGER. *New Bern, NY, USA.* 1790. W.
DETOUCHE—
Francois. *Brighton.* 1839.
& HOUDIN. *Paris.* 1850-62. C.
DETTMANN, Theodore. *London.* 1863.
DEUCONNER, G. *Philadelphia.* 1817.
DEVACHT—
Francois. *Gallipolis, O, USA.* 1792.

DEVACHT—*continued.*
Joseph. *Gallipolis, O, USA.* bro. of Francois.
DEVALL, William Martin. *Colchester.* b.1817-d.1879.
DEVAUX, Edward F. *London.* 1869.
DEVELIN, J. & M. *Philadelphia.* 1848.
DEVENY, Thomas. b.1817. *Liverpool. Lancaster.* 1851.
W.
DEVERALL, John. *Winslow.* mid-18c.
DEVERELL—
John. *Baltimore.* 1789. *Boston.* 1790.
John. *Windsor*(?). ca.1740. C.
DEVERLEY, Hugh. *Perth.* 1843.
DEVEY, William. *Dudley.* (B. 1791-) 1828.
DEVLIN—
Arthur. *Glasgow.* 1860.
John. *Garvagh.* 1846.
John. *Glasgow.* 1860.
P. *Quebec.* 1844.
Patrick. *Greenock.* 1840. W.
DEWAR—
David (& Son). *Doune.* 1837 (-60).
Joseph D. *Worksop.* 1876.
DEWEY—
William. *Ingersoll. Ontario.* 1861.
William. *London.* 1832-39.
William. *London, Canada.* 1862.
William. *Ringwood.* 1859-78.
DEWHURST—
D. Town unknown, *Yorkshire.* 1842. (?*Halifax* area).
Laurence. *Walton, Lancs.* 1722. C.
William. *Ribchester.* 1670. Rep'd. ch. clock.
William Bolton. *Clitheroe.* 1824-58.
DEWICK, Vincent. *Swansea.* 1887.
DEWIN, Joseph. *London.* 1839. Watch case maker.
DEWITT, Garett. *Sparta, Ga, USA.* ca.1840.
DE WOLF, Thomas. *Westtown, Pa, USA.* 1810-15.
DEWSBURY, Samuel. *Manchester.* 1851.
DEXTER—
Charles. *Market Rasen.* 1835.
Dana. *Boston, USA.* ca.1870.
H. *Stockbridge, NY, USA.* ca.1840.
Joseph W. *Providence, RI, USA.* 1824.
William. *Sheepshead.* 1828-35.
William. *Stockbridge, NY, USA.* ca.1830 or later.
DE YOUNG, Michael. *Baltimore, USA.* ca,1830.
DEZERY, Etienne. *Quebec.* 1818.
DEZOUCHE, Isiah. *Dublin.* a.1732.
DIAMOND, Sevrion. *Harwich.* 1874.
DIBDIN, George Michael. *London.* 1881.
DIBLEY, W. *Rye.* 1851-62.
DICHER, Benjamin. *London.* ca.1770. W.
DICK—
John. *Girvan.* 1860.
Robert. *Dailly, Ayrshire.* 1850.
William. *Glasgow.* 1841.
DICKEN—
J. *Newcastle-under-Lyme.* 1868.
John. *Bristol.* mar.1773. C.
William. *Newcastle-under-Lyme.* 1828-42.
DICKENSON—
Charles. *Zanesville, O. USA.* 1815.
Daniel. *Framlingham.* mid-18c.
David. b.1821, *Otley.* 1839, *Skipton.* 1844-58. W.
Richard. *Mount Holly, NJ, USA.* 1760-70.
William. *Philadelphia, USA.* 1843-48.
DICKERSON, John. *Morristown, NJ, USA* and *Philadelphia.* 1755-1828. Later in *Indiana.*
DICKESON, Alfred Thomas. *Weston-super-Mare.* 1883.
DICKINSON—
——. *Egremont.* ca.1780. C.
Frederick Charles. *Bristol.* 1850.
John. *Manchester.* 1817-18. W.
John. *Warrington.* ab.1722.
Matthew. *North Shields.* 1834-58.
Richard. *Liverpool.* 1714 d.1743. W.
Richard. *London.* 1839.
Robert. *York.* 1810-46.
Thomas. Place unknown. ca.1790. C. (?May be next man).
Thomas. *Boston.* 1828.
Thomas. *Lancaster.* Free 1796. W. (B. Succeeded H. Bell, 1801).

DICKINSON—*continued.*
William. *Birmingham.* 1880.
William Henry. *Birmingham.* 1880.
DICKISSON, John. *Nottingham.* 1828.
DICKMAN—
John. *Edinburgh.* 1842.
John. & Son. *Sunderland.* 1827-34.
Jonathan. *Leith.* Late 18c. C.
DICKON, Mary Ann. *Manchester.* 1848.
DICKS, William. *Thrapstone.* 1830.
DICKSON—
James. *Glasgow.* 1860.
John (A?). b.1845, *London,* 1868. *Hitchin,* 1879-85. *Bedford,* 1887-1916, then to *Canada,* where d.1931. W.
Samuel. *Dumfries.* 1860.
WHITE & Co. *Philadelphia.* 1837.
DICTERLE. W. *New Market, Ontario.* 1862.
DIDE, Thomas. *London* and *Enfield.* 1662-d.1687. W. & lant. clock. (Usually DYDE.)
DIDISHEIM (G) & RUEFF BROS. *St. Imer, Switzerland.* 1881.
DIECK, Mrs. Fras von. *London.* 1863.
DIEHL, Jacob. *Reading, Pa, USA.* 1776-1858.
DIERKS, H. C. *Hamilton, Canada.* 1856.
DIESCH—
A. *Melcombe Regis.* 1848-67.
Alfred. *London.* 1881.
DIETERLE, William. *Yorkville, Canada.* 1857.
DIETRICH, J. *Carlisle.* 1858.
DIGBY—
John. *Dublin.* 1772-81. Watch case maker.
Nicholas. *Dublin.* 1744. Engraver.
William. *Dublin.* a.1755-83. W.
William. *London.* 1832.
William. *Welwyn.* (B. 1820). 1828.
DIGHTAM—
Edward. *Dunstable.* 1864-77.
H. *Dunstable.* 1864.
H. A. *Fenny Stratford.* 1869.
DIKE, Nathaniel. *London.* a.1656 to Thomas Wheeler. CC. 1663-64.
DIKEMAN, Edward B. *Grand Rapids, Mich, USA.* ca.1840-60.
DILCHER, C. *Leicester.* 1855(-76). Also DILGER.
DILGER—
Blasius. *London.* 1881.
Charles. *Ballymena.* 1868-80.
Constantine. *Leicester.* (1855-)76.
Emil. *London.* 1875-81.
Ignatius. *Bristol.* 1879.
Isadore. *Drogheda.* 1858.
J. P. *Stepney.* mid-19c. C.
J. & Charles. *Ballymena.* ca.1820-40. C.
J. W. *Sheffield.* 1871.
John. *London.* 1863-81.
Joseph. *Bath.* 1875-83.
Joseph. *London.* 1863-75.
Joseph & Charles. *Ballymena.* 1858.
& KAISER. *London.* 1869-75.
Leo. *London.* 1875-81.
Markus & William. *London.* 1875-81.
P. & D. *Wolverhampton.* 1868.
Paul. *Tipton.* 1876.
WINTERHALDER & CO. *Bradford.* 1853.
DILKE, Edward. *Plymouth.* mar.1760.
DILLON—
& Co. (E. Openshaw), T. E. *New York.* 1869. Succeeded to DILLON & TUTTLE.
Edward. *Waterford.* 1858.
F. C. *Kidsgrove.* 1868.
Foster Crewe. *Tunstall, Staffs.* 1876.
Jonathan. *Manchester.* mar.1804. W.
O. S. *London.* 1869-81.
& TUTTLE. *New York.* 1869.
DIMIER, Bros. & Co. *London.* 1875-81.
DIMMER, George. *Portsea.* 1878.
DIMMICK, James. *West Cowes.* 1878.
DINGLEY—
Anthony. *Tenterden.* mar.1693. Clocksmith.
Richard. *Launceston.* 1823. W.
DINGWALL, John. *Woodstock, Canada.* 1861.

DINKELSPEIL & BUMSEL. *Montreal.* 1862.
DINMORE, Richard. *Tadcaster.* 1822.
DINSDALE, Robert. *London.* 1875-81.
DINZELL, John. *Looe,* 1728. rep'd. ch. clock.
DIPPLE, Francis (& Son). *Warwick.* (B. 1795-)1828.
DISBARROW, John. *Ashen.* Early 18c. lant. clock.
DISBOROUGH, John. *Ashen.* mar.1770. C. May be man above?
DISMORE, Thomas. *Liverpool.* 1848-51.
DISON—
  James. *Potton.* 1830.
  Jeremiah. *Potton.* 1839.
  Jeremiah James. *Cambridge.* 1830-46.
  Joseph. *Whittlesea.* 1840-46.
  Thomas. *Biggleswade.* 1830.
DISSMORE, Richard. *York.* 1818.
DISTIN—
  John. *Totnes.* ca.1750.
  & SIEBE. *London.* 1861.
DISTURNELL, William. *London.* a.1750. CC. 1759-84, than *Newhaven, Conn, USA,* later *Middletown.*
DITCHFIELD, T. *London.* 1851-57.
DITTMYIR, John. *Rochester, NY, USA.* 1879.
DIVERTON, James. *Prescot.* 1751. W.
DIXON—
  Christopher. *Hexham.* 1827. W.
  Edward. *Bristol.* 1870-79.
  Eleanor. *Hexham.* 1834-48.
  George. *Addingham.* ca.1830. C.
  Henry. *Salisbury.* 1848-75.
  Isaac. *Philadelphia.* 1843 and later.
  James. *St. Helens.* 1819. W.
  John. *Iver.* 1798. C.
  John. *Menwith* (nr *Settle*). 1822. W.
  John. *Sowerby Bridge.* 1735-d.1786. C.
  Robert. *Hexham.* 1834-48. C.
  Thomas. *Haddington.* 1837.
  Thomas. *Norwich.* 1858-75.
  William. *London.* 1851-69.
  William. *Pickering.* 1807-40. C.
  William. *Sowerby Bridge.* 1866. W.
  William. *St. Helens.* 1851.
DOBB—
  H. *St. Erth* (*Cornwall*). ca.1830. C. See also next entry.
  Henry. *Helston.* b.1809-d.1852.
  William. a.1639 to James Vautrollier, CC. 1646-62.
DOBBIE—
  Alexander. *Glasgow.* 1860.
  George. *Falkirk.* s. of William. b.1765-d.1836. C.
  George. *Glasgow.* 1860.
  Thomas. *Glasgow.* 1828-48.
  William. *Falkirk.* b.ca.1720-1768 (d.pre-1783). C.
  William. *Falkirk.* s. of George. b.1796-1845. C.
  William. *Falkirk.* 1860. May be same as above.
  William. *Linlithgow.* 1860.
DOBBIN, Jean F. *Quebec.* 1890.
DOBBINGS, William. *Leeds.* 1817-26. C.
DOBBS, Henry M. *New York.* 1794-1802.
DOBBYN—
  G. & W. *Dublin.* 1858.
  George & Son. *Dublin.* 1868-80.
DOBEL(L)—
  Ebenezer. *Hastings.* 1855-78.
  Edward. *Gloucester.* 1850-56.
  Frederick. *Yeovil.* 1872-83.
  Jesse (& E.). *Canterbury.* 1851-74.
  Robert. *Yeovil.* 1840-85.
DOBIE, Richard. *Coventry.* 1880.
DOBSON—
  Frederick Frank. *Driffield.* 1817-58. W.
  George. *Leeds.* 1826.
  John. *Bradford.* 1866.
  John. A. *Baltimore, USA.* ca.1860-70.
  John Butler. *Beverley.* 1858.
  Richard H. *Leeds.* 1871.
  & Son. *Beverley.* 1846-51.
  William. *London* (*High Holborn*). CC. 1670-84. T.C. and lant. clock.
DOCKING, C. *Mildenhall.* 1858.
DODD—
  Abner. *Newark, NJ, USA.* 1830-49.
  Edward. *Eydon.* 1854-77.

DODD—*continued.*
  M. H. *Warwick.* 1860-68.
  Mrs. J. E. *Hartlepool.* 1898.
  Matthew Henry. *Bromsgrove.* 1850.
  Philip George & Son. *London.* (B. a.1816-). 1863-81.
  William. *London.* (B. 1805-21). 1828-32.
DODDINGTON—
  John. *London.* a.1685 to Benjamin Wright. May be next man.
  John. *Kilkenny.* d.1699. May be above man.
DODDS—
  John. *Sunderland.* 1834-51.
  Jonathan. *Sunderland.* 1851.
  Joseph. *Kirton* (*Lincs*). 1828-35.
DODEMARE, G. *Caen.* pre-1693.
DODGE—
  Ezra. *New London, Conn, USA.* 1766. mar.1790-98.
  Ezra. *Providence, RI, USA.* ca.1820
  George, jun. *Salem, Mass.* 1779-1837.
  John. *Catskill. NY. USA.* 1818-19
  Nehemiah. *Providence, RI, USA.* 1796-retired ca.1820.
  Seril. *Providence, RI, USA.* 1788-96. Then *Pomfret, Conn, USA.* 1796-1802.
  & WHITAKER. *Providence, RI, USA.* ca.1800.
  (Nehemiah) & WILLIAMS (Stephen). *Providence, RI, USA.* 1799-1800.
DODSON—
  Thomas. *Beeston. Leeds.* 1795-1828. C. & W.
  W. R. *Norwich.* 1846.
DOD(S)WELL—
  Matthew & Co. *Coventry.* 1880.
  Thomas. *Burwash.* ca.1675. lant. clock (cf. DADSWELL).
  Thomas. *Coventry.* 1880.
DODSWORTH—
  James. *Liverpool.* 1879. W.
  John. *London.* 1646. CC. 1648-70.
DOGGET—
  John. *Boston, USA.* 1780. *Roxbury.* 1812. Retired 1850. Clock cases.
  Samuel. *Roxbury* and *Boston, USA.* bro. of John. Retired 1854.
DOHERTY, Denis. *Londonderry.* 1811-27.
DOKE—
  Richard. *Liverpool.* (B. 1825). 1834. W.
  Sarah. *Liverpool.* 1848. May be wid. of Richard.
  William. *Farnworth.* mar.1820. W.
DOLD—
  A. *Chatham.* 1866.
  A. *Rochester.* 1855.
  Alexander. *Maidstone.* 1847-51.
  Anthony & Co. *Grimsby.* 1876.
  Bros. *Colchester.* ca.1855-ca.1865. (Later DOLD & COFFLER.)
  & Co. *Peterborough.* 1877.
  G. *Colchester.* 1866.
  George. *Brighton.* 1870.
  Joseph. *Inverness.* 1860.
  Joseph & Co. *London.* 1869-75.
  L. *Peterborough.* 1869.
  Liberat. *London.* 1869.
  P. *Maidstone.* 1855.
  P. & Co. *Sudbury.* 1865.
  Thomas. *Merthyr Tydfil.* 1848.
DOLE, H. L. & Co. *Haverhill, Mass, USA.* 1865-77.
DOLL, Joseph. *Lancaster, Pa, USA.* ca.1800-20. Then *Harrisburg.*
DOLLAND, ——. *Burton.* ?early 19c. C.
DOLLERY, J. *Dorking.* 1851-55.
DOLPHIN, Robert. *London.* 1772. W.
DOLT, George. *Newcastle-on-Tyne.* 1844.
DOLTON—
  C. *Arundel.* 1851.
  Charles. *Alresford.* 1839-48.
DOMINION. 1914 and later, product of Pequegnat Clock Co.
DOMINY, Captain Nathaniel. *Sag Harbor* and *East Hampton, NY, USA.* 1732-1812.
DOMMETT, John. *Philadelphia.* 1736.
DON, Alexander. *Albany, USA.* 1815-17.
DONAHAY, William. *Nottingham.* 1876.

DONALD—
Adam E. *Newcastle-on-Tyne.* 1850-56.
James. *Newcastle-on-Tyne.* 1834-36.
John. *Newcastle-on-Tyne.* 1847-52.
& Son. *Newcastle-on-Tyne.* 1856. Chrons.
William. *Rhynie, Aberdeenshire.* 1837-60. Also gas-works manager.
DONALDSON—
Andrew. *Airdrie.* 1836.
David. *Edinburgh.* 1822.
George. *London.* 1844.
J. & P. *Glasgow.* 1860.
James. *Meigle, Scotland.* 1837.
John. *Glasgow.* 1839.
John. *North Shields.* pre-1774. C. Later *South Shields.*
DONCASTER, Thomas. *Wigan.* 1756. Free 1763-98. W. Also silversmith and banker.
DONEGAN—
Edward. *Limerick.* 1858.
John. *Dublin.* 1839. Two shops.
Patrick. *Dublin.* 1838-1924. W. & jewellers.
DONERTY, James. *Clonmel.* 1820.
DONEY—
J. B. *Bath.* 1861-66.
William. *Wincanton.* 1801. C.
DONISTHORPE—
——. *Loughborough.* ca.1780-90. C.
Groves. *Hinckley.* ca.1785 (-B.1797). C.
Groves. *Loughborough.* 1828-35.
Joseph. *Bedale.* 1866. C. & W.
Joseph. *Normanton (Leicestershire).* b.1703-d.1774. W.
DONKIN(G)—
Gerard. *Liverpool.* 1848-51. Imported clocks.
Gerard. *Sheffield.* 1834-37.
James. *Liverpool.* 1828-51. Importer and manufacturer of brass and wood clocks and toys.
John. *Sheffield.* 1834. C.
T. *West Hartlepool.* 1898.
DONNE—
& BOWEN. *Swansea.* ca.1800. (*?London* pre-1768.)
Charles. *London.* 1832.
G. *London.* 1811. Watch case maker.
Lewis. *London.* 1875-81.
DONNELLY, Daniel. *Armagh.* 1820.
DONNER, John. *London.* 1851.
DONNEY, W. J. *Liverpool.* ca.1860. W.
DONO, John. *London.* 1863-69.
DOOLITTLE—
Enos. *Hartford, USA.* bro. of Isaac. 1751-retired 1802.
Isaac, sen. *Newhaven, Conn, USA.* b.1721-1800.
Isaac, jun. *Newhaven, Conn, USA.* s. of Isaac, sen. 1759-1821.
Lucius F. *New York.* 1854.
DORAN & CO. *Montreal.* 1866.
DORCHESTER, Clock & Bell Foundry. *Dorchester, Mass, USA.* 1830s.
DORE, James. *Sandown, Isle of Wight.* 1878.
DORER—
F. *King's Lynn.* 1858-65.
Iver. *London.* 1875.
L. & P. *Greenwich.* 1866.
P. *Greenwich.* 1874.
& Son. *King's Lynn.* 1875.
DOREY, Joseph L. *East Caln, Pa, USA.* 1824. W.
DORFLINGER, Joseph. *Philadelphia.* 1837.
DORION—
Et Cie. *Montreal.* 1879.
L. & Co. *Montreal.* 1866.
DORLEY, William J. *London.* 1857. Clock cases.
DORMAN, Richard. *Milton.* 1823-32.
DORMER, Theodore. *Colchester.* d.1806. W.
DORNHECK, N. *Amsterdam.* Late 18c. W.
DORON, John. *Vincentown, NJ, USA.* ca.1849. W.
DORELL, F. *London.* 1811. Clock engraver.
DORRER—
E. & Co. *Croydon.* 1862.
Eagen & Co. *Dover.* 1845.
DORRINGTON, Theophilus Lutey. *Truro.* 1873. W.
DORSET(T)—
Gregory. *London.* a.1651 to Robert Grinkin, later to William Rogers, CC. 1662-77. *Fleet Street.*
Henry. *Epsom.* 1878.

DORSEY, Philip. *Baltimore, USA.* ca.1800.
DOSER, Antoni. *Ellwengen, Germany.* ?late 17c. C.
DOSSON, William Taylor. *Neath.* 1844-50.
DOTTER—
Charles. *Pentre Rhondda.* 1875-87.
Elias. *Newry.* 1846-58. C.
Joseph & Co. *Pontypool.* 1858.
M. *Newry.* 1840.
DOTY—
George. *Buffalo, NY, USA.* 1835.
John. F. *Rochester, NY, USA.* (From *Albany*). 1844.
DOUGAL(L)—
——. *Glasgow.* 1849. (prob. Alexander).
Alexander. *Glasgow.* 1846.
Alexander. *Strathaven.* 1836.
James. *Glasgow.* 1860.
John (& William). *Kippen, Scotland.* 1836(-60).
DOUGHERTY—
George. *Dublin.* 1780. C.
John Heywood. *Huddersfield.* 1866-71. W.
DOUGHTY—
——. *Derby.* ?18c. br. clock.
Henry. *London.* 1869-81.
Joseph. *Worksop.* 1876.
S. H. *New York.* 1854.
Thomas. *Kidderminster.* 1876.
Tobias. *London.* a.1688 to Henry Child. CC. 1696. (prob. left by 1697).
William. *Birmingham.* 1880.
William. *Ollerton, Notts.* 1876.
DOUGLAS(S)—
Alexander. *Bowmore, Islay.* 1837-60.
Alexander. *Strathaven.* 1860.
Daniel. *Glasgow.* 1860.
Daniel. *Whitehaven.* 1828-34. Also jeweller.
George. *Bonhill, Scotland.* 1837.
George. *Cambusnethan.* 1860.
George. *Dumbarton.* 1860.
George. *Holytown, Scotland.* 1847.
J. *Ryde.* 1867.
J. *West Cowes.* 1859.
J. D. *Havant.* 1848.
James. *Beamsville, Canada.* 1851.
James. *Chertsey.* 1828.
James. *Egham.* 1839-78.
James. *Glasgow.* 1860.
James. *St. Catharines, Canada.* 1857-78.
Robert & Co. *Liverpool.* 1781-1828. W.
Samuel & Robert (jun.). *Liverpool.* 1814.
& Son. *Greenock.* 1842-60.
Thomas. *Niles, Mich, USA.* ca.1870.
W. H. *Stourbridge.* 1868-76.
Walter. *Galston, Scotland.* 1837.
Walter. *Hamilton.* 1860.
William. *Anstruther.* 1860.
DOULD, E. *Derby.* 1852-62.
DOVE—
E. C. *Rochester.* 1874.
John. *Diss.* 1846-75.
John. *London.* ca.1780. W.
Stephen. *Goole.* 1834-37. W.
Stephen. *Hull.* 1846. W.
DOVER—
Herbert James. *Brighton.* 1878.
Thomas. *Dayton, O, USA.* 1830s.
DOVEY, George. *London.* 1875-81.
DOW—
Alexander. *Bromley.* 1845.
Alexander. *Glasgow.* 1837.
Alexander. *London.* 1857-75.
James. *London.* (B. a.1792-)1828-39. Watch case maker.
John. *Glasgow.* 1837.
S. *Martham.* 1846.
William. *Yarmouth.* 1875.
DOWARD—
H. *Widnes.* 1851-58. W.
Henry. *Farnworth.* mar.1786. W.
William. *Farnworth.* 1834-51.
DOWDING, William R. *Andover.* 1878.
DOWDNEY, Burrows. *Philadelphia, USA.* 1768-71.

DOWEL(L)—
James. *Wigtown.* 1869-79.
John. *Wigton.* 1879.
DOWIE, James. *Tain.* 1860.
DOWLE, Robert. *New York.* 1793.
DOWLEY, Robert. *Hexham.* 1781-89.
DOWLING—
G. R. & B. & Co. *Newark, NJ, USA.* 1832-38.
James. *Birmingham.* 1842-68.
Mortagh. *Dublin.* 1700-d.1746. W.
William. *Liverpool.* 1834.
DOWN—
——. *Sheffield.* ca.1800. C.
John. G. *Wrotham.* 1847-55.
DOWNAM, Samuel. b.1723 at *Newton, Suffolk.* Worked
at *Colchester,* then at *Sudbury* 1751-53, then *Colchester* again 1753-68. C.
DOWNE, Theodore. *Colchester.* ca.1735-ca.65. W.
DOWNES—
John. *London.* 1828.
John. *London.* 1875(-81) (or DOWNEY).
John. *Swansea.* Early 19c. C.
Robert. *London.* 1828-44.
DOWNHAM, William. *Dublin.* ca.1800. W.
DOWNING—
Arthur. *Manningtree.* 1866-74.
George R. *Richmond, Va, USA.* ca.1819.
Humfrey. *London, Chancery Lane.* (Not in CC.).
1646-64.
Isaac. *Newcastle-on-Tyne.* 1811. C.
John. *Liverpool.* (B. 1770-)1790-1816. W.
(George) & PHELPS (Silas). *Newark, NJ, USA.*
ca.1815. To *New York.* 1825.
Thomas. *Colchester.* Successor to J. B. Sparke. 1866-
82. W.
W. *Walthamstow.* 1874.
DOWNS—
Anson. *Cincinnati, O* and *Plymouth, Conn, USA.*
1797-1822. To *Bristol.* 1829-76. Then *Waterville.*
1876.
Ephraim. *Bristol, USA.* bro. of Anson. b.1787,
mar.1822. To *Bristol.* 1825. Retired 1842. d.1860.
Joseph. *Rugby.* 1828.
Robert. *Brighton.* 1870-78.
Thomas. *Louth.* ca.1730-40. C.
DOWSETT, Charles. *Margate.* (B. mar.1799)1800-26.
W.
DOWSING, William. *Halesworth.* 1790-1855.
DOWSON—
& ATKINSON. *Bath.* (B. 1791-)1792-95.
John. *London.* 1765(-73). W.
DOYLE—
Henry. *Belfast.* 1843-49.
J. *Birmingham.* 1854.
James. *Newry.* 1784.
James, jun. *Newry.* 1804.
Joseph. *Birmingham.* 1854-60.
Michael. *Cork.* mar.1736-ca.1750. C.
Philip. *Halifax, Nova Scotia.* 1866.
DRAGE—
John. *Leighton Buzzard.* 1839.
John. *Luton.* 1830.
DRAHKRUB, ——. *Paris.* ca.1760. W.
DRAKE—
Edwin. *Halifax.* 1845-71.
George. *Huddersfield.* 1871.
George. *London.* 1875.
Henry Joseph. *London.* 1881.
John. *London.* BC. 1605. Constant rebel against
authority of CC. -1662. *Westminster.*
Richard. *Beaminster.* (B. 1750-)1824-30.
DRAKEFORD—
E. *Congleton.* 1857.
Edward. *Congleton.* 1828-48.
J. *Congleton.* 1857-65.
John. *Congleton.* 1828-48.
DRAPER—
Charles. *London.* 1857.
Henry. *London.* 1844.
John. *Chelmsford.* pre-1760-d.1796.
John. *Youghal, Ireland.* 1780. W.
Mark. *Witham.* 1729-81.

DRAPER—*continued.*
Thomas. *London.* 1875-81.
DRAVIS, Henry. *Halifax, Nova Scotia.* 1866.
DRAWBAUGH, Daniel. *Eberly's Mills, Pa, USA.* 1874.
Electric clocks.
DRAY, Jesse. *London.* 1832.
DRAYCOTT, Samuel. *Burnham Market.* 1830-36.
DRAYSON, Douglas. *Sandwich.* 1862-91.
DRAYTON—
Isaac. *Chardstock.* ca.1800.
James. *Taunton.* 1875.
Mrs. M. *Chard.* 1861-66.
Thomas. *Chard.* 1875-83.
DREHER—
George. *Axbridge.* 1883.
Gottlob. *Bristol.* 1879.
DRENEY, Samuel. *Girvan.* 1850.
DRESCH(L)ER (also spelt DRESCHAR, DRESSLER,
etc.).
George. *Landport.* 1878.
John. *Hull.* 1823-58. C.
K. *Kingston, Surrey.* 1851-66.
Pius. *Hull.* 1838-58.
& ROBOLD. *Paisley.* 1836.
& Son. *Kingston, Surrey.* 1878.
DREW—
——. *Downham.* ca.1860. C.
Edward. *London.* a.1683 to Jeffrey Bayley. CC.
1692-97.
H. *Ventnor.* 1848.
DREYFOUS, Joseph. *Philadelphia, USA.* 1825. W.
DRIELSMA—
Isaac Jones (or Isaac James). *Liverpool.* 1834-51.
Morris. *Liverpool.* 1848-51.
DRILL—
George. *Naas, Ireland.* 1858. W.
William. *Waterford.* d.1726. W.
DRING, Thomas. ?from *London. West Chester, Pa,
USA.* ca.1790. Returned to *England* pre-1800.
DRINKWATER—
J. *Bromsgrove.* 1868.
J. *Stockport.* 1865.
DRIVER, Henry. *Keighley.* 1866-71.
DROCOURT, Pierre & Alfred. *Paris.* ca.1860-89. C.
DROPSIE, M. A. *Philadelphia.* 1842-49.
DROSSADE (or DROSSATE), Samuel. *London.* CC.
1675-97. (Free of the Shipwrights Co.).
DROWN—
Charles Leonard. *Newburyport, Mass, USA.* 1824-60.
John Boardman. *Newburyport, Mass, USA.* 1826-50.
With brother Charles. 1850-60.
Richard. *Newburyport, Mass, USA.* b.ca.1795-1810.
C.
**DROWSET (or DRUSSETT, DROSSUT,
DROESHOUT),** John. *London.* CC. 1635-47.
DROY, Daniel. *Quebec.* 1890.
DROYTON (or DRAYTON), J. *Chardstock.* 1855-67.
DROZ—
Charles & Sons. *Philadelphia.* 1806-14.
Ferdinand Humbert. *Pittsburgh, Pa, USA.* s. of Humbert. ca. 1840. Then *Cleveland, O.* ca.1850.
Hannah. *Philadelphia, USA.* 1842.
Humbert. *Philadelphia.* 1793-d.1813.
John. *Cincinnati, O, USA.* From *Switzerland.* 1824.
DROZ dit BUSSET. *Bordeaux.* ca.1825. W.
DRUGEON, ——. *Paris.* 1878.
DRUKKER—
Mayer. *London.* 1857-69. Swiss.
Simon. *London.* 1863-75.
DRUM(M)OND—
——. *London.* 1830. W.
Alexander. *Perth.* 1860.
Francis. *Shilling Hill, Alloa.* 1837.
Levin J. *Baltimore, USA.* ca.1875.
Robert. *Stirling.* 1860.
DRURY—
Charles William. *Banbury.* 1823-30. C. & W.
Edward George. *London.* 1875-81. Clock cases.
Francis. *London.* 1857-75.
George. *London.* 1832.
George Edward. *London.* 1851-57. Clock cases.
William. *Banbury.* 1773-ca.1810. C. & W.

DRURY —*continued.*
William. *Gainsborough.* ca.1730-40. C.
DRYDEN, George. *London.* 1828-69.
DRYER, C. *Louisville, Ky, USA.* ca.1850. C.
DRYSDALE—
Thomas. *Quebec.* 1844-49.
William. *Falkland, Fife.* ca. 1810-32. C.
William. *Philadelphia.* 1816-51.
William, jun. *Philadelphia.* 1842-45.
DRYSDALL, Daniel. *London.* 1869.
DUBBERLEY—
Sarah. *Monmouth.* 1858-80.
Thomas. *Monmouth.* 1844-52.
DUBEC, ——. *Paris.* ca.1800.
DUBERY, Edward H. *London.* 1875.
DUBOIS—
——. *Quebec and Montreal.* 1730-60.
Edward. *Le Locle.* ca.1860.
& FOLMAR. *New York.* 1816.
Gabriel. *New York.* ca.1810.
P. C. *Alameda, Cal, USA.* 1845-1925.
DUBOURK, Abraham. *London.* 1635.
DUCHESNE, J. B. *Paris.* 18c. br. clock.
DUCKER—
Henry. *London.* 1832-44.
Thomas. *Buckhurst Hill, Essex.* 1874.
DUCKETT, Charles. *Stoke-on-Trent.* 1850-68.
DUCKWORTH—
John. *Halifax.* 1652.
Richard. *Halifax.* 1654-d.1677.
DUCOMMUN, Paul & Co. *Travers (Neuchâtel).* 1889.
DUDLEY—
Benjamin. *Newport, RI, USA.* 1785-1820.
C. *Newport, Isle of Wight.* 1848-67.
S. *Portsmouth.* 1878.
Thomas, *Boston, Mass, USA.* 1807.
Watch Co. *Lancaster, Pa, USA.* 1923-24.
William. *Newport, Isle of Wight.* 1878.
DUDMAN, Thomas. *London.* 1851-63.
DUDSON (sometimes DUTSON), Simon. *London.* a.1647, CC. 1654-56.
DUDUICT or DUDWITT, Jaques or James. *London.*
Prob. from *Blois.* 1622 alien at Mr. Garrett's in St. Martyns.
DUEBER—
-HAMPDEN Watch Co. *Canton, O. USA.* 1888-1927.
John C. *Newport, Ky, USA* and *Canton, O. USA.* 1877-ca.1925.
Watch Case Mfg. Co. *Newport, Ky, USA.* ca.1877-88.
DUERRE, Charles William Rudolph. *London.* 1863-81.
DUESBURY, Samuel. *Manchester.* 1822. Also DEWSBURY.
DUFF—
Daniel. *Paisley.* 1836.
George. *Philadelphia.* 1837.
George C. *New Bern, NC, USA.* 1846.
James. *London.* (B. CC. 1819-)1828.
James (& Son). *London.* 1839-44(51-61).
John. *Turriff.* 1860.
& MILLAR. *Paisley.* 1860.
Thompson. *Donegal.* 1846.
William & Co. *Liverpool.* 1834-48.
DUFFIELD—
Francis. *London.* 1851-69.
William. *Ebbw Vale.* 1850-71.
DUFFT, Louis. *Nicolet, Canada.* 1857.
DUF(F)NER—
Bennet. *Coventry.* 1850-80.
C. *Tadcaster.* 1871.
Constantine. *Leicester.* 1876.
Germanus. *London.* 1881.
Henry. *Newport (Mon).* 1880. Prob. same as next man.
Herman. *Newport (Mon).* 1871-99.
Nikolaus. *Brighton.* 1870.
& Son. *London.* 1881.
Vincent. *Cincinnati, O, USA.* 1850.
DUFOUR, John Moses. *Dublin.* a.1748.
DUFRESNE—
D. P. *Montreal.* 1875-79.
J. B. R. *Montreal.* 1877.
DUGAST, Stephen. *Dublin.* 1683. Watch cases.
DUGGEN, ——. *Bristol.* ca.1800. C.

DUGGIN(S)—
E. *Leamington.* 1880.
George. *London.* 1863.
DUGGLEBY, W. *Kirkby Moorside.* ?mid-19c.
DUGMORE & FOSTER. *Birmingham.* 1816. Dial makers.
DUHME & Co. *Cincinnati, O, USA.* ca.1850.
DUKE—
A. *Southsea.* 1867.
Joseph. *Denbigh.* 1844.
Thomas. *Birmingham.* 1868.
DUKES—
George. *London.* 1839-44.
William. *Thrapstone.* 1824.
DULIN, William. *London.* 1832.
DULLARY, William. *Hartley Witney.* 1830.
DULTY, John. *Zanesville, O, USA.* ca.1810.
DULY—
W. *Hurstmonceaux.* 1862-78.
William. *Hurstmonceaux.* 1839.
DUMAS—
A. *Paris.* 1880.
O. *Paris.* 1855-d.1890.
DUMBELL—
John. *Liverpool.* 1828-34.
John. *Wolverhampton.* 1842-76.
Joseph. *Rochdale.* 1828-34.
Nathaniel. *Wolverhampton.* 1828-35.
Thomas. *Rochdale.* 1822-24. W. and jeweller.
William. *Prescot.* 1828-58.
DUMESNIL, Anthony. *Lexington, Ky, USA.* 1818. (Poss. from *Boston,* where one such B. 1796-1807).
DUMOTET, J. B. *Philadelphia.* Early 19c.
DUMTOT, Joannes. *Duffel, Belgium.* Mid-18c C.
DUMVIL(L)(E)—
James. *St. Ives, Hunts.* 1839-64.
John. *Cambridge.* 1830-75.
John. *Hulme.* s. of Nathaniel. mar.1857.
John. *Olney.* 1798-99. W.
John. *St. Ives, Hunts.* 1830.
Nathaniel. *Hulme.* 1836-57.
Nathaniel. *Stockport.* 1834. May be same as above.
DUN, George. *Douglas, Scotland.* 1860.
DUNBAR—
Butler. b.1791, to *Bristol, USA.* ca.1810, then *Springvale, Pa,* 1815, later returned to *Bristol* till 1852.
Edward Butler. *Bristol, USA.* 1842-ca.1885.
Edward L. (Colonel). *Bristol, USA.* s. of Butler. 1815, mar.1840-72.
George. *Cootehill.* 1805-24.
JACOBS & WARNER. *Bristol, USA.* ca.1850.
(Butler) & MERRIMAN (Titus). *Bristol, USA.* ca.1810. C.
& WARNER. *Bristol, USA.* 1850-52.
DUNCAN—
——. *Glasgow.* 1849.
Archibald. *Armagh.* 1824.
D. *Cupar, Fife.* 1835-50.
Fleming. *Omagh.* 1824.
George. *Banff.* 1827-46.
James. *Aberdeen.* 1860.
James. b.1814. *Scotland* to *Belfast.* mar.1844-d.1864.
James. *Dumfries.* 1855-60. W.
James. *Londonderry.* 1865-68. W.
James. *London.* 1828.
John. *Armagh.* 1820-44.
John. *Killyleagh.* ca.1820-30.
John junior. *Helensburgh.* 1860.
& KLUMP. *Philadelphia.* ca.1910. C.
Patrick. *Strabane.* 1784.
R. *Liverpool.* 1805. W.
Thomas. *Dalbeattie.* ca.1840.
Thomas. *Newcastle-on-Tyne.* 1827-58.
DUNHILL, Joseph. *Doncaster.* 1871.
DUNKLEY—
& FENN. *London.* 1839.
John. *Greenwich.* 1847.
John. *London.* 1828-32.
Thomas. *London.* 1832-44.
William. *London.* (B. 1805-20). 1832.
DUNKS, David. *Accrington.* 1848.
DUNLAP, Archibald. *New York.* 1802.

DUNLOP—
——. *Newcastle-on-Tyne.* Mid-19c. C.
Charles. *Ballynahinch.* 1772, later *Lisburn.* C.
Edward. *Ballynahinch.* ?late 18c. C.
DUNMICK, J. *Sandown.* 1867.
DUNN—
Cary. *Morristown, NJ.* 1776-82. *Newark, NJ.* 1782-88.
Charles, jun. *London.* 1851-81.
Edward. *Dublin.* a.1752-1800. W. & C.
Francis. *Dublin.* 1794-1815. W.
James. *Maguire's Bridge, Co. Fermanagh.* 1846.
John. *Hull.* 1813-34. W.
John. *Howden.* 1844-51.
Mrs. Esther. *Elham.* 1874.
Patrick. *Dublin.* 1858.
Robert. *London.* 1857.
S. E. *Lyminge.* 1847.
Samuel Edward. *Elham.* ca.1819-d.1874.
Thomas. *Bedale.* 1823.
Thomas. *Berwick-on-Tweed.* (B. 1820-) 1827.
William. *London.* 1828-32.
DUNNE, Robert. *New York.* ca.1840.
DUNNING—
(Julius N.) & CRISSEY (Elnathan F.). *Rochester, NY, USA.* 1847.
& CURTIS. *Concord, Mass, USA.* ca.1812-17.
J. L. *Concord, Mass, USA.* 1811-17. C.
John. *Chester.* 1828.
DUNSEATH, David. *Randalstown.* 1846. C.
DUNSFORD, ——. *Plymouth.* ca.1790-ca.1800. br. clock.
DUNSTALL—
J. *Rainham.* 1866.
James. *Gravesend.* 1874.
DUNSTER—
Matthew. *Bodmin.* mar.1724. C.
Roger. *Amsterdam.* Partner with Christopher Clarke. ca.1722. Alone after about 1730. d.1747.
DUNSTONE, James. *Falmouth.* 1782. C. (B. has DUN-STAN 1784).
DUNTON, J. *Dublin.* ca.1770. C.
DUPAN, Marc. *Geneva.* ca.1840. W.
DUPEE, Odrian. *Philadelphia.* 1735. W. *Boston.* 1739-40.
DUPLAIN, Jules. *Montreal.* 1888.
DUPLOCK & WIGGINS. *London.* (B. 1817-25). 1828.
DUPUY, Jean. *Bordeaux.* 1750. Lant. clock.
DU PUY, Jean Pierre. *London.* 1703. W.
DUQUET, Cyrille. *Quebec.* 1841-1922. (Also DERQUET).
DURAND, Asher Brown. *Boston, USA.* 1796-1886.
DURANT—
——. *London.* 1829.
J. P. *Ventnor.* 1859.
John Pope. *Torpoint, Cornwall.* 1873. W.
DURDEN, John. *London.* 1869-81.
DUREN, H. *New York.* ca.1820-40.
DURHAM, William. *Thurso.* 1837.
DUROCHER, Jean-Batiste. *Montreal.* 1875.
DURRAN—
B. P. *Byfield, Northants.* 1847-69.
Benoni Pearson. *Byfield.* 1877. Clock cleaner (prob. above man.).
Eustace. *Banbury.* Mid-19c. C. & W.
Eustace. *Brackley, Oxon.* 1877.
J. *Towcester.* 1864-69.
James Hopkins. *Banbury.* 1832-54. C. & W.
Mrs. Elizabeth. *Towcester.* 1877.
& SMITH. *Brackley, Oxon.* mid-19c. C. & W.
DURRANT—
& ARNOLD. *London.* 1863.
Brothers. *Croydon.* 1878.
G. *Yarmouth.* 1846.
J. *London.* 1863.
John. *London.* 1869-75.
John James (& Son). *London.* 1863-75 (81).
P. *Rouen.* ca.1620. cf. DURAND.
Richard. *Beccles.* 1823-55.
Richard. *London.* 1844-69.
William. *London.* 1875-81.
DURRE, Rudolph. *London.* 1857.
DURSTON, A. *North Petherton, Somerset.* 1875.
DURYEA, W. *New York.* 1854.

DUSGATE—
Francis. *Holt.* 1784-95.
Francis. *Holt.* 1830-36. May be same man as above, or son?
DUTACHE see DESTACHE.
DUTENS, Charles. *Philadelphia* from *London.* 1757.
DUTERREAU, John. *London.* 1828.
DUTTON—
Abraham. *Hanley.* 1835-76. C.
Charles. *Tipton.* 1876.
David. *Mount Vernon, NH, USA.* 1792-1882. C.
David & Sons. *Mount Vernon, USA.* ca.1840. C.
Edward. *Oldbury.* 1868-76.
Edward. *Tipton.* 1842-60.
Fanny. *Walsall.* 1850.
Frederick. *Bromsgrove.* 1828-50.
George Edwin. *Bromsgrove.* 1860.
Matthew. *London.* 1828-32.
Mrs. Charlotte. *London.* 1857. (?wid. of Robert William).
Mrs. S. *Tipton.* 1868.
Reed. *Milford, NH, USA.* ca.1840. C.
Robert William. *London.* 1844-51.
S. *Kidderminster.* 1860-76.
Samuel. *Liverpool.* 1790. W.
T. *Banbury.* 1841-42.
T. *Walsall.* 1860.
T. F. *Longton.* 1860.
Thomas. *Walsall.* 1828-42.
William. *Stourbridge.* 1872-76.
DUVAL, Lewis. *London.* 1851.
DUVERDY & BLOQUEL. *St. Nicolas d'Aliermont.* 1867 and later—
DWERRIHOUSE—
& BELL. *London.* 1839-51.
BELL & Son. *London.* 1857-63.
& BELL. *London.* 1869-81.
CARTON & Co (evidently error for CARTER & Co). *London.* (B. 1825) 1828.
& OGSTON. *London.* (B. 1836)1839-44.
OGSTON & BELL. *London.* 1832.
DWIGHT—
James Adams. *Montreal.* 1818-47. Partner with George Savage, then with Martin Cheney, then one of Twiss Brothers.
& SAVAGE. *Montreal.* 1818.
& Son. *Montreal.* 1842-45.
DYBALL—
E. *Norwich.* 1865.
Edward. *Barnet.* 1874.
DYBORROW, ——. *Ashen.* ca.1715. Lant. clock.
DYDE, Thomas. *London.* See DIDE.
DYE—
J. *Lowestoft.* 1846.
John R. *Lowestoft.* 1839-79.
Robert. *London.* 1851.
William. *Fayetteville, NC, USA.* ca.1805.
DYER—
Ephraim. *Bideford.* ca.1690-ca.1715. C.
Ephraim. *Collingwood, Canada.* 1857-63.
Everall. *Salop.* 1791. W.
Giles. *Boston, USA.* 1673. Clock keeper.
George Jonah. *London.* 1875.
George Wild. *Boston, USA.* Partner with John Sawin till 1828.
Harrison D. *New York.* 1827.
J. *Thame.* ca.1825-30. C.
James. *Sutton.* 1878.
John. *London.* 1832-39. Watch cases.
John Benjamin. *London.* 1869-81.
Joseph. *Concord, Mass, USA.* 1816-20, later at *Middleburg, Vt.*
R. H. *St. Catherines, Canada.* 1877.
WADSWORTH & Co. *Augusta, Ga, USA.* ca.1838-43. C.
Warren. *Lowell, Mass, USA.* 1832.
William. *Barnstaple.* 1740-91. C.
DYFFER, Daniel. *Reading, Pa, USA.* ca.1820. C.
DYKE—
Albert. *New York.* 1926. Chrons.
George. *Southampton.* 1878.
William. *Liverpool.* 1777-90. W.

DYSART—
James. *London.* 1825. W.
James P. *Lancaster, Pa, USA.* ca.1850.
DYSON—
& BROOKHOUSE. *Sheffield.* 1834-37. W.
George. *Manchester.* mar.1847-58. W.

DYSON—*continued.*
Humphrey. *Manchester.* 1824-28.
Jacob. *Farnworth.* mar.1787-95. W.
John. *Leeds.* 1871. (Firm still continues.)
Thomas. *Sheffield.* 1871.
Zacheus. *Sheffield.* (B. 1817-)1822.

E

EACOTT, John. *Bruton.* 1883.
EADE, R. *Steyning.* 1862.
EADEY, Joseph. *Kettering.* 1824.
EAGLE—
  F. *Mildenhall.* ca.1770. C. (Prob. too early dating of next man).
  F. A. *Mildenhall.* 1846.
  James. *Soham.* 1875.
EAGLESFIELD, J. C. *Devizes.* 1848-67.
EAGLESON, James. *Belfast.* 1827.
EAMES—
  Brothers. *Bath.* 1875-83.
  George. *Freshford, Bath.* 1861-83.
  H. *New Alresford.* 1848.
  H. *Petersfield.* 1859-67.
  Henry. *Alresford.* 1839.
  J. *Bath.* 1856.
  Jacob (& Sons). *Bath.* 1861-66(75-83).
  Richard. *Petersfield.* 1813-48. C. & W.
  T. *Chippenham.* 1859.
  Thomas. *Bath.* 1856-66. C.
  Thomas. *London.* 1875-81.
EARDLEY—
  Daniel. *Aberavon.* 1844-48.
  Daniel. *St. Davids.* 1835.
EARL(E)—
  Charles George. *Bristol.* 1879.
  W. *Bristol.* 1856.
  Reginald H. *St. Johns, Newfoundland.* 1871-85.
  Thomas. *Oxford.* a.1796-1812.
  Thomas. *Rye.* 1737. W.
EARLEN. See EARLING.
EARLES, Henry. *London.* 1857-81.
EARLEY, Robert. *London.* 1792. W.
EARLING, T. *Thetford.* ca.1800. C.
EARLY, Henry Thomas. *London.* 1863-81.
EARNSHAW—
  John. *Brighouse.* 1866.
  Thomas, jun. *London.* 1839-69.
EARP—
  Robert. *Philadelphia, USA.* 1811. (May be from *Castle Donnington, Leicestershire,* where one such known late 18c.).
  Thomas. *Kegworth.* ca.1800. C.
EARTHY, Joseph. *Manningtree.* 1855-74.
EASBY—
  R. *Sunderland.* 18c. ?C.
  Robert. *Durham.* 1858-60.
EAST—
  Edward I. Goldsmith & C. & W. b.1602. *Bedfordshire,* a.1618. *London,* CC. 1632-d.1696. Shop in *Fleet St.* Clockmaker to Charles II. Probably a Catholic.
  Edward II. *London.* s. of Edward I. 1688 (not certain if he was a maker).
  Edmund. *London.* s. of Edward II. a.1688, CC. 1696.
  James. *London.* s. of Edward I. Clockmaker to the Queen 1662-88.
  Jeremy (I). *London.* bro. of Edward I. ca.1600. CC. 1640-67. W.

EAST—*continued.*
  Jeremiah (Jeremy). *London.* a.1653 to James Seaburne—may not have completed apprenticeship.
  John. *Dublin.* a.1656 to Daniel Bellingham. (Said to be son of John E. of *London,* W.).
  John. *London.* Evidently working ca.1656, father of above, but nothing known about him.
  John. *London.* a.1654 to Edward East—seems not to have completed apprenticeship.
  Peter. *London.* (B. 1689-)1697.
EASTBURN, James. *Bradford.* d.1751. C.
EASTCOTT, Richard. *Exeter.* Free. 1803. C.
EASTERBROOK, ——. *Torquay.* ca.1830. C.
EASTERLEY, James. *Cork.* 1805-10.
EASTERLY, John. *New Holland, Pa, USA.* 1825-40.
EASTES, William. *Sandgate.* 1845.
EASTGATE, James. *London.* 1875.
EASTHORPE, Thomas. *Birmingham.* 1860-80, also clock cases.
EASTMAN—
  Abel B. *Concord, NH, USA.* ca.1800-06, *Belfast, Me,* ca.1806-ca.21, *Haverhill, Mass.* 1816-21.
  (Robert) & CAREY (James). *Belfast & Brunswick, Me, USA.* 1806-09.
  CLOCK Co. *Chelsea, Mass, USA.* 1886-88.
  Robert. *Belfast* and *Brunswick, Me, USA.* 1805-08.
EASTMOND, William P. *Redhill.* 1878.
EASTON—
  Alexander R. *Aberdeen.* 1860.
  James. *Glasgow.* 1860.
  John. *Petworth.* (B. 1784-) 1828.
EASTWOOD—
  H. *Stockton.* 1898.
  James. *New Glasgow, Nova Scotia.* 1848-80.
EATON—
  Elon. *Grand Rapids, Mich, USA.* ca.1860.
  John H. *Boston, USA.* 1809-10.
  Joseph. *Prescot.* 1824-25. W.
  Samuel A. *Boston, USA.* 1825-42.
EAVES—
  Alfred. *Birmingham.* 1850-80.
  Arthur Frederick. *Birmingham.* 1868-80. Also clock cases.
  Charles. *Birmingham.* 1835-50.
  Charles. *Coventry.* 1880.
  Francis. *Birmingham.* 1880.
  Samuel. *Birmingham.* 1868.
  Samuel Charles. *Birmingham.* 1880.
  Samuel Spilman. *Birmingham.* 1880.
  T. senior. *Birmingham.* 1865. Dial maker.
  T. A. *Birmingham.* 1865. Dial maker.
  Thomas. *Birmingham.* 1854.
  William. *Coventry.* 1854-68, then *Montreal, Canada.* 1868-76.
EAYRE, Thomas. *Kettering.* 1720-d.1757. C. T.C. & bellfounder.
EAYRES, George Walter. *London.* 1869.
EBARALL, John. *Nantwich.* 1828.
EBBEN, William. *London.* (B. 1817-40) 1828-44.

69

EBENSTEIN, Salomon. *London.* 1863.
EBERMAN—
Charles. *Lancaster, Pa, USA.* s. of Joseph. ca.1820.
Joseph II. *Lancaster, Pa, USA.* Poss. s. of Joseph I. 1857-60.
EBERT, Isaac. *Steyr.* ca.1650.
EBORALL—
Edmund. *Llangollen.* 1840-44.
John. *Bromsgrove.* 1872-76.
John. *London.* 1845.
John senior. b. *Coventry.* At *Warwick* ca. 1807 -d. pre-1851
EBSWORTH, John. *London.* a.1657 to Richard Aymes, CC. 1665, d.1699. C. at the Crossed Keys in Lothbury.
EBURN, William. *London.* 1839.
EBY—
Christian. *Manheim, Pa, USA.* ca.1799-1837. C.
George. *Manheim, Pa, USA.* s. of Christian. 1830-60. C.
Jacob. *Manheim, Pa, USA.* s. of Christian. 1830-60. C.
ECALLE, Auguste. *Paris.* 1878-1900. C.
ECCLES, George. *Liverpool.* 1894. C.
ECKEL, Alexander Perry. From *Front Royal, Va, USA* to *Greensboro, NC.* 1821-1906.
ECKERSLEY, Richard. *Chowbent (Lancs).* 1848.
ECKERT, John. *London.* 1863-69.
ECKMULLER, Johann. *Augsburg.* 16c. Table clock.
ECLIPSE. 1914 and later, product of Pequegnat Clock Co.
EDE, John. *Camelford.* 1844-73, then *Stratton.* 1873.
EDER, Franz. *Himberg.* 1825. C.
EDEY—
Edward. *Brixworth.* 1869-77.
James. *Edinburgh.* 1860.
John. *Brixworth.* 1877.
Josiah. *Kettering.* 1830-41.
Stephen. *Kingston-on-Thames.* 1839-55.
EDGAR—
John. *Liverpool, Nova Scotia.* 1864.
John and Robert. *Whitehaven.* 1834.
Joseph. *Halesworth.* pre-1740.
Robert. *Whitehaven.* 1828.
Robert (?II). *Whitehaven.* 1858-69.
S. *Gillingham, Dorset.* 1855-67.
William. *London.* 1857-69.
Worthington. *London.* 1875-81. (Successor to William.)
EDGCOMB, William. *Liskeard.* 1730-ca.1740. C.
EDGECUMBE—
Edwin. *Dartford.* 1866-74.
John G. *Bournemouth.* 1878.
Nathan. *Bristol.* 1840-42.
Thomas. *Coleford.* 1861-83.
EDGERLY, Sylvester. *Roxbury, Mass, USA.* ca.1820-40. C.
EDINGTON, James. *Dunse.* 1860.
EDKINS, James. *London.* 1832-39.
EDMEAD—
Frederick. *London.* 1869.
William. *London.* 1857-81.
EDMISTON, John. *Ballymena.* 1806.
EDMOND, William. *Philadelphia, USA.* 1848.
EDMONDS—
& HAMBLETT. *Boston, USA.* ca.1865. Electric clocks.
John. *London.* (B. 1820-30) 1832.
John. *London.* 1869-81.
T. *Bath.* 1861-75.
Thomas. *Chichester.* a.1742 to John Stripe, C.
W. *Madeley.* ca.1750-ca.1760. C.
EDMONDSON, John. *Bingley.* 1866.
EDMUNDS, E. *Battersea.* 1866.
EDMUNDSON, Thomas. *Warrington, Pa, USA.* 1794-1800. C.
EDSON, Jonah. *Bridgewater, Mass, USA.* 1792-1874.
EDWARD—
George. *Glasgow.* 1860.
George & Son. *London.* 1875-81.
Robert. *Holyhead.* d.1760.
Robert. *Stonehaven.* 1860.
EDWARDS—
——. *Louth.* ca.1815. C.

EDWARDS—continued.
A. (Abraham) & C. (Calvin). *Ashby, Mass, USA.* 1794-d.1797.
Abraham. *Ashby, Mass, USA.* 1761-d.1840.
Alfred. *Darlaston.* 1876.
Benjamin. *Henley-in-Arden.* 1835-50.
Benjamin. *Knowle.* 1835-68. Also gunmaker.
Benjamin. *London.* 1828. d. pre-1869.
Bennet senior. *Bungay.* 1731-51. C.
Bennet junior. *East Dereham.* 1790-1836. C.
Calvin. *Ashby, Mass, USA.* s. of Samuel. 1789-97.
Charles. *Dundee.* 1860.
D. *Liverpool.* pre-1783. W.
David. *Stoke* nr *Coventry.* 1880.
F. H. *Colborne, Canada.* 1862.
Francis. *Liverpool.* 1848. Clock dial enameller.
Frederick. *Cambridge.* 1840.
Gabriel. *Ruthin.* 1856.
Henry Francis. *Norwich.* 1875.
Henry Philip & Co. *London.* 1881.
I. *Willenhall.* 1876.
J. *Coventry.* 1854-60.
J., jun. *Godalming.* 1855.
J. *Southampton.* 1859.
James. *Colchester.* b.1785-1821.
James. *Godalming.* 1828-51.
James. *Guildford.* s. of James of *Artington,* blacksmith. a.1801.
James. *London.* 1828-44.
James. *London.* 1875-81.
James. *Manchester.* mar.1792. C.
James. *Stourbridge.* 1828-60. C. & W.
James Frank. *London.* 1881.
James Thompson. *Dudley.* 1850-76.
John. *Ashby, Mass, USA.* s. of Abraham. 1787-1812. C.
John. *Bristol.* a.1768-80.
John. *Corwen.* d.1862.
John. *Llanidloes.* 1868.
John. *London.* 1869-75.
John. *Mold.* 1868.
John. *Stevenage.* 1828.
John. *York.* 1654-58. C.
John Thomas. *Corwen.* 1868-87.
Joseph Henry. *Cheltenham.* 1870-79.
L. *Oldbury.* 1860-68.
Leonard. *Tamworth.* 1876.
M. H. *Newark.* 1876.
Nathan. *Acton, Mass, USA.* Early 19c.
Owen. *Aberffraw.* 1817-79.
Owen. *Holyhead.* 1817-79.
Philip H. *Burnham.* 1866-75.
Rhys. *Llanerfyl.* ca.1780. C.
Richard. *Liverpool.* 1851.
Robert. *Birmingham.* 1868.
Robert. *Holyhead.* s. of Owen. 1851-90.
Robert. *Ludlow.* 1842-79.
Rowland. *Ludlow.* b. ca. 1842 s. of Robert -1900.W.
Samuel. *Bridgnorth.* (b. ca. 1791). 1825-41.
S. *Dawley, Salop.* 1856.
Samuel. *Llandrindod.* 1887.
Samuel, sen. *Wrentham, Mass, USA.* 1715-83.
Samuel, jun. *Ashby, Mass, USA.* 1804-08, then *Gorham, Me, USA.* 1808-30. C.
T. *Willenhall.* 1860.
Thomas. *Corwen.* 1808-44.
Thomas. *Deddington.* a.1773.
Thomas. *Llangollen.* 1808-76. W.
Thomas. *Llanuwchllym.* Late 18c. C.
Thomas. *Wrexham.* 1798. W.
Timothy. *St. Austell.* ca.1770. C.
W. H. *Birkenhead.* 1878.
W. J. *Penrhyndeudraeth.* Later 19c. W.
William. *Birmingham.* 1868.
William. *Bishopscastle.* b. 1776, s. of Edward. 1851. W.
William. *Cheltenham.* 1850.
William. *Derby.* 1828-49.
William. *Holyhead.* ca.1830-50.
William. *Llandovery.* 1813. W.
William. *Llanidloes.* 1874.
William. *London.* 1844-81.
William. *Tipton.* 1842.
William. *Wednesbury.* 1876.

EDWARDS—continued.
 William & Son. London. 1832-39.
 William Thomas. Ferndale (Wales). 1887.
EDWARDSON, William. Liverpool. 1715-22. W.
EDWARTS, George. London. 1850.
EELLS, Edward, sen. Stonington, Conn, USA. 1739-73.
 Edward, jun. Middlebury, Vt, USA and Medina, NY.
 1773, mar.1802-32. C.
EFFEY, William. Davenport, Ia, USA. ca.1850.
EGAN, Robert. Williamsburg, Va, USA. ca.1772.
EGERTON—
 Matthew, sen. New Brunswick, NJ, USA. 1739-1802.
 Clock cases.
 Matthew, jun. New Brunswick. s. of Matthew, sen.
 ca.1787-1837. Clock cases.
EGG, Edward. Columbia, SC, USA. 1860-ca.1879.
EGGERT, D. New York. 1826-27.
EGGINTON—
 Lewis. London. 1869.
 Thomas & James. Birmingham. 1841. Dial makers.
EGGLESTON—
 J. Salford. Late 19c. C.
 John. Kenton. 1840. W.
 John. Nottingham. 1876.
EGINGTON, Thomas. Margate. 1874.
EGLESE—
 Joseph (& Son). London. 1857-69(-75).
 Joseph (?jun.). London. 1881.
EHNHUUS, Harald. London. 1863.
EKBLOM, Elias. Turku. 1830-55. C.
EKINS, F. G. Greenwich. 1855.
ELBERN, H. Amersham. 1839-69. Also ELBURN.
ELBERT, Albert. London. 1881.
ELBLING, P. Doncaster. 1858.
ELBURN, Henry. Amersham. 1830-69. Also gunsmith
 (and see ELBERN).
ELD, Richard. Manchester. 1702. W.
ELDER—
 William. Forfar. 1860.
 William. Letham, Scotland. 1860.
ELDERSHAW, Thomas. Manchester. mar.1803. C.
ELDRICK, James. Kirkwall. s. of, and successor to Hay
 Eldrick. 1836.
ELDRIDGE—
 A. Tunbridge Wells. 1866.
 Amos. Worthing. 1870-78.
ELECTIME CORP. Brooklyn, NY, USA. 1949.
ELECTRIC—
 CLOCK CO. LTD. London. 1869.
 SELF-WINDING CLOCK CO. Bristol, USA. 1903.
ELEMENT, William. St. Albans. ca.1730. br. clock.
ELEY—
 Hodson, Boston. 1861-76.
 James. Boston. 1876.
 Joseph. Belper. 1846-95.
ELFFROTH, D. Switzerland. 1876. C.
ELFORD—
 Joseph. Axminster. 1714-25.
 Joseph. Lymington. ca.1760. C.
 Joseph Evan. Kingston-on-Thames. 1866-78.
 W. & J. Kingston-on-Thames. 1862.
ELIAS—
 David. Bangor. 1887.
 David. Llanberis. 1874.
 David. Llannerch-y-medd. 1868.
 Druiff & Co. London. 1875.
 Herman. Haverfordwest. 1844.
ELIASOFF, M. & Co. Albany, NY, USA. ca.1888.
ELIOT, William. Baltimore, USA. 1799.
ELISHA, Caleb. London. (B. 1820). 1821-51. W. to Duke
 of York. & chrons.
ELITE. 1914 and later, product of Pequegnat Clock Co.
ELKINGTON—
 J. Leamington. 1868-80.
 Mrs. Mary Ann. Brentwood. 1866-74.
ELKS, James. London. 1828.
ELLAM—
 Thomas. Runcorn. 1834.
 Thomas. Tarvin. 1848-78.
ELLARD, John. Eaton Socon (nr St. Neots). 1854.
ELLEBY—
 H. E. Ashbourne. 1855-64 C.

ELLEBY—continued.
 John senior. Ashbourne. (b. 1768) 1802 -d. 1850.
 & Son. Ashbourne. 1849.
ELLENGER, P. Bishopwearmouth. 1844.
ELLICOT(T)(E)—
 Andrew. Solebury, Pa, USA. 1754-1812. Later a
 professor.
 & COOK. London. 1844.
 Joseph. Bourn. 1868-76.
ELLIDGE—
 John. Bishopwearmouth. 1844-47.
 John. Bradford. 1853.
ELLINGHAM, John. Birmingham. 1880.
ELLINGWORTH—
 Charles. Birmingham. 1860-68.
 Charles. Bromsgrove. 1872-76.
ELLIOT(T)—
 ——. Holywell. 19c. C.
 ——. Moira, Co Down. Early 19c. C.
 ——. Twickenham. Late 18c. C.
 Alfred. Greenwich. 1847-55.
 B. Philadelphia. 1818-22. W.
 B. R. Farmington, Me, USA. ca.1850.
 Benjamin P. Philadelphia. 1843 and later.
 (Zebulon) & BURNHAM. From New York to Salisbury,
 NC, USA. 1821.
 Edward. Lenham, Kent. 1817-47. W.
 George. Ashford. 1866.
 George. Wilmington, Del, USA. Post-1856.
 Hazen. Lowell, Mass, USA. 1832.
 Henry. London. CC. 1688-97. C.
 Henry. London. 1881.
 J. Northwich. 1857-65.
 J. Plymouth, Pa, USA. ca.1770.
 J. Portsea. 1848.
 J. & Sons. Leeds. 1871.
 James. Greenwich. 1720. W.
 James. London. 1881.
 James. Portsea. 1830.
 John. Cork. 1769-95. W.
 John Catchside. Blyth. 1834. C.
 John Catchside. Leeds. 1839-66. C.
 John Henry. Leeds. b.1840-66.
 John (& Son). Ashford. 1802-51.
 Joseph. Cork. 1783-d.1796.
 Joseph. Nottingham. 1876.
 Luther. Amherst, NH, USA. ca.1840.
 R. T. Leeds. 1724. C.
 Richard. Rotherham. 1871.
 Thomas. Greenwich. 1720. W.
 Walter. Plymouth. mar.1747. C.
 William. London. 1863.
 William. Whitby. 1788-92. W.
 William Thomas. Forest Hill. 1874.
 Zebulon. New York. 1814-21. Later Salisbury, NC,
 USA.

ELLIS—
 Ann. Epsom. 1828.
 Benjamin. Philadelphia. USA. 1829-33.
 Benjamin. Woodbridge. 1812-55. C. & W.
 David. St. Asaph. 1835-44.
 Eli. Worcester. 1810.
 Elias. London. 1881. (Succeeded to John & Elias).
 Frank Alexander. Toronto. 1891.
 George. Manchester. mar.1792. C.
 George. Philadelphia. 1850.
 George. Shiffnal. 1842-50.
 Griffith. Cors-y-Gedol. 1855.
 Henry. London. 1844.
 Henry. Windsor. 1854-77.
 J. Southampton. 1859.
 James. Birmingham. 1880.
 James. Leeds. 1822-53.
 James. Quebec. 1820.
 James. Thornhill (Scotland). 1860.
 James E. Liverpool. 1848. Wholesaler.
 James E. Toronto. 1847-77.
 James W. Toronto. 1850-52.
 James Edmund. Birkenhead. 1848.
 John. Charfield (Glos.). 1879.
 John. Liverpool. 1834. Clock dial enamellers.

ELLIS—*continued.*
John. *London.* 1869.
John & Elias. *London.* 1875. (Successor to Mrs. L.)
Joseph. *Clapham (Surrey).* 1878.
Joseph. *Hull.* 1858-72. .
Michael. *London.* 1844-63.
Mrs. L. *London.* 1869. (?wid. of Michael).
Richard. *Melton Mowbray.* 1828-35.
Richard. *North Shields.* 1848.
Robert. *Nottingham.* b.1679-mar.1705. W.
S. *Birmingham.* 1860.
S. *Wellington, Salop.* 1863.
W. *Halesworth.* 1853-58.
William. *London.* 1863-69.
William. *Mold.* 1840-56.
William. *Penrith.* 1879.
William. *Wrexham.* Early 19c. W.
ELLISON—
Robert. *St. Helens.* mar.1771-83. W.
Samuel. *Liverpool.* 1767-77. W.
Thomas. *London.* 1851-69.
Thomas. *York.* 1826-30.
William. *Holywood, Co Down.* 1868-80.
William. *Waltham Abbey.* 1839.
ELLORY, Walter. *Bodmin.* 1647. Rep'd ch. clock.
ELLSON—
David. *London.* CC. 1646-47. d.pre-1650.
William George. *Great Marlow.* 1830. Then *Slough.* 1847.
ELLSUM, William. *Spalding.* 1828-35.
ELLSWORTH, David. *Windsor, Conn, USA.* 1742-1821.
ELLWOOD, E. *Sutton nr Ely.* 1846-58.
ELMORE, M. W. *Ottawa, Kan, USA.* ca.1860.
ELMS, J. *Devonport.* ca.1830. C.
ELPHINGTON(?), W. M. (or William). *London.* ca.1685. C.
ELSON—
Herman N. *Philadelphia.* 1843-48.
Julius. *Philadelphia.* 1842-44.
ELSTON(E)—
George. *London.* 1662. Spring maker.
George Henry. *London.* 1857.
ELSTRAP, J. G. *Lincoln.* 1868.
ELSWORTH, G. *Lymington.* 1859-67.
ELTON—
Francis. *Birmingham.* 1880.
William. *York.* 1846.
ELVINS, William. *Baltimore, USA.* 1796-1808.
ELVIS, Benjamin. *Wingham, Kent.* 1866-74.
ELY—'
Hugh. *Newhope, Pa, USA.* 1799-1803. Then *Trenton, NJ.* 1803-20.
John. *Mifflinburg, Pa, USA.* Early 19c.
EMANUEL—
Brothers. *London.* 1828-32. Wholesalers.
David. *London.* 1851-63.
E. & E. *London.* 1869-81. Jewellers to the Queen and others.
E. & E. *Portsea.* 1859-78.
E. & E. *Portsmouth.* 1859-67.
Emanuel. *Chatham.* 1832.
Emanuel. *Deal.* 1797-1826.
H. M. & Sons. *Portsea.* 1878.
Harry. *London.* 1869.
Joel. *London.* (B. 1824). 1811. W.
Leon. *Southampton.* 1878.
Levi. *Sheffield.* 1837.
Lewis & Son. *London.* 1839-44.
Maurice. *Portsmouth.* 1878.
S. M. *Southampton.* 1859.
EMANUELS, DANZIGER & CO. *Coventry.* Also *London* and *Birmingham.* 1854.
EMBREE, Effingham. *New York.* 1785-94.
EMERICK—
Frederick J. *London.* 1857-63. Clock cases.
John. *London.* 1832-63. Clock cases.
EMERSON—
Dudley. *Lyme, Conn, USA.* 1765-88.
& LLEWELLIN. *Bristol.* 1830.
William G. *Newport, Me, USA.* ca.1830. C.

EMERY—
H. *Coventry.* 1860.
Jesse. *Weare, NH, USA.* 1800.
Samuel. *Salem, Mass, USA.* 1809-64.
EMES, John. *Worcester.* ?ca.1800. C.
EMLER, Albert. *London.* 1881.
EMMERSON—
Joseph & Son. *Leeds.* 1866-71.
William. *Darlington.* 1827.
EMMERTON—
Caleb. *Salford nr Woburn.* 1864-77. C.
James. *Olney.* 1793-1802. W. bro. of John.
John. *Olney.* 1793-1802. W. bro. of James.
EMMERY, John. *New York.* Late 1800s.
EMMET, Edward Tillet. *Boston, USA.* 1764.
EMMONS—
C. G. *Boston, USA.* 1842.
Erastus. *Trenton, NJ, USA.* 1800-20.
EMMOTT, ——. *London.* 1808. W.
EMORY, Jesse. *Weara, NH, USA.* 1759-1838.
EMPIRE CLOCK CO. *Bristol, USA.* 1854.
EMPSALL, Edward. *Brighouse.* 1866.
EMSLEY—
Frederick. *Leeds.* 1866-71. W.
Joseph & Son. *Leeds.* 1866-71. W.
ENCRIS, Joseph. Place unknown. 1758. W.
ENDALL, D. *Great Marlow.* 1869.
ENGARD, Samuel. *Philadelphia.* 1837-42.
ENGEHAM, John. *Yalding.* Free 1789.
ENGEL, C. *Bakewell.* 1855.
ENGELSMAN, ——. *Aylesbury.* 1842. Partner with Frederick Lepmann.
ENGLAND—
Edward Ernest. *London.* 1881.
James. *Baltimore, USA.* 1807-29.
James Robert. *London.* 1851-69.
T. *Chipping Sodbury.* 1856.
T. *Westerleigh.* 1863.
ENGLES, A. *Stratford, Canada.* 1861-63.
ENGLISH—
David. *Manchester.* mar.1803. (B. 1813). W.
J. *London.* 1811. W.
Mary. *Manchester.* 1824.
Thomas. *Hamilton, Ontario.* 1856.
Watch Co. *Birmingham.* 1880.
William. *St. Johns, Newfoundland.* 1877.
ENOCK—
Ezra. From *Sibford Gower (Oxon)* to *London.* (b.1799). 1827-32. Then back to *Sibford Gower.* 1832-d.1860. C. & W.
Henry. *Warwick.* b. 1799, prob. bro. of William, d. 1867. & Son. *Warwick.* 1868.
William (& Son). *Warwick.* 1828-50(54-80).
William. *Warwick.* b. ca. 1801. Working 1827 -d. 1855.Continued as Enock & Son until 1880s.
ENRIGHT—
——. *Edinburgh.* ca.1780. C.
Robert. *London.* 1875.
ENSIGN, Charles. *Troy, NY, USA.* 1842 and later.
ENT, Theobald or Theodore. *Philadelphia.* ca.1742.
ENTWISTLE—
Edmund. *Boston, USA.* 1742. C.
William. *Prescott.* 1824-26. W.
EOFF, Garrett. *New York.* 1785-1858.
ERB, John. *Conestoga Center, Pa, USA.* 1835-60. C.
ERLAM—
Job. *Sutton(nr St. Helens.* mar.1759-63. W.
Percival. *Sutton (nr St. Helens).* (B. 1788-)1795.
ERLING, Jonathan. *Wigan.* 1699. C.
ERNST, ——. *Cooperstown, NY, USA.* ca.1810-40.
ERRICK, James. *Edinburgh.* 1860.
ERRIDGE, Charles. *Rye (from Willingdon).* a.1737 to Thomas Earle. W.
ERRINGTON—
F. *London.* 1844-57.
William M. *Smethwick.* 1876.
William Miller. *Birmingham.* 1880.
ERWIN, Henry. *Philadelphia.* 1817-42. W.
ESAU, Samuel. *Presteign.* 1830.
ESCHLE—
Felix. *Aberdare.* 1868-87.
Fridoline. *Swansea.* 1875.
Mrs. Hannah. *Aberdare.* 1895.
R. *Edinburgh.* 1860.

ESCONBE, Emmanuel. *Toulouse.* 1856. C.
ESDAILE, Andrew. *Bingham.* 1828-49.
ESPLIN, George. *Wigan.* 1831-58. W. & jeweller.
ESSEX—
Leonard. *Youghal (Ireland).* 1611. C.
Model of Pequegnat Clock Co., *Canada.*
Thomas. *Coventry.* 1880.
ESSLINGER, Charles. *Buffalo, USA.* 1840-48.
ESSON, Peter. *Aberdeen.* 1860.
EST (or ENT), John. *New York.* ca.1760.
ESTE, William. *Oxford.* Also *Burford* and *Abingdon.*
1505-26. T.C.
ESTELLE, Samuel. *Germantown, O, USA.* ca.1870.
ESTER, J. H. *Ghent.* ca.1660.
ESTHER—
Henry. *Morpeth.* 1854-58.
Thomas. *Morpeth.* 1848.
ESTON, S. *Ludlow.* 1863.
ETCHE(R)S—
C. *Castleford.* 1871.
G. *Ashbourne.* 1855-64.
John. *Liverpool.* 1828.
Mrs. M. *Nottingham.* 1849.
Samuel Edward. *Middlewich.* 1878.
William. *Altrincham.* 1878.
William. *Canterbury.* 1845.
William. *Leeds.* 1853.
William. *Sheffield.* 1862.
ETCHELLS, Mathew. *Manchester.* mar.1766. C.
ETHERIDGE, George & William Ellis. *Norwich.*
1830-65.
ETHERINGTON—
George. *Montreal.* 1842-63.
John. *Driffield.* d.1790. C.
Rebecca. *York.* dtr. of Thomas I. 1727. W.
Richard. *Tuxford.* 1876.
Thomas I. *York.* 1684-d.1728.
Thomas II. *York.* s. of Thomas I. d.1693. W.
Thomas III. *York.* s. of Thomas II. 1740-d.1741. W.
ETTER—
Arthur. *Halifax, Nova Scotia.* 1870.
Benjamin. *Halifax, Nova Scotia.* 1763-1813.
I. J. *Halifax, Nova Scotia.* 1890-97.
Peter, jun. *Halifax, Nova Scotia.* 1787-98.
ETTRY, John. *Horton.* ca.1750-70. C.
EUBANK—
James. *Glasgow, Ky, USA.* 1834-41. With Joseph.
Alone post-1841.
(Joseph) & JEFFERIES (James). *Glasgow, Ky, USA.*
1820-34.
Joseph. *Glasgow, Ky, USA.* With JEFFERIES 1820-34.
With James 1834-41. Alone 1841-55.
EUNSON, James. *Stromness.* 1836.
EUREKA—
Manufacturing Co. *Boston, USA.* 1860s.
Manufacturing Co. *Bristol, USA.* 1864-68.
Shop, The. *Bristol, USA.* 1837.
EUSTACE—
Edward. *Bristol.* 1850.
J. *Deptford.* 1874.
Richard. *Cork.* b.1765 d.1817. W.
EVA, Richard. *Tregony.* 1734-ca.1780. Then *Falmouth.*
ca.1780-90. C.
EVANS—
A. *Birkenhead.* 1865.
Alfred. *Kirkwood, NY, USA.* ca.1850.
B. *Haverfordwest.* ca.1860. C.
Benjamin. *Aberaron.* 1835.
Benjamin. *Claverdon.* ?early 19c. C.
Benjamin. *Llanpumpsaint.* ca.1800.
& Brown. *Shrewsbury.* 1863-1922. W.
Charles. *London.* 1851-69.
David. *Aberystwyth.* 1887.
David. *Cardigan.* 1887.
David. *Llangynwyd.* ca.1800. C.
David. *?London.* 1699. Sundial.
David. *Newport (Mon).* 1852.
David. *Newquay.* 1868-75.
David. *Pontypool.* 1850-80.
David. *Philadelphia.* 1770-73. *Baltimore.* 1773-84. C.
David. *Swansea.* 1868-75.
David Daniel. *Carmarthen.* Mid-19c. C.

EVANS—continued.
David & Elijah. *Baltimore, USA.* 1789.
E. *Cincinnati, USA.* 1850.
Ebenezer. *Dowlais.* 1868.
Edward. *Aberystwyth.* 1804.
Edward. *Liverpool.* 1848.
Evan. *Aberaeron.* 1868.
Evan. *Aberystwyth.* 18c.
Evan. *Builth.* 1764-91. T.C. & W.
Evan. *Cribyn, Wales.* 1782-1864. C.
Evan (& Son). *Llanfair-Caereinion.* 1835. W. & Tin-
man. (-1844).
Evan. *Llanfihangel-ar-Arth.* 1857.
F. *Coventry.* 1880.
George. *Liverpool.* 1848-51.
George. *Swansea.* 1887.
George Edward. b. ca. 1845 *Maidstone.* To *Oswestry*
ca. 1868-85. W.
George Morgan. *Ystalyfera.* 1887.
Henry. *Llangadog.* 1830-44. C. & W.
Henry. *Newark, NJ, USA.* ca.1850.
Henry. *Stafford.* 1828.
& HUNT. *Ellesmere.* 1863.
J. *Alfreton.* 1849-55.
J. *Clay Cross, Derbyshire.* 1864.
J. *Old Windsor.* 1864.
J. *Wolverhampton.* 1876.
James. *Carmarthen.* 1829-68.
James. *Chenango Point, NY, USA.* 1821.
James. *Coventry.* 1880.
James. *London.* 1869.
James William. *Lewisham.* 1874.
John. *Aberaeron.* 1835-87.
John. *Aberystwyth.* ca.1720. C.
John. *Cardiff.* 1844.
John. *Charles Co, Md, USA.* 1754.
John. *Colwyn Bay.* 1887-90.
John. *Cwmgors.* Mid-19c. C.
John. *Garthbeibio (Wales).* 1874.
John. *Lampeter.* ca.1750-ca.60. C.
John. *Llanarth.* ca.1830.
John. *Llanfyllin.* 1868.
John. *London.* 1869-75.
John. *Pontypool.* 1830-44.
John. *Swansea.* 1875.
John. *Welshpool.* 1856-74. C. & W.
Methuselah. *Carmarthen.* ca.1805-52. C.
Morgan. *Llanarth.* Mid-19c. C.
Moses. *Llanddoged.* 1780-1819. C.
Moses. *Llangernyw.* 1751. C. (Also dresser-clock
known).
Moses. *Llanrwst.* Mid-18c. C.
Mrs. Elizabeth. *Carmarthen.* 1887.
Mrs. M. *Shrewsbury.* 1856.
Mrs. Sarah. *Birkenhead.* 1878. (?wid. of A.).
Nathaniel Corker. *London.* 1881.
R. *Alfreton.* 1855-64.
Rees. *Machynlleth.* 1804-51. C.
Robert. *London.* 1828.
Robert. *London.* 1857.
Rusky. *Merthyr Tydfil.* 1887.
S. *Newport, Del, USA.* ca.1746. C.
Septimus (or Stephen). *Warwick, Pa, USA.* ca. 1810.
& Son. *Godalming.* 1878.
T. *Castleton.* 1864.
T. *Matlock Bath.* 1855.
Thomas. *Aberdare.* 1852-87. C.
Thomas. *Bangor.* 1844-90. C. & W.
Thomas. *Cardigan.* 1830-50. W. & C.
Thomas. *Halifax.* 1837-53.
Thomas. *Llandysul.* 1868.
Thomas. *London.* 1881.
Thomas. *Newcastle Emlyn.* 1801-62. C.
Thomas. *Pontuchel.* 1773. C.
Thomas. *Pontypool.* 1849.
Thomas. *Usk.* ?early 18c.
Thomas W. *Llangadog.* 1875.
W. F. *Birmingham.* 1854.
Walter. *Aurora, Canada.* 1800s.
William. *Great Malvern.* 1842-60.
William. *Machynlleth.* 1835-44. C.
William. *Newport (Mon).* 1852.

EVANS—*continued.*
William. *Pontypool.* s. of Caleb. Post-1780. C.
William. *Upton-on-Severn.* 1835.
William. *Wellington (Salop).* 1850.
William. *Wirksworth.* 1852-95.
William Frederick. *Birmingham.* ca.1850-d.1899. C.
William M. *Cincinnati, O, USA.* 1850.
EVARD, Charles C. *Philadelphia, USA.* 1837.
EVE—
Brothers. *Bristol.* 1879.
Charles. *Maldon, Essex.* 1866-74.
Charles Henry. *Prittlewell (Essex).* 1874.
John. *London.* 1832-44.
EVELEIGH, James. *Gravesend.* 1839.
EVENS—
Evan. *Totnes.* 1850.
Nicholas. *Totnes.* 1823-50.
William, jun. *Kensington, NH, USA.* ca.1770.
EVERALL, John Buckingham. *Montreal.* 1886.
EVERED(S)—
Charles. *London.* 1832.
John. *London.* Early 18c. W.
EVERELL, John. *London.* (B. a.1698-1747). 'Over against the New Church in the *Strand*'. 1731. W.
EVEREST, Edward. *London.* a.1674. to Henry Jones. d.1680. Never a freeman and prob. never made any work of his own.
EVERETT—
Jonas. *Bardwell.* 1855-64.
W. *Norwich.* 1846.
EVERMAN, Jacob. *Lancaster, Pa, USA.* 1773.
EVERS—
Peter. *Chester.* (B. 1791-)1828.
Pieter. *Amsterdam.* ca.1750. C.

EVERSHED, Arthur. *Brighton.* 1878.
EVERT, Nicholas. *Sveaborg.* Early 19c. C.
EVERY, Henry. *London.* 1875-81.
EVISON, Christopher. *Newcastle-on-Tyne.* 1789.
EWALD, F. L. Place unknown. ca.1820. W.
EWBANK—
George. *Elland.* 1740-70. C.
William. *Elland.* 1795.
EWER—
John. *Willesden. Plum Tree Street, Harlesden Green* in 1721. W. (One such a.*London* 1687 to Luke Bird, may be same man.)
R. C. *Ely.* 1846.
EXCELBY, John, jun. *Copmanthorpe.* 1838.
EYETT, John. *Liverpool.* 1864. W.
EYGES, Israel. *London.* 1875-81.
EYLES, John. *Doncaster.* b.1777-d.1849. W.
EYNON—
B. *Maenclochog (Wales).* ca.1850. C.
Benjamin. *Narberth (Wales).* 1887.
John. *St. Davids.* From 1886.
EYRE(S)—
Edwin. *Louth.* 1828-35.
G. *Dudley.* 1868-72.
Henry. *Gateshead.* 1827.
James. *Newtown Limavady, Co Derry.* 1839-46.
Matthias. *Philadelphia, USA.* 1775. Watch springs.
Richard. *London.* Possibly two such. One a.1637 to Oswald Durant. Another still working 1697.
EZEKIEL—
C. *Newcastle-on-Tyne.* 1850.
Moses. *Newcastle-on-Tyne.* 1847.

F

FABER, George. *Sumneytown* and *Reading, Pa, USA.* ca.1780-1805. C.
FABIAN, H. *Chester* and *Lancaster, Pa, USA.* 1853.
FABRAHAM, Joseph. *Waltham Abbey.* ca.1705. Marq. mus. clock.
FABRICE, ——. *Paris.* ca.1720. C.
FACKLER—
  C. & William. *London.* 1857.
  Charles. *London.* 1863-69.
  Charles & Son. *London.* 1851.
  Joseph. *London.* 1875.
  Othmar. *Leeds.* 1850-53.
  William. *London.* 1881.
FADELEY, J. M. *Louisville, Ky, USA.* 1842.
FAFF, Augustus P. *Philadelphia, USA.* 1835.
FAGE, Edward. *London.* Math. instr. mr. Admitted to CC. 1667 though free of another company. -1669.
FAGG—
  ——. *Sandwich.* 1753. C.
  Alfred. *Margate.* 1866-74.
  Jacob. *Folkestone.* 1823.
  Jacob. *Ramsgate.* 1826-28.
  John. *Margate.* (B. ca.1800-)1823-51.
  William. *Ramsgate.* 1847-74.
FAHRENBACH, Pius. *Boston, USA.* ca.1850. See FEHRENBACH.
FAIP, Joseph. *Doncaster.* 1871.
FAIRALL, William. *Ryde.* 1878.
FAIRBAIRN—
  Deacon. *Dumfries.* 1700. Kept town clock.
  John. *Durham.* 1847.
  John. *Wooler.* ?late 18c. C.
FAIRBANKS—
  Joseph O. *Newburyport, Mass, USA.* 1860.
  Levi. *Toronto.* 1837-57. Also gunsmith.
  WHITCOMB, FAIRBANKS & CO. *St. John, New Brunswick.* 1851-77.
FAIRBROTHER—
  Henry James. *Stockport.* 1848-65.
  James. *Witney, Oxon.* 1823. C. & W.
  N. D. *Stockport.* 1878.
  Thomas. *Colchester.* b.ca.1686-d.ca.1750. C.
FAIRCLOUGH—
  Jeffrey. *St. Helens.* 1822-34.
  John. *St. Helens.* 1833. W.
  Samuel. *Halifax.* 1834-50.
FAIRER, Joseph. *London.* 1851-75. T.C.
FAIREY—
  J. *Gravesend.* 1866.
  J. *Melcomb Regis.* 1848-67.
  J. *Portland.* 1855.
  John. *London.* 1811-57.
  John & Son. *London.* 1828-32.
  Joseph. *Lewes.* 1870.
  Joseph Henry. *London.* 1875-81.
  Richard. *London.* 1811-69.
  S. *London.* ca.1810.
FAIRFAX, Thomas. *Halifax.* 1694.

FAIRHAVEN CLOCK CO. *Fairhaven, Vt, USA.* ca.1891-1910.
FAIRLEY, John. *Coventry.* 1868-80.
FAIRMAN, Gideon. With William HOOKER at *Newburyport, Mass, USA,* pre-1810. Then at *Philadelphia* with DRAPER & MURRAY.
FAIRN, James. *Edinburgh.* 1800. Clock cases.
FAIRWEATHER—
  Thomas. *Gateshead.* ca.1780. C.
  Turnbull. *Gateshead.* 1836-51.
  Turnbull. *Newcastle-on-Tyne.* 1838-56.
FAIRY—
  James. *Maidstone.* 1840.
  John. *London.* 1839.
FAITH—
  Alfred John. *Chichester.* 1878.
  George. *Chichester.* 1828-62.
  & Son. *Chichester.* 1870.
FAITHFULL, Frederick. *London.* 1875-81.
FAIVRE, ——. *Trevillers nr Montbeliard.* 1860. C.
FALARDEAU, Joseph I. *Quebec.* 1890.
FALES—
  G. S. *New Bedford, Mass, USA.* ca.1820.
  James. *New Bedford, Mass, USA.* 1810-20.
  James, jun. *New Bedford, Mass, USA.* 1830.
FALK, David & Co. *Manchester.* 1851.
FALKSON, Lewis. *Falmouth.* 1787-1852.
FALLAR, Theodor Kuss. *Manchester.* 1851. Imported clocks.
FALLE, Thomas. *Jersey.* 1832.
FALLER—
  A. *Battersea.* 1878.
  Augustine. *Glasgow.* 1860.
  Benedict. *Leeds.* 1871.
  Frank. *Rhymney (Mon).* 1880.
  Frederick. *Pontlotyn.* 1868.
  Frederick. *Rhymney (Mon).* 1871-75.
  & HETTICH. *London.* 1881. Successor to Joseph.
  & HUMMEL. *Portsea.* 1859.
  J. *Walsingham.* 1858. C.
  J. A. *Landport.* 1859-67.
  John. *Holt.* 1875.
  John. *London.* 1844.
  Joseph. *London.* 1875.
  Lorenz. *London.* 1869-81.
  P. *Coventry.* 1868.
  P. & Co. *Northampton.* 1854.
  Xavier. *Bristol.* 1879.
FALLETT, Samuel. *Gloucester.* 1870-79.
FALLOT, ——. *Paris.* 1750. Lant. clock.
FALLOW(S)—
  Andrew. *Sunderland.* 1851.
  J. & Co. *Manchester.* 1834. German/Dutch clocks.
  Joseph. *Newcastle-on-Tyne.* 1827-36. German clocks.
  Martin. *Newcastle-on-Tyne.* 1836. German clocks.
  Thomas. *Preston.* 1851.
FAR(A)GHER, Thomas. *Toronto.* 1852-77.
FARDOIL, Jean. *Paris.* ca.1720. W.

FAREBROTHER, William T. *London.* 1863-81.
FARIS—
Charles. *?Annapolis, Md, USA.* 1764-1800.
William, sen. s. of William of *London.* To *Philadelphia.* 1729-49. Then *Annapolis.* 1757-92.
William, jun. s. of William, sen. b.1762. *Annapolis.* ca.1790. *Norfolk, Va.* 1792-94. *Havana, Cuba.* 1798. *Edenton, NC, USA.* mar.1803-ca.1815.
FARLEY, Thomas. *Faversham.* (B. 1778-95). 1802. W.
FARLIE, Thomas. *Ballymacarrett.* 1868.
FARLOW, William. *Paris, Canada.* 1851-62.
FARMER—
Charles Henry. *London.* 1869.
John. *London.* (B. CC. 1657). 1662. *Chancery Lane.*
John. *?London.* ca.1740.
John. *Philadelphia.* 1699. W.
M. G. *Salem, Mass, USA.* ca.1849. C.
R. *Maidstone.* 1855-74.
Thomas. *London.* Maker of watch cases and keys. 1640-CC. 1653.
William. *Stockton.* 1820.
William & Thomas. *Stockton.* 1827-51.
FARNDON—
John I. *Deddington (Oxon).* b.ca.1700-d.1743. C. & W.
John II. *Deddington.* b.1736. s. of John I. d.1786. Prob. trained in *London.* C. & W.
John III. *Deddington.* Perhaps s. of John II. 1791-1830.
John. *Woodstock.* 1853. C. & W.
Thomas. *Deddington.* 1791-d.1838. C. & W.
Thomas. *Adderbury.* Prob. same man as at *Deddington.*
Thomas. *Woodstock.* 1823. C. & W. Prob. same man as at *Deddington.*
FARNHAM—
Henry & Rufus, (brothers). *Boston, USA.* 1780.
J. *Bridport.* 1848-67.
James. *Lyme Regis.* 1848-75.
Mrs. M. *Bridport.* 1855.
Samuel H. *Oxford, NY, USA.* 1840-42.
Thomas. *Bridport.* (B. 1780-)1824-48.
W. *Bridport.* 1867.
FARNHILL, William. *Rotherham.* 1822-62.
FARNWORTH—
John. *Liverpool.* 1851.
Thomas. *Blackburn.* 1858.
FARQUHAR—
Andrew. *Peterhead.* 1837-60.
James. *Armagh.* mar.1832. Later to *New York.* ca.1840-51.
John. *Cootehill.* 1846.
Robert. *Clones.* 1846-59.
William. *London.* 1832-81.
William. *Montreal.* 1823-30.
FARQUHARSON—
Charles. *Dunbar.* 1860.
Charles H. *Edinburgh.* 1860.
Robert (W.). *Dundee.* 1847-(60).
William. *Edinburgh.* 1860.
FARR—
Bela. *Norwich, NY, USA.* 1829.
George. *St. Aubin (Jersey).* 1786. W.
John. *Bristol.* (B. ca.1790-1801). 1830.
John. *Utica, NY, USA.* 1834.
John. *Walsall.* 1828.
FARRAN, S. *Ashton-under-Lyne.* 1858. C.
FARRANT—
Elizabeth. *Wrotham.* 1847.
William. *Wrotham.* 1826-39.
William (?II). *Wrotham.* 1874. W.
FARREN, J. *Battersea.* 1878.
FARRER—
Abraham I. *Pontefract.* ca.1695-d.1754.
Abraham (II). *Pontefract.* s. of Abraham I. b.1728-d.1789. C.
Benjamin. *Pontefract.* 1810-66.
Charles. *Pontefract.* b.1761-d.1817. Also at *Doncaster.*
Elizabeth. *Pontefract.* 1866-71.
James. *Pontefract.* 1724-34. C.
John. *Pontefract.* b.1765-d.1817. C.
John. *St. Helens.* mar.1815-19. W.
Jonathan I. *Halifax.* b.1671-1702. C.
Jonathan II. *Halifax.* s. of Jonathan I. b.ca.1700-ca.1745. C.

FARRER—continued.
Joseph. *Pontefract.* d.1765. C.
Joshua. *Doncaster.* b.1771-d.1838. W.
Joshua. *Doncaster.* 1851-71.
Joshua. *Pontefract.* 1804-62.
Samuel. *York.* ca.1625-1648. W.
William. *Heckmondwike.* 1866. W.
William. *Pontefract.* 1716-d.1726. C.
FARRET, ——. *Paris.* 1849. C.
FARRINGTON—
Michael. *Bristol.* 1842.
Thomas. *Bristol.* (B.1819-)1830-50.
FARROW, S. *Ashton-under-Lyne.* 1858. W.
FARRY, William. *Newport.* ca.1740. lant. clock.
FARTHING, Charles J. *London.* 1832. Dialmaker and painter.
FARWELL, T. W. *Weymouth.* 1867.
FASBENDER, John H. *Richmond, Va, USA.* ca.1819.
FASOLDT—
John G. *Albany, NY, USA.* 1869.
Max. *Albany, NY, USA.* 1869.
FATMAN, Brothers. *Philadelphia.* 1843.
FATTON—
& Co. *Philadelphia.* 1840.
Frederick. *Philadelphia.* 1830-39.
Frederick & Co. *London.* 1828.
FATTORINI—
A. & Sons. *Bradford.* ?1826-71. And Later.
Innocent. *Skipton.* 1866-71.
& Son. *Harrogate.* 1866.
FAUCHE(R) & JACOT. *Montreal.* 1857.
FAULDS, Allan. *Kilmarnock.* ca.1830.
FAULKNER—
Alfred. *London.* 1875-81.
J. *Tutbury.* 1868.
James. *New York.* ca.1840.
John. *Burford.* a.1790. C.
Thomas. *London.* 1869-75.
William. *London.* 1875.
FAURE, Louis. *Naples.* ca.1685. W.
FAVER, John. *London.* 1741. W.
FAVILL(E)—
J. N. *Leek.* 1860.
John. *Cheltenham.* 1840-56.
FAVRE, Louis. *London.* 1851.
FAWCETT(E)—
——. *Liverpool.* ca.1715-ca.1725. C.
James. *Richmond (Yorks).* 1830-41. W.
John. *Richmond (Yorks).* 1834-40.
FAWKES—
——. *Plymouth Dock.* ca.1800. C.
F. F. *Ingersoll, Ontario.* 1861.
G. H. *Paris, Canada.* 1861.
FAWSON—
A. *Coventry.* 1860-80.
J. *Coventry.* 1860.
FAY—
Henry. *Albany, NY, USA.* ca.1850.
James & John. *Glasgow.* 1837.
FAYRER, James. *Lancaster.* s. of Thomas. Free 1783. C.
FAZAKERL(E)Y—
Henry. *St. Helens.* mar.1787. W.
James. *Rainford (Lancs).* mar.1824. W.
James. *St. Helens.* mar.1819-30. W.
John. *Liverpool.* 1744-d.1770. W.
John. *St. Helens.* 1780-86. W.
Richard. *Manchester.* b.ca.1785-d.1836.
Thomas. *Prescot.* d.1771. W.
William. *Prescot.* 1742-1829. (Prob. two such.) W.
FEAD, James. *Edinburgh.* 1843.
FEALE—
William. *Newport.* 1848-52.
William M. *Bristol.* 1840-42.
FEAR—
Daniel. *Bristol.* 1879.
Daniel. *Trowbridge.* 1875.
Edwin. *Bristol.* 1850-79.
G. *Reading.* 1877.
FEARN—
Benjamin. *Derby.* 1860-84.
G. *Derby.* 1855-64.

FEARNLEY—
Peter. *Trowbridge.* 18c. C.
Peter. *Wigan.* (b.ca.1749), mar.1776-d.1826. C. &
T.C.
FEATHER(S)—
Edwin. *Haworth.* 1866-71. W.
Peter A. *Dundee.* 1860.
FEATHERSTONE, Francis. *London.* 1828-51.
FEBEN, Walter. *Waterloo, Hants.* 1878.
FEDDERSEN, H. F. *Lancaster, Pa, USA.* ca.1850.
FEDERL, ——. *Vienna.* ca.1815.
FEGAN, James. *Cavan.* 1846-54.
FEHLINGER, Samuel. *Gettysburg, Pa, USA.* Early 19c.
C.
FEHR—
Louis. *Haverfordwest.* 1887.
Louis Phillip. *London.* 1881.
FEHRENBACH (and FEHRENBACK)—
Augustin. *Birmingham.* 1868-80.
B. *London.* 1857-69.
B. *Ramsgate.* 1845.
B. & L. *Bishop's Stortford.* 1859.
& BAURLE. *London.* 1869-75.
Benjamin (or Bennett). *Kidderminster.* 1842-76.
Berthold & Co. *Bishop's Stortford.* 1866-74.
Brothers. *London.* 1881.
Charles. *Bristol.* 1870-79.
& Co. *London.* 1881.
E. *Kendal.* ca.1800?
E. *London.* 1863-69.
Emilian. *Dover.* 1839-51.
Engelbert. *Ipswich.* 1875-79.
F. *Dover.* 1851.
F. & O. *Dover.* 1855.
Gottfried. *London.* 1881.
Gustave. *Wolverhampton.* 1876.
John. *Belfast.* 1843-58.
Joseph. *High Wycombe.* 1842-47.
L. *Reading.* 1854.
Lawrence. *London.* 1857-63.
Leo. *London.* 1881.
Lorenz. *London.* 1869-75.
Pius. *Carlisle.* 1879.
S. & O. *Blackheath Road, London.* 1866.
FEIERTAG, Fabian. *Belfast.* 1880. C.
F(F)EILDER, Thomas. *London.* 1697. (Same as
Fielder.)
FEINTUCH, M. *Toronto.* 1861-67.
FEIRO, Jonathan. *Garvagh.* ca.1820. W.
FEIS, Charles & Co. *London.* 1881. Successor to Adolphe
WOOG.
FELDMAN, Moses. *Birmingham.* 1868-80. Importer.
FELIX, J. *Columbia, Pa, USA.* ca.1840.
FELKIN, John. *Kirton, Lincs.* 1835.
FELL—
Abraham. *Ulverston.* 1733-1800. C. Prob. two such.
Henry K. *Hamilton, Ontario.* 1868.
James. *Kendal.* ca.1770-ca.1800. C.
James. *Lancaster.* Free 1767-1806. W.
John. *Blackburn.* ca.1825-34.
Joseph. *Ulverston.* mar.1779-87. W.
Thomas. *London.* 1863.
FELLOWS—
Abraham. *Montreal.* 1809.
Abraham. *Troy* and *Waterford, NY, USA.* 1810-
1850.
I. W. & J. K. *Lowell, Mass, USA.* 1834.
Ignatius W. *Lowell, Mass, USA.* 1834.
James. *Kensington, NH, USA.* s. of Jeremiah. ca.1800.
C.
James K. *Lowell, Mass, USA.* 1832-34.
Jeremiah. *Kensington, NH, USA.* 1749-1837. C.
Reed & Olcott. *New York.* 1829.
(Louis) & SCHELL. *New York.* ca.1856.
STORM & CURGILL. *New York.* 1832.
FELLS, John. *London.* 1828. Watch case maker.
FELTHAM—
Mrs. M. *Thetford.* 1858-65.
R. D. *Jersey.* 1851. C.
T. *Thetford.* 1846.
William. *Harleston.* (B. 1813-)1830-58.
William. *Stowmarket.* 1839-75.

FELTON—
A. C. *Boston, USA.* ca.1860.
Eli. *Birmingham.* 1815. Dial maker.
R. *Knighton, Wales.* Mid-18c. C.
FENDON, Thomas. *Douglas.* Mid-19c. C.
FENKMAN, William. *London.* CC. 1661.
FENLESTER, Alexander. *Baltimore, USA.* 1807.
FENN—
Charles. *Boughton, Kent.* 1874.
Charles. *Whitstable.* 1866-74.
Charles Walter & Co. *London.* 1881.
G. *Lewisham.* 1851.
J. (& Son). *Greenwich.* 1851-55(-66)-74.
John. *Boughton, Kent.* 1874.
John. *London.* 1844.
Robert. *London.* 1863-69.
William B. *Plymouth, USA.* 1813-mar.1864.
FENNELL, E. *Brighton.* 1862.
FENNER, Thomas. *Rickinghall, Suffolk.* d.1659.
W.
FENNO, James. *Lowell, Mass, USA.* 1834-37. W.
FENNY, James. *St. Helens.* mar.1780-83.
FENNYMORE—
Christopher. *London.* 1746. W. (B. ante-1796.)
William. *London.* 18c. C.
FENON, ——. *Paris.* 1878-89. C.
FENSHAM—
Anthony. *London.* 1863.
& TARRANT. *London.* 1857.
FENTON—
——. *London.* ca.1800. C.
George. *London.* 1863-69.
James. *Tamworth.* 1828-35.
James. *York.* b.1719-40. C.
John. *Congleton.* 1828-34.
John. *Macclesfield.* 1848.
Simeon. *Bury St. Edmunds* 1844-46. then
Mildenhall 1855-74.
Samuel. *London.* 1839-51.
Samuel. *Worksop.* 1828.
FENWICK, Peter, jun. *Crieff.* 1837-60.
FERDARN, John. *Pontefract.* 1817. W.
FEREN—
——. *Dundee.* 1843.
& Co. *Dumfries.* 1837.
FERENBACH—
D. & C. *Edinburgh.* 1850.
Pius. *Edinburgh.* 1860.
FERGUSON—
Alexander I. *Ardtrea, Co Tyrone.* bro. of Andrew.
(b.1779) ca.1800-d.1847.
Alexander. *Belfast.* 1849-50.
Alexander II. *Cookstown.* 1846-68. Also *Ardirea* in
1868.
Andrew. *Ardtrea.* bro. of Alexander. (b.1778) ca.1800-
d.1844.
Archibald. Place unknown. Late 18c. C.
Archibald. *Belfast.* 1843. W.
Archibald. *Johnstone.* 1836.
Charles. *Greenock.* 1860.
Daniel. *Nairn.* 1860.
Elijah. *New Bern, NC, USA.* mar.1833-50.
George. *Philadelphia.* 1820-22.
George Cochran. *London.* 1863-69.
George D. *London, Canada.* 1886.
Hugh. *Carlisle.* 1879.
J. *Dundee.* ca.1770.
James. *Inverness.* 1860.
Mathias Valentine. *Newry.* 1878-92. W.
Montgomery. *Mauchline, Ayrshire.* 1837-50.
Robert. *Gateshead.* 1827.
William Stephen. *Elgin.* 1837-60.
FERMENT, Jean. *Quebec.* 1734-44.
FERNBACH, B. *Bedford.* 1847.
FERNS, Richard. *Prescot.* mar.1768. W.
FERRALL, Edward. *Cranbrook.* 1686-1706. Rep'd ch.
clock.
FERRETT, Thomas. *Gloucester.* 1842.
FERRIER—
John. *Tain.* 1836.
Thornton. *Hull.* 1823.
William Thornton. *Hull.* 1820-51.

FERRIS—
——. *Norwalk, Conn, USA.* 1889. W.
Benjamin. *Waterford, NY, USA.* 1811.
David. *Calne.* 1842-75.
Edward B. *Philadelphia, USA.* 1846-48.
Edward C. *London.* 1875.
James. *Poole.* (B. 1810-)1824. C.
FERRO(O)N, John James. *Lewes.* a.1717 to Thomas
Barrett, C. -1743. C.
FERTIG, Benjamin. b.1778. *Vincent, Pa, USA.* 1802-08.
Philadelphia. 1810. *Pikeland, Pa.* 1815-23.
FESSER—
Frederick. *Coventry.* 1868-80.
Hubert. *Cheltenham.* 1842.
FESSLER—
John, sen. *Frederickton, Md, USA.* 1782-1820.
John, jun. *Frederickton, Md, USA.* s. of John, sen.
1820-40.
& Son (John). *Frederickton, Md, USA.* Early 19c. C.
FEST—
Alfred. *Philadelphia.* post-1815.
& Brother. *Philadelphia.* 1850.
Edward. *Philadelphia.* post-1842.
FETON, J. *Philadelphia.* 1828-40.
FETTER, Ronald. *Thorne.* 1778-84. W.
FETTERS—
Henry, sen.?. *London.* (B. CC. 1632+).
Henry, jun. *London.* CC. 1654-64. *Westminster.*
Nicholas. *London.* CC. 1634-62. *Westminster.*
William. *London.* 1697. (Name uncertain.)
FEVRE, Henry. *London.* 1811. W.
FEWLLER, John. *Liverpool.* 1761-80. C.
FEWTRELL—
J. *Tenbury.* 1868.
John. *Birmingham.* 1880.
FFARR, Robert. *Denbigh.* 1605. TC.
FICKLIN, G. *Birmingham.* 1860.
FIDDAMENT, J. *Stoke Ferry, Norfolk.* 1858. C.
FIDGETT, William. *London.* 1786-1828. W.
FIDLER, Edward S. *Cleator.* 1879.
FIECHTER, Ulrich. *Huddersfield.* d.1944. Musical clock
maker.
FIELD—
Benjamin, sen. *Tunbridge Wells.* 1823-51.
Benjamin, jun. *Tunbridge Wells.* 1851.
Charles Nicholas. *Hitchin.* 1828-51.
Henry Brails. *Shipston.* 1850.
James (& Son). *Hertford.* (B. ca.1800.) 1828(-39).
John. *Bristol.* 1879.
John. *Poughkeepsie, NY, USA.* Early 19c.
John H. *Batavia, NY, USA.* 1811.
Peter, sen. *Hudson, NY, USA.* 1785. *New York.* 1802.
Newburgh, NY. 1807-10.
Peter, jun. *New York.* 1802-25.
R. B. & Co. *Brockville, Canada.* 1860s.
Richard. *London.* ca.1690. W.
Richard Thomas. *Stroud.* 1863-79.
T. & Son. *High Wycombe.* 1877. (Also *Aylesbury.*)
Thomas & Son. *Aylesbury.* 1877.
Thomas. *Bicester.* a.1787. C.
Thomas. *Bristol.* 1781-1812. C.
Thomas White. *Aylesbury.* b.1773. Succeeded Joseph
Quartermain in small d.1832.
Thomas White, jun. *Aylesbury.* s. of Thomas, sen.
b.1805-69.
William. *Cheadle.* 1878.
FIELDER, William. *London.* (B. a.1799.) 1832-39.
Watch cases.
FIELDHOUSE, Benjamin. *Monmouth.* 18c. C. (One
such at *Leominster* 1756.)
FIELDING—
A. *Greenwich.* 1855.
Augustin. *Canterbury.* 1838-59.
Enoch. *Glossop.* 1876-1932.
G. *Glossop.* 1849.
George. *Greenwich.* 1839-51.
James. *Boston.* 1825-68.
Mrs. E. *Glossop.* 1855-64.
Mrs. E. E. *Blackheath.* 1866.
Robert. *Liverpool.* 1784. W.
FIELDSHAW, James. *Leeds.* 1766-87. W.
FIETZEN, Andrew. *Liverpool.* 1834-48.

FIFIELD, John. *Kingston* and *Kensington, NH, USA.*
1758-59.
FILBER, John. *Yorktown* and *Lancaster, Pa, USA.*
1810-25.
FILIPPINI, Frederick. *Pontypridd.* 1887.
FILLANS, William. *Huddersfield.* 1853.
FILLINGER, G. (& Co.). *Ely.* 1846-58(-65).
FILLINGHAM, Robert. *Prescot.* 1794. W.
FINCH—
Benjamin. *Colchester.* b.1820, a.1835. W.
Caleb & Samuel. *Maldon.* 1874.
James. *London.* ca.1770. C.
John. *London.* 1875.
Jonathan. *Prescot.* mar.1819. W.
Mrs. (wid.). *London.* 1713. (B. has Katherine 1719.)
Robert. *London.* 1881.
Stephenson. *London.* 1844.
Thomas. *Steyning.* a.1723 to Benjamin Packem. C.
William. *Halifax.* b.1720-d.1781. C. & W.
FINDING, Samuel. *Bedford.* 1877.
FINDLAY—
John. *Aberdeen.* 1836-60.
John. *Banchory-Ternan* (*Scotland*). 1860.
George Scott. *Ipswich.* mar. 1820-68.
George. *London.* 1832-51.
William George. *London.* 1832.
FINER—
Horatio. *London.* 1844-57.
& NOWLAND. *London.* 1800-39. W.
Thomas. *Chelmsford.* ca.1788-1823.
Thomas. *London.* 1788-1800. (Later partner with
NOWLAND.)
FINLAY—
Andrew. *Gatehouse-of-Fleet.* 1836-60.
George. *Ballymacarrett.* 1868. Watch glasses.
James. *Edinburgh.* 1860.
John & Co. *Glasgow.* 1837.
FINLOW, Ralph. *Prescot.* 1796. W.
FINN—
T. *Whitchurch.* 1856-63.
William. *Wye* (*Kent*). ?1846.
FINNEMORÉ—
William. *Birmingham.* 1815-22. Dial maker.
William & George. *Birmingham.* 1839-41. Dial makers.
William (jun.?). *Birmingham.* 1850. Dial maker.
William & Son. *Birmingham.* 1828-36. Dial maker.
FINNEY—
James. *Prescot.* 1823. W.
John. *Liverpool.* (B. 1754) d.1795. W.
Joseph. *Liverpool.* 1734-d.1772. W. & C.
Richard. *Liverpool.* 1828-34.
Thomas. *Liverpool.* 1848.
FIRDERER—
John. *Birmingham.* 1842-80.
Peter. *London.* 1881.
FIRTH—
D. *Mossley* (*Cheshire*). 1878.
James. *Leeds.* 1871.
John. *Halifax.* 1815. C.
N. C. *Chester.* 1878.
FISH—
George. *Northampton.* 1777-95. W.
Isaac. *Utica, NY, USA.* 1843.
Isac, jun. *Utica, NY, USA.* 1850.
James. *Birmingham.* 1842.
John. *Doncaster.* 1871.
Robert. *London.* 1828-32.
FISHER—
Amos. *Raunds* nr *Thrapston.* 1869-77.
Anthony. *Stafford.* 1860-76.
Charles. *Birmingham.* 1816-42.
Charles. *Rushden.* 1877.
E. *Bath.* 1856-61.
E. *Birmingham.* 1854.
E. *West Bromwich.* 1860.
Ebenezer. *Bath.* 1846-48.
Ebenezer. *Bilston.* 1835.
Ebenezer. *Birmingham.* 1828.
Ebenezer. *Wednesbury.* 1850.
Ebenezer & Margaret. *Ellesmere.* 1839-46.
Edwin. *Bath.* 1819-65. W.
Electric Clock Co. *New York.* 1896.

FISHER—*continued.*
George. *Birmingham.* 1850.
George. *Guelph, Canada.* 1857-63.
George. *Frederick, MD, USA.* 1800.
George. *Troy. USA.* 1837-46.
John Forsdick. *Woodbridge.* 1844-50. W.
John. *Bilston.* 1842-50.
John. *Birmingham.* 1850-60.
John. *Halifax.* ca.1760. C.
John. *Oldbury.* 1842.
John. *Orangeville, Canada.* 1857.
John. *Preston.* ca.1740-ca.50. C.
John. *Tipton.* 1835-42.
John, sen. b. *Germany* to *USA.* 1749. *York, Pa.,*
1759-1808. C.
John. *West Bromwich.* 1828-50.
John D. *Lincoln.* 1876.
John S. *Dudley.* 1860-76.
John S. *West Bromwich.* 1876.
John Scambler. *Walsall.* 1860-76.
John W. *West Bromwich.* 1868-76.
Joseph. *London.* 1828.
& LAGUIRE. *London.* 1857.
Louis. *New York.* ca.1800.
Mrs. M. H. *Woodbridge.* 1853.
Theop. *London.* ca.1770. C.
Thomas. *Birmingham.* 1880.
William. *Huddersfield.* 1853-71. W.
William. *Nuneaton.* 1835.
FISK(E)—
Samuel. *Boston, USA.* bro. of William, mar.1794-97.
Cabinet maker.
Thomas Henry. *Portsmouth.* 1830.
William. *Boston, USA.* Clock cabinet maker. With bro.,
Samuel, till 1797, then alone till 1844.
FITCH—
Eugene. *New York.* 1902.
George. *Gravesend.* 1839-45.
George. *London.* 1828.
George & Co. *London.* 1851-63.
John. *Windsor, Conn, USA.* b.1743, to *Trenton* 1769, to
New York 1782, to *Bardstown, Ky,* ca.1789.
Jonas. *Pepperwell, Mass, USA.* Late 18c.
FITCHBURG Watch Co. *Fitchburg, Mass, USA.*
ca.1874. W.
FITE, John. *Baltimore, USA.* 1783-1818.
FITHIAN see PHYTHIAN.
FITT—
Frederick. *Ottawa.* 1879.
John. *Aylsham.* 1830-65.
S. *Cromer.* 1846.
William Edward. *North Walsham.* 1858-65.
FITTERS, John. *London.* CC. 1685-90. W.
FITTON, Whiteley Samuel. *Sowerby Bridge.* 1866-1871.
FITTS, George. *Bangor, ME, USA.* ca.1830-60.
FITZ—
Thomas. *Salisbury.* (B. 1795-)1830. W.
William. *Portsmouth, NH, USA.* ca.1780. C.
FITZGERALD, G. *St. Albans.* 1866-74.
FITZSIMONS—
——. *St. Catherines, Canada.* 1877.
G. J. *Toronto.* 1877.
FIX—
George. *Reading, Pa, USA.* ca.1802. C.
Joseph. *Reading, Pa, USA.* Early 19c.
FLACH, George W. *Charleston, SC, USA.* From *Hessen,*
Germany. ca.1840-70.
FLACK, George. *London.* (B. 1817-24) 1828-32.
FLAIG—
E. *Danville, Ky, USA.* Early 1860s.
Valentine & Co. *London.* 1839-44.
X. *Leighton Buzzard.* 1854.
FLANAGAN, Michael, jun. *Holmfirth.* 1866. W.
FLASHMAN, George. *Exeter.* mar.1762. C. (B. 1765).
FLATHER—
Robert. *Halifax.* 1850. C.
William I. *Halifax.* b.1786-d.1856. C.
William. II. *Halifax.* s. of William I. b.1810-d.1879.
William. *Sheffield.* 1837-62.
FLECHSELLIN, Anthony. *Vienna.* 1500. May be same
man as Anthony NAGL.

FLEETWOOD—
Henry. *Prescot.* mar.1769. W.
James. *Liverpool.* 1814.
Joseph. *Liverpool.* ca.1750-ca.60. W.
R. ?*London.* 1791. W. (B. 1763-pre 1794).
Robert. *Liverpool.* 1790. W.
Thomas. *Prescot.* 1794. W.
FLEISCHMANN, George. *London.* 1875.
FLEMING—
Ambrose. *South Shields.* 1844.
David. *London.* (B. 1817-24) 1828.
Hugh. *Hexham.* 1757. C.
John. *Glasgow.* 1836.
Richard. *Sevenoaks.* 1793.
Richard. *Liverpool.* (B. 1810) 1814.
Thomas. *London.* 1863.
FLETCHER—
——. *Rampton.* ca.1815. C.
Andrew. *London.* 1869.
Charles. *Barnsley.* 1807-37. s. of Tobias I.
Charles. *London.* 1839-44.
Charles. *Philadelphia.* 1817-33.
Charles. *Retford.* 1835.
Charles Arthur. *London.* 1863-69.
Charles & Thomas. *Philadelphia, USA.* 1830.
D. *Axbridge.* 1861-66.
Edward. *Bradford.* d.1714. C.
George. *Haverfordwest.* 1887.
& HUNTER. *Edinburgh.* 1850.
J. *Newton-in-Makerfield.* 1858.
James S. *Ashton-under-Lyne.* 1851-58. W.
James. *Liverpool.* 1834.
James. *London.* Clock case maker in 1730 at *Long Lane,*
West Smithfield.
James. *Rotherham.* 1790-1822.
John. *Barnsley.* b.1739-d.1778. C.
John. *Barnsley.* 1811-13. s. of Tobias I.
John. *Dewsbury.* ca.1750-ca.60. C.
John. *Halifax.* 1740-70. C.
John. *Haverfordwest.* 1850-75. W.
John. *Leeds.* d.1787. C.
John. *London.* a.1654 or 1655. d.1661. W. (died before
completing apprenticeship).
John. *Manchester.* 1851.
John. *South Norwood.* 1878.
John. b.1820. *Broughton (Lancs). Ulverston* by
1848-51.
John (& Sons). *London.* 1832-51 (57-69).
John Wright. *Birmingham.* 1841-50. Dial makers.
Robert. *Chester.* (B. 1784-) 1828-34.
Robert. *Kirkby Moorside.* 1834.
Robert. *London.* 1869.
Samuel. *Tuxford.* 1842-48.
Samuel. *Cambridge.* 1830.
Samuel. *Dewsbury.* ca.1780. C.
Samuel. *Rampton.* ca.1780-90. C.
Thomas. *Barnsley.* (B. 1802) 1822-62. C.
Thomas. *Kirkby Moorside.* 1840.
Thomas. *Leeds.* 1817-53.
Thomas. *Philadelphia.* 1814-50.
Thomas. *Rotherham.* ?Early 19c. C.
Tobias I. *Barnsley.* 1773-d.1813. C. & W.
Tobias II. *Barnsley.* b.1782, s. of Tobias I.—1837. C.
W. *Croydon.* 1862. Chrons.
W. S. *Picton, Nova Scotia.* 1840.
William. *Leeds.* 1780-1807. Then *York* 1807-09.
William. *Wallace, Nova Scotia.* 1866.
William Frederick. *Ashton-under-Lyne.* 1848-58. W.
William Frederick. *Barnsley.* 1820.
FLEUREAU—
Esaye (or Isaac). *London* from *Orleans, France.*
mar.1694-1703. W. & C.
Jacques. *London.* s. of Esaye. b.1703.
FLICK, Joseph. *Ingersoll, Canada.* 1857.
FLING, Daniel. *Philadelphia.* 1809-22.
FLINN—
E. *Coventry.* 1868-80.
John (& Sons). *Coventry.* 1842-54(60-68).
William & Sons. *Earlsdon, Coventry.* 1880.
FLINT—
——. *Birmingham.* ca.1800. C.
Abraham. *Birmingham.* 1854. Clock case maker.

FLINT—*continued.*
Charles. *Coventry.* 1854-80.
J. *Birmingham.* 1854. Clock case maker.
Mark. *Tunstall (Staffordshire).* 1876.
Mark. *Uppingham.* 1864.
Samuel Ballard. *Driffield.* 1851-58.
William, jun. *Ashford.* d.1813 but business continued till 1851. C. & W.
William James. *Driffield.* d.1846-64. W.
FLOCKHART, Andrew. *London.* 1811-32. W.
FLOOD, William. *Philadelphia.* 1837(-49?).
FLOOKE, John. *London.* (B. 1754) 1760-67. W.
FLOOKS, Charles Henry. *Merthyr Tydfil.* 1887-99.
FLORINGER, Louis. *London.* 1857-81.
FLOUTE, L. F. *Farnham, Surrey.* 1878.
FLOWER—
Arthur. *Shrewsbury.* 1879.
Edward. *Liverpool.* 1851.
George. *London.* a.1670 to James Atkinson (math. inst. mr.). CC. 1682-97. W.
Henry. *Philadelphia.* 1753-75. W. & C.
FLOWERS, John. *Banbury.* 1797-1814. C. & W.
FLOYD—
H. E. *Toronto.* 1862.
Thomas. Pre-1767. London. Later *Charleston, SC, USA.* Also gunsmith.
Thomas. *Truro.* 1783-d.1842.
FLUREAU, Esaye. *London.* See FLEUREAU.
FOCK—
Otto Edvard. *Abo and Turku.* 1823-68. C.
FOGDEN, Henry. *Chichester.* 1839-51.
FOGG—
James. *Liverpool.* mar.1801-28. C.
William. *Farnworth.* 1818. W.
William. *Prescot.* mar.1756. W.
FOGLER, Henry & Son. *Toronto.* 1877.
FOLDERING, Hallam Edward. *? Sheffield.* 1841. Cleaner's label inside a Sheffield clock.
FOLE—
Nathaniel. *Northampton, Mass, USA.* 1819.
Robert. *London.* 1667. CC. math. inst. mr.
FOLEY, William. *Dundee.* 1860.
FOLGER, Walter, jun. *Nantucket, Mass, USA.* b.1765-d.1849. Famous scientist. Complex astron. cl.
FOLIE, Henry. *Merthyr Tydfil.* 1868.
FOLKARD, William. *London.* 1811. W.
FOLKROD, William. *Philadelphia.* 1849.
FOLKS, Robert. *Kirton-in-Lindsey.* 1849-50.
FOLLAND, William. *Exeter.* (B. 1795-)1803. W.
FOLLETT—
Marvill M. *Lowell, Mass, USA.* 1835.
N. M. *Madrid, NY, USA.* 19c.
FOLLOWS—
James. *Birmingham.* 1880.
Thomas. *Newcastle-under-Lyme.* 1828.
FOLWELL, Thomas. *Deal.* 1838-55.
FONNEREAU, James. *London.* 1653. Frenchman. Allowed by CC. to work as journeyman for Mr. Bouquet whilst in England.
James. *St. Gluvias.* 1676. W. (must surely be same as above).
FONTAINE—
John. *London.* b.1693, mar.1728. Later with Moses till ca.1753, then to *Wales* where d.1767.
John & Moses. *London.* ca.1740. (B. pre-1766). C. & W.
Moses. *London.* b.1694, to *Wales.* ca.1753-d.1766.
FONTANA, Charles. *High Wycombe.* 1839-77. W.
FOOLEY, G. *Aylesbury.* 1847.
FOORD, James. *Hastings.* 1862-78.
FOOT(E)—
Charles F. *Bristol, USA.* 1852-55.
John. *Liskeard.* 1769-86. C.
Richard. *Faversham.* 1826-55.
William. *Middletown* and *East Haddam, Conn, USA.* 1772-ca.1836. (Partner with CANFIELD 1795-96).
FORBACH, Joseph. *New York.* 1854.
FORBER, Joseph. *Liverpool.* 1828.
FORBES—
Daniel. *Leith.* 1850.
Edward. *Liverpool* from *London.* 1741. Watch chain maker and gilder.
Francis. *Edinburgh* (B. 1803-) 1814-34. W.

FORBES—*continued.*
John. From *Philadelphia* to *Hartford, USA.* 1770.
John. *London.* 1844-51.
Joshua. *Liverpool.* (B.1825) 1834-51.
& TUCKER. *Concord, NH, USA.* 1841. C.
Wells. *Bristol, NH, USA.* ca.1842. C.
William. *Kintore, Scotland.* 1837.
FORD—
Alfred. *Newtown (Mon).* s. of George Sharratt, 1836-1916.
Andrew. *Norwich.* 1875.
Charles. *London.* 1832.
Daniel. *London.* 1863-75.
George H. *New Haven, Conn, USA.* 1850s. (Partner of Benjamin & Ford).
George Sharratt. *Newtown, Mon.* 1844-68. W.
Hannah. *Wellington, Salop.* 1835.
James (Jacobus). *Bedford.* ca.1670. Lant. clock. ca.1690.
James. *Birmingham.* 1828-35. Also wedding rings.
John. *Oxford.* a.1682-1708. Then Aylesbury till insolvent in 1725.
John. *Treherbert.* 1887.
John Henry. *London.* 1869-75.
Peter. *Lancaster, Pa, USA.* 1783-1820.
Thomas. *Buckingham.* ?ca.1705. C.
Thomas. *Liverpool.* 1834.
William. *Philadelphia.* 1848.
William. *Prescot.* 1824-26. W.
FORDHAM—
John. *Knottingley.* 1819-22.
John Brett. *Knottingley.* b.1819, s. of John Fordham, 1838.
Thomas. *Bocking.* 1828.
Thomas. *Braintree.* 1753. (B. 1779 from *Bishops Stortford*).
FOREL, Abraham. *Nova Scotia.* 1751.
FOREMAN, George Benjamin. *Canterbury.* 1832.
FORESTVILLE—
Clock Co. *Bristol, USA.* 1840s.
Clock Manufactory. *Bristol, USA.* 1849-53 (prop. J. C. Brown). C.
Hardware Co. *Bristol, USA.* 1852. In 1853 became Forestville Hardware & Clock Co.
FORGE, Thomas. *Barking.* 1851-74.
FORMAN, Richard. *London.* 1844.
FORRER—
Christian. *Lampeter* and *Newberry, Pa, USA* from *Switzerland.* 1754-83.
Christian & Daniel. *Newberry* and *Lampeter, Pa, USA.* 1754-74.
Daniel, bro. of Christian. From *Switzerland* to *Lampeter* and *Newberry, Pa, USA.* ca.1755-80. C.
FORREST—
James. *Glasgow.* 1860.
James. *Otley.* b.1833-51.
James. *Pontefract.* 1862-71.
John. *London.* 1857-76. W.
Jonathan. *Launceston.* 1757-ca.1780. C. & W.
Sylvester. *Bradford.* 1853-71.
Sylvester. *Leeds.* Clock dial engraver. 1768-d.1776.
William. *Edinburgh.* 1825-35.
FORRESTER—
Charles Anthony. *Hull.* 1838-40.
P. & F. *Edinburgh.* ca.1780.
Patrick. *Hull.* 1813-40.
Peter & Co. *Edinburgh.* 1783-96. W. (B. has FORREST).
FORSELL, John. *Turku.* 1746-66. C.
FORSTER—
G. *Sittingbourne.* 1802-13. W. & C.
James. *St. Helens.* 1760-67. W.
John. *Newcastle-on-Tyne.* 1834-56.
John. *St. Helens.* 1752-72. W.
John. *Sheerness.* 1790-1823. C.
Ralph. *Farnworth.* mar.1824. W.
Robert. *Corbridge nr Gateshead.* 1848-58 (?d.1884). C.
Thomas. *Carlisle.* 1858.
Thomas. *London.* 1881.
William. *Gravesend.* 1845-47.
William. *St. Helens.* 1755. W.

FORSTER—*continued.*
William Henry. *Gravesend.* (1838-) 1851-66.
William Henry. *Sheerness.* 1832-38.
FORSYTH, T. Morrison. *Turriff, Scotland.* 1835.
FORT—
DEARBORN WATCH & CLOCK Co. *Chicago, USA.* 1918.
John. *London.* a.1660 to Evan Jones. CC. 1672-97. W.
FORTESCUE, J. *New Basford, Nottinghamshire.* 1864.
FORTUNE, John. *Burntisland, Scotland.* 1860.
FOSLER, Charles. *London.* 1844.
FOSTER—
——. *Lincoln (Cornhill).* ca.1830. C. (Prob. William, q.v.)
——. *Sheerness.* ca.1780-90. C. (see John FORSTER).
Edward B. *London.* 1828.
Evan. *Kidderminster.* 1876.
George B. *Boston, USA.* 1842.
Henry. *Ashford, Kent.* 1858-74.
Henry. *Liverpool.* 1848-51.
Henry. *Margate.* 1874.
Henry. *Prescot.* 1826. W.
Horace. *Rye.* 1870-78.
Isiah & Co. *Glasgow.* 1836.
James. *Aberystwyth.* 1832-44. C.
James. *Ashburton.* ca.1790. C.
James. *Leeds.* b.1642-d.1682. W.
James. *Morley* nr *Leeds.* 1869. C.
James. *Peterborough.* (B. 1798) 1830.
John. *Hay.* 1840-44.
John. *Lincoln.* 1813-28 W. succ. to Bunyans.
John. *Manchester.* 1848. Also casemaker.
John. *Market Rasen.* 1850-61.
John. *Prescot.* 1797. W.
John. *Sheffield.* 1871.
John C. *Portland, Me, USA.* 1803.
John & Thomas. *Manchester.* 1851. Imported clocks.
Joseph. *Hull.* 1846.
Joseph. *London.* ca.1800. C.
N. & T. *Newburyport, USA.* Brothers. 1820-86.
Nathaniel. *Newburyport, Mass, USA.* 1797-1893 (with brother, Thomas, 1820-86).
Ralph. *St. Helens.* 1778-d.1786. W.
S. H. *Syston, Leicestershire.* 1855.
Thomas. *Birmingham.* 1868.
Thomas. *Newburyport, Mass, USA.* 1799-1887.
Thomas Wells. *Newburyport, Mass, USA.* 1860.
W. *Boston, Lincs.* 1849.
W. *Southampton.* 1859.
W. *Upwell, Cambs.* 1846.
William. *Kirton, Lincs.* 1850.
William. *Lincoln.* 1828-35.
William. *Lydd, Kent.* 1845.
William. *Manchester.* ca.1750. C. (B. ca.1780-1809).
William John. *Guildford.* 1828.
William Lawrence. *Driffield.* 1858.
FOSTON, J. *Market Rasen.* 1868.
FOTHERGILL—
John. *Knaresborough.* 1807-51. C.
William. *Manchester.* mar.1791. C. (B. 1813).
FOUBISTER, Thomas. *Kirkwall.* 1860.
FOULDS (or FOWLDS)
Allen. *Belfast.* ca.1810-30. C.
James. *Kilmarnock.* Late 18c. C.
FOULKS, Francis. *Bakewell.* 1828-35.
FOUNTAIN—
Thomas. *Sharnbrook* nr *Bedford.* 1864-77.
William. *Streatham (London).* 1878.
FOURNIER—
Bernard Gavin. *Paris.* ca.1770. C.
Louis. *Quebec.* 1863.
Stanislaus. *New Orleans, LA, USA.* 1849. Electric clock.
FOWELL, J. N. *Boston, USA.* 1805-09.
FOWLDS, John. *Kilmarnock.* ca.1810. C.
FOWLE—
(Arthur) Edward. *Westerham.* 1838-d.1879.
Henry. *Uckfield.* 1878.
Henry. *Redhill.* 1862-78.
J. *Sevenoaks.* 1802.
J. H. *Northampton, Mass, USA.* c.1850.
John. *East Grinstead.* 1851-78.

FOWLE—*continued.*
Nathan. *Charleston, SC, USA.* ca.1840.
Richard. *East Grinstead.* 1828-39. (B. has one 1783-95).
Richard. *Lewes* (from *Uckfield*). a.1715 to Richard Turner. C.
Thomas. *Canterbury.* Free 1609. C.
William. *Uckfield.* (B. 1816) 1828-70.
FOWLER—
Abraham. *Amersham.* ca.1760. C.
Daniel. *London.* ca.1770. C.
George. *Horncastle.* 1861-76.
George. *York.* 1819. Then *Doncaster* 1822-51.
J. *Penetangore, Canada.* 1857.
J. F. *Beaconsfield.* 1854-77.
James. *St. Catherines, Canada.* 1861-77.
John. *Chesham.* 1877.
John. *Sheffield.* 1775-94. C.
John C. *Boston, USA.* 1842; *Lynn, Mass,* 1847. Then *Groton, Mass.* 1849.
John William. *London.* 1869.
Joseph. *Bala.* 1868.
Joseph. *Chester.* 1878.
Robert. *York.* 1810, *Leeds.* 1817-37.
Samuel. *Doncaster.* 1862.
Thomas. *Amersham.* ca.1760. C.
William. *Oxford.* a.1667. W.
FOWLES & PARTNER. *London.* 1694. Goldsmith — sold watches.
FOX—
Arthur C. *Cherry Valley, NY, USA.* 19c. chrons.
Asa. *Buffalo, NY, USA.* ca.1810.
August. *Berlin, Canada.* 1851-77 (also FUCHS).
Calvin. *London.* 1851-57.
Charles. *Yeovil.* 1883.
Charles James. *Beverley.* 1834-58.
Edmund. *Ely.* 1846-75.
G. *Castle Donnington.* 1849-55.
Hudson. *Beverley.* 1801-23.
Isaac. *London.* (Minories). Took Isaac Levy as a. in 1757. (B. 1772-94. C.)
J. *Taunton.* ca.1710.
J. *Uppingham.* ca.1775. C.
James. *Beverley.* 1851.
John. *Alverton (Penzance).* 1843. C.
Jonathan. *North Walsham.* 1836.
Mordecai. *London.* (B. CC. 1687)-1697.
Philetus. *Boston, USA.* 1842.
Ralph. *Pontypool.* 1844-58.
Thomas. *Aylsham.* 1875.
Thomas. *York.* b.1748-d.1802. W.
Thomas Henry. *Great Easton (Leicestershire).* 1876.
William. *Liverpool.* 1848.
FOXCROFT—
(James A.) & CLARK (Gabriel). *Baltimore, USA.* ca.1831-39.
James A. *Baltimore, USA.* 1822-39 (partner with CLARK. ca.1831-39).
James Charles. *London.* 1869-81.
William. *London.* 1857-63. Clock cases.
William M. *London.* 1875-81. Clock cases.
FOXTON—
James. *London.* 1828.
James. *Mansfield.* 1828-35.
FOY, Walter. *Watchet.* 1883.
FRAGERCRANS, P. *Princeton, Ill, USA.* 1860s.
FRAME(S)—
George. *Gateshead.* (B. 1811-) 1827-36.
John. *Manchester.* 1848.
Joseph. *Raphoe, co. Donegal.* 1789.
FRAMPTON, Joseph. *London.* 1857-63.
FRANCHERE, J. B. *Montreal.* 1842.
FRANCIES, John. *York.* 1784. W.
FRANCIS—
Andrew. *St. Austell.* mar.1806. W.
Edward. *Carno (Wales).* Later 19c. W.
Joseph. *Stansted.* 1874.
R. *Ipswich.* 1846.
R. *Attleborough.* (B. 1795). ca.1808-ca.1811. C.
Richard Shuckforth. *Ipswich.* 1846-66.
Richard. *Wymondham.* 1830-46.
S. *Birmingham.* 1839. Dial maker.
T. *Coventry.* 1854-60.

FRANCIS—*continued.*
Thomas. *Dumbarton.* 1837.
W. *Forest Hill, Kent.* 1874.
William. *Attleborough.* 1830.
William. *Birmingham.* 1815-30. Dial maker.
William. *London.* 1839. Watch cases.
FRANCISCUS, George. *Baltimore, USA.* ca.1776-91.
FRANCK, Philip. *New Berlin, Pa, USA.* Early 19c.
C.
FRANK—
Benjamin. *Pontypool.* 1880.
Jacob. *Pontypridd.* 1875.
L. *New York.* 1872.
(L.) & LICHTENAUER (M.). *New York.* 1854.
N. *Belfast.* 1835.
FRANKCOM & MOWAT. *Bath.* 1810-35.
FRANKEL, Jacob. *Greenwich.* 1839.
FRANKENSTEIN, Jacob. *London.* 1851.
FRANKFIELD, A. & Co. *New York.* 1866. Cuckoo
clocks.
FRANKLAND, W. *Shipdham, Norfolk.* 1858.
FRANKLIN—
Abraham. *Manchester.* 1824-34. Also jeweller, silver-
smith, etc.
John. *London.* 1857-81.
& MARSHALL. *Seneca Falls, NY, USA.* 1830s.
FRANKS—
——. *Tiverton.* 1619, rep'd. ch. clock.
Abraham. *Paulton.* 1861-75.
E. *Crayford, Kent.* ca.1790. W.
J. & S. *Philadelphia.* 1850.
Jacob. *Philadelphia, USA.* 1845-49.
John. *Londonderry.* 1784.
John Fearn. *Dewsbury.* 1853-71. W.
Mary. *Dewsbury.* 1851.
Richard. *Dewsbury.* 1850.
Richard. *Newark.* 1828.
FRANKSEN, B. *New York.* 1872.
FRARY—
Obediah. *Southampton, Mass, USA.* 1745-75. C.
V. W. *Norwich.* 1858.
FRASER—
Alexander. *Kingsbridge, Devon.* mar.1740.
Alexander. *Menallin Township, Pa, USA.* ca.1840.
Alexander. *Vankleek, Canada.* 1857.
Francis. *Montreal.* 1866-71.
Frederick. *Wrexham.* 1856-90. W. and skel. clock.
H. & J. *Tain.* 1860.
Hugh. *Dunkeld.* 1860.
Hugh. *Tain.* 1836.
Jacob. *Philadelphia.* 1801-77.
James. *Lochgilphead.* 1860.
James. *Windsor, Canada.* 1861-71.
John. *Aberdeen.* 1846-60.
John. *London.* a.1681 to Edward Eyston. Late 17c.
clock signed without town. Believed worked later at
Worcester. Also FRASOR and see John FRAZER.
John. *Perth.* 1843.
FRASIER, Thomas. *London.* 1828.
FRAY, William. *Skipton.* 1784-89. W.
FRAZER—
Alexander. *Comber, Co. Down.* ca.1770-90. C.
H. N. *Vienna, NY, USA.* 1839.
& HAWS. *London.* 1875-81.
James. *Comber.* ca.1790-1810.
J(ohn). *Worcester.* ca.1700. C. (also (FRAZOR) be-
lieved from *London.*
Robert & Alexander. *Philadelphia.* pre-1799; *Paris, Ky,
USA.* 1799; Robert to *Lexington, Ky* 1799-1818.
Alexander went there post-1803.
Samuel. *Baltimore, USA.* 1822-24.
William. *Montreal.* 1819.
FRAZIER, George. *Newry.* 1827.
FREARSON, T. *Coventry.* 1880.
FREDERICK—
Charles. *Aberdeen.* 1860. C.
Leonard. *Bilston.* 1842.
Leonard. *Preston.* 1848.
Leonard. *Wolverhampton.* 1850.
Paul J. *St. Luce, Canada.* 1853. Also tailor.
T. *Wolverhampton.* 1868.
FREDONIA WATCH CO. *Fredonia, NY, USA.* 1884-5.

FREE—
John. *Oxford.* a.1696-d.1726. C. & W.
Penelope. *Oxford.* wid. ot John. 1726-34.
FREEBODY—
Charles. *Ryde.* 1859-78.
T. *Walton-on-Thames.* 1851-66.
FREEDMAN, Mark. *Chipping Ongar.* 1839.
FREEMAN—
Brothers. *Atlanta, Ga, USA.* ca.1850-60.
C. *Coventry.* 1860.
Charles. *Coventry.* 1880.
Charles. *Liverpool.* 1848.
Charles Bailey. *Welwyn.* 1839-74.
D. M. & Co. *Atlanta, Ga, USA.* 1885-87.
Edwin. *Wem.* b. ca. 1838 -61.
H. *Dorchester.* 1854.
Henry B. *West Chester, Pa, USA.* 1857. W. & C.
J. *Birmingham.* 1860.
J. *Highworth.* 1848.
Jacob. *Hull.* 1851-58.
Mark. *London.* 1851-57.
Nathaniel. *London.* 1839-44.
Richard. *Newbury (Thatcham).* apr. 1831 to John
Couderoy. 1864-77.
Richard. *Lechlade.* 1840-50.
S. *Coventry.* 1854. Watch case maker.
T. *Hinckley.* 1864.
T. *Wellington.* 1863.
Thomas. *Nuneaton.* 1880.
W. *Crockerton, Wilts.* 1859.
William. *Baltimore, USA.* ca.1810. Clock cases.
William. *Portsea.* 1859-78.
FREEMANTLE, John G. *Spalding.* 1876.
FREICHER, Leopold. *Salzburg.* ca.1730. C.
FRENCH—
A. W. *Maidstone.* 1812-39. W.
Brooks. *Bedford.* 1839-69.
Brooks. *Upwell.* 1830-36.
Charles. *Brattleboro* and *Rutland, Vt, USA.* pre-1825.
Then to *Ohio* 1825.
& DAVID. *Wenvoe (Wales).* Later 18c. C.
Edward. *Bristol.* ca.1750-60. C.
George. *London.* 1863.
J. M. *London.* 1811. Chrons.
Jacob. *London.* 1857.
James. *Bristol.* (B. 1825) 1830-50.
James Moore. *London.* 1828-44.
John. *Felsted.* 1874.
John. *Hamilton, Canada.* 1851-56.
John. *London.* 1844.
John. *Yalding.* 1832-55.
Lemuel. *Boston, USA.* ca.1790-1820. C.
& PORTHOUSE. *London.* 1863.
& READ. *London.* 1811. W.
Richard Vigor. *Maidstone.* 1823-38.
& Son. *Maidstone.* 1851.
Stephen. *Maidstone.* 1847.
Susanne & APPS, William. *Maidstone.* 1839-55.
Thomas. *Kibworth Harcourt.* 1835.
Thomas. *Pen-Coed.* Early 19c. C.
Thomas Veitch. *Newcastle-on-Tyne.* 1848-58. C.
William. *London.* 1851-81.
William. *Waltham Cross.* 1874.
William Apps. *Maidstone.* 1865-74.
FRENKEL, S. *Toronto.* 1877.
FRESEMEYER, Joseph. *Bryn-mawr.* 1868-87.
FRESER & CO. *Salisbury.* 1842.
FRESLE, Thomas. *Hereford.* 1850-56.
FRESTON—
Albert. *Brightlingsea.* 1874.
Alfred. *Ipswich.* 1879.
FRETWELL, A. W. *Doncaster.* 1871.
FREWEN—
James. *Pontypool.* 1850.
John. *Chipping Sodbury.* 1870-79.
FREWIN—
Ann. *Hereford.* 1850.
J. & S. *Bristol.* 1842.
James. *Abersychan.* 1852-71.
James. *Hereford.* 1830-44.
James. *Sodbury.* 1830-79.
John. *Bristol.* 1879.
John. *Tetbury.* 1850.

FREWIN—*continued.*
Samuel. *Bristol.* b.1754-d.1838.
Samuel. *Newport* (*Mon.*). 1858.
FREYBERG, George. *Tenby.* 1850-75.
FRICK Clock Co. (Frederick). *Waynesboro, Pa, USA.* C. (no date).
FRICKER, John. *Coggeshall.* 1866-74.
FRIDGE, James. *Leeds.* 1794-1802.
FRIDLANDER—
& Co. *Coventry.* 1868-80.
& KLEAN. *Coventry.* 1860.
Moses. *Coventry.* 1842-54.
FRIEDEMANN, J. *London.* 1875.
FRIEDENHEIMER, J. *Brockville, Canada.* 1857.
FRIEDLANDER—
——. *Huddersfield.* mid-19c. C.
J. *London.* 1857.
FRIEDLEY, Emile. *Bradford.* 1866-71. W.
FRIEDRICH, F. *Worcester.* 1868-76.
FRIEND—
Engell. *New York.* mid-19c.
George. *New York.* ca.1820.
John Walter. *Freemantle* (*Hants*). 1867-78.
W. *Newton Abbot.* ca.1810. W.
W. *Southampton.* 1859.
Walter. *London.* 1844.
FRIENDLEY, Ralph. *Prescot.* 1822.
FRIES—
John. *Philadelphia.* 1830-50.
John & P. *Philadelphia.* 1837.
P. *Philadelphia.* 1839 & later.
FRINK, Urban. *Brattleboro, Vt, USA.* ca.1880.
FRISBIE, L. & J. & Co. *Chittenango, NY, USA.* ca.1840.
FRISBY—
—— *Leicester.* Early 19c. W.
John. *London.* 1811. W. (B. 1817-25).
FRISLEY—
——. *London.* 1805. W.
John. *London.* 1828-32.
FRITH—
A. E. *Halifax* (*Yorks*). Prob. late 19c. C.
Richard. *Ashbourn.* 1835.
FRITZ—
C. *Philadelphia.* 1848. Clock repairer.
G. *Bristol.* 1863.
Matthias. *Derby.* 1876.
Matthias & Co. *Birmingham.* 1860-68.
FRIZELL, Louis. *Dungannon, Co. Tyrone.* 1820-24.
FRODSHAM—
& BAKER. *London.* 1857-63.
Charles. *London.* 1863-81. 'By apt. to her Majesty, etc.'
G. & William. *London.* 1851.
George. *London.* 1857-69. Chrons.
George Edward. *London.* 1869-75. Succeeded to Frodsham & Baker.
George William. *London.* 1875-81.
Henry. *Liverpool.* (B. a.1823) 1848-51.
Henry Herbert. *London.* 1881.
John & Son. *London.* 1839-51.
Samuel. *Manchester.* mar.1815. C.
William James. *London.* 1839-44.
FROMANTEEL (also Formantle, etc.). The following data supersedes that in Baillie.
Abraham I. b.ca.1618 Norwich. bro. of Ahasuerus I. To *London* ca.1630, a.1631 in B.C. Not known whether he became a blacksmith or clockmaker, but nothing heard of him after 1631.
Abraham II. s. of Ahasuerus I. b.ca.1646, a. to father in *London* 1662. After a period with father in *Holland* he worked at *Newcastle-on-Tyne.* Returned to *London* where free of CC. 1680. Died *Newcastle* 1731.
Ahasuerus I. b.1607 *Norwich.* To-*London* 1629, BC. 1630, CC. 1632, mar.1st 1631 Maria de Bruijne, mar. 2ndly ca.1660. Sarah Winnock, becoming stepfather of Joshua Winnock, clockmaker. The most famous of all English clockmakers being the first to make pendulum clocks in this country. Took son-in-law, Thomas Loomes, into partnership. Worked at *Mosses Alley, Southwark* and at the *Mermaid* in *Lothbury* (home of Thomas Loomes). d.1693.
Ahasuerus II. b.1640, s. of Ahasuerus I. a.1654 to father. CC. 1663. Went ca.1680 with brother John to

FROMANTEEL—*continued.*
Amsterdam, where he remained. Dtr. married Christopher Clarke there in 1694, giving rise to partnership of Fromanteel & Clarke. Died 1703.
Ahasuerus III. b.1666, s. of John. a.1679 but prob. went to *Holland* with father shortly after. Not known what became of him.
& CLARKE. *Amsterdam* (not *London*). Prob. partnership of Ahasuerus II & Christopher Clarke, his son-in-law. Believed terminated in 1703 when Ahasuerus died. Clarke later a partner with Roger Dunster. Period ca.1694-ca.1722.
Daniel. b.ca.1651, s. of Ahasuerus I. a. to father 1663 but never completed it. Prob. died young. No clocks known by him.
John. b.1638, s. of Ahasuerus I. a.1651 to father, CC. 1663. Studied under Salomon Coster in *Holland* 1657-58, then worked in *London.* Went to *Amsterdam* ca.1680 with brother, and believed died soon after.
Louis. No such person, though some authorities quote this name. Error arose from a mis-reading of a manuscript.
FROMOL, W. B. *Liverpool.* Mid-19c. Wall clock.
FRONMUELLER, Hans. *Augsburg.* 1602.
FROOM, Jonathan. Place unknown. 1813. W.
FROST—
Benjamin. *Reading, Mass, USA.* b.1826-ca.1850. C.
David. *Winchester.* 1867-78.
Edward. *Newport, Mon.* 1835-52.
Henry H. *Bristol.* 1879.
J. *Crich.* 1849.
Jacob. *Kendal.* b.1831-71. W. b. *Renwick* (*Cumberland*), s. of John.
John. *Chichester.* mar.1727-d.1768. C.
John. *Colchester.* b.ca.1745-d.ca.1798. W.
John. *Renwick* (*Cumberland*). 1830-34.
Jonathan. *Exeter.* 1785. W.
Jonathan. *Reading, Mass, USA.* 1798-1881. C.
Merriam & Co. *Bristol, USA.* 1836.
& MUMFORD. *Providence, RI, USA.* 1810.
N. A. *Hanover, NH, USA.* ca.1885.
Oliver. *Providence, RI, USA.* ca.1800.
R. J. *Leighton Buzzard.* 1877.
Samuel. *Burg.* Signature of Samuel F. of *Cheltenham,* q.v.
Samuel. *Cheltenham.*? Early 18c. C.
Samuel. *Newark,* 1853-94. C. From *Nottingham.*
Thomas. *Newport* (*Mon.*). 1858.
William. *Newport* (*Mon.*). 1822-30.
William & Edward. *Newport* (*Mon.*). 1830.
FROWD, William. *Hemel Hempstead.* 1839-74.
FROYNE, George. *Pembroke.* 1868-87.
FRY—
Samuel. *Dublin.* (B. ca.1810) ca.1830 lc. clock with barometer in door.
William. *Odiham.* 1830.
William. *Southampton.* 1839-48.
Young. *New Alresford.* 1859-78.
FRYE—
James. *Haverhill, Mass, USA.* 1853. W.
& SHAW. *New York.* ca.1830.
FRYER—
Edward. *Driffield.* 1851.
Edward. *Pocklington.* b.1821-58. s. of John.
John. *York & Pocklington.* 1812-30. C.
Maria. *Pocklington.* 1830-44. Widow of John.
Michael E. *Wetherby.* 1866-71. W.
Mrs. *York.* 1823. W.
Peter. *Albany, NY, USA.* 1824-25. Then *Norwich, NY,* 1825-40.
William. *York & Pocklington.* 1809-1867. W.
William, jun. *Leeds.* 1817-51. W.
FRYETT, John William. *London.* 1863.
FUBISTER, John. *Edinburgh.* 1836.
FUCHS (or FOX), August. *Berlin, Canada.* 1851-77.
FUCHTER, Fidel. *Maidstone.* 1874.
FUDGE—
——. *Southampton.* Late 19c. C.
Thomas. *Froom* (= *Frome*) ca.1740. C.
FULLER—
——. *Watton* (*Norfolk*). ca.1790. C.
Artemas. *Lowell, Mass, USA.* ca.1840-50.

FULLER—*continued.*
F. A. *Rutland, Vt, USA.* ca.¹830. W.
Henry. *London.* 1839.
& HOLLANDERS. *Landport.* 1878.
& IVES. *Bristol, USA.* 1837.
Joseph. *London.* 1839.
Samuel. *London.* (B. 1810-25) 1828-32. Wholesaler.
Thomas. *Bury St. Edmunds.* 1830-46.
William. *Chelmsford.* pre-1627-32. Locksmith.
FULLERTON—
John. *Larne.* 1865-90. C.
Samuel. *Belfast.* 1854-58. C.
FULLFORD—
James. *London.* ?ca.1740. C.
Richard. *Fareham.* 1830.
FULLWOOD, James. *Newport.* 1875.
FULTON—
David. *Campbeltown.* 1860.
Francis. *Alston.* 1848.
James. *Shelby Co., Ky, USA.* 1835.
John. *Fenwick (Ayrshire).* 1834.
Joseph. *Aughnacloy, co. Tyrone.* mar.1858-92. C.
Watch Co. *Lancaster, Pa, USA.* 1923-24. W.
FUNK, Jacob. *Lebanon, Pa, USA.* ca.1850.
FUNNELL, Edward. *Brighton.* 1870-78.
FURBER, Thomas & Son. *Cheltenham.* 1879. Chrons.
FURDERER, Mrs. P. *Nottingham.* 1855.
FURENBACH, M. O. *Edinburgh.* ca.1800. C.
FERGUSON, John. *Bathgate (Scotland).* 1860.
FURLEY, John. *Huntingdon.* 1839-64.
FURNACE—
Anthony. *Cockermouth.* 1848-58.
Anthony. *Keswick.* 1858-79.
FURNER, Francis. *Rochford.* (B. 1794-)1828-55.

FURNISS—
George. *London.* 1857-81.
Henry. *London.* 1851.
Jacob. *Leeds.* 1834. C.
James William. *London.* 1857-81.
Richard. *London.* 1851.
Samuel. *Newcastle, Md, USA.* s. of William. 1735.
Thomas. *London.* 1839-57.
William. *Newcastle Co., Del, USA.* ca.1740-48. C.
FURNIVAL—
Benjamin. *Liverpool.* 1834.
James. *Marblehead, Mass, USA.* ca.1780.
Thomas. *Sheffield.* ca. 1760. C.
FURRER, John & Co. *London.* 1863-81.
FURSE, William (& Son). *Brighton.* 1839-51 (1855-78)
FURSTENFELDER, Binedict. *Aichen.* ca.1700. W.
FURTER, WINTERHALDER & CO. *London.* 1857.
FURTWANGLER—
Andreas. *London.* 1857-81.
Andrew. *Kilmarnock.* 1860.
F. *London.* 1881.
Michael. *Cardiff.* 1858-71. C.
Michael. *Swansea.* 1852.
FURTWENGLER, Sebastian. *Llanelly.* 1868-87. W.
FURZE, Henry. *London.* 1832.
FUSSELLI, Peter. From *Italy* to *Bowling Green, Ky, USA.* 1850-77. Later back to *Italy.*
FUTVOYE, ——. *Paris.* Pre-1850. C.
FYLER—
——. *Torrington* or *Torringford, Conn, USA.* ca.1830. C.
Orsamus. *Winchester, Conn, USA.* 1831-33. C.
Orsamus Roman. *Chelsea & Bradford, Vt, USA.* b.1793-1833.

G

G. Charlby. Monogram of Thomas Gilkes I of Charlbury.
G, F. X. FXG *Prague.* Early 18c. W.
G, P. PG is mark of C. P. Gontard, q.v.
GABRIEL—
  L. *Quebec.* 1862.
  William. *London.* 1875-81. C.
GADOIS, Jaques. *Montreal.* d.1780.
GADSBY—
  G. *Hathern.* 1849.
  Mrs. A. (Mrs. William). *Lincoln.* 1868-76.
  R. *Leicester.* 1849.
  Sarah. *Leicester.* 1835.
  Thomas. *Leicester.* (B. 1815). ca.1815-28. W. & C.
  Thomas. *Sheffield.* 1837. W.
  Thomas & Son. *Hinckley.* 1828.
  William. *Lincoln.* 1861.
GAFAEL, U. *London.* 1844.
GAGE—
  A. W. *Hamilton, Ontario.* 1875.
  Hiram. *Conn.* or *NH, USA.* ca.1842.
GAGGIS, Valentine. *London.* 1857-69.
GAILLARD, Peter. *Reading, Pa, USA.* From *France.*
  1794-98. C.
GAINES—
  John. *Plymouth, NH, USA.* 1800.
  Richard. *Baltimore, USA.* 1806.
GAINEY, W. B. *Pendleton, SC, USA.* 1859.
GAINSBOROUGH, Humphrey. *Henley-on-Thames.*
  b.1718-96. Engineer and C.
GALBRAITH—
  Alfred. *Georgetown, Canada.* 1857.
  Alfred. *Milton, Canada.* 1857.
  John. *Falkirk.* ca.1770. W.
  Robert H. *Brampton, Canada.* 1857-63.
GALE—
  Ann. *Newport, Isle of Wight.* 1830-39.
  Daniel Jackson. *Bristol, USA.* 1830-1901.
  James. *Salem, Mass, USA.* 1791-1819.
  John. *London.* (B. 1790-)1828-44.
  Joseph. *Fayetteville, NC, USA.* 1792-98.
GALER, John. *St. Albans.* 1828.
GALEWSKI, Israel. *Sunderland.* 1865-81. W.
GALLAGHER, Thomas. *Whitby, Canada.* 1857.
GALLEY, W. *Bury St. Edmunds.* 1858-65.
GALLICHAN, John. *Reading.* 1877.
GALLIE, John. *Dublin.* 1868-74. W.
GALLIMORE—
  Thomas. *Ashbourne.* b. 1817 d. 1874. Clock
  materials & had Horse & Jockey Inn.
GALLINGTON, William. *London.* 1851-57.
GALLOIS, ——. *Paris.* ca.1730. W.
GALLOME, C. *Baltimore, USA.* 1819.
GALLON—
  & CARTER. *Seaham Harbour.* 1834.
  James. *Antrim.* ?Late 18c. or early 19c. C.
  Thomas Carr. *South Shields.* 1847-56.
  William. *South Shields.* 1827-34.
GALLOWAY—
  Henry. *London.* 1869.
  James. *Ballymoney.* ?Late 18c. or early 19c. C.

GALLOWAY—*continued.*
  James. *Leeds.* 1817-50.
  John. *Leeds.* 1817-22.
  John. *Leeds.* 1850-71.
  John G. *Leeds.* 1826-50.
  John & Thomas. *Leeds.* 1834.
  Mary. *Leeds.* 1853-66. W.
  Matthew. *Leeds.* 1817-26.
  Matthew Thomas. *Leeds.* 1850-71.
  Thomas. *Leeds.* 1837.
  William. *Leeds.* 1834-71.
GALOP, James. *Dorchester.* Fore Street in 1730. C.
GALPIN, Moses. *Bethlehem, Conn, USA.* 1821-22. C.
GALT—
  ——. 1914 and later, prod. of Pequegnat Clock Co.
  James. *Williamsburg, Va, USA.* 1766.
  Peter. *Baltimore, USA.* bro. of Stirling. 1777-1830.
  Samuel. *Belfast.* ca.1720-50. C.
  Samuel. *Williamsburg, Pa, USA.* 1751.
  Stirling. *Baltimore, USA.* bro. of Peter. 1802.
  William. *Washington, DC, USA.* 1815-40.
GALWARD, Charles Thomas. *Holywell.* 1887.
GALWAY—
  Brice. *Magherafelt.* 1858-96. W.
  James. *Belfast.* ca.1820-d.1829. C.
  Robert & Sons. *Cookstown.* 1858-68. C.
GAMBELL, Francis. *London.* 1851-57.
GAMBIER—
  George. *Canterbury.* 1847-51.
  Nicholas. *Paris.* 1660.
GAMBEL, Thomas. *Walton by Kimcote (Leicestershire)*
  1704. C.
GAMBLE, Henry. *Farnley (nr Leeds).* 1739-d.1780. C.
  'Died suddenly after eating a hearty dinner.'
GAME—
  G. *Portsea.* 1859-67.
  William. *London.* 1863-75.
  William James. *Stevenage.* 1874.
GAMMACK, James. *Aberchirder, Banff.* 1846.
GAMMAGE—
  B. *Woodbridge.* 1879.
  T. *London.* 1844-63.
GAMMON—
  John. *London.* Early 19c. C.
  William. *Birmingham.* (B. a.1795-)1816.
GAMP, Philip. *London.* 1832.
GANDER, John. *Eastbourne.* 1839.
GANDY, Francis. *Chichester.* 1851-78.
GANE, J. *Frome.* 1861.
GANERON, Stephen. *London.* ca.1735-ca.1740. C.
GANEVAL, Auguste. *London.* 1832-51.
GANNET, Aaron. *Troy, NY, USA.* 1842-44.
GANNEY—
  Henry. *London.* 1875-81.
  William. *London.* 1828.
GANNON—
  S. J. *Springfield.* 1874.
  Sydney James. *Chelmsford.* ca.1870.
GANS, ——. *Paris.* ca.1800. C.

85

GANT—
& DURANT. *London.* 1851.
John. *Farnworth.* mar.1787.
Robert Drake. *Woodbridge.* 1839-46. C.
GANTER—
Adolphe. *Dublin.* 1866-74(-present day).
Bernard. *London.* 1869-75.
Blesi (Blase). *London.* 1857-75.
C. & J. *Ipswich.* 1875-79.
Felix. *Abersychan.* 1868-71.
Francis. *London.* 1839-44.
George. *Newark.* 1840-60. C.
G. *Wolverhampton.* 1860-76.
John & Co. *London.* 1832.
John & Co. *London.* 1869-81.
Joseph. *London.* 1881.
Leopold. *Aldershot.* 1878.
M. *London.* 1863.
Matthew. *Manchester.* 1851. Imported clocks.
Mrs. *Caroline. London.* 1881. (?wid. of Blase).
Mrs. M. A. *Huddersfield.* 1866. W.
P. *Barnsley.* Early 19c. C.
Paul. *Huddersfield.* 1837-53. C.
Thaddaus. *London.* 1875-81.
Thaddens & Co. *Wolverhampton.* 1842-50. German clocks.
GANTLETT—
J. W. *Calne.* 1859-67.
John. *Gt. Sherston* nr *Malmesbury.* 1830.
Quintillian. *Calne.* 1842-59.
Quintillian. *Chippenham.* 1867-75.
GANZ—
Bernard Henry. *London.* 1881.
John Xavier. *Swansea.* s. of Xavier, later 19c.
William. *Llanelly.* 1884 and later.
Xavier. *Swansea.* From *Germany.* 1868-99. C. & W.
GARANDEAN (or GARANDER), *Glande. London.* 1622. Alien. (Prob. GARANDEAU, one such *Paris* 1579-1600.)
GARD—
W. Termain. *St. John, New Brunswick.* 1875.
William. *Exeter.* b.1769-1803. W. & C.
GARDEN—
James. *London.* 1844-75.
Patrick. *Brechin.* 1860.
Peter. *Longside (Aberdeenshire).* 1837.
GARDENER—
A. S. *Coventry.* 1880.
S. H. *Reading.* 1877.
Thomas. *Coventry.* 1880.
William Henry. *London.* 1881.
GARDINER—
——. *Dublin.* 1788-90.
Baldwin. *Philadelphia.* Pre-1827. Then *New York.* 1827-35.
Barzillai. From *Guildford* 1807 to *Charlotte, NC, USA.*
H. *Wrentham.* 1858-65.
Henry. *Belfast.* c.1805-37. C. & W. Also dentist and silversmith.
James P. *Columbia, Pa, USA.* 1843.
John B. *Ansonia, Conn, USA.* 1857.
& KYNOCK. *Edinburgh.* 1846.
(Barzillai) & MCBRIDE (Andrew). *Charlotte, NC, USA.* ca.1810.
Mrs. M. *Wrentham.* 1875.
Peter. *Perth.* 1833.
W. & J. *Perth.* 1837.
William James. *Toronto.* 1906. Watch case maker.
GARDNER—
——. *Ballymoney, co. Antrim.* ca.1780. C.
Alfred. *London.* 1881.
Charles. *Hull.* 1838-58.
& Co. *Canterbury.* 1838.
David. *Tenbury (Worcs.).* 1876-79.
(Henry) & DOWLING (James). *Belfast.* ca.1827. Partnership dissolved 1830.
Edward. *Houghton-le-Spring.* 1851-56.
Edward. b.1826 at *Halton, Lancs. Lancaster.* 1851-69.
George. *London.* 1851-69.
Henry. *Brighton.* 1870.
Henry. *Cardiff.* 1871-87.

GARDNER—*continued.*
J. *Gloucester.* 1863.
James. *Greenock.* 1860.
James. *Liverpool.* 1834.
James King. *London.* 1869.
John. *London.* 1832.
John. *Lydd.* 1826-28.
John. *Painswick* (Glos.) 1840-50.
John. *Sandwich.* 1747-52. C.
Joseph. *London.* 1839. Watch cases.
& LEE. *Belfast.* ca.1818-27. C.
Mrs. Sarah. *London.* 1875. (?wid. of James King).
Obadiah. *London.* ?Late 18c. C.
R. L. from *London* to *Belfast.* 1805-18. W. & dentist.
Robert. *East Sheen.* 1878.
Robert. *Hunmanby.* 1834.
Thomas F. *Kings Lynn.* 1875.
Thomas George. *London.* 1863.
Thomas J. *Dunbar.* 1860.
William. *Birmingham.* 1880.
William. *Brigg.* ?ca.1820.
William. *Harborne.* 1876.
William. *Hull.* (B. 1820-) 1823-40.
William. *Oxford.* a.1755-58. C.
William. *Penrith.* 1828-34. Repairer.
William. *Sandwich.* 1733-d.1758. C.
William. *Woodbridge.* Mar. 1821-30. W.
William Henry. *Coventry.* 1868-80.
William K. *Bridlington.* 1823-40.
GARD'THAUSEN, Christian. *London.* 1869-81.
GARDUS, W. *Whaddon.* 1847.
GARDY, J. *Emsworth.* 1839.
GARGRAVE, John. *London.* 1656.
GARLAN, John R. *Greensboro, NC, USA.* 1843-45. W.
GARLAND—
——. *Plymouth.* mid-19c. C.
——. *Sunderland.* ca.1820. C.
Edward. *Rugby.* 1810. W. (B. pre-1820).
Henry. *London.* 1857.
Thomas. *Dickleburgh.* 1836-46.
William F. *Kingswood (Gloucestershire).* 1879.
GARLICK, T. *Newbury.* 1877.
GARMONSWAY, Charles. *London.* 1844-75.
GARN, Henry. *London.* 1881.
GARNER—
E. *Stratford (Essex).* 1866.
Edwin T. *Utica, NY, USA.* 1842.
John. ?*London.* ca.1675-ca.1690. C. (Perhaps GARDNER.)
William. *Manchester.* mar.1799. C.
GARNETT—
Edmund. *Lancaster.* b.1833-51. W.
Jeremy. *London.* Probably mis-rendering of GAZNETT, q.v.
John. *Bold.* b.1765. s. of William II. mar.1796-1811. W.
Nathan. *Kendal.* 1834-48.
Richard. *Bold.* s. of William II. mar.1785.
Robert I. *Farnworth.* b.1770. s. of William II. mar.1805-d.1843. W.
Robert II. *Farnworth.* s. of Robert I. b.1805-d.1877. W.
Thomas. *Rainhill.* d.1786. W.
Thomas. *Bold.* b.1768. s. of William II. mar.1799, d.1812. W.
William I. *Bold.* b.ca.1711-d.1789. W.
William II. *Cronton.* b.1731. s. of William I, d.1796. W.
William III. *Bold.* b.1760. s. of William II. d.1839. W.
GARNIERE, L. *Montreal.* 1850.
GARNISH, William Henry. *London.* 1881.
GARRARD—
——. *London.* ca.1720. C.
F. *Bury St. Edmunds.* 1865.
G. *Bury St. Edmunds.* 1858.
J. *Bury St. Edmunds.* 1846-58.
Robert. *Bury St. Edmunds.* 1830-39.
Robert. *Bury St. Edmunds.* 1875-79.
GARRATT (& GARRETT)—
Arthur. *Guildford.* 1878.
Benjamin. *Goshen, Pa, USA.* 1771-1856. C. and clock cases.

GARRATT (& GARRETT)—*continued.*
C. *Walthamstow.* 1866.
Charles. *Brighton.* 1878.
Frederick. *Dunstable.* 1877.
& HARTLEY. *Philadelphia.* 1827.
Henry. *Ormskirk.* 1824-28.
James. *Montreal.* 1845-57.
James. *Quebec.* 1852-57.
John. *Peterborough.* 1830.
Mrs. H. *Guildford.* 1862-66.
P. & W. *Peterborough.* 1847-54.
Philip (& Son). *Philadelphia.* 1801-28 (1828-35). C. & W. Importers.
Richard. *London.* 1658. Not admitted to CC.
Samuel Charles. *London.* 1863-81.
& Son. *Hertford.* 1874.
T. & Co. *Montreal.* 1859.
Thomas C. (& Co.). *Philadelphia.* 1829-40(-41). W.
W. *Hertford.* 1859-66.
William. *Peterborough.* 1841.
William. *Pudsey.* 1840.
William. *Thrapstone.* 1830.
GARRICK—
Fergus. *Stranraer.* 1836-60.
John. *Stranraer.* 1836.
GARRISH, D. D. *Boston, USA.* 1854.
GARROD—
Edward. *Saxmundham.* 1839-58.
John Frederick. *Needham Market.* 1858-79.
GARROOD, Gilbert. *Weedon Beck.* 1869-77. Clock cleaner.
GARSIDE, W. H. *Stalybridge.* 1878.
GARSINE, John. *London.* In 1657 was unauthorised journeyman of William Pettit. Dismissal ordered by CC.
GARTH—
John. *Harrogate.* 1822-26.
John. *Knaresborough.* 1822-44. C.
Thornton. *Knaresborough.* 1822-26. Then *Bramley* (*Leeds*) 1834-50.
GARTLEY, John. *Aberdeen.* ca.1780. W. (B. 1799-1810).
GA(E)RTNER—
Charles & Co. *London.* 1863-69.
Jacob. *Philadelphia.* 1790-1800.
GARTON, A. *Lincoln.* 1868.
GARTY—
George. *Dublin.* 1829-31. W. (B. 1795).
George William. *Dublin.* 1824-26.
William. *Dublin.* (B. 1824). 1828-31.
GARVAN & WRIGHT. *Irvine.* 1820.
GASCOIGNE—
Samuel. *London.* a.1668 to Robert Seignior, CC. 1676-97.
William senior. *Newark.* b. 1620 d. 1675. Rep'd TCs & lantern cl.
GASKELL—
Joseph. *Birmingham.* 1880.
Samuel. *Congleton.* 1848.
Thomas. *Knutsford.* (B. 1784-95) 1828.
GASKES—
Samuel Lockett. *Manchester.* 1851.
GASKIN—
John. *Dublin.* a.1725-65.
John. *Dublin.* 1803-80. W.
William. *Wellington* (*Salop*). 1879.
GASSETT, C. *Wandsworth.* 1878.
GASSIZ, A. fils. *St. Imier, Switzerland.* 1881.
GASTELL, ——. *Keighley.* 19c. C.
GATELEY, Michael. *Dublin.* 1868-74.
GATENBY, Robert. *Pateley Bridge.* 1871.
GATES—
Thomas. *Faversham.* Late 17c. lant. clock.
Zacheus. *Harvard* and *Charlestown, Mass, USA.* Early 19c. C.
GATH—
Thomas (I?). *Bristol.* 1830.
Thomas (II?). *Bristol.* 1863-79. C. & W.
GATHERWOOD, Joseph. *London.* 1839.
GATRELL, F. *Fordingbridge.* 1867.
GATTY, Thomas. *Bodmin.* 1844-73. W.

GATWARD—
Benjamin. *Ware.* 1839.
Benjamin. *Welwyn.* 1851-74.
Charles. *Huntingdon.* 1877.
Cornelius W. *Hitchin.* 1859-74. (Also GATWOOD).
E. *Saffron Walden.* 1866.
Henry. *Saffron Walden.* 1866-74.
John. *Hitchin.* 1828-51.
Joseph. *Tonbridge.* 1802.
Samuel. *Hemel Hempstead.* 1828.
Samuel. *Huntingdon.* 1830-64.
Thomas. *Linton.* 1830-46.
GATWOOD, Edmund. *Bangor* (*Wales*). 1874.
GAUCHER, Joseph T. A. *Quebec.* 1914-20. Electric clocks.
GAUDIN, Johan. *Vienna.* 1660.
GAUKROGER, John. *Halifax.* 1820-50. W.
GAUL, Frederick William. *Doncaster.* 1862-71.
GAULD, Harry. *Lumsden* (*Scotland*). 1860. W.
GAULTON, Frederick. *Southampton.* 1867-78.
GAUNT, Benjamin. *Barnsley.* 1871.
GAUTER, Joseph. *London.* 1875. (Perhaps GANTER, q.v.)
GAUTHIER—
A. *Montreal.* 1853.
Laurence. *Quebec.* 1817-54.
GAVREAU, V. *Quebec.* 1822.
GAVIN, Charles. *Bournemouth.* 1878.
GAW, William P. *Philadelphia.* 1816-22.
GAWNE, Charles E. *Barrow-in-Furness.* 1869. W.
GAWTHORP, Nathan. *Skipton.* b.1708-d.1784.
GAY(E)—
Emile. *London.* 1863. French clocks.
Joseph. *Turin.* ca.1785.
LAMAILLE & Co. *London.* 1881. French clocks wholesale.
E. Vicarino & Co. *London.* 1869-75. French clocks.
John. *London.* 1857-69.
GAYDON—
Francis. *Upper Norwood.* 1866-78.
Frederick. *Kingston-on-Thames.* 1866-78.
H. E. *Brentford.* ca.1820. C. (Perhaps same as next man.)
Henry & Edwin. *Richmond, Surrey.* 1862-78. (Late TUTTON.)
Thomas William. *Swansea.* 1887-99.
GAYHART, S. *Camden, NJ, USA.* 1846-49.
GAYLER, John Rogers. *Buntingford* (*Herts*). 1866-74.
GAYLORD—
C. E. *Chicago, USA.* ca.1850.
Homer. *Norfolk, Conn, USA.* 1812. C.
GAZE—
Brothers. *London.* 1863-81.
Samuel Bartholomew. *London.* (B. 1817-24) 1828-57.
GAZNET, Jeremie (sometimes GAZUET). *London.* b.ca.1658, mar.1682-1700. W. & C.
GAZZARD, John. *London.* 1820. W.
GEACH—
George. From *London.* *Liskeard* 1811-18, then *Padstow* 1823, then *Truro.* 1829-d.1834.
John H. *Truro.* s. of George whom he succeeded 1834.
Thomas. *St. Breoke* (*Cornwall*). 1799-1801, then *St. Columb.* 1805-09. C. & W.
GEALL—
J. *Cerne Abbas.* 1855.
James. *Poole.* 1875.
GEARING, Henry Alexander. *Newhaven.* 1878.
GEATER, Henry. *Bristol.* 1825-44. C. & W.
GEBHARD(D)T
John Jacob. *Strasbourg.* s. of Heinrich. b.1631, mar.1661, d.1681.
ROTTMANN & Co. *London.* 1869-75.
GEDDES, Charles. b.1749, from *London* to *Boston, USA* and *New York* 1773-76, then *Halifax, Nova Scotia.* ca.1776-d.1810.
GEDDIE, John. b.ca.1778-pre-1817, *Banff, Scotland,* then *Pictou, Nova Scotia.* 1817-d.1843.
GEE—
Adam. *London.* 1857-69.
Arthur. *Lichfield.* 1868-76.
H. *Derby.* 1849.
James. *Manchester.* mar.1774. C.

GEE—*continued.*
John. *Derby.* 1818-37
John. *Farnworth.* mar.1812. W.
William. *Warrington.* d.1750. W.
GEEHAN, ——. *Stewartstown.* ?Early 19c. C.
GEEN, J. *Epsom.* 1862-66.
GEER, Elihu. *Hartford, Conn, USA.* 1838-87. Printed clock labels.
GEERING, William. *Seaford.* 1851-78.
GEFAELL—
P. & Co. *Leeds.* 1834.
Ulrich. *London.* 1832-39. (See also GAFAEL.)
GEHR, James. *London.* 1857.
GEHRES, Albert. *London.* 1863-81.
GEHRING, John G. *Baltimore.* 1827-31.
GEIGER—
Gottfried. *Lindau.* ca.1690.
Jacob. *Allentown, Pa, USA.* 1787-90. C.
Thomas. *Augsburg.* 1594.
GEISSLER, C. A. *New York.* 1899-1903. Chrons.
GEIST, Simon. *Graz.* ca.1770. C.
GELDART(E)—
John. *York.* b.1648-74. W.
Parkin Hinde. *Darlington.* 1851-66.
GELL—
John I. *York.* 1634-39. W.
John II. *York.* s. of John I. 1663-d.1698. W.
GELLATLY, Robert. *Edinburgh.* 1860.
GELSTON—
George S. *New York.* 1833-37.
Hugh. *Baltimore.* 1832. Importer of watches.
Hugh. bro. of John. *Lisburn.* ca.1752-72, then Lurgan 1772-76, *Belfast* 1776-79, then *Newry* 1779-d.1781. C.
John. b.1736. *Dublin* till ca.1769, then *Newry.* ca.1769-1779, *Ballyrally* 1784-1800, and *Banbridge* d.1804.
William. *Lisburn.* ca.1752-79. C.
William. *Newry.*? Late 18c. Perhaps same as previous man.
GEMMEL, Matthew. *Stewartown.* ca.1760. C. (One such at *New York* 1805, may be same man.)
GEMMIL, John. *Lancaster, Carlisle* and *York, Pa, USA.* ca.1756-60. C.
GENDRON, J. *Prudent, Quebec.* 1851-71.
GENEST, Charles. *Quebec.* 1853.
GENN, James Ditchett. *Falmouth.* 1844-56.
GENT—
George. *Leicester.* 1876.
J. *Congleton.* 1857.
J. A. M. *Penrith.* 1869.
James. *London.* 1869.
Joseph. *Walsall.* (B. 1805-)1828-50.
William. *London.* 1851.
GENTLE, Samuel. *London.* 1863-75.
GENTRY, John. *Shrewsbury.* 1870.
GEORGE—
Arthur. *Swaffham.* 1875.
Henry. *Fishgaurd.* 1840-68.
James. *Fishguard.* 1830-35. C.
John. *Dublin.* 1822-58.
Richard E. *Bishops Castle.* 1879.
Thomas. *London.* ca.1685. C. (B. has pre-1748. W.)
Walter. *Carmarthen.* 1769-d.1781. W.
William. *Perth.* 1848.
GEPP, Frederick Alexander. *Bristol.* 1879.
GERBIE, Charles. *Atlanta, Ga, USA.* 1871.
GERDING & SIMEON. *New York.* 1832. Clock importers.
GERE, Isaac. *Northampton, Mass, USA.* b.1771-1812.
GERHARDT, John. *Montreal.* 1886.
GERMAIN, William. *Bath.* 1875.
GERMAN—
J. *Meon Stoke (Hampshire).* 1859-67.
John. *Bishops Waltham.* 1839.
GERNON, Bernard. a. in *Bristol. London* 1658. Admitted CC. 1659-63.
GEROULD, S. A. *Keene, NH, USA.* ca.1825. W.
GERRARD—
Edward. *St. Helens.* mar.1776-77. W.
John. *Halifax.* 1788. C. (One such in London 1714-ca.1735.)
GERRISH, Oliver. *Portland, Me, USA.* ca.1826-34. C.

GERSON, J. *Sunderland.* 1856. C.
GERTY, Henry. *Dublin.* 1858-80.
GERVIN, Thomas. *Manchester.* mar.1798. C.
GERY, Herman. *Philadelphia.* 1804.
GETTERICK, Ralph. *London.* 1662.
GETTY, William. *Ballymoney.* 1843-58.
GETWOOD, Robert. *Banbridge.* ca.1790. C.
GETZ—
Conrad Emil. *Montreal.* 1902.
George. *Dundas, Canada.* 1865.
George. *Hamilton, Ontario.* 1867-78.
Peter. *Lancaster, Pa, USA.* 1791-c.1820.
GIBB—
James. *Glasgow.* 1846.
James. *Strathaven.* 1860.
William. *Whithorn.* 1836. (*Scotland*).
GIBBS—
——. *London.* 38 Banner St., ca.1840. C.
B. *Burslem.* 1868.
Benjamin. *Newburyport, Mass, USA.* ca.1820.
C. *Bath.* 1856.
Charles. *Bath.* 1861-83.
Frederick. *Nottingham.* 1864-76.
George. *London.* 1832-51.
Henry. *London.* 1857-81.
James. *Philadelphia, USA.* 1847. W.
Joshua. *Deddington* and *Souldern* (*Oxon*). 1805-55.
Richard. *Sibford Gower.* 1852. C. & W.
Thomas. *Haverfordwest.* 1822.
Thomas. *Stratford-on-Avon.* 1828-54.
Walter. *London.* a.1639 to Edward East. CC. 1648-1700, *Fleet Lane.*
William. *Haverfordwest.* 1822-ca.1860. C.
William. *London.* 1863-81.
GIBBONS—
John. *London.* 1828-39.
John. *Manchester.* mar.1806. C.
John. *Neath.* a.1782.
John B. *London.* 1844-69.
Joseph. *Birmingham.* 1880.
Thomas. *Boston, USA.* 1739. Cabinet maker.
William. *London.* 1869-75.
GIBSON—
——. *Ayr.* 1844.
——. *Thetford.* ca.1720. C.
Charles. *London.* 1832.
D. *Great Neston (Cheshire).* 1857.
David. *Holywell.* 1840-44.
G. *Thetford.* Early 18c. C.
George. *Dublin.* 1874-80.
H. *Berwick-on-Tweed.* 1858.
H. *North Shields.* 1858.
Henry. *Berwick-on-Tweed.* 1848-60.
James. *Hebden Bridge.* 1866-71.
James. *Sodbury.* 1840.
John. *Barnard Castle.* 1793.
John. *Beith.* 1837-60.
John. *Dunse.* 1860.
John. *Glasgow.* (B. 1809.) 1841.
John Moffatt. *Alnwick.* 1827-34. C.
John & Thomas. *Alnwick.* 1848-58. C.
R. W. *Leicester.* ca.1850. C.
Richard. *Redcar.* 1866.
Robert. *Dumfries.* 1793-1820.
Thomas. *Berwick-on-Tweed.* 1834-60.
Thomas. *Pocklington.* 1846-58.
Thomas. *Scarborough.* 1846-58.
William. *Ardrossan.* 1860. W.
William. *Aspatria (Cumberland).* 1873. Also optician and agent for sewing machines and bicycles.
William. *Auchinleck.* 1860.
William. *London.* 1875.
William Alexander. *Belford (Northumberland).* 1834.
William (& Co. Ltd). *Belfast.* ca.1860-94. W.
GID, William. *Rotterdam.* 18c. W.
GIDMAN—
Hugh. *Prescot.* 1795. W.
John. *Clay Cross.* 1876.
Thomas. *Knutsford.* 1865-78.
GIDNEY, Robert. *Norwich.* 1830-65.
GIDONAN, Joshua. *Prescot.* 1823-26. W.
GIFFEN, Robert & Son. *Campbeltown.* 1837.

GIFFIN, Thomas. *London.* 1844-75.
GIFFORD—
Benjamin M. *Port Rowan, Canada.* 1871.
S. K. *Camden, SC, USA.* 1836.
GIFFT, Peter. *Kutztown, Pa, USA.* ca.1810.
GIGGINS, William. *Coventry.* 1880.
GILBERT—
——. *Penzance.* ca.1780. C.
A. *Ventnor.* 1878.
Clock Co. (William L.). *Winsted, Conn, USA.* 1871-1934.
Clock Corp. (William L.). *Winsted, Conn and Laconia, NH, USA.* 1934 onwards.
Edward. *Belfast.* (b.1815)1843-d.1853. W. & C.
Eliza. *Belfast.* 1858.
F. *Angers.* ca.1675. C. (One such *London.* CC. 1661.)
Francis. *Chester.* 1878.
George V. *Rugeley.* 1860-76.
H. *Belfast.* ca.1830-40. C.
J. F. *Rochester, NY, USA.* 1830.
James (?sen.). *Lichfield.* 1850.
James (jun.). *Lichfield.* 1868-76.
Jeffrey. *New Romney.* b.1814-74.
Jesse. *Brooklyn, NY, USA.* 1843.
Jesse. *Lydd.* 1839.
JORDAN & SMITH. *New York.* 1832.
Manufacturing Co. *Winsted, Conn, USA.* 1866-71.
(William L.) & MARSH (George) & *Co. Farmington, Conn, USA.* 1830-35. C.
Mrs. Elizabeth May. *Belfast.* (b.1820) wid. of Edward, whom she succeeded. 1853-68.
P. *London.* 1803. W.
RICHARDS & CO. *Chester, Conn, USA.* 1831. C.
& Son. *Belfast.* 1858.
T. *Atlanta, Ga, USA.* 1867.
T. *Hythe.* 1855.
T. *Rugeley.* 1860-68.
Thomas. *Belfast.* ca.1845. C.
Thomas. *Hythe.* 1826-55.
Thomas. *Rugeley.* 1828-42.
Thomas & George. *Rugeley.* 1850.
Thomas William. *Rugeley.* 1876.
William. *Belfast.* Late 19c. C.
William. *Birmingham.* 1854-68.
William. *Greatworth (Northants).* 1747-67.
William. *Rugeley.* 1835-50.
William L. *Winsted, Conn, USA.* ca.1850-66.
William & Son. *Belfast.* (b.1805). ca.1830-d.1890.
GILBERTSON, John. *Ripon.* (B. pre-1764)-d.1793. C.
GILBURD, J. W. *Worthing.* 1862.
GILCHRIST—
Gordan. *St. Andrews, New Brunswick.* 1822.
William. *Haddington (Scotland).* 1860.
GILDER, John. *Whitby.* 1858.
GILES—
A. *Sherborne (Dorset).* 1867.
A. B. *Dronfield.* 1876.
Alfred C. *Reading.* 1877.
George. *Winchester.* 1878.
Henry. *Birmingham.* 1868. Clock cases.
Henry. *Birmingham.* 1880.
Henry. *Oswestry.* 1863-79.
John Scott. *Maidstone.* 1845-55.
Joseph. *Trenton, NJ, USA.* 1800-20.
Nicholas. *Maidstone.* 1807-39. W.
Richard, sen. *Shrewsbury* from *Wellington.* 1842-77.
Richard, jun. *Shrewsbury.* 1863-1908, succeeded by Alcocks.
Richard, sen. *Wellington* (born *London*) 1832-35, later *Shrewsbury.*
William Edward. *Montreal.* 1908. Patented electric alarm clock.
GILFILLAN, James. *Lesmahagow (Scotland).* 1834.
GILHOOLY, Ephraim. *Manchester.* 1848-51.
GILKES—
John. *Burford.* ca.1800. C. & W.
John. *Shipston-on-Stour.* 1740-66. C. & W.
Matthew. *London.* 1839.
Richard. *Adderbury.* b.1715-87. C. & W.
Richard. *Sibford.* ca.1800. C.
Thomas I. *Charlbury.* b.1704-d.1757. C.
Thomas II. *Charlbury.* b.ca.1740-75. C.

GILKES—*continued.*
Thomas. *Chipping Norton.* mar.1758-ca.1770. C.
Thomas. *Shipston-on-Stour.* ca.1760. C.
Thomas. *Sibford Gower.* b.ca.1665-1743. C.
Tobias. *Chipping Norton.* ca.1770. C.
GILL—
Alexander & Son. *Aberdeen.* Early 19c. C.
C. *Shipley.* Later 19c. W.
David. *Aberdeen.* (b.1789)-d.1877.
Edward. *Nottingham.* 1876.
George. *Maidstone.* a.1737.
Henry. *London.* 1869.
Isaac. *Charleston, SC, USA.* 1810.
Isaac. *Manchester.* 1851. Imported clocks.
J. *New Mills (Derbyshire).* 1849-76.
James. *Disley.* 1848.
James. *Huddersfield.* 1871.
John. *Birstwith.* 1838. C.
John. *London.* 1869-81.
John. *Manchester.* 1848.
John Sard. *London.* 1839-44.
Peter & Son. *Aberdeen.* 1846.
R. *Nottingham.* 1864.
& SMITH. *Aberdeen.* 1860.
Walter. *Shipley.* 1840.
William. *Maidstone.* 1737-ca.1770. C.
William. *Truro.* 1853. W.
William. *Winchcomb.* 1863-79.
GILLAM, Thomas. *London.* 1828.
GILLAN—
Andrew. *Quebec.* 1831.
John. *Inverury.* 1837.
GILLEAN, Thomas. *London, Canada.* 1869-90.
GILLETT—
BLAND & Co. *Croydon.* 1862-78. T.C. Also *London* 1862-81, and clock cases.
Thomas. *Oxford.* a.1698-1702.
William. *London.* 1857.
GILLHAM—
W. *Eastbourne.* 1878.
William. *Luton.* 1830-39.
GILLAM, Edward. *Pittsburgh, Pa, USA.* ca.1830.
GILLIES, Robert. *Beith.* (B. 1780-)1802. C.
GILLING—
Sebastian. *Coventry.* 1850-54. Watch case maker.
Thomas. *Altrincham.* 1878.
W. *Coventry.* 1860. Watch case maker.
GILLINGS—
James. *Eye.* 1858-79.
& READ. *Coventry.* 1860.
Thomas. *Coventry.* 1868-80.
GILLIS, Frederick Ludwig. *Liverpool.* 1848-51.
GILLOTT, William. *Tonbridge.* 1845-47.
GILLOWS. *Lancaster/London.* Cabinetmakers and clock case makers, who occasionally sold clocks with their name on. 1740s to early 19c.
GILMAN, B. C. *Exeter, NH, USA.* ca.1790-1830. C.
GILMARTIN, John. *Augusta, Ga, USA.* ca.1820.
GILMORE—
John. *Battle.* Late 17c. lant. clock.
William. *Pittsburgh, Pa, USA.* ca.1830.
GILMOUR—
Henry. *Coleraine.* 1858. C.
James. *Coleraine.* 1839-80. C. (to present day).
John. *?Lisburn.* (b.1769)-1833.
GILMUR, Bryan. *Philadelphia.* ca.1800.
GILSEN, ——. *London, Royal Exchange.* Early 19c. C.
GILSON, Lawrence. *Colchester.* b.1701-a.1715. C.
GIMSON, Samuel. *March.* 1875.
GINCHEREAU, Ulric. *Quebec.* 1890.
GIPFEL, Charles. *Long Stratton.* 1875.
GIRARD—
A. *Mobile, Ala, USA.* 1849.
et JACOT. *Quebec.* 1846.
Peter. *Neuchâtel.* 1851.
GIRAUD, Victor. *New York.* 1847.
GIRDLESTONE, Henry. *Norwich.* 1875.
GIRLING—
B. D. *London.* 1851-57.
Benjamin. *London.* 1875-81. (Successor to Robert.)
J. *Wangford.* 1846.
James. *Wickham Market.* 1846-64.

GIRLING—*continued.*
Robert. *London.* 1869.
GISCARD—
J. *Downham.* 1846-58.
James. *Downham Market.* 1858-75. C. & W.
Jeremiah. *Ely.* 1840. C.
John. *Ely.* 1830.
John. *King's Lynn.* 1830.
John. *Thrapstone.* ?Late 18c. C.
W. *Ely.* 1846.
William. *Downham.* 1830-36.
GITTOS, William. *Bridgnorth.* 1828.
GIVEN—
Alexander. *Newtown Limavady.* ca.1800-46. C.
Andrew. *Newtown Limavady.* mar.1852-80. (& sons—1890.)
James. *Ballymena.* 1843-58. C.
Robert. *Ballymena.* Early 19c. C.
W. B. *Ventnor, Isle of Wight.* 1859.
William B. *Ballymena.* 1843-46.
William James. *Newtown Limavady.* 1846-58. C.
GLADING, F. *Brighton.* 1851-78.
GLADMAN & WILLIAMS. *New York.* 1764.
GLANVILL(E)—
William. *Abingdon.* 1864-77.
William. *St. Austell.* 1847.
GLASCO—
Daniel. *Kilrea.* 1858 (see *Glasgow*).
Robert. *Birr (Ireland).* 1858.
GLASCOCK, Abraham. *Epping.* ca.1765. C.
GLASE, Thomas. *Bridgnorth.* 1828-50. Also gunsmith.
GLASGO, S. *Brixton.* 1878.
GLASGOW—
——. *Cookstown.* ?Early 19c. C.
Daniel. *Kilrea.* mar.1818-46. C.
David. *London.* 1863-81.
H. & Co. *Halifax, Nova Scotia.* 1873-79.
Henry. *Halifax, Nova Scotia.* 1878.
GLASHAM, George. *London.* 1836.
GLASHAN, George. *London.* 1851.
GLASS—
Edward. *Bridgnorth.* ca.1780-90. C.
James. *Bannatyne (Dunbartonshire).* 1837.
James. *Bristol.* (B. 1825-) 1830-50.
James B. *Alexandria (Dunbartonshire).* 1860.
W. *Bristol.* 1856-79.
GLATZ—
Joseph. *Manchester.* 1828-51. German clocks.
& WUNDERLEY. *Manchester.* 1824. Imported clocks.
GLAVE, Albert. *Gloucester.* 1870.
GLAZE, James. *Wednesbury.* 1876.
GLAZEBROOK—
John. *Blidworth.* 1720-d.1766. C.
Thomas. *Nottingham.* Nephew of John. b.1722, mar.1750. Later Mansfield. ca.1750-66. C.
GLEADALL, William. *London.* 1875.
GLEASON, F. A. *Rome, NY, USA.* 1848.
GLEAVE—
John. *London.* 1844.
John. *Liverpool.* (B. 1805-) 1814-51.
Richard. *London.* 1863.
Thomas. *Farnworth.* mar.1805. W.
GLEDHILL—
Lucy. *Huddersfield.* 1822.
Richard. *Manchester.* 1834.
GLEN—
A. J. *Barton, Canada.* 1865.
A. J. *Hamilton, Ontario.* 1856.
James Home. *London.* 1881.
GLENNEY, Alexander & Son. *Arbroath.* 1860.
GLENNY, Joshua. *London.* 1828. Watch case maker.
GLIGEMAN, ——. *Reading, Pa, USA.* 1784 (may be KLINGERMAN, q.v.).
GLOBE CLOCK CO. *Milldale, Conn, USA.* 1883.
GLORE, ——. *NH, USA.* ca.1830.
GLOVER—
Alexander. *Prescot.* mar.1726. W.
Benjamin. *Gildersome.* 1871.
Edwin. *Fayetteville, NC, USA.* 1843. W.
Henry. *Brooklyn, NY, USA.* 1846.
H(enry). *St. Helens.* 1759-ca.1790. W. & C.
Henry. *London.* 1857-63.

GLOVER—*continued.*
J. *Richmond, Surrey.* 1862-78.
James. *Farnworth.* 1834-51.
James. *Sutton (Lancs).* 1806. W.
Job. *Lincoln.* 1868-76.
John. *Bungay.* (b. ca. 1740) ca. 1760 -d. 1810. C.
John. *Prescot.* 1795-d.1801. W.
John. *London.* 1832.
John. *St. Helens.* 1766-85. W.
John. *St. Helens.* 1819. W.
John. *Shrewsbury. (From London).* 1823-34. s. of George Glover whom he succ.
John. *Widnes.* 1851.
Joseph. *Dublin.* 1858-80.
Joseph. *Manchester.* 1824.
Thomas. *Farnworth.* 1848-51.
Thomas, jun. *Dublin.* 1831. W.
Thomas. *Rainhill.* 1760-d.1798. W.
GLUE, D. *Stalybridge.* 1865.
GLYNN, Patrick. *Dublin.* 1858-68.
GOAD—
Alfred (& Sons). *London.* 1857-75(-81).
Thomas Turner. b.1837. *Ulverston; Lancaster.* 1851.
GOATER, John. *Winchester.* 1830.
GOATLEY, Daniel. *Canterbury.* 1823-28.
GOBEL, Henry. *New York.* 1853.
GOBRECHT—
David. *Hanover, Pa, USA.* 1798-1817. C.
Eli. *Pa, USA.* Early 19c. C.
Jacob. *Pa, USA.* Early 19c. C.
GODBALL, J. *Ipswich.* 1865.
GODBY, T. *Scarborough.* 1851.
GODDARD—
Edward. *Hull.* 1834.
Edward. *Sheffield.* 1822-32.
Henry. *Dover.* ca.1730. W.
J. *Glossop.* 1855.
J. *Hastings.* 1851.
John. *Newport, RI, USA.* 1748-85. Clock casemaker.
John. *Newport, RI, USA.* 1789-1843.
Nicholas. *Northampton, Mass, USA.* b.1773-97, then *Rutland, Vt,.* 1797-1823. C.
Robert William. *London.* 1875-81.
Stephen. *Newport, RI, USA.* 1804.
Thomas. *London.* (B. 1817-24-) 1832.
GODDEN—
George. *Harbour Grace, Newfoundland.* 1864.
George. *Tonbridge.* 1847.
Henry. *Town Malling.* 1823-47.
John. *Wingham.* ca.1750. C.
Joseph. *Harbour Grace, Newfoundland.* 1871-85.
Thomas. *Canterbury.* 1832.
Thomas. *Hythe.* 1784-1824.
GODFREY—
Charles. *Kettering.* 1877.
George. *London.* 1828.
Henry. a.1676. *Oxford* to *London.* CC. 1685-1707. C.
& HEWITT. *London.* 1863.
John. *Hull.* 1851.
Joseph. *Sudbury.* b. 1762 d. 1821. C. & W.
Moses. *Cowbridge.* Late 18c.-early 19c. C.
Rowland. *Sudbury.* b. ca. 1798, s. of Joseph, d. 1824. W.
& STORER. *London.* 1875.
Thomas. *Philadelphia, USA.* 1704-49.
Thomas Leigh. *London.* 1881.
William. *London.* 1869.
William. *Philadelphia, USA.* 1750-63.
William. *Winterton (?Lincs).* Prob. early 19c. C.
GODIN, Louis. 1701. W. (Prob from *Paris* where one such 1677-81).
GODRONE, ——. *London.* 1622. Alien journeyman.
GODSALL, Thomas. *Templemore.* 1858. W.
GODSELL, J. *Coventry.* 1880.
GODSHALK, Jacob. *Kowamencin, Pa, USA.* 1771 (also *Philadelphia*).
GODSOE, B. F. *New York.* 1848.
GODWIN—
George. From *London* to *Woodbridge* 1812, to *Ipswich* 1839.
George Alfred. *London.* 1881.
James. *London.* 1881.
William. *Manchester.* 1834.

GOETTLER—
& GIESENHAUS. *London.* 1863-69.
John. *London.* 1875-81.
GOEWEY, P. F. *Albany, NY, USA.* 1855-80.
GOFFE, Thomas. *Maidstone.* a.1709.
William. *Falmouth.* ca.1800-25. C. W. & chrons.
GOITMAN, Louis. *Bristol.* 1850.
GOLAY—
Charles A. *London.* 1857-75.
Eli. *London.* 1851.
GOLD, Isaac. *Lewes.* 1828.
GOLDENBERG, Mendel. *London.* 1881.
GOLDER—
John. *Alloa.* 1830-45. C. & W.
John. *New York.* ca.1810.
Edward. *Hastings.* 1862-78.
George. *Abergavenny.* 1852-68.
John. *Rugby.* 1828-42.
William, jun. *Haverfordwest.* 1844.
William G. *Pontypool.* 1850-68.
GOLDMAN—
L. *Bristol.* 1856.
Lewis (Louis). *London.* 1865-75.
GOLDSBROUGH, George. *Scarborough.* 1790-95.
GOLDSBOROUGH—
John Charles. *Leeds.* 1871.
William. *Scarborough.* ca.1800. C.
GOLDSBURY, ——. *St. Albans* and *Granville, O, USA.* 1828.
GOLDSMITH—
Abraham. *Leeds.* 1871.
Alliance Co. *London.* 1869-81. C.
& Co. *Salem, Mass, USA.* ca.1850.
John. *London.* a.1674 to Robert Williamson. CC. 1681-97.
Thomas. *Douglas, Isle of Man.* 1860.
William. *Douglas, Isle of Man.* 1860.
GOLDSTEIN—
Bernhard. *London.* 1881.
Jacob. *Liverpool.* 1848.
GOLDSTONE—
B. *Philadelphia, USA.* 1839.
Michael. *Manchester.* 1848-51.
GOLDSTRAW, William W. *Leek.* 1876.
GOLDSWORTHY, Joseph. *Redruth.* 1873. W.
GOLETIEL, S. *Lancaster, Pa, USA.* ca.1850-60.
GOLLEDGE—
Richard. *Stratford, Essex.* 1839-55.
Richard. *London.* 1844-57.
GOLLING, Lorenz. *Linz.* 18c. W.
GOLSTON, James. *Bristol.* ca.1740.
GOME, ——. *Aylesbury.* ca.1750. C.
GOMES, William. *Oxford.* 1550-ca.1585. Kept ch. clock.
GOMPERTZ, W. Edmund. *London.* 1851.
GONDIE, Thomas. *Jedburgh.* 1860.
GONTARD, C. P. *Paris.* 1848-55.
GOOCH—
D. *Yarmouth.* 1865.
Ellis. *London.* 1851-63.
& HARPER. *London.* 1811. W.
Horace. *London.* 1828.
John. *London.* 1863.
Thomas. *London.* (B. 1799-1803) 1811. Watch case maker.
Walter. *London.* 1863-75.
William. *London.* 1851.
GOOD, John. *Windsor.* 1877.
GOODACRE, Edwin. *Falkingham.* 1861-76.
GOODALL—
David. *Wetherby.* b.1775-1801. C.
George. b.1731. *Micklefield,* then at *Aberford.* 1766-d.1796. C.
George II. *Tadcaster.* s. of George I. ca.1780-1807.
GOODBAIRN, William. *Leeds.* 1708-ca.1720. C.
GOODCHILD, John. *Bradford.* 1814-22. C.
GOODE—
Francis. *Sleaford.* 1876.
N. *Coventry.* 1880.
William. *Chester.* 1828. C.
William. *Portsea.* 1839-48.
GOODELL, David. *Pomfret, Conn, USA.* Late 18c. C.

GOODFELLOW—
John. *Boston, USA.* 1800-03.
& Son. *Philadelphia.* 1799.
William. *Philadelphia, USA.* 1793-1818.
William, jun. *Philadelphia, USA.* 1799.
GOODGEON, John. *Skipton.* d.1756. C.
GOODGER, Thomas. *Rotherham.* 1862. W.
GOODHALL, William. *London.* (B. a.1751, CC. 1765-72)-1774. W.
GOODHART, Jacob. *Lebanon, Pa, USA.* Early 19c.
GOODHUE, Richard S. *Portland* and *Augusta, Me, USA.* 1830s.
GOODHUGH, Richard. *London.* (B. 1825-40) 1828-57.
GOODING—
Alanson. *New Bedford, Mass, USA.* ca.1810-40.
Henry. *Boston, USA.* (B. 1810-25) 1820-42.
John. *Plymouth, Mass.* Early 19c. W.
Josiah. *Dighton, Mass, USA* and *Bristol, RI* and *Boston.* 1788-1867. C.
GOODMAN—
C. *Chalfont St. Giles.* 1877.
Charles (& Son). *London.* 1851 (1857-81).
David. *Newbridge* (= *Pontypridd*) 1844-68.
David. *Ystrad Rhondda.* 1868.
John. *Cleveland, O, USA.* ca.1850.
John. *Taunton.* 1861-75.
John. *London.* 1832.
J(ohn?) & Son. *London.* 1839-44.
Thomas. *Charleston, SC, USA.* 1733-38. From *London.*
Walter. *Taunton.* 1883.
GOODRICH—
Charles. *Amherstburgh, Canada.* 1861.
Chauncey. *Bristol* and *Forestville, Conn, USA.* ca.1828-57.
Frederick. *Cricklade.* 1867-75.
H. *Bath.* ca.1820. W.
Jared. *Bristol, USA.* 1835-45.
GOODSPEED, Lot. *Middletown, Conn, USA.* ca.1820-30 (clock casemaker).
GOODWILL, Robert. *Newbottle.* 1827-34.
GOODWIN—
Benjamin. *London.* 1857-75. Clock cases.
C. E. H. *Hanley.* 1876.
Charles. *London.* 1869-81. Also clock cases.
& DODD. *Hartford, Conn, USA.* 1816.
E. O. *Bristol, USA.* 1852-55. C.
F. *Sturminster.* 1867.
& FRISBIE. *Unionville, Conn, USA.* ca.1830-50.
George. *Ipswich.* 1839. Clock cases.
H. R. *Macclesfield.* 1857-65.
Henry. *Boston, USA.* 1820.
Henry. *Newark.* 1828-55.
Horace, jun. *Hartford, Conn, USA.* ca.1830-40.
James. *Baltimore, USA.* 1806.
John. *London.* 1832.
Martin. *Manchester.* 1848-51.
S. & Son. *Birmingham.* 1868. Clock cases.
Samuel. *Baltimore, USA.* 1809.
Thomas. *London.* 1662-68. Clerk to CC. (May not be a maker.)
V. C. *Unionville, Conn, USA.* 1830s. Clock cases.
Wallace. *Attleboro, Mass, USA.* 1850s.
Walter. *London.* 1875-81. Clock cases.
William. *Hamilton, Ontario.* 1884.
William. *Harwich.* 1866. C.
William. *Manchester.* 1848-51.
William. *Nottingham.* (B. 1789-1818) 1828-35.
GOODYEAR & Son. *Guildford.* Mid-18c. C.
GOORE—
Giles. *Prescot.* 1825. W.
John. *Prescot.* 1768.
GOOSE, J. *Coltishall.* 1846.
GOOSSEN, CORS & SMOLDERS, Marts. *Geertruydenberg.* ca.1760. C.
GORDON—
——. *Ballynahinch.* ?Late 18c.-early-19c. C.
Albertus S. *Laconia, NH, USA.* 1883-1920.
Alexander. *Insch (Scotland).* 1860.
C. W. *Bromley.* 1866-74.
& FLETCHER. *Dublin.* (B. 1795-) 1805-39. W.
George. *Philadelphia.* 1847.

GORDON—*continued.*
George I. *New York.* 1854.
George H. *Cavan.* 1880-92.
George P. *St. John, New Brunswick.* 1799.
Henry. *Felton.* 1827.
James. *Ballymoney.* ca.1808-46.
James. *Dublin.* 1788-1804.
James. *Midmarr, Scotland.* 1860.
John. *Augher.* ca.1790-1820. C.
John. *Ballymacarrett.* 1858. Watch glasses.
John. *Kirkintilloch.* 1860.
John. *Longton.* 1842.
Joseph I. *Ballymoney.* ca.1770-ca.1820.
Joseph II. *Ballymoney.* ca.1854-90. C.
M. H. *Grimsby.* 1876.
Mrs. *Bromley, Kent.* 1858.
Smyley. *Lowell, Mass, USA.* 1832. Watch cases.
& Son (Joseph & James). *Ballymoney.* 1808. C. & W.
Thomas. *Halifax, Nova Scotia.* 1790-97.
William. *Coventry.* 1880.
William. *Dufftown.* 1836-60.
William. *New Glasgow, Nova Scotia.* 1879.
GORE—
James. *Farnworth.* mar.1798. W.
John. *Liverpool.* 1834.
John. *Prescot.* mar.1819-26. W.
John Thomas. *Eton.* 1869-77.
Richard. *London.* 1869-81.
W. *New Sneinton.* 1855.
William. *Alfreton.* 1876.
GORFIN, Henry. *London.* 1839-63.
GORGAS—
Benjamin. *Ephrate, Pa, USA.* s. of John. Late 18c.
John. *Germantown, Pa, USA.* ca.1720-60.
Joseph. *Ephrata, Pa, USA.* s. of Jacob. 1770-1816.
Solomon. *Ephrata, Pa, USA.* s. of Jacob, 1764-1800, then *Cumberland Co., Pa.* ca.1800-38.
William. *Greensburg, Pa, USA.* ca.1850. C.
GORHAM—
Francis, jun. *London.* 1844-51.
James. *London.* (B. 1820-40) 1832-51 (1857 exors of).
GORMAN—
J. *Lisnaskea.* ca.1790-1800. C.
Thomas. *Dublin.* 1800-20. W.
GORON, Thomas. *Boston, USA.* From *London.* 1759.
GORSHAM, James. *London.* 1839 (see Gorham?).
GORSTIDGE, Samuel. *Liverpool.* (B. 1825) 1834.
GORSUCH, Henry. *Prescot.* mar.1773-d. 1784. W.
GORTHORN, Miss Eliza K. *Royston.* 1874.
GOSHAWK—
John. *East Dereham.* 1830.
John. *N. Elmham.* 1836-46.
GOSLER, George Adam. *York Co, Pa, USA.* Early 19c.
GOSS, Thomas. *Street* (*Somerset*). ca.1770. C.
GOSSELIN—
——. *Paris.* ca.1710. W. (B. ca.1735).
Jaques. *Quebec.* 1757.
N. *Quebec.* 1758.
Pierre. *Quebec.* 1806-84.
GOSTELL, Thomas. *London.* 1839.
GOSTELOWE, Jonathan. *Philadelphia, USA.* 1744-1806.
GOSTLING, Thomas. *Diss.* 1830-36.
GOTHARD—
Thomas. *Dewsbury.* 1866. W.
Thomas. *Ossett.* 1871.
William. *Ossett.* 1871.
GOTOBED, Thomas. *Eton.* 1784.
GOTS(C)HALK—
Henry. *New Britain, Pa, USA.* ca.1760. C.
Jacob. *Philadelphia, USA.* ca.1770. C.
GOTT, William. *Ripon.* 1866-71.
GOTTIER, Francis. *Charleston, SC, USA.* ca.1750.
GOTTLIEB, Andrew William. *Shrewsbury.* 1870-91. W.
GOTZ, George. *London.* 1881.
GOUDY, Andrew. *Lerwick.* 1860.
GOULD—
Abijah. *Nashua, NH, USA.* Early 19c., then *Rochester, NY.* ca.1830, then *Henrietta, NY.* 1830-41.
James. *Baltimore, USA.* 1842.
Lewis. *Wiveliscombe.* 1875.
T. *Wiveliscombe.* 1861.

GOULDAR—
Edward & John. *Gloucester.* 1840-42.
John. *Gloucester.* 1850-79.
GOULDEN—
J. T. *Wilton, Wilts.* 1848-75.
William. *Crewkerne.* 1883.
GOURIET, Victor. *London.* 1857.
GOURLAY—
Alexander. *Kirkcudbright.* 1855. W.
Alexander & William. *Stranraer.* 1860.
James. *Newton-Stewart.* 1836-60.
GOUT—
David Ralph. *London.* 1832-57.
Ralph (?II). *London.* 1863.
GOUTIER, Abraham. *London.* a.1735. To Daniel Matthy.
GOUVERNON, WIDMER & Co. *London and Geneva.* 1869.
GOVE, Richard. *Laconia, NH, USA.* 1833-83.
GOVETT, George. *Philadelphia* 1811-19, then *Norristown, Pa.* 1820-41.
GOW, James. *Dunblane.* 1837.
GOWAN—
George. *London.* Late 18c. C.
James. *Hawick.* 1860.
Peter D. *Charleston, SC, USA.* ca.1820.
GOWANS—
James. *East Linton, Scotland.* 1837-50.
& Son. *Galashiels.* 1860.
GOWER—
George. *Dublin.* Pre-1824. W.
Joseph. *London.* 1869-81.
GOWETH, John. *Oxford.* b.1669-94. C. & W.
GOWING, Stephen. *Lowestoft.* 1830.
GOWLAND—
Clement. *Sunderland.* 1851.
Clement, William & Co. *Sunderland.* 1827-34.
Clement William. *London.* 1881. Regulators (succeeded to James).
G. H. & C. *Sunderland.* 1854-56. W.
George. *Liverpool.* 1851. Chrons.
George. *Toronto.* 1875-77.
James. *London.* 1832-75, C. Also regulators.
Mrs. Ann. *London.* 1863-69. (Successor to Thomas S.)
Thomas. *London.* 1828-57.
Thomas S. *London.* 1863.
William. *Sunderland.* 1851.
GOYER, ——. *Paris.* 1740-1800.
GRACE—
George. *London.* 1881.
John. *Belfast.* Believed from *London.* ca.1767-70.
W. *Haddenham.* 1869.
William. *London.* 1851-69.
GRACEBY, John. *London.* ca.1680. C.
GRACEY, George. *Tanderagee.* 1846-d.61. C.
GRAFF—
Jacob. *Lancaster, Pa, USA.* ca.1775. C.
Joseph. *Allentown, Pa, USA.* ca.1800.
GRAFTON—
Edward. *London.* 1851.
John. *London.* 1851.
John & Edward. *London.* 1832-44.
Samuel. *London.* 1863.
GRAHAM—
A. *Bridgwater.* ca.1820-25. C.
Alexander. *Berwick-on-Tweed.* 1848-60.
Benjamin. *Wakefield.* 1862-71. W.
Charles W. *Middlesbrough.* 1866. W.
Edmund. *Middlesbrough.* 1866. W.
G. H. *Cokstown.* 1865.
Henry. *London.* 1851.
J. *Keswick.* 1858.
James. *Girvan.* 1837.
James. *Glasgow.* 1860.
James. *Spalding.* 1861-76.
James. *Whitburn.* (d?)1833. C.
John. *Airdrie.* 1836.
John? *Belfast.* ca.1800. W.
John. *Hawick.* 1860.
John. *Kilrea.* ca.1800. C.
John. *Langholm.* 1837.

GRAHAM—*continued.*
John. *Liverpool.* 1851.
John. *Moffat.* 1837-60.
Jonathan. *Lockerbie.* 1860.
Joseph. *Farnworth.* mar.1834. W.
Joseph. *Glasgow.* 1837.
Joseph. *Prescot.* 1825. W.
Mitchell. *Philadelphia.* 1837.
Robert. *Monaghan.* 1858-65.
Thomas. *Carrickfergus.* ca.1800-20. C.
Thomas. *Hawick.* 1837-60.
Thomas. *St. Helens.* 1777. W.
Timothy. *Cockermouth.* 1858-79.
W. *Darwen.* ca.1750-60. C.
W. B. *Dublin.* 1813. W.
William John. *Toronto.* 1887.
CRAIG, William. *Stewartfield (Aberdeenshire).* 1836.
GRAINGE, L. *Bradford.* ca.1840-50. C.
GRAINGER, Thomas. *London.* 1863-69.
GRAISBERRY—
John. *Belfast.* Late 18c. (one such at *Larne.* ca.1780).
(?John). *Larne.* ca.1775-ca.80. C.
GRAIZELY, Arthur. *Chaux de Fonds.* 1881.
GRANAGE, Alfred. *Stow on Wold.* 1879.
GRANGER—
Francis. *Nottingham.* 1864-76.
James. *London.* 1857.
Richard. *London.* 1665.
GRANT—
——. *Cookstown.* ?Early 19c. C.
Alexander. *London.* 1881.
Alexander. *Newburgh.* 1834.
Alexander. *Stirling.* 1825-d.1875.
Alfred. *New Haven, Conn, USA.* ca.1850.
B. *Bridgwater.* 1861-66. Repairer.
Edward. *Keady.* 1846.
Edward. *Ottawa.* 1862.
F. & J. *Ottawa.* 1875.
G. *Southampton.* 1839-67.
George. *Edinburgh.* (B. a.1776)-83. C.
Henry. *Belleville, Canada.* From *England.* 1862.
Henry. *Cardiff. Christchurch, Hants* pre-1844, then *Cardiff* 1844-55, later emigrated to *Canada.*
Henry. *London.* 1832.
Henry. *Montreal.* 1866-76.
Henry. *Portsmouth.* Pre-1844. Later *Christchurch.*
Henry. *Staplehurst (Kent).* 1874.
Henry. *Warbleton.* 1862-78.
Israel. *St. Louis, Mo, USA.* ca.1820.
James. *Albany, NY, USA.* 1789.
James. *Gosport.* 1859-78.
James. From *London* to *Hartford, Conn, USA.* 1794-96. Then *Wethersfield.*
James. *London.* 1875.
James. *Quebec.* 1783.
John. *Clapham.* 1839.
John. *Fraserburgh.* 1860. Also compasses.
John. *Fyvie.* 1846.
John. *Greenock.* 1860.
John. *London.* 1828-44.
Joseph. *Boston.* 1876.
Joseph. *Helensburgh.* 1837.
Joseph. *Spalding.* 1868.
& MATTHEWS. *Bristol, USA.* 1841.
& NICHOLLS. *London.* 1875.
Richard. *East Hoathly (Sussex).* 1862-78.
S. *Leicester.* 1876.
William. *Boston, USA.* 1815-30. C.
William. *Edinburgh.* a.1821 to William Drysdale.
William. *Gosport.* 1830-48.
William. *London.* 1832-57.
GRANTHAM, John. *London.* 1762. W.
GRATLE (?GRATTE), Henry. *London.* ca.1785. C.
GRATTON—
J. *Bakewell.* 1864.
Joseph. *Mansfield.* 1876.
GRAV ELECTRIC CLOCK CO. *New York.* 1899.
GRAVE—
G. *London.* 1811.
George. *London.* (B. 1820). 1828-51.
GRAVELEY, John. *Alton.* Mid-18c. C.
GRAVELL, William & Son. *London.* 1828-63.

GRAVELLE, ——. *Paris.* ca.1810. C.
GRAVES—
Alfred. *Willow Grove, Pa, USA.* 1845.
Henry. *St. Ives (Hunts).* 1830.
Henry A. *London.* 1863.
William. *Kettering.* mar.1753.
GRAY—
——. *Elgin.* Post-1820.
A. *Ottawa.* 1846.
(De Mory) & ALDER (W. D.). *New York.* 1868.
Andrew. *Aberdeen.* ca.1750. C.
& BALDWIN. *Loughborough.* 1849.
Benjamin. *Loughborough.* Late 18c.-1835.
Benjamin & William. *Leicester.* 1835.
& Co. *Leicester.* 1876.
& CONSTABLE. *London.* 1794. W.
Daniel. *Hamilton, Ontario.* 1856.
Daniel. *Leamington.* 1850.
Daniel. *Montreal.* 1853-54.
David. *Banbridge.* 1870-72. W.
George. *Melksham.* 1867-75.
Henry. *Inverkeithing (Scotland).* 1834.
James. *Barrhead (Scotland).* 1860.
James. *Macduff.* 1846-60.
James. *Manchester.* mar.1797. C.
James. *New York.* ca.1840.
James. *St. Neots.* 1877.
John. b.ca.1792. *Milltown.* To *Belfast.* 1815-d.1860. W. (Partner ca.1819-ca.50 with Robert Gray.)
John. *Dungannon.* 1820.
John. *Trowbridge.* 1867-75.
Joseph. *Dungannon.* 1824.
Joseph Henry. *Birmingham.* 1854-80. Clock cases.
Peter & Co. *Edinburgh.* 1850.
R. *Glasgow.* ?late-18c. C.
Robert. *Belfast.* b.ca.1795, d.1875. (With John, q.v.)
Robert. *Edinburgh.* 1844. C.
Robert & Son. *Glasgow.* 1837.
Thomas. *Dunse.* a. of Davidson there in 1808.
Thomas. *Leicester.* 1876.
W. *Leicester.* 1849-64.
William. *Belfast.* s. of John. b.1823, d.1854. W.
William. *Calder.* 1860.
GRAYDON, George. *Dublin.* a.1753 to Charles Gillespy. d.1805. C. & W.
GRAYHURST, Harvey & Co. *London.* 1828.
GRAYLING, Thomas. *Rye.* From *Westfield.* a.1718 to Peter Roberts. W.
GRAYSON—
John. *Henley-on-Thames.* 1823-50. C. & W.
William. *Henley-on-Thames.* 1839-60. C. & W.
William. *Kendal.* 1851.
GREADER, Henry. *Willenhall.* 1842-50. W.
GREATBA(T)CH—
John. *Birmingham.* 1850-68.
Richard. *Birmingham.* (B. 1808-)1816-35.
Thomas. *Birmingham.* 1880.
William. *Birmingham.* 1842-54.
GREATHEAD, Nathan. *Leicester.* 1855-76.
GREATREX—
Edward. *Birmingham.* ca.1760. C.
James. *Bristol.* 1850.
GREAVES—
Henry. *London.* 1869-81.
James. *Cork.* 1858.
William. *Bromsgrove.* ?mid-18c. C.
William. *Cork.* d.1810. W.
William. *Leicester.* ca.1760-1791. W. & C.
GREBANT, ——. *London.* 1622. Alien with two apprentices in *St. Bartholomews.*
GREBEL, ——. *Berlin, Ontario.* 1856.
GRECIAN. 1914 and later, product of Pequegnat Clock Co.
GREEN—
C. *London.* 1851-57.
Charles. *Bath.* 1875-83.
Charles E. *Grantham.* 1868-76.
E. *Leicester.* 1864.
E. (late Thomas). *Liverpool.* 1822.
Edwin. *Carshalton.* 1878.
Elijah. *Yaxley, Suffolk.* 1875-79.
Elizabeth. *Tenbury. Worcestershire.* 1828.
Enoch. *Whitwick, Leicestershire.* 1876.

GREEN—*continued.*
G. *Southsea.* 1848-67.
George (alias Smith). Early life in *Leicester,* then 1750 to *Cirencester,* then *Ross-on-Wye.* Died at *Oxford.* 1762. C.
George. *Tenbury.* 1835-60.
George. b. ca. 1803 *Manchester. To Liverpool* by 1838, *Bilston* by 1841, *Bridgnorth* by 1870-71. W. & C.
Henry. *Bradford.* 1853-66.
Henry. *Landport.* 1878.
Henry. *London.* 1832-44.
Henry Edward. *Castle Rising (Norfolk).* 1865.
I. *Stourbridge.* 1868.
Isaac. *Smethwick.* 1876.
Isaac. *Stourbridge.* 1842.
J. *Albany, NY, USA.* 1797.
J. *Leicester.* 1849.
J. *Nottingham.* 1864.
James. *Easingwold.* 1788. d.pre-1801.
James. *London.* a.1678 to Henry Jones. CC. 1685-97.
James. *Richmond, Yorks.* 1746-57. C.
Job. *London.* 1857-63.
John. *Carlisle, Pa, USA.* Early 19c.
John. *Dartmouth, Nova Scotia.* 1876-81.
John. *Grantham.* 1868-76.
John. *Halifax.* 1830.
John. *London.* 1839. Watch cases.
John. *Oxford.* 1794-1823. C. & W.
John. *Plymouth.* mar.1776.
John. *Skipton.* 1704-d.1742. C.
John W. *Hamilton, Ontario.* 1867-69.
John William. *Leeds.* 1866. W.
Josiah. *Newcastle-on-Tyne.* 1850-52.
Mrs. A. *Grantham.* 1861.
Peter. *Bristol.* b.1800-d.1843.
R. *Castle Rising.* 1846.
Richard. *London.* 1863-81.
Robert. *Bristol.* 1850.
Robert. *Norwich.* 1875.
Samuel. *Bradford.* 1822.
Samuel. *London.* 1832.
Samuel. *Witney (Oxfordshire).* From *London.* ca.1818-23.
T. *Bristol.* 1856.
Thomas. b.1843. *Wigan.* At *Kendal.* 1871.
Thomas. *London.* 1828-44.
W. *Cromer.* 1846-65.
W. E. *Reigate.* 1866.
W. H. *Bath.* 1856.
W. H. *Gloucester.* 1863.
W. H. *Reigate.* 1862.
William. *Bristol.* 1838-42.
William. *Cley.* 1830-36.
William. *Doncaster.* b.1759, d.1824. W.
William. *Grantham.* 1828-76.
William. *Hamilton, Canada.* 1865.
William. *Milton-under-Wychwood.* ca.1770. C.
William. *Prescot.* 1819-34. C.
William. *Wigan.* 1834-51.
GREENALL, William. *Parr (Lancashire).* ca.1770-ca.80. C.
GREENAWALT, William. *Halifax, NC, USA.* 1826.
GREENAWAY—
George. *Bath.* 1798.
John. *London.* 1844.
William. *London.* 1832-44.
GREENBANK & SON. *Sedbergh (Yorks).* 1871.
GREENBAUM, Philip. *London.* 1881.
GREENBIRT, I. *Halstead (Essex).* 1866.
GREENER—
Charles. *London.* 1863. German clocks.
C. & Francis. *Liverpool.* 1822-34.
William. *London.* 1863.
GREENFIELD—
J. *Kidderminster.* 1872.
Joseph. *Ironbridge.* 1870.
GREENHALGH—
Henry. *Manchester.* 1851.
J. *Bury.* ca.1840. W.
John. *Accrington.* 1848-58.
John. *Manchester.* 1828-51. German clocks.
John & Son. *Manchester.* 1848.

GRENHALL, David. *Pontyeats.* Mid-19c. C.
GRENHELGH, T. *Brampton, Canada.* 1862.
GREENHILL(S)—
John. *Ashford.* 1650. Lant. clock.
Richard. *Ashford.* ca.1700-late-18c. C.
Richard. *Maidstone.* ca.1700. C.
Samuel. *Canterbury.* 1716. *Winesheap Street.* Bankrupt 1723. W.
GREENHOW, Joseph (S.). *Chelmsford.* 1839-74.
GREENING—
Ann. *Chepstow.* 1858.
Benjamin. *Bristol.* 1850.
Benjamin. *Chepstow.* 1822-50.
Benjamin (?jun). *Chepstow.* 1868-80.
Benjamin. *Clapham.* 1878.
Benjamin. *Coleford.* 1830.
Benjamin. *Newnham (Gloucestershire).* 1870.
Charles. *Bristol.* (B. 1830-)1840.
George. *Bristol.* 1830.
John. *Bristol.* a.1828.
Josiah. *Frampton.* ?later 18c. C.
Nathan. *Cinderford.* 1879.
Thomas. *London.* 1863-75.
GREENLAND, John. *Ashford.* 1675. C.
GREENLEAF—
David, jun. *Hartford, Conn, USA.* 1765-1835. C.
Orlando. *Sterling, Canada.* 1857.
GREENLEES, Daniel. *Galston (Scotland).* 1860.
GREENOUGH—
Norman Cummings. *Newburyport, Mass, USA.* 1820-66. W. & chrons.
Samuel & Sons. *Bolton.* ca.1900.
GREENTREE, J. *East Meon.* 1867.
GREENWAY—
George. *Chippenham.* 1830.
Jabez. *Birmingham.* 1880.
James, jun. *Tiverton.* 1725. C.
William. *Oldbury.* 1868-76.
GREENWELL—
George. *Alston.* 1828.
George. *Newcastle-on-Tyne.* 1844-58.
GREENWICH CLOCK & INSTRUMENT CO. *New York, USA.* 1888-1935.
GREENWOOD—
Charles. *Liverpool.* 1848.
J. Hamer. *Whitby, Canada.* President of Canada Clock Co.
James. *Hebden Bridge.* ca.1835. C.
John. *Canterbury.* 1832-40. W.
John. *Dover.* 1823-28.
John. *London.* 1844.
John (& Sons). *London.* 1857(63-81).
John. *Rochester.* 1793-95. W. 'Opposite the Rump in High Street'.
Michael. *Skipton.* 1797-98. C.
Samuel N. *Halifax, Nova Scotia.* 1815.
Thomas (& Co.). *London.* 1851(-63-81). Importers of American clocks.
W. *Asbury.* 1864.
William. *Birstall.* 1866-71. W.
William. *Rochester.* 1823-47.
GREGG, Jacob. *Alexandria, Va, USA.* ca.1810-20.
GREGGS, Thomas. *Bishop's Stortford.* ?late 18c.-ca.1820.
GREGO—
Charles. *London.* 1863-75.
Charles. *Perth.* 1860. Also optician.
GREGORY—
George. *London.* 1869-75.
George A. *Swansea.* 1887-99.
Henry. *Gloucester.* 1870.
James. *Basingstoke.* ca.1790-1813. C.
James. *London.* a.1650 to Benjamin Hill. CC. 1657-60.
James. *Ormskirk.* 1824.
Ja.1es. *Southport.* 1828-34. W.
John. *Ashton-under-Lyne.* 1848.
John. *Keady (Ireland).* 1858.
John. *London.* 1839.
Mrs. S. A. *Basingstoke.* 1878.
P. *West Bromwich.* 1868.
Robert. *London.* a.1670 to Henry Crumpe. CC. 1678-97.

GREGORY—*continued.*
W. *Maidenhead.* 1854-64.
William. *Basingstoke.* 1830-67.
William. *Haddenham.* 1847.
William Tucker. *Gloucester.* 1850.
GREGSON—
John. *Lancaster.* 1811-30. W.
John. *Liverpool.* ca.1825-ca.30. W.
John. *Ulverstone.* 1822-25. C. and silversmith.
T. *Newcastle-on-Tyne.* 1856-58.
GREIG—
David. *Perth.* 1810-60.
David. *Stonehaven.* 1835-60.
GREINER, Charles. *Charleston, SC, USA.* ca.1780.
GREISHABER, E. *Louisville, Ky, USA.* ca.1840.
GREMELS—
Joseph. *London.* 1863.
Paul. *London.* 1869-81.
& SCHWAR. *Greenwich.* 1874.
GRENDON, Henry (also GRINDON). *London.* CC. 1633-38.
GRENER & Co. *Manchester.* 1822.
GRENFELL, John. *Sutton Coldfield.* 1880.
GREN(N)ON—
Edward. *London.* 1851.
Michael. *Dublin.* 1858.
Paul. *Paris.* 1912. C.
GRENVOIZE, George. *London.* 1832.
GRESHAM, John. *Hull.* 1838-46.
GRETTON, William. *London.* ?late-17c. Lant. clock.
GREVILLE, Henry. *Dublin.* ca.1780. C.
GREY—
Ernest. *Aberdeen.* Late of *Calcutta.* 1848. C.
George. *Hull.* 1840.
Thomas. *London.* ca.1800. W.
GRIBBEN—
Edward, sen. b.ca.1793. *Coleraine.* Then *Belfast.* mar.ca.1815, d.1841. C. & W. (With John Wallace 1815-20.)
Edward, jun. *Belfast.* s. of Edward, sen., whom he succeeded. ca.1840-68.
& WALLACE. *Belfast.* 1815-20. Succeeded to Thomas Singlehurst. ca.1816. Partnership dissolved ca.1820.
William. *Belfast.* 1892. W.
GRICE—
Job. *Ormskirk.* Mid-18c. C.
Thomas. Town not stated. 1705. Sundial. (Prob. *London,* where one such a.1667 to Richard Ames.)
GRIDLEY—
L. P. & C. E. *Logansport, Ind, USA.* ca.1860.
Martin. *Logansport, Ind, USA.* ca.1840.
Timothy. *Sanbornton, NH, USA.* ca.1808. C.
GRIEBEL, Christopher. *Strasbourg.* ca.1750. (B. ca.1770).
GRIESHABER—
A. & W. *Ebbw Vale.* 1880.
Augustus. *Pontypool.* 1868-99.
Weibert. *Neath.* 1887.
GRIESSELICH, Henry & Co. *London.* 1875.
GRIEVE—
David. *Carlisle.* 1879.
Hay. *Belfast.* 1849-68. W.
GRIFFEN—
Henry. *New York* and *Brooklyn.* 1793-1818.
& HOYT. *New York.* ca.1820-30.
Nathan. *Lamborne.* 1877.
Peter. *New York.* ca.1810.
W. *Atlanta, Ga, USA.* 1871.
GRIFFENBERG, J. *Norwich.* 1858.
GRIFFIN—
B. *Wisbech.* 1846.
Francis. *London.* 1857.
George. *Cork.* ca.1694-d.1726. W.
George. *Tamworth.* 1876.
Richard. *Burford.* 1853.
Samuel. *Stewkley.* 1781-83. W.
Thomas. *London.* 1839.
Walter. *London.* 1869.
William. *London.* 1844.
GRIFFIS, Paul. *Southminster.* 1839-55.
GRIFFITH—
Benjamin. *Newcastle-under-Lyme.* 1828.

GRIFFITH—*continued.*
Benjamin. *Tunstall* (*Staffordshire*). 1835.
David. *Caernarvon.* s. of Richard of *Denbigh.* 1800-94. C.
David. *Pwllheli.* 1887-95.
David Lloyd. *Denbigh.* 1887-90. W.
Edward. *Bangor* (*Wales*). 1844.
Edward. From *London.* To *Litchfield, Conn, USA.* 1790. Then *Savannah, Ga, USA.* 1796.
F. W. *Croydon.* 1878.
John. *Bethesda.* 1856. C.
John. *Blenau Ffestiniog.* Same man as at *Caernarvon.* q.v.
John. *Caernarvon.* 1868-d.ca.1903. C. & W.
John Widlake A. *London.* 1844-81.
L. *Philadelphia.* 1842.
Mrs. Jane. *London.* 1863. (?wid. of William.)
Richard. *Bangor.* 1887.
Richard. *Denbigh.* 1791-1840. C. & W.
Robert. *Denbigh.* 1764-. Painted dial of ch. clock.
Robert. *Gravesend.* 1855-66.
Samuel. *Philadelphia.* 1847.
Thomas. *Beddgelert.* ca.1760-d.1784. C.
W. R. *Llannerch-y-medd.* 1884. C.
William. *Cwmwysg* (*Llywel*). 1707-86. C.
William. *Defynnog.* Prob. mid-18c. C.
William. *London.* 1851-57.
William. *London.* 1869-81.
William (& Co.). *London.* 1828(-32).
GRIFFITHS—
Benjamin. *Dudley.* (B. 1790-)1842.
Benjamin. *Tipton.* 1850.
Charles. *Newport* (*Pemb.*). 1844.
David. *Liverpool.* 1851.
E. *Chester.* 1865.
E. *Ton-y-Pandy.* 1875.
Edward. *Abergele.* 1839-93.
George Jeffrey. *Pontypool.* 1858.
J. *Clapham* (*Surrey*). 1878.
James. *Glens Falls, NY, USA.* 1836.
Jeremiah. *Monmouth.* 1850-80.
John. *Builth.* 1875.
John. *Greenville, SC, USA.* From *Europe.* 1855.
John. *London.* 1839.
John. *Ross.* 1856-79.
John. *Tremadoc.* 1868.
Nehamiah. *Chester.* 1834-57.
Rowland. *Llanrwst.* 1835.
Samuel. *Birkenhead.* 1878.
Sarah. *Millom.* 1879.
W. *Upper Tooting.* 1878.
W. & R. *Birmingham.* 1858-65. Dial makers. (Successor to Finnemores.)
William. *Birmingham.* 1842-60.
William. *Bridgend.* 1830-71. W.
William. *Hereford.* 1830-35.
William Henry. b. ca. 1819 *Bishopscastle.* 1851, then *Ludlow* from 1856-75. W. & C.
GRIFFT, Peter. *Kutztown, Pa, USA.* ca.1810. C.
GRIGG, ——. *Bishop's Stortford.* ca.1680. C. (John Grigg a. in *London* 1684 to Edward Hunt — may be him?).
GRIGGS—
Ebenezer. *Bristol, USA.* 1810. C.
Solomon. *Bristol, USA.* 1810. C.
GRIGNION, Daniel. *Topsham.* ca.1760. C. (One such at *London.* b.1684-d.1763.)
GRIGSON, Philip. *London.* a.1663 to Ahasuerus Fromanteel, jun. (II).
GRILLEY, Silas. *Waterbury, Conn, USA.* 1808.
GRIM, George. *Orwigsburg, Pa, USA.* ca.1820. C.
GRIMALDE—
——. *London.* 1794. W.
& JOHNSON. *London.* (B. 1809-25)-1828.
GRIMANDE, Samuel. *London.* 1832.
GRIMES—
William. *London.* CC. 1682-97. C.
William. *Northampton.* 1655-56. Rep'd ch. clock. (Also GRYMES).
GRIMSHAW—
James. *London.* 1832-44.
Thomas. *Liverpool.* mar.1834. C.

GRIMSLEY—
——. *Leicester.* 1849.
Edward. *Leicester.* 1876.
GRINBERG & REICHMAN. *Brighton.* 1878.
GRINDALL, John, jun. *Dumfries.* 1789.
GRINDON—
William. *Brighton.* 1828.
William. *Olney.* 1798. W.
GRINKIN—
Edmund. *London.* s. of Robert I. Alive 1626. B. Co. 1637. Warned to join CC. in 1640. Died pre-1660.
Edmund. *The Hague.* Reputedly 1650-80, which I cannot confirm.
Robert I. *London.* B. Co. 1598-d.1626. Father of Robert II and Edmund.
Robert II. *London.* s. of Robert I. b.post-1605. B. Co. 1636. CC. 1637. d.1661. *W. Fleet Street.*
GRINLAW—
Alexander. *Danse.* 1837.
Andrew. *Danse.* 1860.
GRISDALE, G. *London.* 1811. W.
GRISHABER, John. *Pontlotyn.* 1875.
GRISWOLD—
A. B. & Co. *New Orleans, LA, USA.* ca.1850-60.
Chauncey D. *Troy, NY, USA.* 1838.
Daniel White. *East Hartford, Conn, USA.* 1767-1844. C.
Joab. *Buffalo, NY, USA.* 1835.
GROCE, George. *Leicester.* 1828-49.
GROCOTT, ——. *Holywell.* Late-18c. or early 19c. C.
GROFF, Amos. *Rawlinsville, Pa, USA.* ca.1850.
GROHE, James. *London.* 1832-81.
GROKE, ——. *London.* Early 19c. C.
GROOM(E)—
George. *Croydon.* 1878.
H. *Thornage (Norfolk).* 1865.
John. *Colchester.* ca.1658-80. C.
GROOMS, J. *Coventry.* 1868.
GROOT, J. J. *Wormerveer (Holland).* 1721. C.
GROPENGEISSER, J. L. *Philadelphia.* 1841. C.
GROSH, Peter Lehn. *Lancaster Co, Pa, USA.* 1830. Clock dial painter.
GROSREY—
Celestin. *London.* 1839-44.
Eugene. *London.* 1875-81.
Peter C. *London.* 1857-69. (Also *Geneva.*)
GROSSE, Anthony. *London. Cornhill.* 1662-66.
GROSSMAN—
FURRER & CO. *London.* 1869.
H. J. *Dover.* 1858.
John & Co. *London.* 1875-81.
GROSVENOR—
Robert. b. ca. 1792 *Market Drayton* - 1856, then *Ellesmere* 1856-79.
Thomas. *Knowle.* 1880.
GROTZ, Isaac. *Easton* and *Bethlehem, Pa, USA.* 1810-35.
GROUNDS, Johnson. *Wigan.* 1858-ca.1890. W.
GROUNDWATER, Robert Paterson. *Kirkwall.* d.1850.
GROUT—
John, sen. *London.* 1844-63.
John, jun. *London.* 1857-81.
William. *London.* a.1648 to Joseph Quash. CC. 1660-68.
GROVE—
Christian. *Heidelberg Township, Pa, USA.* ca.1800. C.
Joseph. *London.* 1828.
Richard. *London.* 1632. CC.
Richard. *London.* 1811.
William. *Hanover, Pa, USA.* 1830-57. C.
GROVES—
John. *Leeds.* 1850-53.
William. *Leeds.* 1823-67. T.C.
GRUAR, Thomas J. *Newport (Mon).* 1875.
GRUBB, H. *Taunton.* Early 19c.?
GRUBY, Edward L. *Portland, Me, USA.* 1834.
GRUCHY, Jean. *Jersey.* 1763. C.
GRUEN—
D. & Sons. *Cincinnati, O, USA.* 1890-98.
& SAVAGE. *Columbus, O, USA.* 1879-82.
GRUEZ, John. *New York.* 1821. Clock case maker.
GRUIER, Armand. *London.* 1857.

GRUMBINE, Daniel. *Hanover* and *East Berlin, Pa, USA.* 1824-50.
GRUMBLE, Joseph. *Leicester.* 1876.
GRUNAUER, Erasmus. *Vienna.* 1447. C.
GRUNDY—
Benjamin. *Dudley.* 1876.
John. *Whalley.* ca.1775-80. C.
GRUNTHAL, Julius. *Leeds.* 1871.
GRYMES, William. *Northampton.* 1655-56. Rep'd ch. clocks.
GRZIMISH, Philip. *London.* 1857-69.
GUANELIA, A. *Bristol.* ca.1830. C.
GUARIN, Jacques. *Lyons.* ca.1660. C. (Also GUERIN.)
GUARNERIO—
G. *London.* 1863.
G. & Co. *Peckham.* 1878.
Gaspar. *Scarborough.* 1866.
Peter. *Huntingdon.* 1839-54.
GUAY, L. E. B. & Cie. *Quebec.* 1890.
GUBBIN, John. *Truro.* 1823-d.1827.
GUDGEON—
George. *Bury St. Edmunds.* s. of John 1817-39.
John. *Bury St. Edmunds.* Partner with Mary Lumley as Lumley & Gudgeon 1785-1800. Then alone until d. 1835. C. & W. (but also a John junior recorded?).
GUELPH, B. 1914 and later, product of Pequegnat Clock Co.
GUEPIN—
Abraham. *London.* 1702. W.
David. *London.* ca.1690. C.
Isaac. *Lewes.* 1712-18. W.
Peter. *London.* ca.1700. C.
GUEST—
Robert. *Yarmouth, Nova Scotia.* 1833-37.
T. *Birmingham.* 1854.
Thomas. *Yarmouth, Nova Scotia.* 1866.
GUIBAUDET, Gustave E. *Paris.* Succeeded to Brocot. 1889.
GUIBLET—
Jules. *London.* 1869.
& RAMBAL. *London.* 1863.
GUIDE, German. *London.* 1881.
GUIGNON, Julien Louis. *St. Nicolas d'Aliermont.* 1889. (Also *Paris.*)
GUILD—
Jeremiah. *Cincinnati, O, USA.* ca.1831.
John. *Philadelphia, USA.* 1818-24.
GUILLAUME—
A. *London.* 1851-69.
Edward & Charles. *London.* 1857-63. (Successor to George).
Frances Ami. *London.* 1832.
George. *London.* 1844-51.
& MOORE. *London.* 1875-81.
GUILMET, ——. *Paris.* 1878.
GUIMERIN, T. J. *Atlanta, Ga, USA.* 1871.
GUINAND—
Albert & Co. *London.* 1863.
Celestine. *London.* 1857-63.
GUINARD, F. E. *Baltimore, USA.* ca.1810.
GUINNESS—
B. *Brighton.* 1855.
John. *Brighton.* 1839-55.
GULDBRANSDEN, Paul. *Belfast.* From *Glasgow.* 1860-62. W.
GULICK—
Nathan. *Easton, Pa, USA.* 1777-1818. Then *Maysville, Ky.* 1818-26.
Samuel. *Northampton, Pa, USA.* 1756-1825.
GULLIAM, William. *Falmouth.* 1791-95.
GULLIFORD—
Robert. *Cutcombe (Somerset).* 1861-75.
William. *Luccombe (Somerset).* 1875-83.
William. *Porlock.* 1866-83.
GULLOCK—
John. *Rochford.* a.1761-96. W. & C.
Philip Howe. *Rochford.* s. of John. a.1783-1839.
Robert. *Canewdon (Essex).* s. of John. a.1781-88. W.
GUNN—
Charles. *Coventry.* 1880.
William. *Wallingford.* mar.1714-ca.1740. C.
GUNNELL, William. *Newbury.* 1823-31.

GUNNING—
John William. *Cheltenham.* 1870-79.
Thomas. *Somerton.* 1861-83.
GUNTER—
E. & A. *Toronto.* 1877.
Henry. *Norwich.* 1846. W.
James. *Bridport.* 1875.
GUNTHER—
Egmund. *Toronto.* 1857.
F. & E. *Toronto.* 1859-67.
GUNTON—
H. R. *Norwich.* 1858-75.
Henry. *Norwich.* 1836-58.
GURD, Joseph L. *Montreal.* 1875.
GURDEN, Joseph. *Oxford.* mar.1755, d.1772. C. & W.
GURDON—
G. *Hadleigh.* 1846-53.
G. *Wells.* 1858.
George. *Woodbridge.* Mar. 1811-22. then *Hadleigh* 1822-46.
W. *Burnham Market.* 1858.
GURLY, Daniel. *London.* 1839.
GURNER, Thomas. *Ickleton (Cambridgeshire).* 1875.
GURNEY—
Frederick Kingston. *London.* 1863.
L. F. *Bridgeport, Conn, USA.* ca.1870.
Thomas. *Luton.* ca.1740. C.
GURR—
Edward. *Ewhurst.* 1870.
George. *Ewhurst.* 1855-62.
J. T. *Robertsbridge (Sussex).* 1878.
GURTLER, Hans Georgen. *Uppsala.* ca.1750. C.
GURTON, Richard. *London.* 1839.
GUTHRIE, John. *Kelso.* 1860.
GUTKAES, Franz. *Vienna.* ca.1830. C.
GUTMAN, John. *London.* 1857.
GUTWEIN, ——. *Warzburg.* ca. 1765. C.

GUTTERIDGE—
C. *Bath.* 1856-66.
Charles. *Hulme.* (b.1820). mar.1851. W.
Job. *Wellington (Shropshire).* 1870.
John. *London.* 1832.
GUTTMAN—
Isaac. *Sheffield.* 1871.
Tobias. *Sheffield.* 1862-71.
GUY—
——. *Merthyr Tydfil.* 1824. C.
——. *Shaston.* ca.1770-ca.1800. C.
E. E. *Devizes.* 1848-59.
Edward, sen. *London.* 1832-81.
Edward, jun. *London.* 1844.
George. *Sherborne (Dorset).* 1875.
Henry. *Great Malvern.* 1876.
James. *Bristol.* a.1816-30.
Jasper. *Shaftesbury.* 1830.
Jessie. *Shaftesbury.* 1824.
John. *Liverpool.* (B. 1780-)d.1799. W.
P. A. *Quebec.* 1851.
Peter. *Liverpool.* 1689-d.1741. C.
Richard. *Millbrook, Canada.* 1862.
Thomas. *Hurstbourne Tarrant.* 1878.
W. *Hurstbourne Tarrant.* 1859-67.
William E. *London.* 1851-57.
GUYE, Philippe & August. *London.* 1863-81. Wholesalers.
GUYER, Benjamin. *Philadelphia.* 1848.
GUYERDET, ——. *Paris.* 1830. C.
GWILLEAM, John. *Camborne (Cornwall).* 1787. W.
GWINN, Thomas. *Ripley (Surrey).* 1839-55.
GWINNETT, Henry James. *Cardiff.* 1887.
GWYNN, Henry. *Carmarthen.* 1764. C.
GYLES, Daniel. *Bristol.* 1781.
GYMER—
Robert. *Norwich.* 1740-47. W. (B. d.1751).
Robert. *Yarmouth.* ca.1730. C.

H

H, H. Monogram of Hieronymus Heintzelman.
H & H. Trade mark of Harris & Harrington.
HR see R, H.
H, W, WH is monogram of Walker & Hughes.
H, Z, ZH-ca.1580. C. (Tabernacle clock.)
HAAS—
  B. & Co. *Geneva* and *Besancon.* ca.1830. W.
  & GOETZ. *New York.* ca.1820-30.
  Gottlieb. *Red Hook, NY, USA.* ca.1830.
  James A. *Philadelphia, USA.* 1846-49.
  John & Co. *New York.* ca.1825. C.
  Joseph Adolph. *Taunton.* 1861-75.
  N. *Philadelphia, USA.* 1796-1846.
  (Ph.) & Soehne (Baden). *London.* 1875. Adolph
    Boesch, agent.
HABEL, G. *Gunst.* 17c. C.
HABERSTRAW, Rudolph. *Leeds.* 1871.
HABERSTROH—
  G. *Taunton.* 1861-75.
  Jacob. *Dewsbury.* 1853.
  Jacob. *Cork.* 1862. C.
HABRECHT, Joachim. *Schaffhausen.* (B. 1537-)1567.
  C.
HACK, Gerrard. *Kendal.* b.1855-71. W.
HACKER, Michael. *Tewksbury, NJ, USA* and *German-*
  town, Pa. ca.1757-96. C.
HACKET—
  Cornelius. *London.* 1697.
  James. *Cork.* 1856-70. W.
  Richard. *Harringworth (Northants).* mar.1754.
  Simon. *London.* CC. 1632-67. Cornhill.
HACKNEY, Thomas Henry. *Gravesend.* 1866.
HADDACK & LANSDOWN. *Bath.* Pre-1832.
HADDEN, David. *Hunmanby (nr Filey).* 1834.
HADDER, William. *Philadelphia, USA.* 1837.
HADDON—
  C. *Taunton.* 1866.
  & FRANKLIN. *Taunton.* 1875.
  G. H. *Northampton.* 1877.
HADE, J. G. *Leamington.* 1860.
HADEN, Henry. *Dublin.* 1809. W.
HADERER—
  G. & D. *West Bromwich.* 1860.
  G. & E. *Birmingham.* 1854.
  George. *West Bromwich.* 1868-76.
HADFIELD—
  James. *Bristol.* Clock engraver, insolvent 1774.
  John. *Manchester.* 1738. C.
  Thomas. *Hyde.* 1834.
  W. *Hinckley.* 1855.
  William. *Leamington.* 1842-60.
HADIDA, Joseph. *London.* 1881.
HADINGTON, William. *London.* 1759. W.
HADLEY—
  Brothers & ESTELL. *Chicago, USA.* 1870.
  Thomas. *Birmingham.* (B. 1780-90) 1828-42.
HADROT & AVRIL. *London.* 1875.
HADWEN—
  Hannah. *Liverpool.* 1767-69.

HADWEN—*continued.*
  Isaac I. *Kendal.* b.1687. d.1737. C.
  Isaac II. *Liverpool.* b.1723, s. of Isaac of *Kendal.* To
    Liverpool ca.1737—d.1767. W.
  Isaac III. *Liverpool.* 1777-84 W. (B. 1800).
  Joseph. *Liverpool.* 1761-69. C.
HAETTICH, Andrew. *Cleveland, O, USA.* c.1850.
HAFFNER—
  Andrew. *Nottingham.* 1876.
  Charles. *London.* 1869-81.
HAGEN, Olive. *London.* 1851.
HAGER, Dietrich Peter. *Wolfenbuttel.* ca.1720. W.
HAGERTY, Bernard. *Ballyshannon.* 1858-90.
HAGEY—
  George. *Trappe, Pa, USA* and *Sterling, O, USA.*
    1850.
  Jacob. *Lower Salford Township, Pa, USA.* ca.1790-
    1831. C.
  John. *Philadelphia* and *Germantown, Pa, USA.*
    ca.1820.
  Jonas. *Springtown* and *Hallertown, Pa, USA.* Mid-19c.
    C.
  Samuel. *Franconia* and *Germantown, Pa, USA.* 1820-
    40s.
HAGGARD—
  John. *Wootton Bassett.* 1848-75.
  William. *Swindon.* 1842-59.
HAGGART, James. *Callander.* 1860.
HAGGERTY, James. *Cork.* 1767. C.
HAGUE—
  B. *New York.* 1854.
  S. *Belper.* 1855.
HAHL, Manufacturing Co. *Baltimore, USA.* ca.1875.
HAHN—
  C. G. *Philadelphia, USA.* 1789.
  Henry. *Reading, Pa, USA.* 1754-1843. C.
HAIGH—
  John. *Honley (Yorks).* 1871.
  T. Jefferson. *Baltimore, USA.* 1829-31.
  Thomas. *Mossley* nr *Manchester.* 1878.
  Thomas Henry. *Huddersfield.* 1866. W.
HAIN, John. *London.* 1839.
HAINES—
  Charles. *Swindon.* 1830-59. Also silversmith.
  F. *Sherborne.* 1855.
  Henry. *Knowle.* 1880.
  J. *Bedford.* 1847-77.
  James. *London.* 1832.
  N. *Knowle.* 1860-68.
  Nathan. *Knowle.* 1842-54.
  W. *Matlock Bath.* 1864.
  William. *Bedford.* 1830-54. C.
HAIR—
  George Bush. *London.* 1839-51.
  J. *Stratford (Essex).* 1851.
  William. *Birmingham.* (B. ca.1780-1808) 1816.
  William. *Worcester.* 1784-86. C.
  William E. *Boston.* 1849-50.
HAIRE, James. *Limavady.* 1858.

HAIS(Z)—
Mathias. *Birmingham.* 1868-80. Clock cases.
Mathias & George. *Walsall.* 1850.
& ISELE. *Dublin.* 1858-80 (-present day).
HAIZMAN, ——. *Canterbury.* 1865.
HAKES, A. H. *Norwich, Conn, USA.* ca.1860.
HALDANE, James. *Edinburgh.* mar.1811. W.
HALDEN, William. *Colwich (Staffordshire).* 1835-42.
HALE—
Joshua. *Lowell, Mass, USA.* ca.1840.
Nathan. *Windsor, Vt, USA.* 1772-ca.1800, then *Chelsea, Vt, USA.* ca.1800-1810.
Thomas & Co. *Bristol.* 1850- T.C.
Thomas & Sons. *Bristol.* 1863- T.C.
William. *Holyhead.* 1856-68.
William C. *Salem, Mass, USA.* ca.1850.
HALES, Edward. *Cork.* 1773. W.
HALEY—
Charles. *London.* (B. a.1762, CC. 1781-1825) 1828-32.
Michael. *Halifax.* 1651.
William. *London.* ca.1790. C.
HALFHIDE—
Alfred C. *Southampton.* 1878.
Alfred Charles. *London.* 1863.
Charles. *Ross.* 1870.
Edward. *Hereford.* 1844-50.
George William. *London.* 1863-81.
William. *Hereford.* 1835.
HALFORD—
John & William. *London.* 1844.
Robert Hosier. *London.* 1881.
William. *Northleach.* 1830-63.
HAL(L)IFAX—
——. 1914 and later prod. of Pequegnat Clock Co.
John I. *Barnsley.* b.1695. d.1750. C.
George. *Doncaster.* b.1725, s. of John I. d.1811. C.
Joseph. *Barnsley.* b.1728, s. of John I. d.1762.
Thomas. *Barnsley.* b.1721, s. of John I. Later *London.*
HALKSWORTH, William. *London.* 1839-57.
HALL—
A. B. *Ohio City* and *Cleveland, O, USA.* 1820-30.
Amasa. *Atlanta, Ga. USA.* 1859.
Andrew. *Hatfield.* (B. from *London* 1802-04) 1828-39.
Asaph. *Goshen* and *Hart Hollow, Conn, USA.* 1797-1842.
& BLISS. *Albany, NY, USA.* 1816-18.
Charles. *London.* (B. a.1810-20) 1832-51.
Charles P. *Ingersoll, Canada.* 1851-71.
D. *Ripley (Derbyshire).* 1864.
D. G. *Lewiston, Me, USA.* ca.1850.
David. *Philadelphia & Burlington, NJ, USA.* 1777.
Francis. *Whitby.* 1866. W.
Frederick Charles. *Yoxford.* 1879.
Frederick Richard. *London.* 1851.
George. *Wooler.* 1827-34.
George & John. *Wakefield.* 1862.
Henry. *Birmingham.* 1860-68. Watch repairer.
Henry. *Workington.* 1873-79.
Henry S. *London.* 1863.
Henry William. *Lititz, Pa, USA.* ca.1830-40.
I. N. *Stratford, Canada.* 1857-63.
J. *Ipswich.* 1858-65.
J. *Northampton.* 1864
J. *Stourbridge.* 1860.
J. F. *Yoxford.* 1858-65.
James. *Prescot.* 1773. W.
Jesse. *Wainfleet.* 1850-68.
John. *Beverley.* ca.1765-1813. C. Several mus. clocks known.
John. *Birmingham.* 1860-68.
John. *Geneva, NY, USA.* ca.1810. C.
John. *Grimsby.* 1828-50.
John. *London.* 1832-39.
John (& Co) *Manchester.* 1848-51. Regulator.
John. *Philadelphia.* 1806-40.
John. *Ross.* 1870.
John. *Sleaford.* 1828-35.
John. *Southampton.* 1878.
John. *West Chester, Pa, USA.* 1793-1867. W.
John. *York.* 1853. W.
John H. *New Haven, Conn, USA.* ca.1780. W. & C.
John Joshua. *Belfast.* 1805-09. C.

HALL—*continued.*
Jonas G. *Montpelier* and *Roxbury, Vt, USA.* ca.1870. W.
Joseph. *Alford.* 1835-50.
Joseph. *Birmingham.* 1816.
Joseph. *Grimsby.* 1828-50.
Joseph. *Holmfirth.* 1822.
Joseph. *Skipton.* b.1828-66. W.
Joseph. *Wakefield.* ca.1800? C.
Joseph. *Wigton.* 1834.
Joseph & Elizabeth. *Alston.* 1797. (These names on a clock assumed to be owners, but B. has Joseph as 1810 C.)
Joseph Starling. *Beaumaris.* 1887.
K. *Dover.* 1866.
Kennett. *Dover.* 1838-47.
Martin. *Yarmouth.* (B. 1752-)62. C. & W.
Mrs. M. *Northampton.* 1869.
Peter. *Philadelphia, USA.* 1818-24.
R. *Worksop.* 1864.
Robert. *Oswestry.* Mar. 1800-38.
Robert. *Wells.* 1830-36.
Samuel. *Bickerstaffe (Lancashire).* d.1737. W.
Seymour & Co. *Unionville, Conn, USA.* ca.1830.
T. J., jun. *Northampton.* 1854.
Thomas. *Desford (Leicestershire).* ca.1750. C.
Thomas. *Edinburgh.* Free 1729, d.pre-1784. W. Previously in *London* and *Paris.*
Thomas. *Farnworth.* 1805. W.
Thomas. *Liverpool.* 1834. W.
Thomas. *Maidstone.* Early 18c. C.
Thomas. *Prescot.* 1778. W.
Thomas. *Southampton.* 1878.
Thomas. *Wells.* ca.1770. C.
Thomas, sen. *Northampton.* 1841-54.
& WADE (Nathaniel). *Newfield, Conn, USA.* ca.1796.
William. *Birmingham.* 1868-80.
William. *Coventry.* 1828.
William. *Eyemouth.* 1837.
William. *Hammersmith.* ca.1770? C.
William. *Ipswich.* 1864-85.
William. *Liverpool.* 1827-34. W.
William. *North Shields.* ca.1790. C.
William. *Rothbury.* 1834.
William & Son. *Dunse.* 1860.
HALLAGAN, John. *Dublin.* 1819. W. Also goldsmith and jeweller.
HALLAM—
——. *Wymeswold.* early 19c. C.
Edward. *London.* 1857.
Henry. *London.* 1832.
James. *Leicester.* ca.1770-91. C.
James. *London.* 1828.
John. *London.* (B. 1809-11) 1828.
John. *Lutterworth.* ca.1740. C.
& Son. *Nottingham.* 1849-55.
T. *Nottingham.* 1864-76.
Thomas. *Lutterworth.* 1835-76.
Thomas. *Melton Mowbray.* 1828-35.
Thomas. *Nottingham.* succ. John Hallam 1819-44. then & Son (Thomas junior) 1844-91.
HALLE, A. *Louisville, Ky, USA.* ca.1840.
HALLER (or HELLER), John. *Berlin, Canada.* 1851-63.
HALLETT—
A. F. *Hamilton, Ontario.* 1875-78.
F. H. *Melksham.* 1848.
Jehu. *Bere Regis.* 1875.
John. *Lyme Regis.* 1824-30. C.
Richard D. *Beaminster.* 1867-75.
William G. *Bradford (Wilts).* 1842-59.
William George. *Hastings.* 1870-78.
HAL(L)IDAY—
Elias H. *Philadelphia, USA.* 1828-33.
G. J. *London.* 1851.
Hiram. *Albany, NY, USA.* 1834-44.
Jacob. *Banbridge.* 1865-68, then *Belfast* till 1892. Importer.
Peter. *Wigtown.* 1837-60.
Robert. *Kirkcudbright.* 1832-60.
HALLIWELL—
David. *Warrington.* 1834.

HALLIWELL—*continued.*
John. *Wakefield.* ca.1800. C.
John. *Warrington.* 1822. C.
Thomas. *Dewsbury.* 1822-71. W.
HALLMUTH, A. *Dudley.* 1868 (cf. Hellmuth).
HALLOWELL, John. *Montreal.* 1831-d.1853.
HALLOWS—
John. *Liverpool.* 1851.
Jonathan. *Liverpool.* 1824.
HALLS, John. *Nayland.* 1853-79.
HALLUM, Timothy. *Lutterworth.* 1828 (cf. Hallam).
HALLYWOOD, Job. *Belfast.* 1865.
HALPERN, Brothers. *Manchester.* ca.1870. W.
HALSALL—
Edward. *Bristol.* (B. 1825) 1830-42.
Robert. *Ormskirk.* 1828.
HALS(H)EY—
Henry. *Lancaster.* d.1683. W.
John. *Norwich.* 1725-54. W.
HALSTED, Thomas. *Ryde, Isle of Wight.* 1859-78.
HALTON—
John. *Manchester.* 1803. C.
& SLOCOMBE. *London.* 1875-81.
HAM—
G. *Torpoint (Cornwall).* 1856.
George. *Portsmouth, NH, USA.* ca.1810.
Henry. *Bodmin.* ca.1810. C.
Jabez. *Winterton (Lincolnshire).* 1849-68.
James. *New York.* 1754, math. inst. mr.
John. *Kingsand (Cornwall).* 1844. W.
John. s. of Joseph. *Liskeard* ca.1800, then *Millbrook (Cornwall)* ca.1820-23. C.
Joseph. *Liskeard.* ca.1775-1806. C.
S. T. *Frimley.* 1866-78.
Supply. *Portsmouth, NH, USA.* 1788-1862.
T. *Frimley.* 1862-66.
Thomas. *London.* 1844.
William Pote. *Liskeard.* ca.1820-73. C.
HAMBLET, William Walter. *London.* 1832.
HAMBLETON, Thomas. *Birkenhead.* 1857-78. Skel. clock.
HAMBLING & MURRAY. *Cheltenham.* 1879.
HAMBLY, William. *London.* Late 18c. W. (One such at *Falmouth* 1784-95, may be same man?).
HAMBLYN, Zachariah. *Southport (Portsea).* 1830.
HAMER, John. *Llanidloes.* 1868.
HAMILL, James. *Ballymoney.* 1858. C.
HAMILTON—
——. 1914 and later, prod. of Pequegnat Clock Co.
& ADAMS. *Elmira, NY, USA.* 1837-42.
Clock Co. *Hamilton, Ontario.* 1877-84. Later called Canada Clock Co., q.v.
& Co. *Drogheda.* 1858.
Daniel S. *Elmira, NY, USA.* 1848.
J. *Bristol.* 1856.
J. H. *Portsmouth.* 1867.
James. *Belfast.* 1865-68. C. (?also *Londonderry*).
James. *Hamilton.* 1860.
James. *Philadelphia, USA.* 1848.
John. *Bradford.* 1871.
John. *Glasgow.* 1783-84. C.
R. J. *Philadelphia.* 1837-46.
Robert. *Hexham.* 1854-58.
S. P. *Savannah, Ga, USA.* ca.1860.
Samuel. *Belfast.* 1880-87.
Samuel. *Philadelphia, USA.* 1837.
Sangamo Corp. *Springfield, Ill, USA.* 1928-30. Electric clock.
W. *Woodstock, Canada.* 1851.
Watch Co. *Lancaster, Pa, USA.* 1892-1909.
HAMLEN—
Nathaniel. *Augusta, Me, USA.* 1790-1820. C.
William John. *Marlborough.* 1875.
HAMLET—
John. *Ringwood.* 1839.
William. *Liverpool.* 1851.
HAMLIN, William. *Providence, RI, USA.* 1797.
HAMLYN, Thomas. *Ashburton.* 1790-1800. C. & W.
HAMMAN, Peter. *Philadelphia, USA.* 1817.
HAMMERSLEY, John. *London.* 1869.
HAMMERTON, Robert. *Birmingham.* 1880.
HAMMETT, Edgar. *Swansea.* 1887.

HAMMON—
G. J. *Coventry.* 1860. Watch case maker.
George. *London.* 1832-39. Watch cases.
J. *Coventry.* 1860. Watch cases.
J. W. *Coventry* and *Clerkenwell.* 1854. Watch cases.
James (& Sons). *London.* 1844(51-69).
John William. *London.* 1828-39. Watch cases.
HAMMOND—
Ann. *Clare.* 1839.
Charles B. *London.* 1857-81.
Clock Co. *Chicago, USA.* 1936.
Edward. *Hoddesdon.* 1839-59.
Frederick Samuel. *Hunstanton St. Edmunds (Norfolk).* 1875.
George & Thomas. *Manchester.* 1848-51.
John. *Clare.* Mar. 1790-1830. Also silversmith.
John. *Colchester.* 1810.
John. *Hoddesdon.* 1828.
L. *Cambridge.* 1858.
Miss H. *Clare.* 1846.
Richard. *Guildford.* s. of William of Compton, a.1790, free 1797.
Samuel. *Battle.* b.1668-d.1736. C.
Samuel. *New York.* ca.1840-60. Chrons.
Samuel Moss. *Witham (from Maldon)* 1798-1812. W.
T. C. *Portsea.* 1859.
Thomas. *Anglesey.* ca.1860-75.
Thomas Cundall. *York* 1825, *Richmond* 1829.
HAMNETT—
James. *Glossop.* 1876-1916.
S. *Stockport.* 1865.
HAMON, George. *Jersey.* 1834.
HAMPDEN Watch Co. *Springfield, Mass, USA.* 1877-88.
HAMPSHIRE—
Edward. *Upper Mill (nr Manchester).* 1871.
William. *Dewsbury.* 1866. W.
HAMPSON—
——. *Bolton.* Late 18c.
Albert. *Aberystwyth.* b.1838-d.1867.
Henry. *London.* 1857. Successor to William.
Robert. *Bury.* ca.1775-(B. 1787). C.
Robert. *Dublin.* 1826-58. W.
Robert. *Manchester.* 1828. From *Warrington.*
(Robert) & THELWELL. *Manchester.* 1824.
Robert. *Warrington.* 1758 -ca. 1790. C. Later at *Manchester* qv.
Thomas. *Liverpool.* ca.1780. C.
W. jun. *London.* 1836. C.
William. *Place not stated.* ca.1800 regulator (perhaps next man).
William. *London.* 1828. C. & clock hand maker.
William. *London.* 1851-69.
William (& Sons). *London.* 1832-39 (44-51).
HAMPSTON—
John (& Co.). *York.* b.1739-d.1805. C.
PRINCE & CATTLE. *York.* Partnership ca.1777- ca.1810.
HAMPTON—
James Brandon. *Salisbury, NC, USA.* 1801-32.
(James B) & PALMER (John C.) *Salisbury, NC, USA.* 1830-32.
Samuel. *Birmingham.* 1842.
Samuel. *Chelsea, Mass, USA.* 1840.
HAMSON, Clark. *Waterbury, Conn,* 1812.
HAMY, William. *Dublin.* 1802-19.
HANBURY—
J. *West Haddon.* ca.1780. C.
Valentine. *Watford (Northants).* 1777. C.
HANBY, E. *Bradford.* 1866-71.
HANCOCK—
Abraham. *Wetherby.* 1844 (?error for Anthony).
Abraham. *Winchester.* 1830.
Anthony. *Otley.* 1822-34. Then *Wetherby* 1837 and later. C.
C. F. (?*London*). ca.1850. C.
C. J. *Midsomer Norton.* 1875.
& COX (& Co.). *Yeovil.* 1861-66 (75-94).
Daniel. *Bradford.* 1866-71.
Daniel. *Manchester.* 1846-55. W.
Frederick. *Aberystwyth.* 1830.
George. *London.* 1832-39. Watch dial maker.

HANCOCK—*continued.*
George. *Midsomer Norton.* ca.1800. C.
I. *Midsomer Norton.* 1861-66.
James. *Yeovil.* 1840.
John. *Castle Carey.* ca.1800. C.
John. *Leeds.* 1773-80. Then *Bradford* 1780 and later. W.
John. *Yeovil.* (B. 1791-)1840.
John E. *Risca* (Mon.). 1880.
Luke. *London.* 1857-81.
Thomas C. *Blandford.* 1867-75.
William H. *Knaresborough.* 1851-71. W.
HAND, Patrick Bernard. *Dublin.* 1858.
HANDCOCK, Edward. *London.* 1844.
HANDEMAN, E. S. *Bridge.* Late 18c. C.
HANDFORD—
Thomas. *Loughborough.* 1864-76.
Thomas. *Ross.* 1879.
HANDLEY—
Alfred. *Bristol.* 1879.
Alfred. *Stroud.* 1870.
Fredrick Herbert. 1881.
George. *London.* ca.1780. C.
John. *Runcorn.* 1828-65.
William. *Sedbergh.* 1822.
HANDS—
C. *Clapham.* 1862.
Daniel. *Bilston.* 1850.
Frederick Adolphus. *Ramsgate.* 1866-74.
Henry. *Norwich.* 1875.
Joseph. *Coventry.* 1860-80.
HANDSCOMB—
E. *Woburn.* 1854.
Ebenezer. *Ampthill.* 1830.
Samuel & E. (son). *Woburn.* 1830-47.
HANDY, Henry. *London.* 1851-57.
HANEN, Thomas. *London.* 1662 (May be HANDE).
HANET—
Catherine. *Paris.* Widow, at Le Tabernacle. 1687-1705. C.
Nicolas. *The Hague.* 1657-60. *Paris.* ca.1660-70.
HANEYE, Nathaniel. *Bridgewater, Mass, USA.* ?Late 18c. C.
HANKINS, Jacob. *Farningham.* 1826-45.
HANKS—
Benjamin, *Windham, Conn, USA.* 1755-79; *Litchfield* 1779-90, *Mansfield* 1790. Later *Troy, NY,* where d.1824.
Henry Garlick. *Malmesbury.* 1830-59. Also stamp distributor.
HANLIN—
John. *Belfast.* 1815.
Robert. *Dublin.* 1858.
HANLON—
James. *London.* 1881.
William. *Dublin.* 1809-39. W.
HANMAN, Thomas. *Gloucester.* 1863-79. Wall clocks.
HANNA—
Hugh. *Wabash, Ind, USA.* 1834.
Hugh White. *Rathfriland.* ca.1840. C.
James. *Dublin* to *Quebec* 1763-1807.
James Godfrey. s. of James. *Quebec.* 1807-18. Partner with Francois De Lagrave 1816-18.
HANNAFORD, Mrs F. *London.* 1869.
HANNAM—
Edward. *London.* 1869-81.
Joseph. *Bristol.* a.1805-12. C.
George. *Workington.* 1834.
HANNUM, John. *Northampton, Mass, USA.* 1837-49.
HANNY—
James. *Shrewsbury.* 1835-79. C.
William Stourton. *Shrewsbury.* b. ca. 1836 s. of James Hanny, clockmaker -1875.
HANSCOMBE, Ebenezer. *Newport Pagnell.* ca.1730-40. C.
HANSELL—
Henry. *Chatham.* 1839.
James. *Philadelphia, USA.* 1816-50. C.
HANSEN Manufacturing Co. *Princeton, Ind, USA.* ca.1936.
HANSFORD, John. *Ilminster.* 1861-83.
HANSLAPP—
Robert. *London.* a.1646 to Edward East. CC. 1653-63.

HANSLAPP—*continued.*
William (alias William WILLIAMS). *London.* a.1654 to Robert Hanslapp. CC. 1663. d.1690. C.
HANSON—
Charles. *Batley.* 1866. W.
Charles. *Huddersfield.* 1834-71. W.
Charles. *London.* 1839.
Richard. *Leeds.* 1834. W.
William. *Windsor.* 1830-64.
HANTSCHKE, G. J. *Toronto.* 1861-63.
HANWORTH, W. *Harleston.* 1858.
HARAN, Henry. *Glasgow.* 1860.
HARBEN—
Thomas, sen. *Lewes.* b.1712, a.1726, d.1766. C.
Thomas, jun. *Lewes.* s. of Thomas, sen. b.1739, mar.1766, d.1803. Successor in 1778 to Richard Comber. W. Also goldsmith and banker.
HARBINSON, John Francis (& Co). *Belfast.* 1865-87 (?-1896). W.
HARCOURT, Alfred. *Norwich.* 1875.
HARDACRE, John. *Bramley* (*Leeds*). 1807. *Rodley* (*Leeds*). 1822.
HARDAKER, James. *Bradford.* 1853-71. C.
HARDCASTLE—
Henry. *Birmingham.* 1860-80.
Jacob. *Pateley Bridge.* 1834.
HARDEMAN—
Edwin Samuel. *Canterbury.* 1838-55.
Samuel. *Bridge* (*Kent*). 1794-1839. Previously in partnership with William NASH.
William Henry. *Bridge.* 1848-74.
HARDEN—
Charles. *Hythe.* b.1793-1861. W.
James. *Philadelphia, USA.* 1818-24. Clock dial maker.
John. *Birmingham.* 1816. Dial maker.
HARDER, W. A. *New York.* 1848.
HARDEY, John. *Hull.* Late 18c. C.
HARDIE—
& CHRISTIE. *Manchester.* 1824. Clock dial makers.
James. *Aberdeen.* 1846.
John. *Morpeth.* (B. 1820-)1827. C.
HARDING(E)—
C. *Forton* (*Hants*). 1878.
Charles. *Ashburton.* 1790-1800.
E. H. *Canterbury.* 1865.
Edmund John. *Hitchin.* 1874.
Henry. *London.* 1844-69.
James. *London.* ca.1760(?).
James. *London.* 1839.
John. *London.* b.ca.1662, mar.1693-97. C. *St. Martin-in-the-Fields.*
John. *Middleton* (*Isle of Wight*). 1878.
John. *Pen-y-groes.* 1874.
John. *Portmadoc.* 1856-68. C.
John Martin. *London.* 1828-32. Watch cases.
Joseph. *Timsbury.* 1861-75.
Mary. *London.* 1839.
Newell. *Haverhill, Mass, USA.* 1796, then *Boston* till 1862.
Thomas. *Bristol.* a.1797-1805.
W. *Newport* (*Isle of Wight*). 1867.
Walter. *St. Neots.* 1877.
HARDMAN—
Edmund John. *Birmingham.* 1880.
Garrard. *Liverpool.* s. of John of *Wavertree.* 1768-d.1771. W.
Gerrard. *Farnworth* (*Lancs*). 1784. W.
J. *Lebanon, O, USA.* From *Va.* 1832.
John. *Wavertree* (*Lancs*). 1748-d.1773. W.
Samuel. *Prescot.* 1819. W.
Thomas. *St. Helens.* 1787. W.
William. *Manchester* 1809. C.
HARDON, John. *Workington.* 1879. Also musical boxes.
HARDT, Harry. *London.* 1881.
HARDWICK—
——. *Ridgmount.* ca.1750. C.
John. *Malton.* mar.1735. W.
John. *Whitby.* d.1752. W.
Leslie. *Annapolis Royal, Nova Scotia.* 1866.
Robert. *Huntingdon.* 1864-77.
William. *Neath.* 1887.

HARDY—
Albert. *Purewell* (*Hants*). 1848-78.
Edward. *Halifax.* ca.1780.
Edwin Elvey. *Wells.* 1861-83.
George. *Kingston, Canada.* 1850-65.
George. *London.* 1857.
Henry. *London.* 1832-39. Watch cases.
J. *Nottingham.* 1864-76.
James. *Dorchester.* 1830-67.
John. *Aberdeen.* 1837.
John. *Birmingham.* 1880.
John. *St. Ives* (*Hunts*). 1864-77.
John Noah. *Leeds.* 1866-71. W.
Richard. *Newark.* 1813-44. W. & C.
Samuel. *St. Ives* (*Hunts*). 1839-54.
Thomas. *Bulwell.* 1876.
Thomas. *London.* 1828. Watch cases.
Thomas. *Nottingham.* ca.1770-ca.1790. C.
Thomas. *Nottingham.* ca. 1770-1810. C.
William. *Charleston, SC, USA.* From *London.* 1773-75,
    when returned to England.
William. *London.* 1832.
HARE—
Abraham. *London.* 1811 (B. 1815). W.
Alfred. *Kidderminster.* 1868-76.
HARFLEET, Cornelius. *Sandwich.* ca.1750. C.
HARFORD—
John. *Bath.* 1658. Lant. clock.
John. *Bath.* 1730-70.
William. *Bath.* 1795-d.1797.
HARGRAVE(S)—
J. *Sleaford.* ca.1790. C.
James. *Bawtry.* b.1768-d.1835. W.
John. *Hull.* 1846-58.
Robert. *Doncaster.* 1785-d.1795. W.
Robert. *Skipton.* ca.1800.
HARGREAVE(S)
James. *Blyth.* 1828.
John. *Settle.* 1837.
Thomas. *Burnley.* ca.1780-90. C.
Thomas. *Settle.* 1790-1834. C. & W.
William. *Leeds.* 1871.
William. *Settle.* 1853-67. C.
HARINGTON, Thomas. *Henley-on-Thames.* 1494-1515.
    Kept ch. clock.
HARKER—
H. *Bradford.* 1866.
James. *Liverpool.* b.1689-1725. W.
John. *Askrigg.* 1807-12. C.
Nathaniel. *Bishop Auckland.* 1834.
Nathaniel. *Pateley Bridge.* 1837. W.
Richard. *Abingdon.* 1877.
Richard. *Birkenshaw.* 1866. W.
Richard. *Bradford.* 1853.
William. *Pontefract.* 1871.
HARKHAM?, Thomas. *London.* 1697 (spelling
    uncertain).
HARLACHER, Benjamin. *Washington, Pa, USA.* Early
    19c.
HARLAND—
Christopher. *Ramsgate.* 1858.
E. *Ramsgate.* 1855. Chrons.
Henry. *Croydon.* 1878.
John. *Hull.* 1813.
HARLE—
Charles. *London.* 1828-39. Watch case maker.
Charles. *London.* 1875-81.
Joseph. *Rothbury.* 1741-52. C.
Robert. *London.* 1869.
HARLEY—
Joseph. *Wingrave* (*Bucks*). 18c. C.
W. *London.* Late 18c.
William. *Shrewsbury.* apr. 1717 to Thomas
    Gorsuch. d. 1764. C. & W.
HARLING—
Mrs. J. *London.* 1857.
William. *London.* 1844.
HARLOCK—
James. *London.* 1844.
Robert. *Ely.* 1840.
HARLOW—
B. *Hanley.* 1793. W.
Benjamin. *Bath.* 1883.

HARLOW—*continued.*
Benjamin. *Lane End* (*Staffs*). 1828. C.
Benjamin. *Longton.* 1842.
Benjamin. *Macclesfield.* 1834.
Benjamin Wyatt. *Ashbourne.* s. of Robert Harlow
    senior whom he succ. ca. 1828-45.
Mrs. & Son. *Ashbourne.* 1849.
HARMAN—
George. *High Wycombe.* 1790-1838. Chime-maker.
H. *Clevedon.* 1861.
Samuel. *Cork.* ca.1754-d.1784. W.
Thomas. *Cork.* s. of Samuel. ca.1770. W.
William. *Cork.* b.ca.1769-d.1837. W.
HARMER (& HARMAR)—
J. *Beccles.* 1858-65.
Jasper. *Islington.* 1730. (*London* 1685-1716.)
Joseph. *Peterborough.* 1877.
Joseph. *Whittlesey.* 1875.
Joseph Norman & Co. *Fletton* nr *Peterborough.* 1877.
Joseph N. *London.* 1881.
Thomas. *Montreal.* 1875.
HARMSON, ——. *Newport, RI, USA.* 1720s.
HARMSWORTH, John. *Cambridge.* 1830-46. C.
HARNSWORTH, John. *London.* 1828.
HAROV, Anselm. *Quebec.* 1851-57.
HARPER—
A. *Burton-on-Trent.* 1876.
Alexander. *Halifax, Nova Scotia.* 1813.
Alfred. *Ashbourne.* 1876-91. Also at *Derby* 1891.
Alfred. *Louth.* 1849-50.
Benjamin. *Philadelphia, USA.* 1843.
C. *London.* 1816. W.
D. A. *Montreal.* 1866.
F. *Darlaston.* 1868.
Francis. *Bilston.* 1868-76.
John George. *Bristol.* 1879.
John M. *Philadelphia, USA.* 1841.
W. E. *Philadelphia, USA.* 1839. W.
William. *Antrim.* ca.1770-92. W.
William. *Halifax.* 1755.
William. *Montreal.* 1842.
William. *Pontefract.* 1866.
William. *Prescot.* 1797. W.
HARRIDGE—
G. *Richmond* (*Surrey*). 1862.
J. *Gosport.* 1848-59.
HARRILD, Robert. *London.* 1851-57.
HARRINGTON—
Charles. *Brattleboro, Vt, USA.* ca.1830. W.
Henry. *Salem, Mass, USA.* 1832-55.
Michael. *Cork.* 1856-58.
R. *Aldershot.* 1867.
R. *Farnham.* 1851.
R. *Witham.* 1851.
R. M. *Farnham.* 1851-66.
Robert. *Farnham.* 1839.
Samuel. *Amherst, Mass, USA.* 1842-45. W.
Thomas. *Great Baddow* (*Essex*). 1839.
William. *Philadelphia, USA.* 1849.
HARRIS—
Aaron. *Morwenstow* (*Cornwall*). ca.1830. C.
Alexander. *Paisley.* 1834-41. W.
& ANDREWS. *Coventry.* 1880.
Arthur. *Coventry.* 1868.
Arthur William. *Montgomery.* 1887.
Asher. *London.* 1863-81.
C. D. *Coventry.* 1880.
Charles. *High Wycombe.* 1877.
Charles. *Liverpool.* (B. 1825-)1834.
Charles. *Reading.* 1837.
Charles. *Tunbridge Wells.* 1874.
Clement. *London.* (B. a.1820) 1828-44.
& Co. *Philadelphia, USA.* 1830s. C.
David. *Louth.* 1876.
David. *Swansea.* 1840.
Dinah. *Witney.* wid. of William, 1823-39. C. & W.
E. *King Swinford.* 1860.
Edward. *Bromyard.* 1830.
Edward. *Ipswich.* 1875.
Edward B. *Montreal.* 1844.
Edward B. *Quebec.* 1848-57.
Elias & Co. *London.* 1875.
Elizabeth. *Bromyard.* 1844.

HARRIS—*continued.*
& ELLIOTT. *Coventry.* 1860. Watch case maker.
Frederick. *Manchester.* 1851.
Frederick Warne. *London.* 1832.
G. *Wellingborough.* 1864-69.
George. *Fritwell.* b.1614-d.1694.
George. *London.* 1875.
George. *Pembroke Dock.* 1850-68.
George. *Redruth.* 1818-ca.1820. C. & W.
George. *Truro.* 1822. W.
H. *Mildenhall.* 1846.
H. *Westbury.* 1848.
H. & Co. *Birmingham.* 1860.
& HARRINGTON. *New York.* 1880-1919.
Henry. *Grimsby.* 1876.
Henry. *Mildenhall.* 1839.
Henry. *Mynydd Islwyn.* Late 18c.-early 19c. C.
Henry. *Thetford.* 1858-75.
Henry. *Truro.* 1823-47.
Henry & Co. *Birmingham.* 1880.
Henry James. *Manchester.* 1828-51.
Israel. *Truro.* 1844.
J. *Arlington* (*Gloucestershire*). 1856.
J. *London 'Old Bailey'.* ca.1680. W.
J. & A. *Liverpool.* 1863. W.
Jacob. *Paulton.* 1861-75.
James, sen. *Brixton.* 1862-78.
James J., jun. *Brixton.* 1866-78.
James. *Grimsby.* 1861-68.
James. *London.* 1857.
James. *Louth.* Same man as at *Grimsby.* 1861.
James. *Witney.* mar.1795-1852. C.
James Robert. *London.* 1875-81.
John. *Charleston, SC, USA.* 1729-39.
John. *Coventry.* 1880.
John. *London.* 1832-39. Watch cases.
John. *London.* 1857-81.
John. *Mildenhall.* 1830.
John. *Oxford.* a.1668-?96. W.
John. *Sevenoaks.* 1855-66.
John. *Sutton Valence.* 1847-51.
John. *Truro.* 1847.
John. *Witney.* mid-19c. C.
John C. *Wellingborough.* 1877.
John James. *London.* 1811-57. W.
Joseph. *Sheffield.* 1862.
Joseph. *Swansea.* 1887.
Joshua. *Burford.* ca.1760-1810. C. & W.
Joshua. *Witney.* s. of William. 1809-31. C.
L. *Manchester.* 1881. W.
Lazarus. *Aberavon.* 1848.
Lazarus. *Liverpool.* 1827-51 (one such *London* 1802-1824).
Lipman. *Worthing.* 1839.
Louis. *London.* 1851.
Morris Hart. *Penzance.* 1844-49. C.
Mrs. Annie Louisa. *Rhyl.* 1887.
Mrs. E. *Paulton.* 1883 (wid. of Jacob?). Clock dealer.
Mrs. M. *Belfast.* 1865-68. Trade parts warehouse.
Nathan. *Crediton.* a.1737 to John Tichell.
Nicholas. *Fritwell.* b.1657. s. of George. d.1738. C.
Robert. *London.* ca.1700. C.
Robert. *Oxford.* 1777-ca.1800. C. & W.
Robert. *Paisley.* (B. 1820)-60.
Robert. *King's Lynn.* 1875.
S. *Cambridge.* 1865.
Samuel. *Falmouth.* 1823. W.
Samuel. *Gillingham.* 1830.
Samuel. *Sheffield.* 1862-71.
& Sons. *Swansea.* 1875. Successor to David H.
& STANWOOD. *Boston, USA.* 1842.
Stephen. *Tonbridge.* ca.1690-ca.1730. C.
T. *Brackley.* 1854-69.
Thomas. *Barnstaple.* 1781-95.
Thomas. *Bromyard.* 1850-63.
Thomas. *Canterbury.* 1823-28.
Thomas. *Deddington.* b.1732-d.1797. C.
Thomas. *Felton.* 1834.
Thomas. *Hawkhurst* (*Kent*). 1847-66.
Thomas. *Lutterworth.* 1849-76.
Thomas. *Redruth.* 1823. W.
Thomas Robert. *London.* 1863-81.

HARRIS—*continued.*
W. *Rye.* 1851.
W. *Chippenham.* 1830-58.
W. *Newmarket.* 1846.
Walter. *Maidstone.* 1704.
William. *Aldershot.* 1878.
William. *Knighton.* 1887.
William. *London.* 1828-32. Watch cases.
William, jun. *London.* 1875-81.
William, sen. *London.* 1863-69(75-81).
William. *Newmarket.* 1858-75.
William. *Paisley.* 1860.
William. *Westbury* (*Wilts*). 1859-75.
William. *Witney.* 1794-1823. C.
William Henry. *East Dereham.* 1875.

HARRISON—
——. *Highland Co., O, USA.* 1792.
Adam. *Dumfries.* 1887.
David. *Leeds.* 1871.
E. *London.* Early 19c. C.
Edward. *Warrington.* Late 18c. C.
Francis. *Hexham.* 1834-58. C.
Francis. *London.* 1828-57.
Frederick. *Derby.* 1876. C.
Frederick. *Wellesbourne* (*Warwickshire*). 1880.
G. *Ollerton.* 1855.
J. *Congleton.* 1857.
J. *Warkworth.* 1858. C.
J. *West Hartlepool.* 1898.
James. *Birmingham.* 1828.
James. *Birmingham.* 1868.
James. *Bristol.* 1830.
James. *Farnworth.* 1834. W.
James. *Hull.* 1834-58. T.C. (also *Barton, Lincs*).
James. *Waterbury, Conn, USA.* 1767-1814, then Boston.
John. *Darlington.* 1827-d.1876. ?From *London.*
John. *Market Drayton* 1822-36.
John. *Edinburgh.* 1822. May be same as John David.
John. *Liverpool.* 1824-51.
John. *Mold.* 1835.
John David. *Edinburgh.* 1821.
Lemuel. *Waterbury, Conn, USA.* bro. of James. 1800-11.
Mrs. A. *Southwell.* 1849.
Richard. *Mold.* 1840-56.
Richard. *Ormskirk.* ca.1790-1800. W.
Robert. *Bradford.* 1871.
Robert. *Exeter.* ca.1780-90. C.
S. (& Sons). *Warwick.* 1854-60(-68).
& Son. *Darlington.* Successor to John, mid-19c.-1898, and later.
Susannah. *New York.* ca.1855.
Thomas. *Liverpool.* 1774-d.1814. W. & C.
Thomas. *Morpeth.* 1834.
William Stephenson. *Southwell* (*Notts*). 1844-79. W.
W. E. *Stockton.* 1898.
William. *Charlbury.* ca.1770-92. C.
William. *Farnworth.* 1833. W.
William. *Tadcaster.* 1822-26.
William Stephenson. *Tadcaster.* 1834.
Wooster. bro. of James and Lemuel. *Trumbull, Conn, USA.* 1772-1800, then *Newfield.*
HARROCKS, Joshua. *Lancaster.* 1748-64, then *Eamont Bridge.*
HARROWER, John. *Aberfeldy* (*Scotland*). 1860.
HARRY, Samuel. *Hertford.* 1874.
HARRYS, William. *Cardiff.* 1734.
HARSCHER, Ferdinand. *Macclesfield.* 1848-78.
HART—
Alpha. bro. of Henry. *Goshen, Conn, USA.* ca.1820-36.
& BREWER. *Middletown, Conn, USA.* 1800-03.
David. *Chippenham.* 1867-75.
David. *Malmesbury.* 1842-59.
Eliphaz. *Norwich, Conn, USA.* ca.1810.
G. *Bridgeport, Pa, USA.* ca.1850.
George. *Derrock, co. Antrim.* mar.1836. C.
George. *London.* 1875.
H. *Painswick.* 1879.
H. *Woolwich.* 1847.
Henry. *Goshen, Conn, USA.* bro. of Alpha. ca.1830-35. C.

HART—*continued.*
J. & Co. *London.* 1839.
Jacob. *Hull.* 1823.
James. *Canterbury.* mar.1752-61.
John. *Yarmouth.* (B. 1759-)d.1773.
John George. *London.* 1875.
John N. & Son. *London.* 1839.
Josiah. *North Walsham.* 1830-36.
Judah. *Norwich, Conn, USA.* 1805-16.
Lewis. *London.* 1863.
M. *Bury St. Edmunds.* 1846-53.
M. *Pittsburgh, Pa, USA.* ca.1820 (of Morgan & Hart).
Moses. *Liverpool.* 1848.
N. & Co. (& Son). *London.* 1828(-32).
Napthall & Son. *London.* 1844.
Napthali. *Liverpool.* 1848-51.
Orrin. *Bristol, USA.* 1824-33. C.
Philip. *London.* 1881.
Robert. *Bridgewater.* 1866-75. Cleaner.
Robert. *York.* 1858-66. W.
Sampson. *Hull.* 1851.
Sampson. *London.* 1869.
Samuel. *Devizes.* 1830-75.
Samuel. *Malmesbury.* 1875.
Thomas. *Ballycastle.* ca.1824-ca.1840. C.
Thomas Henry. *London.* 1863.
W. *Christchurch.* 1848-59. Fuzee and chain makers.
(Judah) & WILLCOX (Alvin). *Norwich, Conn, USA.* 1805-07.
William & Co. *Christchurch.* 1867-78. Fuzee and chain makers.
HARTFORD—
Alexander. *Bushmills, co. Derry.* 1846.
William. *Bushmills, co. Derry.* 1854-58. C.
HARTH, H. C. *New York.* ca.1850.
HARTHILL, John. *Birmingham.* 1854-68.
HARTLEY—
Jeremy. *Norwich.* 1680-1717.
Jeremiah. *Philadelphia.* 1837.
John. Town not stated. Same man as at *Halifax,* q.v.
John. *Halifax.* ca.1765-ca.1780. C. (Usually signed clocks without place name.)
John. *London.* 1832-39. Watch cases.
John. *York, Pa, USA.* Early 19c.
Sarah. *Bradford.* 1866.
Thomas. *Snaith.* ca.1770-d.1784. C.
HARTMAN—
Emil. *New York.* ca.1840.
Emil. *San Francisco, Cal, USA.* 1875.
Otto. *London.* 1869-81.
& SCHULTZ. *London.* 1863.
HARTNALL, Thomas. *Cirencester.* 1830-50.
HARTSHORN—
Edmund. *London.* 1869-81.
Thomas. *Coleshill.* 1835.
William. *London.* 1875.
HARTUNG, Charles. *London.* 1844.
HARTZLER, Joseph. *Bear Town, Pa, USA.* ca.1850.
HARVARD—
Clock Co. *Boston, USA.* 1880.
John C. *Pontlotyn.* 1868.
HARVEY—
Albert. *London.* 1869-81.
Charles. *London.* 1875.
Christopher. *Colchester.* b.1702. a.1723.
Daniel. *Hull.* 1851-58.
David. *London.* 1863-69.
Emanuel. *Launceston.* mar.1775-d.1818.
Francis. *Hayle (Cornwall).* 1786. W.
G. *Walthamstow.* 1866.
George. *London.* 1832-57.
Hannah. *Pontypool.* 1835.
Henry & Co. *London.* 1828-32.
Henry Thomas. *Wandsworth.* 1878.
James. *Abergavenny.* 1777-1822. W.
James. *Chesterfield.* 1876.
James. *London.* 1844.
James. *London.* 1863-81.
James. *Pontypool.* 1844.
James Tracey. *Abergavenny.* 1830-58.
& JAMISON. *Belfast.* 1858.
John. *Newark.* 1864-76.

HARVEY—*continued.*
John. *London.* 1851-81.
John. *Pontypool.* Believed from *London.* ca.1822-30.
John. *Wellington.* 1863-79.
Joseph. *Axminster.* 1790-1800.
Manfred. *Birmingham.* 1880.
Moses. *St. Marys, Canada.* 1862.
Mrs. Charlotte. *London.* 1881 (successor to Samuel James).
Richard Andrew. *Penzance.* 1823-35. W.
Robert. *Oxford.* 1588-ca.1600 and later.
Samuel James. *London.* 1863-75.
Thomas. *Baldock.* 1839-74.
Thomas. *Colchester.* b.ca.1618-d.1679. C.
Thomas. *Penzance.* 1840-49. When went to *India.* C.
Thomas. *York.* a.1801-30.
William. *Stirling.* s. of George. b.1808-d.1883. W.
HARVIE—
Robertus. Prob. Robert Harvey of *Oxford.*
William. *Wigan.* 1651, rep'd. ch. clock.
HARWICH or **HARWICK,** William. *Eton.* 1830-39.
HARWOOD—
Edwin. *London.* 1844.
F. *Hoddesdon.* 1859-66.
Frederick G. *Waltham Abbey.* 1874.
George. *Rochester, NY, USA.* ca.1830.
James. *Halifax.* 1714-21.
Joshua. *Halifax.* ca.1768. C.
HASCY—
Alexander R. s. of Samuel. *Albany, NY, USA.* 1831.
Samuel. *Albany, NY, USA.* ca.1825.
Samuel & Son (Alexander). *Albany, NY, USA* 1829-31.
HASELDINE, Percy & George. *London.* 1881.
HASELTINE & WENTWORTH. *Lowell, Mass, USA.* 1832.
HASIE, Mark. *New York.* ca.1855.
HASKELL—
James. *Salisbury.* 1842-75.
James Albert. *Ipswich.* 1875-79.
HASKINS, James. *Wolverhampton.* 1876.
HASLAM, Samuel. *Bakewell.* 1876.
HASLEDEN, Richard. *London.* 1875.
HASLEHURST—
Samuel. *London.* 1851-69.
Samuel. *Yeovil.* 1875.
HASLER—
John. *London.* ca.1760. C.
Ulrich. *Birmingham.* 1860-80.
HASLEWOOD, G. *Steeple Bumpstead.* 1851-66.
HASLOCK, John. *Tunbridge Wells.* 1874.
HASLUCK, Paul Nooncree. *London.* 1875.
HASSAM, Stephen. From *Boston* to *Charlestown, NH, USA.* ca.1787.
HASSELL, John. *Coventry.* 1842.
HASSFIELD Brothers. *London.* 1881.
HASTINGS—
——. *Aurora, Canada.* 1800s.
B. B. *Cleveland, O, USA.* 1837.
George W. *Workington.* 1858-73.
Hercules. *Burford.* ca.1700-ca.1730. C.
Joseph W. *Barrie, Canada.* 1862-71.
R. *King's Lynn.* 1846.
T. D. *Boston, USA.* 1854.
HASWELL—
——. *Lutterworth.* ca.1840. C.
Archibald. *London.* 1839-75.
James. *London.* 1881.
HATCH—
John B. *Attleboro, Mass, USA.* ca.1880.
John Curtis. *Yarmouth.* 1875.
Jonathan. *Westtown, Pa, USA* from *Conn.* 1810-15.
William. *London.* 1839.
HATFIELD—
D. *Husbands Bosworth.* ca.1815-ca.1830. C.
& HALL. *Manchester.* 1834.
John. *Kibworth Harcourt.* 1835. C.
John. *Wigston.* 1828.
HATHAWAY, J. *Odiham.* 1859.
HATHERLEY, William H. *Portsea.* 1878.
HATHORN, John. *Newcastle-on-Tyne.* See Haythornthwaite.

HATHORNTHWAITE, John. *Newcastle-on-Tyne.*
From *Kirkby Lonsdale.* d.1779. Signed clocks as
HAWTHORN.
HATTER, Thomas. *Wigan.* 1821. W.
HATTERSLEY—
Isaac. *Hedon.* 1846-51.
Isaac. *Rotherham.* mar.1827-37.
HATTICK, Konrad. *London.* 1863.
HATTON—
George. *St. Albans.* 1866-74.
George Cooper. *Lancaster.* 1826-69. W.
James. *St. Helens.* 1768. W.
John. *London.* 1646. CC.
John. *London.* 1832.
Joseph York. *London.* 1828-32.
William. *Liverpool.* (B. 1825) 1827-36. W.
William. *London.* 1828-32.
HAUGHTON—
J. & Son. *Hyde.* 1878.
James. *London.* 1828.
John. *New Castleton (Scotland).* 1836.
S. *New Haven, Conn, USA.* 1810. C.
HAUGHWOUT, E. V. & Co. *New York.* ca.1850. C.
HAUSBURG, E. O. *New York.* 1897-1900. C.
HAUSHALL, John. *Philadelphia, USA.* 1816.
HAVARD, John C. *Pontlotyn.* 1868.
HAVELOCK, George. *Guisborough.* (B. 1700?) 1737-99.
C.
HAVEN, Thomas. *London.* CC. 1652. Had worked for
past 10 years at *Chelmsford.*
HAVERS, John Henry. *Birmingham.* 1880.
HAWDON, Robert. *Hexham.* 1757. C.
HAWES—
J. H. *Ithaca, NY, USA.* 1853-62. C.
William. *London.* 1869-81.
HAWGOOD—
William. *London.* 1832.
William. *Pocklington.* 1851.
HAWKE—
C. *Wickham Market.* (B. 1810-)1846.
Charles. *Aldborough.* 1839.
Charles Henry. *Wickham Market.* 1830-75.
Christopher. *Bodmin.* mar.1661-73. C. (Also HOCKE.)
Daniel. *Woodbridge.* 1839.
John. *Stalbridge (Dorset).* 1824.
HAWKEN, Thomas Edward. *Lostwithiel.* 1873.
HAWKER, Charles. *Daventry.* 1864-77. Cleaner.
HAWKES—
Alfred. *Birmingham.* 1880.
Alfred. *Erdington.* 1880.
T. F. *London, Canada.* 1875.
Thomas. *London.* 1869.
William. *Birmingham.* 1868-80.
HAWKESWORTH—
——. *Cork.* b.1782-d.1856. W.
Edward. *Cork.* b.1794-1858.
Walter. *York.* b.1745-1772. C.
HAWKINGS, Richard. *London.* 1863.
HAWKINS—
Ambrose. *Exeter.* 6, *The Close* in 1695. d.1705. C.
(Month clock known.)
Ann. *Exeter.* Late-17c. C. (May be error for
Ambrose?).
Charles. *Cheltenham.* 1850-63.
E. *Coventry.* 1860-68.
Frederick. *Birmingham.* 1880.
& FURNISS. *London.* 1851.
John. *Cheltenham.* 1870.
John. *London.* a.1695 to Cornelius Harbert.
John. *Southampton.* West side of *High Street,* corner of
Butcher Row. 1723. W. (May well be the *London* man
above).
Joseph. *Axminster.* 1790-1800.
Joseph. *Southampton.* 1830.
Mark senior. *Bury St. Edmunds.* b. ca. 1674, mar.
1701, d. 1750. W. & C.
Mark junior. *Bury St. Edmunds.* s. of Mark sen.
b. 1707 d. 1767. Also at *Newmarket.* C.
Philip. *Royston.* 1828-30.
Richard. *London.* 1857.
Richard. *Oxford.* 1637-81. Church dial painter.
S. *Luton.* 1864.
S. *Wootton (Isle of Wight).* 1848.
Samuel. *Chesham.* 1830-54.

HAWKINS—continued.
T. *Taunton.* 1866.
William Henry & Co. *London.* 1875.
HAWKSWELL, Thomas. *Brompton (nr Pickering).*
1801-23. C.
HAWLEY—
Frederick. *London.* 1863.
J. *Coventry.* 1860.
James. *Birmingham.* 1868-80.
John. *London.* 1844-51.
John & Charles. *London.* (B. 1820-)1828-39.
John & Son. *Coventry.* 1880.
Thomas & Co. *London.* (B.1795-)1828-39.
HAWMAN, John. *Stockton.* 1827-34.
HAWORTH—
Richard. *Liverpool.* 1834-51.
Thomas. *Halifax.* 1671.
HAWSON, James. *Liverpool.* 1834. Dial enamellers.
HAWTHORN—
Frederick Alexander. *London.* 1863.
John. *Newcastle-on-Tyne.* See Hathornthwaite.
Thomas. *Burslem.* 1842. C.
Thomas. *Hanley.* 1828-35. C.
Thomas. *Shelton.* 1850.
W. *New Sneinton (Nottinghamshire).* 1849-55.
HAWTING—
John. *Oxford.* a.1745-d.1791. C. & W.
William. *Oxford.* s. of John H. a.1770. C.
HAWXHURST—
& DE MILT. *New York.* ca.1790.
Nathaniel. *New York.* 1786-93.
HAY—
Dunham John. *Norwich.* 1834. W.
George. *Buckie (Scotland).* 1860.
George. *Callington (Cornwall).* mar.1785-ca.1800. C.
Mrs. Celia. *Shrewsbury.* 1863-70.
Mrs. C. & Son. *Woolwich.* 1866-74.
Patrick. *Stewardfield (Scotland).* 1860.
Peter. *Edinburgh.* 1850.
Peter. *London.* 1832-39.
Robert. *Belfast.* 1784. C.
Thomas. *Peterhead.* 1860.
Thomas William. *Shrewsbury.* 1828-56.
William. *Wolverhampton.* 1828-50. Also T.C. and
jeweller.
HAYCOCK—
John & Thomas. *Ashbourne.* Successor to S. B. Harlow,
1826-49.
Silas Henry. *Ramsgate.* 1874-94.
Thomas. *Ashbourne.* Successor to John & Thomas.
1826-d.1868.
Thomas, jun. *Ashbourne.* s. of Thomas, sen. 1868-
d.1906. C.
William. *Ashbourne.* s. of Thomas, sen. 1868-d.1904.
HAYDEN—
David. *Waterbury, Conn, USA.* ca.1808.
& FREEMAN. *New York.* 1788.
S. & Son. *Boston, USA.* 1803.
Stephen. *Butler, O, USA.* ca.1804. C.
HAYDOCK, C. G. *Philadelphia, USA.* 1785-98.
HAYDON—
John. *Croydon.* b.ca.1661, mar.1687. W.
William. *London.* 1697. (B. has one *Croydon.* 1687-91.
Prob. same man.)
William. *Birmingham.* 1868.
HAYES—
Christopher Naylor. *Oldham.* ca. 1811 -d. 1856.
Edward. *Leigh.* 1848-58.
& EVANS. *London.* 1875.
George. *London.* 1881.
J. *Grimsby.* 1868-76.
James Henry. *Oldham.* s. of Christopher Naylor.
1848 -d. 1869.
John. *Grimsby.* 1861.
Josiah. *Birmingham.* 1800-30. Dialmaker.
Josiah & Son. *Birmingham.* 1835. Dialmakers.
M. *Birmingham.* 1854-60.
Peter B. *USA.* 1788-1842.
Robert. *Kirkby Lonsdale.* 1828. C.
Samuel. *Ellesmere.* 1879.
Thomas. *London.* 1832.
W. *Oldham.* 1858.
William. *Liverpool.* 1851.

HAYFORD—
Henry. *London.* 1844-51.
William. *London.* (B. 1820-24) 1828.
HAYHURST—
John. *Hastings.* 1870.
John. *Preston.* 1851-58.
HAYLER—
Benjamin. *Chatham.* Pre-1769-1800. W.
William. *Chatham.* ca.1770-1851. C. Perhaps two such.
HAYMAN—
George. *London.* 1875-81.
Robert. *Canterbury.* Free 1722. W.
HAYNES—
Alexander. *London.* 1802. W.
Edward. *Birmingham.* 1880.
George. *London.* ca.1735.
Henry. *Daventry.* 1753.
J. *Coventry.* 1868.
John. *London.* Prosecuted in 1666 by CC. for practising without having been apprenticed.
Joseph. *Brinklow.* 1880.
Joseph. *Ilkeston.* 1864.
Lafayette. *Troy, NY, USA.* 1836.
Robert. *Stamford (Lincs).* 1850.
Robert Broughton. *Stamford (Lincs).* 1841-50.
Samuel. *Cahir (Ireland).* s. of Samuel of *Cork.* ca.1840-52.
Samuel & Son. *Cork.* 1820-58.
Thomas (& Son). *Stamford (Lincs).* 1822-28(29).
William. *Buckingham.* 1702-04. C. Perhaps from *London,* where one such free 1703.
HAYS—
Michael S. *New York.* 1769.
Thomas. *Shrewsbury.* 1870.
HAYTER, Samuel. *Mere (Wilts).* 1842. Also 'organist'.
HAYTHORNTHWAITE, Peter. *Kirkby Lonsdale.* ca.1710-20. C.
HAYTON—
Daniel. *Hereford.* 1879.
Daniel. *London.* 1869.
HAYWARD—
E. *Deal.* 1866.
E. *Eastbourne.* 1878.
Edward. *Ashford.* 1847-74.
Edward. *Folkestone.* 1866-74.
Francis. *London.* 1869.
Frank Walter. *Norwich.* 1875.
G. *Winchester.* 1859.
J. *Bridgnorth.* 1879.
J. *Winchester.* 1867.
J. H. *Winchester.* 1839.
James. *Bath.* 1875-83.
James. *Hertford.* 1866-74.
James. *Hoddesdon.* 1874.
James. J. *Norwich.* 1846-75.
John. *Gloucester.* 1870.
Mrs. E. & Sons. *Winchester.* 1859.
R. *Needham (Norfolk).* 1846.
Robert Hunt. *London.* 1832-39.
T. *Winchester.* 1848.
Thomas. *Stockbridge.* 1830.
Thomas. *Winchester.* 1830.
Thomas Scott. *Littlehampton.* 1870.
W. *Farnham (Surrey).* 1866.
William. *Christchurch.* 1878.
William. *Edenbridge.* 1858-66.
William. *Evesham.* ca.1680. Lant. clock.
HAYWOOD, David. *Manchester.* 1843-51. W.
HAYWORTH, Richard. *Liverpool.* 1848.
HAZELWOOD, Henry. *Woodbridge.* 1875-79.
HAZEN, N. S. *Cincinnati, O, USA.* ca.1840.
HAZLETON, Hammersley. *Armagh.* 1840-90.
HAZLETT, Francis. *Bradford.* 1866.
HEAD—
J. *Winchester.* 1859-67.
John. *Bingham (Norfolk).* 1836-46.
Samuel. *London.* 1875.
HEADMAN, William. *Philadelphia, USA.* 1828-50.
HEADRICK, James. *Forfar.* 1860.
HEADWORTH—
A. *Ware.* 1851-66.
T. *London.* 1844-57.

HEAL, J. H. *Landport.* 1867.
HEALD—
Alfred. *Wisbech.* 1840-75.
T. *Chorley.* 1858.
HEALE, John. Place unknown. Lant. clock dated 1672.
HEALES, William. *London.* s. of Benjamin. Bap.1788, mar.1812. Lived at *Shoreditch.* -1815.
HEALY—
C. *Potterspury (Northants).* 1864.
Charles W. *Syracuse, NY, USA.* ca.1850.
John W. *Worcester, Mass, USA.* ca.1850.
Thomas. *London.* 1832.
Thomas. *Manchester.* 1822-24.
HEAP—
Charles. *London.* 1832-39.
Charles. *London.* 1875-81.
John. *Burnley.* 1848-58.
Thomas. *Northwich.* 1878.
HEAR, Joseph. *Appleby Magna (Leicestershire).* ca.1812. W.
HEARD, S. *Oxford.* ?late-19c. C.
HEARFIELD, John. *Leeds.* 1871.
HEARN(E)—
E. *Edinburgh.* 1850.
William & Co. *Toronto.* 1853. Successor to George Savage.
HEARSEY, Charles. *London.* 1869-81.
HEARSOM, D. *Thorpe-le-Soken.* 1855-66.
HEARTWELL, John. *Bourton-on-Water.* 1830-50.
HEATER, George. *Wantage.* 1854-77.
HEATH—
John. *Oxford.* a.1756.
Reuben. From *Vermont* to *Scottsville, NY, USA.* 1791-1818.
Roger. *London.* ca.1760. C.
Stevens. *Chillicothe, O, USA.* 1815.
T. *Great Berkhampsted.* 1851-74.
Willard B. *Bangor, Me, USA.* ca.1830-60.
HEATHER—
George. *Melcombe Regis.* 1875.
James. *Belfast.* (b.1799 at *Magheralin, co Down.*) d.1860. W.
HEATON—
G. *Leeds.* 1835-61.
John. *Bierley nr Bradford.* ca.1765. C.
John. *Leeds.* 1822-61. W.
Michael I. *Haworth.* ca.1760-d.1774. C. Often omitted placename.
Michael II. *Haworth.* 1774. Grandson of Michael I.
Robert. *Leeds.* 1853-71. W.
Thomas. *Sheffield.* (B.1817-)1822-37. W.
HEBDEN—
Francis. *Halifax.* 1837-66. C.
George. *Halifax.* 1816.
James. *Halifax.* 1822-37. C.
Jeffrey. *Hawes.* b. 1823 -d. 1912. C.
HEBDITCH—
John. *Chatteris.* 1875.
John. *Haxey (Lincolnshire).* 1876.
Thomas Hooker. *London.* 1875.
HERBERT, Anthoine (Anthony). *London. Horse Shoe Alley, Shoreditch.* 1701.
HEBTING, Frederick. *London.* 1828-32. C.
HECHT, William. *London.* 1869.
HECKETT—
John. *Stratton (Cornwall).* 1584. Kept ch. clock.
Thomas. *Stratton (Cornwall).* 1529-d.1572. Rep'd ch. clocks.
HECKLE, Henry Harrison. *Liverpool.* 1848-51.
HEDDERLEY, Charles. *Philadelphia, USA.* ca.1790.
HEDDON, —. *Stratton (Cornwall).* 1603. Rep'd. ch. clocks.
HEDGE—
J. *Soham.* 1865.
Jacob. *Colchester.* s. of Thomas II. b.1769-d.ca.1794. C.
James. *Cambridge.* 1875.
John. *Colchester.* b.1737-d.1778. C.
Nathaniel III. *Colchester.* b.1710-d.1795. C. & W.
Nathaniel IV. *Colchester.* b.1735-d.1821. C. & W.
Nathaniel VI. *Colchester.* s. of Thomas IV. b.1803-39.
Thomas II. *Colchester.* b.1744-d.1814. C.

HEDGER—
E. *Coventry.* 1854-68.
G. *Leamington.* 1880.
George. *London.* (B. 1820-)1828-32.
George. *London.* 1851.
H. *Coventry.* 1868.
Henry & James. *Coventry.* 1842.
James. *Coventry.* 1850.
HEDGES—
George. *Waterford, NY, USA.* ca.1825. *Buffalo, NY, USA.* 1828-48.
John & Son. *Maidstone.* 1832.
HEDLEY, John. *Durham.* 1834-56.
HEELEY, T. *Wilmslow.* 1857.
HEELIS, Emanuel. *Skipton.* b.1726-d.1767. C.
HEENAN, P. *Belfast.* 1827.
HEFFER, William. *London.* 1828-32.
HEFFERNAN, Edward Francis. *New York.* pre-1885. Then *Toronto.* 1885-89.
HEFFLEY—
Ananias. *Berlin, Pa, USA.* ca.1825-60.
Daniel. *Berlin, Pa, USA.* ca.1831-49.
HEIGHTON, T. *Kettering.* 1854.
HEILBURN, Michael. *Baltimore, USA.* ca.1840.
HEILIG—
Herman. *Germantown, Pa, USA.* ca.1850.
Jacob. *Philadelphia* and *Lancaster, Pa, USA.* 1770-1824. W.
John. *Germantown, Pa, USA.* 1801-50. W.
HEINE, Anton. *Woodstock, Canada.* 1857.
HEINEKEY, Robert. *Liverpool.* 1834.
HEINEMAN—
George. *Philadelphia, USA.* 1847-49.
L. C. *Philadelphia, USA.* 1849.
HEINITSCH, Charles. *Lancaster, Pa, USA.* Late 18c.
HEINTZ, Frederick William. *London.* 1875.
HEINTZELMAN—
Hieronymus. *Lampeter, Pa, USA.* b.*Switzerland.* ca.1750. Used monogram HH.
John Conrad. *Manheim, Pa, USA.* s. of Hieronymus. 1787-1805. C.
Peter. *Manheim, Pa, USA.* s. of John. ca.1800.
HEINY, Clements. *New York.* 1842.
HEIRE, ——. *London.* ca.1760. C.
HEISELY, Frederick. *Frederick, Md, USA.* 1759-93. *Lancaster, Pa.* 1793-1801. *Harrisburg.* 1801-20s. *Pittsburgh.* 1820s-39.
HEISMAN, William. *London.* 1844.
HEISS, James P. *Philadelphia, USA.* 1849.
HEI(T)ZMAN—
——. *Faversham.* 1858.
Anthony. *Rochester.* 1838-74.
Brothers. *London.* 1869-81.
Charles. *Canterbury.* 1832-47.
Charles. *London.* 1863-75.
Charles. *Maidstone.* 1839.
Ferdinand. *London.* 1869-81.
Fidel & Co. *London.* 1839-51.
Frederick J. *Walsall.* 1876.
Frederick Joseph. *Cheadle.* 1860.
G. *Bloxwich.* 1868-76.
George. *Hastings.* 1839. German clocks.
George. *Maidstone.* 1832-47.
James. *Walsall.* 1868-76.
John. *Canterbury.* 1840.
John. *Kirkcaldy.* 1837.
John. *London.* 1857.
Joseph. *Walsall.* 1868.
Joseph Celestin. *London.* 1863.
M. *March.* 1846.
M. G. & B. *Canterbury.* 1838-40.
M. & J. *Canterbury.* 1847-51.
Matthew. *London.* 1839.
Mathias. *Canterbury.* 1855-74.
Mathias. b.1802. *Germany.* Worked at *Wisbech.* c.1830-40. Also at *March.* Then at *Cheadle.* From ca.1845-1851.
Pius. *Newport* (*Mon*). 1858-81. W. & C.
Pius. *Pontypool.* 1868-81. C. & W.
Raymond. *Cardiff.* 1852-87.
Rochuz. *London.* 1851-81.
Tobias. *London.* 1857.

HEI(T)ZMAN—*continued.*
William. *Blandford.* 1875.
& WILLIAM. *Dublin.* 1880.
HELDEN (sometimes HILDEN)—
Cornelius. *London.* b.ca.1670. a.1686 to Nathaniel Delander (not Daniel Delander). 1692. W.
Onisephorus. *London.* 1630-1656.
HELDER, Peter. *Haslemere.* 1829-39.
HELDYARD, J. *Woodbridge.* 1846.
HELGE, August Myhre. *London.* 1863.
HELIWELL, Richard. *Buxton.* 1876.
HELLABY, W. *Derby.* 1849.
HELLER (or HALLER), John. *Berlin, Canada.* 1851-63.
HELLERING, Joseph. *Swindon.* 1875.
HELLEWELL—
J. *Leeds.* 1832. C.
Richard. *Halifax.* 1626-36. Rep'd. ch. clocks.
HELLIER—
Charles. *London.* 1857-69.
Henry. *Bristol.* a.1717-27.
HELLIWELL, William. *Leeds.* 1815-53. C. Also supplied the trade.
HEL(L)MUTH—
A. *Kidderminster.* 1872.
Andrew. *Dudley.* 1868-76.
Brothers. *Dudley.* 1860.
HELM—
——. *Ormskirk.* ca.1780. C.
Christian. *Philadelphia.* 1802-04.
HELMSCHMIDT, Hans. *Augsburg.* 1570.
HELSBY—
Edward. *St. Helens.* 1822.
Edward. *Wigan.* 1824. W.
James. *Farnworth.* 1821. W.
James. *St. Helens.* 1786. W.
James Gooden. *Liverpool.* 1834.
John. *Liverpool.* 1834.
John. *St. Helens.* 1820. W.
Richard. *St. Helens.* 1810-32. W.
Thomas. *Prescot.* 1817-24. W.
W. *St. Helens.* 1858. W.
William. *St. Helens.* 1769-87. W.
HEMBRY, George. *North Curry* (*Somerset*). 1822. C.
HEMINGTON, George. *London.* 1704. W.
HEMINGWAY—
A. *Chicago, Ill, USA.* 1856.
David. *Leeds.* 1763. C.
John. *Manchester.* 1822-34.
HEMLEY, Nathaniel. *Kensington, NH, USA.* ca.1760. Blacksmith and perhaps C.
HEMMING, Charles. *London.* a.1679 to James Wolverston, prob. never completed it.
HEM(M)INS—
Edward. *Bicester.* 1720-d.44. C. and bellfounder.
Joseph. *Banbury.* 1741-44. T.C.
HEMPHILL, Thomas J. *Philadelphia, USA.* 1836-41.
HEMPSALL, G. *East Retford.* 1864.
HEMPSON, Robert. *New York.* ca.1825.
HEMPSTED, Daniel Booth. *New London, Conn, USA.* 1784-1852. Cabinet maker.
HEMSLEY, Richard. *Montreal.* 1869-80.
HEMSWORTH, Henry. *Leeds.* 1834.
HENCE, William Henry. *Daventry.* 1864-77.
HENDEL—
Bernard. *Carlisle, Pa, USA.* ca.1800.
Jacob. *Carlisle, Pa, USA.* ca.1810.
HENDEN, Harry, jun. *Dartford.* 1874.
HENDER—
Edmund. *Bodmin.* ca.1820-47. C. & W.
Nathaniel. *Camelford.* ca.1760. C.
Thomas Row. *Looe.* 1844.
William. *Callington* (*Cornwall*). 1814. Then *Altarnun* where d.1848. C. & W.
HENDERICK, John. *Droitwich.* 1835.
HENDERSON—
A. *Ripon.* 1866. W.
Adam. *Poughkeepsie, NY, USA.* ca.1831-46.
Alexander. *Bury.* 1851.
Alexander. *Richmond Hill, Canada.* 1862.
C. *Leicester.* 1864.
D. S. *Belfast.* 1880.

HENDERSON—*continued.*
George Henry & Co. *Southampton.* 1867.
Herbert. *London.* 1863.
James. *Armagh.* 1854. C.
James. *Dublin.* (B. 1795) 1807-28. W. and goldsmith and jeweller.
Jeremiah. *Scarborough.* b.1718. s. of Robert. ca.1765. C.
John. *Brigg.* 1849-76.
John. *London.* 1662. (Almost certainly an error for John Hilderson, q.v.).
& LOSSING. *Poughkeepsie, NY, USA.* ca.1835.
Miles. *Penrith.* 1769-78. C.
Robert. *Scarborough.* b.1678-d.1756. C.
Thomas. *Houghton-le-Spring.* 1834-56.
William. *Appoquinimink, New Castle, Del, USA.* 1770.
William. *Dundee.* 1850.
HENDERY, R. & Co. *Montreal.* 1847-50. From *Scotland.*
HENDRICK—
Barnes & Co. *Forestville, Conn, USA.* 1849-52.
Charles. *Dartford.* 1851-55.
Charles S. *Oswestry.* 1879.
(Ebenezer M.) & CHURCHILL (John). *Bristol, USA.* 1847.
Ebenezer M. *Bristol, USA.* 1850. C.
HUBBELL & BEACH (Levi). *Bristol, USA.* 1854. C.
(E. M.), HUBBELL (Laporte) & Co. *Bristol, USA.* 1848-53. C.
John & Peter. *Liverpool.* 1822-40. W.
T. J. (or T. A.). *London.* 1844-51.
Thomas. *London.* 1857-63.
HENDRICKSON, John. *London.* 1652. Journeyman to John Champion for six months.
HENDRIE—
——. *Wigton.* ca.1760. C.
Robert. *Forres (Scotland).* 1860.
William A. *Chicago, Ill, USA.* 1880-82.
HENDRIX, Uriah W. *New York.* 1756.
HENDRY, John. *Keith (Scotland).* 1860.
HENEBERGER, Peter. *Harrisonburg, Va, USA.* 1784-1869. C.
HENEY, Patrick. *Dublin.* 1798-1831.
HENFREY, James. *Leicester.* 1828-35(-46?).
HENN—
J. W. *Cradley Heath.* 1876.
S. *Darlaston.* 1860.
S. *Tipton.* 1868.
Silas. *Tipton.* 1842-50.
Silas. *Dudley.* 1872-76.
HENNESSY, Bernard. *Swansea.* Believed from *London.* 1848-75. C. & W.
HENNING, Robert. *Lymington, Hants.* 1830-48.
HENNINGHAM, G. B. *London.* 1851.
HENRET, John. *Prescot.* 1818. W.
HENRY—
Francis. *Douay (France).* ca.1820. W.
I. *Great White (Hunts).* 1847.
James. *Hamilton, Canada.* 1861-65.
James. *Keith.* 1837.
James. *Maysville, Ky, USA.* 1820. C.
John. *Goderich, Canada.* 1857-62.
John. *Ryde.* 1859-78.
Joseph. *Cheltenham.* 1879.
R. *Bristol.* ca.1780. C.
William. *London.* (B. 1820) 1828.
HENSHALL, H. *Whitchurch.* 1863.
HENSON—
Charles. *Market Deeping.* 1828.
Mrs. E. *Wansford (Northants).* 1869.
Robert. *London.* 1851-63. Marble clock cases.
Samuel. *London.* 1875-81. Marble clock cases.
T. *Kingscliffe* nr *Wansford.* 1847-64.
HENTSHEL, Paul & Carl. *Waterloo, Canada.* 1890 to present day.
HENWOOD, Digory. *Bodmin.* ca.1770. Then *Fowey.* 1765. C.
HENY—
P. *Calne.* 1859.
Pius. *Fyfield* nr *Marlborough.* 1867-75.
HEPPLEMAN, John. *Manheim, Pa, USA.* 1790s-ca.1810.

HEPTING—
Alios. *London* 1828. C.
& FURDERER. *London.* 1881.
HEPTON—
Frederick. *Philadelphia, USA.* 1785.
John. *Northallerton.* 1840.
Robert Henry. *Wellingborough.* 1877.
William (& Sons). *Northallerton.* 1822-40(-51).
HEPWORTH, William. *York.* 1866. W.
HEQUEMBOURG, C. *New Haven, Conn, USA.* ca.1818. W.
HERANCOURT, (G.) & DRESBACHE (C. T.). *Columbus, O, USA.* ca.1830.
HERBERT—
David. *Aberystwyth.* 1815-65. C.
Edward. *London.* a.1657 to Humphrey Downing. CC. 1664.
Evan. *London.* a.1680 to Thomas Herbert. CC. 1691-97.
James K. *Reading.* 1877.
John. *London.* b.ca.1657, a.1672 to Nicholas Payne. CC. 1682, mar.1686-1713. C.
John. *Oxford.* mar.1742-94. C.
R. *Walsall.* 1860
William. *Bristol.* 1870-79.
William. *Ludlow.* 1828-50.
HERBIN—
John. *Halifax, Nova Scotia.* 1871-74.
John. *Windsor, Nova Scotia.* 1866-71.
HERBSTREET (& HERBSTREIT)—
J. *Handsworth.* 1868.
Joseph. *Birmingham.* 1868-80.
Joseph. *Smethwick.* 1876.
HERDMAN—
Charles. *Armagh.* Free 1744-65. C.
Robert. *Belfast.* ca.1700, lant. clock. (B. 1728.)
HERITAGE, Ephraim. *Cork.* b.1756-d.1793. W.
HERMAN(N)—
Anthony. *London.* 1875-81.
& BEURLE. *London.* 1839.
George. *London.* 1857.
& HUMMEL. *King's Lynn.* 1846.
I. *East Dulwich.* 1878.
Ignatz. *London.* 1844.
James. *Manchester.* 1831-51.
Joseph. *Bradford.* 1822.
Joseph. *Dartford.* 1874.
Joseph. *Leeds.* 1834-53. C.
Joseph. *Manchester.* 1824-28.
Peter. *Leeds.* 1817-26. C. & T.C.
HERON—
——. *Downpatrick.* Late 18c.-early 19c. C.
Alexander. *Newtownards.* 1818-54. C.
David. *The Ards.* ca.1740-ca.1760. C.
Erskine. *Charleston, SC, USA.* From *London.* 1762-65 (one such at *Edinburgh.* a.1752).
James. *Greenock.* 1836.
James. *Newtown, Pa, USA.* ca.1780. C.
James I. *Newtownards.* ca.1757-d.1784. C.
James. *Newtownards.* ca.1794. C.
John. *Lisburn.* ca.1770-85. C.
John. *Portavoe.* ca.1775-85. C.
& Son. *Greenock.* 1836.
Ursula. *Newtownards.* wid. of James I. Ca.1784-d.1811.
William. *Donaghadee.* ca.1770-80. C.
William. *Newtownards* ca.1778-ca.1795, then *Liverpool* c.1795-d.1800. C.
HERR—
J. *Bolton.* 1858.
Joseph. *Bradford.* 1866-71.
Joseph. *Goderich, Canada.* 1857.
William, jun. *Providence, RI, USA.* 1849.
HERRICAN, William. *Philadelphia, USA.* 1850.
HERRICK, Edwin. *London.* 1869.
HERRITT, W. *Southampton.* 1839.
HERRMANN—
& COBB. *King's Lynn.* 1858.
Ignaz. *London.* 1869-75.
HERRON, John. *Blyth.* 1848-58.
HERTZ, Jacob. *Lancaster, Pa, USA.* Early 19c. C.
HERVEY, M. *Wellingborough.* 1847-54.

HERWICK, Jacob. *Carlisle, Pa, USA.* 1779.
HERZOG—
Carl & Co. *London.* 1881.
Joachim. *Wyll.* 18c. C.
HESELDIN, George. *Sheffield.* 1834-37. W.
HESELGRAVE, Robert. *York.* 1866. W.
HESELTON—
George. *Bridlington.* 1834-38.
James C. *Beverley.* 1858.
HESELWOOD—
John. *York.* b.1810-d.1873.
Richard. *York.* 1851-66. W.
HESKETH, Thomas & John. *Farnworth.* 1848-51.
HESLOP—
John. *Newcastle-on-Tyne.* 1801. Watch glass maker.
Richard. *Huddersfield.* 1850-71. W.
William. *London.* 1869-81
HESMONDHALGH, Edward. *Ribchester.* 1813. Ch.
clock dial.
HESS—
George. *Zurich, Canada.* 1860-d.1891 (b. *Germany*
1838).
Ralph & Co. *Liverpool.* 1848-51.
Rosetta. *Liverpool.* 1834.
HESSAY, George. *Hovingham (Yorks N.R.).* 1823. W.
HETH, Robert. *Thame.* 1573. T.C.
HETHERINGTON, J. *Kendal.* 1926.
HETTICH—
Clement. *London.* 1828-44.
Emanuel. *London.* 1875.
Julius. *Cardiff.* 1875-99.
Mrs. Mary. *London.* 1881. (Successor to Emanuel.)
William. *London.* 1851-57.
HETTICK, C. *King's Lynn.* 1846-65.
HETZEL(L)—
——. *Milford, Pa, USA.* ca.1834. C.
John. *Newtown, NJ, USA.* 1795.
HEUER, Lambelet & Co. *London.* 1881.
HEURTIN, William, jun. *New York.* 1703-65.
HEUWEILER, ——. *Germany.* 1838. T.C.
HEWARD, John. *London.* b.ca.1672, a.1689 to John
Miller. CC. 1694-97. C.
HEWARDINE, George. *Stilton (Hunts).* 1839-64.
?HEWENER, George. *London.* 1697 (writing bad).
HEWES—
Peter. *London.* 1646-66 (see also HUES/HUGHES etc.).
Robert. *Chelmsford.* Same man as at *Colchester,*
ca.1735-ca.1765. C.
Robert. *Colchester.* b.1711-d.1769. C.
HEWETSON—
——. *London.* Early 18c. C.
James. *St. Austell.* mar.1826. W.
HEWETT—
E. *Norwich.* 1865.
Richard. *Fowey.* ca.1790-d.1846. W. & C.
HEWITSON, Richard. *Liverpool.* 1817. W.
HEWIT(T)—
A. E. *North Bridgewater, Mass, USA.* ca.1860.
Alexander. *London.* a.1685 to Benjamin Bell -1691.
David, *London.* 1857.
George. *Marlborough.* ca. 1740-90. C.
George. *Prescot.* 1823-25. W.
J. *Coventry.* 1880.
J. *Rushden.* 1847-64.
James. *Edinburgh.* a.1816.
John. *Burnham (Bucks).* 1842-54.
John. *Coventry.* 1860-80.
John. *Lincoln.* 1876.
John. *Prescot.* 1819-24. W.
Joseph. *Prescot.* 1796. W.
Joshua. *Huyton (Liverpool).* 1794. W.
Joshua. *Prescot.* 1788-d.1802. W.
Joshua. *Prescot.* 1818. W.
L. *Higham Ferrers.* 1854-69.
Owen. *Watlington.* 1823. C. & W.
Samuel. *London.* 1857-81.
Thomas. *Coventry.* 1880.
Thomas (& Son). *London.* 1832-44(51-75).
William. *Gainsborough.* 1861-76.
HEWKINS, ——. *Tenterden.* Till 1914. C.
HEWSON—
W. *Rodmersham.* 1866-74.

HEWSON—*continued*
William. *Lincoln.* 1828-61.
HEXT—
George. *Ringwood.* 1878.
Giles. *Christchurch.* 1859-78.
William. *St. Austell* from *London* 1743.
HEYDON, Thomas. *Farnborough.* ca.1780-ca.1800. C.
HEYDORN—
C. *Hartford, Conn, USA.* 1808-11.
(C.) & IMLAY (R.). *Hartford, Conn, USA.* 1808-11.
HEYES—
Christopher. *Oldham.* 1822-34.
Thomas. *London.* 1863.
W. *South Wimbledon.* 1878.
William. *London.* 1869.
HEYLAND, Charles. *Cork.* 1844.
HEYLING (also HILLINGS), Bernard. *London.* CC.
1652, from *Antwerp.*
HEYMAN, John. *Canterbury.* Free 1714. W.
HEYMUYS, Adam. *Amsterdam.* ca.1760. C.
HEYRICKE, Samuel. Believed *Leicestershire.* 1687
sundial.
HEYS, William. *Liverpool.* 1780. W.
HEYTEN, Robert. *Oxford.* a.1656. W.
HEYWOOD—
——. *Carnarvon.* 1832.
John. *Birkenhead.* 1848.
Thomas. *Bangor (Wales).* 1818-36. C. & W.
Thomas. *Birkenhead.* 1848-57.
Thomas. *Wrexham.* 1840-87. W.
William. *Bradford.* 1853.
William. *London.* 1811 (B. 1840).
William. *Northwich.* 1828.
HIATT, Henry. *Prescot.* 1848-51.
HIBBARD—
Caleb, jun. *Williston, Pa, USA.* 1781-1809, then *Ohio*
ca.1818-19.
Richard. *Alford.* 1835.
HIBBEN—
Andrew. *Charleston, SC, USA.* 1765-84.
HIBBERD, Thomas. *Sheffield.* 1837. W.
HIBBERT—
John. *Haxey (Lincs).* ca.1800. C.
Thomas. *Manchester.* 1848-51.
HIBBINS, John. *Stamford (Lincs.).* 1876.
HICK—
Charles F. *Rochester.* 1845.
Lambert. *York.* 1866-80. W.
HICKCOX, Samuel R. *Humphreysville, Conn. USA.*
ca.1830.
HICKLING—
John. *London.* 1839-44.
Samuel. *Long Eaton (Derbyshire).* 1876.
HICKMAN—
Edward. *Oxford.* 1818-26. C. & W.
M. *Leeds.* 1853.
Sampson. *Leeds.* 1871.
William. *London.* 1839.
William. *Stamford (Lincs).* 1828-35.
HICKMOTT, C. *Maidstone.* 1865.
HICKOX, A. *Brighton.* 1851.
HICKS—
Charles. *Liverpool.* 1822-28.
J. *London.* 1736. W.
James. *London.* (B. 1802-15. ?Same man.) 1832.
John. *Stamford (Lincs).* 1861-76.
HIDE—
Charles. *Banbury.* 1757-d.1773. C. & W.
J. *Sleaford.* See Hyde.
William B. *London.* 1857.
HIGBY, S. S. *New Hartford, Conn, USA.* ca.1830. C.
HIGDON—
Panteness. *Brewham (Somerset).* 1744-ca.1760.
Thomas. *East Pennard (Somerset).* 1717-93. C.
HIGGINBOTHAM & HIGGINBOTTOM—
Joseph. *Long Eaton (Derbyshire).* 1876.
Joseph. *Wexford.* (B. 1824)-58.
Thomas. *Birmingham.* 1868 d. 1892.
HIGGINS—
Charles. *Canterbury.* 1865-74.
George. *Alcester.* 1828-42.
J. *Huddersfield.* 1814.

HIGGINS—*continued.*
John. *Belfast.* ca.1800. C.
John Simms. *Canterbury.* 1859-74.
T. *Shipston-on-Stour.* 1854-76.
T. *Ashby-de-la-Zouch.* 1849-55.
Thomas. *Burford.* 1823. C. & W.
Thomas. *Gloucester.* 1840-50. Also optician.
Thomas. *Shipston (Worcestershire).* 1842-50.
William. *Birmingham.* 1839. Dial maker.
William. *London.* 1839-51.
HIGGINSON—
Charles. *Prescot.* 1796. W.
& FAIRCLOUGH. *Liverpool.* ca.1775. W.
Henry. *Liverpool.* 1675-d.1694. W. (One such CC. 1662 *London*, might be same man.)
John. *London.* post-1690-98.
Nicholas. *London.* CC. 1646-62. *Chancery Lane.*
HIGGS—
E. *Newbury.* 1864.
Ernest W. *Sandown.* 1878.
& EVANS. *London.* (B. 1775-1825) 1828-32.
Henry. *Newbury.* 1877-1903.
& Son. *Rugby.* 1868.
W. *Southampton.* 1859.
W. W. *Newbury.* 1854.
William. *Millbrook (Hants).* 1878.
William. *Ruabon.* 1835.
William. *Rugby.* 1842-80.
William. *Wandsworth.* 1866-78.
HIGH—
W. *Bishop's Stortford.* 1851.
William. *London.* 1863-81.
HIGHAM—
Mary Ann. *Brentwood.* 1839.
Nathaniel. *Brentwood.* 1828.
Robert. *Manchester.* 1848.
Thomas. *Liverpool.* 1799-1821. W.
HIGHFIELD—
Alfred. *Ipswich.* 1875-79.
Edward. *London.* 1857-63.
Nathan. *Liverpool.* 1780-90. W.
Samuel. *London.* 1811.
William. *London.* 1832-39.
William & Sons. *London.* 1869.
HIGHLEY, John. *Sheerness.* 1861-74. W.
HIGHO, James A. *London.* 1844-69.
HIGMAN, Jacob. (b.1763.) From *London* ca.1806 to *St. Austell* where d.1841. C.
HILBIRT, I. *Haxey.* ca.1810-20. C.
HILBURN, John Jacob. *Bowling Green, Ky, USA.* 1836-77.
HILDEBURN—
& Brothers. *Philadelphia, USA.* 1849.
Samuel. *Philadelphia, USA.* 1810-37.
(Samuel) & WATSON. *Philadelphia, USA.* 1833. W.
WOOLWORTH. *Philadelphia, USA.* 1816-19.
HILDERSON, John. *London.* 1657-62. (Also appears as HINDERSON & HENDERSON in error.)
HILDRETH—
Jonas. *Salisbury, Vt, USA.* 1805.
W. *Cork.* 1720. W.
HILDYARD—
Brothers. *Woodbridge* and *Wickham Market.* 1879.
David. *Caterham Valley.* 1878.
David. *Westerham.* 1874.
John. *Woodbridge.* 1839-75.
John. *St. Ives (Huntingdonshire).* 1877.
HILES, Thomas. *Stroud.* 1830.
HILL—
——. *Wells.* 1710-11. C.
B. *Penrith.* ca.1815. C.
Baron. *Crewe.* 1865-78.
Benjamin. *London.* From B. Co. Brother in CC. 1640-d.1670. W. *Fleet Street.*
Benjamin. *London.* 1881.
Benjamin Morris, jun. *Richmond, Pa, USA.* 1771.
Brothers. *Coventry.* 1842-50.
C. *Preston.* 1858.
Charles. *Steubenville, O, USA.* 1823-25.
Charles. *Zanesville, O, USA.* 1815.
Charles John. *Coventry.* 1868-80.
D. *Reading, Pa, USA.* 1820-40.

HILL—*continued.*
Daniel. *Colchester.* 1843-d.1884.
E. J. *Albion, NY, USA.* ca.1850.
Frederick. *Lamborn (Berks).* 1837-64.
Frederick A. *Southsea.* 1878.
G. *Shrewsbury.* 1856.
George. *Borrowstounness.* 1860.
George. *Bowness (Scotland).* 1844.
George. *Great Bedwin (Wilts).* 1842-48.
George. *Whiteburn (Scotland).* 1836.
Henry. *New York.* ca.1750. C.
Henry Vincent. *Sandwich.* 1891-95.
J. *Bakewell (Derbyshire).* 1849-76. C.
J. *Sudbury.* Late 18c. C.
J. *Yeovil.* 1861.
J. J. *Castle Cary.* 1866.
J. J. *Ilminster.* 1861.
J. W. *Kansas, USA.* 1873.
James. *Penrith.* 1848.
Joachim. *Flemington, NJ, USA.* 1783-1869. C.
John. *Aylesbury.* 1771-80. C. & W.
John. *Boxford.* 1839.
John. *Bradford.* 1853-71.
John. *Chertsey.* 1878.
John. *Evesham.* 1828-42.
John. *Grimsby.* 1876.
John. *Hastings.* 1878.
John. *London.* 1839.
John. *Lydd (Kent).* 1847-55.
John. *Nottingham.* 1876. Clock cases.
John. *Risborough (Bucks).* Late 17c. marq. l.c. clock.
John. *Sheffield.* 1822. C.
John. *Whitby.* 1840.
John. *Wirksworth.* 1835.
John A. *Northampton.* 1877.
John B. *Beverley, Mass, USA.* ca.1850.
John Spencer. *Burslem.* 1842-50.
Jonas. *York.* 1770-c.1775.
Joseph. *Bristol.* 1840-42.
Joseph. *Freemantle (Southampton).* 1878.
Joseph. *Newcastle-under-Lyme.* 1860-76.
Joseph. *Romsey.* 1839-78.
Mrs. E. A. *Southampton.* 1878.
N. (& Son & Co.). *Coventry.* 1854(-60).
N. *Coventry.* 1880.
Peter. *Mount Holly and Burlington, USA.* 1796-1821.
Richard. *Liverpool.* (B. 1807-14) 1824.
Robert. *Penrith.* 1828-34.
Robert (II?). *Penrith.* 1858-79.
& ROSS. *Zanesville, O, USA.* ca.1830.
Samuel. *Hamburg, Pa. USA.* ca.1800.
Samuel. *London.* 1844.
Samuel. *Totley.* ca.1770. C.
Stephen. *London.* 1839.
T. & G. *Lambourn.* 1854.
Thomas. *Barton-on-Humber.* 1828-50.
Thomas. *Farnworth.* 1800. W.
Thomas. *Kilbride.* 1836.
Thomas. *Little Heywood (Staffordshire).* 1835.
Thomas. *London.* 1632-35. W.
Thomas. *Rochdale.* 1828-34.
Thomas. b. ca. 1792. *Prees, Shropshire. Malpas* ca.1832. *Wem* 1841-51. W.
Thomas S & Co. *Coventry.* 1854-80.
Thomas S., jun. and sen. *Hamilton, Canada.* 1856-75
W. *Southampton.* 1867.
W. H. & Sons. *Coventry.* 1860.
William. *Coventry.* 1835.
William. *Farnworth.* mar.1798. W.
William. *London.* 1851.
William. *New York.* ca.1810.
William, sen. *Walsingham.* 1733.
William. *Whitby.* 1851.
William Buley. *Birmingham.* 1828.
William F. *Boston, USA.* 1810.
William Henry. *Coventry.* 1850-60.
William Henry. *London.* 1863-81.
William Rowland. *Birmingham.* 1850.
HILLABY, Richard. *Sheffield.* 1871.
HILLARD—
Christopher. *Hagerstown, Md, USA.* ca.1825.
George W. *Fayetteville, NC, USA.* 1823.
James. *Charleston, SC, USA.* ca.1730-49.

HILLARD—*continued.*
William. *Fayetteville, NC, USA.* 1801-05.
HILLARY, Edmund. *Rhos-y-medre.* 1874-76.
HILLDROP, Thomas. *Hartford, Conn, USA.* 1773.
HILLDRUP, Thomas. *Hartford, Conn, USA.* From *London* ca.1775-90.
HILLER, Joseph. *Boston* and *Salem, USA.* 1770.
HILLERY, John. *London.* a.1681 to Richard Jarret. Prob. never completed it.
HILLIARD—
George. *Dublin.* 1858.
William. *Tralee.* 1858. W.
HILLIER—
G. L. *Basingstoke.* 1859-67.
George. *Basingstoke.* 1839-48.
John. *Huxley.* ca.1810. C.
William. *London.* a.1669 to Thomas Creed. CC. 1679-1733.
HILLMAN—
Alfred. *Chepstow.* 1868.
F. *New York.* 1854.
Walter. *London.* 1881.
William. *St. Johns, Canada.* 1864.
HILLS—
Amariah. *New York.* 1845.
Benjamin A. Quaker. b. 1807 *Coggeshall (Essex).* a. there 1821 to John Knight. Then *Sudbury* where succ. G. Dawson 1830. Mar. 1831 d. 1859. Also at *Bures. Nayland* and *Boxford.* W.
Brothers. *Sudbury.* 1879.
(William), BROWN (J. C.) & Co. *Bristol, USA.* 1840.
Charles C. *Haverhill, Mass, USA.* 1853.
D. B. *Plainville, NJ, USA.* 1888.
Dwight B. *Bristol, USA.* 1895. C.
George & Son. *Plainville, Conn, USA.* ca.1870. C.
(William), GOODRICH (Jared) & Co. *Plainville, Conn, USA.* 1841-45. C.
GODERICH & Co. *Farmington, Conn, USA.* 1841-45.
John. *East Grinstead.* 1828-62. W.
Joseph Francis. b. 1833 s. of Benjamin of *Sudbury* whom he succ. 1859 -d. 1915. Also postmaster.
W. & G. *Rochester.* 1874.
William. *Farmington, Conn, USA.* 1835-45.
William. *Rochester.* 1847-74.
HILLWORTH, Frederick. *Philadelphia, USA.* 1844-49.
HILLYARTINER, Philip. *Philadelphia, USA.* 1844.
HILLYER, Thomas. *Stony Stratford.* 1830.
HILTON—
Amos. *Halifax.* 1853.
Evan. *Wigan.* 1667-d.1699. W.
John. *Halifax.* 1834-37. C.
John. *Liverpool.* 1848-51.
& Son. *Halifax.* 1850.
HIMELE, James. *New York.* 1786.
HIMELY, John James. *Philadelphia, USA* ca.1786 and *Charleston, SC, USA* 1796.
HIMLI, ——. *Colmar.* ca.1695. W.
HINCH(C)LIFF(E)—
J. *Oldham.* 1858.
James. *Holmfirth.* 1853.
Joe Lee. *Huddersfield.* 1866-71. W.
HINCKLEY, William. *Ironbridge.* 1850.
HIND(E)—
Benjamin William. *London.* 1839-57.
George. *Edinburgh.* a.1823.
Thomas. *Bolsover.* b.1817-d.1836. C.
William. *Oldham.* 1851.
HINDELANG, Emil. *Dudley.* 1860-72.
HINDERSON, John. *London.* 1662. CC. (Wrongly recorded as Henderson, almost certainly slip for HILDERSON, q.v.)
HINDERWELL (also **HINDERWILL**)—
——. *Stockton.* ca.1810. C.
John. *Norton nr Stockton.* 1827-34. C.
Matthew. *Stockton.* 1827. C.
Thomas. *Norton nr Stockton.* 1851-56.
HINDLE, ——. *Leeds.* ca.1840-50. C.
HINDLEY—
——. *York.* Same as Henry Hindley, q.v.
Henry. *Wigan.* b.ca.1701-1730, then *York* 1731-d.1771. C. Very fine maker.
Joseph. *York.* s. of Henry. a.1742-d.1774. C.
Roger. *London.* bro. of Henry. Watch cap maker.

HIND(E)S—
George. *London.* 1857-81.
Henry. *Ampthill.* 1839.
Henry. *Market Street (Bedfordshire).* 1830.
Henry. *St. Albans.* 1859-74.
Joseph. *Colchester.* b.ca.1820-45. W.
Joseph. *Stamford (Lincs).* 1828-35.
Richard. *London.* 1881.
William Henry. *Aylesbury.* 1842.
HINDMORE, ——. *London.* Early 18c. C.
HINES, Frederick. *London.* 1878.
HINKLE, John P. *Philadelphia, USA.* 1824.
HINKLEY—
John. *London.* ca.1760. C.
William. *Newcastle-under-Lyme.* 1860-76.
HINKS & RADCLIFFE. *London.* 1881.
HINKSMAN, ——. *Madeley.* ca.1750. C.
HINMAN, Alonzo. *New Haven, Conn, USA.* ca.1840.
HINSDALE, Epaphras. *Newark, NJ, USA.* 1769-1810.
HINSHON, J. P. *Terre Haute, Ind, USA.* ca.1850.
HINTON—
G. *Clapham.* 1862.
James. *London.* 1828.
Josiah. *London.* 1832-39.
William. *London.* 1832.
William Isaac. *London.* ca.1820.
HIPKISS—
& HARROLD. *Birmingham.* Japanners and dial makers, 1800-c.1820.
R. *Birmingham.* ca.1805-ca.1811. Dial maker.
HIPP, George. *Kempton.* 1619. W.
HIPPACH, Karl. *London.* 1875-81.
HIRD—
Alexander & Co. *Bradford.* 1837.
Edward. *Ambleside.* 1858-69.
Edward. *Ulverston.* ca.1880.
Edward & William. bros. of Henry II. *Barrow-in-Furness* ca.1864, then *Ulverston* ca.1880.
Henry. *Ambleside.* mid-19c. C.
Henry I. *Broughton (Lancs).* a.1801-ca.1815, then to *Thornhill, Scotland.* d.1854.
Henry II. *Ulverston.* Nephew of Henry I. 1851-69.
& Sons. *Ambleside.* 1879.
HIRSCH, PRITCHARD & CO. *London.* 1875-81, French clocks.
HIRSCHFIELD Adolph. *London.* 1851-81.
HIRST—
Brothers. (George K. & John T.). *Leeds.* Successor to George K. 1851-71.
George Koester. *Leeds.* b.1787. mar.1810-37. W. Successor to Samuel & Son.
H., jun. *Beaconsfield.* 1847.
Henry. *Beaconsfield.* 1847.
James. *Beaconsfield.* 1823-47.
James. *Great Marlow.* 1838.
John. *Hull.* 1846.
John. *London.* 1839. Watch cases.
Samuel & Son. *Leeds.* 1817-26.
Walter. *Sheffield.* 1871.
William. *Heckmondwike.* 1871.
HIRT, Martin. *Colchester.* ca.1855-ca.1865. C.
HISCOCKS, Zachariah. London. 1839-51.
HISLOP—
Alexander. *Glasgow.* 1823.
Alexander. *Liverpool.* (B. 1825-)1828-51.
Alexander. *Stockport.* 1834.
Charles. *London.* 1844.
J. *East Dereham.* 1846-65.
Richard. *London.* 1832-51.
Robert. *Peebles.* 1860.
William. *Biggar.* 1860.
William. *Tunbridge Wells.* 1874.
HITCHCOCK—
H. *Lodi, NY, USA.* Early 19c.
Henry. *Chard.* ca.1800.
R. *USA.* 1874.
S. R. *Humphreysville, NY, USA.* c.1810.
William. *Oxford.* a.1675.
HITCHEN(S)—
Joseph. *Birmingham.* 1850-54.
Robert. *Wakefield.* 1822.
Samuel. *Leeds.* 1826.

HITCHIN—
James. *Wednesbury.* 1835.
Samuel. *Kidderminster.* 1842.
Samuel. *Wednesbury.* 1850-76.
HITCHINS—
Alfred. *London.* 1875-81.
Frederick. *Beauharnois, Canada.* 1857.
Thomas. *Birmingham.* 1858-80. Dial makers.
HITCHINSON, J. *Walsall.* 1860.
HITCHMAN, Alfred. *London.* 1863.
HITT, Peter. *Liskeard.* 1748-d. pre-1803. C.
HITZMAN—
Celestin & Co. *London.* 1857.
Henry. *Cambridge.* 1865-75.
Joseph. *Cambridge.* 1840-65.
HIX, James. *Bristol.* 1826.
HOAD, Henry. *West Malling.* 1858-74.
HOADLEY—
Ammi. *Bethany* and *Plymouth, Conn, USA.* 1762-1834.
Luther. *Winsted, Conn, USA.* bro. of Samuel. 1807-13.
Samuel. *Winsted, Conn, USA.* bro. of Luther. 1807-13.
Samuel & Co. *Winchester, Conn, USA.* ca.1809.
HOBART—
Aaron. *Abington, Mass, USA.* 1770.
Gabriel. *York.* 1750-54. W.
HOBBINS—
Edward. *Solihull.* 1880.
Frederick. *Birmingham.* 1880.
Joseph. *Feckenham.* 1820. C.
HOBBIS, John. *Usk.* 1880.
HOBBORT, William. *Bury St. Edmunds.* 1830.
HOBBS—
Charles. *Eastover (Somerset).* 1866-75.
Elias. *Dulverton.* 1875.
James. *Baltimore, USA.* ca.1800.
John. *London.* 1828.
Nathan. *Boston, USA.* 1842.
Samuel. *Canterbury.* Free 1743. W.
HOBDAY—
Edgar. *Clevedon.* 1875-83.
L. N. *Birmingham.* 1880.
HOBDELL, Henry Banshard. *Oxford.* 1832-44. C.
HOBSON—
L. *Birmingham.* ca.1830. Dial maker.
Thomas Charles. *London.* 1869-81.
HOCH—
C. *London.* 1844.
John. *Brixton.* 1878.
Nicholas. *London.* 1851-63.
HOCHENADEL, ——. *London.* ca.1700. W.
HOCHSTETTER—
Joseph. *London.* 1697 (prob. same as HECKSTETTER, q.v.).
William. *London.* 1857-69.
HOCHWALD, J. *Toronto, Canada.* 1861-63.
HOCK—
Cattfried. *London.* 1875-81.
Raymund. *Coleford.* 1870-79.
HOCKEN—
Henry. *Camelford.* 1823. W.
John. *Camelford.* 1856-73.
John. *Stratton.* 1612. Rep'd ch. clocks.
William. *Camelford.* 1847.
HOCKEY—
Charles. *Stalbridge (Dorset).* (B. 1790-) 1824-30.
& Son. *Stalbridge.* 1848.
HOCHFIELD, David. *London.* 1875-81.
HOCKIN(G)—
James. *Chacewater.* ca.1830-44. C.
James. *Truro.* 1856-73.
John I. *Camborne (Cornwall).* 1791-d.1805. C.
John II. *Camborne.* 1805-23. C.
John. *Redruth.* 1844-47.
Samuel. *Camborne.* 1856-73.
Samuel. *Redruth.* ca.1850. W.
Samuel. *St. Agnes.* ca.1830. C.
Thomas. *Lostwithiel.* 1824. C.
William. *Camborne.* 1873. W.
William. *Redruth.* 1809.
HOCKLEY, Thomas. *Philadelphia, USA.* ca.1800.
HODDELL & HODDILL—
James. *Coventry* and *London.* 1842-54.

HODDELL & HODDILL—*continued.*
James & Co. *Coventry* and *London.* 1860-80.
James & Co. *London.* 1857-81.
HODGE—
J. *Helston (Cornwall).* 1844-60. C.
John. *Liskeard.* ca.1800. C.
John. *Newport (Isle of Wight).* 1839.
John. *West Cowes.* 1848.
Jonathan. *Helston.* 1844. Bankrupt 1860.
William. *Fowey.* 1873. W.
HODGES—
Anthony. *Oxford.* a.1664-72. W.
Benjamin. *Hereford.* 1870.
Charles Henry. *Newtown (Wales).* 1886-90.
Erastus. *Torrington, Conn, USA.* 1831-42. C.
Frederick. *Dublin.* 1796-1839.
James. *London.* 1832-51.
John & Sons. *Maidstone.* 1839.
(Erastus) & NORTH (Norris). *Torrington Hollow, Conn, USA.* ca.1830. C.
HODGESKYNE, ——. *Maidstone.* ca.1578.
HODGESON, Charles. *Newcastle-on-Tyne.* 1856-58.
HODGETTS, Mrs. Jane. *Llandudno.* 1887.
HODGKIN, Robert. *London.* ca.1705-ca.1720. C.
HODGKINS—
E. A. *London.* 1846. W.
& BOOTH. *Birmingham.* 1842-60. Wholesalers.
George. *Toronto.* 1865-77.
HODGSON—
G. *Gateshead.* 1858.
John. *Annan.* 1837.
John. *Lancaster.* 1834-51. W.
John. *Skipton.* d.1729. C.
John. *York.* 1716-d.1719. s. of Marcus.
Jonathan. *Hexham.* 1760 (B. 1784).
Robert. *Hexham.* 1796-1834.
Thomas. *Durham.* 1827-51.
Thomas. *Preston.* Late 18c. C.
William. *Philadelphia, USA.* 1785.
William Batty. *Lancaster.* 1820-51. C. & W.
HODIERNE, William Harvey. *West Ham (Essex).* 1839.
HODNETT—
John. *West Bromwich.* 1876.
T. *West Bromwich.* 1860-68.
HODSOLL—
Joseph. *Quebec.* 1857-63.
William. *London.* (B. 1805-08) 1844.
HODSON—
Charles. *Dudley.* 1835-42.
Charles. *Worcester.* 1850-60.
Mrs. Sylvia. *Bristol.* 1870-79.
S. *Bristol.* 1863.
Thomas. *Chorley.* 1690-d.1756. C.
William. *Bristol.* 1850-56.
HOEBER, Louis. *London.* 1857-81. Wall dial clocks.
HOEFLER, William. *Birmingham.* 1868-80.
HOEN, Benjamin. *West Zaandam.* ca.1775.
HOENDSCHKE, ——. *Dresden.* ca.1744.
HOER—
Ernest. *London.* 1881.
Oskar. *London.* 1881.
HOERNLIN, Peter. *Vienna.* Pre-1660. W.
HOETZER, ——. *Munchen.* ca.1660.
HOEY, James. *Moneymore.* 1858.
HOFER, Charles. *Macon, Miss, USA.* ca.1850.
HOFF—
George, jun. *Lancaster, Pa, USA.* 1765-1818.
George Frederick. *Lancaster, Pa, USA.* 1810.
John. *Lancaster, Penn, USA.* s. of John George, b.1776-d.1818. C.
John George. *Lancaster, Penn, USA.* b.1733-d.1816. C.
John George jun. *Lancaster, Penn, USA.* s. of John George sen. b.1788-d.1822. C.
(John) & HEISELEY (Frederick). *Lancaster, Pa, USA.* 1793-1801.
HOFFARD, Samuel. *Berlin, Pa, USA.* 1825-50.
HOFFMAN—
C. M. *Lebanon, NH, USA.* ca.1850.
Carl G. *Tenby.* 1887.
HOFFMAYER (also HOFFMEYER, HOFFMEIR, etc.)
A. J. *Liverpool.* 1856. W.

HOFFMAYER—*continued.*
Alexander. *Liverpool.* 1851.
Bernard. *Londonderry.* 1858. C.
Brothers. *London.* 1875-81.
& CLEY. *Liverpool.* 1848.
Edward. *London.* 1875.
George. b. ca. 1821 at *Baden, Germany. Leamington.* 1854. & HERDSTREET. *Leamington.* 1850.
M. *Birmingham.* 1860.
M. *Chelmsford.* See Huffmyer.
Martin. *Liverpool.* 1834.
Matthew W. *Birmingham.* 1868-80.
WUNDERLE & Co. *Londonderry.* 1846. C.
HOFFNER, Henry. *Philadelphia, USA.* 1791.
HOFFNUNG, Abraham. *Montreal.* 1857-63.
HOFFONYER & TRITSIKLER. *Chelmsford.* ca.1839
HOGARTH, John. *North Shields.* 1834.
HOGBEN, Thomas. *Smarden (Kent).* ca.1740-92.
HOGG—
Alois. *London.* 1875.
George. *London.* 1857.
Isaac. *London, Canada.* 1875.
James. *Gifford (Scotland).* 1837.
James. *Haddington.* 1860.
James. *Strabane.* ca.1800. C.
James. *Tranent (Scotland).* 1860.
Peter. *Felton.* Late 18c. C.
Thomas. *Skipton.* ca.1830-40. C.
HOGUET—
Augustus. *Birmingham.* 1835.
Augustus. *Philadelphia, USA.* 1814-33.
HOLBURN—
Charles William. *South Cave.* b.1759-d.1811. C.
Edmund. *Norwich.* 1875.
HOLBOROUGH—
Thomas. *Colchester.* b.1676-1706, then Ipswich where d.1727. C.
Thomas II. *Ipswich.* b. 1706, s. of Thomas I -1732 C.
HOLBROOK—
H. *Medway, Mass, USA.* ca.1830. T.C.
John. *Dublin.* 1829.
Major George. *Wrentham* and *Brookfield* and *Medway, Mass, USA.* 1767-1846. C. and bellfounder.
William. *Worcester.* 1868.
HOLCROFT, John. *Prescot.* 1794-97. W.
HOLDEN—
Benjamin John. *Witham.* 1839.
Eli. *Philadelphia, USA.* 1843. C.
John. Place unknown, prob. *Lincolnshire.* ca.1790. C.
George. *Sheffield.* (B. 1825) 1834-62.
George. *Stowmarket.* 1830.
Henry. *New Haven, Conn, USA.* ca.1840.
J. *Boston.* ca.1830. W.
J. *Great Marlow.* 1854.
James. *Sheffield.* 1871.
John. *Lincoln.* ca.1770. C.
Joseph. *Dayton, O, USA.* 1836.
Joseph. *Liverpool.* 1848-51.
HOLDER—
Charles. *Colchester.* s. of Robert. b.1788-1828. C.
George. *Colchester.* s. of Robert. b.1777-d.1818. W.
Peter. *Haselmere.* 1829-39.
Robert. *Colchester.* ca.1775-d.1801. W.
William. *Bristol.* 1781.
William. *Cheltenham.* 1879.
HOLDRED, Theophilus. *London.* (B. 1820-29) 1828-1832.
HOLDREDGE, A. A. *Glens Falls, NY, USA.* 1841.
HOLDSWORTH—
Henry. *Tadcaster.* 1844.
Jonas. *Halifax.* 1816-37.
William. *Bingley.* 1822.
HOLE, Henry. *London.* 1839.
HOLFORD, J. *Birmingham.* 1854.
HOLGATE, William. *Wigan.* (B. ca.1780-1814) 1822. C.
HOLINGUE, ——. *St. Nicolas d'Aliermont.* 1859-76.
HOLISON, ——. *Liverpool.* 1831. W.
HOLL—
Frederick Richard. *London.* 1857-63.
G. *Thetford.* 1858.
George T. *Old Buckenham.* 1830-58.

HOLLAND—
——. *Nottingham.* ca.1800. C.
Charles. *London.* 1857.
Charles. *Oakham.* ?Late 18c. C.
Charles. *Walsall.* 1876.
Gabriel. *Coventry.* ca.1740. C.
Gabriel. *Swannington (Leicestershire).* ca.1750. C.
Hamnet. *Kings Lynn.* 1875.
James. *Birkenhead.* 1878.
James. *Manchester.* mar.1791. W.
John. *London.* 1828.
Samuel. *Nottingham.* 1835-55.
Thomas. *Atherstone.* 1842.
Thomas. *Brighton.* (B. 1822-)1828.
William. *Birkenhead.* 1878.
William. *Chester.* 1782-1828. C.
HOLLANDER(S)—
David. *Birmingham.* 1868-80.
George. *Birmingham.* 1868.
Emil. *Portsea.* 1867-78.
J. J. *Portsea.* 1867.
HOLLENBACH, David. *Reading, Pa, USA.* With Daniel Oyster 1822-26, alone 1826-40.
HOLLER, Henry. *Cenier Co., Pa, USA.* 1818. C.
HOLLEYHEAD, Armet. *Sheffield.* 1871. Clock dealer.
HOLLIDAY—
Cornelius Brooks. *London.* 1839. Watch cases.
Edward. *London.* CC. 1650-62.
HOLLIDGE, Samuel. *London.* 1881.
HOLLINGER, A. *Philadelphia.* 1839.
HOLLING(S)HEAD—
George. *Olney.* s. of James. b.1818-77.
James. *Olney.* 1808-d.1863.
HOLLINGSWORTH, George. *Liverpool.* 1780-84. W.
HOLLINRAKE—
J. *Mytholmroyd.* 1871.
James. *Todmorden.* 1866-71.
HOLLINS—
Edwin John. *Coventry.* 1880.
H. *Stoke.* 1860.
Richard H. *Newcastle-under-Lyme.* 1850-76.
HOLLINSHEAD—
Hugh. *Mount Holly* and *Moorestown, NJ, USA.* mar.1775-86.
George. s. of Morgan. *Woodstown, NJ, USA.* 1775-79.
Jacob. *Salem, NJ, USA.* 1768-78.
Job. *Haddonfield, NJ, USA.* Pre-1821.
John. *Mount Holly* and *Burlington, NJ, USA.* 1745-ca.1780.
John, jun. *Burlington, NJ, USA.* T.C.
Joseph. *Burlington* and *Moorestown, NJ, USA.* 1740-75. C.
Morgan. *Moorestown* and *Chester, NJ, USA.* 1775-1832.
HOLLINSON(E)—
Alexander. *Liverpool.* ca.1790-1807. W.
William. *Liverpool.* ca.1800. W.
HOLLIS—
J. M. *Chester, Pa, USA.* 1821.
Walter. *Castle Donington.* 1876.
HOLLISTER, George. *London.* 1863-81.
HOLLIWELL—
John. *Warrington.* 1828. C.
Thomas. *Farnworth.* mar.1812. C.
William. *Alfreton.* 1828.
HOLLMAN, John. *Lewes.* 1828.
HOLLOME, G. *Boston.* 1868.
HOLLOWAY—
Edward. *Banbury.* 1833-42. Clock case maker.
J. *Ramsgate.* 1838.
Nelson J. & Co. *London.* 1863-81. American clock importers.
Robert. *Baltimore.* ca.1820.
Thomas. *Great Haseley.* 1736-d.1764. C. & T.C.
Thomas. *Youghal (Ireland).* 1778-84. C.
William. *Great Heasly (Haseley?).* Late 18c. C.
William. *Stroud.* b. ca. 1638. mar. 1664 -d. 1695. W. & C.
HOLLOWELL—
John. *Montreal.* 1842.
Mrs. C. *Northampton.* 1854.
HOLLYOAK—
Joseph. *Hinckley.* 1835.

HOLLYOAK—*continued.*
Joseph. *Smockington (Leicestershire).* ca.1830. C.
HOLLYWOOD, John. *Dudley.* 1835. German clocks.
HOLMAN—
D. *Baltimore, USA.* ca.1810.
H. J. *Lewes.* 1851.
John. *Lewes.* 1797-1803. W.
Mrs. M. *Lewes.* 1855.
Salem. *Hartford, Conn, USA.* ca.1820. C.
& Son. *Lewes.* 1839.
Walter George H. *Burnham (Somerset).* 1883.
HOLMDEN—
John George. *London.* (B. a.1781, CC. 1788-)1844. W.
and watch cases.
R. *Bognor.* 1851.
R. *Lower Norwood.* 1851-55.
R. *Ringwood.* 1848.
HOLME—
D. *Derby.* 1849-55.
James. *Manchester.* mar.1665. W.
John. *Broughton (Lancs)* from *Cockermouth.* b.ca.1760-88. W.
Joseph. *Prescot.* 1825. W.
Lawrence. *Liverpool.* ca.1758-90.
Lawrence. *Rochdale.* mar.1754-ca.1770.
& SMITHARD. *Derby.* 1835. C.
Thomas. *Hexham.* ca.1750-60. C.
Thomas. *Liverpool.* 1767-69. W.
Thomas. *Prescot.* 1807-26. W.
HOLMER, John. *Belfast.* 1804.
HOLMES—
Aaron. *Boston, USA.* 1842.
Andrew. *Lane End (Staffordshire).* 1828-35.
C. *London.* 1851.
Charles Joseph. *Cheadle.* 1828-42.
Christopher. *Dublin.* 1805-24.
Conway. *High Wycombe.* 1823-30.
Frederick & Co. *London.* 1857.
Henry. *Longton.* 1842.
J. *Philadelphia, USA.* 1842.
James. *Belfast.* ca.1815. C.
James A. *London.* 1863.
John. *Guildford.* 1828-39.
John. *London.* 1811. W.
Joseph. *Penrith.* 1879.
Matthew. *London.* 1828-39.
Matthew. *Sculcoates.* 1813. W.
Matthew S. & Son. *London.* 1844.
Peter. *Liverpool.* 1834.
Richard Henry. *Hull.* 1823. C.
Robert. *Dublin.* ca.1720-87. W. (Also free of Belfast).
Robert & Samuel. *Dublin.* Partners till 1769. Robert then alone till d.1787. Samuel alone 1769-d.1795.
S. A. *Eastbourne.* 1878.
Samuel. *Wellington.* 1828.
Thomas. *Daddry Shields.* ca.1850-60. C.
Thomas. *Ilkley.* 1871.
W. E. *Cirencester.* 1863.
William. *Norwich.* 1865-75.
HOLMGRE(E)N
Charles. *Hamilton, NY, USA.* ca.1850.
Charles J. *Quebec.* 1862-64.
HOLMIAE, Wiedeman. *Vienna.* Late 17c. W.
HOLROYD—
Benjamin, *Mansfield.* mar.1706. C.
John. *Wakefield.* 1730-75. C. (B. 1814).
Joseph. *Leeds.* 1759-d.1786? W.
Richard. *Chester.* 1834.
Richard. *Halifax.* 1814-22.
Richard. *Stockport.* 1828.
HOLT—
Alfred. *Northampton.* 1877.
George. *London.* 1863.
George. *Manchester.* 1851.
H. *Towcester.* 1864-69.
Henry B. *Leicester.* 1835.
J. *Altrincham.* 1857.
J. *Coventry.* 1860.
J. *Stockport.* 1857.
John. *Goole.* 1844.
John. *Rochdale.* ca.1790-1808. C. (B. ca.1820).
John. *Titchfield.* ca.1740. C.

HOLT—*continued.*
Matthew. *London.* 1832.
Mathew I. *Wigan.* ca.1745-d.ca.1782. W.
Mathew II. *Wigan.* s. of Mathew I. b. ca.1745-d.1805. W.
Robert. *Birmingham.* 1835-42.
Robert. *Haslingden.* 1851.
T. *Rochdale.* 1858.
Valentine. *Rochdale.* b.1781-d.1846.
William. *Lancaster.* 1767. Free at *Lancaster* but working in *London.* W.
William. *Petworth.* 1839.
William. *Wigan.* b.ca.1735-d.1780. W.
HOLTBY, Abel. *Driffield.* 1840-58. C.
HOLTON, Henry. *Wells River, Vt, USA.* ca.1850.
HOLVERSHIELD, R. *Stratford, Canada.* 1857.
HOLWAY, Philip. *Falmouth, Mass, USA.* ca.1800. C.
HOLYOAK(E)—
——. *Sharnford (Leicestershire).* Early 19c. C.
Joseph. *Lutterworth.* 1835.
HOLZMAN—
FERNBACH & Co. *Kings Langley.* 1874.
& FERNBACH. *London.* 1875.
Johannes. *Vienna.* 1775.
L. *London.* 1863-69.
HOME—
Robert. *Edinburgh.* 1766. W.
Watch Co. used by American Watch Co., *Waltham.* ca.1868.
Watch Co. *Boston, USA.* ca.1890. W.
HOMER—
A. *Ballybay.* 1802.
William. *Moreland, Pa, USA.* 1849.
HOMERSHAM, John. *Canterbury.* 1838.
HOMERTON, W. Place unknown. 1824. W.
HONE, Edward Stephen. *London.* 1875.
HONEY, Thomas. *Launceston.* 1844-84.
HONEYBONE—
Amy. *Nottingham.* 1876.
Charles John. *Fairford.* 1870-79.
George. *Fairford.* 1840-63.
R. *Rugby.* 1860.
Richard. *Fairford.* 1830.
Richard. *Luton.* 1877.
Richard. *Nottingham.* 1835-64.
Richard. *Wanborrough.* ca.1740-50?
T. *Brentford.* 1851-55.
Thomas. *Bexley Heath.* 1874.
HONNY, Thomas. *Dublin.* ca.1720. C.
HOOD—
A. *Bridgewater, Nova Scotia.* 1864.
Alexander. *Dunfermline.* 1860.
George. *Colinsburgh (Scotland).* 1840-55.
J. *Melcombe Regis.* 1855-67.
James, jun. *Melcombe Regis.* 1875.
John. *Cupar, Fife.* 1840-d.1888. W. (pupil of R. S. Rentzsch).
Robert. *Blandford.* (B. 1780-)1830-55.
Robert, jun. *Blandford.* 1867-75.
W. *Hinckley.* 1855-64.
William. *Atherstone.* 1828-42. C.
William. *Tarbolton (Scotland).* 1843. W.
HOOK—
Emil. *Bristol.* 1863-70.
John. *Midsomer Norton.* 1883.
Michael. *Lancaster, Pa, USA.* ca.1708.
HOOKER—
& GOODENOUGH. *Bristol, USA.* ca.1840.
Mary. *Lewes.* Successor to William. ca.1833-39. C.
& MORGAN. *Pine Plains, NY, USA.* ca.1815-20.
Thomas. *London.* ca.1750. C.
William. *Lewes.* 1803-32. C.
HOOLEY, Richard. *Flemington, NJ, USA.* From England. 1796-1840. C.
HOOPER—
& ALLEN. *Cardiff.* 1875.
E. M. *Luton.* 1869.
George. *Carlton.* 1876.
George. *London.* 1875.
John. *Dursley.* 1870-79.
Samuel. *Mere (Wiltshire).* 1875.
Thomas. *Northampton.* 1830.

HOOPER—*continued.*
W. *Tewkesbury.* 1856.
William. *Wednesbury.* 1876.
William. *Wotton-under-Edge.* 1863-70.
HOOPS, Adam. *Somerset, Pa, USA.* ca.1800.
HOOTEN, Peter. *Liverpool.* 1834. W.
HOOTON—
G. *Kirton-in-Lindsey.* 1861.
George. *Brigg.* 1861-76.
HOOVER, John. *Emmetsburg, MD, USA.* ca.1850.
HOPE—
F. M. *Sag Harbor, NY, USA.* 1870. C.
Thomas. *York.* Pre-1778-85. W.
W. D. *Sevenoaks.* 1874.
HOPKIN—
Abraham. *Bourne (Lincs).* 1835.
Lewis. *Caerlan (Tonyrefail).* ?1708-71. Sundials.
Rees. *Llanarthne.* a.1814.
William. *Llandeilo.* 1847-1941.
William. *Tuxford.* 1828.
William & Edward. *Llandeilo.* 1887.
HOPKINS—
Abraham. *Burton-on-Stather.* 1828.
(Edward) & ALFRED (Augustus). *Harwinton, Conn, USA.* 1820-27. C. (1831-42?).
Asa. *Northfield and Fluteville, Conn, USA.* 1799-1813. C.
E. *Charlton, Brackley (Northants).* 1854-69.
E. & C. & Co. *Coventry.* 1880.
Henry P. *Philadelphia, USA.* 1831-33. Dealer.
Jason R. *Washington, DC. USA.* 1875. W.
John. *Bristol.* a.1649-57.
John. *Llandovery.* 1830-d.1842. C.
John. *London.* CC. 1640-54.
Joseph. *Waterbury, Conn, USA.* ca.1750.
& LEWIS. *Litchfield, Conn, USA.* ca.1825. C.
Maria. *Dublin.* 1858.
Orange. *Litchfield and Campville and Terryville, Conn, USA.* 1791-1830s. C.
R. *Brentford.* 1851-55.
Robert. *Philadelphia, USA.* 1833.
& SCHWAR. *London.* 1881.
Thomas. *Wootton-under-Edge.* 1850-56.
William R. *Geneva, NY, USA.* 1841.
HOPKINSON, T. *Hunslet (Leeds).* Early 19c. C.
HOPLEY—
George Henry. *Dover.* 1851.
W. F. *Dover.* 1855.
William. *Dover.* Free 1818-55.
HOPPER—
B. C. *Philadelphia, USA.* 1844-48.
Benjamin. *Philadelphia.* 1850.
Christopher James. *London.* 1869.
James. *Worsborough.* mar.1735. C.
Joseph M. *Philadelphia.* 1816-22.
William. *Durham.* 1851-56.
HOPPERTON, Emanuel. *Leeds.* b.1705-d.1753. C.
HOPPIN—
Matthew. *Exeter.* Locksmith. 1626.
William. *Exeter.* Locksmith. Free 1626. d. pre-1648 (a. of Mathew).
HOPTON—
James. *Edinburgh.* 1826-50. German clocks.
Richard. *Methley.* 1871.
HOPWOOD—
J. C. & Son. *Colchester.* ca.1882-1914.
Joseph Cooke. *Colchester.* 1845-ca.82. W. & C.
R. *Gloucester.* 1879.
Robert. b. ca. 1837 *Cirencester.* To *Bridgnorth* by1861-75. W.
Son & Payne. *Colchester.* 1914-39.
HORAH, James. *Salisbury, NC, USA.* 1826-64.
HORDEN—
John. *Kettering.* 1877.
W. P. *Kidderminster.* 1860.
HORDERN—
John. *Leamington (from Coventry).* b. ca.1789. 1828-50.
Joseph. *Manchester.* 1848.
Samuel. b. ca. 1823 at *Coventry,* s. of John. *Leamington* 1854.
HORLEY, Alfred. *London.* 1869-78.
HORMER, J. *Tonbridge (Kent).* 1802. W.
HORN—
Alexander. *Fyvie (Scotland).* (B. ca.1825-)60.

HORN—*continued.*
Alfred. *New Haven, Conn, USA.* ca.1840.
E. B. *Boston, USA.* 1847-60.
Eliphalet. *Lowell, Mass, USA.* 1832-47. C.
Henry Bishop. *Akron, O, USA.* (b.1819) 1845-68. C.
& PECK (Timothy). *Litchfield, Conn, USA.* 1808. C.
HORNBLOWER—
Jonathan. *Penryn.* ca.1780. C.
Mrs. A. *London.* 1844.
HORNBOSFELL, Abraham. *Bruxelles.* ca.1660.
HORNBY—
Garrard (& Son). *Liverpool.* (B. 1780-) 1795-1824 (-1837). W.
George. *Liverpool.* 1848-51.
Henry. *Liverpool.* 1814-22.
James. *Liverpool.* 1824-48.
John. *Liverpool.* 1707-24. Watch spring maker.
John. *Liverpool.* 1781-1834.
Richard. *Liverpool.* 1814-51. W.
Thomas. *Liverpool.* 1784-1822. W.
Thomas. *Prescot.* ca.1790-1800. W.
HORNCY, William. *Hetton-le-Hole.* 1851. (See HORNER).
HORNE—
John. *Canterbury.* mar.1784-1828.
John. *Waltham Abbey.* 1828.
Samuel. *London.* a.1647. CC. 1654-d.1685. *Chancery Lane.*
William. b.1837-pre-1860. *Oldham.* 1860. *Askrigg.* 1866. *Leyburn.*
William. *London.* 1828-44.
HORNER—
Charles. *Halifax.* 1871.
David. *Monaghan.* 1809-27.
James. *Ripon.* 1721-d.1761. C.
John. *Ballybea.* 1802.
William. *Hetton-le-Hole.* 1856.
HORNSEY—
Thomas Edward. *York.* 1826-58.
William. *Exeter.* mar.1767. W.
HORSFALL—
J. H. *Coventry.* 1860.
Matthew. *Leeds.* mar.1765-75. Hour glass maker.
William. *York.* 1846.
HORSTMAN—
Gustav. *Bath.* 1861-83. (Also *Weston-super-Mare* 1861.)
Henry. *Weston-super-Mare.* 1875-83.
HORSWILL, A. *Leamington.* 1880.
HORTON—
A. *London.* 1869-81.
& BURGI. *New York.* ca.1776.
Richard. *Oversley Green (Warwickshire).* 18c. C.
Thomas. *London.* 1875.
HORWITZ—
Daniel & Co. *London.* 1869.
Max & Co. *London.* 1881.
HORWOOD—
H. *Chesham.* 1869.
J. *Chesham.* 1847.
Moses. *Bristol.* 1759-74.
William. *Bristol.* a.1829-39.
HOSKING—
Henry. *Camelford.* 1814. W.
John. *Camborne.* 1844-d.1853. W.
John. *Penzance.* ca.1747-ca.1760. C.
Luce. *Ventnor.* 1878.
Richard. *Callington.* 1844-47.
W. L. *Ventnor.* 1859-67.
William Luce. *London.* 1881.
HOSKINS—
& BIRD. *London.* (B. 1825). 1828.
Jonah. *Bristol.* a.1806-23.
Jonah. *London.* (B. 1822-24) 1832-39.
Joseph & Son. *London.* 1844.
HOSTERMAN, Thomas. *Halifax, Nova Scotia.* 1802-16. (Partner with Benjamin Etter 1802-13).
HOSTETLER, Lafayette. *Toronto.* 1915.
HOSTETTER—
Jacob, jun. *Hanover, Pa, USA.* To *Ohio.* 1822(-ca.31).
Jacob, sen. *Hanover, Pa, USA.* 1786. Then *Ohio.* 1825-31.

HOSTETTER—*continued.*
S. *Hanover, Pa, USA.* ca.1820. W.

HOTCHKISS—
Alva. *Poughkeepsie, NY, USA.* 1825-35. Then *New York.* 1835-45.
& BENEDICT. *Auburn, NY, USA.* ca.1830. C.
& Co. (A. S.). *New York.* ca.1869-77.
Elisha. *Bristol* and *Burlington, Conn, USA.* ca.1820-37. C.
& FIELD. *Burlington, Conn, USA.* ca.1825.
Hezekiah. *New Haven, Conn, USA.* 1721-61.
John. *Rochester, NY, USA.* 1845.
& PIERPOINT. *Plymouth, Conn, USA.* ca.1810. C.
Robert & Henry. *Plymouth, Conn, USA.* ca.1840.
SPENCER & CO. *Salem Bridge, Conn, USA.* ca.1830. C. and buttons.
William. '*Md*', *USA.* 1775.

HOTHAM, Henry. *London.* a.1665 to John Pennock. CC. 1673. At 'ye Black Spread Eagle at ye west end of St. Paules'. Lantern clock.

HOTTEN, John Thomas. *Penzance.* 1852. Bankrupt. W.

HOTTON, William. *Probus (Cornwall).* 1827-56. W.

HOUGH—
Frederick. *Coventry.* 1868-80.
John. *Newport, Ind, USA.* 1820s-30s.

HOUGHTON—
——. Place unknown. ca.1830. C. (*?Cornwall*).
Edmund. *Norwich.* b.1726-ca.1750. Then *Colchester* where d.ca.1778.
J. *Beccles.* 1853.
James. *Burtonwood (Lancashire).* 1783-97. W.
James. *London.* 1857.
James. *Ormskirk.* 1806-55. W.
James. *Prescot.* 1818-26. W.
James. *St. Helens.* 1848-58.
John. *Birmingham.* 1797-1843. Succeeded by William Frederick Evans.
John. *Prescot.* mar.1779. W.
John. *Prescot.* 1824-26. W.
John. *St. Helens.* 1778. W.
John. *St. Helens.* 1824. W.
John. *Uppingham.* 1828-49.
Michael. *Liverpool.* 1819. W.
Richard. *Liverpool.* 1715-61. W.
Richard. *Wigan.* 1781-1806. C.
S. *Wigan.* 1858.
Stephen. *St. Helens.* 1848-51.
Stephen & Son. *Ormskirk.* 1824.
T. *Wigan.* 1858.
Thomas. *Kirkham.* 1851.
Thomas. *Prescot.* 1794. W.
William. *St. Helens.* mar. 1782-85. W.
William. *St. Helens.* 1821-28. W.

HOUGUET, Augustus. *Philadelphia, USA.* 1819-25.

HOUISON—
James. *York.* 1832-38. W. & C.
John. *London.* (B. 1805-11). 1828-32.

HOULAHAN, Edward. *Thurles (Ireland).* 1858. W.

HOULGREAVE, Charles. *Childwall (Lancs).* 1795. W.

HOUNSOM, James. *Liss (Hants).* 1848-78.

HOURSTON, William. *Kirkwall.* 1845-60.

HOUSE & ROBINSON. *Bristol, USA.* 1851.

HOUSEDON, J. *Johnstone.* ca.1800. C.

HOUSER, Matthias. *Oakham.* 1876.

HOUSIAUX, F. *Wolverhampton* 1868.

HOUSMAN, Jacob. *Lancaster.* 1732-80. C.

HOUSTON—
James. *Johnstone.* 1836.
Sean. *Dublin.* 1775-88. (B. 1795). W.
William. *Philadelphia, USA.* ca.1800. C.

HOUTON, Richard. *Oversley Green.* ca.1710. C.

HOVEY—
Cyrus. *Lowell, Mass, USA.* ca.1870.
J. R. *Norwich, NY, USA.* 1817.

HOVIL, John. *London.* (B. 1820-)1828-32.

HOW—
Edward. *Holford (Somerset).* ca.1790. C.
Ephraim. *Northampton.* 1755. Prob. from *London*, where one such a.1716. CC. 1729.
George. *London.* ca.1720. C.
J. *Carlisle.* 1858.
James I. *Bromley.* 1797, mar. ca. 1800 Martha Scott. d. 1843.

HOW—*continued.*
James II Nicholas. *Bermondsey. London* 1822-30, then *St. Louis, USA* ca. 1846.
Martha, Thomas & John. *Bromley.* 1847-66.
Peter. bro. of James I. *Eltham* 1827-40, then *Woolwich* 1847-66
Richard. *Market Harborough.* 1774. W.
Robert. *Holford (Somerset).* 1787.
Silas Samuel. *Tonbridge.* 1826-66.
Thomas. *London.* a.1670 to Nathaniel Delander. C 1677-97.

HOWARD—
——. *Kirkdale.* 18c.W.
Albert. *Boston, USA.* ca.1878-93.
Alfred. *London.* 1857-75.
Christopher. *Boston (Lincs).* 1828.
Christopher. *Donnington (Lincs).* 1835.
Clock Co. (E.). *Roxbury, Mass, USA.* 1903-34. W.
Clock Products Co. *Waltham, Mass, USA.* 1934.
Clock & Watch Co., The. 1861-63. Formerly E. Howard & Co.
& DAVIS. *Boston, USA.* 1842-59.
DAVIS & DENNISON. *Roxbury, Mass, USA.* ca.1850
E. & Co. *Roxbury, Mass, USA.* 1857-61. C.
E. Watch & Clock Co. 1881-1903.
Edward. *Prescot.* mar.1808-19. W.
George. *London.* ca.1750. W.
Henry. *Liverpool.* 1784-90. (B. 1796). W.
Henry. *Rockton, Canada.* 1865-69. Also dentist.
Henry. *Wigan.* 1796.
J. *Boston, USA.* 1870s.
James. *Manchester.* 1848-51.
John. *Liverpool.* 1851.
John. *London.* 1689-97. W.
John. *Prescot.* mar.1776. W.
Nicholas. *Yarmouth.* 1705.
Thomas. *Farnworth.* mar.1833. W.
Thomas. *Liverpool.* 1848-51.
Thomas. *London.* Early 19c. W.
Thomas. *Rainhill.* 1818. W.
Thomas Wardle. *Liverpool.* 1851.
Watch Co. *Waltham, Mass, USA.* 1910-27.
Watch & Clock Co. *Boston, USA.* 1863-81.
William. *Boston, USA.* 1813-21.
William. *Dublin.* 1767-94. W.

HOWARTH—
John H. *Perth, Canada.* 1871.
Peter. *Ribchester.* 1719. Rep'd ch. clocks.
Squire. *Ramsbottom.* 1848.

HOWATSON, James. *Sanquhar.* 1860.

HOWCOTT, Nathaniel. *Edenton, NC, USA.* mar. 1828.

HOWDEN—
Charles. *Pennycuik (Scotland).* 1860.
& Son. *Edinburgh.* 1860.

HOWE—
——. *Downham.* ca.1790. C.
Brothers. *Hull.* ca.1880. C.
Elias. *Boston, USA.* 1843.
F. *Hereford.* 1863.
F. & Son. *Burslem.* 1868.
George. *Rochdale.* 1848-58.
John. *Alresford.* 1830.
John. *Wigton.* (B. 1820). 1828-48.
Jubal. *Boston, USA.* 1830.
Robert. *Bristol.* 1801.
W. H. *Clare (Norfolk).* 1846.
William. *Weston-super-Mare.* 1875.
William H. *Sudbury.* 1853-79.

HOWELL—
Amy. *Carmarthen.* 1824-30.
& COOKE. *Birmingham.* 1849-65. Dial makers.
David. *Carmarthen.* 19c. (pre-1824?). C.
& HALL. *Albany, NY, USA.* 1801.
John. *Carmarthen.* 1822.
John. *Colchester.* 1794. W.
Nathan. *New Haven, Conn, USA.* 1741-84.
Richard. *Bristol.* 1762-81.
Silas White. *New Brunswick, NJ, USA.* 1770. *Albany, NY.* 1797.

HOWELLS—
——. *Carmarthen.* 1818.
Peter. *Whitland.* 1875.

HOWELLS—*continued.*
Thomas. *Hay (Brecon).* b.1749, d.1821. Later in life was a woollen manufacturer.
William. *Tenby.* 1887.
William. *Falmouth.* 1790-1812. C., W. and chrons.
HOWES—
B. *Downham Market.* 1858-65.
Benjamin. *Downham.* 1830-46.
H. *Sudbury.* 1846.
Henry. *Ballington* nr *Sudbury.* 1830-39.
James. *London.* 1697.
John P. *Wallingford.* 1837-64.
M. *Peterborough.* 1864.
Martin. *Stoke Ferry (Norfolk).* 1875.
Suckling. *Downham Market.* 1787-(B. 1791).
HOWIE, Adam. Place not stated. ca.1765. C. (May be same as Allan Howie of *Irvine.*)
HOWGILL, John. *Masham.* 1732. Clock repairer.
HOWITT (or HOWETT), Richard. *Falkingham.* 1849-50.
HOWLETT, John. *Cheltenham.* 1830-79. C.
HOWLEY, J. *Birmingham.* 1860.
HOW(S)—
Edward. *Holford.* 1768. C.
Thomas. *London.* a.1670. CC. 1677-97.
HOWSE—
——. *Swindon.* ca.1790-ca.1800. C.
Richard. *Marlborough.* 1842-75.
Thomas. *Marlborough.* 1830.
HOY—
Dunham John & John. *Mattishal (Norfolk).* 1836-1865.
E. *Elmham.* 1865.
Edward John. *Mattishall.* 1875.
Edward William. *Whissonsett (Norfolk).* 1836-58.
Elizabeth. *York.* 1822.
George. *York.* mar.1808-pre-1822.
John. *Hingham.* 1865-75.
Thomas. *Kelso.* ca.1778.
HOYLE—
A. *Huddersfield.* 1871.
F. *Prescott, Canada.* 1862.
Frank. Place unknown *(Yorkshire).* 1891. Repairer.
James. *Halifax.* 1693. Rep'd ch. clocks.
Joseph. *Stalybridge.* 1848.
William. *Bolton.* 1824.
William. *Rochdale.* (B. 1820-)1824-28.
HOYLES—
C. *Basingstoke.* 1859.
Clater. *Sleaford.* 1849-50.
HOYT—
Freeman. *Sumter, SC, USA.* 1805-69.
George A. *Albany, NY, USA.* 1822-44.
George A. & Co. *Albany, NY, USA.* 1829-33.
George A. & Son. *Albany, NY, USA.* 1845. W.
George B. *Albany, NY, USA.* 1846.
Henry. *New York.* ca.1810. *Albany, NY, USA.* 1828-36.
James H. *Troy, NY, USA.* 1838. W.
S. & Co. *New York.* ca.1836.
Seymour. *Brooklyn, NY, USA.* ca.1840.
HUBAND—
& BEVIS. *Evesham.* 1876.
Charles James. *Evesham.* 1850-72.
HUBASHEK, Andrew. *Toronto.* 1861-65.
HUBBARD—
Brothers. *New York.* 1862. W.
C. K. *Hartford, Conn, USA.* ca.1860.
Charles. *London.* 1832-39. Watch cases.
Daniel. *Medfield, Mass, USA.* ca.1820.
Edward. *London.* 1832-44. Also imports Dutch clocks.
Edward. *Oakham.* ca.1770-ca.1780. C.
Gilbert & Co. *Bristol, USA.* 1857.
& HITCHCOCK. *Buckland, Mass, USA.* Early 19c.
James M. *Dorking.* 1866-78.
John. *Kilrea.* ca.1760-75. C. & W.
Mrs. J. *Dorking.* 1851-55.
S. *Leatherhead.* 1851.
Thomas, sen. *Dorking.* 1828-39.
Thomas, jun. *Dorking.* 1839.
HUBBELL—
(Laporte) & BEACH (Levi). *Bristol, USA.* 1859-63.

HUBBELL—*continued.*
& BOARDMAN. *New Haven, Conn, USA.* ca.1860-70. C.
L. & Son. *Bristol, USA.* 1874-79. C.
Laporte. *Bristol, USA.* 1849-89.
HUBBLE, Daniel. *East Malling.* 1847-74.
HUBER—
Andreas. *Munich.* ca.1900. C.
C. *Bedford.* 1864.
& Co. *Shrewsbury.* 1842.
& FESSON. *Wellington (Shropshire).* 1842. German clocks.
G. *Gloucester.* 1856.
George. *Stroud.* 1879.
Joseph. *Cheltenham.* 1850-70.
Raymond. *London.* 1863-81.
HUBERT, Estienne. *Rouen.* ca.1620-ca.1657. W.
HUBRY, Gregory. *Stourbridge.* 1850.
HUCK, Robert. *Epsom.* 1839.
HUCKER, Charles. *Slough.* 1854-77.
HUDDY, William Hotton. *Tregony (Cornwall).* From *Newlyn East.* 1856-73.
HUDON, John. *London.* ca.1755. W.
HUDSON—
Charles. *East Retford.* 1876.
David (& Son). *Otley.* 1822-26. C. (1834-37).
Edward. *Mount Holly, NJ, USA.* 1810-14.
George. *Hastings.* 1878.
H. *Longton.* 1868.
I. *Bradford.* 1866.
J. C. *San Francisco, Calif, USA.* ca.1860.
James. *Wetherby.* 1866. W.
John. *Otley.* b.1807. (?s. of David). -1871.
Mrs. Ann. *Wetherby.* 1871.
William. *Mount Holly, NJ, USA.* ca.1810-16.
William Henry. *Hereford.* 1856-79.
William Henry. *London.* 1839.
HUDSPETH, ——. *Newcastle.* ca.1830-40. C.
HUE—
James, jun. *Edinburgh.* 1741. Gilder and japanner of clock cases.
Pierry (Peter). *London.* CC. 1632-35.
HUESTIS, G. A. *Windsor, Nova Scotia.* 1890-97.
HUEY (or HOEY), James. *Cookstown.* From *Kilrea.* 1865.
HUFFMYER (also HOFFMEIER), Martin. *Chelmsford.* 1839-51.
HUG—
Constantine. *Merthyr Tydfyl.* 1865-75.
Lawrence. *Birmingham.* 1842-80.
Sebastian. *Tredegar.* 1850-58.
William. *Dublin.* 1858-80.
HUGGETT—
Frederick & Thomas. *Ramsgate.* 1832-47.
John. *Hawkhurst.* 1832-39. C.
HUGGINS—
——. *London.* ca.1830. C.
John. *West Rudham (Norfolk).* 1875.
Joseph. *Huddersfield.* 1822. W.
T. *Wolverhampton.* 1860.
Thomas. *Gloucester.* 1850-63.
HUGHES—
Ann. *Caernarvon.* 1835.
Christopher. *Dublin.* 1831-38.
David. *Ffestiniog.* 1874. C.
David. *London.* 1832-44.
Edmund. *Hampton* and *Middletown, Conn, USA.* 1788.
Edward Evan. *Caernarvon.* 1887-90.
Ezekiel. *Machynlleth.* a.1786. From *Newtown.* Later to *Llanbryn Mair.* Then to USA in 1795. (*Fort Washington* now *Cincinnati.*)
Frederick J. *Watford, Canada.* 1907.
G. W. *Birmingham.* 1854-60.
George. *Kington (Hereford).* 1844.
George W. *Philadelphia, USA.* 1829-33.
Henry E. *London.* 1851-69.
Hugh. *Gaerwen.* Also *Llangefni.* Also *Llanrhyddlad.* Also *Pwllheli.* These are probably all the same man as at *Caernarvon.* 1794-1806.
James. *Ashton-under-Lyne.* 1851.
James. *London.* 1869.
James. *Swindon.* 1842-48.

HUGHES—*continued.*
John. *Caernarvon.* s. of Hugh. 1844-d.1863. W.
John & Co. *Caernarvon.* wid. of John H. 1876. W.
John. *Ffestiniog.* 1887. W.
John. *Llanfair.* Mid-18c. C.
John. *Llanrwchllyn.* b.1767. Wooden clocks.
John. *London.* bro. of Morris. a.1689-d.1695.
John. *Maidstone.* 1784.
John. *Nefyn.* 1874 and later (?to 1910). W.
John. *Pwllheli.* 1886. C.
John. *Swansea.* 1844.
John William. *Llanelly.* Later 19c. to about 1895.
Joseph. *Machynlleth.* 1856-87. C.
Lewis. *Liverpool.* 1848.
Mrs. Jane. *London.* 1851. (?wid. of David).
Mrs. Mary. *Holyhead.* 1887.
Owen. *Bala.* Late 19c.
Owen Henry. *Aberdyfi.* 1887.
Patrick. *Dublin.* 1820-25.
R. Ll. *Llangollen.* s. of Robert. 1895-1905. W.
Richard. *Caernarvon.* 1730-48. Rep'd. ch. clocks.
Richard P. *Llanrwst.* 1887.
Robert. *Llangollen.* From *Rhos-y-Medre.* ca.1873-d.1895. W.
Robert. *Rhos-y-Medre.* 1864-ca.1873. W. Later *Llangollen.*
Samuel. *Llangollen.* 1887.
T. H. *Wrexham.* s. of Robert of Llangollen. 1895-1905. W.
Thomas. *Caernarvon.* ca.1795. C.
Thomas. *Ruabon.* 1887.
Thomas Roger. *Manchester.* 1848-51.
William. *Birmingham.* 1880.
William. *London.* At the Dial near *King Street, High Holborn* till 1784. (CC. 1781). Later to *Llanfflewin.*
William. *London.* 1869-75.
William. *Maidstone.* Free 1732.
William. *Rhos-llannerch-rugog.* 1868.
William Charles. *Bethesda.* 1887. C.
HUGUENAIL, Charles T. *Philadelphia, USA.* 1799-1828.
HUGUENIN—
A. & Sons. *London.* 1881.
Aime. *Liverpool.* 1824.
Auguste. *London.* 1828-32. Also musical clocks.
Charles Frederick. From *Switzerland.* 1797-1802. To *Philadelphia.* Also *Halifax, Wilmington* and *Fayetteville, SC, USA.*
Sarah. *Liverpool.* (B. 1825-)1828.
Son & HALL. *London.* 1881. (Successor to John M. Badollet & Co.)
HUGUS—
Jacob. *Greensburg, Pa, USA.* 1768-1835. C.
Michael. *Berlin, Pa, USA.* bro. of Jacob. Early 1800s. C.
HUKINS—
G. N. *Hawkhurst.* 1851.
George Hopper. *Tenterden.* b.1824-d.1910 in *Guernsey.*
J. *Alcester.* 1868.
James. *Tenterden.* b.1800-d.1882.
HULBERT (also HULBURT)—
Benjamin. *Marshfield (Gloucester).* 1840-50.
Charles. *Marshfield.* 1840-79.
Charlotte. *Marshfield.* 1830.
Horace. *New Haven, Conn, USA.* ca.1850.
S. *Bath.* 1856.
Samuel. *Widcombe* nr *Bath.* 1875.
HULK, J. *Coventry.* 1860.
HULL—
Charles. *Dublin.* 1774-83.
Charles. *Hillsborough.* ca.1775-85. C.
Edwin. *Quebec.* 1844-51.
Wid. of Edwin. *Quebec.* 1851-57.
John K. *London.* 1857-81.
N. F. *Dublin.* 1804. W.
Nesbitt. *Downpatrick.* ca.1790.
R. *Newmarket.* 1846-65.
HULME—
F. *Leamington.* 1868.
James. *Altrincham.* 1848.
James. *Stretford.* 1834. W.

HULME—*continued.*
Richard. *London.* 1857.
Thomas. *Manchester.* mar.1844-48. C.
HULSE—
Henry. *Manchester.* 1822-28.
Isaac. *London.* 1662.
HULSEMOORE, Charles. *Birkenhead.* 1878.
HUMBERDROZ, ——. *Philadelphia, USA.* 1799.
HUMBERT—
Charles. *New York.* ca.1810.
Frederick. *London.* 1828. Watch case maker.
HUMBY—
M. *Bournemouth.* 1867.
M. *Salisbury.* 1859.
HUME—
John. *Kelso.* 1836.
R. *Glanton* nr *Wooler.* 1834-58.
Richard. *Workington.* 1873.
HUMEL—
D. *Brighton.* 1851-55.
& EGAN. *Aberaron.* 1865.
Sigismund. *Swansea.* 1868-75. C.
Valentine. *Stafford.* 1850.
HUMMEL(L)—
Brothers. *London.* 1881.
Frederick. *London.* 1875-81.
& HOLBROOK. *Worcester.* 1860.
J. *Kings Lynn.* 1858-65.
J. & Co. *Portsea.* 1867.
John. *Derby.* 1855-76.
John. *Worcester.* 1850.
Joseph. *Lynn.* 1855. W.
Joseph. *Peterborough.* 1847. C.
Joseph. *Swansea.* 1840.
Silvester. *Carmarthen.* 1847.
Silvester. *Cross Inn* (= *Ammanford*). Prob. same man as at *Carmarthen.*
Silvester. *Ystradgynlais.* 1868.
HUMMELY & THOMAS. *Piqua, O, USA.* 1829. C.
HUMPHREY(S) (also HUMPHRIES)—
Andrew. *Runcorn.* 1848-57.
David. *Chester.* 1878.
David. *Lexington, Ky, USA.* 1789-93.
Henry. *Trowbridge.* 1830.
John. *Thrapston.* 1777-95. W.
Joseph. *Beeston (Nottinghamshire).* 1876.
Joseph. *Bewdley.* 1876.
Joseph. *Liverpool.* 1848-51.
Joshua. *Charlestown, Pa, USA.* ca.1744. C.
Norman. *New York.* 1840s.
Samuel. *Tarporley.* 1834-78.
Thomas. *Barnard Castle.* 1812-48. C.
Thomas. *Caernarvon.* pre-1880. C.
Thomas. *Hartlepool.* 1851.
Thomas. *Hurstpierpoint.* 1855-78.
William. *Barnard Castle.* b.1812. s. of Thomas. -1837. Then to *Hartlepool* where d.1887.
William. *Brighton.* 1870-78.
William. *Deal.* 1832-40.
William. *Thame.* ca.1850. C.
William. *West Auckland.* 1834.
William Gilby. *Northampton.* 1877.
HUMPHRY—
Thomas. *Guildford.* 1790-d.pre-1815.
William. *London.* 1832.
William Hammond. s. of Thomas of *Guildford.* Free at *Guildford.* 1815. But prob. worked at *Deal* only.
HUNDSON, Thomas H. *Chelmsford.* 1792. W. (near *Shire Hall*). (B. 1784.)
HUNGATE, William. *Chelmsford.* a.1761-68.
HUNNS, John Andrew. *Peterborough.* 1877.
HUNOT, Samuel. M. *London.* 1844-63.
HUNSDON—
Edward. *Chelmsford.* ca.1730-82.
Henry. *Chelmsford.* ca.1759-d.1817 (Henry H. there in 1728 as goldsmith).
Thomas Hinde. *Chelmsford.* ca.1784-98 (see HUNDSON).
HUNT—
& BELL. *Toronto.* 1871.
Charles. *Headcorn.* 1826-66.
E. *New York.* 1789.

HUNT—*continued.*
Edward. *Farnworth.* 1818. W.
Edward. *Hurstpierpoint.* 1878.
Elizabeth. *Overton* nr *Basingstoke.* ca.1770. C.
G. *Pewsey* (*Wilts*). 1848-75.
George. *Amesbury.* 1830-48.
George (I?). *Bristol.* 1840-50.
George (II?). *Bristol.* 1870.
George. *Northampton.* 1854-77.
George. *Portsea.* 1830-48.
George Henry. *Worcester.* 1876.
Henry. *Farnworth.* 1818. W.
Henry. *Farnworth.* mar.1831. W.
Henry. *Rayleigh.* 1874.
Henry. *Yarmouth.* 1846-75.
Henry & Son. *Salisbury.* 1830.
Hiram. *Robbinston* and *Bangor, Me, USA.* 1806-66. C.
Isaac. *Farnworth.* 1827. W.
J. *Upwell* (*Cambridgeshire*). 1846-65.
James. *Bristol.* a.1805-12.
James. *Farnworth.* 1834-51.
James. *Oxford.* 1851.
James. *Soham* (*Cambridgeshire*). 1840.
John. *Farmington* and *Plainville, Conn, USA.* ca.1830-1840. C.
John. *Farnworth.* 1818. W.
John. *London.* CC. 1671-97.
John. *Oxford.* mar.1813-23. C.
Noe. *London.* 1697 (may be HURT, q.v.).
Peter. *Prescot.* mar.1819. W.
R. *Coventry.* 1860.
Richard. *Bristol.* ca.1700. C.
Richard. *Manchester.* (B. 1808-14)-1822.
& ROSKELL. *Manchester.* 1851.
Samuel. *London.* 1869.
Thomas. *Farnworth.* 1848-51. W.
Thomas. *London.* (B. 1822-)1832-57.
Thomas. *Oxford.* 1777-ca.1800. W.
Thomas. *Prescot.* mar.1775. W.
W. *Farnworth.* 1858. W.
William. *London.* 1832.
William George. *London.* 1881.
William & Son. *Yarmouth.* 1830-36.
HUNTBACH, T. *Congleton.* 1865.
HUNTER—
A. *Halifax, Nova Scotia.* 1819.
Alexander. *New Cumnock* (*Scotland*). 1837.
Arthur L. *London, Ontario.* 1875.
Charles (& Co). *London.* 1857.
Charles Patrey. *Long Sutton* (*Lincolnshire*). 1850.
Cuthbert. *Newcastle-on-Tyne.* Late 18c. C.
& EDWARDES. *London.* 1839-44.
Henry. *Prescot.* 1824. W.
J. T. & Son. *Clapham.* 1878.
James. *Belfast* from *Dublin.* 1789-91. W. (Late partner of McCabe & Hunter).
James. *Dunfermline.* 1860.
James. *Liverpool.* 1824.
John. *Edinburgh.* a.1824.
John. *Farnworth.* mar.1824.
John & Co. *London.* 1851.
Mrs. M. *Brixton.* 1862.
Mrs. Mary. *London.* 1857.
Nathan. *Port Glasgow.* (B. 1820-)-36. Also postmaster.
Peter. *Edinburgh.* 1846.
Peter. *London.* 1881.
R. *Hartlepool.* 1898.
Richard. *Leeds.* 1834-37. W.
Robert. *Fochabers* (*Scotland*). 1860.
S. *Clapham.* 1851-78.
Samuel, sen. *London.* 1857.
Samuel (& Son). *Clapham.* 1828-39.
Thomas. *Hull.* 1858-72.
Thomas. *Liverpool.* 1828.
Thomas, sen. *Quebec* from *London, England.* 1813-d.1826.
Thomas, jun. *Quebec.* 1822-30.
Thomas. *Ramsgate.* 1823-51.
Thomas Stephen. *Bridlington.* b.1818. s. of Richard. W.
William. *Brixton.* 1828-51.
William. jun. *Campbeltown.* 1860.

HUNTER—*continued.*
William. *Driffield.* 1846.
William. *London.* 1857.
William. *Wakefield.* 1834-37.
William D. *Ramsgate.* 1855.
HUNTINGDON, H. M. J. *Chester.* ca.1780. W.
HUNTINGTON—
Gordon. *Windham, Conn, USA.* 1763-1804.
M. P. & Co. *Milton, NC, USA.* 1819.
& PLATTO. From *Plymouth, Conn USA* to *Ithaca, NY, USA* 1855-68, later Ithaca Calendar Clock Co. till 1919.
HUNTLEY—
John Jolley. *Barnsley.* 1862-71. W.
W. *Frome.* 1861.
William. *Bradford* (*Wiltshire*). 1842-48.
HUNTSMAN, Benjamin. b.1704-mar. 1729. *Mansfield,* then *Doncaster* 1733-42, then *Sheffield* 1742-d.1776.
HUNZINGER, A. *London.* 1904.
HUPTON—
Ezra. *Thurton* (*Norfolk*). 1875.
H. E. *Lowestoft.* 1875.
HURDEL, ——. *Vienna, Canada.* 1853.
HURDERWILL, Thomas. *Stockton,* prob. error for *Hinderwill.* q.v.
HURDNELL, Thomas. *Fenny Stratford.* b.1733, mar.1763-67. C.
HURDUS, Allen. *Cincinnati, O, USA* from *England.* 1806. C.
HURLEY & ELLIOTT. *Hemel Hempstead.* ca.1800. C.
HURLOW, Thomas L. *Pembroke Dock.* 1868-87.
HURLSTON, Alfred & Co. *Coventry.* 1880.
HURREN, Simon Carew. *Yoxford* (*Suffolk*). 1853-79.
HURRY—
Jeremiah. *Whittlesey.* 1875.
John, sen. *Whittlesey.* 1830-65.
John, jun. *Whittlesey.* 1830-65.
HURST—
David. *Lancaster, Pa, USA.* ca.1850.
Edward. *Howdon.* 1827-34. C.
Edward. *North Shields.* 1848-58.
Edward. *South Shields.* 1856.
Frederick. *London.* 1851-57. (Also clock cases.)
George. *Farnworth.* mar.1834. W.
Isaac. *Hinckley.* 1876.
J. *Weston-super-Mare.* 1866.
James. *Beaconsfield.* 1830-47.
James. *Prescot.* mar.1778. W.
James. *London.* 1885.
M. *Middleton* (*Manchester*). 1858.
William. *London.* 1832.
William. *London.* 1857. (Successor to Frederick.)
William Joseph. *London.* 1839-51.
HURSTFIELD, John. *Farnworth.* mar.1823. W.
HURSTHOUSE, Joseph. Place unknown. mid-18c. C.
HURT—
Josiah. *Bristol.* 1759 dissolved partnership with Jordan.
& Son. *Birmingham.* 1880.
Thomas. *London.* 1828.
& WRAY. *Birmingham.* 1850-60.
HURTIN—
Christian. *Goshen, NY, USA.* 1792.
Joshua. *New York.* bro. of William III. 1738-80.
William III. *New York.* 1776.
HUSBAND—
Charles James. *Evesham.* 1828.
James. *Evesham.* 1828.
Thomas. *Hull.* ca.1760-d.1812. C. & W.
Walter. *Truro.* 1784. C.
HUSSEY—
Francis. *London.* 1685.
Henry. *Brighton.* 1870-78.
Henry. *London.* 1857-63.
John. *Birmingham.* 1816.
William. *London.* 1851-57.
HUSTON—
James. *Trenton, NJ, USA.* 1761-74. C.
Joseph. *Albany, NY, USA.* ca.1850.
William. *Philadelphia, USA.* 1754-71. C.
HUTCHIN(G)S—
Abel. *Roxbury, Mass, USA.* 1763-86, then *Concord, NH, USA.* 1788-1821.

HUTCHIN(G)S—*continued.*
Charles. *Sturminster.* 1875.
Levi. *Abington, Conn, USA.* 1786 to *Concord, NH, USA.* C.
Nicholas. *Baltimore, USA.* ca.1810.
Thomas. *Shepton Mallet.* 1866-83.
HUTCHINSON—
Anthony. *Leeds.* ca.1770-78. C.
Charles. *Swansea.* 1830-52.
Cuthbert. *Sunderland.* 1851-56.
Edward. *London.* ca.1710. C.
G. & Co. *Clapham.* 1862-78.
G. G. *St. John, New Brunswick.* 1819-60.
George. *Clapham.* 1828-51.
George. *London.* 1857.
George, jun. *St. John, New Brunswick.* 1840-64.
Harwood E. *Escrick.* 1840. C.
Isaac. *Keswick.* 1848-73.
Isaac. *Maryport.* 1869-79.
John. *Appleby.* 1820-34.
John. *Birstall.* Late 18c. C.
John. *Bradford.* d.1741. C.
John. *Chester-le-Street.* 1847.
John. *Clitheroe.* 1803-24. W.
John. *Darlington.* 1850.
John. *Worksop.* 1828.
R. *Londonderry.* ca.1800. C.
Richard. *Colchester.* b.1676-d.1746. C.
Richard. *Huddersfield.* 1853.
Richardson. *Norwich.* 1854. W.
Robert. *Bellshill, Hamilton.* 1860.
Samuel. *Philadelphia, USA.* 1828-39.
Thomas. *Colchester.* b.1711-d.1764. W.
Thomas W. *Kirkby Stephen.* 1879.
William. *Sedbergh.* 1822.
William. *St. John, New Brunswick.* 1835.
HUTCHISON—
George. *Edinburgh.* a.1770-76.
George. *Stirling.* 1782. W.
Robert. *Douglas* (*Lanarkshire*). 1836.
HUTINSON, William. *Philadelphia, USA.* post-1800.
HUTSON, W. *Chelmsford.* ca.1859.
HUTTING, Thomas. *Syston* (*Leicestershire*). 1876.
HUTTON—
A. *Market Rasen.* 1861.
Charles. *Burslem.* 1828-50.
Charles. *Doncaster.* mar.1828-34. From *Burslem.*
& IMRAY. *London.* 1857.
J. *London.* 1851.
John. *London.* 1863-69.

HUTTON—*continued.*
T. *Birmingham.* 1860.
T. *Mildenhall.* 1865.
Thomas. *Tewkesbury.* 1879 (later White).
William. *Belfast.* 1784.
HUX—
James. *London.* 1851-63.
John (& Sons). *London.* 1828-39 (1844-57).
(Richard Rodwell) & Sons. *London.* 1863-75.
Thomas James. *London.* 1875.
HUXLEY, Thomas. *Tunbridge Wells.* 1832.
HUXTABLE—
E. *Newton Abbot.* ca.1820. C.
J. *Trowbridge* (also *Frome*). 1848.
John. *London.* 1857.
William Rendle. *London.* 1881.
HUYGENS, Steven. *Amsterdam.* ca.1695. C.
HYAM(S)—
& Co. *London.* 1869.
David & Co. *London.* 1869-81.
John. *Sunderland.* 1847.
Joseph. *London.* 1857.
Joshua. *London.* 1832-57.
N. *Brighton.* 1855.
P. *Norwich.* 1865.
HYATT—
Edward (& Son). *Wolverhampton.* 1850-68(-76).
George. *Birkenhead.* 1865-78.
John. *Wednesbury.* 1868-76.
HYDE—
J. O. *New York.* 1854.
John. *Barnsley.* b.1757-d.1787. C.
John. *Sleaford.* 1828-61.
John E. *New York.* 1885.
John I. *Sleaford.* 1825 -d. 1853. C.
HYETT, Louis. *Swansea.* 1871-87.
HYLAND—
John. *London.* 1863-81.
William C. *New York.* 1848.
HYMAN—
Brothers. *Cardiff.* 1899.
Henry. *Lexington, Ky, USA.* 1799.
Joseph. *Bangor.* First half 19c. C.
Philip. *Chatham.* 1866-74.
S. *Montreal.* 1842-44.
Samuel. *Baltimore.* 1799, then *Philadelphia.* ca.1800.
HYNDMAN—
B. *Newtown Limavady.* ca.1820-40. W.
George. *Belfast.* 1827.
HYVER, G. A. *New Orleans, La, USA.* ca.1850-60.

I

IBACH, Alexander. *Edinburgh.* From *Geneva* via *Paris.* 1831. W.
IBESON, Henry. *Buxton.* 1864-76.
IDDISON—
James. *Kirkby Malzeard.* 1871 (sometimes Addison).
Thomas. *Bedale.* ca.1870-d.1911. C.
IDEAL. 1914 and later, product of Pequegnat Clock Co.
IDESON, William. *Bradford.* 1853-71.
IGGLESDEN—
John. *Dover.* Free 1818. W.
John. *Chatham.* 1823-32. W.
G. R. *Dover.* 1866-74. (Also Iggleston.)
IHRIE, Edward. *Easton, Pa, USA.* Early 19c.
ILDERTON, Robert. *Newcastle-on-Tyne.* 1708-d.1745.
ILES—
Jonas. *Dudley.* 1868-72.
Thomas. *London.* 1769. W.
ILIFFE—
C. G. *Coventry.* 1880.
Henry T. *Ashton-under-Lyne.* 1851-58.
Samuel. *Leicestershire.* 1828.
ILLINGWORTH—
Benjamin. *Leeds.* 1850. C.
William. *Kirkby Lonsdale.* 1828-34.
William. *Leeds.* 1838-50.
ILLINOIS—
Springfield Watch Co. *Springfield, Ill, USA.* 1869-79.
Watch Co., The successor to Illinois Springfield Watch Co., 1885-1927.
ILLMAN, Charles (& Son). *Greenwich.* 1839-55(-74).
IMBERY—
Brothers. *Wolverhampton.* 1876.
J. & A. *New York.* 1840.
John. *New York.* ca.1820-30.
IMBSER, Philip. *Tubingen.* 1555.
IMHAUSER & Co. *New York.* 1873.
IMHOF—
Jacob. *New York.* ca.1840.
& MUKLE. *London.* 1863-75.
Nicolaus. *London.* 1844-63.
IMISON (John). *Mossley* (nr *Manchester*). ca.1785. C.
IMLAY, ——. *New Haven* and *Hartford, Conn, USA.* ca.1801-07.
IMPERIAL CLOCK CO. *Hyland, Ill, USA.* ca.1936.
IMPORT, Joshua. *London.* 1863-69.
INCH, John. *Annapolis, Md, USA.* 1745-49.
INDEPENDENT WATCH CO. *Fredonia, NY, USA.* Successor to Cornell Watch Co. 1877-85.
INDGE, Thomas Coventry (& Sons). *Chard.* 1861-66 (1875-83).
INGERSOLL—
Robert H. & Brother. *Waterbury, Conn, USA.* Also *Trenton, NJ, USA.* 1892-1922.
Robert Hawley. *New York.* 1859-1928.
INGHAM—
F. *Darlington.* 1898.
Henry. *Todmorden.* 1851.
Joshua. *Cockermouth.* 1869-79.
Richard. *Wakefield.* 1862-66.

INGHAM—*continued.*
S. *Bury.* 1858.
Samuel. *Ripon.* 1822-26. C.
T. H. *Wakefield.* 1869.
William. *Todmorden.* 1848-71.
INGLEBY, Clement. *London.* 1881.
INGLIS—
——. *Quebec.* 1819.
Adam. *London.* 1839-51.
& ANDERSON. *London.* 1863-69. (Successor to David.)
David. *London.* 1844-57.
David. *Manchester.* 1822.
Mary. *Manchester.* 1824-28.
INGRAHAM—
& BARTHOLOMEW. *Bristol, USA.* 1831.
& Co., E. *Bristol, USA.* 1857-80.
& Co., E. *Bristol, USA.* 1880-84.
Co., The E. *Bristol, USA.* 1884-present day.
E(lias) & A(ndrew), Co. *Bristol, USA.* 1852-56.
Elias. *Bristol* and *Ansonia, USA.* 1805-85. C.
(Elias) & GOODERICH (Chauncey). *Bristol, USA.* 1832.
Henry. *Philadelphia, USA.* 1829-33.
Reuben. *Preston* and *Plainville, Conn, USA.* 1745-1811. C.
& STEADMAN. *New York* and *Connecticut, USA.* 1850.
INGRAM—
Alexander. *Abergavenny.* 1852.
Alexander. *Wotton-under-Edge.* 1830.
David. *Bourne* (*Lincolnshire*). 1868-76.
James. *Dumfries.* 1812.
James Oliver. *Cardiff.* ca.1830-68.
John, jun. *Abergavenny.* 1822-44.
John, sen. *Abergavenny.* 1830.
John. *Antrim.* ?Late 18c. C.
John. *Belfast.* ca.1780-1800. C.
John. *Cardiff.* 1840-87.
John. *Cowbitt.* Early 18c. C.
John. *Hayle.* 1844-56.
John. *Penarth.* 1887-99.
Robert. *Abergavenny.* 1830-44.
Thomas. *London.* a.1686 to Alexander Warfield, CC. 1695-ca.1730. W.
Thomas Charles. *Abergavenny.* 1835-68. C.
W. *Gloucester.* 1856.
William. *Catrine* (*Scotland*). 1850.
William. *London.* 1839-44.
William (& Son). *Ayr.* 1836 (1850-60).
INKPEN, John. *Horsham.* 1726-45. C.
INKSTER, Henry. *Stromness.* 1836.
INMAN—
James. *Colne.* 1824.
James. *Stockport.* 1834-48.
John A. *Liverpool.* 1859. W.
Sophia. *Liverpool.* 1851.
William. *Liverpool.* 1851.
William. *London.* 1869.

INNES—
David. *London.* 1881. (To the trade.)
George. *Glasgow.* 1828-41.
William. *Quebec.* 1848-64.
INNOCENT, Robert. *London.* 1832.
INSKEEP, Joseph. *Philadelphia, USA.* ca.1800. C.
INSKIP—
Alfred T. *Shefford (Bedfordshire).* 1877.
H. *Shefford.* 1854-69.
Thomas. *Shefford.* 1830-47.
INTERNATIONAL Time Recorders. *New York.* 1936.
INVERSIN, Louis. *London.* 1869.
IONES (see JONES).
IORNS, Thomas. *Alcester.* 1828-68.
IREDALE, Charles. *London.* 1863-81.
IRELAND—
Henry. *Jersey.* 1832.
Henry. *London.* 1653, CC. 1654-63 (B. -1675). *Lothbury.* Lant. clock.
William. *Prescot.* mar.1781. W.
IRESON, Oakley. *Nassington (Northants).* 1877.
IRISH—
A. *Gosport.* 1848.
James. *Brighton.* 1792. W. and silversmith.
John. *Plymouth.* ca.1690.
Michael. *Lewes.* 1803-15. W.
IRVIN (or IRWIN), Henry. *London.* 1844-51.
IRVINE, John Watt. Place unknown. 1788. C. (May be John WATT of Irvine?.)
IRVING (also ERVIN), Alexander. *London.* a.1688 to John Wells, CC. 1695-97. C.
IRWIN—
Edward. *?Co. Down.* Early 19c. C.
Henry. *London.* 1844-51.
John. *Rathkeale.* 1858.
William. *Dublin.* 1802. W.
ISAAC—
A. W. *West Hartlepool.* 1898.
Abraham. *Newport (Monmouth).* 1868.
Brothers. *South Shields.* 1856.
Henry Patrick. *London.* 1869-75.
James. *Carmarthen.* 1810-d.1820. C.
Richard. *Carmarthen.* ca.1800. (See ISSAC.)
Richard James. *Llanelly.* 1830-50.
Robert. *Llanelly.* Second half 18c.
William. *Llanelly.* 1835-75. C.
ISAACS—
Abraham. *Newport (Monmouth).* 1868.
& Co. *Abergavenny.* 1850-58. Also pawnbrokers.
E. *Leicester.* 1849-64.
& HARRIS. *Abergavenny.* 1868-80.
Isaac. *London.* a.1756-1768.
James. *Braintree.* a.1753 to Thomas Fordham.
John. *Chatham.* 1839-47.
John. *London.* 1828.
John. *London.* 1851.
Lewis. *London.* 1828-51.
Michael. *London.* 1832.
Ralph. *Liverpool.* (B. 1814-)1822-34.
Solomon. *Liverpool.* 1828-34.
ISELE, Bernard. *Sheffield.* 1862. C.
ISHERWOOD—
J. *Bolton.* 1851-58.
W. *Bolton.* 1858.

ISMAY, John. *Oulton* nr *Wigton.* b.1668, mar.1698 (-d.1706?). C.
ISRAEL, Moses. *London.* a.1749.
ISSAC, Richard. *Carmarthen.* Early 19c. C. ca.1800. (?Error for ISAAC.)
ISSOT, John Clough. *Beverley.* 1834-51.
ITHACA—
Calendar Clock Co. *Ithaca, NY, USA.* 1865-1919. C.
Clock Co. *Toronto.* 1875.
IVE—
George Henry. *London.* 1839-63.
Robert. *Worthing.* 1851-70.
Trumplett. *London.* 1869.
William. *London.* 1832.
IVERS, Christopher. *Kilrush.* 1858.
IVERSON, Christian Henry. *London.* 1863-75.
IVES—
Amasa. *Bristol, USA.* 1777-1817.
Amasa & Chauncey. *Bristol, USA.* 1811.
BLAKESLEE & Co. *Bridgport, Conn, USA.* ca.1870-80. C.
(Shaylor) & BREWSTER (Elisha C.). *Bristol, USA.* 1840-43. C.
Brothers. (Ira, Amasa, Philo, Joseph, Shaylor and Chauncey). *Bristol, USA.* 1809-59.
C. (Chauncey) & L. (Lawson) C. *Bristol, USA.* 1830-38. C.
Charles Graneson. *Bristol, USA.* 1816-24.
Chauncey. *Bristol, USA.* 1787-1857. C.
Enos. *New York.* 1848.
Ira. *Bristol, USA.* 1775-1848.
James S. *Bristol, USA.* 1836.
John. *Bristol, USA.* 1822.
Joseph. *Bristol, USA.* 1782-1862. C.
Joseph & Co. *Bristol, USA.* 1821-22. C.
Joseph Shaylor. *Bristol, USA.* s. of Ira. 1887.
Lawson C. & Co. *Bristol, USA.* s. of Philo. 1839-43. C.
& LEWIS. *Bristol, USA.* 1819-23. C.
Philo. *Bristol, USA.* 1780-1822.
Porteus R. *Bristol, USA.* s. of Joseph. 1805-1867.
Rollin. *Bristol, USA.* s. of Porteus. 1861-65.
Shaylor. *Bristol, USA.* 1785-ca.1840. C.
T. *Colchester.* 1866.
Zacharia. *London.* CC. 1682-97. C.
IVESON—
Richard. *Kirkby Stephen.* 1828-69.
Richard, jun. *Kirkby Lonsdale.* 1879.
IVEY—
Henry. *St. Allen (Cornwall).* 1854.
Nicholas. *Redruth.* 1775-ca.1795. C.
Nicholas. *Truro.* 1791. W.
IVEY—
Thomas. *Merthyr Tydfil.* 1887.
William. *Marazion (Cornwall).* 1873.
IVISON—
——. Place not stated. Prob. *Yorks.* ca.1860. C.
Henry. *Liverpool.* 1834.
John. *Jeristown* (nr *Carlisle).* ca.1740. C.
John. *Carlisle.* (B. 1810-)1828-34.
IVORY, Charles. *London.* 1857-69.
IZOD, William. *London.* a.1641 to John Burges, CC. 1649.

J

J, H, HI, trademark of Henri Jacot, q.v.
JACCARD—
 David (& Son). *London*. (B. 1820-) 1828-69(-81).
 Samuel. *London*. 1828, repeating watches.
JACK, Miss Susan. *Toronto*. 1877.
JACKMAN—
 A. *Faversham*. 1866.
 Alfred & Son. *Bath*. 1883.
 Charles. *Fareham*. 1867-87.
 F. *Hambledon (Hampshire)*. 1867.
 Frederick. *Emsworth (Hampshire)*. 1878.
 G. *Bishops Waltham*. 1848.
 G. *Havant*. 1859.
 George. *Tunbridge Wells*. 1855-74.
 Joseph. *London*. 1683. Lant. clocks. (B. 1708-16. W.).
 Morris. *Newport (Isle of Wight)*. 1859-78.
 William. *Fareham*. 1830-59.
JACKS—
 (James) & GIBSON. *Charleston, SC, USA*. ca.1780-
  dissolved 1800.
 James. *Jaimaica, West Indies*. ca.1777; *Charleston, SC,
  USA*. 1784-97; *Philadelphia* 1797-99; *Charleston*
  1822.
JACKSON—
 Abraham. *Liverpool*. 1814-34.
 Alfred T. *Merthyr Tydfil*. 1868-99.
 Alfred T. *Swansea*. 1899.
 Allan. *Lochgilphead*. 1836-60.
 & BRAHAM. *Bristol*. 1840.
 C. *Burton-on-Trent*. 1868.
 C. *Earl Shilton*. 1855.
 C. *Measham* nr *Atherstone*. 1849-64.
 Charles. *Bristol*. 1842.
 Charles. *Dunchurch*. 1835.
 Charles. *Kenilworth*. 1842-80.
 Charles. *Schenectady, NY, USA*. 1816.
 Charles John. *Birmingham*. 1880.
 Charles John. *Saltley (Warwickshire)*. 1880.
 David. *Hull*. 1846.
 & DEACON. *Hinckley*. 1876.
 E. *Wisbech*. 1865.
 Edward. *Great Marlow*. 1823.
 Edward. *London*. a.1672, CC. 1680-97.
 Edward. *York*. b.1786-d.1859.
 Edwin W. *Coventry*. 1880.
 F. L. *Middlesbrough*. 1898.
 Francis. *Cork*. 1856-58.
 Frederick. *Liverpool*. 1848-51.
 G. *Battersea*. 1878.
 George. *Dolgellau*. 1805.
 George. *Gainsborough*. 1828.
 George. *Lancaster*. 1839-58.
 George. *Llanrwst*. Prob. same man as at Dolgellau.
 George, jun. *East Marlborough, Pa, USA*. ca.1778-
  1836. C.
 Henry. *Newcastle-on-Tyne*. 1836-48.
 Henry. *Toronto*. 1837-65.
 Henry & Son. *London*. 1839-44.
 Henry & William. *London*. 1832. Watch cases.

JACKSON—*continued*.
 Herbert. *Polesworth (Warwickshire)*. 1860-80.
 Isaac. *London Grove* and *New Garden, Pa, USA*.
  1734-1807. C.
 Isaac. *Wylam*. b.1796-d.1862.
 Israel. *Goole*. 1858-71.
 J. *Dublin*. 1797-1822. (Later jewellery only.)
 J. *Rochester*. 1866.
 J. *St. Helens*. 1858.
 J. H. *Tunstall (Staffordshire)*. 1868. C.
 J. J. *Hockley (Nottinghamshire)*. 1864.
 J. W. *St. Catharines, Canada*. 1861-64.
 James. *Accrington*. 1851.
 James. *London*. 1828-39. Watch cases.
 James. *Worthing*. 1828.
 James H. *Perth*. 1828-36.
 John. *Barrow-on-Humber*. 1835-76.
 John. *Barton-on-Humber* 1828.
 John. *Beaconsfield*. 1798-1838.
 John. *Boroughbridge*. 1822-37. C.
 John. *Brompton*. 1826-28.
 John. *Gravesend*. 1832.
 John. *Hanley (Staffordshire)*. 1876.
 John. *Henley-on-Thames*. 1786. W.
 John. *Killinchy* and *Barnathaghery*. Early 19c. C.
 John. *Kirkham*. 1834.
 John. *Lancaster*. (From *Shropshire*). b.1797-1869.
 John. *Liverpool*. 1799-1848 (May be two such?).
 John. *London*. 1839. Watch cases.
 John. *Marlborough, Pa, USA*. 1746-77.
 John. *Oakham*. 1846.
 John. *Nova Scotia*. 1778.
 John. *Scarborough*. 1846-58.
 John. *Stafford*. 1842-60.
 John. *York*. mar.1789. W.
 John, jun. *London*. 1828-32.
 John Smith. *London*. 1844.
 Jonathan. *Beaconsfield*. ca.1740. C.
 Joseph. *Bradford*. d.1716. C.
 Joseph. *Liverpool*. 1834.
 Joseph. *Nottingham*. 1876.
 Joseph. *Tunstall (Staffordshire)*. 1860.
 Joseph. *Warton (Lancashire)*. 1730. C.
 Joseph A. (born *England*). *Millcreek, Del, USA* 1790,
  then *Philadelphia* 1802-10. C.
 Joseph Maidens. *Spilsby*. 1850.
 Matthew, jun. *Hull*. 1838. C.
 Mrs. J. *Stafford*. 1868.
 N. *Hammersmith*. 1851-55.
 Owen. *Cranbrook*. ca.1760. C.
 Richard. *East Springfield, O, USA*. 1801-12. C.
 Richard. *Hexham*. From *York* 1762-d.1779. C.
 Richard. *Maidstone*. ca.1760. C.
 Samuel. *Lancaster*. 1851.
 T. *Coventry*. 1868.
 T. R. *Birkinhead*. 1865.
 Thomas. *Farnworth*. mar.1787. W.
 Thomas. *Nottingham*. 1876.
 Thomas. Born *England*. To *Portsmouth, NH, USA*,

123

JACKSON—*continued.*
   Kittery, ME, Boston and *Preston, Conn.* ca.1727-1806. C.
   Thomas G. *Rugby.* 1868-80.
   Thomas Maiden. *Horncastle.* 1849-50.
   W. *Birmingham.* 1854. Clock case maker.
   W. *Frodsham.* 1857-65.
   W. *Spilsby.* 1849.
   Walter. *Measham (Derbyshire).* 1876.
   William. *Cork.* 1846-58.
   William. *Easingwold.* 1790-mar.1809.
   William. *Frodsham.* 1828-48.
   William. *Liverpool.* 1848-51. C.
   William. *London.* 1839. Watch cases.
   William. *London.* 1832-81.
   William. *Loughborough.* ca.1730-ca.40. C.
   William. *Lutterworth.* Mid-18c. C.
   William. *Maidstone.* 1796-1823.
   William. *Northwich.* 1834. W.
   William. *Runcorn.* 1828.
   William & Henry. *London.* 1828. Watch cases.
   William Henry and Samuel. *London.* 1851-81.
   William Richard Radley. *Maidstone.* 1823-39.

JACOB—
   Aime. *Paris* and *St. Nicolas d'Aliermont.* 1855-72.
   Celestin. *Philadelphia, USA.* 1840. W.
   Charles. *Annapolis, MD, USA.* 1773-75; *Port Tobacco, Baltimore,* 1778.
   Charles W. *Colchester.* 1866-74.
   (Charles) & CLAUDE. *Annapolis, MD, USA.* 1775.
   Emile. *St. Roch, Quebec.* 1890.
   G. *Woodbridge.* 1846.
   Garrard. *Dedham.* 1866-74.
   Garrard. *Stradbrook.* 1839.
   George P. *Winchester.* 1848-78.
   Isaac. *South Shields.* 1851.
   John. *London.* Alien 1622. (Prob. an apprentice then.)
   Moses. *Redruth.* 1769-d.1807. C.
   Moses Jacob. *Falmouth.* 1813-60. C. & W.
   Paule. *London.* Alien 1622. (Prob. an apprentice then.)
   Robert. *Halstead.* 1839-51.
   Robert. *Sudbury.* 1830.
   Sarah. *Redruth.* wid. of Moses. post-1807.

JACOBS—
   A. J. & E. *Newport (Monmouth).* 1880.
   Alfred M. & Co. *London.* 1863-81.
   Augusta. *Swansea.* 1868.
   B. *Quebec.* 1852.
   Barnet. *Dukes Place, London.* Took Isaac Isaacs a.1756.
   Bethel. *Hull.* 1838-51.
   Edmund. *London.* 1832-39.
   Edward. *London.* 1828.
   Emanuel. *Hull.* 1834.
   Ephraim. *London.* (B. 1820-24) 1828-32.
   Frederick. *Sheffield.* 1862-71. C.
   G. *Long Melford.* 1853.
   Isaac. *Poole (Dorset).* 1824-30.
   Israel. *Scarborough.* 1834-51. W.
   Israel & Son. *Hull.* 1813-46.
   Jacob. *New Market, Canada.* 1853.
   Jacob. *Montreal.* 1819.
   James. *Bristol.* 1863.
   Joseph. *(London?).* a.1756 to D. Vauginon.
   Lazarus. *Swansea.* 1868.
   Levi. *Penzance.* 1823. W.
   Louis. *Swansea.* 1868.
   & LUCAS. *Hull.* 1851-58.
   Morris. *Newport (Monmouth).* 1880.
   S(olomon?). *Falmouth.* ca.1800, see also *Penzance.*
   Solomon. *Penzance.* d.pre-1803.

JACOBSON—
   Israel. *North Shields.* 1848.
   Nathan W. *London.* 1851.

JACOT—
   A. *Baltimore, USA.* 1842.
   Ami Francis. *London.* 1869.
   Descombes, jun. *London.* 1863.
   Emile. *Quebec.* 1863.
   H. L. & Co. *Montreal.* 1842-63.
   Henri. *Paris.* 1855-1900. C.
   & SANDOZ. *London.* 1857.

JACQUES (see also JAQUES)—
   Abraham. *Eamont Bridge.* ca.1753.
   Clock Co. (Charles). *New York.* ca.1880.
   John. *Wakefield.* ca.1760-1822. C.
   & Son. *Wakefield.* ca.1825-30. C.
   Thomas. From *Midhurst* to *Petersfield.* a.1740 to Daniel Tribs. C.

JAEGER Watch Co. *New York.* 20c.

JAFFREY—
   Alexander. *New Pitsligo (Scotland).* 1860.
   William. *Glasgow.* 1841-60.

JAGAR/JAGER—
   George. *Kingston-on-Thames.* 1839.
   George. *London.* 1844.
   Hannah. *Canterbury.* 1823 (May be JAGGER).

JAGEL (or JAGLE), Magnus. *London.* 1832-44.

JAGGER—
   Aaron. *Ashton-under-Lyne.* 1834.
   Richard. *Manchester.* 1848-51.
   Richard. *Oldham.* 1834.

JAMES—
   B. *Norwich.* 1846.
   Benjamin. *Brighton.* 1870.
   Benjamin. *London.* ca.1730. C.
   Benjamin. *Shaston.* ca.1760. C.
   Charles. *St. John, New Brunswick.* 1825.
   Daniel. *Llandovery.* mar.1841-44.
   David. *Aberavon.* 1865-75.
   David. *Briton Ferry (Wales).* 1875.
   David. *Talgarth.* 1875.
   Edward. *Farnham.* 1828-78.
   Edward. *Odiham.* ca.1800-ca.1810. C.
   Edward. *Philadelphia.* Late 18c. C.
   Evan. *Dolgellau.* ca.1750-60.
   F. W. *Tredegar.* 1850.
   Francis. *London.* a.1655 to Henry Ireland, CC. 1662.
   H. *Petersfield.* 1867.
   H. *Portsmouth.* 1859.
   Henry. *Godalming.* 1862-78.
   J. *Aldershot.* 1859.
   J. *Bury.* 1858.
   J. H. *Staines.* 1851-55.
   J. J. *Augusta, Ga, USA.* ca.1820.
   James. *Haverfordwest.* 1844.
   John. *Edinburgh.* 1846.
   John. *Egham.* 1828-62.
   John. *Hull.* 1840-46. C.
   John. *Liverpool.* 1848-51.
   John. *London.* a.1653, CC. 1661-62.
   John. *London.* 1869.
   John. *Newcastle Emlyn.* 1868-87. C.
   John. *Saffron Walden.* 1874.
   John Henry. *Egham.* 1866-78.
   John & Son. *Petersfield.* 1850.
   Joshua. *Boston, USA.* 1810-20.
   R. *Saxmundham.* 1865.
   Robert. *London.* 1832.
   Robert. *Shaftesbury (Dorset).* 1824-30.
   Samuel. *Stevenage.* 1859-74.
   Solomon C. *New Quay (Wales).* 1887.
   Stephen. *Bristol.* a.1813-30.
   Thomas. *Pen-y-Bont (Radnor).* 1875.
   Thomas. *Petersfield.* 1839-59.
   Thomas. *Pontypool.* 1880. Clock cleaner.
   Thomas. *Woolwich.* 1847.
   Thomas William. *Tredegar.* 1858-68.
   William. *Bath.* 1742-ca.1770. W.
   William. *Bishopsworth (nr Bristol).* 1866-83.
   William. *Brecon.* 1835.
   William. *Bristol.* a.1807-26.
   William. *Portsmouth, RI, USA.* Early 18c.
   William. *Rhymney.* 1844-48. C.
   William B. *Golspie (Scotland).* 1860.
   William Charles. *London.* 1881.

JAMESON—
   George. *Hamilton.* 1729. Sundials.
   Jacob. To *Dayton, O, USA.* 1823, then *Columbia, Pa, USA.* ca.1830.
   James. *Stranraer.* 1836.
   John. *Dublin.* 1858-80.
   Thomas. *London.* 1875-81.
   William. *Leigh.* 1851.

JAMIESON—
James. *Stranraer.* 1836.
Robert. *Glasgow.* 1838.
Thomas. *Ayr.* 1836-60.
JAMIN, Jean Baptiste. *Baltimore, USA.* ca.1800.
JAMISON—
Alexander. *Banbridge.* Early 19c. C.
George. *Cork.* 1813, from *London,* where one such 1786-1810.
George John. *Portsmouth.* 1830-59.
James. *Newtown Stewart.* ?Early 19c. C.
JANS, Josephus. *Passeau.* Early 17c. C.
JANTZ, Philip. *London.* 1875.
JAPY—
Freres & Co. *London.* 1881.
Louis. *Berne.* 1849.
Marti, Rous. *Paris.* 1863.
JAQUES (see also JACQUES)—
Augustus. *London.* 1844.
François. *London.* 1869.
John. *London.* 1832-63.
Thomas. *Loudon.* 1875-81.
Victor. *London.* 1828.
William. Place unknown (probably *Yorkshire).* 1825. Sundial.
JARDINE, Robert. *Bathgate (Scotland).* 1836-60.
JARMAN, John. *London.* 1811. W.
JARRAD, William A. *Southampton.* 1878.
JARRET, Sebastian. *Germantown, Pa, USA.* 1772. C.
JARREY, Stephen (Etienne). *London.* Journeyman 1655.
JARVIS—
George Weedon. *London.* 1875-81.
Henry. *Prescot.* 1795-1826. W.
John. b. ca. 1821 *Whitchurch (Shropshire).* mar. 1842-79. W.
John Jackson. *Boston, USA.* From *London* 1787. C.
Matthew. Place unknown. (?*Northern England).* ca.1730. C.
William. *Horncastle.* 1868-76.
JAVAN, Henry (& Son). *London.* 1844-57(-75).
JAY—
Alfred. *Chichester.* 1839.
Francis. *Horncastle.* 1876.
Joseph. *London.* 1857.
Thomas. *Norwich.* 1818. W.
JAYE—
Elizabeth. *Ipswich.* 1839.
John. *Ipswich.* 1830.
Mrs. E. *Ipswich.* 1858 (May be Elizabeth above.)
JEAN, Louis. *Ashby de la Zouche.* 1828.
JEANES, Thomas. *Philadelphia, USA.* 1835-37.
JEANNERET—
Henry William. *London.* 1869.
J. *London, Canada.* 1851.
Olirrer. *London.* 1828.
Theophilus H. *Philadelphia, USA.* 1818.
JEANNIN, Alphonse. *London.* 1839-44.
JEARY, J. *Blakeney.* 1846.
JEEVES, Anthony. *Edinburgh.* From *Oxford* 1744. C.
JEFFARD, ——. *Newton Abbot.* 1790. W.
JEFFERIES, also JEFFREYS, JEFFEREYS, etc.
Benjamin. *Leeds.* 1866-71.
Benjamin & Co. *Leeds.* 1871. Clock case makers.
CURTIS G. *Glasgow, Ky, USA.* 1785-1834. W.
Edward. *Andover.* ca.1750. C.
Edward. *Chatham.* s. of George. mar.1789.
Frederick. *London.* 1857.
G. F. *Camberwell.* 1866.
& HAMS. *London.* (B. 1810-25) 1832.
J. J. *Devizes.* 1867.
J. & Son. *Epworth.* 1861.
James. *Canterbury.* 1823-65.
James. *Wingham.* 1847-55.
James. *Witney.* 1823-53.
John. *Biggleswade.* ca.1830. C.
John. *Bristol.* (B. 1780-)1801. Partner with McCarthy.
John. *Stow on the Wold.* 1870.
John B. *London.* 1863-81.
John R. *Biggleswade.* 1830-77.
Joseph. *Leeds.* 1853-71. W.
M. *Birstwith.* ca.1830-ca.50. C.

JEFFERIES—*continued.*
Miss Elizabeth. *Maldon.* 1828-55.
Rev. George. *Franklintown, O, USA.* 1814. Clock cases.
Robert. *London.* 1851-57.
Samuel. *Farnham.* 1862-78.
Samuel Fuller. *Philadelphia, USA.* 1759-76. Later to *England.* W.
Thomas. *Liverpool.* ca.1815. W.
William. *Liverpool.* b.1809-36. Then *Lancaster* ca.1836-58. W.
William. *London.* 1881.
William. *Maldon.* From *Witham.* a.1774-1807. W.
William. *Scarborough.* 1864. C.
William R. *Ashford.* 1874.
JEFFERSON—
E. & J. *Hull.* 1858.
John. *Driffield.* 1823-58.
John. *Pocklington.* 1834.
Matthew. *London.* 1832-44.
Samuel. *London.* (B. 1809-) 1828-51.
Samuel Henry. *London.* 1857.
JEFFORD, Robert. *Cork.* 1783.
JEFFS—
John. *Oxford.* 1640-50. T.C.
John Thomas. *Luton.* 1877.
Thomas. *Birmingham.* 1880.
W., jun. *Cheltenham.* 1863.
JEKYLL, John. *Hedon.* 1851.
JELLINGS, F. & A. *Stockton.* 1898.
JELL(E)Y—
Charles. *Coventry.* b.1786-1828. W.
James C. *London.* 1869.
Jonathan. *Coventry.* b.1784-1830. W.
Richard. *Coventry.* b.1796-1834. W.
William. *Coventry.* b.1782-1832. W.
JEMMETT, Samuel. *Battle.* From *Eastbourne.* a.1711 to Samuel Hammond. C.
JENCKS, John E. *Providence, RI, USA.* ca.1800.
JENKINS—
Daniel. *Cardigan.* 1840-44.
Daniel. *Newcastle Emlyn.* 1835. C.
David Owen. *London.* 1857-81.
Evan. *Crickhowell.* 1835.
Frank. *Bristol.* 1879.
George, jun. *Hythe.* 1845.
& GREEN. *London.* 1851.
H. T. *Christchurch.* Fuzee and chain makers.
Harry. *Bristol.* 1879.
Henry. *Portishead.* 1883.
Henry, jun. *Christchurch.* 1830. Fuzee, chain and hook makers.
Henry T. *Christchurch.* 1848-67.
I. & H. *Albany, NY, USA.* 1815.
Ira. *Albany, NY, USA.* 1813.
Isaac. *Tredegar.* 1844-50.
J. & JEARS. *Christchurch.* 1839. Fuzee, chain and hook makers.
James. *Mold.* 1887-90.
John. *Cardigan.* 1835.
John. *Swansea.* 1830-52. C. & chrons.
John. *Tenby.* 1850-71.
John Bartholomew. *Newcastle Emlyn* pre-1832, then *Crickhowell* 1832-75.
Newth. *Cork* 1763, then *Bandon* 1787.
Osmore. *Lowell, Mass, USA.* 1837. W.
Thomas. *Ballymacarrett.* 1835.
Thomas. *Dowlais.* 1835-52.
Thomas Hopwood. *Swansea* 1840, then *Llanelly* 1844, then *Carmarthen* 1847-87. C.
Thomas Jenkin. *Dowlais* 1844, then *Merthyr Tydfil* 1865-87. W.
W. *Richmond, Ind, USA.* ca.1820.
William. *Bandon.* ca.1820-ca.30. W.
William. *Bristol.* 1863-79.
William. *Cardigan.* 1887.
William. *London.* 1863-81.
William. *Narberth.* 1830.
JENKINSON—
Edward. *Bawtry.* 1834-71.
Edward. *Blyth.* 1828.
Edward. *Nottingham.* mar.1718. W.

JENKINSON—continued.
Edward. Swinton nr Rotherham. 1871.
John. East Retford. 1855-76.
Thomas. Sandwich. b.1696, free 1719-d.1755. C. & W.
JENNE & ANDERSONS. Big Rapids, Mich, USA. ca.1860.
JENNENS—
John Creed (& Son). London. 1863-75(-81).
Joseph. London. 1881.
JENNER & KNEWSTUB. London. 1881.
JENNERER, Charles F. St. Louis, Mo, USA. ca.1820.
JENNINGS—
Charles. Idle. 1871.
Clement Heeley. Swansea. 1887.
Francis Thomas. London. 1869.
Hugh. Rathfriland 1843-54, Castlewillan 1886-d.1888. C. (In army in India 1854-86.)
John William. Birmingham. 1842-68.
Mountifort. Dublin. 1858-80.
Robert. Fritwell (Oxfordshire). b.1722-d.1773. W.
William. Colchester. b.1797-1850. W.
William. Fritwell. bro. of Thomas. b.1716-d.1780. W. & C.
JENNIS, F. S. Barnstead Parade, NH, USA. 1884.
JENSEN, Frederick & Son. New York. 20c.
JEPHCOTT—
T. Coventry. 1860.
Thomas. Llandudno. 1874-87.
William. Birmingham. 1868-80.
JEPSON, William. Boston, USA. 1830.
JERGER, William. London. 1881.
JERGLE, Jonathan. Birmingham. 1835.
JEROME—
Andrew. New Haven, Conn, USA. ca.1850.
& BARNES. Bristol, USA. 1833-37. C.
C(Chauncey). Philadelphia, USA. 1846-49.
C(Chauncey) & N(Noble). Bristol, USA. 1834-39.
Charles. New Haven, Conn, USA. ca.1850.
Chauncey. Morristown, NJ, USA. ca.1812.
Chauncey. b.1793, worked Plymouth (USA) 1816-22, Bristol 1822-45, New Haven 1845-55, d.1868.
Chauncey & Noble. Richmond, Va, USA and Hamburg, SC, USA. 1835-37.
& Co. Liverpool. 1851, branch of American Co.
& Co. New Haven, Conn, USA. ca.1850. C.
& Co. Philadelphia, USA. 1852. C.
& DARROW, Bristol, USA. 1824-33.
DARROW & Co. Bristol, USA. ca.1825.
GILBERT (W. L.) & GRANT (Zelotes) & Co. Bristol, USA. 1839. C.
& GRANT. Bristol, USA. 1842-43.
Manufacturing Co. Boston, USA. 1852-54.
Manufacturing Co. New Haven, Conn, USA. 1845-55. C.
Noble. bro. of Chauncey. Bristol, USA. 1824-49.
S. B. & Co. New Haven, Conn, USA. 1856-78. C.
THOMPSON & Co. Bristol, Conn, USA. 1827. C.
JESSOP(S)—
George. Manchester. 1851.
& JACKSON. Tavistock. ca.1820-ca.40. C.
JESSUP—
Jonathan. York, Pa, USA. ca.1787-ca.1850.
Joseph U. York, Pa, USA. s. of Jonathan. ca.1820-50.
JEUCHNER, Julius. London. 1869-81.
JEWEL(L)—
——. 1914 & later, product of Pequegnat Clock Co.
Jerome & Co. Bristol, USA. 1847-49.
(Lyman), MATTHEWS (Daniel) & Co. Bristol, USA. 1847-53.
Moss. London. 1828.
W. St. Columb. ca.1820. C.
& WARNER. Bristol, USA. 1846.
JEWER, Stephen. Farnham. ?Early 18c. Lant. clock.
JEWETT—
Amos. New Lebanon, NY, USA. 1753-1834. C.
Augustine. Newburyport, Mass, USA. 1860.
JEX—
Johnson. Letheringsett. 1779-1852.
Johnson. Norwich. 1811-20. W.
JEYES, James. Northampton. 1745-53. Prob. from London.
JEZEPH, James. London. 1828.

JIBB—
Joseph. Whitchurch (Hampshire). 1830.
Joseph. Whithorn (Scotland). 1860.
JILLARD, Brothers. Harbor Grace, Newfoundland. 1871-85.
JOB—
Frederick. Birmingham. 1868-80.
Frederick. London. 1839-57.
Henry. London. 1857-63.
Henry William. London. 1844.
James H. Hastings. 1878.
John Pentecost. Truro. 1805-d.1821. W.
Joseph. Hastings. 1828-70.
Joseph H. St. Leonards. 1839-70.
R. H. St. Leonards. 1878.
Robert. London. (B. 1804-08) 1828-44.
JOBSON, Frederick William. Marden (Kent). 1866-74.
JOCELIN, Simeon. New Haven. 1746-1823. C.
JOCELYN—
Albert Higley. New Haven, Conn, USA. s. of Nathan. 1827-1900. Wood engraver.
Nathaniel. New Haven, Conn, USA. bro. of Simeon Smith, 1796-1891.
Simeon Smith. New Haven, Conn, USA. bro. of Nathan. 1825-79. Engraver.
JOEL—
Isaac. Preston. 1824.
Jacob. Manchester. 1848-51.
L. Coventry. 1868.
Son & DEAL. Coventry. 1880.
T. Newcastle-on-Tyne. 1844.
JOHAN(N)SEN—
Asmus. London. 1863-69.
Hans C. London. 1869.
JOHN, Lewis. Michaelston (Wales). d.1784.
JOHNS—
John William. Falmouth. 1856-73.
Richard. Exeter. Free 1628. C.
Thomas. Cerne Abbas. 1824-48.
William. Newbridge (Pontypridd). Early 19c.
JOHNSON—
——. Colchester. 1836. W.
——. Newtown Limavady. 1829. W.
——. St. Andrews, Canada. 1853.
Aaron Wood. Newcastle-on-Tyne. 1848.
Alwyne E. Ryde. 1878.
Andrew. Boston, USA. ca.1840.
B. Greenwich. 1851.
Benjamin. Batley. 1866-71.
Benjamin. Morley. 1866.
Brothers. Sanbornton, NH, USA. Second quarter 19c.
Brothers. Toronto. 1877.
Charles. Darlington. 1866.
Charles. Peckham. 1878.
Charles. Stratford (Essex). 1851-57.
Charles. West Ham. 1866-74.
Charles. Wigan. 1848-58.
Charles F. Owego, NY, USA. 1846. T.C.
Chauncey. Albany, NY, USA. 1824-41. C.
Christopher. Knaresborough. 1759-1822. C. & W.
Christopher. York. 1771. C.
& Co. Bolton. 1824.
& CROWLEY. Philadelphia, USA. 1830-33.
D. Three Rivers, Canada. 1853.
Daniel B. Utica, NY, USA. ca.1834-43.
David. Industry Village, Canada. 1857.
David. London. 1832.
David. Sheffield. 1834-62. W.
E. Hinkley. 1849.
E. D. & Co. London. 1844.
Edward. Derby. 1855-76.
Edward Daniel (& Son). London. 1857-75(-81).
Edwin. Winchcombe. 1850-79.
Eli. Boston, USA. 1821.
Elisha. Greensboro, NC, USA. 1841.
F. Belmont, NH, USA. 1884.
F. T. Dublin. 1820-31.
Frederick. Lincoln. 1849-50.
Frederick. London. 1875-81.
G. C. Lincoln. 1861.
George. Grantham. 1828.

JOHNSON—*continued.*
George. *London.* a.1641 to John Nicasius. CC. 1649 (-1664?).
George. *London.* 1839.
George. *Settle.* mid-18c. C.
George & William. *London.* 1832.
& GRANT. *Carlisle.* 1869.
H. J. *Devizes.* 1875.
Hannah & Mary. *Knaresborough.* 1834-37.
Henry. *Cowbridge.* 1852-87. C.
Henry. *Salford.* Early 19c. W.
Isaac. *Yarmouth.* ca.1740-50. C.
Israel H. *Easton, Md, USA.* 1793.
J. *Bristol.* 1856.
J. *Clifton (Gloucester).* 1840.
J. *Ecton (Northants).* 1847.
J. *Halesworth.* 1846.
J. B. *Middlesbrough.* 1898.
J. J. *New York.* ca.1860.
Jabez. *Charleston, SC, USA.* 1795.
James. *Kingston, Canada.* 1857-65.
James. *London.* 1828-57.
James. *Montreal.* 1849.
James. *Prescot.* 1823-25.
James. *Whitby, Canada.* 1862.
James & Son. *London.* 1851.
Jeremiah. *Deptford.* a.1770-1808.
John. *Ayr.* 1819.
John. *Halesworth.* 1839.
John. *London.* a.1673 to Barlow Rookes. CC. 1680-97.
John. *Manchester.* mar.1792. C.
John. *Rothbury.* ca.1770. C.
John. *Swansea.* 1835.
John. *Sydenham.* 1847.
Joseph. *Liverpool.* 1814-51.
Leon. *London.* 1828.
Leonard. *Newtown Limmervady.* Early 19c. C.
& LEWIS. *Philadelphia, USA.* 1837-42.
M. B. *Watertown, NY, USA.* 1838.
Mary. *Knaresborough.* 1844.
Michael. *Barnard Castle.* b.1782-d.1843. C.
Michael. *Yellow Creek, O, USA.* 1816. C.
Nathaniel. *London.* 1857-81.
Owen. *Manchester.* mar.1803. C.
Mrs. R. *Halesowen.* 1876.
P. *London.* 1783. W.
Peter. *London.* 1857.
R. L. *Eastbourne.* 1878.
R. S. & R. D. *Sanbornton, NH, USA.* 1844.
R. & T. *Darlington.* 1898.
Richard. *London.* 1832.
Richard. *Oldham.* 1848-58.
Richard. *Ripon.* b.1759-d.1844. s. of Christopher.
Richard. *Sheffield.* 1871.
Richard. *St. Helens.* mar.1751-72. W.
Richard James. *Watton (Norfolk).* 1875.
Robert. *Darlington.* 1851-66.
Robert. *Linlithgow.* 1835.
Robert. *Philadelphia, USA.* 1832-50.
S. *Harrow.* 1851-55.
Samuel. *Liverpool.* 1848.
& Son. *Darlington.* 1834.
T. *Coventry.* 1854.
T. C. & Sons. *Halifax, Nova Scotia.* 1874 to present.
Thomas. *Derby.* 1835-64.
Thomas. *London.* 1828.
Thomas. *St. Helens.* 1763-70. W.
Thomas C. *Halifax, Nova Scotia.* 1853-1923.
Thomas & Isaac. *Prescot.* 1834.
W. *Clapham.* 1862.
W. *High Wycombe.* 1847-54.
WALKER & TOLHURST. *London.* 1869-81. Wholesalers.
William. *Bristol.* a.1792-1812.
(Alias ROBERTS). William. *Flixton (Lancashire).* See Roberts.
William. *High Wycombe.* 1842-54.
William. *Kingston.* 1878.
William. *Leighton Buzzard.* 1839-69.
William. *London.* 1832-81.
William. *Montreal.* 1842.
William. *Pwllheli.* 1835.

JOHNSON—*continued.*
William. *Winchcombe.* 1830-42.
William. *York.* b.1642-1713. W.
William H. *London.* 1875-81.
William S. *New York.* 1841-61. C.
William Henry. *Llantrisant.* 1871-87.
JOHNSTON(E)—
——. *Kilrea.* ca.1800? C.
A. (Arthur) & W. *Hagerstown, Md, USA.* ca.1785-1815. C.
Adam. *Downpatrick.* ca.1775-95. C.
Arthur. *Hagerstown, Md, USA.* 1785-ca.1820. C. (Later bank director, 1837.)
David. *Arbroath.* Late 18c. C.
David Graeme. *London.* 1839.
Edmond. *Dublin.* 1874-80.
Edward (Neddy). *Ballymoney.* 1820-30. C. & W.
& FISCHER. *Hagerstown, Md, USA.* ca.1846.
George. *Comber.* Early 19c. C.
Hugh. *Barrhead.* 1836.
J. H. & Co. *New York.* ca.1820.
J. P. *Toronto.* 1861-63.
James. *Ballinderry.* Early 19c. C.
James. *Kingston, Canada.* 1857-65.
James. *Linlithgow.* 1830.
James. *Portsoy (Scotland).* ca.1825-37.
John. *Carlisle.* 1879.
John. *London.* 1828-32.
John. *Peterhead.* 1837.
John A. From *Belfast.* d.1841. At *Brandon, USA.*
John F. *Dublin.* 1820-27. W. and jeweller.
Matthew. *Edinburgh.* 1820.
& MELHORNE. *Hagerstown* and *Boonsboro, Md, USA.* ca.1785-1818.
Mrs. H. *London.* 1844-81. (Wid. of David Graeme?)
Peter. *Mitcham.* 1828-51.
R. *Bangor (Co Down).* 1785-95. C.
Richard. *Alston.* 1873.
Richard. *Carlisle.* 1879.
Robert. From *Downpatrick.* b. 1769. *Ballymoney.* ca.1795-d.1825. C.
Robert. *Cincinnati, O, USA.* 1851.
Robert. *Comber.* 1785-1810. C.
Robert. *Crumlin.* 1868-80.
Robert. *Lurgan.* ?early 19c. C.
Samuel. *Langholm (Scotland).* 1837.
Thomas. *Belfast.* ca.1755-57. W.
Thomas. *Belfast.* From *Ballyclare.* 1819. W.
Thomas. *Comber.* ca.1800. C.
Thomas. *Dublin.* 1765-1806. W.
William. *Glasgow.* 1847.
JOICE, Samuel. *Minehead.* ca.1790. C.
JOLIE, Julien. *Paris.* ca.1760. C.
JOLLIFFE—
Charles. *Birmingham.* 1868.
J. *Portsea.* 1848.
J. *Southsea.* 1848-59.
James L. *Southsea.* 1867-78.
Josiah. *Landport.* 1867-78.
W. & Son. *Northampton.* 1854-69.
William. *Northampton.* 1841-47.
JOLL(E)Y—
Henry. *Alfreton.* 1876.
J. W. *Colchester.* 1866.
John. *Belfast.* 1657.
Richard. *London.* 1832.
T. *Chesterfield.* 1864.
Thomas. *Coventry.* d.1836. W. bro. of William of *Leicester.*
Thomas. *Loughborough.* 1835-55. C.
William Higginson. *Mansfield.* 1853-94. W.
William. *Leicester.* 1828-49. W. bro. of Thomas of *Coventry.*
William. *Rugby.* 1842.
JONAS—
Bros. *London* and *Geneva.* 1875.
Isaac Aaron. *Liverpool.* 1848.
Joseph. From *England.* To *Cincinnati, O, USA.* 1817-25.
Joseph. *Philadelphia.* 1817.
JONCKHEERE, Francis. *Baltimore, Md, USA.* 1807-1824.

JONES—

——. *Denbigh.* 1791.
——. *Denbigh.* 1876. C.
A. *Grantham.* 1868.
Abner. *Weare, NH, USA.* 1780. C.
Alexander. *Great Marlow.* 1726. C.
Anthony Charles. *London.* 1875-81. Clock cases.
& ARNELL. *London.* 1857.
Arthur Edward. *London.* 1881.
& ASPINWALL. *Boston, USA.* 1809.
BALL & CO. *Boston, USA.* 1840-46.
BALL & POOR. *Boston, USA.* See SHREVE, CRUMP & LOW.
& BAMFORD. *Neath.* ca.1776. C.
Benjamin. *Aberavon.* 1865-87.
Benjamin. *Yorkville, Canada.* 1861-65.
Charles. *London.* 1875. Clock cases.
Charles F. *Whitchurch.* 1878.
Christopher. *Coventry.* 1850-54. Watch case maker.
& Co. (George A.). *Bristol* and *New York, USA.* 1870-72. C.
D. *Derby.* 1849-55.
D. W. *Neath.* Early 19c.
Daniel. *Aberavon.* 1844.
Daniel. *Amlwch.* 1874-87. W. & C.
Daniel. *Neath.* 1791-1811.
Daniel. *Steubenville, O, USA.* 1808.
David. *Aberavon.* 1868.
David. *Aberystwyth.* 1816.
David. *Bala.* 1840-76.
David. *Bangor.* 1835.
David. *Bethesda.* 1844-87. C.
David. *Brighton.* 1870.
David. *Caeo (Wales).* ca.1778. C.
David. *Cowbridge (Wales).* 1830-35.
David. *Gwynfe.* 1799.
David. *Lampeter.* 1831-67.
David. *Llandovery.* 1830-35.
David. *Llandysul.* 1822-d.1846.
David. *Llandysul.* 1875-87.
David. *Llangefni.* 1890. W.
David. *Llannerch-y-medd.* 1856-68.
David. *London.* 1851-81.
David. *Merthyr Tydfil.* b.1763-d.1842. C. & W.
David, jun. *Merthyr Tydfil.* 1835-52.
David. *Mountain Ash.* 1868.
David. *Narberth.* 1830-50.
David. *Newtown.* 1856-90. C.
David. *Pentre Rhondda.* 1875.
David. *Rhydowen.* Prob. same man as at *Lampeter.*
David. *Ruthin.* 1822-44. C.
David Price. *Denbigh.* 1887-90.
David & Andrew. *Pwllheli.* 1887.
Edward. *Bristol.* 1830. Chrons.
Edward. *Holywell.* 1822-74. C.
Edward. *London.* 1851-81.
Edward. *Welshpool.* 1728.
Edward K. *Bristol, USA.* 1837.
Eli. *London.* 1863.
Elias E. *Bethesda.* 1868-90. C.
Elias E. *Bryngwran.* Prob. same man as at *Bethesda.*
Erasmus. *Aberystwyth.* 1868.
Evan. *Abergele.* 1856. C.
Evan. *Dolgellau.* 1868-74. C.
Evan. *Llandovery.* 1840-44.
Evan. *London. Fleet Street.* 1646. CC. 1647-(B. 1690).
Evan. *Machynlleth.* 1835.
Evan. *Mold.* 1874.
Evan. *Shrewsbury.* 1879.
F. *Bristol.* 1863.
& FORRESTER. *Hull.* 1813-23. W.
Francis. *Bristol.* 1850-56.
Francis. *London.* (B. CC. 1825.) 1832.
Francis. *Mold.* 1874-87.
Frank. *Presteign.* 1887.
Frank Albert. *London.* 1881.
Frederick (I?). *Bristol.* 1842.
Frederick (II?). *Bristol.* 1870-79.
& FRISBEE. *New Hartford, Conn, USA.* ca.1830-40. C.
Garwood. *Beccles.* 1823. mar. 1830, ret'd. 1860
George. *London.* 1875-81.
George James. *London.* 1857-63.

JONES—continued.

Griffith. *Caernarvon.* 1874-87. C.
Griffith. *Ruthin.* Prob. same man as at *Caernarvon.*
Griffith D. *Baltimore, USA.* 1824-27.
H. S. *Leicester.* 1864.
Harlow. *Canandaigua, NY, USA.* 1811.
Hart. *King's Lynn.* 1830-36.
Henry. *Bristol.* 1850.
Henry. *Clonmel.* 1858.
Henry. *London.* 1869-81.
Henry W. *Llandysul.* 1819-80. C. and poet.
Herbert. *Newtown (Mon).* 1887.
Hugh. *Bangor (Wales).* 1835-44.
Hugh. *Caernarvon.* 1895.
Hugh. *Conway.* 1868-74.
Hugh. *Corwen.* d.1832.
Hugh. *Machynlleth.* Late 18c. C.
Humphrey. b. ca. 1762. *Oswestry.* 1801 -d. 1848.
J. *London.* 1857.
J. *Northampton.* 1847.
J. *Reading.* 1864.
J. *Warboys.* 1847.
J. B. *Boston, USA.* 1822.
J. D. *Pwllheli.* 1874. W.
Jacob. *Baltimore, USA.* 1817.
Jacob. *Pittsfield, NH, USA.* ca.1800. C.
James. *Birmingham.* 1860-68.
James. *London.* 1806-11. W.
James Horatio. *Kingston-on-Thames.* 1878.
Jenkin. *Glyn Nedd.* 1795-1883.
Johannes. Place not stated. Same as John Jones of *Abergavenny* who d.1789.
John. *Abergele.* 1856. W.
John. *Aberystwyth.* 1825-d.1877. W. & C. Prob. more than one.
John. *Barmouth.* 1868.
John. *Bethesda.* s. of Thomas of *Denbigh.* 1803-79. W. & C.
John. *Birmingham.* 1880.
John. *Caernarvon.* 1835-44. C.
John. *Carmarthen.* 1844. C.
John. *Chalfont.* 1783. W.
John. *Chester.* 1828-34.
John. *Colwyn Bay.* 1874.
John. *Coventry.* 1880.
John. *Dolgellau.* 1840-68. C.
John. *Four Crosses nr Pwllheli.* 1856.
John. *Kington (Hereford).* 1835.
John. *Liverpool.* Several such 1822-51.
John. *Llandudno.* 1874-90.
John. *Llanertgoyland.* Same as *Llannerch-coedlan.* q.v.
John. *Llanidloes.* 1835-44.
John. *London.* 1851-81.
John. *Manchester.* mar.1796-1834. W.
John (& Son). *Mold.* 1840-56(-68).
John. *Narberth.* 1871-75.
John. *Oxford.* 1662-87. Later *Llandaff.* Invented a clock driven by bellows.
John. *Paris.* ca.1850.
John. *Philadelphia.* 1772.
John. *Prescot.* mar.1771. W.
John. *Pwllheli.* 1856-78. W. & C.
John. *Ruthin.* ?18c. C.
John. *St. Clears (Wales).* 1835.
John. *Tenby.* 1830-40. W.
John. *Wrexham.* 1878-82. W.
John A. *Glyn Neath.* 1875.
John Edward. b. 1817 *Liverpool.* To *Aldborough (E. Riding).* by 1851-72. W.
John Emlyn. From *Crickhowell.* 1820-73.
John Parry. *Aberystwyth.* 1868-87. C.
John Richard. *Ramoth (Wales).* 1765-1822.
John & Thomas. *Conway.* 1868.
John William. *Tredegar.* 1850.
Jonathan. *London.* 1863-69.
Joseph. *Coleford.* 1830-42.
Joseph. b. ca. 1816 *Berriew (Montgomery). Oswestry.* 1854-85.
Lewis. *Tregaron.* 1868.
LOWS & BALL. *Boston, USA.* ca.1835-39.
Luther. *Llanstephan.* 1875.
Matthew. *Coleford.* 1863-79.
Morres. *St. Helens.* 1763. W.
Mrs. Jane. *Bethesda.* 1887.

JONES—*continued.*
Noel. *Hudson* or *Troy, NY, USA.* ca.1800. C.
& OLNEY. *Newark, NY, USA.* ca.1835.
& PARDMORE. *Troy, NY, USA.* ca.1887.
Peter. *Liverpool.* 1834.
Peter. *Prescot.* 1818-24. W.
Rice. *Monmouth.* Later 18c. C.
Richard. *Aberdare.* 1862-87.
Richard. *Aberystwyth.* b.1852-d.1872.
Richard. *Hereford.* 1830-35.
Richard. *Llanelly.* 1871-87.
Richard. *Llanidloes.* 1860-74.
Richard. *London.* 1875.
Richard. *Ton Pentre.* 1887.
Robert. *Amlwch.* ca.1830-56.
Robert. *Conway.* 1835-44.
Robert. *Dyffryn* (*Wales*). Later 19c.
Robert. *Gloverstone.* d.1747. W.
Robert. *Liverpool.* ca.1830-50. W.
Robert & Son. *Liverpool.* 1814.
Robert. *Pembroke.* 19c. C.
Robert. *Ruthin.* 1809-ca.1820. C. & W.
Robert Henry. *London.* 1875-81.
Robert Kent. *Birkenhead.* 1865-78.
Roland. *Utica, NY, USA.* 1837.
Samuel. *Aberystwyth.* 1851.
Samuel. *Bath.* 1756-92.
Samuel. *Llanidloes.* 1868-90.
Samuel. *London.* 1832.
Samuel. *Sevenoaks.* 1826-51.
Samuel. *Towyn.* 1844.
Samuel G. *Baltimore, USA.* 1815-29.
Samuel G. *Philadelphia.* 1799.
Sarah. *Bristol.* 1830.
SHREVE, BROWN & CO. *Boston, USA.* 1852.
Simon. *London.* 1857.
Simon. *Luton.* 1869-77.
Stephen. *Lampeter.* Early 19c. C.
T. *Coventry.* 1860. Clock cleaner.
T. *Derby.* 1855.
T. *West Bromwich.* 1868.
T. *Whitstable.* 1855.
T. W. *Tredegar.* 1852.
Thomas. *Bangor.* 1844.
Thomas. *Colwyn Bay.* ca.1830-ca.1850. W.
Thomas. *Conway.* 1874-87.
Thomas. *Denbigh.* Early 19c.
Thomas. *Dublin.* a.1738 to Thomas Blundell. C.
Thomas. *Liverpool.* 1848-51.
Thomas. *London.* 1844-81.
Thomas. *Prescot.* mar.1777-97. W.
Thomas. *Pwllheli.* 19c. W.
Thomas. *Swansea.* 1835.
Thomas Alfred. *London.* 1881.
Timothy. *London.* 1832-39.
W. *North Shields.* 1858.
& WALLY. *Liverpool.* 1788. W.
W. & Son. *Aberystwyth.* ca.1850. C.
William. *Abergele.* 1874-d.1895. W.
William. *Aberystwyth.* 1816-39.
William. *Birmingham.* 1816.
William. *Bristol.* a.1735-42. Then *Brecon.* 1754.
William. *Caernarvon.* 1886-95.
William. *Denbigh.* 1856.
William. *Flint.* 1874.
William. *Llanfyllin.* 1796-1844. C.
William. *London.* 1875-81.
William. *Ludlow.* 1828.
William. *Manchester.* 1824-48.
William. *Merthyr Tydfil.* 1830-48.
William, jun. *Merthyr Tydfil.* 1844-48. C.
William. *New York.* 1846. Chrons.
William. *Rhyl.* 1868-74.
William. *Ross.* 1830-56.
William. *Southampton.* 1839.
William. *Tredegar.* d.1841. W.
William. *Tredegar.* Same man as at *Merthyr Tydfil.* 1844-48.
William Gurner. *London.* 1851-81.
William H. *Charleston, SC, USA.* 1837-ca.1841.
William Lloyd. *Pontypool.* ca.1840. W.
William Rees. *Aberystwyth.* 1887.

JONES—*continued.*
William Williams. *Caernarvon.* 1887.
& WOOD. *Syracuse, NY, USA.* 1846.
JONSON & ABRAHAM. *Bodmin.* 1844. W.
JOPLING, Philip. *Chester-le-Street.* 1834. C.
JORDAN/JORDON—
A. *Montreal.* 1849.
Daniel. *Chesham.* 1842-77. Also gunsmith and bell-hanger.
Dominic. *Birmingham.* 1854-68.
James. *Chatham.* ca.1740. C.
James. *Stadhampton.* b.ca.1751-mar.1776. C.
John. *Manchester.* 1834-51.
Mrs. Elizabeth. *London.* 1839-44.
R. *Richmond, Va, USA.* ca.1820-30.
Roe. *Cork.* a.1735. d.1762. C.
Samuel. *Chesham.* 1877.
Thomas. *Blaenavon* (*Mon*). 1868.
Thomas. *Chesham.* mar.1804-39.
Thomas. *Leicester.* 1835.
Thomas. *Stadhampton.* 1770-90. C. & W.
William. *Chesham.* mar.1806. C.
William. *Rickmansworth.* 1828-51.
JOSCELYNE, LAMBERT & CO. *Canterbury.* 1832.
JOSEPH—
——. *London.* 1786. W.
——. *Ratcliffe.* ca.1740. C. (May be same as above)
Aaron. *Seaham Harbour.* 1856.
Abraham. *Falmouth.* 1827.
B. L. *Liverpool.* 1834-51. W.
Benjamin. *Swansea.* 1830-52. C.
Brothers. *Liverpool.* 1851.
Charles P. H. *Paris.* 1889-1900. Carr. clocks.
David. *Birmingham.* 1868.
David. *Bishopwearmouth.* 1851-56.
David. *Tregaron.* 1875.
Elias. *Liverpool.* 1790-1800. W.
Isaac. *Boston, USA.* 1823.
J. *Sunderland.* 1856.
J. *St. Austell.* ca.1820. C.
J. G. *Cincinnati, O, USA.* ca.1830.
J. G. & Co. *Toronto.* 1846-77.
Jacob. *Sunderland.* 1834. C.
Joseph. *Falmouth.* 1823. W.
Joseph. *London.* 1863.
Joseph. *Redruth.* 1823-47.
M. *Birmingham.* 1860.
M. *Bishopwearmouth.* 1856.
Moss. *London.* 1857.
Myers & Co. *London.* 1863.
Nesham. *Sunderland.* 1851-56.
JOSEPHS, ——. *Jersey.* 1834.
JOSLIN, Gilman. *Boston, USA.* 1855-76.
JOY—
Alfred. *Yarmouth.* 1830.
James. *Huddersfield.* 1871.
Julius. *Manchester.* (B. 1804-14) 1822.
Robert. *Elland.* 1866. W.
Robert. *Halifax* (*Greenland*). 1871.
Thomas. *Brighouse.* 1866-71. W.
JOYCE—
——. *Cockshill.* ca.1775. C.
——. *Tanderagee.* pre-1770. C.
C. *Hungerford.* 1864.
C. *Ramsbury* (*Wilts*). 1859-67.
Caleb. *Newbury.* 1877.
Henry. *Denbigh.* 1868-90. W.
J. *Bradford* (*Wilts*). 1848.
James. *Ruthin.* 1835-68.
James. *Whitchurch.* b. 1821, d. 1883.
Jeremiah. *London.* 1832.
John. *Exeter.* mar.1757. (Said to be *London* a.1754, but doubtful).
John. *London.* 1881.
John. *Newbury.* 1833-66. C.
John & Robert. *Ruthin.* 1835-44. C.
Robert G. *Ruthin.* 1856-87. W.
Samuel & Conway. *London.* 1832.
& Son. *Ruthin.* 1868.
Thomas. *Whitchurch.* b. 1793 s. of James Joyce, clockmaker. d. 1861. C.
Thomas Price. *Manchester.* 1856-70. W.

JOYCE—*continued.*
Walter C. *Ruthin.* 1874.
JOYNE, John. *London.* a.1660. CC. 1687-97.
JOZEAU, Jeans. *London.* 1863.
JUDD—
Henry G. *New York.* 1846.
Thomas William. *East Dereham.* 1875.
JUDEN—
Stephen. *London.* 1828.
William. *London.* 1832.
JUDGE—
C. *Toddington.* 1864-69.
Charles. *Luton.* 1877.
Mark & Son. *Potton.* 1869-77.
Matthews. *Luton.* 1864-77.
T. *Canterbury.* 1866.
T. *Hampstead.* 1851.
Thomas. *Hythe.* 1874.
Thomas. *London.* 1881.
W. *Ampthill.* 1864-69.
William. *Harpenden.* 1874.
JUKES—
John. *Birmingham.* 1839. Dial maker.
J(ohn) (jun.?). *Birmingham.* 1850-65. Dial maker.
Josiah. *Birmingham.* 1839-42. Dial maker.
JULER—
——. *North Walsham.* ca.1825. C. (Prob. George q.v.)
G. *Snettisham.* 1858.

JULER—*continued.*
George. *Burnham Market.* 1875.
George. *North Walsham.* 1830-58.
George. *Southwold.* 1879.
Matthew. *Norwich.* 18c. C.
Miles. *Yarmouth.* 1858-75.
William. *Burnham Market.* 1830-58.
JULES, ——. *Paris.* ca.1840. Carr. clocks.
JULIAN, Thomas. *Helston (Cornwall).* 1873. W.
JULLION, Francis. *London.* ca.1780. C.
JUMP—
I. & A. *London.* Carr. clock. mid-19c. (Prob. Jos. & Alf., q.v.)
Joseph & Alfred. *London.* 1857-69.
Joseph & Henry. *London.* 1875-81.
Thomas. *Prescot.* 1795. W.
JUNG, Herman. *London.* 1869-81.
JUPP & BARBER. *London.* 1832.
JURY, John. *Charlottetown, New Brunswick.* 1830.
JUST—
George. *London.* 1839.
Leonard. *London.* (B. 1790-1825) 1832-44. W.
Mrs. R. *London.* 1863.
William. *London.* (B. a.1797) 1832.
JUSTICE, Joseph (J. J.). *Philadelphia.* 1844-48.
JUTSUM, William. *London.* 1863.
JUVET, Louis Paul. *Glens Falls, NY, USA.* 1879-86.

K

K, H, HK, mark on table clock. ca.1610.
KABEL, Joseph. *New York.* ca.1850.
KAESER, John James. *Birmingham.* 1860-68.
KAIL—
Ambrose. *Blandford.* 1824-30.
Henry. *Blandford.* 1830.
KAISER—
Bros. *Dublin.* 1874.
Constantine. *London.* 1881.
Dilger & Co. *London.* 1863.
Dominick. *Birmingham.* 1868.
Elias. *Cardiff* from *Germany.* ca.1860-87. W.
F. *Dublin.* 1874-80.
G. & D. *Birmingham.* 1860.
John. *London.* 1881.
Joseph. *London.* 1851-57.
Kleyser & Co. *London.* 1844.
Martin. *London.* 1875-81.
Matthias. *Cardiff.* bro. of Elias. 1859-68.
Mrs. Anne Maria. *Cardiff.* 1887.
Mrs. B. *Landport.* 1867.
Paul. *Dublin.* 1858-68.
S. *Landport.* 1859.
William. *London.* 1857-75.
KALABERGO, John. *Banbury.* 1832-d.52. C. & W.
KALLENDAR, Orlando. *Exeter.* s. of Thomas. Free 1784. W.
KALLMAN, Charles. *Newbury, NY, USA.* 1869. Patented clock combined with fly trap.
KALTENBACH/KALTENBACK—
Adrian. *Cardiff.* 1875-87.
Andrew. *London.* 1844.
Bertin. *Maesteg.* 1887.
E. & A. *Cardiff.* 1868.
Edward. *Cardiff.* 1875-87.
Francis Joseph. *Pontypridd.* 1868-75.
J. *Fratton (Hampshire).* 1867.
Maximilian. *Abergavenny.* 1850-52.
Peter. *London.* 1857-81.
Philip & Co. *London.* 1828-39. C.
Samuel. *Neath.* 1875-87.
KAM(M)ERER—
Joseph & Co. *London.* 1832-51.
Mathias. *Llanelly.* 1887.
KANE, Thomas. *Liverpool.* 1848-51.
KARKEEK, George see CARKEEK.
KARN, A. L. *Philadelphia, USA.* 1809.
KARNER, C. *Philadelphia, USA.* 1809.
KATES, James. *Newmarket.* 1830-46.
KATTERNS, Daniel. *Thrapston.* ca.1765-ca.90. (B. 1795).
KAUF(F)MAN—
A. *Clapham.* 1878.
Abraham. *London.* 1869-81.
Barnhard. *London.* 1863.
Carl. *London.* 1844-57.
I. *Hartlepool.* 1898.
Joseph. *London.* 1875-81.
Mrs. Jessie. *London.* 1851.

KAUL—
A. J. *London.* 1832.
Adolphus. *London.* 1851.
Alexander. *London.* 1857.
Mrs. Elizabeth. *London.* 1863 (?wid. of Alexander).
KAUP, Francis Joseph. *Liskeard.* 1873-86. W.
KAVANAGH, Charles. *Dublin.* 1788.
KAY(E)—
J. *North Shields.* 1858.
James. *Sheffield.* 1871.
John. *Carlisle.* 1848.
John. *Galt, Canada.* 1857-62.
John. *Manchester.* 1848-51.
John. *Paris, Canada.* 1857.
Samuel. *Manchester.* 1822-24. W.
& Son. *Worcester.* Late 19c. wall dial.
Thomas. *Southport.* 1851.
Walter. *London.* From *Ewhurst.* a.1725 to Henry Neve. C.
Will. Place not stated. *Northern England,* mid-18c. C. (May be William of *Manchester?*).
William. *Doncaster.* 1851-71.
William. *London.* 1828.
William. *London.* 1781(-B. 1804).
William. *Manchester.* 1742, d. pre-1745. C.
KAYSER, Joseph. *Sheerness.* 1839-47.
KEALY, Charles. *Richhill (co. Armagh).* ca.1775-85 (?-ca.1810). C.
KEANE, Francis. *Newry.* 1824-46. C.
KEARLY, R. *Brompton (Kent).* ca.1800. C.
KEARN, Felix. *New Haven, Conn, USA.* ca.1840-50.
KEARNEY, Hugh. *Wolcottville, Conn, USA.* 1831-35. C.
KEAT—
& Co. *London.* 1875.
Henry. *London.* 1857.
Mrs. Sophia. *London.* 1844-51 (?wid. of Edward).
William. *London.* 1857-69.
KEAT(E)S—
Charles. *Belper.* 1855-76.
George. *Cheadle.* 1842-50.
KEDZIE, John. *Rochester, NY, USA.* 1838-46.
KEEFF, Thomas. *London.* 1832. Watch cases.
KEEL, John. *Philadelphia, USA.* 1835-37.
KEELER, Joseph. *Norwalk, Conn, USA.* ca.1870.
KEELING—
George. *London.* 1839-51.
George A. *London.* 1857-75.
Thomas. *Birmingham.* 1800-15. Dialmaker.
KEEN(E)—
C. *Coventry.* 1868.
Charles. *Abergavenny.* 1899. W.
Charles. *Birmingham.* 1880.
James. *Coventry.* 1828-54.
James. *Digby, Nova Scotia.* 1866.
Mark. *Birmingham.* 1868-80.
KEENAN—
Henry. *Belfast.* 1846-50.
Henry. *Lurgan.* ca.1795-1810. C.

131

KEEPFER—
John. *Denbigh* from *Neustadt*. 1850-56.
Wilhelm. *Denbigh.* 1862-90. W.
KEEPING—
R. *Ryde.* 1848.
R. *Southampton.* 1839.
KEER, ——. *Diss.* 1836.
KEET, James. *Ryde.* 1839.
KEETCH, S. C. *Aurora, Canada.* 1800s.
KEEY, William Henry. *Bristol.* 1870-79.
KEHEW, William Henry. *Salem, Mass, USA.* 1829-64.
KEHLE, J. *Norwich.* 1858-65.
KEIGHLEY, William. *Keighley.* 1822-37.
KEIM, John. *Reading, Pa, USA.* 1745-1819. C.
KEISER, John & Co. *Wolverhampton.* 1842.
KEITH—
David. *Inverness.* 1850-60.
James, jun. *Coleraine.* 1846.
William. *Inverness.* 1837-60.
KELHAM—
Mary. *Chelmsford.* ca.1823.
Simon. *Holbeach.* 1850.
Thomas. *Grimston.* ca.1760. C.
KELLAM, Thomas. Place not stated. 1752, clock with
slate dial. (Prob. Thomas Kelham of *Grimston,*
q.v.)
KELLETT/KELLITT—
Henry. *Bredbury.* 1865-78.
John. *Thorne.* 1814-51. W.
Thomas. *Thorne.* 1775-86. C.
Thomas, jun. *Thorne.* 1846-71.
KELLEY—
Allen. *Sandwich, Mass, USA.* 1810-30.
Benjamin. *Atlanta, Ga, USA.* 1874.
David. *Philadelphia, USA.* 1806-16.
KELLING, J. G. *Gnadenfrey.* ca.1790. W.
KELLOGG, Daniel. *Hebron, Conn, USA.* 1766,
mar.1794-1811, then *Colchester, Conn.* 1811-33, then
Hartford 1833-55.
KELLS, R. *Preston.* 1858.
KELLY—
Andrew. *Glasgow.* 1835.
Charles Aylmer. *Dublin.* 1784-1815. W.
Hezekiah. *Norwich, Conn, USA.* 1793.
Hugh. *Nenagh (Ireland).* 1858.
J. *Bristol.* 1856.
J. *Richmond (Surrey).* 1862-78.
James. *Monaghan.* 1784.
Jane. *Drogheda.* 1858.
John. *Bagnalstown (Ireland).* 1858. W.
John. *Bristol.* 1840-42.
John. *Liverpool.* (B. 1813-) 1822-34.
John. *New Bedford, Mass, USA.* 1836.
Michael. *New York.* 1854.
Peter. *Edinburgh.* a.1770.
Philip. *London.* 1857-69.
Thomas. *London.* 1875-81.
Walter. *Cork.* a.1745-57.
William. *Macclesfield.* 1878.
KELMAN, Charles. *New Pitsligo (Scotland).* 1860.
KELSEY—
Benjamin. *London.* 1869.
O. *Gainsborough.* C. Date unknown. (May be C.
Kelvey?)
KELSO—
——. *Antrim.* 1849. W.
Edward. *Donaghadee.* 1854-58. W.
KELVER (or KELVEY), C. G. *Birkenhead.* 1857-65.
KELVEY—
Ann. *Gainsborough.* 1828.
Charles Grant. *Gainsborough.* 1850.
Ebenezer. *Nottingham.* 1835.
Robert. *Dover.* Late 18c. C.
& Son. *Gainsborough.* 1835.
KELYNACK, Richard. *Newlyn (Cornwall).* 1805. W.
KEMBER—
Joseph (or Josiah?). *Shaw, nr. Newbury.* 1743-
ca. 1775. C.
William Henry. *Surbiton.* 1862-66.
KEMLO, Francis. *Chelsea, Mass, USA.* ca.1840.
KEMP—
Alfred. *Tunbridge Wells.* 1851-74.

KEMP—*continued.*
Benjamin. *Chester.* 1878.
Charles. *Barnet.* 1874.
Charles. *Hackney.* 1874.
Charles. *London.* 1857-75.
Edwin. *Battle.* Successor to Henry 1870-78. W.
George. *London.* 1863-81.
George Edmund. *London.* 1857-63.
George Firman. *Rayleigh.* 1828.
H. *Mayfield.* 1855.
Henry. *Battle.* 1862-70 (in 1867. Successor to John
Noakes). W.
Henry. *Seaford.* 1828-39.
Henry. *Wokingham.* 1864-77.
John. *Wolverhampton.* 1868-76.
Levi. *Norwich.* 1875.
Richard Sampson. *Cork.* 1818-28.
Thomas Firmin. *London.* 1844.
V. *Rotherfield.* 1878.
William. *Lewes.* (B. pre-1756) 1759-91. W.
William. *Newhaven.* 1828-39.
William. *Syston (Leicestershire).* ca.1820. C.
William Henry. *London.* (B. CC. 1808) 1839. W.
William Henry. *Peckham.* 1845-51.
KEMPIE, Andrew. *Perth.* 1837.
KEMPS—
Anthony. *London.* 1663.
Henry Joseph. *Brighton.* 1870.
KEMPTHORNE, William Worth. *Neath.* 1830.
KEMSHEAD—
Harvey (& Son). *Manchester.* 1822-34.
Robert. *Manchester.* 1834.
Robert & Son. *Manchester.* 1828.
KENDAL(L)—
——. *Canterbury.* Early 19c. C. (Prob. Richard, q.v.)
Caleb. *South Woodstock, Vt, USA.* ca.1811. W.
D. C. *Boston, USA.* 1842.
& DENT. *London.* 1881-84.
Edward. *Falmouth.* 1753. W.
John. *Romsey.* 1839.
Joseph. *Sedbergh.* 1838. W.
Richard. *Canterbury.* 1838-40.
S. *Downton (Wiltshire).* 1859.
Samuel W. *Portland.* 1867-75.
Thomas. *Bingley.* 1853.
KENDRICK—
Charles. *Alcester.* 1842-50.
F. H. *Alcester.* 1860.
Frank B. *Lebanon, NH, USA.* 1845-retired 1910,
d.1936. W.
George. *Droitwich.* 1850-76.
KENLEY, K. *Ottawa.* 1875.
KENMUIR—
Alexander. *Lisburn.* mar.1852-92.
Henry. *Lisburn.* 1888. W.
James. *Lisburn.* 1865.
William. *Lisburn.* 1854-68. C.
William (& Sons). *Ballynahinch.* 1843-80. W.
KENNARD, John. *Newfields, NH, USA.* ca.1820.
KENNEDY—
Alexander. *Coleraine.* 1858.
Alexander. *Dungannon.* 1840.
Alexander. *Enniskillen.* 1831-46. C.
Edward. *Dublin.* 1827-28.
Electric Clock Co. *New York.* ca.1860.
Elisha. *Middletown, Conn, USA.* 1766-88.
Hugh. *Ballycahan.* mar.1833, *Coleraine,* 1858.
Hugh. *Belfast.* 1825-27. W.
John, jun. *Maybole (Scotland).* 1837.
Mark. *Macclesfield.* 1848-65.
Patrick. *Philadelphia, USA.* 1795-1801.
Roger I. *Dublin.* 1784-1814.
(?Roger II. *Dublin.* 1820-25).
Samuel. *Letterkenny.* mar.1828-58.
Samuel I. *Rathfriland.* Early 19c.
Samuel II. *Rathfriland.* s. of Samuel I, 1854-68.
Thomas. *Armagh.* 1763 and perhaps later. C.
Thomas. *Kilmarnock.* 1837.
Thomas. *Rathfriland.* ca.1835-58, *Castle Wellan*
1858-80.
Thomas. *Wigan.* (b.1727). ca.1754-d.1806. W.
William. *Tanderagee.* (b.1768) 1781-d.1834. C. Blind.

KENNEDY—*continued.*
Also bagpipe maker.
William I. *Antrim.* 1784-ca.1810. C.
William II. *Antrim.* (b.1802) 1828-d.1847. W.
KENNELL—
John. *London.* 1881.
Joseph. *Birmingham.* 1880.
Samuel. *Taunton.* 1875.
KENNETON, James. *Exeter.* mar.1764.
KENNETT—
John. *London.* 1651. (Poor journeyman.)
W. R. *Peckham.* 1855-78.
William. *London.* 1851.
William. *Warwick.* 1880.
KENNEY, Asa. *West Milbury, Mass, USA.* ca.1800.
KENNING, William. *Banbury.* b.1648-ca.1675, then *London* where CC. 1685-87. Lant. clock.
KENNON, Henry. *Killyleagh.* Late 18c. clock label printer.
KENNY, Robert. *Ballyshannon.* 1824-46.
KENT—
Brothers. *Toronto.* 1867-77.
C. D. *Forest Hill.* 1866.
Christopher David. *London.* 1857-75.
John. *Manchester.* 1769-1825. W. & C.
John. *Saffron Walden.* (B. ca.1764-90) 1828-39.
John. *Salford (Manchester).* 1804-(B. 1808-13).
John. *Shrewsbury.* 1870.
& JONES. *London.* 1869.
Joseph. *Saffron Walden.* 1828.
Luke. *Cincinnati, O, USA.* ca.1820-ca.40.
Mary Ann. *London.* 1832-39.
R. *Widnes.* 1858.
Thomas. *Cincinnati, O, USA.* s. of Luke. 1821-44.
William. *Saffron Waldon.* ca.1760-ca.70. C.
William Worsley. *Manchester.* s. of John. b.1804, mar.1825-1851. W.
KENVIN, William J. *Southampton.* 1878.
KENWORTHY, John. *Oldham.* 1848.
KENYON—
James. *Liverpool.* 1715-43. C.
Lawrence (& Son). *Stockport.* 1848-65(-78).
William. *Liverpool.* (?b.1667) 1708-20. C.
KEPLINGER—
John. *Baltimore, USA.* Early 19c.
Samuel. *Baltimore, USA* (from *Gettysburg*). 1770-1849.
William. *Baltimore, USA.* 1829.
KERBY—
Benjamin. *Bristol.* 1830.
Benjamin. *Coventry.* 1850-60. Watch case maker.
Edward. *Coventry.* 1868.
Edward Rosser. *Coventry.* 1850-60.
Francis. *Jersey.* 1832-53.
Francis. *Warminster.* 1859-75.
Isaac. *London.* 1807. W.
Isaac Blanch. *Coleford.* 1842.
M. *Jersey.* 1833.
KERFOOT—
Robert (?I). *Liverpool.* 1754-d.1774. W.
Robert (?II). *Liverpool.* 1761-96. W.
Robert (?III). *Liverpool.* mar.1822. W.
KERKELO(O), Jacob. *Amsterdam.* ca.1730-ca.40 (B. ca.1750). C.
KERKHAM, Hugh. *St. Helens.* 1786. W.
KERN—
(Joseph) & Co. *Swansea.* 1850. C. & W.
Herman. *Lancaster, Pa, USA.* ca.1850.
Joseph. *Lymington.* 1878.
Joseph. *Swansea.* 1844-48. C. & W.
KERNER—
& DOLD. *London.* 1869.
Louis. *Muskingum County, O, USA.* ca.1850-ca.70.
Nicholas. *Marietta, Pa, USA.* ca.1850.
P. *Nottingham.* 1849-64.
& PAFF. *New York.* 1796. C.
KERNICKE, Christopher. *St. Neots (Cornwall).* 1622. Rep'd. ch. clocks.
KERNOR, I. *Poole (Dorset).* 1848-67.
KERR—
——. *Lurgan.* ca.1780-ca.1800. C.
Alexander. *Dumfries.* 1796.

KERR—*continued.*
Andrew (?or Alexander). *Belfast.* 1854-59.
Francis. *Monaghan.* ca.1770-80. C.
George. *Monaghan.* 1784.
Henry. *Dundee.* 1863.
Henry. *Edinburgh.* 1857.
Henry. *Loanhead (Scotland).* 1850.
Henry M. *Halifax, Nova Scotia.* 1863-81.
Hugh. *Raphoe (co. Donegal).* 1787.
Joseph. *Bolton.* 1851.
Joseph. *Monaghan.* 1785-1800. C.
W. *Basingstoke.* 1848.
W. S. *Chertsey.* 1851.
William, jun. *Providence, RI, USA.* 1850s.
KERRICK—
George Townley. *Chipping Sodbury.* 1879.
George Townley. *Penarth.* 1887.
KERRIDGE, B. J. *Wimborne.* 1855-75.
KERRISON—
James. *Norwich.* 1858-75.
Robert M. *Philadelphia, USA.* 1842.
KERRY, John. *Liverpool.* mar.1821-51. W.
KERSEY—
Frederick. *Romsey.* 1878.
Robert. *Easton, MD, USA.* 1793.
T. *Stowmarket.* 1846-53.
James. *Stockport.* 1878.
KERSHAW, John. *Oldham.* 1851-58. W.
KERSWELL, G. *Portsea.* 1859.
KESLER, John. *Haverfordwest.* 1868-87.
KESSELMEIR, Frederick. *Wooster, O, USA.* 1844.
KETCHAM & HITCHCOCK. *New York.* ca.1810.
KETCHING, William. *Edinburgh.* 1850.
KETTERER—
——. *Portsea.* ca.1830. C. (see Ketterer & Co.).
A. *Grimsby.* 1861.
Anthony. *Worcester.* 1872-76.
Bartholomew. *London.* 1857-63.
Brothers. *Ware.* 1874.
& BRUGGER. *Leeds.* 1853.
Charles. *Greenwich.* 1847.
& Co. *Greenock.* 1860.
& Co. *Oldham.* 1834. German clocks.
& Co. *Portsea.* 1839-48.
Crispin. *Greenwich.* 1847.
David & Co. *Worcester.* 1868-76.
J. *Ware.* 1859-66.
J. B. *St. Albans.* 1874.
M. *Southampton.* 1859.
Matthew. *Frome.* 1875-83.
Mrs. F. *Bedford.* 1877.
John. *Ripon.* 1834, then *Leeds* 1842-77. W.
John & Kaberry. *Leeds.* 1837.
KEUFFMAN, Charles. *London.* 1851.
KEVITT, Richard. *London.* 1834. W.
KEW—
George. *Philadelphia, USA.* 1840.
William. *Ashford.* 1874.
KEWLETT, Joseph/James(?). *Bristol.* 1773. W.
KEY—
John. *Warrington.* mar.1761. C.
Richard. *London.* 1851.
William. *St. Columb (Cornwall).* 1809.
William. *Swansea.* 1830-99.
KEY(E)S—
David. *London.* 1857-81. Wholesaler.
Elias. *Canning Town.* 1874.
Elias. *London.* 1863-69.
Joseph. *Exeter.* (b. ca. 1789) 1822 -d. 1865. C.
Joseph. *London.* 1857.
Rufus. *Lowell, Mass, USA.* 1833.
William. *Exeter.* 1803. C.
KEYSER/KEYSOR—
Benjamin & Co. *London.* 1881.
Joseph. *Philadelphia, USA.* 1828-33.
KEYSTONE—
Standard Watch Co. *Lancaster, Pa, USA.* 1886-90.
Watch Case Co., The. *Philadelphia, USA.* 1899-ca.1903. W.
Watch Corp. *Philadelphia, USA.* 1927.
KEYTE—
George. *Wolston (Warwickshire).* 1868.

KEYTE—*continued.*
Richard. *Witney.* 1770. C. & W.
KEYWORTH—
Robert. *Washington, DC, USA.* ca.1820.
T. *York, Pa, USA.* ca.1850.
KIBBLE—
Richard. *Greenwich.* 1847-74.
William. *London.* 1863-81.
KIDD(S)—
——. *Malton.* Prob. combination of Gilbert & Thomas. Late 18c. C.
Gilbert (sen.?). *Malton.* d.1778. W.
Gilbert & Nephew. *Malton.* 1786. W.
John. *Birkenhead.* 1878.
John. *Stratford-on-Avon.* 1842-50.
Thomas. *Malton.* 1807-d.1823. C.
KIDDLE, George. *Chelmsford.* 1855-78.
KIDDY—
Frederick. *Wethersfield* nr *Braintree.* 1851-74.
J. *Wirksworth.* 1849-55.
KIENZLE, A. *Peckham.* 1878.
KILB(O)URN—
——. *Epworth.* ca.1810-ca.20. C.
Henry. *New Haven, Conn, USA.* ca.1840-ca.50.
Hiram. *Lowell, Mass, USA.* 1837.
KILES, William. *Cork.* a.1784-92, then to *Waterford.*
KILGOUR—
George. *London.* 1881.
Thomas. *Inverness.* ca.1700. C.
KILHAM, S. *Holbeach.* 1849.
KILLAM—
George Roland. *Providence, RI, USA.* 1873-1930. C.
Guy. *Pawtucket, RI, USA.* 1939. C.
KILLICK—
Charles. *London.* 1869.
& WEBB. *London.* 1843. W.
KILLINGTON, E. *Yarmouth.* 1865.
KILLINGWORTH—
John. *Olney.* b.1769. mar.1796-1830. W.
William. *Olney.* s. of John. b.1800-d.1863.
KILLMAN—
John. *Deptford.* 1847.
John William. *London.* 1875.
KILLY—
Albin. *Landport.* 1878.
J. B. *Southampton.* 1859.
KILNER, Samuel. *Ulverston.* 1760. C. (?d.1787).
KILPIN—
Edmund B. *Ryde.* 1830-48.
Edmund Burke. *Llandudno.* 1874.
Edward. *London.* 1869.
Edward. *Midhurst.* 1828.
George. *Southport.* 1830.
J. *Landport.* 1848.
Joseph. *Southsea.* 1839.
Joseph George. *London.* 1881.
KILSHAW, Nehemiah. *Prescot.* mar.1759.
KILWICK, James. *Colchester.* b.1708-1727. C.
KILZINGER, Joseph. *London.* 1861.
KIMBALL/KIMBELL—
F. A. Used by Otay Watch Co. ca.1889. (*USA.*)
& GOULD. *Haverhill, Mass, USA.* 1865.
Porter. *Stanstead, Canada.* Pre-1841.
Thomas. *London.* 1844-51.
KIMBERLEY—
George. *Coventry.* 1850-68.
Thomas. *Mevagissey (Cornwall).* 1856-73. W.
KIMICH/KIMICK—
Bernard. *Newport (Mon.).* 1852-58.
E. *Hemel Hempstead.* 1866-74.
E. & Co. *Luton.* 1859. W.
J. *Dunstable.* 1847.
KIMMUIR, William. *Ballynahinch.* 1858.
KIMPTON, Benjamin. *Eltham.* 1874.
KIND, John. *Liverpool.* 1834-48.
KINDER, T. & Co. *Coventry.* 1860-80.
KINEHAM, William. *West Farnham, Canada.* 1871.
KINEHAN—
James, jun. *Acton Vale, Canada.* 1862.
William. *St. Hyacinthe, Canada.* 1862.
KING—
——. *Loddon (Norfolk).* ca.1800. C. (Prob. Robert

KING—*continued.*
q.v.)
Alexander. *Liverpool.* Formerly of *Ringwood* and *Romsey.* Insolvent 1772.
Alexander. *Peterhead.* 1836.
Alfred. *Chippenham.* 1830-59. Also jeweller.
Alfred. *London.* 1881.
Andrew. *Hull.* ca.1850-1920. (Alias Andrius Konig).
Benjamin. *Peterhead.* 1846.
Charles. *Winslow.* 1798. C.
Christian. *Leicester.* 1846. C.
Christian. *Yarmouth.* 1836.
Edward. ('King Edward'), 1914 and later, product of Pequegnat Clock Co.
Frederick. *Windsor.* 1877.
George Henry. *London.* 1881.
George Trafalgar. *Kings Lynn.* 1865-75.
Henry. *Bristol.* 1768-81.
Henry. *Hamburg, Pa, USA.* ca.1800 (Real name Henry ROI).
Henry N. *New York.* ca.1810.
Horace. *Kings Lynn.* 1865-75.
J. *Kings Lynn.* 1846.
J. *New Zealand.* ca.1790. W.
James. *Devizes.* 1875.
James. *London.* 1839-81. (Succeeded Thomas.)
John. *Ampthill.* 1830.
John. *Colchester.* b.1784-d.1809. W. & C.
John. *Loughborough.* 1835-76 and later.
John. *Montrose.* 1840.
John. *Southwold.* 1839.
John. *Yarmouth.* 1830.
John Bartholomew. *London.* 1881.
John James. *London.* 1851-69.
Jonathan. *London.* (b.ca.1666) a.1682 to Richard Watts, CC. 1689-97. W.
Joseph. *Birmingham.* 1880.
Joseph. *Skipton.* 1871.
Joseph. *Winslow.* 1798-1830.
Josiah. *Stroud.* 1879.
Robert. *Leeds.* 1866-71.
Robert. *Loddon (Norfolk).* 1836.
Samuel. *Newport, RI, USA.* 1748-1819. Math. insts.
Thomas. *Alnwick.* 1795. C.
Thomas. *Baltimore, USA.* ca.1820.
Thomas. *Birmingham.* 1868-80.
Thomas. *Coventry.* 1880.
Thomas. *Huddersfield.* 1837.
Thomas. *London.* 1844.
Thomas. *Titchfield (Hampshire).* 1867-78.
Thomas & Benjamin. *London.* 1828-39.
W. *Billericay.* 1866.
William. *Chelmsford.* ca.1870.
William. *Birmingham.* 1868.
William, jun. *Birmingham.* 1868.
William. *London.* 1881.
William Finch. *Chelmsford.* 1866-74.
William & Harry. *Ipswich.* 1879.
William Johnson. *Kings Lynn.* 1875.
William Robert. *Witham.* 1874.
KINGBURY, William. *Polstead (Suffolk).* 1879.
KINGDON, Henry. *Ipswich.* b. 1837 d. 1895.
KINGS, James. *Cradley (Hereford).* 1879.
KINGSMILL (alias WOGAN), George. *London.* CC. 1667, former a. of William Smith.
KINGSTON—
1914 and later, product of Pequegnat Clock Co.
J. T. *Ramsgate.* 1855-74.
Joseph. *Coventry.* 1880.
Joseph. *Tetsworth.* 1786. W.
Richard. *Brinklow (Warwickshire).* 1868-80.
KINKAID, George. *Cincinnati, O, USA.* ca.1830.
KINKEAD—
——. *Strabane.* ?Early 19c. C.
Alexander. *Christiana Bridge, Del, USA.* ca.1788.
James. *Philadelphia, USA.* 1765. (B. 1774) (from *Strabane, Ireland* but late of *Salisbury, Pa.*).
Joseph. *Christiana Bridge, Del, USA.* 1778-96. C. (B. 1820).
KINLAN—
Thomas. *Dublin.* 1858. (B. has one such at *Granard* 1824).

KINLAN—*continued.*
Thomas. *Tunbridge Wells.* 1866.
KINMOUTH, Hugh. *Cork.* 1844.
KINNEAR—
Alexander. *Ballibay, co. Monaghan.* 1868.
C. D. *Portobello.* 1836.
Conrad, sen. *Glasgow.* 1836-d.1840.
Conrad, jun. *Glasgow.* 1840.
J. *Edinburgh.* 1850.
John. *St. John, New Brunswick.* 1838.
M. *Newcastle-on-Tyne.* 1856.
KINNER, Raymond. *Belfast.* 1858. C.
KINSEY, David P. *Newtown (Mon.).* 1868.
KINSMAN, Daniel. *Toronto.* 1861.
KINSTLEY—
Joseph. *Pen-y-Graig* and *Ton-y-Pandy.* 1887-99.
& LEHMAN. *Ton-y-Pandy.* 1875.
KIPLING, John. *Stafford.* 1828-35.
KIPPAX, John. *East Retford.* 1849-76.
KIPPIN, George. *Bridgeport, Conn, USA.* 1822.
KIRBY—
George. *Canterbury.* 1851-74.
H. *Great White* nr *Ramsey (Hunts).* 1854.
H. *Whittlesey.* 1858.
H. *Woburn.* 1864.
J. *Grantham.* 1868.
James. *St. Neots (Hunts).* 1864-77.
John. *Greenwich.* 1866-74.
John. *London.* 1875.
John. *St. Neots (Hunts.)* 1847-64.
Mrs. Elizabeth. *Woburn.* 1869-77.
R. B. *Ware.* 1851.
Richard. *St. Neots (Hunts).* 1830.
William. *Reading.* 1877.
KIRCHOFF, J. H. *Philadelphia, USA.* 1805.
KIRCKHAFF, E. H. *Philadelphia, USA.* 1803.
KIRK—
——. *Dumfries.* ca.1800.
——. *Humberstone (Leicestershire).* ca.1790. C.
Aquilla. *York, Pa, USA.* ca.1795, later *Baltimore.*
Charles. *Bristol, USA.* 1828-37, *Wolcott* 1837-43, then
 New Haven 1847. C.
Elisha. *Yorktown, Pa, USA.* 1780-90. C.
J. *Coventry.* 1854-60. Watch case maker.
James. *Edinburgh.* a.1648. C.
John. *Bristol, USA.* 1831.
Joseph. *Skegby.* ca.1730. C.
Kezia & Thomas. *Hull.* 1840.
Nathaniel. *Kibworth (Leicestershire).* Later 18c. C.
Robert. *Ballymoney (co. Antrim).* a.1767-70.
Samuel. *Ballymoney.* ca.1790-1800. C.
Thomas. *Hull.* 1846-74. (succ. to B. Jacobs).
Timothy. *Hanover, Pa, USA.* ca.1783.
(Charles) & TODD. *Wolcott, Conn, USA.* Early 1840s.
 C.
W. *Stroud.* 1856.
William. *Manchester.* (b.ca.1752) mar.1778-94, later
 Stockport where d.1834. C.
William. *Northampton.* 1841-64.
William. *Stowmarket.* Partner with John Ablitt
 1810-14, then alone 1814-30. C.
KIRKALTHORPE, William S. *London.* ca.1730. C.
 (dubious?).
KIRKBY, Robert B. *Malton.* 1851-66.
KIRKE, I. *Harstoft.* C. Date unknown.
KIRKHALL, Thomas. *Bolton.* mar.1625-37. T.C. (?not
 1673).
KIRKHAM, ——. *Holywell.* 18c.
KIRKLAND—
John & Joseph. *Tickhill* (nr *Doncaster).* 1862-71. W.
Samuel. *Northampton, Mass, USA.* ca.1835.
KIRKPATRICK—
Martin. *Dublin.* a.1712-d.1769. W.
Robert. *Farringdon.* 1837.
Thomas. *New York.* 1846.
William. *Lincoln.* 1861.
KIRKWOOD—
Alexander. *Paisley.* 1820-24. W.
John. Place unknown. ca.1730. C.
John. From *London. Charleston, SC, USA.* 1761.
KIRMER, Anselm. *London.* 1881.
KIRNER—
Andrew. *London.* 1869.

KIRNER—*continued.*
Augustine & George. *Cleator Moor (Cumberland).*
 1873.
Brothers. *London.* 1857.
D. *Wolverhampton.* 1868-76.
George. *Cleator.* 1879.
George. *Whitehaven.* 1869-73.
Joseph. *London.* 1875-81.
Peter. *London.* 1869.
Raymond. *Belfast.* 1835-58. C.
KIRTON—
——. *London.* ca.1785. C.
G. *Stockton.* 1898.
G. T. *Sunderland.* 1856.
George. *Bishopwearmouth.* 1851-56.
George. *South Shields.* 1827-34. C.
James. *Alston.* d.1729. C.
William. *Newcastle-on-Tyne.* 1827-58. W.
KISLER, Anthony. *Dublin.* (B. 1795) 1820-31. C.
KISTLER—
Andreas. *Romford.* 1839-66.
Dominic. *London.* 1869.
George. *Penzance.* 1856.
George & Mathias. *Penzance.* 1873.
Matthew. *Penzance.* 1844-47.
Mrs. Hannah. *Romford.* 1874.
Thomas Andrew. *Redruth.* 1873. W.
KITCHEN—
Benjamin. *London.* 1839-51.
Samuel. *Pudsey* (nr *Leeds).* 1834. W.
KITSON, ——. *Nuneaton.* C. Dates unknown.
KITTLER, J. *Cheltenham.* 1863.
KITTS, John. *Louisville, Ky, USA.* 1838-78.
KITZ, Louis. *London.* 1851.
KLAFTENBERGER, Charles I. *London.* 1863-81.
KLAUSMANN, Roman. *London.* 1875.
KLEAN, Michael & Co. *London.* 1863-81.
KLECKNER, Solomon. *Mifflinburg, Pa, USA.* ca.
 1830.
KLEEMAN, Edward. *Hull.* 1858.
KLEIN—
George S. *Toronto.* 1894.
John. *Philadelphia, USA.* 1838.
KLEISER, also **KLEYSER, KLEISIER,** etc.—
A. & Co. *Reading.* 1864-99.
Albert. *Toronto.* 1875-77.
Andrew. *York.* 1851. C.
Andrew & Joseph. *York.* 1866-ca.1885. W.
& BERKLEY. *Gloucester.* 1863.
& BROTHERS. *London.* 1844-57.
Charles & Co. *London.* 1863.
& Co. *London.* 1828. C.
& co. *London.* 1851-57.
Constantine. *Hereford.* 1870.
Constantine Bernard. *Mumbles.* 1875-87.
& DILGER. *London.* 1869.
Felix. *Hereford.* 1850-79.
Felix. *London.* (B. 1817-24) 1828. C.
J. *Unionville, Canada.* 1851.
J. & Co. *London.* (B. 1820?) 1875-81.
Jacob. *Philadelphia. USA.* 1822-24.
James. *London.* 1863-81.
Joachim. *London.* 1828-32. Imported clocks.
John (& Co.). *London.* 1832 (39-69). Importer of
 German clocks. C. & W.
John. *Toronto.* 1856.
John & Son. *London.* (B. 1820) 1828. C.
M. & Co. *Doncaster.* 1862.
Matthias. *Guildford.* 1862-78.
Ottmar. *York.* 1866.
Richard. *Holyhead.* 1887-95. Successor to J. Bader,
 q.v.
Simon P. *Toronto.* 1865-77.
T. *Gloucester.* 1879.
& TRITSCHLER. *London.* 1839-44.
W. *Portsea.* 1848-59.
KLINE—
B. *Philadelphia. USA.* 1841.
and Co. *New York.* 1862.
John. *Lancaster, Pa, USA.* ca.1800, *Philadelphia* 1812-
 20, then *Reading, Pa.* 1820-30.
KLING, Jacob. *Reading, Pa, USA.* 1758-1806. C.

KILNGELE—
Cornelius. *Newcastle-on-Tyne.* 1848-52.
D. & Co. *Newcastle-on-Tyne.* 1858.
KLINGMAN, Daniel. *York, Pa, USA.* ca.1820.
KLISER, Jacob. *Toronto.* 1837.
KLOCK, Jan. *Amsterdam.* Early 18c. C.
KNAGGS, William. *Whitby.* 1866. W.
KNAPP—
Charles G. Used by U.S. Watch Co. ca.1870.
Jesse. *Boston, USA.* 1825-42.
John. *Cork.* 1723-d.1757.
John. *London.* ca.1690. C. (-Early 18c.)
John. *Reading.* Early 18c. Lant. clock.
Louis. *Wandsworth.* 1878.
Peter. *Leicester.* 1828.
William. *Cork.* a.ca.1753-ca.1760. (One such at Annapolis 1764.)
KNEEDLER, Jacob. *Horsham, Pa, USA.* 1791.
KNEELAND—
& ADAMS. *Hartford, Conn, USA.* 1792. C.
Samuel. *Hartford, Conn, USA.* 1788-93.
KNEESHAW—
Robert. *Stokesley.* 1823. C.
William. *Pickering.* 1851-66. W.
KNELL—
E. *Gosport.* 1848-59.
Edwin. *Gosport.* 1878.
Edwin. *Winchester.* 1878.
Henry. *Berlin, Canada.* 1862.
KNIBB—
John. *Oxford.* bro. of Joseph. b.1650-d.1722. C. & W.
Joseph. *Oxford.* bro. of John. b.1640-ca.1670, then *London* till ca.1697, then *Hanslope* till d. 1703.
Samuel. (b.1625). Cousin of John & Joseph. *Newport Pagnell* till ca.1662, then *London, Westminster* 1662, CC. 1663-67, d.ca.1670. C.
Sebastian. *Newcastle-on-Tyne.* 1821.
KNICKERBOCKER Watch Co. *New York.* ca.1891. Imported Swiss watches.
KNIGHT—
——. *Norwich.* ca.1800. C.
A. *Bath.* 1861.
Alfred. *Birmingham.* 1854.
Alfred. *London.* 1863-69.
Benjamin. *Statesville, RI, USA.* 1840. C.
Charles. *Great Dunmow.* 1839-66.
Charles. *Thaxted.* 1839.
& FLETCHER. *London.* 1881.
Frederick. *Devizes.* 1867-75.
G. *Grimsby.* 1876.
George. *Holywell.* 1874.
George. *Warminster.* 1848-75.
George. *Yeovil.* 1875.
Harry. *Dunmow.* 1874.
Henry. *Birmingham.* 1828-50. T.C. and bellfounder.
J. *Earls Barton.* 1854-69.
J. *Wickham.* 1848-67.
J. & Son. *Northampton.* 1877.
James. *Fareham.* 1878.
James. *London.* 1844.
John. *Coggeshall.* 1839.
John. *Halifax.* 1769-74. Bankrupt. Successor to Thomas Ogden.
John. *Thaxted.* 1828.
John. *Wickham.* 1830.
Joseph & Son. *Coggeshall.* 1828.
Joshua. *Bristol.* 1830.
Levi. *Salem, Ind, USA.* ca.1845. C.
Peter Mongo. *London.* 1828-32.
R. *Odiham.* 1839.
Richard. *Norwich.* ca. 1800 -ret'd. 1815. C.
Richard James. *Basingstoke.* 1859-78.
Thomas. *Bristol.* 1836-50.
Thomas. *Newport (Isle of Wight).* 1830-67.
Thomas and Mary. *Denmead.* 1761. C. (May be owners' names?).
Stephen Wintruph. *Manchester.* mar.1827. C.
Thomas. *Manchester.* 1822-51.
W. *Marden (Kent).* ca.1760. C.
W. *Newport (Isle of Wight).* 1867.
W. *Portsea.* 1848.
William. *Bristol.* 1830.

KNIGHT—*continued.*
William. *Halstead.* 1839-66.
William. *London.* 1839.
William. *Petersfield.* ?Mid-18c. C.
KNIGHTS, George. *Beccles.* 1879.
KNOLLES, Richard Hayes. *Cork.* b.1774-96.
KNOPFF—
A. L. R. *Buxton.* 1864.
Rudolph. *London.* 1875.
KNORR & Co. *London.* 1869.
KNOTT, John. *Silloth (Cumberland).* 1879. W.
KNOTTESFORD, William. *London.* a.1656 to Henry Child, CC. 1663-97.
KNOWER, Daniel. *Roxbury, Mass, USA.* ca.1800. T.C.
KNOWLES—
Andrew. *Bolton.* 1724-ca.1740. C.
David. *Bradford.* 1853.
Edward. *Backbarrow.* 1813. C.
George Henry. *Birmingham.* 1850-60.
George J. *Winchester.* 1859-78.
J. *Winchester.* 1848.
James. *London.* 1832-57.
John. *Farnworth.* mar.1802. W.
John. *Huntly (Scotland).* 1860.
John. *Kincardine O'Neil.* 1860.
John. *London.* 1839.
Joseph. *Birmingham.* 1850.
Joseph. *Brixton.* 1862.
Joseph. *London.* 1844.
Joseph H. *Birmingham.* 1868.
Mrs. Rosannah. *Birmingham.* 1868.
Robert. *Bangor, Me, USA.* ca.1830-40.
T. *Birmingham.* 1854.
Thomas. *Belfast.* 1865.
Thomas. *Bolton.* ca.1780-d.1787. C.
Thomas. *Cartmel.* mar.1784. C.
KNOWLTON, Luke. *Stanstead, Canada.* 1841.
KNOX—
James. *Tralee.* (B. 1824-)1858. W.
John I. *Belfast.* Free 1729-33. C., W. and goldsmith.
John II. *Belfast.* 1758-d.1783. C.
John III. Believed at *Larne* 1769-83, then *Belfast* 1783-retired 1816. C. & W. (Nephew of John II.)
John. *Guildford.* 1828.
KOCH—
——. *Vienna.* Sundial, first half 18c.
C. B. *London.* 1875.
Richard. *York, Pa, USA.* 1805-36. C.
KOCKSPERGER, Henry. *Philadelphia.* 1837.
KOEBER, G. F. *Frankfort.* ca.1750.
KOHL, Nicholas. *Willow Grove, Pa, USA.* 1830.
KOOS—
George. *Aberavon.* 1887-99.
Leander. *Merthyr Tydfil.* 1868-99. C.
KOPLIN—
T. *Norristown, Pa, USA.* ca.1850.
Washington. *Norristown, Pa, USA.* 1850.
KORFHAGE, Charles. *Brooklyn, NY, USA.* 'Estd. 1877'.
KORNMAN, Jo. ?*London.* ca.1700. W.
KOSTER, J. *St. Catherines, Canada.* 1853.
KRAEMER, F. *New York.* 1848.
KRAFT, Jacob. *Shepherdstown, W. Va. USA.* ca.1800.
KRAHACHER, Franz X. *Salzburg.* ca.1730. C.
KRAHE, William. *San Francisco, Cal, USA.* 1861.
KRAKAUER, S. *Barnsley.* ca.1820-ca.30. C.
KRAMER, M. & Co. *Boston, USA.* 1830-41.
KRAMM, Brothers. *London.* 1863.
KRAUTH—
& Co. (Adolphe). *Quebec.* 1857.
F. T. *Birmingham.* 1880.
G. A. *Montreal.* 1856-63.
KREMER, Adam. *Paris.* 1889-1900. Carr. clock.
KREUTZ—
Andrew. *London.* 1875.
G. *West Hartlepool.* 1898.
George. *London.* 1863.
Jacob M. G. & Co. *London.* 1857-63.
Joseph. *London.* 1869.
Leopold. *London.* 1875. (Successor to Joseph.)
KREUTZBERG, Julius. *London.* 1875-81.
KRINGE, Jacob. *New Market, Va, USA.* ca.1790.

KROEBER—
  Clock Co., F. *New York*. 1880-99. C.
  F. *New York*. 1872-78.
  Frederick J. Clock Co. *New York*. 1887. C.
KROHN—
  Brothers (& Co.). *London*. 1875(-81).
  Henry. *London*. 1869.
KROUSE, John J. *Northampton, Pa, USA*. ca.1830.
KROUT, Jacob. *Plumstead, Pa, USA*. ca.1830.
KRUEGER, Adolph. *Camden, NJ, USA*. 1850-62.
KRUGER, L. *Cleveland, O, USA*. ca.1850.
KRZECZKOWSKI—
  Stanislaus. *London*. 1881.
  Stanislaus. *Peckham*. 1878.
KUHN, Adam. *London*. 1875.
KULP, William. *Lower Salford, Pa, USA*. ca.1800. C.
KUNDERT, Fritz. *London*. 1875.
KUNER—
  Aloys. *Coggeshall*. 1866-74.
  & CASY. *Sheerness*. 1866.
  Isidor. *Cowbridge* (*Wales*). 1875.
  Isidor. *Pontypridd*. 1868-87.

KUNER—*continued.*
  Isidor. *Porth*. 1887.
  S. *Canterbury*. 1865.
  S. & Co. *Bristol*. 1879.
KUNKLE, John. *Ephrata, Pa, USA*. 1830-40. (May be
    GUNKLE.)
KUNSMAN, Henry. *Raleigh, NC, USA*. 1820-23.
KUNUYN, James. *Place unknown*. ca.1600. W.
KURCZYN, N. P. M. *Montreal*. 1819.
KUSS—
  Berhard & Co. *Bishopwearmouth*. 1851. C.
  G. & A. *Newcastle-on-Tyne*. 1858.
  G. & Co. *Newcastle-on-Tyne*. 1852-56.
  Germann. *Newcastle-on-Tyne*. 1848.
  German & Co. *Hull*. 1846.
  Xaver. *London*. 1863-81.
KUTCH, J. W. *New Market, Canada*. 1862.
KYEZOR—
  Henry. *London*. 1851.
  Louis (sen.?). *London*. 1839-63.
  Louis, jun. *London*. 1851-81.

L

L, EG, EGL, trademark of E. G. Lamaile, q.v.
L, G, GL, trademark of E. G. Lamaile, q.v.
LABHART, W. L. New York. ca.1810.
LABROW, Thomas Scott. *Manchester.* 1845-58. W.
LACEY—
James. *Laxton (Notts).* 1855-76.
Paul. *Bristol.* a.1779-1840.
W. *Ollerton.* 1864.
LACHANCE, Aristide. *Quebec.* 1887.
LACHE, J. A. *Montreal.* 1862-63.
LACHLIS(T)ON—
J. *Whitehaven.* 1858.
Margaret. *Whitehaven.* 1869.
LACKER, Michael. *Manchester.* 1828-34. Dutch and German clocks.
LACKEY, Henry. *Philadelphia.* 1808-11.
LACY, E. *Bristol.* 1840.
LADBROOK, Daniel. *Maidstone.* 1858.
LADD, ——. *Amherst, Nova Scotia.* 1835-ca.1840.
LADE, Michael. *Canterbury.* mar.1723-d.1737.
LADOMUS—
Charles. A. *Chester, Pa, USA.* ca.1825-55.
Jacob. *Philadelphia, USA.* 1843.
Joseph. *Chester, Pa, USA.* ca.1850.
Louis. *Philadelphia, USA.* 1845.
LADSON—
H. *Wrotham.* 1851.
Henry James. *London.* 1863-69.
LADUE, S. P. *Rockford, Ia, USA.* 1859.
LADYMAN, ——. *Hawkshead.* Late 18c. C.
LAFAYETTE, George G. *Brockville, Canada.* 1871.
LAFEUER & BEARY. New York. 1854. T.C.
LAFLEUR, Thomas. *Quebec.* 1871.
LAFLIN, Matthew. Name used by National Watch Co. of *Chicago,* ca.1870.
LAFOY, Theodore. *Newark, NJ, USA.* 1848-52.
LAGARDE, ——. *Paris.* 1860. C.
LAGES, Ernest. *London.* 1863-81.
LAGGATT, Henry. *Montreal.* 1844-70.
LAGGENHAGER, John George. *London.* 1863.
LAIDET, Peter. *London.* 1863-69.
LAIDLAW—
George. *Greenlaw (Scotland).* 1860. Also photographer.
Thomas. *Hetton-le-Hole.* 1827.
Thomas. *Newcastle-on-Tyne.* (B. 1820) 1836). (Also see Laidler.)
LAIDLER, Thomas. *Newcastle-on-Tyne.* 1834. (See Laidlaw.)
LAIGHT, John. *London.* 1734. W.
LAING—
Andrew. *Stamfordham.* 1827.
George. *Aberfeldy.* 1837.
George. *Kelso.* 1860.
James. *Keith.* 1837-60.
William. *Elgin.* 1860.
William. *Fort William.* 1837.
LAIRD—
David White. *Leith.* 1836-50.

LAIRD—*continued.*
John and Andrew. *Glasgow.* 1837.
LAIT, Edward. *Hull.* 1846. W.
LAITHWAITE—
Robert. *Liverpool.* 1715-34. W. and watch cases.
William. *Liverpool.* (B. 1701-) 1710. Watch cases.
LAKE—
Bryan. *London.* a.1667 to John Ebsworth, CC. 1674.
Edmund Edye. *Cheadle.* 1868.
Edward John. *Norwich.* 1875.
F. & C. *Taunton.* 1861-66.
Isaac. *Braintree.* 1828-39.
James Arnold. *London.* 1869-81.
Thomas. *Taunton.* ca.1770-95. C. & W.
William Bentall. *Romford.* 1851-77.
LAKEMAN—
Ebenezer Knowlton. *Salem, Mass, USA.* 1819.
John. *Exeter.* mar.1766.
LAKIN—
——. *Leicester.* ca.1760-ca.1790. C.
William. *Tamworth.* 1850-68.
LAMAILLE, E. G. *London* and *Paris.* 1880-1900.
LAMB—
Anthony. *New York.* 1749. Mathematical instruments.
Charles King. *Yarmouth.* 1865-75.
Cyrus. *Oxford, Mass, USA.* ca.1830.
Edward. *London.* 1832.
H. T. *Woolwich.* 1855.
Henry. *London.* 1875-81.
Henry, jun. *London.* 1875.
Henry Thomas. *London.* 1857-81.
J. *Leigh (Essex).* 1828.
(J. &) H. *Windsor.* (1847)-1854.
James. *Bicester.* mar.1818-53. W.
John. *Leighton Buzzard.* 1864-77.
John. *London.* 1828-32. C.
John. *London.* 1875.
John. *Sunderland.* 1827-34.
Joseph. *London.* 1832.
Mrs. M. *Yarmouth.* 1858-65.
Richard. *Stratford (Essex).* 1828-39.
Sarah (Miss). *London.* 1832-44.
Thomas William. *Yarmouth.* 1846-75.
W. *Bootle.* 1858.
LAMBELL, Samuel. *Northampton.* 1750.
LAMBER, Henry. *London.* 1839.
LAMBERT—
Charles. *Brighton.* 1862-78.
& CRAGG. *Colchester.* See Cragg, Edwin Newton.
Henry. *Bristol.* 1863-70.
J. *Colchester.* 1866.
John. *London.* 1863-81.
John William. *Hereford.* 1863.
L. C. *New York.* 1872.
Mrs. Eliza. *Hereford.* 1850-56.
Norris. From *Cambridge* to *Chelmsford.* 1755.
Peter. *Berwick-on-Tweed.* 1827-37.
S. *London.* 1851.
William. *Birmingham.* 1835-44.

138

LAMBERTOZ, D. *Wilmington, NC, USA.* 1797.
LAMBIE, Hugh. *Liverpool.* 1851.
LAMMERS, Charles. *London.* 1857.
LAMMIE, W. *Worcester.* 1868-72.
LAMOND, J. & Co. *Leith.* 1850.
LAMONT—
  George. *London.* 1869.
  John Harrison (& Co.) (b.1788 at *Bloomfield*) *Belfast.* 1843-d.1844.
LAMONTAGNE—
  Elzear. *Quebec.* 1857-65.
  George. *Quebec.* 1862.
  Michael. *Quebec.* 1833-65.
LAMOYNE, Augustus. *Philadelphia, USA.* 1816.
LAMPARD—
  J. *Bruton (Somerset).* 1861.
  J. *Mere (Wiltshire).* 1859.
  John. *Bristol.* 1870-79.
LAMPLOUGH—
  John Frank. *Bridlington.* 1823. d.1882. C.
  L. E. *Bridlington.* ca.1820-30. C. (Prob. error for above).
LAMPMAN, George. *Thorold, Canada.* 1861.
LAMPREY—
  B. *Burford.* ca.1760. C.
  Benjamin. *Banbury.* mar.1696-d.1721. C.
  John. *Banbury.* s. of Benjamin. b.1704-d.1759. C.
  John II. *Banbury.* s. of John I. b.1734-71. C.
LAMSON, Charles. *Salem, Mass.* 1817-42.
LAMUDE, John. *London.* s. of Reuben of *Chard.* a.1749 to George Wentworth.
LAMVINE, Augustus. *Philadelphia, USA.* 1811-16.
LAMY, Charles. *London.* 1863-69.
LANCASTER—
  ——. *Leeds.* Mid-19c. C.
  Abraham. *Bradford.* 1822. C.
  Francis. *Liverpool.* (B. 1820) 1824. Also chrons.
  James. *Liverpool.* (B. 1825-)1834. Also chrons.
  John. *Prescot.* 1760- (B. 1795). W.
  Richard. *London.* b.ca.1664, a.1684 to Henry Merryman, mar.1687. W.
  Pennsylvania Watch Co. *Lancaster, Pa, USA.* 1877. Formerly Adams & Perry Watch Co.
  Pennsylvania Watch Co. Ltd. 1878-79.
  Watch Co. *Lancaster, Pa, USA.* 1879-92.
LAND, William Richard. *Bradford.* 1871.
LANDAH, John. Place unknown, *USA.* ca.1792. C.
LANDENBERGER & Co. *London.* Late 19c.
LANDIS—
  Isaac. *Coatesville, Pa, USA.* ca.1845. W.
  Isaac C. *Coatesville, Pa, USA.* 1898-1914.
LANDON, G. *Much Wenlock.* 1879.
LANDREAU, ——. *Bordeaux.* ca.1685. W.
LANDRY, Alexander. *Philadelphia, USA.* ca.1790. Imported European clocks.
LANDSBERT, Henry. *London.* 1881.
LANE—
  Aaron. *Elizabethtown, NJ, USA.* 1753-1819. C.
  Charles. *Helston.* 1823-56.
  Charles Benson. *London.* 1869-81. (Successor to James. W.)
  George. *Bristol.* a.1810-30.
  George. *Swansea.* 1848-75.
  James. *Birmingham.* 1850.
  James. *Bradford.* ca.1830. C.
  James. *Philadelphia.* 1803-18.
  James Nall. *London.* 1869.
  James W. *London.* 1863.
  John. *Gloucester.* 1879.
  John. *London.* 1863.
  Lyman J. *New Haven, Conn, USA.* ca.1840-50.
  Mark. *Elizabethtown, NJ, USA.* 1835-37. C.
  Mark. *Southington, Conn, USA.* 1828-33. C.
  N. *New York.* 1848.
  Samuel Edward. *Berkley (Gloucestershire).* 1863-70.
  & STURGEON. *Lancaster, O, USA.* 1827. C.
  William. *Colchester.* 1784-87. W.
  William Puckey. *Helston.* 1873.
  Wright. a.*Oxford* 1676, then *London* where CC. 1687, then *Oxford* again 1689.
LANG (or LONG), John. *Moretonhampstead.* Mid-18c. C.

LANGBRIDGE, John. *Do(u)lton (Devonshire).* ca.1740-d.1775. C.
LANGDON—
  (William) & JONES (George A.). *Bristol, USA.* 1845-55.
  (Edward) & ROOT (Samuel E.). *Bristol, USA.* 1851-54, then Root alone till 1896.
L'ANGE, Augustus. *London.* 1832-44.
LANGE—
  Christian. *London.* 1863-81.
  Hermann. *London.* 1863-75.
  & WYLEYS. *Charleston, SC, USA.* ca.1785.
LANGELAAN, James. *Godalming.* 1878.
LANGENBACH—
  Michael. *London.* 1863-81.
  William. *London.* 1863-69.
LANGFORD—
  Edwin. *Bristol.* 1879.
  Joseph. *London.* 1881.
  Thomas. *London.* 1869-81.
  William (& Sons). *Bristol.* 1825-70(-79).
  William. *London.* 1875.
LANGILLE, J. A. *Acadia Mines, Nova Scotia.* 1890-97.
LANGLANDS & ROBERTSON. *Newcastle-on-Tyne.* 1784.
LANGLEY—
  Charles. *River (Kent).* 1851-74.
  John. *Burton-on-Trent.* 1835.
  John. *Sutton Coldfield.* 1850-80.
  Thomas. *Liverpool.* 1848.
  Thomas. *Oxford.* a.1673-87. W.
LANGLOIS, F. *London.* 1851.
LANGLYE, William. *Liskeard.* 1605. Made town clock.
LANGMACK, H. *Davenport, Ia, USA.* ca.1850.
LANGMAN, William Henry. *Derby.* 1876.
LANGMEAD, George. *St. Johns, Newfoundland.* 1846-77.
LANGSHAW, Hugh. *Prescot.* 1795-97. W.
LANGSTAFF, Thomas. *Middlesbrough.* 1840-47. W.
LANGSTON, Charles Frederick. *London.* 1869-81.
LANGTON—
  Edward William. *London.* 1875-81.
  Francis. *Liverpool.* 1767-68. W.
LANGWORTHY, William Andrews. *Saratoga Springs, NY, USA.* 1822.
LANNY, David. *Boston, USA* from *Paris,* 1789-90, then *New York.* ca.1793-1802.
LANSDOWN—
  George. *Bath.* 1832-61.
  Mrs. M. *Bath.* 1866 (?wid. of George).
LANSING, Jacob H. *Rochester, NY, USA.* 1847-48.
LA PLACE, Charles. *Wilmington, NC, USA.* 1795. W., later at *Philadelphia.*
LAPORTE, S. *Ottawa.* 1875.
LA QUAIN, M. *Philadelphia.* 1794.
LARARD—
  J. *Clapham.* 1862-66.
  James. *London.* 1839-57.
LARCOMB(E)—
  James. *London.* 1857.
  James B(e?)dford. *London.* 1863.
LARGE—
  ——. *Billesdon (Leicestershire).* Early 19c. C.
  John. *Farnworth.* mar.1803. W.
  John. *Manchester.* mar.1792. C.
  'Squire'. *Putnam, O, USA.* ca.1810. Clock cases.
  Thomas. *Melton Mowbray.* 1855-76.
  Thomas. *Syston (Leicestershire).* 1849.
LARGEN, Robert. *Philadelphia, USA.* 1840s. C.
LARKIN—
  John. *Tamworth.* 1761.
  Joseph. *Boston, USA.* 1841-48.
LARKINS, Henry. *London.* 1881.
LARNER—
  A. *Camberwell.* 1878.
  Alfred Thomas. *London.* 1844-63.
LAROCH, John. *London.* (B. a.1785-1825) 1828. W. & C.
LAROCHELLE, Etienne. *Quebec.* 1890.
LARPENT—
  Isaac. b.1711 *Norway.* d.1788 *Copenhagen.*
  & JURGENSEN. *Copenhagen.* ca.1785. W.

LARRARD, Thomas (& Sons). *Hull.* 1813-40(46-58).
LARSEN, Ole. *Frohang (Norway).* 1760. C. and perhaps later.
LARTER, James O. *Kenninghall (Norfolk).* 1830.
LASH—
  & Co. *Toronto.* 1864-76.
  John F. *Hamilton, Ontario.* 1881.
LASHMAR, Richard. *Henfield.* 1839-51.
LASHMER, Richard Marshall. *Brighton.* (B. 1822) 1828.
LASHMORE—
  Alfred. *Llangollen.* 1895.
  Edward. *Oswestry.* 1863-79.
  John. *Southampton.* 1878.
  Thomas. *Southampton.* 1830-67.
LASKEY, Samuel. *Moreton (Hampstead).* *(Devon).* ca.1760-70. C.
LASSELL, William. *Toxteth Park, Liverpool.* 1758-90 (B. -1807). C.
LASSETER—
  Charles. *Steyning.* 1828-51. Also brazier.
  John. *Arundel.* 1828-39.
  W. *Arundel.* 1851-62.
  Walter. *Havant.* 1878.
  William. *Havant.* 1839.
  William J. *Arundel.* 1870-78. C.
LAST—
  ——. *Acle (Norfolk).* ca.1800. C.
  Frederick Edward. *Abergavenny.* 1858-99. Also optician.
  George Clifford. *Yarmouth.* 1875.
  James. *Stalham.* (B. 1796) 1836. (May be two such?)
  W. B. *Bury St. Edmunds.* 1858.
  Walter. *Sudbury.* 1875-79.
  William. *Walsham le Willows (Suffolk).* 1853-79.
  William Bradbury. *Shrewsbury.* 1870.
  William Nelson. *Bury St. Edmunds.* 1839-79.
  William Roberts. *Yarmouth.* 1846-65.
LATCH, William. *Newport (Mon).* 1830-71.
LATCHOW, John. *Baltimore, USA.* 1829.
LATHAM/LATHOM—
  & CLARK. *Charleston, SC, USA.* 1790.
  George. *St. Helens.* 1837. W.
  James. *Albany, NY, USA.* 1795.
  John. *Hyde.* 1878.
  John. *Wigan.* Free 1749-57. C.
LATIMER—
  David. *Ecclefechan (Scotland).* 1860. Clock cleaner.
  William. *Hastings.* 1878.
LATOURNAU, John B. *Baltimore, USA.* 1796-1853.
LATROBE, Samuel H. *Bristol.* 1870-79.
LATSHAR, John. *York, Pa, USA.* 1780-84.
LATTA, A. *Philadelphia.* 1837.
LATTEN—
  R. *Kings Lynn.* 1846.
  Robert James. *Kings Lynn.* 1836.
LAUDER, Andrew. *Belleville, Canada.* 1851-58.
LAUFORD & Co. *Birmingham.* 1880. Importers.
LAUGHENBACH, Michael. *London.* 1857.
LAUGHLIN, A. S. *Barnett, Vt, USA.* ca.1860.
LAUGHTON—
  Valentine R. *Larne.* 1843-54.
  William. *London.* CC. 1683-97.
LAULE, Fidely. *Edinburgh.* 1850-60.
LAUNT, A. *Chesterfield.* 1849-55.
LAUNY, David F. *New York.* 1823.
LAURENCE—
  F. *Weston-super-Mare.* 1866.
  G. *Cheltenham.* 1856.
  Isaac. *Cheltenham.* 1850.
  J. *Southampton.* 1848.
  Josiah. *Chelmsford.* ca.1848.
  Richard. *Bath* and *Warminster.* 1729-73.
  S. *London.* 1844.
  Teft. *Louth.* 1850-61.
LAUSSINE, Esaius. *Edinburgh.* 1595 not LAUSSME-B.).
LAUTIER, Charlotte. *Bath.* 1848-54.
LAVATT, Joseph see LOVATT.
LAVERACK, David. *Rotherham.* mar.1826-34.
LAVOINE, Jules A. *London.* 1869.
LAW—
  Abraham. *Goldington* nr *Bedford.* Late 18c. C.
  Anthony. *London.* 1839.

LAW—*continued.*
  Brothers. *Clapham.* 1878.
  Brothers. *London.* 1869-81.
  David. *Kilmarnock.* 1837.
  James. *Castle Douglas.* s. of Robert. post-1830-36.
  James II. *Castle Douglas.* s.of James I, post-1830.
  Patrick. *Belfast.* 1820-39. Brass caster.
  Robert. *Castle Douglas.* ca. 1790-1860. C. (may be two such?).
  Robert. *Dumfries.* 1887.
  Robert. *Kirkcudbright.* 1723.
  Samuel. *Dumfries.* 1848-60.
  Samuel. *Rochdale.* 1824. C.
  Thomas. *Castle Douglas.* ca.1723.
  Thomas (& Son). *London.* 1863-69(75-81).
  William. *Burnley.* 1848. Clock repairer.
  William. *Kirkcudbright.* (B. 1820) 1836-60.
  William. *London.* 1869.
  William. *Philadelphia, USA.* 1839-41.
LAWING, Samuel. *Charlotte, NC, USA.* 1807-65 (as LAWING & BREWER 1842-43).
LAWLER, Ralph. *Hertford.* 1866-74.
LAWLEY—
  Albert B. *London.* 1875.
  B. *London.* 1851-69.
  Bernard. *London.* 1839.
  J. *Bath.* 1856-66.
  John. *Bristol.* 1830.
  Joseph. *Wellington* b. ca. 1780, mar. 1831 d. 1854. W.
  Theodore. *London.* 1857-75.
  Theodore. *Manchester.* 1828. German and Dutch clocks.
  Thomas Henry. *Birmingham.* 1880.
LAWLOR, James John. *St. John, New Brunswick.* 1853-64.
LA WRAY PIERE, William. *London.* Alien 1622.
LAWRENCE—
  ——. *Kingston-on-Thames.* 1741. C.
  Edmund (or Edward). *London.* 1875-81.
  George. *Keith.* 1837.
  George. *London.* (B. 1820) 1832-39.
  George. *Lowell, Mass, USA.* 1832.
  George. *Ramsgate.* 1847.
  Henry. *Bristol.* 1773.
  J. *Eltham (Kent).* 1838.
  James. *Eltham.* 1847-74.
  James. *London.* 1828-39.
  John. *Philadelphia, USA.* 1798.
  Josiah. *London.* 1832-39.
  Mrs. Ellen. *Louth.* 1868-76.
  Mrs. M. A. *Midhurst.* 1878.
  Richard. *Godalming.* 1878.
  S. *London.* 1851.
  Silas H. *New York.* ca.1840.
  T. *Louth.* 1849.
  Thomas. *Thame.* a.1759.
  Thomas Vigurs. *Penzance.* mar.1811-17. W.
  William. b.1762. *Lancaster.* mar. *Cartmel* 1787-1811. C.
  William. *Cudsdon.* Same man as at *Thame.*
  William. *Halifax.* mar.1788. C.; then *Manchester.*
  William. *Mount Pleasant, NY, USA.* ca.1850.
  William. *Thame.* 1744-64. C.
  William. *Todmorden.* ca.1820.
  William Edward. *London.* 1875-81.
LAWRENSON, James. *Prescot.* 1819. W.
LAWRIE, John. *Birmingham.* 1868-80.
LAWS, Michael Graham. *Berwick-on-Tweed.* 1845.
LAWSE, Jonathan. *Amwell, NJ, USA.* 1778-1809.
LAWSHE, John. *Amwell, NJ, USA.* ca.1750. C.
LAWSON (and LAUSON).
  Benjamin. *Askrigg.* mar.1786. C.
  & GODDARD. *London.* 1875
  Henry. *Hindley* nr *Wigan.* 1834.
  James. *London.* (B. a.1778?) 1806. W.
  John. Not town stated. Prob. the *Bingley* man, q.v.
  John. *Bingley.* ca.1775-d.1789. C. Also at *Bradford.*
  John. *Bradford.* ca.1775-d.1789. C. Also at *Bingley,* where died.
  John. *Bradford.* 1822-37. C.
  John. *Huddersfield.* 1850-53.
  John. *London.* 1851-63.

LAWSON—*continued.*
John. *Ince* nr *Wigan.* 1851.
Peter. *London.* 1881.
Robert. *Bridlington.* 1858-88.
Robert. *Manchester.* mar.1776. C.
Samuel. *Keighley.* b.1721-d.1770. C.
Thomas (& Son). *Brighton.* 1839-62(70-78).
Thomas. *Kirkoswald.* 1848-79. Also glazier.
Thomas. *South Shields.* ca.1790. C.
William. *Bradford.* 1814-ca.1825. C.
William. *Keighley.* ca.1750-1822. C.
William. *Newton le Willows (Lancashire).* (b.1738)-
   d.1805. C.
William. *Todmorden.* 1824-66.
LAWTON, John. *Sheffield.* 1871.
LAWYER, Loring. *Newark, NY, USA.* ca.1835. C.
LAXTON—
W. H. *Coventry.* 1854-68.
William Henry. *Coventry.* 1880.
LAY, Asa, jun. *Hartford, Conn, USA.* ca.1784.
LAYBOURNE, Mark. *Driffield.* mar.1793-94. C.
LAYCOCK—
A. *Otley.* 1834. T.C.
Henry. *Rotherham.* 1862. W.
Henry John. *Eastbourne.* 1870.
J. *Leeds.* ca.1800. C.
John. *Bedminster.* Early 17c. Sundial (?ca.1620).
John Henry. *Nottingham.* 1876.
Thomas. *Bradford.* 1871.
LAYES, E. *Birmingham.* ca.1800. C.
LAYZELL, John. *London.* 1857.
LAZARUS—
Elizabeth & Son. *London.* 1839.
Henry. *London.* 1844-57.
I. *London.* 1851-57.
Isaac. *London.* 1828-57.
Jacob. *London.* 1857.
John. *London.* 1875-81.
Joseph & Jacob. *London.* 1863-69.
Joseph & John. *London.* 1875.
& LAWRENCE. *London.* (B. 1825-) 1828-32.
   Wholesalers.
Lewis. *Carmarthen.* 1835.
Peter. *Vienna.* ca.1800. C.
Samuel. *London.* 1869-81.
LEA—
Alfred. *Aberdare.* b.1868-1924.
Alfred. *Leeds.* 1871. Also clock case makers.
Charles. *Pembroke Dock.* 1887.
Henry. *St. Helens.* 1821-25. W.
John. *Lutterworth.* 1602. Rep'd ch. clocks.
Richard. *Ringwood.* Early 19c. C.
LEACH—
& BRADLEY. *Utica, NY, USA.* 1832-34.
Caleb. *Plymouth, Mass, USA.* 1776-90.
Charles. *London.* 1851-81.
Charles B. *Utica, NY, USA.* 1843-47.
E. *Romsey.* 1859.
Edwin. *Bristol.* 1870-79.
George. *Redhill.* 1878.
George. *Salisbury.* 1830-48. C.
Henry. *Chorley.* 1828-34.
Henry. *Preston.* 1851.
J. *Preston.* 1858.
John. *Bingley.* 1871.
John. *Romsey.* 1830-48.
Joseph Britton. *Brighton.* 1828.
Reuben. *Romsey.* 1867-78.
Richard. *London.* marq lc. clock, supposedly ca.1700.
Richard (& Son). *London.* 1832-63(69-81).
Thomas. *London.* 1851-69.
Thomas, jun. *London.* 1881.
LEADBEATER (also LEADBETTER, etc).—
Charles. *Wigan.* 1824-29. W.
H. *Macclesfield.* 1878.
Henry. *Congleton.* 1848-57. C.
J. *Rotherham.* 1871.
J. *Kidsgrove.* 1860.
Peter. *Congleton.* 1828-48.
Peter John. *Congleton.* 1848.
& Son. *Sandbach.* 1832.
T. *Sandbach.* 1857.

LEADBEATER—*continued.*
Thomas. *Manchester.* mar.1830. W.
Thomas. *Prescot.* mar.1802-19. W.
Thomas. *Wrexham.* 1856-76.
Thomas, sen. *Sandbach.* 1828-34.
Thomas, jun. *Sandbach.* 1834-48.
Timothy. *Wigan.* Post-1834.
W. *Kidsgrove.* 1860.
William. *Sandbach.* 1865-78.
William. *Wigan.* 1828-51.
LEADER, 1914 and later, product of Pequegnat Clock
   Co.
LEADINGHAM, Adam. *Premnay (Scotland).* 1860.
LEAH—
Henry. *Stalybridge.* 1848.
Samuel. *?London.* 1777. W.
Samuel Henry. *London.* (B. 1820-) 1828-44.
LEAPMAN—
Isaac. *Bristol.* 1830.
Lewis. *London.* 1875-81.
Moss. *London.* 1869-75.
LEAR—
Henry. *Simcoe, Canada.* 1857-74.
James. *Birmingham.* 1868.
Richard. *Plymouth.* ca.1770. C.
LEARMONT, William. *Montreal.* 1841-70.
LEATHAM, John. *Waterford.* ca.1780. C.
LEATHER, Richard. *Prescot.* 1823-25. W.
LEATHERBARROW, Charles. *Liverpool.* 1784. W.
LEAVENWORTH—
Mark. *Waterbury, Conn, USA.* 1774-1849. C.
William (& Son). *Waterbury, Conn, USA.* ca.1802-10
   (*Albany, NY, USA.* 1817-23).
LEAVER—
& BREEZE. *London.* 1875.
Charles. *Birmingham.* 1880.
LEAVIT(T)—
Dr. Josiah. *Hingham, Mass, USA.* ca.1772. C.
M. F. *Kalamazoo, Michigan, USA.* 1870.
LE CHEMINANT—
& Co. *London.* 1881.
John, sen. *London.* 1857-75.
J(ohn), jun. *London.* 1875.
LECK—
Robert. *Jedburgh.* 1837.
Robert. *Mauchline Tower (Scotland).* 1850.
William. *Jedburgh.* 1837.
LE CLAIRE, Augustus. *London.* 1881.
LECLUSE, Felix. *London.* 1844-81.
LECOMBER, John. *London.* 1881.
LE COMPTE, John. *London.* 1851-75.
LE COUNT (also LE COMPTE), Daniel. *London.* CC.
   1676-97. W. & C.
LEDDERER, F. *Norwich.* 1846.
LEDER(S), John. *Liverpool.* 1834-51.
LEDGER, Eldred. *Cobham.* 1878.
LEDOUX, François. *Amiens.* Early 18c. C.
LEE—
Absolom. *West Rudham.* 1836-75.
Cuthbert. *London.* a.1668 to Robert Williamson, CC.
   1676-97. (B. -1718). C.
George. *Askrigg.* b.1784. To Skipton by 1806. C.
George. *Galt, Canada.* 1851.
George. *Owen Sound, Canada.* 1857.
George. *Skipton* from *Askrigg.* 1806-53. C.
H. *Jersey.* 1832-34.
H. *Worthing.* 1851.
Henry. *Bolton.* 1851-58.
Henry. *Burslem.* 1860-76.
Henry. *London.* 1869.
I. J. *Woolwich.* 1851.
Isaac. *Liverpool.* 1834.
Isaac John. *London.* 1839-51.
J. *Bolton.* 1858.
James. *Bristol.* 1850.
James. *Farnworth.* mar.1785. W.
James Griffith. *London.* 1875-81.
Jane. *Milford Haven.* 1844.
John. *Coventry.* 1842-54.
John. *High Wycombe.* 1767-1823. T.C.
John. *Liverpool.* 1824.
John. *Owen Sound, Canada.* 1857.

LEE—*continued.*
John Henry. *London.* 1875.
John Stanford. *Brighton.* 1855-78.
Joseph I. *Belfast.* 1818-49 (as Gardiner & Lee 1818-27).
Joseph II. *Belfast.* s. of Joseph I. 1868.
M. *Brighton.* 1851.
Michael. *Milford Haven.* 1830-35.
Peter. *Prescot.* 1819. W.
Peter. *Skipton.* b.1817, s. of George. -1871. W.
Richard. *Great Marlow.* 1768-70. W.
Richard. *Halifax.* 1665. Rep'd ch. clocks.
Roger. *Leicester.* Free 1691-ca.1700. W. & C. and lant. clock.
Samuel. *London.* a.1687 to Jeffrey Bailey, CC. 1694 (B. -1719).
(Joseph) & Son (Joseph). *Belfast.* 1849-66 (Josephs I and II above).
Susan. *Milford Haven.* 1840.
Thomas. *Bury.* 1851.
Thomas. *Farnworth.* mar.1760. W.
William. *Great Marlow.* Succeeded brother, Richard, in 1770-71. W.
William. *Leicester.* s. of Roger. ca.1715-d.1744. C.
William. *Northampton.* 1777-1830. C.
William. *West Haddon (Northants).* 1864-77.
LEECH—
Henry. *Bristol.* 1879.
John. *Kirkham.* 1822-29. W. and silversmith.
Thomas. *Prescot.* mar.1768.
William. *London.* 1851-57.
William. *Maidstone.* 1832.
LEEDHAM—
J. *Halifax, Nova Scotia.* 1863.
J. B. *Brierley Hill.* 1860.
William. *Sherburn nr Aberford.* 1866-71. W.
LEEDS—
Gideon. *Philadelphia, USA.* 1841.
Howard G. *Philadelphia.* 1840.
Paul Colley. *Sittingbourne.* 1838-74.
LEEFE, Benjamin. *Malton.* 1866-86. W.
LEEMING, Thomas. *Settle (Yorks).* 1866-71. W.
LEES—
Charles. *Nottingham.* 1828-64.
G. H. & Co. *Hamilton, Ontario.* 1844-95.
J. & Co. *Ashton-under-Lyne.* 1858.
J. H. *Bury.* 1858. W.
J. H. *Wolverhampton.* 1860.
James. *Ashton-under-Lyne.* 1697-1706. C.
James. *Manchester.* d.1797. C.
James. *Middleton (Manchester).* 1848.
John. *Manchester.* mar.1802. C.
John. *Middleton (Manchester).* d.1785. C.
John. *Middleton.* 1824-34. W. -
Jonathan. *Bury.* ca.1730. Died *Middleton* 1785. C.
Mark. *Walsall.* 1860-76.
Samuel. *Ashton-under-Lyne.* 1824-48. Also jeweller.
Thomas. *Bury.* ca.1790-1848. C.
Thomas. *Hamilton, Ontario.* 1862-95.
Thomas. *Middleton (Manchester).* 1848-58.
William. *Accrington.* 1848-58.
William. *Haslingden.* 1822-58. W.
William. *Middleton (Manchester).* 1808-13. W.
William. *Walsall.* 1876.
LEESON—
Henry. *Birmingham.* 1880.
Henry. *Coleshill.* b.1844, s. of William I. -1900. C.
William I. *Coleshill.* b.1809, a. in *London* but ran away. Worked firstly at *Minworth,* then *Curdworth,* then *Coleshill* 1831-d.1886.
William Tansley. *Coleshill.* b.1842, s. of William I. mar.1868-78. C. Lived at the Clock Inn.
LEETE, Thomas. *London.* 1869-81.
LEFEVER, George. *Wisbeach.* 1830-40.
LEFEVRE, N. *Montreal.* 1875.
LEFFERTS—
Charles. *Ovid, NY, USA.* 1827.
Charles. *Philadelphia, USA.* 1818-22.
& HALL. *Philadelphia.* 1818-22.
LEFFI, S. *Newport (Isle of Wight).* 1859.
LEFFLER—
Aaron. *Lydney.* 1879.
Adolf. *Pontarddulais.* 1887.

LEFFLER—*continued.*
Charles. *Bristol.* 1870-79.
Charles. *Portishead.* 1875.
I. *Dudley.* 1868.
J. M. & Co. *Tipton.* 1860-76.
John. *Nuneaton.* 1842.
Martin. *Tipton.* 1842-50.
N. *Birmingham.* 1854-60.
LE FORTIER, ——. *Jersey.* 1834-37.
LEFRANC, ——. *Paris.* 1878-80. C.
LE GALLAIS, John. *Jersey.* 1853.
LEGARE—
Alphonse. *Quebec.* 1890.
Phillipe. *Quebec.* 1826-d.1843.
LEGE, Annibal. *London.* 1875-81.
LEGER—
Anseline Marcel. *Shediac, New Brunswick.* 1887.
Leon C. *Quebec.* 1871.
Zoel Marcel. *Shediac, New Brunswick.* 1887.
LEGG, John. *Blechingley.* (B. 1787-)1828-39.
LE GOUPILLOT—
E., jun. *Jersey.* 1835.
& Son. *Jersey.* 1837-53.
LE GOUX, J. F. *Charleston, SC, USA.* Late 18c.
LEGRAND—
E. (?*Paris*). ca.1840. Carriage clocks.
James (Jacques) (sen.). *London.* CC. 1640-64, Frenchman.
LE GRAVE, Samuel Christopher. *London.* 1828-32.
LE GROS—
John F. *Baltimore, USA.* 1793.
Peter Joseph. *Belfast.* 1835-39.
Peter Joseph. *London.* 1828. Also mechanical model maker.
LEHMAN(N)—
Frederick F. *Aylesbury.* 1877.
Germanus. *London.* 1869-75.
John. *Swansea.* 1868-87.
Lewis. *London.* 1875.
Severin. *Mold.* 1868-74.
LE HURAY—
Nicholas, sen. *Philadelphia* from *Germany.* 1809-31. C. (?B. has from Guernsey ca.1780).
Nicholas, jun. *Ogletown, Del, USA.* s. of Nicholas, sen. 1834.
Nicholas, jun. *Philadelphia, USA.* 1800-46.
LEIBERT, Henry. *Norristown, Pa, USA.* ca.1850.
LEIF—
George. *Helmsley.* 1810. C.
George Turbutt. *Sheffield.* 1822-37.
LEIGE, Daniel. *London.* ca.1710. C.
LEIGH—
David. *Pottsdown, Pa, USA.* 1849.
James. *?Lancashire* or *Cheshire.* ca.1700. C. (Might be JAMES of *Leigh, Lancs?*)
James. *Bold.* b.1762. s. of Thomas. mar.1785-1808. W.
Joshua. *Liverpool.* (B. 1825-)1834.
Peter. *Wigan.* 1793. C.
Thomas. *Sutton/Farnworth.* a.1752, mar.1760, d.1795. C.
William. *Eccleshall.* 1868-76.
William. *Liverpool.* 1780. C.
William. *Liverpool.* 1851.
William. *Newton-le-Willows (Lancashire).* 1822. C.
LEIGHTON—
John. *Lancaster.* 1838.
John. *London.* CC. 1653.
Thomas. *Birmingham* and *Paris.* 1868.
Thomas & Sons. *Birmingham.* 1880. Dial makers.
Walter. *Montrose.* 1830-60.
LEINHARDT, Christian. *Carlyle, Pa, USA.* 1782.
LEIPNIK, Isaac M. *London.* 1869.
LEIST, M. *Halifax, Nova Scotia.* 1877-79.
LEITCH, Daniel. *Kincardine-on-Forth.* 1836.
LEITH, Angus. *Tain.* 1860.
LEITHEAD—
James. *Galashiels.* 1836.
James. *Moffat.* 1835-60. W.
LE JEUNE & PERKEN. *London.* 1881. Ships clocks.
LELLI, S. *Newport (Isle of Wight).* 1839.
LEMAIRE, ——. *London.* ca.1700. W.

LEMAITRE—
& BERGMANN. *London.* 1863-69.
Henry. *Dublin.* 1788.
Jules. *Canterbury.* From *Paris.* 1865.
P. T. *London.* 1811. Watch case maker. (B. has Paul Thomas. CC. 1815-24.)
LEMAN, T. *Ditchingham (Norfolk).* 1858.
LEMIST—
(William) & TAPPAN (W. P.) *Philadelphia, USA.* 1816-19.
William King. *Philadelphia, USA.* 1791-1820. C. (d.1840.)
LEMM, J. *Esher.* 1851-66.
LEMMON (see also LEMON)—
Charles. *Brentwood.* 1874.
Charles. *Croydon.* 1851-78.
Charles. *London.* 1844.
Henry. *Horsham.* 1870.
Henry. *London.* 1839-81.
Obadiah Orange (really!). b.1791 at *Christchurch.* *Battle* 1820-27, then *Hampton Wick* 1827-61. C., W. and gunmaker.
LEMOINE, A. *Philadelphia.* 1810-17.
LEMON (see also LEMMON)—
Abraham & Morris. *Douglas, IoM.* 1860.
J. J. *Louisville, Ky, USA.* 1845.
William. *Lynden, Ontario.* 1868. Also gunsmith.
LEMREP, Thomas. *London.* 18c. W.
LENDON, Miss H. *Taunton.* 1861.
LENGGENHAGER, John G. *London.* 1851-57.
LENHARDT, Godfrey. *York, Pa, USA.* 1779-1819.
LENNON, Edward Francis. *London.* 1881.
LENTON—
——. *Congleton.* ca.1780. C.
Ebenezer. *Clapham.* 1878.
LEOFFLER—
Bernard. *Copetown, Canada.* 1857.
& Co. *Greenwich.* 1849.
Joseph. *Pontypool.* 1880.
Peter. *Greenwich.* 1839.
LEOMBERG, Benjamin. *Swansea.* 1868-75
LEON—
Brothers. *London.* 1851.
George Isaac. *London.* 1844-51.
Jacob. *Aberystwyth.* 1835-44.
Maria. *Aberystwyth.* 1868.
LEONARD—
Charles. *Camberwell.* 1828.
Charles. *London.* 1832.
F. *Wolverhampton.* 1860.
J. *Botley (Southampton).* 1848.
Joseph. b.1772. *London.* a. of William Mills there, then *Winchelsea* 1804. C. & W.
Thomas. *London.* 1851-69.
Watch Co. *Boston, USA.* 1911. W.
LEONI—
& Co. *New York.* ca.1840.
Frederick. *Andover.* 1839-48.
LEOWE, Lewis. *London.* 1869.
LEPINE—
Charles. *Deal.* 1839-45.
Edwin. *London.* 1851-57.
Henry. *Canterbury.* bro. of Charles. 1823-38. W.
Robert T. *Nova Scotia.* 1884.
LE PLASTRIER (also LE PLASTERER)—
I. *London.* 1851.
Isaac. *London.* (B. CC. 1813-20) 1828-44.
Isaac & Son. *London.* 1828-32.
Louis. *London.* 1832-44.
Mrs. Sophia. *London.* 1857. (Succeeded William & Co.)
Robert. *Dover.* Free 1810-40.
Robert Louis. *Ramsgate.* 1832.
William. *Dover.* Free 1800-1802. W.
William I. *London.* 1857.
William L. & Co. *London.* (B. s. of Isaac, a.1821) 1844-51.
LEPMANN, Frederick Fidelio. b.1820 *Germany. Aylesbury* 1842, where partner with Engelsman. Alone from 1845-91.
LEPPER—
Francis. *Belfast.* ca.1778-retired 1835. W. and dentist.

LEPPER—*continued.*
George. *Belfast.* (B. 1778) ca.1790-1810. C. & W.
LEPPLEMAN, Edward. *Buffalo, NY, USA.* 1836-39.
LEPROHON, Alfred. *Montreal.* 1875.
LERICHE, Charles L. *London.* 1857.
LE ROUX—
Edward. *Jersey.* 1832.
M. *Jersey.* pre-1832.
LEROY—
& Son. *London.* 1857-81.
Theodore Marie. *Paris.* b.1827-d.1899.
LERWILL, William. *Birmingham.* 1880. Clock cases.
LESCHEY, Thomas. *Middletown, Pa, USA.* Early 19c.
LESCHOT, Louis A. *Charlottesville, Va, USA.* d.1838.
LESCOTT, Lambert. From *Paris* to *Providence, RI, USA.* 1769, then *Hartford, Conn, USA.* Later to *London, England.*
LESIEUR, ——. *Paris.* 1819. C. (B. 1822-25).
LESLIE—
Alexander. *London.* 1875.
David. *Montreal.* 1820.
John. *Ottawa.* 1848-76.
Peter. *Burntisland (Scotland).* 1837.
Robert. From *London, England,* to *Philadelphia, USA.* 1793-d.1803.
Robert & Co. *Baltimore, USA.* 1796.
W. F. *Oakville, Canada.* 1857.
William J. ca.1791 *New Brunswick, NJ, USA;* 1799-1817 at *Trenton.*
(William J.) & WILLIAMS. *New Brunswick, NJ, USA.* 1791-1806.
LESOURGEON, Samuel. *London.* a.1735 to James Hubert.
LESQUEREUX, L. & Son. *Columbus, O, USA.* 1804.
LESTER—
J. U. *Oswego, NY, USA.* 1843-45.
James. *London.* 1881.
W. *Dunstable.* 1877.
LETCHER, Walter. *St. Agnes (Cornwall).* 1856-73.
LETELLIER—
Charles. *Basseville, Canada.* 1845.
Charles. *Beaumont, Canada.* 1857.
John. *Philadelphia, USA.* 1770-93. Later a dentist at *Wilmington, Del, USA.*
LEUBA—
Henry. *Lexington, Ky, USA.* 1818.
Henry. *Basel.* (B. 1823) 1851. C.
LEUBA-PRINCE, ——. *Geneva.* 1881.
LEVASSEUR—
Firmin. *London.* 1851-57.
Joseph. *Quebec.* 1890.
LEVER, Peter. *St. Helens.* 1831-37. W.
LE VERROUX, Louis. *Paris.* ca.1670. C.
LEVERSHA, J. *Stogursey (Somerset).* 1861-66.
LEVERTON, W. *St. Columb (Cornwall).* d.1837. W.
LEVEY—
Brothers. *Toronto.* 1861.
Emanuel. *Dover.* Free 1818-51.
& MYERS. *Portsea.* 1867.
S. G. *Toronto.* 1862.
LEVI—
——. *Ramsgate.* 1802 (B. has Noah 1789).
Abraham. *Wednesbury.* 1842.
Alfred. *London.* 1869-81.
Barnet. *Liverpool.* 1828-51. Also jeweller.
Colman, *Ross.* 1850.
Emanuel. *Birmingham.* 1850.
Garretson. *Philadelphia, USA.* 1840-43.
Henry. *London.* 1881.
Isaac. *Philadelphia, USA.* (From *London*) 1790. W. (Later with Michael.)
Isaac. *Ross.* 1835-44.
Israel Morris. *Oxford.* 1853. C. & W.
John. *London.* 1857.
John & Alfred. *London.* 1863.
Lewis. *Bedford.* 1839-64.
Michael, *Philadelphia, USA.* ca.1750; *Baltimore* 1802-16.
Morris. *Birmingham.* 1868.
Moses. *Woolwich.* 1823.
Solomon. *Canterbury.* 1788-93.
LEVICK, George. *Market Rasen.* 19c. C.

LEVIEN, Louis W. *Chatham.* 1845.
LEVIN—
  Alexander. *Penzance.* 1847.
  & FERGUSON. *Alexandria, La, USA.* ca.1850.
LEVINBERG, B. *Bristol.* 1863.
LEVINE, Moses. *Norwich.* 1875.
LEVINSON, Michael. *Sheffield.* 1862. W.
LEVIS, John. *Montreal.* 1875.
LEVITT—
  Isaac & Morris Tobias. *London.* 1851-57.
  Lewis. *London.* (B. 1825) 1828-32.
  Solomon. *London.* 1832.
  Stephen. *Chelmsford.* Quaker. b. 1689. mar.
    1708-1782. C.
LEVY—
  A. *Gosport.* 1867.
  A. & Son. *Slough.* 1854.
  Aaron. *Chatham.* 1826-32.
  Aaron Alexander. *Newcastle-on-Tyne.* 1848-52.
  Abraham. *Romford.* 1874.
  Abraham. *Ross.* 1830-44.
  Alexander. *Eton.* 1827-54 (also at *Slough* in 1854).
  Alexander William. *Eton.* 1837.
  Brothers. *Hamilton, Ontario.* 1895.
  Brothers & SCHEUER. *Hamilton, Ontario.* 1875-85.
  Charles. *Briton Ferry (Wales).* 1868.
  Charles. *Truro.* 1844.
  & COHEN. *Toronto.* 1877.
  Daniel. *Carmarthen.* 1840-44. C.
  & DENZIGER. *Liverpool.* 1848-51.
  Emanuel. *Exeter.* ca.1780. C.
  Frederick. *Birmingham.* 1860-80. Wholesaler.
  H. & A. *Hamilton, Ontario.* 1868.
  Henry (H. A.). *Philadelphia, USA.* 1841.
  Henry & Co. *Manchester.* ?19c. W.
  Isaac. *London.* a.1757 to Isaac Fox.
  Isiah. *Gainsborough.* 1828.
  Jacob. *Truro.* 1822-47. C. & W.
  Jonas (& Son). *London.* 1811-51 (1857-81). W.
  Joseph. *Hull.* 1813. W.
  Joseph. *London.* 1828.
  Joseph Solomon & Co. *London.* 1881.
  Josiah. *Gainsborough.* 1835.
  Lawrence. *Birmingham.* 1850-60.
  Lawrence. *London.* 1832-39.
  Lewis B. *Philadelphia, USA.* 1841-45.
  M. & Co. *Philadelphia, USA.* 1816.
  Michael. *Hull.* 1790-1813. W.
  & MICHAELS. *Halifax, Nova Scotia.* 1876-81.
  Morris. *Middlesbrough.* 1866. W.
  Morris L. *Birmingham.* 1868.
  Moses. *Ipswich.* 1839.
  Moses & Mark. *Toronto.* 1856.
  & MOSS. *London.* 1844.
  R. & Co. *London.* 1851-63.
  Reuben. *London.* 1869.
  Robert. *London.* 1863-81.
  Solomon H. *London.* 1851.
  Wolff & Co. *London.* 1869.
LEWIN, Joseph Hodge. *Leicester.* 1835. C.
LEWIS—
  ——. *Philadelphia, USA.* 1796.
  ——. *Tunis, NY, USA.* post-1800.
  Alfred. *Liverpool.* b.1830, to *Kendal* ca.1853-60. W.
  Benjamin B. *Bristol, USA.* 1864-70. C.
  Benjamin B. & Son. *Bristol, USA.* 1871. C.
  Charles. *St. Albans, O, USA.* 1823-25.
  Charles. *Wincanton.* ca.1720-40. C.
  & Co. (C. S.). *Bristol, USA.* 1875.
  Curtis. *Reading, Pa, USA.* 1770-1847. C.
  Erastus. *New Britain* and *Waterbury, Conn, USA.*
    ca.1825.
  Evan. *Ferndale.* 1887-99.
  Frederick H. *Rochester, NH, USA.* 1849.
  G. *Coventry.* 1860-80.
  George. *Cannonsburg, Pa, USA.* ca.1830. Clock case
    maker.
  George. *Colchester.* b.1784-1805, then *London* till
    1809, then back to *Colchester* 1809-12.
  George. *Kingston-on-Thames.* 1878.
  George. *Manchester.* 1824.
  George. *Narberth.* 1835-50.
  George. *Pembroke Dock.* 1830.

LEWIS—*continued.*
  George Samuel. *Newcastle-on-Tyne.* 1827-48.
  Henry. *Dudley.* 1855. C.
  Isaac. *Newark, NJ, USA.* 1782.
  & IVES. *Colebrook, Conn, USA.* pre-1849.
  (Lewis) & IVES (Joseph). *Bristol, USA.* 1819-23. C.
  Jackson. *San Jose, Cal, USA.* ca.1870.
  James. *Swansea.* 1844.
  John. *Brixton.* 1878.
  John. *Ewell.* 1866-78.
  John. *London.* 1857.
  John. *Tredegar.* 1850-99.
  John Bateman. *Merthyr Tydfil.* 1830.
  John (John M.). *Philadelphia, USA.* 1830 and later.
  John Philip. *Bilston.* 1842-76.
  John Walter. *Knighton (Scotland).* 1887.
  Joseph. *Bristol.* 1726-40.
  Joseph. *Chepstow.* 18c. C.
  Joseph. *Hove.* 1862-78.
  Levi. *Bristol, USA.* 1809-23.
  Morris. *Liverpool.* mar.1834. W.
  Peter. *Liverpool.* 1695-d.1699. W.
  Rees. *Newbridge* (= *Pontypridd*). 1830-40. C.
  Richard. *Wincanton.* 1760. C. & W.
  Robert. *Liverpool.* mar.1834. W.
  Robert. *London.* 1832.
  & Sons. *Brighton.* 1862-70.
  Thomas. *Bala.* b.1790-1830. C.
  Walter. *Bristol.* 1879.
  William. *London.* 1851-57.
LEWISON, Lewis. *Toronto.* 1913.
LEWITT—
  A. *Leicester.* 1849.
  Benjamin. *Leicester.* (B. 1815-) 1828-35.
  George. *Leicester.* 1876.
LEWNS, Charles Frederick. *Rye.* 1839-62.
LEWSE, John. *London. Chancery Lane* 1662.
LEWTHWAITE—
  Edwin. *Halifax.* 1850-71.
  William. *Liverpool.* 1696-1710. W. and watch case
    maker.
LEWTON—
  George. *Kingswood.* 1830-79.
  George. *Watley's End, Winterburn (Bristol).* 1856-79.
  R. *Stapleton (Bristol).* 1856.
  Richard. *Kingswood.* 1863-79.
LEWTY, James. b. 1823 *Nottingham. Bottesford.*
  Leicestershire. 1841-99.
LEY—
  John. *St. Ives (Cornwall).* mar.1805-15. W.
  Thomas Stephen. *Hayle.* 1844.
  William. *Hayle.* 1847.
LEYLAND, Thomas. *Prescot.* 1823-51. W.
LEYNS, John. *London.* d.1603.
LEYOAK, William. *London.* mid-18c. W.
LEZARD & Sons. *London.* 1875-81.
LIBBY, R. S. *Port Hope, Canada.* 1857.
LICENSE, Paul. *Ipswich.* 1853-74.
LICHTY, Jacob. *Cracow.* ca.1745. C.
LICKERT—
  H. *Cambridge.* 1846.
  John. *Reading.* 1837-54.
  Matthew. *London.* 1869.
  & WINTERHALDER. *Reading.* 1864.
LIDBROOK (also **LADBROOK**), Thomas. *Oxford.*
  a.1679. C.
LIDDELL—
  A. *Folkestone.* 1866.
  Adam. *Liverpool.* 1848-51.
  Adam. *Newry.* 1784-1840. C. (With Woodhouse 1824.)
  James. *Bathgate.* 1825.
  William. *Portobello.* 1839.
LIDDLE—
  James. *Kirkwall.* 1860.
  & Sons. *Edinburgh.* 1830-50.
LIDDY—
  & McCORRY. *Newry.* 1805.
  Richard. *Newry.* 1816-28.
LIDELL, Thomas. *Frederick, Ind, USA.* ca.1860.
LIEBERT, Henry. *Norristown, Pa, USA.* 1849.
LIECHTI, Erhard. *Wintertur (Switzerland).* 1561-1604.
  T.C.
LIEUTAUD, B. *Paris.* d.1780. C.

LIGGINS, Mrs. M. A. *Coventry.* 1854. Watch case maker.
LIGHT—
G. *Lymington.* 1848.
John. *Bristol.* 1826-50.
LIGHTFOOT, Roger. *St. Helens.* 1770-79. W.
LILIENTHAL, J. *New Orleans, La, USA.* ca.1850-ca.1860.
LILL(E)Y—
Edward. *Mossley (nr Manchester).* 1871-78.
Mrs. M. *Poole.* 1867.
W. C. *Poole.* 1848-55.
LILLIE, William. *Fraserburgh.* ca.1790. C.
LIMB(ER)RY—
John. *Curry Mallet.* d.1646.
John. *Curry Mallet.* d.1676.
John. *Olney.* b.1811, mar.1845-68.
LIMEBURNER, John. *Philadelphia, USA.* 1790.
LINACRE/LINAKER—
Henry. *Liverpool.* 1834 (B. 1796-1829).
William. *Wellingborough.* pre-1742.
LINCOLN—
Name used by Auburndale Watch Co. *USA.* 1879 (not 1789).
B. *Saxmundham.* 1846.
Benjamin Corbould. *Aldborough (Suffolk).* 1823-64. Also library.
& REED. *Boston, USA.* 1842.
LIND, John. *Philadelphia.* 1775-1805.
LINDBERG—
Charles J. *St. Johns, Newfoundland.* 1871.
John. *St. Johns, Newfoundland.* 1864-85.
LINDER, Charles. *Geneva, NY, USA.* ca.1810.
LINDERBY, Francis Heyward. *London.* 1828.
LINDEY, Thomas. *Bristol.* (B.1812) 1830-40.
LINDLEY, James. *London.* b.ca.1777, mar. ca.1803-13. W. Later jewellers. d.1846.
LINDON, Mrs. E. *Coventry.*1880.
LINDORFF, Henry William. *London.* 1863-69.
LINDSAY—
——. *Canada.* Model of Pequegnat Clock Co.
Alexander. *Dungannon.* 1840-68. C.
James. *London.* (?B. a.1793-)1828.
Thomas. *Frankford, Pa, USA.* ca.1810. C.
W. K. *Pittsburgh, Pa, USA.* ca.1825.
W. R. *Davenport, Ia, USA.* ca.1850.
LINDSL(E)Y—
Timothy. *Reading, Pa, USA.* 1815-25.
William. *Portsmouth, O, USA.* 1838-41.
LINEBAUGH, H. W. *Keokuk, Ia, USA.* ca.1860-70.
LINEHAM, Francis, *Newark.* 1876.
LINERD, John. *Philadelphia, USA.* 1816.
LINES—
George. *London.* 1832. Clock case maker.
William. *Warwick.* b. 1794, working 1818 -d. 1829. C.
LINESON, C. *Newent.* 1856.
LINFORD—
Edward. *London.* (B.1805-20) 1828-32.
John Tagg. *Ipswich.* 1839.
Robert. *Norwich.* 1858-75.
LINGHAM, H. *Maidstone.* 1855.
LINK—
Henry. *Bristol.* a.1756-77. C.
William. *Cleobury Mortimer.* 1879.
LINLEY, Charles Pugh. *London.* 1875-81.
LINNARD, John. *Swansea.* 1830-75.
LINNEL, Knowles. *St. Albans, O, USA.* ca.1825. C.
LINSELL, Henry James. *Little Hadham (Hertfordshire).* 1859-74.
LINSEY—
John Henry. *London.* 1851.
Thomas Busby (& Co.). *London.* 1844-69 (1875-81).
LINSLEY, William. *London.* 1828. Watch case maker.
LINTELL, C. *High Wycombe.* 1854.
LINTON,
Isaac. *Whitehaven.* 1873.
John. *Liverpool.* 1824.
LION—
Charles. *York.* 1787. C.
Isaac. *Ramsgate.* 1826-40.
Robert. *Carnwath.* 1836.
LIPPOLD, Ernest Alfred. *London.* 1881.

LIPSETT, Joseph. *Ballyshannon.* 1789.
LISETER, William. *Ironbridge.* 1828-35.
LISNEY—
William. *London.* 1844.
William. *New York.* ca.1840 (?perhaps same man as above, post-1844).
LISSETT, Richard. *London.* (B. a.1785-)1832.
LISTER—
James. *Halifax.* 1830-71.
James. *Horsforth (nr Leeds).* 1807. C.
John. *Halifax.* 1830-50.
Joseph. *Halifax.* 1822-37.
Nicholas. *Sheffield.* 1797 -d. 1809. W.
Samuel. *Bradford.* ca.1800. C.
Samuel. *Midgley nr Halifax.* 1775-d.1830.
Thomas I. *Luddenden nr Halifax* (and later in *Halifax*). b.1717. s. of William I. d.1779. C.
Thomas II. *Halifax.* b.1745. s. of Thomas I. d.1814. C. Made several world time dials.
Timothy. *Bingley.* 1866-71. W.
William I. *Keighley.* mar.1715-d.1731. C.
William II. *Keighley.* b.1721. s. of William I. d.ca.1786. C.
William. *Midgley nr Halifax.* b.1748. s. of Thomas I. d.1811. C.
William. *Newcastle-on-Tyne.* (B. 1815-)1827-56.
Wooldridge. *Birmingham.* 1868.
LISTON, William James. *London.* 1875-81.
LITCHFIELD MANUFACTURING CO. *Litchfield, Conn, USA.* ca.1850.
LITHERLAND—
Ann. *Liverpool.* (B. 1825) 1828.
& Co. *Liverpool.* (B. 1799-)1800-07. W.
DAVIES & Co. *Liverpool.* 1814-51. 'Inventors, patentees and manufacturers of the lever watch', also chrons and imported clocks.
John. *Liverpool.* 1677-d.1687. W.
John. *Warrington.* mar.1755. W.
Peter. *Liverpool.* 1831. W.
Richard. *Liverpool.* 1814-48. 'Inventors and patentees of the new chronometer.'
LITHGOE/LITHGOW—
& BENTLEY. *Stafford.* 1850-60.
Joseph. *Farnworth.* mar.1826. W.
Thomas. *Coventry.* 1837. W.
William. *Halifax, Nova Scotia.* 1838-52.
LITTLE—
Archibald. *Reading, Pa, USA.* ca.1810-20, 1839 *Camden, NJ,* 1840 *Philadelphia.*
David. *Carlisle.* 1858-73.
& ELMER. *Bridgetown, NJ, USA.* ca.1830. W.
J. *London.* 1869.
James. *Dumfries.* 1788-89.
James. *London.* 1881.
John. *Annan.* 1836-60.
John. *Dumfries.* 1887.
LITTLEFIELD, James. *London.* 1832.
LITTLEFORD, John. *London.* 1828. (B. 1817-24.)
LITTLEJOHN—
Wilson. *Peterhead.* 1846.
Wilson. *Turriff.* 1860.
LITTLER, H. *Tutbury (Staffs).* 1868.
LITTLEWOOD—
B. *Woolwich.* 1866-74.
Benjamin, jun. *Woolwich.* 1874.
George. *London.* 1839-44.
John. *London.* 1839-44.
LITTLEWORTH, George. *London.* (B. a.1789-CC. 1822) 1828-51.
LIVERSAY, George. *Prescot.* 1666 W.
LIVESEY—
Christopher. *Leeds.* 1798. Engraver.
John. *Bolton.* ca.1730-ca.1750. C.
& WARD. *Halifax.* 1840-50.
LLEWELLIN—
——. *Llansadwrn.* mid-19c. C.
David. *Llansamlet.* ca.1815. C.
Peter & Co. *Bristol.* 1840-50.
LLEWELLINS & JAMES. *Bristol.* 1863.
LLOYD—
Andrew J. *Halifax, Nova Scotia.* 1866.
David. *Leominster.* 1830-56. Also gunsmith.

LLOYD—*continued.*
David. *Welshpool.* 1887-90. W.
Edward. *Ffestiniog.* 1874-87.
Edward. *Holywell.* 1844.
Edward. *Llanfyllin.* 1868-86.
Edward. *Ruthin.* 1856-68.
Henry. *Farnworth.* mar.1834. W.
J. *Horsham.* 1862.
James. *Guildford.* 1878.
James. *Prescot.* 1824. W.
John. *Brecon.* 1802-40. W.
John. *Builth.* 1830-44.
John. *Hereford.* 1850-56.
Joseph. *Toronto.* 1891.
Joseph. *Wigan.* 1811-16 (B. 1816-24).
Samuel. *London.* 1857.
Thomas. *Farnworth.* mar.1798. W.
Thomas J. D. *Coventry.* 1880.
Thomas Nathaniel. *Watford.* 1866.
William. *Manchester.* mar.1802. W.
William. *Springfield, Mass, USA.* 1779-1845. Clock cases.
William John. *London.* (B. CC. 1825.) 1839-81.
LOADER (or **LODDER**), Frederick. *Basingstoke.* 1867-78.
LOAF(F)LER—
A. & Co. *London.* 1863-69.
Leopold & Co. *London.* 1857-81.
LOCH, John. *Tillicoultry (Scotland).* 1860.
LOCHARD, Robert. *London.* a.1647 to John Matchet. CC. 1656.
LOCK(E)—
Anne. *Bath.* 1848.
Edward. *Oxford.* b.1729, d.1813. W.
George. *Bristol.* s. of James. 1812.
Henry. *Oxford.* 1802. W.
& HUTTON. *Dunfermline.* 1825. (*Not* LOCKE & MULTON-B.).
J. C. *Camberwell.* 1878.
John. *Cincinnati and Newark, O, USA.* 1840.
John. *London.* 1869-81.
John. *Whitburn.* ca.1820. C.
Mrs. A. *Bath.* 1856.
Richard. *London.* 1875.
Robert. *Wallingford.* 1837.
Walter. *Curry Rivel.* 1875-83.
William. *Cheltenham.* 1830.
William New. *Bristol.* s. of James. 1830.
LOCKHART—
Andrew. *London.* 1828.
Andrew. *Whitehaven.* 1879.
Edward. *Belfast.* 1880-94.
LOCKIER, Mrs. *Nailsworth.* 1840.
LOCKWOOD—
Benjamin. *Swaffham.* b.1737(-95).
Frederick. *New York.* ca.1831.
John Wiggins. *London.* 1832-39.
Robert. *Coddenham (Suffolk).* 1875-79.
& SCRIBNER. *New York.* 1847.
William. *Huddersfield.* 1871.
William & COOKE, John. *Huddersfield.* 1850-53.
LOCKYER, Charles. *Sandhurst.* 1847-74.
LODDER, William Alfred. *Southend.* 1874.
LODGE—
George. *Barnsley.* 1871.
Thomas. *London.* b.ca.1663, mar.1688-ca.1695. W. & C.
LOE, Robert. *Brading, Isle of Wight.* 1839-78.
LOEFFLER—
Lawrence. *Greenwich.* 1847-51.
U. *Greenwich.* 1855.
LOFFLER—
E. *Colchester.* 1866.
Ferdinand & Joseph. *Winslow.* 1869-77.
George. *London.* 1857-81.
& ROMBACH. *London.* 1863.
LOFTER, John. *Sculcoates.* 1814. C.
LOFTUS—
James. *London.* a.1726. C.
John. *Liverpool.* mar.1822. W.
LOGAN—
A. Sidney. *Goshenville, Pa, USA.* 1849-1925.

LOGAN—*continued.*
Robert. *St. Louis, Mo, USA.* ca.1820.
T. *Dorchester.* 1855.
Thomas. *Dorchester.* (B. 1795-)1830.
Thomas. *Maybole (Scotland).* 1820-37.
William. *Ballater.* 1846.
LOGGINS, John. *Rochdale.* 1791. W.
LOHE & KEYSER. *Philadelphia, USA.* 1831-35. W.
LOHMAN, Axel. *London.* 1881.
LOHRMAN, H. *Smithville, Canada.* 1857.
LOMAS—
J. *Alfreton.* 1855-64.
J. *New Mills (Derbyshire).* 1864.
J. *Stockport.* 1864-76.
Joseph. *Sheffield.* 1822-34. C.
Richard. Place unknown. 1717. Believed *North Lancashire.*
Samuel. *Poulton (Lancashire).* mar.1744. C.
Thomas Edward. *Sheffield.* 1837-62. C.
LOMAX—
James. *Blackburn.* b.1749-d.1814. C. & W.
Joseph. *Prescot.* mar.1802-1819. W.
Samuel. *Blackburn.* 1749-1781. And perhaps later. C.
William. *Prescot.* 1795-97. W.
William. *Prescot.* mar.1795-97. W.
LOMBARD, Daniel, jun. *Boston, USA.* 1830.
LOMES, William. *New York.* ca.1840.
LONDON—
—— (and LONDON B.). 1914 and later, product of Pequegnat Clock Co.
James. *London.* (B. 1820-)1832-44.
John. *Bristol.* ca.1690. Lant. clock.
& RYDER. *London.* 1869-81.
LONDRAPEAR, William. *London.* 1662. (Probably LA WRAY PIER, q.v.)
LONG—
Charles. *Boreham.* 1855-78.
George. *Hanover, Pa, USA.* ca.1800-11. C.
Henry. *London.* 1869.
J. *Kings Lynn.* 1846.
J. *Shipton-on-Stour.* 1860.
John. *Chipping Campden.* 1870.
John. *London.* 1828.
John. *St. Austell.* 1847.
Moses. *Armagh.* 1846.
R. *Gunnislake (Cornwall).* 1856.
Robert. *St. Austell.* 1844.
S. *Kingston-on-Thames.* 1862-66.
S. *Putney.* 1851-55.
Samuel (S. R.). *Philadelphia, USA.* 1842-46.
Samuel. *Rayleigh.* 1839.
Theodore. *Newcastle-on-Tyne.* 1834.
William. *Bristol.* 1870.
William Henry. *Cardiff.* 1875.
LONGBOTTOM, Joseph. *Leeds.* 1853.
LONGHURST, John. *Reigate.* 1828.
LONGIN, R. *USA.* 1876.
LONGLEY, T. *Kirkby Moorside.* ca.1820.
LONGMORE, William. *Manchester.* 1834-51.
LONGSTAFF—
George. *Darlington.* 1648. Kept ch. clock.
John. *Bishop Auckland.* 1834-56.
LONG(S)WORTH, Peter. *Liverpool.* (B. 1824-)1828-34.
LOOF—
Edward Fry. *Tunbridge Wells.* 1855-74.
William, sen. *Tunbridge Wells.* 1823-55.
William, jun. *Tunbridge Wells.* 1845-47.
LOOMES, Thomas. *London.* At the Mermaid in *Lothbury* nr *Bartholomew Lane End.* CC. 1649. Succeeded to John Selwood 1651. Partner from 1658 with Ahasuerus Fromanteel I, whose daughter he married. died ca.1665. Lantern clocks.
LOOMIS—
Henry. *Bristol, Conn, USA.* 1832-33. C.
Henry. *Frankfort, NY, USA.* ca.1830. C.
William B. *Middletown* and *Wethersfield, Conn, USA.* ca.1825-31. C.
LOOSE, A. A. *Peterborough.* 1864.
LOOSLEY—
John. *London.* 1875.
Richard. *London.* 1875-81.

LOPRESTI, Antonie Joseph. *London.* 1875-81.
LORAINE, James. *Newcastle-on-Tyne.* 1827-34. (B 1820).
LORD—
& GALE (Joseph). *Fayetteville, NC, USA.* 1792.
George William. *Olney.* 1869-77.
& GODDARD. *Rutland, Vt, USA.* 1797-1823. C.
Henry. *Liverpool.* 1848-51.
James. *Woodbury, NJ, USA.* 1821-35. C.
John. *Farringdon.* (B. 1784-97) 1830.
R. *Coventry.* 1880.
William. *London.* 1638. CC.
LORIMER—
David. *London.* 1811. (B. 1815-20).
William. *London.* 1828-39.
LORING—
Henry William. *Boston, USA.* ca.1812-40.
Joseph. *Sterling, Mass, USA.* 1791-1812. C.
LORMIER—
& EDWARDS. *London.* 1828.
William. *London.* 1844-63.
LORRIMAN, John. *London.* 1875-81.
LORTON, William B. *New York.* 1810-54. C.
LOSADA, Jose R. de. *London.* 1839-81. 'By appointment to several courts'.
LOSEBY—
Amos (late Henry). *Leicester.* 1876.
Edward. *Coventry.* 1806. To *Leicester.* ca.1817-76. Succeeded by son, Edward Thomas. b.1817-d.1890.
Edward. *Shiffnal.* ?early 19c. W.
LOSS—
Augustus. *Pittsburgh, Pa, USA.* ca.1840.
P. *Germantown, Pa, USA.* ca.1850.
LOTHERINGTON—
John Flint. *Hull.* Later 19c. Succeeded to William.
William. *Hull.* 1846-51.
LOUARTH, Jasper. *London.* CC. 1641-d.pre-1650.
LOUBEAU, Pierre. *Lyon.* ca.1600.
LOUDAN, William. *London.* (B. 1825) 1828-51.
LOUDON, David. *Kilwinning.* 1843.
LOUGHBOROUGH, John. *Durham.* (B. 1820) 1827-1834. C.
LOUGHOR, John. *Bridgend.* 18c. C.
LOUIS—
Abraham. *Birmingham.* 1835.
Benjamin. *Windsor, Nova Scotia.* 1864.
Joseph. *Montreal.* 1864.
LOVATT (sometimes **LAVATT**)—
J. *Audley (Staffordshire).* 1860.
Joseph. *Newcastle-under-Lyme.* 1828-35.
Joseph, sen. *Tunstall (Staffordshire).* 1842-68. Also bellhanger.
Joseph, jun. *Tunstall (Staffordshire).* 1868-76.
LOVE—
B. *Rugby.* 1854.
F. *Nottingham.* 1855.
John. *Glasgow.* 1828.
LOVEDAY—
James. *Wakefield.* 1851-71.
William. *Chelmsford.* 1874.
LOVEGROVE—
H. *Colnbrook.* 1854.
Henry. *Slough.* 1877.
& ROBERTS. *Slough.* 1869.
LOVELACE—
James. *Bridport.* ca.1700. C.
Jacob. *Exeter.* Free 1721-d.1755. C.
LOVELL—
A. *Nottingham.* 1864.
A. E. *Philadelphia, USA.* 1841-49.
& Co. (G.). *Philadelphia, USA.* ca.1880.
Manufacturing Co. Ltd. *Erie, Pa, USA.* 1893. C.
& SMITH. *Philadelphia, USA.* 1841-43(-ca.1880).
LOVERIDGE, S. & Sons. *Birmingham* and *London.* 1868.
LOVETT—
James. *Mendon, Mass, USA.* 1728-1814. C.
William. *Stoke-on-Trent.* 1842.
LOVIS—
Captain Joseph. *Hingham, Mass, USA.* 1775-1804.
Charles Dupley Silvain. *Montreal.* 1824-43.
J. & A. *Montreal.* 1857-67.

LOVIS—continued.
Joseph. *Montreal.* 1809.
& Sons. *Montreal.* 1819.
LOW—
& Co. (John J.). *Boston, USA.* 1832. Imported watches.
James. *Errol (Scotland).* 1860.
John. *Kirriemuir.* 1837.
Thomas. *Dundee.* 1828.
Thomas. *Perth.* 1843.
LOWDEN—
William. *London.* 1811. W.
William John. *London.* 1832.
LOWE—
——. *Dartford.* 1777-1830. W. & C. (See Thomas Lowe).
& BENNETT. *Coventry.* 1860.
Charles. *London.* 1851. Clock cases.
Edward. *Chester.* (B. 1793) 1828-34.
Edward. *Liverpool.* 1761-90. W.
G. *Chester.* 1857.
George. *Gloucester.* 1830-42.
George. *Llandudno.* 1868-76.
George Cliff. *Manchester.* 1851.
George & Sons. *Chester.* 1834.
Henry. *London.* 1869.
Henry. *Norwood.* 1878.
J. *Chester.* 1857.
J. B. *Barmouth.* 1887-90.
James. *Over Darwen (Lancashire).* 1858. W.
James. *Prescot.* mar.1767.
Jesse William. *Over Darwen.* 1848. (May be same as James above).
John. *Chester.* 1848.
John. *Over Darwen.* 1851.
John. *Stockport.* 1878.
Joseph. *Bawtry.* 1834-37.
Joseph. *Blyth (Nottinghamshire).* 1828.
M. *Wragby (Lincolnshire).* 1849.
Richard. *London.* b.ca.1665, mar.1690. Watch case maker.
Robert. *Preston.* 1834-51.
& Sons. *Chester.* 1878.
Thomas. *Dartford.* 1823-32.
Thomas. *London.* 1851.
Thomas. *Philadelphia.* 1772.
W. *Portslade.* 1878.
LOWELL—
George. *London.* 1656.
Paul, sen. *London.* BC. 1628-1662. *Show Lane.*
Paul, jun. *London.* s. of Paul, sen. a.1646 to father. CC. 1653-1662. *Show Lane.*
LOWLEY (or **LOWRY**), William. *Ballymoney.* 1806-24. W.
LOWMAN, Jeremiah. *Ramsgate.* (B. from *London.* Pre-1818-32. C. & W.
LOWNDS, John. *London.* 1844-63.
LOWNES—
Davis. *Philadelphia, USA.* 1785-1810.
Hyatt. *Hagerstown, Md, USA.* ca.1792.
LOWREY, David. *Newington, Conn, USA.* 1740-1819. C. Also guns.
LOWRIE, Ebenezer. *London.* (B. a.1814, CC. 1828) 1828. Watch cases.
LOWRY—
George. *Leeds.* 1834. W.
John. *Belfast.* 1849-80. C. & W. and chrons.
Morgan. *Leeds.* b.1682-d.1757. C. Maker of repute.
Samuel. *London.* 1857.
William. *Antrim.* ca.1850-80. W.
William. *Ballymoney.* 1792-1806. W.
LOWS, BALL & CO. *Boston, USA.* 1842.
LOWTHER—
John. *Clerkenwell.* 1832. Clock case maker.
Robert. *Castleford.* 1866. W.
Robert. *Knottingley.* 1862.
& Son. *Clerkenwell.* 1839-51. Clock case makers.
Thomas Nicholas. *Clerkenwell.* 1828-32. Clock case maker.
LOYES, William. *Wincanton.* 1820.
LUBERT, Florens Gerard. *Amsterdam.* (B. ca.1730) ca.1765. C.

LUCAS—
Alexander. *Glasgow.* 1849.
Charles. *London.* 1851.
Henry George Duffett. *Bristol.* a.1835-46.
John. *London.* 1851.
John Ferdinand. *London.* 1881.
Joseph. *Leicester.* 1864-76.
Joseph. *London.* 1857-81.
Joseph. *Norwood.* 1878.
Thomas. *Huddersfield.* 1866. W.
Thomas. *Quorndon (Leicestershire).* 1876.
William. *Coventry.* 1880.
LUCE, David. *London.* ca.1690-ca.1700. W.
LUCIEN, ——. *Paris.* Pre-1850. C.
LUCKE, John P. *Philadelphia, USA.* 1849.
LUCKETT, Joseph. *Northleach.* 1870-79.
LUCKHURST, William (& Son). *Dover.* 1832-66.
LUCKMAN, John. *Birmingham.* 1850.
LUCY, D. E. *Houlton, Me, USA.* ca.1860.
LUDBROOK—
D. W. *Oundle.* 1864.
Daniel. *Braintree.* 1874.
Daniel. *Chelmsford.* ca.1848.
Daniel. *Debenham.* 1839.
John. *Soham (Cambridgeshire).* 1840-75.
LUDWIG—
John. *Philadelphia, USA.* 1790.
Louis. *Huddersfield.* 1866. W.
LUFKIN & JOHNSON. *Boston, USA.* 1800-10.
LUGG, Jasper. *Gloucester.* ca.1656-ca.1685. Lant. clock.
LUKE—
Edward. *Belfast.* 1865-68. C. & W.
James. *Alnwick.* 1834.
James. *Stockton.* 1856.
LUKENS—
Isiah. *Horsham* and *Philadelphia, USA.* 1779-1837.
J. *Philadelphia, USA.* 1837.
Seneca. *Horsham, Pa, USA.* 1751-d.1829. C.
LULHAM, W. H. *Montreal.* 1871.
LULMAN, Alfred. *London.* 1857-69.
LUM, John. *Ripponden.* ca.1840. C.
LUMB, William. *Kippax.* b.1737-d.1803. C.
LUMLEY, David. *Canning Town.* 1874.
LUMS—
Joseph. *Chelmsford.* ca.1726-31.
Stephen. *Chelmsford.* ca.1740. C.
LUMSDEN—
David. *Anstruther.* (b.1827). 1850-retired 1896 (d.1909).
George, jun. *Pittenweem (Scotland).* (b.1832). s. of George, sen., whom he succeeded 1849-d.1899.
John. *Montreal.* Pre-1804. When he sold out to R. Irish.
LUND—
& BLOCKLEY. *London.* 1875-81.
Brothers. *London.* 1881.
John Richard. *London.* 1832.
LUNDIE—
John. *Dundee.* (B. 1809-37)-1860.
William. *Aberdeen.* ca.1800. C.
LUNDY—
& FARROW. *Grimsby.* 1868.
J. F. *Grimsby.* 1876.
W. *Darlington.* 1898.
LUNN, Charles. *Edinburgh.* a.1799. Free 1806.
LUNT, Samuel. *St. Helens.* 1851.
LUPP—
Charles. *New Brunswick, NJ, USA.* bro. of Lawrence. 1788-1825.
Harvey. *New Brunswick, NJ, USA.* 1809-15.
Henry. *New Brunswick, NJ, USA.* s. of Peter. 1760. mar.1788-1816.
John. *New Brunswick, NJ, USA.* 1734-1804. C.
John H. *New Brunswick, NJ, USA.* s. of William. ca.1830.
Laurence. *New Brunswick, NJ, USA.* bro. of William for whom he worked 1806.
Peter. *New Brunswick, NJ, USA.* 1760-d.1807. C.
Samuel Vickers. *New Brunswick, NJ, USA.* s. of Henry. 1789-1809.
William. *New Brunswick, NJ, USA.* s. of John. 1766-1845. C.

LUPPIE & SOLCHA. *Hull.* 1840.
LUPTON—
Charles. *London.* 1828. Watch case maker.
Clifford. *London.* 1869-81.
George. *Altrincham.* (B. 1772-95. C.) 1828.
Henry B. *Newcastle-on-Tyne.* 1852-56.
John. *Altrincham.* 1834.
Thomas. *Harrogate.* 1871.
Thomas. *Leeds.* 1853.
William. *Keswick.* 1879.
William II. *York.* 1681. s. of William I. d.1689. C.
LUSCOMB(E)—
Richard. *Totnes.* ca.1730-ca.40. C.
Samuel. *Salem, Mass, USA.* 1773.
LUTIGER, James Xavier. *London.* 1863.
LUTMAN, John. *Lewes.* 1870-78.
LUTZ, Fidel. *London.* 1857.
LUX—
Clock Manufacturing Co. *Waterbury, Conn, USA.* 1917.
Paul. *Waterbury, Conn, USA.* 1914-47. Founder of Lux Clock Manufacturing Co.
LYCETT, Henry Finer. *London.* 1863-69.
LYLE, Thomas. *Belfast.* 1784.
LYMAN—
G. E. *Providence, RI, USA.* ca.1840.
Roland. *Lowell, Mass, USA.* 1832-37.
Thomas. *Windsor, Conn, USA.* 1770. *Marietta.* ca.1791.
LYNCH—
Abraham. *Baltimore, USA.* ca.1800.
Anthony. *Newbury.* 1755-96. C.
LYNDALL, William. *Philadelphia, USA.* 1844.
LYON—
Charles. *Bridlington.* 1858-88. Nephew and successor to Craven L. C.
Craven. *Bridlington.* (b.1793) 1822-retired 1868. Died 1888. C.
Edward. *Prescot.* 1796. W.
Edward. *Prescot.* 1816-19. W.
Edward W. *Harbour Grace, Newfoundland.* 1871.
George. *Liverpool.* 1691-1700. W.
George. *Wilmington, NC, USA.* 1819-44.
Henry. *Farnworth.* mar.1819. W.
Henry. *Toronto.* 1861.
Henry. *Trowbridge.* 1875.
Isaac. *Ramsgate.* 1768-1840. (Prob. two such?). W.
J. *Greenwich.* 1874.
J. *Peckham.* 1866.
James Walter & Co. *Edinburgh.* 1842.
John. *Prescot.* 1796. W.
John. *Warrington.* 1666-72. C. (B. 1685).
Joseph. *Prescot.* mar.1797. W.
Judah. *Swansea.* 1840-44.
L. & S. *Chatham.* 1845.
& LEVI. *Birmingham.* 1868.
Lewis. *London.* 1839-69.
Mary Ann. *Doncaster.* 1858-68. Successor to Sherwood Lyon.
Maurice & Samuel. *London.* 1875-81.
Ridley John. *Romford.* 1866.
Sherwood. *Bradford.* 1871.
Sherwood. *Doncaster.* mar.1827-51. C.
Simon. *Rochester.* 1866-74.
Solomon. *Birmingham.* 1880.
Thomas. *London.* 1881.
William. *Farnworth.* mar.1804. W.
William. *London.* 1857-81.
LYONS—
Edward. *Bristol.* 1879.
Edward. *London.* 1881.
Emma. *Ripon.* 1866. W.
Francis Henry. *Cardiff.* 1887-99.
H. B. *St. Johns, Newfoundland.* 1864-77.
Henry. *London.* 1875-81.
J. *Bristol.* 1856.
James. *Folkestone.* 1845.
John. *Bristol.* 1850-70.
John, jun. *New Haven, USA.* ca.1840.
LYTHE, John Smith. *York.* 1841. W.
LYTHGOE, John. *St. Helens.* 1851.

M

M, A, Trademark of F. A. Margaine.
M, E. & Co., Trademark of E. Maurice & Co.

MAC, MC and **M'**
All names with these prefixes are taken together here, and
written **Mc**.

McADAM—
James. *Newcastle-on-Tyne.* 1827.
Robert, jun. *Dumfries.* 1840-67.
Walter. *Bathgate.* 1840-50.

McALLISTER—
A. L. *USA.* 1821.
George. *Halifax, Nova Scotia.* 1866-81.

McALPIN(E)—
George. *Edinburgh.* 1836-46.
James. *London.* 1869.

McAULEY, John. *Armagh.* 1820.

McBEAN, James. *Inverness.* 1837.

McBEATH, Alexander. *Fraserburgh.* 1837.

McBRIDE—
(A.) & GARDNER (B.). *Charlotte, NC, USA.* 1807.
James. *Lisburn.* 1880-94. W.

McCABE—
& Co. *London.* 1875.
(Robert) & Co. *London.* 1839-75.
(Thomas) & HUNTER (James). *Belfast.* 1789. Part-
ners.
James. Brother of Thomas. *Lurgan.* ca.1760-ca.1770.
Then *Belfast.* Then ca.1778. *London.* Died ca.1811.
John. *Newry.* Then to *Dublin.* 1767. *Baltimore, USA.*
In 1774.
Patrick. *Lurgan.* ca.1730-d.1766. C.
& STRACHAN. *London.* (B. 1825) 1828-32.
Thomas. s. of Patrick. *Lurgan.* Where b.ca.1740. Then
ca.1766. To *Lisburn.* Later *Belfast.* ca.1767-ca.1798.
Died 1820. W. & C.
William I. *Newry.* 1772-1785. C. & W.
William II. *Newry.* ca.1785-ca.1800. Later *London.* C.
& W.
William. *Lisburn.* ca.1766-72. Then *Newry.*
William. *Richmond, Va, USA.* ca.1790-ca.1820.

McCALL—
John. *Dalkeith.* 1829.
John. *Edinburgh.* 1836.

McCALLION, S. *Eton.* 1869.

McCALLUM—
Archibald. *Quebec.* 1855-65.
Henry. *Belfast.* 1868.
Malcolm. *Barrhead (Scotland).* 1860.

McCALVEY, Charles. *Enniskillen.* 1784.

McCANNA, John. *Montreal.* 1831.

McCARDLE, William. *Strabane.* 1839.

McCARTEN, John. *Downpatrick.* 1843-46.

McCARTER, John. *New York.* 1854.

McCARTNEY—
Arthur. (b.ca.1792 at *Bangor*). *Belfast.* ca.1820-
d.1837. C.
J. G. *Dartmouth, Nova Scotia.* 1878-81.
John N. *Belfast.* 1849-53. Then went to *USA.*

McCARTNEY—continued.
John Neill (& Co.). *Armagh.* 1840-46. C. (?later
Belfast).

McCARTY, William. *Aurora, Ontario.* 1800s.

McCASKY, Joseph Warden. *Cookstown.* 1843-46.

McCAUL, John. *Tillicoultry (Scotland).* 1860.

McCAULEY—
Henry. *Rathfriland.* 1846-68.
J. *St. Hyacinthe, Canada.* 1851.
William Hugh. *Rathfriland.* 1840.

McCELLEM, Samuel. *Greenwich.* 1847.

McCHESNEY, Joseph. *Belfast.* 1880-94. W.

McCLAHARTY, Alexander. *Kingston.* 1878.

McCLARY—
Samuel II. bro. of Thomas. *Wilmington, Del, USA.*
Early 19c.
Thomas. bro. of Samuel. *Wilmington, Del, USA.* Early
19c.

McCLEAN—
Edward. *Hyde.* 1878.
John. *Belfast.* 1783-89. W.
John. *Lisburn.* ca.1772-85.

McCLEAVE—
Brothers. *Belfast.* 1865-74. W.
James. *Belfast.* (b.1818). 1843-d.1847.
Robert. *Belfast.* (b.1831)-d.1874. W.
Stewart R. *Belfast.* (b.ca.1828-d.1874). Engraver.

McCLEERY, John. *Monaghan.* 1846.

McCLELLAN—
——. *Deptford.* 1839.
Alexander. *London.* 1869.
& Co. *London.* 1857.
S. *Greenwich.* 1851-55.
Samuel. *London.* 1851-63.
Samuel Hughes (& Son). *London.* 1863-75(-81).

McCLELLAND, James. *Ballanahone (Co Antrim).*
ca.1790-ca.1800. C.

McCLENAGHAN, James. *Strabane.* 1788.

McCLENNAN (or McCLELLAN), Alexander John.
Greenwich. 1866-74.

McCLINTOCK, O. B. *Minneapolis, Minn, USA.* 1945.
Electric clocks.

McCLOSKEY, F. *Philadelphia, USA.* 1850.

McCLUER, Heman. *Hamburg, NY, USA.* 1834. Sundials.

McCLUNG—
S. *Three Rivers, Canada.* 1857-65. (Must surely be the
man below).
Samuel. *Carlisle.* 1848. (Presumably same man as
above).

McCLURE—
David. *Boston, USA.* 1810.
Samuel. *Montreal.* 1819-20.

McCLURG, Joseph. *Coleraine.* 1795.

McCLUSKY, Francis. *Ballycastle.* 1846.

McCOLLIN, Thomas. *Philadelphia, USA.* 1824-33.

McCOLLUM, James Harry Keighly. *Toronto.* 1907.

McCOMBE, James. *Ottawa.* 1862.

McCONEGALL, Patrick. *Londonderry.* 1784.

McCONNELL—
J. Clemens. *Waynesburg, Pa, USA.* ca.1845.
John. *Belfast.* 1880-94.
McCONVILL, Edward. *Liverpool.* 1848. German clocks.
McCORMACK, Henry. *Philadelphia, USA.* 1833.
McCORMIC(K)—
Andrew. *Ballibay, co. Monaghan.* 1846-54.
Hugh. *Belfast.* 1843-68. W.
Samuel. *Belfast.* 1854.
McCOURT, John. *Dudley.* 1842. German clocks.
McCOURTIE, George. *Laurencekirk (Scotland).* 1860.
McCOY, George W. *Philadelphia, USA.* 1837.
McCRACKEN—
Francis. *Portaferry, co. Down.* 1824-58.
John. *Cootehill, co. Cavan.* 1854-68. C.
William. *Glasgow.* 1841.
McCREARY, Edward. *London.* 1828.
McCREDIE, Thomas. *Stranraer.* 1836-60.
McCREEVY, Robert. *Carrickfergus, co. Antrim.* 1843-68. W.
McCROSSAN, Joseph & Co. *Glasgow.* 1860. American clocks.
McCROW, Thomas. From *Edinburgh* to *Annapolis, Md, USA.* 1767.
McCULLAGH/McCULLOCH/McCULLOUGH—
Andrew. *Belfast.* 1660-66. Goldsmith.
David. *Bangor.* ca.1780-90. C.
Hugh. *Belfast.* (?1774) 1784. W.
James. *Kingswear.* ca.1740. C.
James. *Stirling.* 1860.
John. *Halifax, Nova Scotia.* From *Glasgow* 1837-d.1875.
John (b.1809). *Newry* ca.1835, to *Rodney, Mississippi, USA,* where d.1837.
William. *Belfast.* ca.1830. C.
William. *Philadelphia, USA.* 1841.
McCUNE—
Henry. *Liverpool.* 1851.
Thomas. *Belfast.* 1670-79. Goldsmith and believed W.
McCUTCHEON—
David. *Belfast.* 1854.
David. *Downpatrick.* 1858-90. C.
H. *Belfast.* ca.1800-ca.1830. C.
Hugh I. *Holywood, co. Down.* ca.1780-95. C.
Hugh II. *Newtownards.* ca.1805-d.1849. C.
Hugh III. *Holywood, co. Down.* ca.1835-d.1876. C.
McDANIEL, William H. *Philadelphia, USA.* 1825.
McDONALD—
A. *Toronto.* 1877.
Bernard. *Malton.* 1858.
Charles. *Lexington, Ky, USA.* 1818.
David. *Glasgow.* 1836.
G. B. *Gloucester.* 1879.
H. S. *Stirling.* 1860.
James. *Aberdeen.* 1846.
John. *Chippewa, Ontario.* 1852.
Malcolm. *Elgin.* 1860.
William. *Invergordon.* 1836-60.
William. *Nairn.* 1836.
McDONNELL, David. *Tallycross* nr *Portaferry, co. Down.* (b.1793)-d.1814.
McDONOUGH, John. *Dublin.* 1786-88.
McDOWELL (and variants)
C. & J. *Pontefract.* 1826.
Charles. *Boroughbridge.* 1822. W.
Charles. s. of William of *Pontefract.* b.1790. *Wakefield* and *Leeds* till ca.1839, then *London* where d.1873. C.
F. *Philadelphia, USA.* ca.1800.
John. *Belfast.* 1767-70.
John. *Philadelphia, USA.* 1817.
Mrs. Mary. *London.* 1875. (Succeeded/wid. of Charles).
William. *Thirsk.* mar.1780. W.
William. *Pontefract.* 1807-37. C. & W.
McELMON, David R. *Halifax, Nova Scotia.* 1866.
McELWAIN—
David & George. *Rochester, NY, USA.* ca.1840-ca.1850.
William. *Coleraine.* (b.1794)-d.1816.
McELWEE—
James. *Londonderry.* 1774-d.1778.

McELWEE—*continued.*
James. *Philadelphia, USA.* 1813.
McENTIRE, James. *Monaghan* 1785. *Enniskillen* 1824.
McEVILLE, William. *Roxton Falls, Quebec.* 1871.
McEWAN—
James. *Crieff.* 1846.
John. *Crieff.* 1837-60.
William. *Auchterarder.* 1840-60. W.
William. *Edinburgh.* 1849.
McFADDEN—
J. B. *Pittsburgh, Pa, USA.* ca.1835.
John B. & Co. *Pittsburgh, Pa, USA.* ca.1840.
McFARLAND—
John. *Omagh.* 1846.
John & Co. *Greenock.* ca.1780-ca.1790. C.
McFARLANE—
A. P. *Glasgow.* 1841.
D. *Glasgow.* 1841.
D. & Son. *Glasgow.* 1837.
James. *Auchterarder.* 1860.
James. *Perth.* 1842.
John. *Belfast.* 1854-68. C.
John. *Coleraine.* mar.1831-39. C.
Peter. *Glasgow.* 1841.
Robert. *Perth.* 1833.
& Son. *Glasgow.* 1828.
Thomas. *Ballater (Scotland).* 1860.
William. *Philadelphia, USA.* 1805.
McFEE—
Angus. *Belleville, Canada.* 1858-63.
D. *Whitehouse, co. Antrim.* ca.1810-ca.30. C.
McFERRAN, William. *Manchester.* 1851.
McGAW—
John. *Cookstown* and *Stewartstown, co. Tyrone.* 1817-54.
Peter. *Houghton le Spring.* 1827.
McGEORGE—
David. *Castle Douglas.* 1837.
Ebenezer. *Dumfries.* 1755-93.
John. *Kirkcudbright.* 1836.
McGHAN, R. *Greenwich.* 1851-66.
McGHIE, ——. *Abbey Green, Lesmahagow (Scotland).* 1860.
McGIBBON—
Maxwell. *Dumfries.* 1790.
Richard F. *Paisley.* 'est'd 1866'. C.
McGILCHRIST, John. *Barrhead.* 1830.
McGILLIS, Donald. *Williamstown, Canada.* 1851-58.
McGILLVRAY, John. *Stornaway.* 1860. W.
McGLOCHLIN, ——. *Newberry, Canada.* 1861.
McGLOGHLON, W. D. *London, Ontario.* 1875-90.
McGONAGILL, James. *Belfast.* 1843-65. W. (?d.1869).
McGOWAN, William. *Wigtown.* 1860.
McGRATH—
Edward. *Belfast.* 1858-80. W.
Hugh. *Portaferry.* 1863-68.
McGREGOR—
Alexander. *Lasswade (Scotland).* 1860.
Anthony. *Manchester.* mar.1792. C.
Archibald. *Dumfries.* 1882-87.
D. W. *Glasgow.* 1848.
Duncan. *Comrie.* 1837.
Duncan. *Greenock.* 1860.
Forrest. *St. Ninians.* 1830-80.
J. *San Francisco, Cal, USA.* ca.1850-ca.60. C.
James & Son. *Edinburgh.* 1836-60.
John. *Invergordon.* 1860.
John. *Stornoway.* 1836.
John. *Wick.* 1837.
Peter. *Perth.* 1840.
Thomas. *Eyemouth.* 1860.
Thomas. a.1808 at *Dunse,* then *Ayton* 1837.
McGREW—
Alexander. *Cincinnati, O, USA.* 1829.
Wilson. *Cincinnati, O, USA.* 1829-64.
McHARG—
Alexander. *Albany, NY, USA.* 1817-ca.1850.
(Alexander) & SELKIRK. *Albany, NY, USA.* 1815.
James. *Donaghadee, co. Down.* 1779. W.
McHINCH—
Robert. *Belfast.* ca.1790-ca.1820. W.
William. *Donaghadee.* ca.1805-24. C.

McHUGH, James. *Lowell, Mass, USA.* ca.1860.
McILHENY (J. E.) & WEST. *Philadelphia, USA.* 1818-22.
McILWAIN(E)—
Alexander. *Belfast.* 1880-87.
James. *Rathfriland.* 1820-24.
John & Sarah. *Rathfriland.* ca.1775. W.
Nathaniel. *Belfast.* 1839.
McILWRATH, Alexander. *Belfast.* 1784.
McINNES—
Neal. *Lochgilphead.* 1837.
William. *Glasgow.* 1834-60.
McINTOSH—
Alexander. *Pitlochry.* 1860.
Robert. *Dingwall.* 1860.
McINTYRE—
Joseph. *Crieff.* 1837-60.
Watch Co. *Kankakee, Ill, USA.* 1905.
McIVOR—
James. *Cookstown.?* ca.1780-ca.1800. C.
Murdo. *Dingwall.* 1836. (Prob. the man below).
Murdo. *Rochester, NY, USA.* 1844. (Prob. the above).
McKANNA—see McKenna.
McKAY—
Andrew. *Fraserburgh.* 1860.
C. from *London* to *Charleston, SC, USA.* 1790-1799. W.
& CHISHOLM. *Edinburgh.* 1835-1908.
David. *Arbroath.* 1837.
H. *Haverhill, Mass, USA.* 20th century.
Henry. *Birmingham.* 1880. Importers.
James Thomson. *Aberdeen.* 1836.
John. *Dorking.* 1828.
John R. *Elgin.* 1860.
SPEAR & BROWN. *Boston, USA.* 1854.
William P. & Co. *Boston, USA.* 1842-54.
McKECHNIE, William. *Paisley.* 1860.
McKEE—
Hugh. *Portaferry.* ca.1794-d.1834. W.
James. *Lurgan.* 1865.
John. *Chester.* 1816. C.
McKEEN, Henry. *Philadelphia, USA.* 1823-50. W.
McKELLOW—
John. *Guildford.* 1828.
John. *Maidstone.* 1823-51.
(Miss) J. *Maidstone.* 1851-55.
McKENNA, James. *Hull.* 1838-40. C.
McKENNY—
G. *London.* 1844.
Patrick. *Belfast.* 1784.
Theophilus. *London.* 1839.
McKENSIE, J. A. *Quebec.* 1824.
McKENZIE—
——. *Dundonald, co. Down.* ?Early 19c. C.
Alexander. *Barrie, Canada.* 1857-65.
Alexander. *Brompton (Kent).* 1847.
James. *Cumineston (Scotland).* 1860.
John. *Downham.* 1875.
John. *Dumfries.* 1848.
John. *Tain.* 1860. Also hairdresser and perfumer.
Malcolm. *Dingwall.* 1860. Also jeweller.
Robert. *Inverness.* 1860.
William. *Aberchirder.* ca.1840.
William. *Tain.* 1860.
McKEOWN—
James. *Coleraine.* ca.1780-1824. C.
James. *Cookstown.* ca.1770-ca.1790. W.
Thomas. *London.* 1851-69.
Thomas. *Maghera, co. Donegal.* 1846-90.
McKERROA/McKERROW, James. *Pathhead (Scotland).* 1836-60.
McKEY, William. *Londonderry.* ca.1830. W.
McKIBBIN—
John. *Lisburn* (?from *Newtownards*). ca.1775-d.1784. C.
John. *Newtownards.* ca.1771-80.
McKIE—
Andrew. *Fraserburgh.* 1837.
George. *London.* 1851.
James, jun. *London.* 1828-81.
James, sen. *London.* 1828.
James & George. *London.* 1828-44.

McKIE—*continued.*
John. *Broughshane, co. Antrim.* 1783. Journeyman to John Creighton.
John. *Ellon.* 1837.
Samuel. *Richhill, co. Armagh.* ca.1730-71. C.
William. *Aberdeen.* 1837.
McKINDER—
Townson. *Spilsby.* 1828-35.
William. *Spilsby.* 1835.
McKINLAY, Peter. *Edinburgh.* 1836.
McKINLEY, Edward. *Philadelphia, USA.* 1830-37.
McKINNEY—
Frederick William. *London.* 1863-81.
Herbert Jones. *London.* 1851-81.
Robert. *Wilmington, Del, USA.* 1845.
McKIRDIE, John. *Aberdeen.* 1846-60.
McKIRDY—
Hugh. *Birkenhead.* 1848.
Hugh. *Glasgow.* 1836.
McKNIGHT—
H. *Leamington.* 1880.
& Son. *Portaferry.* 1846.
McKUTCHEON, ——. *Bangor.* 19c. W.
McLACHLAN—
Charles. *London.* 1857.
Hugh (& Son). *London.* 1828-44(51).
John. *Dumfries.* 1820-28.
McLAREN, L. *Edinburgh.* 1850.
McLAUGHLIN, Samuel. *Quebec.* 1848-52.
McLEAN—
George. *Glasgow.* 1837-61.
John. *Dublin.* 1786.
John. *Dumfries.* 1856-83.
Patrick Gordon. *London.* 1875.
McLEISH, William. *London.* 1875.
McLELLAN—
John. *Aberfeldy (Scotland).* 1860.
R. N. C. *Halifax, Nova Scotia.* 1847.
John. *Peterborough, Canada.* 1876.
John. b.1814 in *Dingwall,* d.1886 in *London.*
Kenneth. *London.* 1828.
William. *Inverness.* 1835.
William. *London.* 1828.
McLEOD—
A. *York, Canada.* 1833.
Alexander. *Inverness.* 1860.
J. & Co. *Aberdeen.* 1846.
Peter. *New Lancaster, Canada.* 1871.
McLOGAN, John. *Portadown.* 1854.
McMAHON—
Brothers. (Arthur & Albert). *Belfast.* 1868. Late of Messrs Gilbert & Son.
Joseph. *Montreal.* 1842.
Owen. *Carrickmacross.* 1784.
Patrick. *Strabane.* 1865-90.
Robert J. *Belfast.* 1865-90.
William F. *Hamilton, Ontario.* 1867-78.
McMANUS, John. *Philadelphia, USA.* 1840.
McMASTER(S)—
David. *Sarnia, Canada.* 1871.
Hugh A. *Philadelphia, USA.* 1839.
Samuel. *Ballymena.* 1824. C.
Mrs. M. (Mrs. William). *Quebec.* 1855-64.
William. *Quebec.* 1844-55.
McMILLAN—
Andrew. *Glasgow.* 1832-61.
James. *Ardrossan.* 1850.
John. *Sowerby Bridge (Yorks WR).* 1805.
Peter. *Aberdeen.* 1836.
Richard. *Manchester.* mar.1801. C.
McMINN, John M. *Manchester.* 1848-51.
McMORDIE, Hans. *Ardmillan, co. Down.* ca.1750. C.
McMULLEN, Edward. *Philadelphia, USA.* 1846-48. C.
McMULLIN—
——. *Workington.* 1858.
Robert. *St. Catharines, Canada.* 1863.
McMURRAY, Thomas. *Toronto.* 1833-57.
McMYERS, John. *Baltimore, USA.* 1799.
McNAB, J. & A. *Perth.* 1837-49.
McNAGHTEN, Francis. *Dromore, co. Down.* 1843-46. C.
McNAIR, John. *Nantwich.* 1865-78.

McNALLY, Charles. *York, Canada.* 1833.
McNAMARA—
——. *Philadelphia, USA.* 1765. W.
M. *Perth, Canada.* 1862.
Patrick. *Belfast.* 1817.
McNAUGHTON—
Alexander. *Montreal.* 1819.
James. *Galt, Canada.* 1861.
James. *Tanderagee.* Early 19c. C. (Prob. McNeaghton, q.v.)
James. *Toronto.* 1861.
McNEAGHTEN, James. *Portadown.* 1840-46. C.
McNEE, William. *Edinburgh.* 1850.
McNEICE, John. *Newry.* 1854-92. C. & W.
McNEIL(L)—
——. *Lurgan.* ca.1795-1800.
E. *Binghampton, NY, USA.* 1813.
Thomas. *Greenock.* 1860.
William. *Greenwich.* 1874.
McNEILLY—
Alexander. *Armagh.* (B. 1824) 1846.
Brothers. *Belfast.* 1858.
McNEICE, John. *Bishopscastle (Scotland)* 1861-75.
McNEISH, John. *New York.* ca.1810. (One such at *Falkirk* ca.1805).
McOURT, P. *Dudley.* 1860.
McOWAN—
James. *Crieff.* 1860.
Joseph. *Crieff.* 1860.
McPARLIN, William. *Annapolis, Md, USA.* 1800.
McPHAIL, Charles. *London.* 1832.
McPHERSON—
——. *Beauly (Scotland).* 1860.
A. *Quebec.* 1848.
& Co. *Montreal.* 1862.
D. M. *Montreal.* 1849-65.
John. *Nairn.* 1836.
John. *North Shields.* 1834.
John. *Stornoway.* 1860.
Robert. *Philadelphia, USA.* 1837.
Sweeney Eugene. *Boston, USA.* 1830.
McQUIBAN, William. *Forres.* 1860. C.
McRAE—
Alexander. *Inverness.* 1837.
John. *Inverness.* 1837.
John. *Pollokshaws.* 1860.
McREADY, William. *Ryde (Isle of Wight).* 1859-78.
McROBB, William. *Strichen (Scotland).* 1860.
McSTOCKER, Francis. *Philadelphia, USA.* 1831.
McVEY, William. *Portobello (Scotland).* 1860.
McVICAR, Archibald. *Lundie Mill (Fife, Scotland).* 1830-42.
McWALTER, Moses. *Balfron (Scotland).* 1836.
McWATERS, George. *Glasgow.* 1836.
McWILLIAM, James. *Keith.* 1860.
MABBETT, Charles. *Salisbury.* 1875.
MABY, E. *Ashburn.* 1806. W.
MABYN—
Davye. *Stratton.* 1601. Kept ch. clock.
John. *Stratton.* 1563. Kept ch. clock.
MACE—
Andrew. *Stokesley.* 1866. W.
John. *Ditchley.* ca.1770-ca.1775. C.
John. *Taston (Oxon).* Early 19c. C.
Robert Harris. *Taston (Oxon).* 1853.
MACHAM, Samuel. *London.* ca.1710. C.
MACHEN, Thomas W. *New Bern, NC, USA.* 1812-30.
MACK—
James. *Cromer.* 1836.
& Son. *Cromer.* 1875.
Thomas. *Cromer.* 1875.
Thomas. *Jedburgh.* ca. 1780-1816. C. & W.
MACKLIN, Thomas. *Weymouth.* 1830.
MACKY, George. *Londonderry.* (b.1777). ca.1824-d.1827. W. & C.
MACOMB Co., The. *Macomb, Ill, USA.* ca.1885. C.
MACOUN & Co. *Belfast.* 'Est'd 1833' -1865. Clock importers.
MADDEFORD, Charles. *Theale (Berkshire).* 1837-64.
MADDEN—
Mathew. *Bath.* 1740. W.
W. H. *Belleville, Canada.* 1851.

MADDEY, John. *Nottingham.* b.1689—mar.1739. W.
MADDINSON, Robert. *Rugeley.* 1828.
MAD(D)OCK(S)—
Charles. *Over.* 1865-78.
G. *Tarporley.* 1857-65.
George. *Middlewich.* 1848.
John. *Talagarth.* ca.1750. C.
Joseph. *Tarporley.* 1848.
Samuel. *Over Budworth (Cheshire).* 1834. W.
Thomas. *Frodsham.* 1848-78.
MADDOX—
Charles. *Liverpool.* (B. 1818-21) 1824-28. W.
James. *Hereford.* 1830-56.
MAERS, Angel Jacob. *London.* 1869-81.
MAFFIA—
Angelo. *Hertford.* 1851-74.
Dominic (& Son). *Monmouth.* 1868-71 (-1880).
Edward, *Monmouth.* 1880.
Peter. *Monmouth.* From *Italy.* 1841-71.
Peter & Dominic. *Monmouth.* 1868.
MAGANN—
Patrick. *Belfast* 1792, then *Charleston, SC, USA,* 1792-94. C.
MAGILL—
Daniel. *Crumlin, co. Antrim.* 1818.
Edward. *Hamilton, Canada.* 1851.
MAGGS—
Charles. *Axbridge.* ?mid-19c. C.
Charles. *Swansea.* 1887.
Joseph. *Calne.* 1875.
R. *Wells.* 1861.
William. *Clutton* nr *Bristol.* 1875-83.
MAGNUS—
Simeon. *Chatham.* 1838.
Simon. *Swansea.* 1844.
MAGRATH—
Charles Frederick. *London.* 1851-81.
James. *London.* 1851-81.
MAGUIRE—
John. *Carrickmacross.* 1846-58.
S. & Co. *Belfast.* 1868. Importers.
MAHLENDORFF, Edward F. *London.* 1863-81.
MAHOLLAND, Robert. *Philadelphia, USA.* 1850.
MAHON, Thomas. *Brighton.* 1870-78.
MAHOOD—
Samuel & Son. *London.* 1881.
& Son. *Beckenham.* 1874.
MAHR, Wendel. *London.* 1881.
MAIER, Adolph. *London.* 1875-81.
MAILLARD, Osmond Charles. *Gloucester.* 1850.
MAIN, James. *Dunbar.* 1860.
MA(I)NWARING, ——. *Burwash.* 1804. C.
MAIRE, Charles. *Louisville, Ky, USA.* ca.1840.
MAIRS, R. *Birkenhead.* 1865.
MAISEY—
Christopher. *Cricklade.* 1830-42.
Christopher. *Highworth.* 1848.
MAISON, Aristide Poterel. *Bryn-Mawr.* 1868.
MAITLAND—
James. *Neilston (Scotland).* 1830.
John. *Lochwinnoch (Scotland).* 1836.
MAJOR—
Augustus. *Evesham.* 1860-72.
George. *Brighton.* 1878.
W. & A. *Cheltenham.* 1856.
William. *Cheltenham.* 1830-63.
William. *Stourbridge.* 1872.
MAKEPEACE, George. *London.* 1857-63.
MAKER, Matthew. *Charleston, SC, USA.* Pre-1776.
MALCOLM—
John. *Donaghadee.* 1772. C.
John. *Windsor, Nova Scotia.* 1791-95.
Walter. *Inverary.* 1860. Also agent for steamships.
William. *Callander.* 1827-37.
MALCOLMSON—
J. *Cavan.* ca.1825-40. W.
Rachel. *Lurgan.* 1785.
Thomas. *Lurgan.* 1800.
MALDIN, James. *Raine.* Late 17c. lant. clock.
MALE, J. *Cambridge.* 1865.
MALI, Charles. *Wisbech.* 1875.
MALKIN, John. *Kirby-in-Ashfield.* 1771-1780. C.

MALLARD, E. P. *Dawley (Shropshire).* 1879.
MALLER, Christian. *Radkersburg.* ca.1780.
MALLERY, John. *Chatham.* 1832.
MALLET(T)—
  Henry. *Birmingham.* 1880.
  Henry. *Woodbridge.* 1839. Also silversmith.
  J. *Barum (?Barnstaple).* ca.1840. C.
  John. *Barnstaple.* 1811-45, then *Bath.* ca.1866-1935 (several generations).
  Peter. *London.* Late 17c. C. (One such at *Barnstaple* 1705.)
  W. *Tipton.* 1868.
  William. *Weston-super-Mare.* 1883.
MALLEY, John. *Lancaster.* b.1823-51. W. and jeweller.
MALLINSON, Thomas. *Holmfirth.* 1871.
MALLOCH, William. *Edinburgh.* 1806. Clock case maker.
MALLORY—
  George. *New York.* 1846. Dealer.
  & MERRIMAN. *Bristol, USA.* 1831.
  Ransom (1792) to *Bristol, USA.* 1821-53.
MALLOY, George. *Lowell, Mass, USA.* 1859.
MALLS, Philip. *Washington, DC, USA.* ca.1800. C.
MALONE, William. *Guelph, Canada.* 1851.
MALT—
  James. *Wisbech.* 1858-75.
  Robert. *Spalding.* 1849.
MALTBY, Henry D. *York.* 1811-23. C.
MALYON—
  Elizabeth. *London.* 1832 (?wid. of Isaac).
  Isaac. *London.* 1828.
MAN, William Charles. *Gloucester.* 1879.
MANBY, John. *Skipton.* (b.1793)-1853. C.
MANCHESTER—
  Cyril B. *Pawtucket, RI, USA.* 1867. Toy clock banks.
  G. D. *Plainfield, Conn, USA.* ca.1850-60.
  John. *London.* a.1691 to William Hayden, CC. 1700.
  Thomas. *Bolton.* 1834-51. Clock cleaner.
MANCOR, James—see Mencor.
MANDALE, John. *Workington.* 1879.
MANDER—
  C. E. *Clewer (Berkshire).* 1877.
  T. *Windsor.* 1854-77.
MANFREDINI, Orolors. *Milan.* 18c. C.
MANGER, John (& Co.). *London.* 1875(-81).
MANGIE, Edward. *York.* 1659-67. W.
MANGIN, S. H. *London.* 1793. W.
MANHAM, Samuel. *Leeds.* 1871.
MANHATTAN—
  Clock Co. *New York.* 1899.
  Watch Co, The. *New York.* 1883-ca.1892.
MANIGLIER, John. *London.* 1832-44.
MANILL, Vincent. *London.* ca.1700. C.
MANLEY—
  Caleb. *Beccles.* s. of Daniel senior. mar. 1704 d. 1739. C. & W.
  Cornelius. *Norwich.* 1702-14. W. & C.
  James. *London.* ca.1780.
  John. From *London* to *Chatham.* 1777-1828. W.
  William. *Cardiff.* 1852.
MANN—
  John. *London.* 1668. CC.
  John. *Norwich.* 1830.
  Matilda. *Norwich.* 1836.
  Robert. *Norwich.* 1761-1831. C.
  & STEPHENS. *Coventry.* 1828-35.
  Thomas. Place unknown (believed *Yorkshire*) ring sundial dated 1673. Sundial dated 1681.
  Thomas. *Halifax.* 1743. Rep'd ch. clocks.
  William. *Gloucester.* 1840-79. Also chronometers.
  William. *London.* ca.1770. C.
MANNING—
  ——. *Southampton.* ?Early 19c. (Prob. James, q.v.)
  Benjamin. *London.* 1851-75.
  BOWMAN Co. *Meriden, Conn, USA.* 20c. Electric clocks.
  & Co. *London.* 1881.
  & Co. *Worcester.* 1876.
  Frederick. *Worcester.* 1868-72.
  James. *Southampton.* 1830-59.
  James. *Wellingborough.* d.1723. C.
  & MANNING. *Great Malvern.* 1860.
  Richard. *Ipswich, Mass, USA.* ca.1748-67. C.

MANNING—*continued.*
  Thomas. *Salem, Mass, USA.* 1685. C. and gunsmith.
  W. H. *Tunbridge Wells.* 1874.
  William. *Worcester.* 1823-50. W.
MANNINGS—
  James. *Chichester.* 1828-39.
  John. *Steyning.* 1878.
  William. *Steyning.* 1839-55.
MANROSS—
  Brothers. *Bristol, USA.* 1856-69.
  E. (Elijah) & C. (Charles) H. *Bristol, USA.* 1854-56. Succeeded by Manross Brothers.
  Elijah. *Bristol, USA.* s. of Elisha, 1827-1911.
  Elijah, jun. *Bristol, USA.* 1862-70.
  Elisha. *Bristol, USA.* (1792) 1813-56. C.
  (Elisha) & NORTON. *Bristol, USA.* 1839.
  PRITCHARD & Co. *Bristol, USA.* 1841.
  & WILCOX. *Bristol, Conn, USA.* 1831-39. C.
MANS, John. *Columbia, Pa, USA.* Pre-1818.
MANSELL—
  & ALLEN. *Birmingham.* 1865. Dialmakers.
  Mrs. Marv. *Boughton (Kent).* 1847.
  Thomas. *Fakenham.* 1865-75. C.
  William (& Son). *London.* (B. a.1816) 1828(-32-39). Watch cases.
MANSER & ASHLEY. *London.* (B. 1820) 1839.
MANSFIELD—
  Adrian. *London.* 1875-81.
  J. *Newbury.* 1877.
  J. *Shaftesbury (Dorset).* 1848-67.
  John. Place unknown. ?ca.1810. C.
  John Martin. *Shepton Mallet.* 1875.
  Samuel A. *Philadelphia, USA.* 1848.
  Thomas. *Shaftesbury (Dorset).* 1824-30.
  Thomas James. *Shaftesbury (Dorset).* 1875.
MANSIR, Robert. *London.* (B. a.1808) 1828. Watch cases.
MANSON—
  Alexander. *Thurso.* 1837.
  Cornelius. *Peterborough.* 1864-77.
  Johannes. *Sweden.* ca.1800. C.
  William Bailie. *London.* 1851.
MANSWORTH, Michael. *London.* 1632. CC.
MANTEGANI—
  A. *Wisbech.* 1846.
  J. *Wisbech.* 1858.
  John. *Wisbech.* 1875.
MANTUA—
  P. & J. *Luton.* 1854.
  Peter. *Luton.* 1839.
MANUAL & BRILLMAN. *London.* 1851-63.
MANUEL, Jules. *Philadelphia, USA.* 1849.
MANWARING—
  Hezekiah. *Wandsworth.* 1878.
  Richard. *Maidstone.* 1838-55.
MAPLE LEAF, 1914 and later, product of Pequegnat Clock Co.
MAPPLE, D. D. *London.* 1851. C.
MAPSON—
  J. *Cricklade.* 1848.
  Joseph. *Highworth.* 1867-75.
  Joseph. *Wootton Bassett.* 1830-42.
  Thomas. *Melton Mowbray.* 1876.
  William. *Tetbury.* 1863-79.
  William. *Wootton Bassett.* 1848-75.
MARACHE, Solomon. *New York.* 1759.
MARAND, Joseph. *Baltimore, USA.* 1804.
MARANVILLE, Alusha. *Winsted, Conn, USA.* 1861.
MARBLE, Simeon. *New Haven, Conn, USA.* 1801-07.
MARCER, William. *Liverpool.* mar.1726-28. W.
MARCH, Charles. *London.* 1857-75.
MARCHAND—
  ——. *France.* 1878. Carriage clocks.
  Louis H. *St. Johns, Canada.* 1871.
  Lucien. *London.* 1844-51.
  Thomas. *London.* 1857
  Samuel. *London.* a.1670 to Edward Eyston, CC. 1677.
  William. *Huddersfield.* 1866-71.
MARCHISI, Joseph. *Utica, NY, USA.* 1845.
MARCHMENT, ——. *St. Albans.* Late 17c. C.
MARCOU, J. P. *Amsterdam.* Early 18c. W.

MARCUS—
Gerrit. *Amsterdam.* ca.1750. C.
Henry. *Hay.* 1830.
R. *Jersey.* 1837-75.
MARFIELD, John. *London.* 1662. Perhaps error for Warfield.
MARGAINE, François-Arsene. *Paris.* 1869-1914.
MARGETTS, George. *Old Woodstock (Oxfordshire).* b.1748-pre-1779. C. & W. Later *London.*
MARGON, Thomas. *Manchester.* 1848.
MARGOSCHIS, Samuel. b. *Poland.* ca. 1807. At *Cheltenham* ca. 1840-46, then *Leamington* ca. 1846-68.
MARGOT, David. *London.* 1828.
MARGRAF, Christopher. *Rome.* 1596.
MARGRETT—
S. *Leamington.* 1868.
William. *Gloucester.* 1870.
William. *Hereford.* 1879.
MARGUERITE, ——. *Paris.* ca.1685. Br. clock.
MARIAS, Samuel. *London.* 1881.
MARIE—
John. *Hartford.* 17c. C. & W.
M. *Charleston, SC, USA.* 1796.
MARIEN, John. *New York.* 1848.
MARINE Clock Co. *New Haven, Conn, USA.* ca.1847. C.
MARINER, William. *Longtown.* 1873-79.
MARION Watch Co. *Marion (New Jersey City), NJ, USA.* 1872-75.
MARJORAM—
H. *North Walsham.* 1865-75.
Walter William. *London.* 1881.
MARK—
J. & A. *Peterborough.* 1854.
J. & Co. *Peterborough.* 1864.
J. & L. *Peterborough.* 1864.
MARKHAM—
& CASE. *Columbia, SC, USA.* ca.1845.
R. *Gainsborough.* 1861-68.
MARKS—
A. J. *Sunderland.* 1856.
Aaron. *Sheffield.* 1862-71. W.
E. *Hertford.* 1859.
Isaac. *Philadelphia, USA.* 1795-99. Watch dealer.
Jacob. *Merthyr Tydfil.* 1875.
L. *Portsea.* 1859.
Levi. *Bridgend.* 1838.
Lewis. *Houghton le Spring.* 1834.
Lewis. *Hull.* 1846-72.
Lewis. *London.* 1828-44.
Lyon/Leon. *Liverpool.* 1848-51.
Mark. *Cardiff.* 1829-52. C.
Michael. *Swansea.* 1830-44.
Samuel. *Bridgend.* 1848-68. C.
Samuel. *Cardiff.* 1835.
Samuel. *Cowbridge.* 1835-44. C.
MARLOW—
——. *Cranbrook.* 1732-47. W.
Charles J. *Coventry.* 1868-80.
& Co. *York, Pa, USA.* 20c. C.
M. & Son. *Bath.* ca.1820. C.
MARPLES, Robert Moffatt. *London.* 1875-81.
MARQUAND—
& Brothers. *New York.* 1823.
Frederick. *New York.* 1825.
Isaac. From *London* 1791-96, *Edentown, NC, USA,* then 1804 *New York.*
James. *Chatham, Canada.* 1857-62.
MARR—
Christopher. *London.* 1832-63.
J. & A. *Falkirk.* 1860.
John Wilkinson. *Backbarrow.* 1838. W.
William. *Ulverston.* 1837. W.
MARRAT, William. *Oundle.* 1777. W.
MARRIAN, John H. *New York.* 1847.
MARRINER, William. *Oxford.* a.1774-82.
MARRIOT(T)—
Alfred. *Wallingford.* 1877.
Benjamin. *London.* 1851-69.
G. E. *Lymington.* 1867.
George Edward. *London.* 1875.
James. *Bishops Waltham (Hampshire).* 1830.
John. *London.* a.1641 to John Midnall.

MARRIOT(T)—*continued.*
John. *London.* a.1690 to Richard Jarratt, CC. 1715.
John. *Lymington.* 1830-59.
William. *Cambridge.* 1830.
William, sen. *Northampton.* 1824-41.
William, jun. *Northampton.* 1841-69.
William Hampson. *London.* 1875.
MARRIS—
Charles. *Hull.* 1813-23. W.
John. *Market Deeping.* 1850. (Cf. Morris.)
MARROW, Richard. *Liverpool.* Lost watch 1796.
MARSDEN—
James. *Prescot.* 1824. W.
Samuel. *London.* (B. 1820) 1832-44.
Thomas. *Hensingham (Cumberland).* 1834.
MARSH—
——. *Coggeshall.* Early 19c. C.
Arthur Frederick. *Diss.* 1858-75.
Charles Hollands. *Dover.* (?s. of John, jun.). a.1834-66.
Edward. *Prescot.* mar.1768. W.
George. *Bristol* and *Farmington, Conn, USA.* 1829-33. C.
George C. *Winchester* and *Winsted* and *Bristol* and *Wolcottville* and *Farmington* and *Dayton, USA.* 1828-31.
George & Co. *Farmington, Conn, USA.* 1830s.
Gilbert & Co. *Bristol* and *Farmington, Conn, USA.* 1828. C.
Henry. *Hvyton (Liverpool).* ?Late 18c. C.
Henry. *London.* 1839-44.
& HOFFMAN. *Albany, NY, USA.* ca.1890.
J. T. *St. Helens.* 1858. W.
James. *Farnworth.* mar.1800. W.
James. *London.* 1828.
John, jun. *Dover.* s. of John, sen., a.1825-47.
John, sen. *Eastry (Kent).* 1771-d.1788. C.
John. *Highworth.* ca. 1730. C. (?1728). (B. has Ono Marsh d.1733, which is probably an error for Jno.)
John. *London.* 1828-32. Watch cases.
Joseph. *Wolverhampton.* 1868-76.
Richard Charles. *Birmingham.* 1880.
T. *Mortlake.* 1866.
T. *Southampton.* 1859-67.
T. K. *Paris, Ky, USA.* ca.1804.
Thomas. *Bold.* 1790-95. W.
Thomas. *Dorking.* 1855-78.
Thomas. *Farnworth.* mar.1790. W.
Thomas. *Lymington.* 1830.
Walter. *London.* 1857-69.
William. *Diss.* 1836-46.
William B. *Cambridge.* 1840-46.
William B. *London.* 1851.
WILLIAMS & Co. *Dayton, O. USA.* ca.1830. C.
WILLIAMS, HAYDEN — Co. *Dayton, O, USA.* Post-1833. C.
MARSHALL—
——. *Highworth.* 1728. C. (May be error for Marsh (John), q.v.).
——. *South Emsall.* Early 19c. C. (Prob. Francis, q.v.).
& ADAMS, *Seneca Falls, NY, USA.* ca.1825. C.
Benjamin. *London.* a.1672 to Abraham Prime, CC. 1680 (B.-1732).
Charles. *Greenwich.* 1839-47.
Christopher. *Halifax.* 1701. Rep'd ch. clocks.
& Co. *Gosport.* 1878.
Edward. *London.* 1832.
Edward, jun. *London.* 1851-57.
Edward. *Yarmouth.* 1836.
Ernest. *Durham.* Late 18c. C.
Francis. *South Emsall.* 1838. C.
Francis. *Wath on Dearne (Yorks).* 1838. C.
G. *London.* 1858. W.
Henry. *London.* 1655.
James & Walter. *Edinburgh.* 1816-60. (Not James and Waller -B. 1816).
John. *Bishopwearmouth.* 1841-56.
John. *Chulmleigh.* 1740. C.
John. *Doncaster.* 1862.
John. *Moorthorpe (nr South Kirkby, Yorks).* 1838-71. C. & W.
John. *Newark.* b.1725-mar.1747. C.

MARSHALL—*continued.*
John. *Newry.* 1846-90. W.
John. *St. Columb (Cornwall).* 1617. Rep'd ch. clocks.
Joseph. *Leicester.* (B. 1751) -ca.1780.
Joseph. *Lincoln.* 1828-35.
Joshua. *Lincoln.* ca.1800. C. (?May be error for Joseph.)
Miss Mary. *Elora, Canada.* 1857.
Peter. *South Queensferry (Scotland).* 1830-60. W.
R. *Clay Cross.* 1876-1922.
Richard. *London.* 1857-81.
Robert. *Greenside* nr *Newcastle-on-Tyne.* ca.1775-1856. C. (Perhaps two of same name.)
Robert. *Ryton (County Durham).* 1827.
Robert. *Shotley Bridge* (nr *Gateshead).* 1848-58.
Samuel. *London.* a.1682 to Cornelius Jenkins, CC. 1689-1719.
Thomas. *Haley Hill* (nr *Halifax).* 1837-53. C.
Thomas. *Halifax.* 1659-63. Rep'd ch. clocks.
Thomas. *Halifax.* 1810-22. C.
Thomas. *London.* 1869.
Thomas R. *Oxford.* 1852. C. & W.
W. H. *Huddersfield.* 1871.
& WHITE. *Petersburg, Va, USA.* ca.1825.
William. *Bellie (Scotland).* (b.1747)-d.1833. C.
William. *Grantham.* 1868-76.
William. *Halifax.* 1810-50.
William. *London.* (B. 1817-) 1828-32.
William. *Newry.* 1840. C.
William. *Market Rasen.* 1828-35.
William H. *Boston.* 1876.
William James. *London.* 1844-69.
William (& Son). *London.* 1844-51(-57).
MARSHMAN—
Frederick. *Northampton.* 1877.
Robert. *Walthamstow.* 1874.
MARSLAND, W. *Stockport.* 1865.
MARSON, Edward. *Rugby.* 1828-54.
MARSTON—
A. E. *Coventry.* 1880.
Charles. *Birmingham.* 1880.
James and Henry. *Coventry.* 1850-80.
John. *London.* 1844.
Thomas, sen. *Coventry.* 1828.
Thomas, jun. *Coventry.* 1835-42.
Thomas. *Queniborough* nr *Leicester.* 1876.
William. *London.* a.1659 to Joseph Munday, later to Edward Whitfield, CC. 1669. (B.-1685).
MARTEN—
& BISHOP. *London.* 1857-69 (Successor to William Marten).
Henry & William. *London.* 1839-44.
William. *London.* 1851.
MARTI—
J. & Cie. *France.* ca.1860. C.
S. & Cie. *Le Pays de Montbeliard.* 1867. C.
MARTIN—
——. *Atherstone.* ca.1840. C. (See George.)
——. *Crosby* (nr *Liverpool).* 1750. C.
A. *Canterbury.* 1846. W.
Abraham. *London.* CC. 1682-97.
Alfred. *Brighton.* 1839-62.
Alfred. *Greenwich.* 1866-74.
Andrew. *Swansea.* 1875-99.
& BALCHIN, *London.* 1875-81.
& BALCHIN. *Wandsworth.* 1878.
BASKET & MARTIN. *Cheltenham.* 1840-56.
Celestin. *Sandwich.* 1862-82.
& Co. *Cheltenham.* 1863-70.
Edward. *London.* 1839.
Emile. *St. Nicolas d'Aliermont.* 1859-67.
Emile & SAUTER Brothers. *London.* 1869.
Felix. *Swansea.* Mid-19c.
Felizian. *Swansea.* 1887.
Frank. *London.* 1881.
George. *Atherstone.* 1842-80.
George. *Lancaster, Pa, USA.* Makers of catgut and ropes for clocks, 1780-1830.
George. *London.* 1828.
George. *St. John, New Brunswick.* 1864.
George. *Swansea.* 1868-75.
George A. *Bethel, Me, USA.* ca.1870.

MARTIN—*continued.*
J. *Maidstone.* 1865.
J. B. *Hamilton, Ontario.* 1852.
James. *Faversham.* 1847-55. W.
James. *Maidstone.* 1845-74.
James. *Midhurst.* 1839.
James. *Ventnor.* 1878.
John. *Comber, co. Down.* Late 18c.-early 19c. C.
John. *Faversham.* 1818-23. W.
John. *Gatehouse (Scotland).* 1860.
John. *Kincardine-on-Forth.* 1836.
John. *Liverpool.* 1750-ca.1790. C.
John. *Maidstone.* 1832-47.
John. *New York.* 1831. W.
John. *Worksop.* 1828.
John C. *London.* 1839-63.
John J. *Philadelphia, USA.* 1844.
(Thomas) & MULLIN (Robert). *Baltimore, USA* 1764.
Patrick. *Philadelphia, USA.* 1820-50.
Peter. *Glasgow.* 1840-60. W.
Peter. *New York.* ca.1810.
Richard. *Eton.* 1854-77.
Richard. *Helston (Cornwall).* ca.1780. C.
Richard. *Northampton.* ca.1695.
Robert. *Dumfries.* 1730.
Robert. *Worksop.* 1835.
Samuel. *Cheltenham.* 1830.
Samuel. *Dublin.* ca.1780-ca.90. W.
T. *Spratton (Northamptonshire).* 1854-69.
Theophilus. *London.* 1857-81.
Thomas. *London.* b.1666, a.1685, mar.1688. W.
Thomas. *Thirsk.* 1780-1823. C. & W.
Thomas I. *Wigan.* Free 1675-d.1716. C. & W.
Thomas II. *Wigan.* b.1688-1744.
Valentine. *Boston, USA.* 1842.
W. *Guildford.* 1866.
William. *Darlington.* 1828-50. C.
William. *London.* 1828-44.
William. *Wrexham.* 1887-90.
William Day. *Sheffield.* 1862. W.
William George. *London.* 1875.
William Nicholas. *Shrewsbury.* b. ca. 1815-75. W.
Wilkinson Day. *Holmfirth.* 1838-50. W.
MARTINEAU, Joseph, jun. ?*London.* ca.1750. W.
MARTINOT, Gilles. *Paris.* ca.1715. W.
MARTINS, ——. *Bath.* 1810. C. Also at Cheltenham.
MARTINSTEIN, George. *London.* 1839.
MARTLOCK, Martin. *Clare.* 1875. Prob. error for Mortlock, q.v.
MARTYN—
J. *Looe.* ca.1780. C.
J. N. *Hamilton, Canada.* 1851.
John (sen. and jun.). *Quebec.* 1826-49. W. Also naturalist.
John Nicholls. *Falmouth.* ca.1800. W. & C.
Mrs. John. *Quebec.* 1851-52.
William. *St. Columb.* 1800-d. 1809. W. & C.
MASCALL, A. G. *Middlesbrough.* 1898.
MASCHER, John F. *Philadelphia, USA.* 1845.
MASHAM, Samuel. *Wiltshire, Md, USA.* 1774.
MASHFORD, James C. *Grimsby.* 1876.
MASI—
& Co. *Washington, DC, USA.* 1833.
Seraphim. *Washington, DC, USA.* ca.1830. Dealer.
MASKEL, ——. *Salisbury.* Early 19c. C.
MASKELYN, J. *Cheltenham.* 1863.
MASKENS, Henry. *Dover.* 1839.
MASON—
——. *Bryn-mawr.* 1844.
Used by Illinois Springfield Watch Co. 1870-75.
Alexander. *Liverpool.* 1792 lost watch.
C. *Canterbury.* 1851.
Charles. *Canterbury.* 1838-40.
Edward. *Aberystwyth.* 1864-68.
George. *Waseca, Minn. USA.* ca.1870. C.
H. G. *Boston, USA.* 1844-49.
Henry. *Dewsbury.* 1875-1901. W.
Henry. *London.* a.1647 to Jeremy Gregory.
J. *Birmingham.* 1854.
J. *Worcester.* 1860.
James. *London.* 1844-63.

MASON—*continued.*
John. *Bawtry.* Perhaps same man as at *Doncaster.*
John. *Bristol.* 1673.
John. *Doncaster.* ca.1740-1771. C.
John. *Rotherham.* 1822-71. W.
John & Edward. *Worcester.* 1828-50.
Joseph. *Buckingham.* 1704-ca.1750. C.
Joseph. *London.* 1832-39. Musical clocks.
P. *Somerville, NJ, USA.* Early 19c.
Richard. *London.* (B. BC. 1615)-CC. 1635.
Robert. *Worksop.* 1835.
S. *Burton Latimer.* 1854.
Samuel Taylor. *Olney.* 1811-30.
& Son. *Canterbury.* 1874.
Thomas. *Blackburn.* 1858. W.
Thomas. *Colchester.* Successor to John Cooper, 1835-53. C. & W.
Thomas. *Preston.* 1848-51. (*Bamber Bridge* nr *Preston.*)
Thomas James. *London.* 1869-75.
Timothy. *Gainsborough. Lords Street.* 1727. C.
Timothy B. *Boston, USA.* 1830.
William. *Bristol.* 1870-79.
William. *Daventry.* 1802.
William. *London.* a.1680 to Mary Harris, CC. 1688(-B. 1713).
William. *Walsall.* 1876.
William. *Warwick.* 1811-1830s, then *Leamington.* ca. 1835-50. C.
William H. *Mount Holly, NJ, USA.* 1834-61. C.
MASOTH, Charles. *Canterbury.* 1847.
MASPERO, William Angelo. *Leatherhead.* 1878.
MASPOLI—
Augustus. *Hull.* 1838-55. Also barometers.
James. *Hull.* 1851. W.
Monti & Co. *Sandwich.* 1838-40.
Peter. *Sandwich.* 1845-47.
Vittore. *Sandwich.* b. abroad ca.1810-41.
MASSAM, Robert. *Kirkby Lonsdale.* ca.1760-70. C.
MASSERON, Thp. *Paris.* ca.1650. W.
MASSEY—
Alexander. *Sunderland.* ca.1810. C.
Charles. *London.* 1832.
Charles R. *Philadelphia, USA.* 1837-39. W.
Edmund. *London.* a.1674 to Joseph Knibb, CC. 1682 (B. -1690).
Edmund. *London.* 1839-51.
Edward James, jun. *London.* 1839-51.
Edward John. *Liverpool.* 1848 (B. 1838). Chron. maker.
Francis. *Liverpool.* 1824.
Francis F. *London.* 1839.
Francis Joseph. *London.* 1839-44.
John. *Charleston, SC, USA.* ca.1736.
John. *London.* 1828.
Joseph. *Charleston, SC, USA.* 1722-36.
Thomas. *London.* 1839-69.
William. *Nantwich.* 1828-34.
& WINDHAM/WYNDHAM. *London.* 1833-39.
MASSIE, Edward. *Liverpool.* Lost watch 1785.
MASSINGHAM—
G. P. *Boston.* 1868.
John. *Fakenham.* 1738-1809. C.
P. *Kings Lynn.* 1846.
W. *Boston.* 1861.
William. *Kings Lynn.* 1836.
MASSOT, Horace. *Charleston, SC, USA.* ca.1780.
MASTERMAN, G. *London.* 1795. W.
MASTERS—
A. *Coventry.* 1880.
Alexander Francis. *London.* 1875-81.
H. *Coventry.* 1880.
H. & A. *Coventry.* 1868.
Henry. *London.* 1881.
Henry. *Woodsetton* (*Staffordshire*). 1876.
J. *Gosport.* 1859.
J. N. *Rye.* 1870-78.
J. R. *Tewkesbury.* 1863.
John I. *Bristol.* a.1699 -insolvent. 1739.
John. *Bristol.* 1780.
John. *London.* 1875-81.
John. (b. *England* 1770). *Newfoundland* 1786-1818, then *Boston, USA,* 1820, then *Bath, Me, USA* -1846. C.

MASTERS—*continued.*
John. *Tenterden.* b.1818. d.1887.
John Neve. *Rye.* b.1846. d.1928.
John Neve. *Tenterden.* 1855.
John Robert. *Birmingham.* 1880.
Richard. *Halifax, Nova Scotia.* 1817-38.
Samuel. *Kingston-on-Thames.* 1722. W.
Thomas. *Littlehampton.* 1870.
Thomas. *Southampton.* 1830.
William. *Bath, Me, USA.* s. of John, b.1806-1854.
William. *Cheltenham.* 1840.
William. *London.* a.1672 to Henry Young, CC. 1701.
MASTERSON, Patrick. *Killyshandra, co. Cavan.* 1846.
MASTERTON—
A. E. *Penge.* 1878.
Alfred Ernest. *London.* 1881.
MASTON—
& GATH. *Bristol.* 1850. (Prob. Muston & Gath).
John. *London.* 1661-62, *Westminster.* C.
Thomas. *London.* (B. pre-1777-) 1788-1822. W.
MATCHAM, Samuel. *London.* ca.1700. C.
MATEER, Samuel. *Belfast.* ca.1775-1800. W.
MATHENS—
David. *Carmarthen.* Later 18c. C.
David. *Talley.* Later 18c. C.
MATHER—
C., sen. *Nottingham.* 1864.
C., jun. *Nottingham.* 1864.
Eli. *West Bradford, Pa, USA.* 1828.
Nathaniel. *Liverpool.* mar.1707. W.
Robert. *Nottingham.* 1835-76.
William. *Nottingham.* (B. 1818-)1828-64.
MATHERS—
Adam Andrew. *Coleraine.* ca.1810-59. C. & W.
George. *Peterhead.* 1837.
J. ?*Coleraine.* 1855.
Samuel. *Ballycastle.* ca.1835-59. C.
William. *Ballycastle.* ca.1760. W.
MATHESON, John S. *Leith.* 1880.
MATHEWS & MATTHEWS—
A. L. *St. Catherines, Canada.* 1853.
Aaron. *Oakville, Canada.* 1870-75.
Alfred. *Leighton Buzzard.* 1847-77.
Andrew. *Peterhead.* 1860.
Barnaby. Place unknown. Mid-18c. C.
Charles. *Rhyl.* 1887-90.
Christopher. *Oxford.* 1667-88. Turret clock dial painter.
David & Edwin. *Bristol, USA.* 1832.
Edward. *Leeds.* 1853-66. W.
Edward. *London.* s. of Frederick Samuel. b.1853. a.1871. d.1936. Gold watch case maker.
Edward. *Welshpool.* 1822-44.
Edward, jun. *Welshpool.* 1844-68. W.
Edwin. *Welshpool.* 1874.
Frederick Samuel. *London.* Goldsmith Co. 1851. Successor to Louis Comtesse. Gold watch case maker.
Gilbert Edward. *Knighton.* 1887.
J. *Rochdale.* 1858. C.
James Howell. *Oswestry.* b. ca. 1839. s. of Richard Matthews, watchmaker, -1861. W. & jeweller.
Jacob. *Bristol.* 1842.
James. *London.* 1828-32.
(David), JEWEL (Lyman) & Co. *Bristol, USA.* 1851-53. C.
John. Place unknown. ca.1720. C.
John. *Ballycastle, co. Antrim.* ca.1800. C.
John. b. ca. 1797 *Meifod (Montgomery).* To *Bishopscastle.* by 1828-79. W.
John. *Leighton Buzzard.* 1830-47.
John. *London.* 1839.
John Francis. *London.* 1869-81.
John Turner. *Cardigan.* 1844 (see Matthias).
Joseph John. *London.* 1881.
Mrs Ann. *Penrith.* 1869-73.
Peter. b. *Blandford.* At *Battle* 1837, then *Uckfield* ca. 1840-70.
& PENNOYER. *Bristol, USA.* 1830.
Richard R. *Bishopscastle.* b. ca. 1842. s. of John Matthews, watchmaker. At *Builth Wells.* ca. 1864-67, then back to *Bishopscastle* -1875. W.
Richard. b. ca. 1814 *Welshpool. Oswestry* by 1836-56. W.

MATHEWS & MATTHEWS—*continued.*
Robert. *Ely.* 1830-46. C.
Thomas. *Knighton.* 1868-75.
Thomas. *Leicester.* 1876.
Thomas & Richard. *Oswestry.* 1835.
& THORP. *London.* 1844.
W. *Wigton.* 1858-69. Also music seller.
William. *Coventry.* 1880.
William. *Egremont.* 1873-79.
William. *Leighton Buzzard.* 1839.
William. *London.* 1881.
William. *Penrith.* 1834-58.
William L. *Bridport.* 1867-75.
MATHEWSON—
& HARRIS. *New Hartford, Conn, USA.* ca.1830 C.
J. *Providence, RI, USA.* 1849.
James. *Kilconquhar.* s. of Andrew. b.1813-d.1882.
MAT(T)HEY—
Henry. *London.* 1857-81.
Lewis. *Philadelphia, USA.* 1797-1803.
Philibert. *London.* 1832. Watch cases.
MATHEY-TISSOT. *Les Ponts-de-Martel, Switzerland.*
ca.1900.
MAT(T)HIAS—
John Turner. *Cardigan.* 1850-87 (see Matthews).
Thomas. *Haverfordwest.* Later 18c. C.
MATHIESON, ——. *London.* mid-18c. C.
MATHIEU, Gaston. *New York.* 1845.
MATLACK—
William. *New York.* 1769-1777, then later *Philadelphia.*
William. *Philadelphia, USA* from *Charleston, SC.*
1787-1828.
MATT, Mathew. *Cardiff.* 1865-71. C.
MATTHY, Daniel. *London. St. Giles in the Fields.* Took
A. Goutier apr. in 1735.
MATTOCKS—
M. & Co. *Coventry.* 1864.
M. & Co. *Bakewell.* 1864.
Mark & Co. *Matlock.* 1864-76.
William. *Bournemouth.* 1878.
MAUCKENMULLNER, Jacob. *Vienna.* 1509.
MAUD—
Charles. *Daventry.* 1877.
E. B. & C. *Daventry.* 1847-69.
Edward. *Daventry.* 1841-47.
Edward & Charles. *Daventry.* 1830.
William. *Ryde.* 1878.
MAUDS, B. E. *Daventry.* 1780.
MAUDSLEY, George. *Wakefield.* 1775-d.1780. W.
MAUGER, Charles Martin. *Jersey.* 1810. W.
MAUGHAM, Joseph Heppell. *Newcastle-on-Tyne.* 1834-
56. C.
MAULE—
Alexander. *Cheltenham.* 1879.
Andrew. *Wooler.* 1827.
James. *Berwick-on-Tweed.* 1827.
John. *Wooler.* 1834-48. C.
William. *Coldstream.* 1837-60.
MAUNDER—
Aaron. *Launceston.* 1856-73. C. & W.
Michael. *Bodmin.* 1844-56.
MAUNTON, G. *Woodbridge.* 1853.
MAURER, Hans. Place unknown. ca.1640.
MAURICE—
E. & Co. *Paris.* 1889. carr. clock.
John. *Haverfordwest.* 1791-ca.1800. C. & W.
MAUS—
Frederick. *Philadelphia, USA.* 1782-93. C.
Jacob. *Philadelphia* and *Trenton, USA.* 1780-90.
Philip. *Lebanon, Pa, USA.* ca.1800. C.
Samuel. *Pottstown, Pa, USA.* ca.1790. C.
William. *Quakertown, Pa, USA.* ca.1810.
MAUSON, Charles. *London.* 18c. C., mid to late.
MAUTZ, John. *Philadelphia, USA.* 1841.
MAVER, Francis. *Fochabers (Scotland).* 1837.
MAVIN, John. *Cleator.* 1879.
MAW, L. *Darlington.* 1898.
MAWDSLEY—
Hargreaves. *Southport.* 1848-58. W. Jeweller and opti-
cian.
John. *Manchester.* 1851.
John. *Philadelphia, USA.* 1846.
MAWLEY. Robert. *London.* 1766. W.

MAWMAN—
George. *Beverley.* 1813-23. W.
George, sen. *Patrington.* 1840-58.
George junior. *Patrington (E. Yorks).* 1858-92.
MAWSON, William. *Leeds.* 1834-71. W.
MAXANT, E. M. L. *USA.* 1879.
MAXEY, Charles. see MAZEY.
MAXFIELD, Robert. *Wetherby.* 1866. W.
MAXIMILIAN & KALTENBACH. *Brecon.* 1844.
MAXWELL—
F. *Battersea.* 1878.
Henry. *Edinburgh.* a.1822.
James. *Boston, USA.* Early 1700s, hourglasses.
William. *London.* 1857-81.
MAY—
David. *Portadown* 1854, then *Downpatrick* 1858.
Edward. *Lyndhurst.* 1878.
Edward. *Oxford.* 1725. May be same man as at *Witney.*
Edward. *Witney.* ca.1725-ca.1760. C.
FIRNHABER & Co. *London.* 1881.
George. *Belfast.* 1843-46. C.
George. *Bidford (Warwickshire).* 1880.
George. *Lurgan.* 1854-65. (May be from *Belfast.*)
J. *Bath.* 1856.
James. *Landport.* 1867.
John. *Bath.* 1740.
John. *Bath.* 1848.
John. *Witney.* 1725-95. C.
John B. *Bath.* 1875-83.
Mrs. M. *Bath.* 1861-66.
& PAYSON. *Baltimore, USA.* 1789.
Robert. *Deptford.* 1839.
Samuel. *London.* 1844-57.
Samuel. *Philadelphia, USA.* 1765.
Thomas. *Witney.* s. of John. 1772.
W. *Penge.* 1866.
William. *Brighton.* 1839.
William. *East Grinstead.* 1878.
William. *Portadown.* ca.1800-59. C.
William Alfred. *Dover.* 1874.
MAYALL, Alfred Henry. *Hay.* 1887.
MAYBOM, F. C. *Paris.* 1640. W.
MAYBON, E. *Paris.* ca.1670.
MAYELL—
Edwin. *Liskeard.* 1844-56.
Joseph. *Lostwithiel.* 1823-28.
MAYER—
Anthony (& Son). *London.* 1857-61 (1869-81).
Dominick. *Birmingham.* 1868.
Elias. *Philadelphia, USA.* 1831.
F. *Reading.* 1854.
Henry. *Peterborough.* 1841(-1847).
Henry. *Tunstall (Staffordshire).* 1828.
John B. *Niagara Falls, Canada.* 1861.
Joseph. *Liverpool.* 1848. Also electroplate goods dealer.
Kasper. *Newcastle-on-Tyne.* 1848-56.
Martin. *London.* 1881.
N. & Son. *London.* 1869-81.
Saul. *Manchester.* 1848-51.
MAYERS, J. *Luton.* 1864.
MAYES—
G. *Southwold.* 1846.
G. *Stowmarket.* 1858.
John. *London.* 1844-57.
Patrick. *Killyshandra, co. Cavan.* Early 19c. C.
MAYHEW—
George Frederick. *London.* 1863-81.
Henry. *Bury St. Edmunds.* 1879.
John. *Rendham.* mid-18c. C.
Joseph. *Rochester.* 1866-74.
MAYLAN, Louis. *Moseley (Worcestershire).* 1873.
MAYLAND or **MAYLARD,** Thomas. *London.* a.1682 to
Henry Reeve. CC. 1698.
MAYLARD, Charles. *Birmingham.* 1880.
MAYLER, William. *Llandudno.* 1868-76.
MAYN, Benjamin. *Kelvedon.* 1839-51. W.
MAYNARD—
——. *Turin.* ca.1760. C.
Christopher. *London.* a.1660 to Simon Hackett, CC.
1667-92, 'Royal Exchange'.
Edward James. *London.* 1875.
George. *New York.* 1703-30.

MAYNARD—*continued.*
J. *Teignmouth.* ca.1800. C.
John senior. *Long Melford.* b. 1636 d. 1689.
Blacksmith & TC.
W. *Coventry.* 1854.
MAYNE, William. *Llanelly.* 1868.
MAYO—
Charles & Co. *London.* 1828.
John. *Coventry.* 1828-54.
William. *Coventry.* (B. 1817-) 1828-50.
William. *Rochdale.* 1851. Also jeweller.
William Edwin. *Gloucester.* 1840-42.
William & Son. *Manchester.* 1834-51.
MAYOR, John. *Manchester.* 1834. German and Dutch
clocks.
MAYSENHOELDER—
& BADLEY. *Montreal.* 1862.
& BOHLE. *Montreal.* 1849-54.
David. *Montreal.* 1840-80.
John & MEVES, Otto. *Montreal.* 1858-60.
MAVES & BADLEY. *Montreal.* 1862.
MAYSMOR(E)—
Humphrey. *Wrexham.* ca.1716-30. W. (?From *London*
where one such CC. 1692).
William. *Wrexham.* ca.1720. W. (?From *London* where
one such a.1693 to John Higgs).
MAYSON, William. *Whitehaven.* 1879.
MAZEY—
Charles. *Witney.* 1771.
Charles. *Woodstock.* 1795.
MAZZI, Jean Baptiste. *Florence.* ca.1730. C.
MEAD—
Adriance & Co. *Ithaca NY. USA.* 1831.
Benjamin. *Castine, Me, USA.* Early 19c. C.
MEADER—
Charles. *Stalbridge (Dorset).* 1875.
J. *Sturminster.* 1875.
John William. *London.* 1881.
MEADLEY, Richard. *Beverley.* 1713. Clock keeper.
MEAGEAR, Thomas J. *Wilmington, Del, USA.* 1833-
ca.1850.
MEAGER, George S. *Yarmouth, Isle of Wight.* 1867-78.
MEAKIN—
William. *Dublin.* 1699.
William. *Ellesmere.* 1828.
MEAL, J. *Slaithwaite (nr Huddersfield).* Early 19c. C.
MEALE, William. *London.* 1839. Watch cases.
MEAR, Thomas. *Dursley.* (B. 1789-93) 1830.
MEARNS—
& Co. *Belfast.* 1868. W.
John. *Aberdeen.* 1792-1846.
John. *Belfast.* 1865.
MEARS—
Alonzo. *London.* 1869-81.
Charles. *Philadelphia, USA.* 1828-35.
George. *London.* 1881.
H. *Wigan.* 1858. W.
James John. *Eastbourne.* 1870.
Thomas Joseph. *London.* 1881.
William. *Reading, Pa, USA.* 1785. C.
MEASURE—
Leanne. *London.* 1828.
S. *Merton.* 1866.
W. *Ashby de la Zouche.* 1864-76.
MEATH & FRAZER. *USA.* 1831. W.
MEBERT, Isaac. *London.* ca.1656. Diamond cutter.
MECANLEY, William. *Liverpool.* 1848.
MECHIN, Henry. *London.* 1881.
MECHLIN, Jacob. *Reading, Pa, USA.* ca.1759.
MECKE, John. *Philadelphia, USA.* 1837.
MECKEVELT, Johannes. *Vorden, Holland.* ca.1760.
C.
MECOM, John. *New York.* 1763-70. W.
MEDCALF(E)—
——. *Woodstock.* 1800.
G. *Brightlingsea.* 1866.
George. *Haverhill.* 1875-79.
Joseph. *London.* 1869-81.
William. *Liverpool.* 1795. W.
William. *Preston.* b.1838-63, then *Kendal* 1863-77. W.
MEDICI, Mark. *Middlesbrough.* 1866. W.
MEDLEY, Robert. *Colne.* 1791-1802. C.
MEDLOCK, Charles. *London.* 1869-81.

MEE—
Francis. *Higham Ferrers.* ca.1780-1799. C.
H. *Chesterfield.* 1855-64.
MEECH, John. *Fordingbridge.* 1878.
MEEK—
John. *Edinburgh.* mid-19c. C.
John. *London.* 1828-32. Also musical clocks.
Peter Joseph. *London.* 1857-81.
Samuel. *London.* ca.1740. W.
MEEKS, ——. *Basingstoke.* Early 19c. C.
MEER Brothers. *Frankfort, Ky, USA.* ca.1835.
MEERS, Stephen. *Hempnall (Norfolk).* ca.1760. C.
MEGARY, Alexander. *New York, USA.* ca.1820-30.
MEGONEGAL, W. H. *Philadelphia, USA.* 1844.
MEIER, Felix. *New York.* ca.1880.
MEILY, Emanuel. *Lebanon, Pa, USA.* ca.1810.
MEIN, William. *Newcastle-on-Tyne.* b. ca.1765-left 1801.
Prob. went to *London.*
MELANDER, John. *Toronto.* 1864-77.
MELCHER, ——. *Plymouth Hollow, Conn, USA.*
ca.1790.
MELCHIONA, Antonio. *Geneva.* ca.1790. W.
MELHOM, Michael. *Boonsboro, Ind, USA.* ca.1830.
MELICK, James Godfrey. *St. John, New Brunswick.*
1824-64.
MELLANBY, J. *Stockton.* 1898.
MELLIN, Cornelius. *London.* Alien. 1622. Blackfriars.
Master with four apprentices and six journeymen.
MELLING & Co. *Liverpool.* 1848-51. Chrons.
MELLOR—
E. G. *Montreal.* 1866-76.
John. *Liverpool.* 1821-25. W.
John. *Manchester.* mar.1790-1824. W.
W. *Matlock.* 1855.
MELLUISH—
J. *Bath.* 1861-75.
J. *Walcot nr Bath.* 1856.
MELLY Brothers. *New York.* ca.1829.
MELOCHE, Charles. *Montreal.* 1862-67.
MELROSE—
George. *Yarmouth, Nova Scotia.* 1837.
James. *Edinburgh.* 1826.
MELSON, M. *Aldershot.* 1867.
MELUN, Michael. *Falmouth.* ?Late 18c. W.
MELVILL(E)—
Henry. *Wilmington, NC, USA.* From *London.* 1798.
Robert. *London.* 1832-39.
William. *London.* 1832. Watch cases.
MELVIN, Richard. *Belfast* (from *London*) and also
Glasgow. Sundial maker ca. 1790-1864. (esp. slate dials).
MEMESS—
Robert. *London.* (B. 1817-CC. 1825) 1828-44.
Robert. *Woolwich.* 1826-28. W.
MEMPSTEAD, John. *Ash.* 1782-91. C.
MENADUE, William. *Camborne* and *St. Agnes.* 1844.
MENAGE—
Ernest. *London.* 1881.
William. *London.* 1881.
MENCOR (sometimes MANCOR), James. *Bolton.* 1834-
58. W.
MENDALL, Benjamin. *Holt.* 1830-46.
MENDE, W. *Bristol.* 1856.
MENDELSON, Henry. *Manchester.* 1834-51.
MENDS—
Benjamin. *Philadelphia, USA.* ca.1800.
James. *Philadelphia, USA.* 1796.
John. *Plymouth.* mar.1767.
MENEFY—
W. T. *Whitchurch.* 1859-67.
William. *Whitchurch.* 1839-48.
MENTHA, Fritz. *Manchester.* 1848.
MENTZ, Lawrence. *Trenton, Canada.* 1857.
MENZIES—
James. *Philadelphia, USA.* ca.1800.
John. *London.* 1839.
John. *Philadelphia, USA.* 1804-51.
John, jun. *Philadelphia, USA.* 1835 and later.
Robert. *Alloa.* 1830.
Robert. *Crieff.* 1827.
MERCER—
Hay. *Aberdeen.* 1837.
John. *Farnworth.* mar.1830. W.
John, sen. *Hythe.* pre-1737. W.

MERCER—*continued.*
John, jun. *Hythe.* s. of John sen. pre-1796. W.
John. *Lewes.* a.1713 to Thomas Barrett, C.
Joseph Barnes. *Coventry.* 1860-80.
Joseph H. *Appleby.* 1869-79.
Miss Elizabeth. *London.* 1857.
Thomas. *London.* 1863-69.
Thomas & Son. *London.* 1881.
Thomas & Son. *Warwick.* 1827-30, then
Coventry. 1828-60.
Thomas J. & Sons. *Coventry.* 1868-80.
William. *Hythe.* 1781. W.
William. *Liverpool.* (B. 1734-)d.1754. W.
William. *London.* 1832.
William. *Prescot.* mar.1803-26. W.
MERCHANT—
Augustin C. *Bristol.* 1879.
Benjamin Hyatt. *Shepton Mallet.* 1861-83.
Nathaniel. *Eastbourne* from *Mayfield.* a.1718 to
Richard Weller. C.
Thomas. *Banbury.* 1847-51. C. & W.
MERCKEL, Emil. *London.* 1881.
MEREDITH—
B. *Havant.* 1848.
John. *London.* s. of Lancelot, a.1654 to father, CC.
1664.
Joseph P. *Baltimore, USA.* 1824-28.
Lancelot. *London.* CC. 1637-65.
Thomas. *Llanfyllin.* 1868-90.
William. *Chepstow.* ca.1775 (-B. 1791). C. & W.
William. *Merthyr Tydfil.* 1852-99. C.
William. *Tredegar.* 1852.
MEREDYDD, Robert. *Corwen.* Early 19c.
MEREJOTT (or MERIGEOT), John. *London.* 1731.
Ropemaker's Alley, Little Moorfields. C.
MERGENTHALER, Ottmar. (1854 Germany.)
Washington, DC, USA. 1872-76. *Baltimore* 1876.
MERGIN, W. J. *Upper Norwood.* 1862-66.
MERIFIELD & GENN. *Falmouth.* 1823. W.
MERIMEE, William. *USA.* 1796.
MERITO, William. *Canterbury.* 1847.
MERITT, William. *Canterbury.* 1838-74. C.
MERKEY, Thomas. *Paris.* 1673.
MERLIN, James. *London.* 1776. C.
MERREDAY, John. *London. Chancery Lane.* 1662.
MERRELL—
A. *Vienna, O, USA.* ca.1828. C.
F. *Winchcomb (Gloucestershire).* 1856.
William. *Winchcombe.* 1850.
MERRICK—
Alfred. *Eton.* 1847-77.
Joseph. *Eton* and *Windsor.* 1842-53.
Joseph. *London.* 1828-44.
Joseph. *Windsor.* 1837-54.
Mrs. M. *Windsor.* 1864-77.
Thomas. *London.* 1857-63.
MERRIDEW—
A. W. *Coventry* pre-1874, then *Rhyl* 1874-76. Prob.
same as below.
A. W. *Tunstall (Staffordshire).* 1868. Prob. same as
above.
Frederick H. *Caernarvon.* 1886-90.
MERRIE, John P. *Utica, NY, USA.* 1833.
MERRILL—
Henry. *Richmond (Surrey).* 1839-66.
William. *Richmond (Surrey).* 1828.
MERRIMAN—
Benjamin. *London.* (B. CC. 1682-1734) of *St. Martins
Ludgate,* C. & W.
(Dr. Titus), BIRGE (John) — Co. *Bristol, USA.*
1819-22. C.
& BRADLEY. *New Haven, Conn, USA.* 1825.
BRADLEY & Co. *New Haven, Conn, USA.* ca.1810.
& Co. *New Haven, Conn, USA.* post-1805. C.
Henry. *London.* a.1667 to Richard Bowen, CC. 1674-
1711. W.
(Titus) & DUNBAR (Butler). *Bristol, USA.* pre-1812.
C.
Marcus. *New Haven, Conn, USA.* s. of Silas, ca.1805.
R. & Co. *Bristol, USA.* Early 1820s.
Reuben. *Cheshire* and *Litchfield, Conn, USA.* 1842. C.
Samuel. *New Haven, Conn, USA.* s. of Silas, ca.1805.

MERRIMAN—*continued.*
Silas. *New Haven, Conn, USA.* 1739. mar.1760-1805.
C.
Titus. *Bristol, USA.* Early 1800s. C.
MERRIN, Henry. *London.* 1839-75.
MERRISON—
J. *Norwich.* 1846.
J. A. *Loddon.* 1846.
James. *Wymondham.* 1830.
James A. *Norwich.* 1830-36.
MERRY—
F. *Philadelphia, USA.* 1799.
J. *Nottingham.* 1849.
MERSON, James. *Huntly.* 1837-60.
MERTON & Co. *Liverpool.* ca.1785. C.
MERTZ—
Franz Xavier. *Treherbert.* 1875-89.
John. *Cavan.* 1868.
MERYETT, John. *London.* 1857.
MERVILLE, Charles. *London.* Took L. Chaune apr. in
1765.
MERYETT, W. *Ripley, Surrey.* 1851-66.
MERZ—
C. *Paris.* ca.1875.
Laurence. *London.* 1857-75.
Matthew. *London.* 1881.
MERZBACH, LAND & FELLHEIMER. *London.*
1881.
MESSER—
Henry. *Chertsey.* 1839.
Henry. *Colnbrook (Buckinghamshire).* 1839.
MESTIER, B. *Philadelphia, USA.* 1817.
MESURE—
H. *Shaftesbury.* 1848.
Henry. *Gillingham.* 1848.
Lianna & Co. *London.* 1857.
Lionel. *Eltham.* 1845.
Lyonel. *Maidstone.* 1845.
METAR, Henry. *London.* ca.1710. C.
METCALF(E)—
Edmund. *Halifax.* 1837.
F. *Hopkinton, Mass, USA.* ca.1825.
George. *London.* 1863.
George. *Millom.* 1879.
J. *London.* 1851.
James. *Halifax.* ca.1780, then *Askrigg* 1788. C.
John. *Halifax.* 1850-53.
John. *Liverpool.* 1790-1824.
Joseph. *London.* 1832.
Luther. *Medway, Mass, USA.* ca.1800. Clock case
maker.
Mark. *Askrigg.* b.1693-d.1776. C.
Mary. *Maryport.* 1828.
& NICHOLL. *Halifax.* ca.1800. C.
T. *Birkenhead.* 1857.
Thomas. *Easington.* 1856.
Thomas. *Seaham Harbour.* 1856.
Thomas. *Stockton.* 1827-51.
William. *Preston.* b.ca.1833-*Kendal.* ca.1863. W.
METTEN, Laurens. *St. Louis, Mo, USA.* ca.1850.
MEURON & COMP. *Paris.* ca.1820. W.
MEVES, Otto. *Montreal.* 1862.
MEVREY, F. *Philadelphia, USA.* ca.1799.
MEWES, William August. *London.* 1863.
MEYER—
Albert. *Cincinnati, O, USA.* ca.1850.
Felix. *New York.* 1880.
George. *London.* 1857-63.
J. H. *New York.* 1803.
Jacob. *London.* 1863-69.
Johann Georg. *Munich.* ca.1660. C.
John. *London.* 1869.
Joseph. *London.* 1875.
L. *Bristol.* 1805.
L. J. *Bristol.* 1863.
Leodegari John. *Bristol.* 1850-79.
M. T. (& Co.). *Bristol.* 1840(-42).
Nicholas. *London.* 1844-63.
MEYERS—
Frederick. *Barnsley.* 1871.
George. *Belfast.* 1819.
H. L. *Ottawa.* 1875.

MEYERS—*continued.*
John. *Fredericktown, Md, USA.* 1793-1825.
Samuel. *Montreal.* 1875.
MEYLAN, Louis. *Birmingham.* 1880.
MEYLER, ——. *Haverfordwest.* Early 19c. C.
MICABIUS, John. *London.* 1632. Must be error for John Nicasius, q.v.
MICHAEL—
A. (& Son). *Chesham.* 1838(-39).
Benjamin. *Mevagissey.* b.1768-d.1853. W. Also *St. Austell.*
Joseph (& Samuel). *Chesham.* s. of A., whom he succeeded 1842-47 in partnership with brother, Samuel.
Levi. *Swansea.* 1830.
& MARKS. *Sunderland.* 1851.
Maurice. *Bristol.* 1840-79.
Maurice. *Coventry.* 1880.
& MICHELL. *St. Austell.* ca.1840. C.
Samuel. *Chesham.* 1842-47 with brother, Joseph. Alone 1854.
MICHAELS, David. *London.* 1828.
MICHALORITZ, T. Z. *Manchester.* 1851.
MICHAL(L)—
Benjamin. *Mevagissey.* mar.1785-ca.1800. C
Benjamin & Co. *St. Austell.* 1847. prob. error for Michael, q.v.
Frederick Barron. *Truro.* 1873. W.
James. *London.* ca.1700. C.
Joseph. *Penzance.* 1844-d.1847.
William. *St. Austell.* 1843-44.
MICHIE—
J. E., jun. *Norwood.* 1866.
James. *Brechin.* 1837.
James Carey. *Brixton.* 1862-78.
James Carey. *London.* 1857.
James Carey. *Norwood.* 1878.
MICKLEWRIGHT—
Erasmus. *London.* a.1666 to Richard Bestwick. Springmaker, CC. 1673-97.
Francis. *Hastings.* 1878.
MIDDLEDITCH, John. *London.* 1832-51.
MIDDLEMISS—
M. *Newcastle-on-Tyne.* 1850-58.
Robert. *Hull.* 1822. W.
MIDDLETON—
Aaron. *Burlington, NJ, USA.* 1732.
J. D. *Grimsby.* 1876.
James. *Edzell* nr *Brechin.* 1860. Also photographer.
James. *London.* 1828.
M. *London.* Early 18c. W.
Richard. *Swansea.* 1835-50.
William. *Darlington.* 1716. W.
William. *London.* 1844.
William T. *London.* 1844-57.
MIDGET. 1914 and later, product of Pequegnat Clock Co.
MIDGLEY, Richard. *Halifax.* 1745-63. C.
MIDLAND Counties Watch Co. *Birmingham.* 1880.
MIDNALL (sometimes MEDNALL), John. *London.* CC. 1632-35. Shortly after he was arrested for debt. W.
MIDWINTER—
Charles. *London.* 1863-69.
Charles. *Woking.* 1878.
John. *Oxford.* mar.1764-72. W.
MILBURN—
John. *Leeds.* 1853-71.
William. *Darlington.* d.1854.
MILES—
George. *London.* (B. 1799-1824) 1828.
Henry & Co. *Bristol.* 1879.
Henry Thompson. *London.* 1869.
James. *Leamington.* 1868-80.
Robert. *Sheffield.* 1871.
Robert. *Stalham.* 1858-75.
Samuel George. *Coltishall.* 1865-75.
Stephen. *London.* ca.1760. C. (?one such at *Kidderminster* 1753-72).
William. *Poole* (*Dorset*). 1875.
MILETTE, François. *Montreal.* 1819.
MILFORD, John. *Belfast.* 1784.

MILK, Thomas. *Md, USA.* 1775.
MILL—
——. *St. Ninians.* 1784. W.
John. *Kilkhampton* (*Cornwall*). 1856-73.
& JONES. *Swansea.* 1887. Successor to William Henry Mill.
William Henry. *Neath.* 1887.
William Henry. *Swansea.* 1871-75. Later Mill & Jones, q.v.
MILLAN, Richard. *Shoreham.* 1878.
MILLAR (see also MILLER)—
Andrew. *Moore, Canada.* 1857.
Archibald. *Castle Douglas.* ?Late 18c.
Archibald. *Kirkcudbright.* 1820.
David. *London.* 1844.
George. *Carluke.* 1836.
George. *Lessuden* (*Scotland*). 1860.
Hugh. *Kilwinning.* 1860. Also ironmonger.
James W. *Toronto.* 1856-63.
John. *Bedford.* ca.1840. C.
John. *Birkenhead.* 1878.
John. *London.* 1839.
R. & Son. *Edinburgh.* 1850.
Robert. *Edinburgh.* 1826-38.
MILLARD—
Charles. *Yarmouth.* 1875.
H. R. *Montreal.* 1871.
T. *Stratford* (*Essex*). 1851-74.
T. W. *Gloucester.* 1856-63.
Thomas. *London.* 1851-57.
Thomas Henry. *Bristol.* 1879.
William. *Leeds.* 1866-71. W.
MILLER—
——. *London.* 1656.
——. *USA.* ca.1776. C.
Aaron. *Elizabethtown, NJ, USA.* 1747. C. and bells.
Abraham. *Easton, Pa, USA.* 1810-30.
Alexander. *Perth.* 1820.
Alfred. *Brighton.* 1870-78.
Andrew. *Edinburgh.* ca.1827. Successor to Robert Logie. Son continued till 1903.
Archibald and William. *Airdrie.* 1835.
Brothers. *London.* 1857-69. Successor to Matthew.
Catherine. *Caernarvon.* 1874.
Charles. *Reigate.* 1828.
(Herman) Clock Co. *Zeeland, Mich, USA.* 1929. C.
& Co. *Dudley.* 1860.
Cornelius. *New Jersey.* Late 18c.
David. *Montreal.* 1875.
Edward F. *Providence, RI, USA.* 1824.
Elizabeth. *Reigate.* 1839.
F. *London.* 1779.
Ferdinand. *Vienna.* ca.1700.
Fidel. *Dudley.* 1868-76.
George. *Batley.* 1871.
George. *Germantown, Pa, USA.* ca.1771. C.
George. *London.* 1881.
George. *Philadelphia, USA.* 1809.
George II. *Philadelphia, USA.* 1828-33.
George. *Toronto.* 1877.
George & Co. *London.* 1832.
Gordon Hanway. *Hereford.* 1870.
Henry. *East Hanover, Pa, USA.* 1825.
Henry A. *Southington, Conn, USA.* ca.1830. C.
Hugh. *Farnworth.* mar.1793. C.
Isaac. *Newport* (*Shropshire*). 1828.
J. *Dumfries.* 1810.
J. *Preston, Ontario.* 1861.
J. B. *Portland, Ore, USA.* 1860s.
J. W. & Co. *Toronto.* 1853-77.
James. *Blairgowrie* (*Scotland*). ca.1785. C.
James. *Blairgowrie.* 1860.
James. *Brighton.* 1870.
James. *London.* 1839-75.
James. *Perth.* 1845-48.
James. *Selkirk.* 1860.
James. *Worthing.* 1878.
James A. *Penge.* 1878.
James Robert. *Brighton.* 1870.
John W. *Ballyclare.* 1858.
John. *Bedford.* 1830-54.
John. *Germantown, Pa, USA.* '1735'. C.

MILLER—*continued.*
John. *Kingsclere.* 1830.
John. *Lincoln.* Early 18c. C.
John. *Liverpool.* 1747. W.
John. *London.* a.1666 to Samuel Knibb, later to Joseph Knibb, CC. 1674-97. *Charing Cross.* C.
John. *Prescot.* 1823-26. C.
John. *Preston.* 1798. C. & W.
John. *Selkirk.* 1836.
John. *Walton le Dale (Lancashire).* mar.1741. C.
Joseph. *Closeburn (Scotland).* 1860.
Kennedy. *Elizabethtown, NJ, USA.* 1830-33. C.
Lawrence. *St. Neots (Huntingdonshire).* 1730. W. and gunsmith.
Matthew. *Kirkintilloch.* 1860.
Matthew. *London.* 1851.
Mrs. *M. London.* 1851 (?wid. of Robert).
Mrs. Priscilla. *London.* 1857.
Pardon. *Providence* and *Newport, RI, USA.* ca.1820-ca.1840.
Peter. *Lynn* and *Ephrata, Pa, USA.* (b.1772) 1793-1837 (d.1855).
Philip. *New York.* 1763-69.
Philip. *Tredegar.* 1868-71.
Q. *Dumfries.* 1820.
Ralph. *London.* a.1690 to Robert Halstead, CC. 1697.
Richard William. *London.* 1857-81.
Robert. *Belfast.* 1865.
Robert. *London.* 1832-44.
Robert. *Perth.* 1841.
Robert. *Prescot.* 1752. W.
Robert. *Newcastle-on-Tyne.* 1834.
S. *Brighton.* 1862.
S. W. *Philadelphia, USA.* 1843.
Samuel. *London.* 1851-57.
Samuel Henry. *Redhill.* 1866-78.
& Sons. *London.* 1869-81.
& Stewart. *Hereford.* Late 19c. C.
Thomas. *Ballyclare.* 1858-80.
Thomas. *Philadelphia, USA.* 1819-41.
W. H. C. & Co. *Chicago, USA.* 1860s.
William. *Aberdeen.* 1846.
William. *Airdrie.* 1860.
William. *Farnworth.* 1824. W.
William. *Hythe.* 1826.
William. *Lurgan.* 1756-62. C.
William. *Wellingborough.* 1735-43.
William Frederick. *Dover.* 1823.
William Frederick. *Sandgate.* 1838.
William S. *Philadelphia, USA.* 1844-48.
& WILLIAMS. *Cincinnati, O, USA.* 1830-40. C.
MILLERCHIP, Mrs. D. *Coventry.* 1868-80.
MILLERET—
George. *London.* 1863.
Paul (& Co.) *London.* 1869(75-81).
MILLET—
Edward. *London.* a.1672 to Nicholas Beck, CC. 1680-1684. Later at *Hogsdon, Middlesex.*
John. *Bristol.* 1879.
John. *London.* a.1677 to William Clement, 1684 at *Hogsdon, Middsx.* Prob. brother of Edward above.
Thomas. *Ware.* 1722. W.
MILLICHAMP, Charles. *Presteign.* 1871-87.
MILLIGAN, James. *Dumfries* 1849.
MILLIKEN, Thomas. *Carrickfergus.* 1802. W.
MILLINGER, Joseph. *Chatham.* 1823.
MILLINGTON—
Isaac. *Lancaster, Pa, USA.* ca.1850.
John James. *London.* 1881.
S. *Nottingham.* 1864.
MILLIS—
& CHASE. *New York.* 1853.
J. R. & Co. *New York.* 1845.
Joseph. *USA (Pennsylvania?).* 1700-1799. C. (One such at *London* pre-1760).
William. *New York.* ca.1840.
MILLOT, ——. *Paris.* 1877.
MILLROSE, George. *Chester, Nova Scotia.* 1838.
MILLROY, John. *Dumfries.* 1794.
MILLS—
Alfred J. *Toronto.* 1874-77.
Andrew. *Halifax, Nova Scotia.* 1816.

MILLS—*continued.*
E. *Bungay.* 1846. (Prob. Edward, q.v.)
E. *Yarmouth.* 1858-65.
Edward. *Bungay.* 1830-53.
George. *Basingstoke.* 1878.
George. *London.* 1869.
George. *Ripon.* 1724. Made ch. clocks (B. 1750).
George William. *Birmingham.* 1868.
Henry. *Caerleon.* 1868-80.
Henry. *London.* 1851-69.
J. *Birkenhead.* 1857.
J. *Birmingham.* 1860.
John. *Bristol.* ?Late 18c. C.
John. *Oldham.* d.1790. C.
John. *Burtonwood (Lancashire).* mar.1773-1780. W.
Robert. *Taunton.* ca.1710. C.
Thomas. *Bytown, Canada.* 1851-53.
Thomas. *Caerleon.* 1822
Thomas II. *Caerleon.* s. of Thomas I. 1844-58.
Thomas. *Caerphilly.* 1852-75.
Thomas. *Gloucester.* 1830.
Thomas. *London.* 1648-1662. *Westminster.*
Thomas. *Newport (Monmouth).* 1835.
Thomas. *Peterborough, Canada.* 1857.
W. *Tamworth.* 1860.
W. N. *Pictou, Nova Scotia.* 1866-80.
William. *Birmingham.* 1868.
William. *Gloucester.* ca.1790. C.
William. *London.* ca.1786.
William. *Montgomery.* 1844. C.
MILLSOM, William. *Bristol.* ca.1775-80. C.
MILLUM, Moses. *Baltimore, USA.* 1819.
MILLWARD—
A. *Redbridge (Southampton).* 1867.
Henry James. *London.* 1863-69.
MILLWARD & MARTELL. *Southampton.* 1878.
MILNE—
George. *Aberdeen.* 1846.
James. *Montrose.* 1784.
Mrs. Ann. *Crayford.* 1874.
Robert. *Aberdeen.* 1821.
Robert (& Son). *Montrose.* 1837(-60).
Robert. *New York.* 1798-1802; *Philadelphia, USA.* 1817.
William. *Dunfermline.* 1842.
MILNER—
Charles. *Pocklington.* b. 1815. s. of Reuben,whom he succ. 1851-72.
Joseph. *Leyburn.* b.1808-30. C.
Joseph. *Sunderland.* 1834.
R. *Wigan.* 1858. W.
Reuben. *Pocklington.* 1813-46. C. & W.
Robert. *Pudsey.* 1871.
Thomas. *Wigan.* 1848-58. C.
MILNES, George. *Huddersfield.* 1834.
MILSOM, Daniel. *Bristol.* a.1813-30.
MILSTEAD, Thomas. *Bristol.* a.1815-26.
MILTON—
——. Place unknown. ?Later 18c. C. English.
——. 1914 and later, product of Pequegnat Clock Co.
MILWARD—
John. *Kendal.* b.1833-71. W.
John. *London.* (B. 1805-20) 1828.
Thomas. *London.* 1807. W.
MINAS, Louis. *London.* 1863-69.
MINCHIN—
John. *Moreton in the Marsh.* 1830.
John J. *Stow in the Wold.* 1840-63.
MINDEL, Gustavus. *Philadelphia, USA.* 1850.
MINERS, William B. *Plas-Marl.* 1887.
MINORS, G. *Woolwich.* 1855-74.
MINNES, James Henry. *Tunbridge Wells.* 1874.
MIN(N)ETT—
George. *Risca (Monmouth).* 1868.
George. *Ross.* 1850-63.
MINNIECE, James A. *Londonderry.* 1854-80.
MINOR—
E. C. *Jonesville, Mich, USA.* ca.1870.
Richardson. *Stratford, Conn, USA.* 1736. mar.1764-1797.
MINOT, J. *Boston, USA.* ca.1800-10. C.
MINSHULL—
——. *Birmingham.* 1816-30. Dialmakers.

MINSHULL—*continued.*
John. *Glossop.* 1846-57.
John. *Ashton-under-Lyne.* 1825-28.
John. *Denbigh.* 1835-44. C.
MIROY—
Brothers. (& Sons). *London.* 1857-63(-69).
Ernest. *London.* 1881. Succeeded to Miroy-Requier & Co.
Freres. *Paris.* 1851-ca.1860.
—REQUIER & Co. *London.* 1875.
MITCHEL(L)—
——. *Chard.* 19c. C.
Alexander. *Ballymena.* 1784.
Alexander. *Glasgow.* 1822-47.
Alexander & Son. *Glasgow.* 1798-1861.
& ATKINS. *Bristol, Conn, USA.* 1831. C.
BAILEY & Co. *New York.* 1854-60. C.
Barwise. *Cockermouth.* 1828-34.
Charles. *London.* 1881.
David. *Belfast.* 1827.
Frank. *Stamford.* 1880.
George. *Bishopwearmouth.* 1851-56.
George. *Bristol, USA.* 1774-1852. C.
George. *London.* 1839-51.
Henry. *Halifax.* 1866.
Henry. *New York.* 1786-1802. C.
& HINMAN. *Bristol, USA.* 1828-30.
Hiram. *Hull.* 1858.
James. *Crawley.* 1862-78.
James. *Jersey.* 1792. W.
James. *Keith.* 1860.
Jesse C. *Buffalo, NY, USA.* 1835.
John. *Bishopwearmouth.* 1851-56.
John. *Mevagissey.* Grandson of Benjamin. 1819. W.
John. *Wakefield.* 1853-71. W.
John & William. *Glasgow.* 1836-61.
John and William. *London.* 1851-63.
Joseph. *Colchester.* b.1765-d.1843. C. & W.
Mary. *Cockermouth.* 1848.
(Henry) & MOTT (Jordan). *New York.* ca.1790-1809. C. & W.
Phineas. *Boston, USA.* 1809-30.
Prof. *Cincinnati, O, USA.* ca.1850.
Richard. *Lane End.* 1835.
Robert. *Birkenhead.* 1878.
Robert. *Sunderland.* 1827-56.
& Son. *Colchester.* (Joseph & son, William). ca.1820-21. W.
& Son. *Gosport.* 1867.
Thomas. *London.* 1844-63.
Vance & Co. *New York.* 1860-80.
W. C. *Arnprior, Canada.* 1871.
W. H. *Toronto.* 1877.
Walter. *Plymouth.* mar.1746. C.
William. *Aberdeen.* (B. 1820)-1860.
William. *Colchester.* s. of Joseph. b.1798-ca.1821, then *London* ca.1821-d.1835.
William. *Strabane.* 1865.
William A. *Brampton, Canada.* 1871.
William John. *London.* 1863-69.
William, jun. *Richmond, Va, USA.* 1843-49.
William Smith. *London.* 1869.
MITCHELSON, David. *Boston, USA.* 1774. (One such at *Kingston, Jamaica,* 1795.)
MITSCHKE, A. *West Hartlepool.* 1898.
MITTEN, Francis. *Chichester.* 1711. C.
MITTLEBERGER—
James F. *Montreal.* 1806-20.
James Frederick. *Quebec.* 1826.
John T. *St. Catherines, Canada.* 1851.
MITTON, George. *Worcester.* 1876.
MIX—
Brothers. *Ithaca, NY, USA.* 1854-63.
Elisha. *New Haven, USA.* 1840-50.
MOAT—
Isaac. *Thorne.* 1851-62.
William. *Sandwich.* 1867.
MODD, James. *Donnington nr Spalding.* 1849-68.
MOFFAT, James. *Musselburgh.* Son of Alexander. 1831 and later. W.
MOGER, James. *Royston.* 1866-74.

MOGG—
James. *Basingstoke.* ?mid-18c. C.
William George. *London.* 1881.
William George. *Peckham.* 1878.
MOGGS, Harry. *Wincanton.* ca.1820.
MOGINIE—
C. *London.* 1857-69.
Joseph. *London.* 1839.
Samuel. *London.* (B. 1820-42) 1828-51.
MOHLER, Jacob. *Baltimore, USA.* 1744-73. W.
MOIR—
R. *West Cowes.* 1859-67.
Robert. *Cowes.* 1830-39.
Robert. *Landport.* 1878.
William. *New York.* 1878. C.
MOJON, MONTANDON & Co. *London.* 1881.
MOLE—
Charles. *Birmingham.* 1868.
J. & L. *Wolverhampton* also *Birmingham.* 1876.
John and L. *Birmingham.* 1860-68.
W. J. T. *St. Neots (Huntingdonshire).* 1864.
MOLEE, Peter. *London.* 1828.
MOLENKAMP, Douwe. *London.* 1875.
MOLINS, C. *London.* ca.1725. W.
MOLL, John. *Wilmington, Del, USA.* ca.1680. C.
MOLLER, Jens Peter. *London.* 1857-63.
MOLLINGER—
Henry. *Philadelphia, USA.* 1794-1804.
Jacob. *Neustadt.* mid-18c.
MOLYNEUX (and variant spellings).
John. *Enniskillen.* 1846-80.
John. *Liverpool.* 1699-(B. 1734). W.
Robert. *Place unknown.* ca.1800. C.
Robert. *London.* 1844-51.
Robert & H. *London.* 1839.
Robert (& Sons). *London.* 1828(-32).
MOLZ, Gottlieb. *Sheffield.* 1862. C.
MONCAS—
John. *Liverpool.* 1822-28.
Thomas. 1834. Chrons.
MONCRIEF(F)—
John. *London.* ca.1700 W. (B. a.1688).
Mitchell. *South Shields.* 1856.
MONCTON. 1914 and later, product of Pequegnat Clock Co.
MONDAY (also MUNDAY), Joseph. *London.* a.1647 to Isaac Plovier, CC. 1654-d.1663.
MONDEHARE, Johannes. *London.* 18c. C.
MONER, Don Alberto de Badaxi. ca.1820. W.
MONEY—
& BASSOLD. *London.* 1863-69.
Henry (& Son). *London.* 1875(-81).
James Charles. *London.* 1839.
Samuel. *Lowestoft.* 1864-85.
MONK—
Francis Albert. *Shoreham.* 1870.
George. *London.* 1857.
John. *Bolton.* 1848-51. (-?58).
Samuel. *Bolton.* 1834.
Thomas & Son. *Stamford.* 1798-1857. C. & Gun-smith.
MONKHOUSE—
Thomas. *London.* ca.1760. C.
MONKS—
Charles. *Prescot.* s. of George I. b.1777-d.1842.
F. A. *London.* 1863.
George I. *Prescot.* b.1750-d.1815. C.
George II. *Prescot.* s. of George I. b.1775-d.1827. C.
Louis. *Sutton.* 1878.
T. *Camberwell.* 1862.
T. W. *Brixton.* 1862-66.
Thomas. *London.* 1839-57.
Thomas W. *Camberwell.* 1866-78.
W. *Peckham.* 1878.
Walter. *London.* 1881.
MONNIER, Daniel. *Philadelphia, USA.* 1825-50.
MONNIN, ——. *Paris.* ca.1848-ca.1860. C.
MONNOA, M. A. *Farringdon.* 1837.
MONRO(E)—
Benjamin (James). *London.* 1832-44(-51).
Charles. *Bangor, Me, USA.* 1840.

MONRO(E)—*continued.*
E. & C. H. & Co. *Bristol, USA.* ca.1850.
Hector. *Leith.* 1836-d.1847.
Hugh. *Edinburgh.* 1825.
MONS, William. *Harrietsham (Kent).* 1838-55.
MONTANDON—
Hannah (Mrs. H. L.). *Lancaster, Pa, USA.* wid. of Henry Lewis, 1802-10.
Henry Lewis. *Lancaster, Pa, USA.* to 1802. C.
(Hannah) & ROBERTS (Oliver). *Lancaster, Pa, USA.* 1802.
MONTCASTLE, William R. *Warrenton, NC, USA.* 1844-56.
MONTEITH—
Benjamin. *Philadelphia, USA.* 1818.
Charles. *Philadelphia, USA.* 1847.
& Co. *Philadelphia, USA.* 1845.
& SHIPPEN. *Philadelphia, USA.* 1817.
MONTGOMERY—
Andrew. *Baltimore, USA.* 1822-24.
Henry. *Belfast.* 1805-18.
Hugh. *Omagh, co. Tyrone.* 1784.
James. *Belfast.* ca.1810-20. C.
Robert. *Ballyrogan.* ca.1775-85. C.
Robert. *Belfast.* From *Lisburn.* 1781-1800. C.
Robert. *Lisburn.* 1781. Later *Belfast* (?from *Ballyrogan*).
Robert. *New York.* 1786.
MONTI—
Anthony. *Canterbury.* 1845-65.
Antonio. *Ramsgate.* 1874.
John. *Canterbury.* 1847.
Joseph. *Canterbury.* 1838.
Maspoli & Co. *Sandwich.* 1839. C. & W.
Peter. *Sandwich.* (b.1817 in *Italy*.) 1849-95.
MONTJOY, John D. *York,* Canada, 1833-53.
MONTREAL—
1914 and later. Product of Pequegnat Clock Co.
Watch Case Co. *Montreal.* 1896.
Watch Co. *Montreal.* 1890.
MOOAR(?), Lot. *Nashua, NH, USA.* ca.1830.
MOODY—
David. *London.* a. to William Partridge, CC. 1649-62. Fleet Street.
George. *Broadstairs.* 1832. W.
Leonard. *Ripponden.* d.1791.
M. *Cambridge.* 1846.
MOON—
———. *Brenchley.* 1845.
George. *Tunbridge Wells.* 1839.
Henry. *Birmingham.* 1868-80. Dial makers.
& HOLMES. *London.* 1851. (Successor to William Moon.)
J. *Radford (Nottinghamshire).* 1864.
R. *Liskeard.* 1856.
Richard. *Brighton.* 1878.
Robert. *Philadelphia, USA.* 1768. Casemaker.
T. *Coventry.* 1854.
T. *Sutton.* 1866.
MOONEY, John. *Ballyshannon.* 1784.
MOONLINGER, Henry. *Philadelphia, USA.* 1794-1804.
MOOR(E)—
Alexander. *London.* 1844-75.
Alfred. *London.* 1869-81.
Ambrose. *Dublin.* 1781. W.
Arthur. *Cardiff.* 1887-99.
Arthur. *London.* 1857-75.
Benjamin R. and J. *London.* 1839.
C. *Norwich.* 1858-65.
C. H. *Birkenhead.* 1865-78.
Charles. *London.* 1828.
Charles Alexander. *London.* 1832-81.
Daniel. *Coltishall (Norfolk).* 1836.
Daniel. *Coventry.* 1828-50.
David. *Randalstown, co. Antrim.* 1782-1824.
Edmund Thomas. *London.* 1832-39.
Edward. *Basingstoke.* 1830.
Edward. *Dublin.* 1768.
Edward. *Hastings.* 1870.
Edward. *Kendal.* s. of Joseph. b.1849-92.
Edward. *Leeds.* 1741. 'Engraver of clock faces'.

MOOR(E)—*continued.*
Edward I. *Oxford.* 1714-d.1774. C. (Parish of All Saints.)
Edward II. *Oxford.* s. of Edward I. 1751-1772. C. & W. May be also at *London*
Edwin Lock. *London.* 1869-81.
Elizabeth. *Ferrybridge (W. Yorks).* Prob. widow of Francis 18c C.
Francis. *Ferrybridge.* ca.1750-ca.1775. C.
Francis M. *Belfast.* 1854-94. s. of James of *Belfast.*
Frederick. *London.* 1844-63.
Frederick. *New Haven, Conn, USA.* ca.1840.
George. *Cumnock.* 1837-60.
George. *London.* 1839-75.
George H. *Lynn, Mass, USA.* ca.1860.
George & Son. *London.* 1881.
Henry. *Leicester.* 1864-76.
Henry. *London.* 1851-63.
I. T. *Blackburn.* W. (?19th century).
J. *Norwich.* ?ca.1838.
J. & E. *Hastings.* 1862.
J. G. *Birmingham.* 1860.
James. *Belfast.* 1814-24.
James. *Birmingham.* 1880.
James Newey. *Birmingham.* 1868.
John. *Church Stretton.* 1850.
John. *Keighley.* 1822. C.
John (& Son). *London.* 1828(-32). T.C.
John. *Norwich.* 1830-46.
John. *Putney.* 1862-78.
John. *Toronto.* 1851-63.
John. *Warminster.* 1795-d.1796. W.
John Curtis. *Worthing.* 1828-39.
John Hassall. *Birmingham.* 1828-60.
John & Son. *London.* 1831-68. T.C.
John & Son. *London.* 1844-81. Chiming clocks.
John & Sons. *London.* 1880.
Joseph. *Coventry.* 1880.
Joseph I. *Kendal.* b.1821-d. pre-1861. W.
Joseph II. *Kendal.* s. of Joseph I. b.1844-79. W.
M. *Hilsborough (Norfolk).* 1858-65.
Mrs. Ann *London.* 1881 (?wid. of Alexander).
Mrs. Mary A. *London.* 1869.
Nelson A. *Newark, NJ, USA.* ca.1850.
Robert. *Belfast.* 1783.
Robert. *Bolney (Sussex).* a.1735 to Thomas Stockler, C.
Robert. *Philadelphia, USA.* 1798.
& Sons. *Coventry.* 1842.
T. *Ashbourne.* 1849.
Thomas. *Derby.* 1849-1912.
Thomas. *Ipswich.* ca. 1714-62. C. & W.
Thomas. *Salisbury.* ca.1780-ca.1790. C.
W. B. *Worthing.* 1855.
William. *Kettering.* 1824.
William. *Walthamstow.* 1874.
William. *Worthing.* 1870.
MOORFIELD, James. *Prescot.* mar.-1785-95. W.
MOORHOUSE—
Robert. *Padiham* nr *Burnley.* 1848-58. W.
Thomas. *Sheffield.* 1811.
W. *Preston.* 1851-58. W.
William. *Liverpool.* ca.1820-30. C.
William. *Wetherby.* 1826-34. C.
MOORS, Samuel. *London.* ca.1790. C.
MORAT (or MORAL), Joseph. *Leeds.* 1866-71, T.C.
MORATH—
Fedele & Brothers. *Liverpool.* 1848-51. Importers.
Michael. *London.* 1832-81.
MORDAN. (or Mordant). William. *Nottingham.* mar. 1768 -ca. 1780. C.
MORDEN—
———. *Nottingham.* ca.1780. C.
Samuel. *Colchester.* b.1746-75.
MORDIKE, John Frederick. *Bridgnorth.* 1879.
MORECOMBE, John. *Barnstaple.* 1622-58. Made ch. clocks.
MOREHEAD, Robert. *Plymouth.* mar.1691.
MOREL, Jacques. *London.* mid-17c. W. Prob. James Morrell.
MORELAND—
Samuel J. *Gloucester.* 1879.
T. *Chester.* 1857-65.
Thomas. *Chester.* 1834-48.

MORENCY, Charles. *Levis, Quebec.* 1871.
MORETTI, Rocco. *Cardiff.* 1865-87.
MOREY—
J. *Ryde (Isle of Wight).* 1848.
Josiah. *Kirkby Stephen.* 1869. W.
Michael. *Orton (Westmorland).* 1879. W.
Robert. *London.* ca.1730.
MORGAN—
Arthur J. *Birmingham.* 1880.
Charles P. *Truro, Nova Scotia.* 1865-97.
Daniel C. *Ystalyfera.* 1875.
David. *Llandovery.* 1875-87. C. & W.
Donald. *Kirkwall.* 1845-60.
Edward. *Newtown (Mon.).* 1822.
Elijah, jun. *Poughkeepsie, NY, USA.* 1783-1857.
George. *Cowbridge.* 1835-48.
George. *Halifax, Nova Scotia.* 1866-72.
George. *London.* 1875-81.
Gideon. *Pittsburgh, Pa, USA.* 1819-26. C.
& HART. *Pittsburgh, PA, USA.* 1819.
Henry W. *Landport.* 1878.
Humphrey. *Abergavenny.* Said to be late 17c, but prob. later.
James. *Dromore.* 1712-29. C.
James. *London.* 1869-75.
John. *Aberystwyth.* 1816-35. W.
John. *Cambridge.* 1830-40.
John. *Halifax, Nova Scotia.* 1869-72.
John. *Merthyr Tydfil.* 1822.
John. *Monmouth.* 1753-88. W. & C.
John. *Narberth.* 1830-35. C.
John. *Wolverhampton.* 1876.
John, jun. *Bristol.* (B. 1825) 1830-42. Also mail timekeeper.
John Lettey. *Newent.* 1870.
John Varley. *Liverpool.* 1848-51.
Luther S. *Salem, Mass.* 1842-46.
Mrs. C. *Cambridge.* 1846.
Richard. *London.* CC. 1630-56. CC. Spring-maker.
Richard. *Minehead.* 1875.
Robert I. *London.* CC. 1637.
Robert. *London.* 1869-81.
T. *Minehead.* 1861-66.
Theodore. *Salem, Mass, USA.* (1778.) mar.1806-45.
Thomas. *Bristol.* 1830.
Thomas, jun. *Bristol.* 1830.
Thomas. *Caerphilly.* 1875.
Thomas. *London.* CC. 1659-97.
Thomas. *Philadelphia, USA.* pre-1772. *Baltimore.* 1772-79. *Philadelphia* again 1779-93.
Thomas. *Swansea.* 1744-91. C.
W. *Bridport.* 1867.
W. *Chesham.* 1847.
W. *Fortuneswell (Dorset).* 1855.
W. *Pontypool.* 1850.
W. L. *Hereford.* 1856.
William. *London.* a.1650 to Simon Bartram. CC. 1658.
William. *Pontypool.* 1858. Dealer.
William S. *Poughkeepsie, NY, USA.* s. of Elijah. 1807-36.
MORGANTI, John Baptist. *Brighton.* 1839.
MORGENSTERN, Jacob. *Tenby.* 1880-87.
MORICE, David & William. *London.* (B. 1805-25). 1828-32. Same as David & Son.
MORIN—
——. *Paris.* ca.1720. W.
Augustus. *Philadelphia, USA.* 1835.
MORISON—
Alexander & Arthur. *London.* 1857.
George. *Aberdeen.* 1761-ca.1770. C. & W.
MORLAND—
John. *Richmond (Yorkshire).* 1829. C.
R. No town. Same man as at *Kirkby Malzeard.* q.v.
Richard. *Kirkby Malzeard.* ca.1800-22. C.
William. *Boroughbridge.* 1871.
William. *Durham.* 1851.
MORLEY—
Edward. *Norwich.* 1875.
George. *Boston.* 1876.
Henry. *Mildenhall.* 1864-92.
James. *Brighton.* 1878.
James Bailey. *Henfield.* 1870-78.

MORLEY—*continued.*
Martin. *Halifax.* 1871.
Richard. *Idlicot (Cotswolds).* ca.1680. Lant. clock.
MORPHY—
Andrew. *Ingersoll, Canada.* 1861.
Andrew. *London, Ontario.* 1862-76.
Brothers. *Toronto.* 1851-61.
Edward M. *Toronto.* 1843-77.
S. *Brantford, Canada.* 1861.
MORRALL, William. *Birmingham.* 1868. Repairer.
MORRELL—
John. *Baltimore.* 1822.
John. *Whitby.* 1823-34. C.
& MITCHELL. *New York.* 1816-20.
Vincent. *London.* ca.1700. C.
MORREY, John. *Manchester.* mar.1801. C.
MORRICE—
G. M. *Great Berkhampstead.* 1859.
George. *London.* 1811. W.
MORRILL—
Benjamin. *Boscawen, NH, USA.* (1794). 1816-45. C.
H. C. *Baltimore.* 1835-ca.1840. C.
MORRIS—
A. *Stone.* 1868.
Abel. *Reading, Pa, USA.* 1774. C.
Benjamin. *Hilltown* and *New Britton, Pa, USA.* (1748). 1768-1830(-33). C.
& BRODIE. *St. Albans.* 1859.
David. *London.* ca.1810.
E. H. *Ipswich.* 1865.
Edward. *Stone.* 1828-60.
Edward Henry. *Hove.* 1870.
Elijah. *Canton, Mass, USA.* ca.1820.
Enos. *Hilltown, Pa, USA.* ca.1780.
George. *Liverpool.* (B. 1813). 1824.
George Henry. *London.* 1881.
Harris. *Walsall.* 1842-50.
Henry. *Canton, Mass, USA.* ca.1820.
Henry. *London.* 1875.
James W. *Faversham.* 1866-74.
John. *Bristol.* 1842.
John. *Chipping Norton.* ca.1770. C.
John. *Derby.* 1730. W.
John. *Eye.* 1864-77.
John. *London.* 1851.
John. *Market Deeping.* 1835-49. See Marris.
John. *Pershore.* 1750.
John. *Peterborough.* 1830.
John William. *London.* 1863-69.
Louis. *London.* 1875.
Mary. *Bolton-le-Moors.* 1824-28.
Mrs. Sarah M. *London.* 1857.
P. *Newport (Monmouth).* 1822.
Richard. *Sawbridgeworth (Hertfordshire).* 1874.
Robert. *Bala.* 1840-44.
Robert. *Stone.* 1850-76.
S. *Southampton.* 1848.
Samuel. *London.* 1668. Maker of measures.
Stephen. *Bledlow Ridge (Buckinghamshire).* Early 18c. C.
Thomas. *Albrighton.* 1856-79.
Thomas. *Church Aston.* 1879.
W. (& Son). *Eastbourne.* 1851-62.
(William) & WILLARD (Simon). *Grafton, Mass, USA.* ca.1770. C.
William. *Cardiff.* 1887-99.
William. *Eastbourne.* 1828-39. Also auctioneer.
William. *Liverpool.* ca.1810. W.
William. *Philadelphia, USA.* 1837.
William. *Princes Risborough (Buckinghamshire).* 1869-77.
William. *Utica, NY, USA.* 1832.
William, jun. *Philadelphia, USA.* 1844.
William Robert. *Caernarvon.* b.1861-99.
Wollaston. *Md, USA.* 1774.
MORRIS-JONES, Sir John. *Llanfair-pwll.* 1864-1929. Made electronome master clock working three impulse dials.
MORRISON—
Alexander. *Glasgow.* 1837.
Archibald. *Glasgow.* ca.1770-80. W.
& DENNING. *London.* 1839.

MORRISON—*continued.*
J. *Derby.* 1855.
John. *London.* 1851-81.
& MCEWEN. *Edinburgh.* 1849.
Robert. *Belfast.* ca.1785-1800.
Theodore. *Bridge of Dee, Aberdeen.* 1846.
Thomas. *Ayton* nr *Gateshead.* 1875. W.
Thomas. *Toronto.* 1856.
William. *Alloa.* 1841.
William. *London.* 1828.
William. *London.* 1863-81.
William C. *Toronto.* 1843-87.
MORRISSEY, C. R. *Philadelphia, USA.* 1837.
MORROW—
John. *Ardmillan, co. Down.* 1750. Dial maker and
engraver.
John. *Liverpool.* 1823. W.
John. *Stratford (Essex).* 1866-74.
MORSE—
——. *Mathry.* mid-19c.
Andrew, jun. *Bloomfield, Me USA.* 1835.
& BLAKESLEE. *Plymouth, Conn, USA.* 1841-55.
Celebrated Perpetual Calendar Clock Co. *Chicago, Ill,
USA.* ca.1893.
Chain Co. *Bridgeport, Conn, USA.* 20c.
& Co. *London.* 1839.
(Miles) & Co. *Plymouth Hollow, Conn, USA.* 1846-55.
C.
J. *Rochester.* 1866.
James. *St. Davids.* 1868-75.
John. *Malmesbury.* 1830. W. & C. Also silversmith and
cutler.
Miles. *Plymouth, Conn, USA.* 1841-55. C.
Moses L. *Boston, USA.* 1813.
Richard. *London.* 1844-69.
MORSS—
D. *Hitchin.* 1859.
Daniel. *London.* 1869.
Henry. *London.* 1869-81.
Samuel. *London.* 1851.
MORT—
Jordan. *New York.* 1802-25.
& MITCHELL. *New York.* Successor to Jordan Mort.
MORTEN, Robert. *Dundee.* ca.1800. C.
MORTER, Frederick Robert. *South Heighingham
(Norfolk).* 1875.
MORTIMER—
David. *Hereford.* 1830-44.
W. *Heckmondwike.* 1871.
William. *Cullen.* 1837.
William. *Portsoy.* 1837-60.
MORTLOCK—
James. *Clare.* 1844-64
Mrs. M. A. *Clare.* 1865.
Martin. *Clare.* 1879.
Thomas. *Clare.* mar. 1768. C.
W. *Clare.* 1846-58.
William I *Clare.* mar. 1801-55.
William II. *Clare.* 1864-75.
MORTON—
Alexander. *Armagh.* ca.1717-50. C.
E. Daniel & Co. *Bristol, USA.* 1855.
G. *Birkenhead.* 1878.
George. *Keighley.* 1853. C.
George. *London.* 1869-75.
James. *Dunbar.* d.1846.
Joseph. *Aberford.* b.ca.1724-d.1774. C.
Peter. *Herne Bay.* 1838.
Robert S. *Dunbar.* 1837.
Thomas. *Manchester.* 1848-51.
Thomas. *St. Helens.* mar.1805-48. W.
MORTWICH, Abraham. *Bristol.* 1850.
MOSELEY and variants—
Alfred Edward & Co. *London.* 1881.
Braham P. *London.* 1851.
Braham Phillip. *Swansea.* 1848.
& Co. *London.* 1875.
Ephraim & Moses. *London.* 1839-44.
& FIENBURG. *Swansea.* 1844.
Henry P. *Southampton.* 1839-78.
J. *Holbeach.* 1861.
Jacob. *Neath.* 1830-52. C.

MOSELEY—*continued.*
Jacob. *Swansea.* 1822-44. C. & W.
Joshua. *Penistone.* 1757. C.
Maurice. *Brighton.* 1839-70.
Moses. *London.* 1828-44.
Robert E. *Newburyport, Mass, USA.* ca.1848.
Robert Ellis. *Swansea.* 1835-40.
Robert & Sons. *London.* 1828-44.
Thomas Henry. *Neath.* 1846. W.
William. *London.* b.ca.1659. a. to Benjamin Graves.
CC. 1680. mar.1684. C.
William. *Penistone.* 1822-pre-1834. C. & W.
MOSER—
——. *Paris.* 1851. C.
Christian. *Llandeilo.* 1840-44.
MOSES—
A. & L. *Portsea.* 1878.
Abraham. *Brighton.* 1839.
Abraham. *Frome.* ?early 19c.
D. L. *Brynaman.* 19c.
E. *London.* 1851.
Edward. *London.* 1875-81.
Emanuel F. *Swansea.* 1868.
J. Moses. *Swansea.* 1830-48.
John. *Bowness.* 1879. W.
Joseph. *Gloucester.* 1850.
Joseph. *West Auckland.* 1851.
Michael & Emanuel. *Swansea.* 1844.
Morris. *London.* 1857-69.
Moses. *Dover.* 1801-28. W.
Moses. *London.* 1828-32.
& Sons. *Swansea.* 1868.
Thomas. *Wolcottville, Conn, USA.* 1831-36. C.
MOSHER—
& DAVIS (S.). *Hamilton, NY, USA.* 1834.
S. *Hamilton, NY, USA.* 1830.
MOSLEY (see also **MOSELEY**)—
——. *Mount.* Early 19c.
——. *Penistone.* ca.1740-60. C.
Charles. *London.* 1857-63.
John. *Tideswell (Derbyshire).* 1876.
Joshua. *Dronfield.* 1835. C.
MOSS—
——. *Rochdale, Mass, USA.* 1818. W.
Benjamin. *London.* 1832-39.
Charles. *Liverpool.* 1848.
E. *Clack* nr *Chippenham.* 1867-75.
George. *Carlisle.* (B. 1814-20). 1828.
George. *Heskett-Newmarket (Cumberland).* 1834-48.
Isaac. *Sheffield.* 1837. W.
J. *Clack (Wiltshire).* 1848-59.
James. *Liverpool.* 1834.
James Dennett. *Liverpool.* 1834.
John. *Rochdale.* 1822-34.
Michael. *Preston.* 1848.
Myer. *Truro, Nova Scotia.* 1866.
& MYERS. *Sheffield.* 1862. W.
Nathaniel M. *Barking.* 1828-39.
Richard. *Ulverston.* 1774-d.1776. W.
Tristram. *Londonderry.* 1622. C.
William Selby. *Manchester.* 1824-51.
MOSSMAN & Son. *Edinburgh.* 1813-1906.
MOTE, Garett. *London.* Foreign journeyman to Isaac
Plovier, 1655.
MOTHERSOLE—
Michael. *Harleston.* 1865-75.
Thomas. *Bury St. Edmunds.* 1858-79.
MOTT—
Charles. *Halifax, Nova Scotia.* 1880.
& MORRELL. *New York.* 1802-10.
& MOURNE. *New York.* 1790(-1805). W.
William. *London.* 1832-39.
MOTTEUX, Samuel. *London.* a.1686 to Philip Cor-
deroy. CC. 1697.
MOTTRAM, P. *Newcastle-on-Tyne.* 1856.
MOTTU—
Brothers. *London.* 1844-75.
& Son. *London.* 1839.
MOULAM, G. *Bath.* (B. 1819?). 1856-66.
MOULAND—
J. *Mitcham.* 1855-66.
James. *London.* 1863.

MOULD, William. *Towyn.* 1874-87.
MOULANIE/MOULINIE—
Aine & Co. *London.* 1839.
John F. (& Co.). *London.* 1844(-51-57).
& LEGRANDROY. *London* and *Geneva.* 1863-75. W.
(Christian Lange, agent).
MOULTON—
(William) & CARR (James). *Lowell, Mass, USA.* 1834.
Edward (E. G.). *Rochester, NH, USA* and *Saco, Me, USA.* 1807-25. C.
Francis E. *Lowell, Mass, USA.* 1832-35.
Joseph. *Newbury, Mass, USA.* 1694-ca.1756.
Thomas. *Rochester, NH, USA.* Early 19c.
Thomas M. *Dunbarton, NH, USA.* ca.1800. C.
William. *Birkenhead.* 1878.
MOUNT—
Vernon Watch Co. *Mount Vernon, NY, USA.* ca.1936. W.
W. Henry. *Ramsgate.* 1847-66.
William. *London.* a.1682 to Withers Cheney. CC. 1692-97.
MOUNTAIN—
Alfred. *London.* 1881.
Samuel P. *Philadelphia, USA.* 1842.
MOUNTFORD—
Edward Lewis. *Worcester.* 1842.
John. *Philadelphia, USA.* 1818. Watch importer.
Thomas Harvey. *Worcester.* 1850-60.
Zachariah. *London.* a.1676 to William Speakman. CC. 1684-86.
MOUNTJOY—
John. *New York.* ca.1810.
William. *New York.* ca.1805.
MOURET, ——. *Caen.* 18c. C.
MOUTOUX, Carl. *Brooklyn, NY, USA.* ca.1850.
MOWAT, A. *Bath.* 1856-66.
MOWBRAY—
William. *Doncaster.* b.ca.1730-d.1793. W.
William. *Hartlepool.* 1856. C.
William. *London.* ca.1770. C.
William. *Pontefract.* 1764. W.
MOWCAR, James. *Bolton.* 1828.
MOWDAY—
E. *Great Berkhampstead.* 1874.
S. *Great Berkhampstead.* 1851-66.
MOWGROVE—
Francis. *New York, USA.* 1816.
Peter. *Charleston, SC, USA.* ca.1735-39.
MOXHAM, James A. *Coleford.* 1852-58. Later ironmonger.
MOXON—
Joseph. *Bristol.* Late 18c. C.
Josh. *London.* ca.1830. C.
MOYLE, Edwin Charles. *Chichester.* 1878.
MOYS, John. *Croydon.* 1730. W.
MOYSTON, John Hugan. *Schenectady, NY, USA.* 1772-1844.
MOZART—
——. *Boston, USA.* 1823-30.
Don J. *Boston, USA.* 1820-23. mar.1854. To *Bristol.* 1863. *Providence, RI.* 1864. *Ann Arbor.* 1868-d.1877.
MUCHENBERGER, Karl. *Brighton.* 1878.
MUCKARSIE, George James. *London.* (B. a.1801, CC. 1814-24). 1828.
MUDGE, George. *London.* 1869-81.
MUDIE—
Thomas. *Halifax, Nova Scotia.* 1807-16.
Thomas. *Pictou, Nova Scotia.* 1828-67.
MUDON, John. *London.* 1881.
MUELLER, see **MULLER—**
MUGNIER, Julius James. *London.* 1863.
MUIR—
John. *Stromness.* 1860.
Robert. *Dalry.* 1850-60.
MUIRHEAD—
——. *Glasgow.* 1838.
Henry. *Glasgow.* 1839.
Henry. *Rothesay.* 1836.
& Son. *Glasgow.* 1860.
MUKLE, Nicklaus. *London.* 1863-81.
MULES, William. *Kilkhampton (Cornwall).* ca.1830. Later *Stratton.* 1844.

MULFORD, John H. *Albany, NY, USA.* 1842. W.
MULITORE, Alfred. *Halifax, Nova Scotia.* 1866.
MULLAN, Thomas. *Dungannon.* 1820-24.
MULLEN, Thomas. *Belfast.* 1784.
MULLENEUX, ——. *Derby.* ca.1820. C.
MULLER/MUELLER—
Brothers. *London* and *Geneva.* 1851-81.
Charles. *London.* 1875-81.
Frederick. *Savannah, Ga, USA.* 1736-47. C.
George. *London.* 1869.
Henry. *London.* 1857.
Hieronymus. *Schottwein.* 1565.
Johann Heinrich. *London.* 1863-81.
Marx Leopold. *London.* 1857-63.
Nicholas. *New York.* 1850-72. C.
Oscar. *London.* 1875.
's Sons (Nicholas). *New York.* ca.1880. C.
Theodor. *London.* 1869-75.
MULLIGAN—
Thomas. *Enniskillen.* ca.1840-d.1841.
W. *Kingston, Canada.* 1851.
MULLIKEN—
Benjamin. *Bradford, Mass, USA.* ca.1740.
John. *Bradford, Mass, USA.* s. of Robert. 1690 and later.
Jonathan. *Bradford, Mass, USA.* s. of Robert. mar.1742-ca.1810.
Jonathan. *Newburyport, Mass, USA.* s. of Samuel I. 1746-1782.
Joseph. *Concord, Mass, USA.* s. of Nathaniel. ca.1777.
Joseph. *Newburyport, Mass, USA.* 1804. C.
Joseph. *Salem, Mass, USA.* 1771-95.
Nathaniel. *Newburyport, Mass, USA.* s. of Jonathan of *Newburyport.* 1776. C.
Nathaniel I. *Lexington, Mass, USA.* s. of John. d.1767. C.
Nathaniel II. *Lexington, USA.* s. of Nathaniel I. 1767-77.
Robert. 1665. From *Scotland.* To *Boston, USA.* 1683. To *Bradford, Mass, USA.* ca.1683-88.
Samuel. *Newburyport, Mass.* s. of Jonathan. 1769. Later to *Hallowell, Me, USA.*
Samuel I. s. of John. To *Newburyport, USA.* 1750. d.1756. C.
Samuel II. s. of Samuel I. 1761. mar.1783. 1790-96. *Salem, USA.* Then 1803-07. *Lynn, Mass, USA.* -1847. C.
MULLINEX, ——. *Derby.* ?late 18c. To ca.1820. C.
MULTER, Peter A. *New York.* 1854.
MUMFORD—
——. *Providence, RI, USA.* ca.1810. W.
William. *Helmsley.* ca.1750. C.
MUMMERY—
Charles. *Folkestone.* 1839.
Thomas, sen. *Dover.* Free 1808-55.
Thomas, jun. *Dover.* a.1814-47.
MUNBY, Joseph. *Great Aycliffe.* 1851-56. W.
MUNCASTER—
C. J. *London.* 1863.
George. b. ca. 1824. *Leamington* 1851. *Douglas (Isle of Man)* 1860.
John. *Manchester.* mar.1809. At *Dalton.* W.
John. *Pontefract.* b.1748-d.1809. C.
John. *Ulverstone.* 1824-28.
John. *Whitehaven.* 1834.
William. *Douglas (Isle of Man).* 1860.
MUNCEY—
Edward. *London.* 1869.
James. *London.* 1869.
P. *Clapham.* 1866.
MUNCHIN, M. *USA.* 1874.
MUNDAY (also sometimes MONDAY)—
Benjamin. *London.* See Harvey, alias Munday.
Joseph. *London.* 1656-d.1663.
MUNDEN, Francis. *London.* a.1662 to John Savill. CC. 1670.
MUNFORD—
Edward. *Yarmouth.* 1836-46.
J. *Crewkerne.* 1861.
John. *Crewkerne.* 1883.
John & Edward H. *Crewkerne.* 1866-75.
MUNGER—
A. & Son. *Auburn, NY, USA.* ca.1839-47.

MUNGER—*continued.*
Asa. *Auburn, NY, USA.* bro. of Sylvester. 1778-1851.
  C.
Austin E. *Auburn, NY, USA.* 1811-47. Then *Syracuse.*
  1847-92.
(Asa) & BENEDICT (J. H.). *Auburn, NY, USA.*
  1826-33.
(S.) & DODGE (Abraham, jun). *Ithaca, NY, USA.*
  1824.
(S.) & PRATT (Daniel). *Ithaca, NY, USA.* 1826-32. C.
Sylvester. 1790-1857. bro. of Asa. mar.1816. *Clinton,*
  NY. Then 1822 *Onondaga.* Then 1823 *Ithaca.*
MUNGHAM, John. *Smarden (Kent).* 1704. Kept ch.
  clocks.
MUNN—
Neil. *Rothsay.* 1860.
Thomas. *Maidstone.* 1858.
William. *Sheerness.* 1838-47.
William E. *Sheerness.* 1851-74.
MUNNS—
J. *St. Albans.* 1851.
William. *St. Albans.* 1839.
MUNRO(E)—
Alexander. *St. John, New Brunswick.* 1795.
& Co. *Charleston, SC, USA.* 1795.
Daniel. *Concord and Boston, USA.* 1775-1859. C.
Daniel & Nathaniel. *Concord, Mass, USA.* 1800-1807.
  C.
Deacon Nehemiah. *Boston, USA.* Late 18c. Clock
  casemaker.
Donald. *Dingwall.* 1860.
Dugald. *Aberfeldy.* 1837.
Hugh. *Dollar.* 1837.
John. From *Edinburgh,* via *London.* To *Charleston, SC,*
  USA. 1785-1809.
(Daniel) & JONES (Ezekiel). *Boston, USA.* 1807-09.
Nathan. bro. of Daniel. *Concord, Mass, USA.* 1777-
  1817. *Baltimore.* 1817-61.
Thomas. *New Glasgow, Nova Scotia.* 1879.
(Nathaniel) & WHITING (Samuel). *Concord, Mass,*
  USA. 1808-17. C.
MUNT—
Bisley Henry. *Haverfordwest.* 1884-87. C.
Bisley Henry. *Milford Haven.* Same as at *Haverford-*
  west.
Frederick Joseph. *Buckingham.* 1869-77.
MUNYAN—
A. H. *Northampton, Mass, USA.* 1848.
Brothers. *Pittsfield, Mass, USA.* ca.1860.
MURAT, M. *Nottingham.* 1864.
MURCH—
Ierom (= Jeremy). *Honiton.* mid-18c. C.
John. *Exeter.* s. of William, Lanternmaker. Free 1698
  W.
Mathew. *Honiton.* ca.1817-ca.1825. C.
MURDEN & SON. *Dover.* 1874.
MURDOCH—
Andrew. *Glasgow.* 1828.
J. G. *Ramsgate.* 1874.
James. *Newton, Ayr.* 1837-60.
James. *Tarbolton.* 1850.
John. *Belfast.* 1784.
MURDOCK—
& Co., J. *Utica, NY, USA.* ca.1820.
John. *Woodbury, NJ, USA.* 1777-86.
MURFITT—
John. *Cottenham.* 1846-75.
R. *Littleport (Cambridgeshire).* 1846-65.
W. *Sutton (Cambridgeshire).* 1846.
William. *Littleport.* 1840.
MURPHY—
Hugh. *Dungannon.* 1840-80.
John. *Antrim.* 1853-1904. C.
John. *Charlestown, Pa, USA.* ca.1790. C.
John. *Northampton.* 1755.
John. *Northampton* and *Allentown, Pa, USA.* 'From
  Ireland'. 1787. C.
John J. *Halifax, Nova Scotia.* 1876-81.
Patrick. *Aughnacloy, co. Tyrone.* 1868-80.
Robert (R. E.). *Philadelphia, USA.* 1848.
Thomas. *Allentown, Pa, USA.* Late 1830s.
Thomas. b.1809. *Dublin.* At *Lancaster.* 1851. C.

MURRAY—
A. S. *London, Ontario.* 1863-90.
Alexander. *Cromarty.* 1860.
Frederick. *London.* 1869-81.
George. *Lochgilphead.* 1837.
J., jun. *Hastings.* 1855-62.
J. W. *Workington.* 1873.
James (& Co.). *London.* (B. 1810) 1828 (1851-
  1881).
John. *Aberdeen.* 1843.
John. *Cavan.* 1827-46.
John. *Dumfries.* 1887.
John. *Lockerbie.* 1860.
John. *Longtown.* 1848.
John. *Nairn.* 1860.
John. *St. Leonards.* 1870-78.
Joshua Joyce. *London.* 1857-81.
Miss. *Cavan.* 1854-59.
Robert. *Lauder.* 1837-60.
Robert. *Whiteburn (Scotland).* 1860.
& Son. *Fredericksburg, Va, USA.* 1805.
William. *Bellingham (Northumberland).* 1848-58.
William. *Liskeard.* 1823-d.1883.
William. *Longtown.* 1834.
William. *Rothbury.* From *Edinburgh.* ca.1828-1903.
MURRELL, William. *Bolney (Sussex).* a.1747 to
  Thomas Stickler, C.
MURREY, Robert. *Richmond.* mid-19c. s. of William of
  Rothbury.
MURRY, Robert. *Bowmore (Islay).* 1860.
MUSCROFT, G. *Sheffield.* 1821. C.
MUSGRAVE—
John *Keswick.* 1834-48.
& KELLY. *Buffalo, NY, USA.* 1812.
MUSGROVE, James. *Burton-on-Trent.* 1828-42.
MUSKET(T)—
Frederick Thomas. *London.* 1881.
Iohn. *Prescot.* mar.1766. W.
MUSSARD, Daniel. *London.* (B.C.C. 1686)-1697.
MUSSELWHITE, William. *Bicester.* 1787-91.
MUSSEN, Thomas. *Downpatrick.* ca.1790.
MUSSON—
Baldus (Balthazar). *Louth.* 1835-68.
George Balthazar. *Louth.* 1876.
Samuel. *Hinckley.* mid-18c. C.
MUSTON—
——. *Bristol.* 1832. Chrons.
Charles. *London.* 1828-39. Watch cases.
Charles. *London.* 1851-63. (Successor to George.)
& GARTH. *Bristol.* 1856. Chrons.
George. *Bristol.* 1840-42. Also chrons.
George. *London.* 1832-44.
MUTCH, Alexander. *Ellon (Aberdeenshire).* 1860.
MUTTON, Samuel. *Lostwithiel.* 1823. Later *Liskeard.*
  1847.
MUZZELL, Cornelius. *Horsham.* (B. 1793-97
  clocksmith) 1828-39.
MYALL, Charles. *Framlingham.* 1879.
MYATT, Samuel. *Newcastle-under-Lyme.* 1848-50. C.
MYCOCK, Thomas. *Waterhouses (Staffordshire).*
  1860-76.
MYER—
A. & Son. *Hereford.* 1879.
George. *New York.* 1846.
Sydney. *Southampton.* 1878.
MYERS—
——. *Leeds.* ca.1860. C.
Abraham. *London.* 1839-57.
& Co. (S. F.). *New York.* 1889.
Daniel. *London.* 1857.
David. *London.* 1844.
Mrs. Elizabeth. *London.* 1863 (?wid. of Abraham).
F. W. Hockley (*Nottinghamshire*). 1864.
Frederick. *Md, USA.* Late 18c. C.
George. *Darlington.* 1850-d.1886. C.
George. *Huddersfield.* 1834. C.
Henry. *Ramsey (Isle of Man).* 1860.
Isaac. *London.* 1844.
Israel. *Hull.* 1858.
John. *Darlington.* 1802. C.
John. *London.* Early 18c. C.
Iohn (& Co.). *London.* 1857-69 (1875-81).

MYERS—*continued.*
M. (& W.). *Stony Stratford.* Partnership dissolved 1840.
Moses. *Poughkeepsie, NY, USA.* 1840s.
Richard. *Darlington.* Early 19c. C.
Richard. *Knaresborough.* 1822-26.
Solomon. *London.* 1857-81.
Thomas. *Sheffield.* Successor to Moss & Myers. 1871.
Thomas Henry. *Birkenhead.* 1848.

MYGATT, Comfort Starr. *Danbury, Conn, USA.* 1763-1804, *Canfield, Ohio.* 1807.
MYLE, Samuel. *Lebanon, Pa, USA.* ca.1810.
MYLES (or MILES), William. *Stockport.* 1848-57. C.
MYLNE—
George E. *London.* 1851-69.
James. *London.* ca.1740. W.
MYLREA, Basil. *Peel (Isle of Man).* 1860.

N

N, W, WN, monogram of William North of *London*, q.v.
NADAL, Jean. *London*. 1875-81.
NADAULD—
  & JACKSON. *London*. 1828-32.
  Miss Mary Ann. *London*. 1844.
NADROW, Thomas. Place unknown. ca.1772. W.
NAGELE—
  J. B. *Wolverhampton*. 1868-76.
  P. & J. *Wolverhampton*. 1860.
NAGL, Anthony. *Vienna*. 1500-18.
NAISH, ——. *Clutton* (*Somerset*). ?19c. C.
NANCOLAS, Anthony. *Falmouth*. 1783-ca.1785. W. &
  C.
NAPIER—
  David & Son. *London*. 1869. T.C.
  Thomas. *Glasgow*. 1789-1803. C. (not *Edinburgh*).
NAPTON, William. *London*. a.1688 to Edward Enys,
  sen., CC. 1695.
NARNET, Nicholas. *Paris*. 18c. C.
NARRAWAY, W. *Chesham*. 1877.
NASH—
  Frederick Francis. *London*. 1881.
  George. *London*. 1863-81.
  John. *London*. CC. 1667-97.
  John. *London*. ca.1770. C.
  Thomas. *New Haven, Conn, USA*. 1638.
  William. *London*. 1828-32. Clock case maker.
NASHUA, Watch Co. *Nashua, NH, USA*. 1859-62.
NATHAN—
  Benjamin & Co. *Birmingham*. 1860-80. Wholesalers.
  David. *Canterbury*. 1838-55.
  David. *Liverpool*. 1848-51.
  H. *Birmingham*. 1854-60.
  Jacob. *Manchester*. 1808-11.
  James. *Hyde*. 1878.
  John. *Liverpool*. 1822-29.
  Lemon & Jacob. *Manchester*. 1824-34.
  Mosley. *Liverpool*. 1851.
  Philip. *Liverpool*. 1824-34.
  Phineas. *London*. 1839-51.
  Rosina. *Liverpool*. 1848-51.
  S. *Hyde*. 1865.
NATIONAL—
  Self Winding Clock Co. *Bristol, USA*. 1903-04.
  Watch Co. *Elgin, Ill, USA*. 1864-74.
NATIVE SON. Mark of Otay Watch Co., Cal, USA.
  1889.
NAU, Richard. *London*. a.1653 to Peter Bellon, CC.
  1661-68.
NAUL, W. *Coventry*. 1860. Watch case maker.
NAUNTON—
  Charles. *Woodbridge*. 1858-92.
  George. *Rochford*. 1855-74.
NAV(E)Y—
  Joseph. *Leeds*. 1850-53.
  William. *Cleckheaton* (also *Brighouse* and *Hipper-
  holme*). 1834-66.
NAWTHROP, William. *Batley*. 1793. C. (see also
  Northrop).

NAYLOR—
  Andrew. *Scotter* (*Lincolnshire*). 1868-76.
  J. *Lowestoft*. 1858-65.
  J. F. *Penge*. 1878.
  James. *Pudsey*. 1834. C.
  John. *London*. 1857-81.
  Joseph. *Cleckheaton*. 1834. W.
  Thomas. *Childwall*. mar.1763. W.
  Thomas. *Holbeach*. 1828.
  Thomas. *Lowestoft*. 1839-53.
  W. *Owston* (*Lincolnshire*). 1861.
  W. & T. *Lower Bebington* (*Cheshire*). 1857.
  William. *Owston Ferry* (*Lincolnshire*). 1868-76.
NAYSBUT, William. *Barnard Castle*. 1832-33. W.
NEALE(E)—
  Daniel. *Philadelphia, USA*. 1823-33.
  Elisha. *New Hartford, Conn, USA*. 1829-32. C.
  Euclid. *Aylesbury*. 1724-d.1736. C.
  Francis. *Aylesbury*. 1750-75. C. (B. ca.1715?.)
  & FREEMAN. *Coventry*. 1860. Watch case maker.
  H. J. *Grimsby*. 1868.
  Henry. *Bourne*. 1876.
  Henry. *London*. 1857.
  Henry. *Thorp Arch* nr *Wetherby*. 1844.
  Henry. *Wandsworth*. 1862-78.
  James. *Aylesbury*. 1737-59. C.
  James. *London*. (B. a.1816) 1828.
  John. *London*. 1863-81.
  Michael. *London*. 1839. Watch cases.
  Richard. *Hunton* (*Kent*). 1866-74.
  T. *Beccles*. 1865.
  T. *Peckham*. 1874.
  Thomas. *Birmingham*. 1791-1803. Dial maker
  (?watches).
  William. *Brighton*. 1828-62.
NEAT, George Thomas. *London*. 1875-81.
NEED—
  Robert. *London*. 1857-75.
  W., jun. *Hornet* (*Sussex*). 1851.
NEEDHAM—
  James. *Rotherham*. 1862-71. W.
  Frederick. *Birmingham*. 1868-80. Clock cases.
  Jonathan. *Coventry*. 1860.
  Joseph. *Coventry*. 1850-54.
  Woodford. *Birmingham*. 1868. Clock cases.
NEEDS, Peter. *Leeds*. 1756-d.1760. W.
NEELY, Thomas. *Maghera*. 1846.
NEGUS, T. S. & J. D. *New York*. 1845.
NEHRING, Hinrich. *Reval*. ca.1670. C.
NEIGHBOUR, William. *London*. a.1677. CC. 1685-97.
NEIL(L)—
  Brothers (John R. & James). *Belfast*. 1849-59. C. & W.
  James (& Co). *Belfast*. s. of Robert. ca.1845-83. C. &
  W.
  John R. *Belfast*. 1845-82. C. & W.
  John R. & Co. *Londonderry*. 1839.
  Robert. *Belfast*. (b. ca.1775, a.1791) 1803-ca.1845.
  (d.1857.) Spent some time in *London* pre-1803. C. &
  W.

NEIL(L)—*continued.*
Robert, jun. *Belfast.* s. of Robert, sen. Later 19c. Died in *Naples.*
Robert. *Quebec.* 1848-57.
Robert & Sons. *Londonderry.* 1839-59.
NEILSON—
——. *Annapolis, Md, USA.* 1734.
——. *Dromore.* 1862.
Alexander. *London.* 1869.
Cornelius. *Perth, Canada.* 1851-63.
George. *Boston, USA.* 1830.
George. *Dumfries.* 1810-20.
& LAWRENCE. *London.* 1869.
Thomas. *London.* 1863-81.
Thomas Walker. *London.* 1881.
William. *Halifax, Nova Scotia.* 1838.
William H. *Belfast.* 1800-1818. C.
NEINE(N)GER/NEINI(N)GER—
Anthony. *Bristol.* 1850.
Anthony. *Stonehouse.* 1856-79.
J. *Gloucester.* 1856-63.
Joseph. *Gloucester.* 1879.
William. *Brighton.* 1839-62.
NEISSER, Augustin. 1736 *Georgia, USA,* 1739 *Philadelphia,* 1772-80 *Germantown.*
NELKEN, Jules. *London.* 1875.
NELMES, Robert. *London.* 1828-44.
NELSON—
——. 1914 and later, product of Pequegnat Clock Co.
——. *Jersey.* 1832.
——. *Lurgan.* Early 19c. C.
——. *Tobermore, co. Derry.* ca.1800. C.
Alexander. *Dungannon.* 1865-68.
Alexander. *West Chester, Pa, USA.* 1820.
Bernard. *Liverpool.* 1848.
& BUTTERS. *Montreal.* 1857.
H. A. & Sons. *Montreal* and *Toronto.* 1875.
J. & R. *Dromore.* ca.1825-35. C.
James I. *Banbridge, co. Down.* ca.1783-1820.
James II. *Banbridge.* (b.1783). ca.1805-d.1845.
James III. *Banbridge.* ca.1843-68. (With Robert N. in 1843).
John. *Bury.* 1858. W.
John. *London.* 1844-57.
John A. *Boston, USA.* ca.1825.
Joseph I. *Banbridge.* 1820-24. C.
Joseph II. *Banbridge.* s. of James, ca.1850, when went to *New York.*
Joseph & Robert. *Dromore, co. Down.* 1846-59. C.
Lawrence. *Stafford.* 1876.
& LEFFORT. *Montreal.* 1875.
R. *Enniskillen, co. Fermanagh.* 1868-80.
R. J. *Davenport, Iowa, USA.* ca.1850.
Robert. *Derby.* 1876.
Robert. *Omagh.* 1865-90.
Robert. *Prescot.* 1796. W.
Robert. *St. Thomas, Canada.* 1851-58.
Robert I. *Dromore.* ca.1825-46.
Robert II. *Banbridge.* ca.1845-50, when went to *New York.*
Samuel. *Banbridge.* 1865-68.
T. *London.* 1851.
Thomas. *Birmingham.* 1842-60.
Thomas. *Hinckley.* 1828-35.
Thomas. *Liverpool.* (B. 1807-) 1823-34.
Thomas. *Manchester.* mar.1790. C.
Thomas. *Philadelphia.* ca.1800.
Thomas. *Southwell.* ?Late 18c.
WOOD & Co. *Toronto* and *Montreal.* 1842-71.
NELTHORPE, A. *Boston.* 1876.
NEMERT, Gottlieb Christian. *Reading, Pa, USA.* 1841.
NEMSER, Joseph. *Birmingham.* 1880.
NESBITT—
George. *Sunderland.* (B. 1820-)1827. C.
James. *Downpatrick.* 1846.
NESS—
George. *Kirkby Moorside.* s. of and successor, to Thomas. b.ca.1830-1912.
Peacock. *Kirkby Moorside.* 1840-66. W.
Thomas. *Kirkby Moorside.* 1840.
William. *Kirkby Moorside.* 1823-34. C.
NESSL, Joh. *Christ. Laybach.* ca.1730. W.

NESTMAN, Julius. *Bradford, Canada.* 1862.
NETHERCOT(T)—
——. *Tredington (Oxfordshire).* 1707. Rep'd. ch. clock.
George. *Wantage.* ca.1770. C.
John. *Long Compton (Oxfordshire).* ca.1750. C.
John. *Standlake (Oxfordshire).* ca.1750. C.
William. *Long Compton (Oxfordshire).* ca.1750. C.
NETHERCROFT, George. *Wantage.* 1763.
NETTLETON—
Heath & Co. *Scottsville, NY, USA.* ca.1820. C.
W. K. *Rochester, NY, USA.* ca.1830.
Wilfred H. *Bristol, USA.* 1850-70. C.
NEUCHATEL Exportation Co. *London.* Adolphe Vicarino agent, 1863.
NEUENS, Peter. See NEWENS.
NEUHOFF, Theodore. *London.* 1881.
NEUMANN, Charles. *London.* 1881.
NEVAY, William. *Forfar.* 1837.
NEVE, Henry. *London.* (B. 1700-52) *St. Clement Danes* parish.
NEVILL, A. *Coventry.* 1854.
NEVI(N)SON, Robert. *Kendal.* b.ca.1719-mar.1745. C. & W.
NEVITT, Thomas. *Bristol.* ca.1780. C.
NEWALL—
John. *Wigan.* 1666. W.
Thomas. *Sheffield, Mass, USA.* 1809.
NEWARK Watch Co. *New York.* 1863-69.
NEWBALD, E. *Ryde (Isle of Wight).* 1859-67.
NEWBERY—
George Frederick. *Brighton.* 1862-70.
J. & R. *Philadelphia, USA.* 1816.
James W. *Philadelphia, USA.* 1819-ca.1850.
NEWBOLD—
James. b. 1794 *York,* mar. 1817 *Notts,* then *Snaith* 1817-22, then *York* 1830 -d. 1848.
John. *Rotherham.* 1871.
NEWBURY—
& BIRELY. *Hamilton, Ontario.* 1857-65.
H. *Hanslope (Buckinghamshire).* 1869-77.
NEWBY—
James. *London.* 1851.
James. *Malton.* 1866. W.
James & William. *Kendal.* ca.1765-1800. C.
NEWCOMB(E)—
Henry. *Lowell, Mass, USA.* 1837.
Joseph. *London.* (B. a.1798, CC. 1807) 1828.
Robert. *Henley-in-Arden.* 1880.
Thomas. *Boston, USA.* 1784-ca.1808.
NEWELL—
——. *Leicester.* 1795-ca.1810. C.
A. *Boston, USA.* ca.1785.
James J. *Utica, NY, USA.* 1834.
Lott. *Bristol, USA.* 1818.
Norman. *Rochester, NY, USA.* 1844.
Sextus. *Bristol, USA.* 1809-11.
Theodore. *Poultney, Vt, USA.* 1820.
Thomas. *Sheffield, Mass, USA.* 1810-20.
NEW ENGLAND CLOCK CO. *Bristol, USA.* 1851. C.
NEWENS (sometimes **NEUENS),** Peter. *London.* 1839-44.
NEW ERA. Used by Lancaster (Pa) Watch Co. ca.1878.
NEWEY—
Joshua. *Birmingham.* 1880.
R. *York.* ca.1770-ca.1780. C.
NEW(H)ALL—
Frederick Augustus. *Salem, Mass, USA.* 1818-59.
Henry. *Cleobury Mortimer (Shropshire).* 1828-42.
William. *Boston, USA.* ca.1850.
NEWHART, ——. *Lebanon, Pa, USA.* ca.1840.
NEW HAVEN—
Clock Co. *Brantford, Ontario.* 1906.
Clock Co. *New Haven, Conn, USA.* 1853-1950 (? and later).
Watch Co. *New Haven, Conn, USA.* 1883-1908.
NEWHOUSE, Edward. *Halifax.* 1775. C.
NEWINGTON—
H. *Mayfield.* 1862.
Horace. *Wadhurst.* 1862-78.
J. O. *Lamberhurst.* 1845.
J. O. *Mayfield.* 1878.
J. O. *Wadhurst.* 1851-55.

NEWINGTON—*continued.*
O. *Worthing.* 1878.
NEW JERSEY WATCH CO. *New York, USA.* 1888.
NEWLAND(S)—
A. *London.* 1810. W.
Christopher. *Newport (Monmouth).* 1880.
George. *Kilmarnock.* 1860.
J. *Alton.* 1839-48.
J. *Birkenhead.* 1857.
J. *Farnham.* 1862-66.
John. *York.* 1526. Made minster clock.
John B. *Paisley.* 1860.
Joseph. *Farnham.* 1828.
NEWLIN, Edward G. *Philadelphia, USA.* 1848.
NEWLOVE—
John. *Scarborough.* 1840.
John. *York.* 1823-34. C.
NEWLYN, Alfred. *London.* 1881.
NEWMAN—
Albert Walter. *Cardiff.* 1887. Later *Barry Dock.* 1899.
Alfred James. *Guildford.* 1851-78.
Charles. *Kings Lynn.* ca.1760. C.
Clock Co. *Chicago, Ill, USA.* 1878-91.
Edward. *Wisbech.* 1830-46.
J. *Chatham.* 1855.
John. *Belvedere.* 1874.
John. *Bexley Heath.* 1866-74.
John. *Boston, USA.* 1764.
John. *London.* (B. a.1804-25) 1832.
John. *London.* 1863.
Robert. *London.* 1832.
Samuel. *Norwich.* 1875.
Thomas. *Cavan* 1824. *Letterkenny* 1846-66.
W. *Birmingham.* 1854.
William. *London.* (B. 1817-20) 1828-32.
William Henry. *Halifax, Nova Scotia.* 1826-97.
NEWNHAM—
Frederick. *London.* 1839. Watch cases.
John. *Horsham.* a.1726 to John Inkpen, C.
NEWNTEN, George. *?Seene.* ca.1640. Lant. clock.
NEWSAM, Bartholomew. (B. 1568) appointed Clockmaker to Queen Elizabeth in 1590. Died 1593.
NEWSHAM, Richard. *Liverpool.* 1790-96. W.
NEWSOME—
& Co. *Coventry.* 1880.
Mrs. Selina. *Coventry.* 1880.
T. *Coventry.* 1860-68.
& YEOMANS. *Coventry.* 1868.
NEWSON—
D. D. *London.* ca.1760. C.
F. B. *Badingham (Suffolk).* 1846-58.
N. W. *Halesworth.* 1853-79.
NEWSTEAD—
Christopher. *York.* b.1730-d.1801. C.
William. *Melbourne.* 1835.
William. *Repton.* 1828.
NEWTH, William. *Schenectady, NY, USA.* 1837-42.
NEWTON—
Alexander L. & Co. *London.* 1832. Wholesaler.
Brothers. *Bradford.* 1871.
Francis. *Norwich.* Partner with John Cooper, 1835-36. C.
Francis. *Truro (Kenwyn).* 1787. W.
George. *Manchester.* 1848. Imported clocks.
George. *Millom.* 1851.
HART & CO. *London.* 1828.
Isaac. *Manchester.* 1851.
Isaac L. *Salem, Mass, USA.* 1796.
J. L. *Trenton, NJ, USA.* From *London.* 1804-20. W.
James. *London. Red Lyon Street.* ca.1760.
Joseph. *Liverpool.* 1781-1834. W.
Ottis R. *Niagara Falls, Canada.* 1861.
William. *London.* 1863-81.
William. *New York.* 1840s.
NEW WATCH CO. *Waterbury, Conn, USA.* 1898-1912.
NEW YORK—
City Watch Co. *New York.* 1897. W.
Standard Watch Co. *Jersey City, USA.* 1885-1902.
Watch Co. *Providence, RI, USA.* 1866-1927.
Watch Manufacturing Co. *Springfield, Mass, USA.* 1875.
NEYSSER, William. *Germantown, Pa, USA.* Late 18c.

NIBLETT, Henry. *London.* 1828.
NICAISE, Gabriel. *Nauvoo, Ill, USA.* 1848-93. C.
NICASIUS—
John, sen. *London.* CC. 1632- ?1679. W.
John, jun. *London.* a. ca.1638. Free in BC. 1647. W.
NICHOL(L)—
Isaac. *London.* a.1674 to *Withers Cheney.* CC. 1681. (B. -1712.)
John. *Halifax.* 1837.
John L. *Belvedere, NJ, USA.* 1790-1818. C.
Thomas. *Camborne.* 1873. W.
NICHOLAS—
Caleb. *Birmingham.* 1787-1842. C.
Charles. *Birmingham.* 1790. W.
David. *Fishguard.* 1830-50. C.
John. *Daventry.* 1808-77. W.
John. *Ystalyfera.* 1887.
Joseph. *Redruth.* mar.1821. W.
Robert. *Sheerness.* 1839-47.
Samuel. *London.* 1839-51.
Stephen. *London.* 1857.
Thomas James. *Fishguard.* 1887.
William. *London.* (B. 1820-25) 1828.
William C. *Winchendon, Mass, USA.* ca.1840. W.
NICHOL(L)S—
——. *Walton-on-Naze.* Late 19c. C.
A. L. *Streatham.* 1866.
Alexander. *London.* 1863-75.
C. R. *Fulton, NY, USA.* 1860s.
Edward. *London.* 1881.
Hammond. *Canterbury.* 1789-1810. W.
Henry. *Liverpool.* 1848-51. Also watch materials dealer.
I. *Fullneck* (nr *Pudsey*). 1796. C.
Isaac. *Wells.* ca.1740. C. May be from *London* where one such 1697.
J. *Basingstoke.* 1859.
James. *Bristol.* b.1775, a.1788-d.1825.
James. *Newport (Isle of Wight).* (B. 1814. W.) 1830-39.
James. *Ryde (Isle of Wight).* 1830-39.
James Henry. *Redruth.* 1856-73.
John. *Birkenhead.* 1828.
John. *London.* 1811 clock case maker, *6 Redlion Street, Clerkenwell.*
John. *Margam (Wales).* ca.1700. C.
Joseph Henry Penrose. *Cambridge.* 1873. W.
Mrs. P. *Basingstoke.* 1867.
Richard. *Leominster.* 1850.
Richard Crowe. *Leominster.* 1856-70.
Roger. *London.* a.1659 to John Savill, CC. 1667 (B.-1708).
Thomas. *Oundle.* 18c. C. (?mid-century).
W. W. *Streatham.* 1862.
Walter. *Newport, RI, USA.* 1849.
William. *Cheltenham.* 1842-79.
NICHOLSON—
Charles. *Brighton.* (B. 1822-)1828.
Charles. *London.* 1839. Clock case maker.
George. *Wandsworth.* 1862-78.
H. *Brighton.* 1851-70.
Henry Howard. *London.* 1839. Watch cases.
J. *Andover.* 1867.
J. *Fareham.* 1848-59.
J. A. W. *Barwick.* 1855 (?may be *Berwick-on-Tweed*).
James. *Brighton.* 1828.
John. *Berwick-on-Tweed.* 1834-37.
John. *London.* 1832.
John. *Stockbridge.* 1839-48.
John & William. *Berwick-on-Tweed.* 1844-60. C.
Joseph. *Halifax.* 1871.
Richard. *Berwick-on-Tweed.* (B. 1806-22)-1827. C.
Robert. *Ainthorpe* nr *Danby in Cleveland.* 1840.
Robert. *Hull.* 1813. Engraver.
Thomas. *London.* 1857-81. Clock cases.
William. *Whitehaven.* ca. 1720-50. C.
NICKALLS—
& KNIGHT. *Reigate.* 1878.
T. *Reigate.* 1851-66.
NICKEL, William. *Londonderry.* 1839-46.
NICKERSON, Jesse. *Hamilton, Canada.* 1853-66.
NICKISSON—
John. *Newcastle-under-Lyme.* 1828-42.

NICKISSON—*continued.*
Sampson. *London.* (B. 1808-) 1832-44.
NICKOLDS—
Edward. *Brewood (Staffordshire).* 1842-76.
Philip. *Birmingham.* 1868.
Thomas. *Albrighton.* 1856-70.
Thomas. *Shiffnal.* 1842.
NICOL—
James. *Edinburgh.* ca.1780. C.
James. *Kilmarnock.* 1850-60.
NICOLE—
& CAPT. *London.* 1844-75.
& Co. *London.* 1881.
NIELSEN & Co. *London.* 1881.
NICOLET, Julian. *Baltimore, USA.* 1819-31 to
Pittsburgh, Pa, USA.
NICOLL—
James. *Canongate (Edinburgh).* mid-18c. C. (B. a.1721-
60).
John. *London.* 1839-51.
William. *London.* 1828-57.
NICOLLET—
Joseph M. *Philadelphia, USA.* 1797.
Joseph W. *Philadelphia, USA.* 1798.
Mary. *Philadelphia, USA.* 1793-99. W.
NICOLLS—
Henry. *Launceston* 1791, *St. Austell* 1793. C. & W.
William. *Little Billing (Northamptonshire).* pre-1688.
NICOLSON, Alexander. *London.* 1875.
NIDDIE, John. *Markinch (Scotland).* 1860.
NIEBERGALL, Frederick. *Rondout, NY, USA.*
ca.1850.
NIEILLY, Emanuel. *Lebanon, Pa, USA.* 1848.
NIELD, Daniel. *Rochdale.* 1834. Clock dealer.
NIMMO, N. *Kirkcaldy.* ca.1780. C.
NINDE, James. *Baltimore, USA.* 1799-1835.
NINNES—
J. W. *Tunbridge Wells.* 1855-66.
& LOOF. *Tunbridge Wells.* 1851.
NINNIS, Isaac. *St. Ives (Cornwall).* ca.1800. C.
NISBET, George Hunter. *Tunstall (Staffordshire).*
1876.
NIVEN, James. *Glasgow.* 1860. C.
NIX, John. *Eastbourne.* 1878.
NIXON—
George. *Dumfries.* 1860.
George. *Farnworth.* mar.1832. W.
John. *Cobourg, Canada.* 1851-58.
NOAKES—
Henry. *Burwash.* 1878.
J. *Battle.* 1855.
J. F. *Faversham.* 1874-84.
John. *Burwash.* 1833-d.1870s. W.
John (Thomas). *Battle.* b.1833. s. of John N. of
Burwash, W. -1859. C.
John Thomas. *Tonbridge.* 1874.
NOBBS, W. *Norwich.* 1875.
NOBLATT, Edward. *Preston.* Late 18c. C.
NOBLE—
——. *Peterborough.* ca.1790. C.
Charles. *London.* 1839.
J. *Brentwood.* 1855.
James. *Lancaster.* Free 1733, d.pre-1784. W.
& MURRAY. *St. Catharines, Canada.* 1877.
Philander. *Pittsfield, Mass, USA.* ca.1830.
T. & Son. *Bath.* 1856.
Thomas. *Orillia, Canada.* 1871.
NOE, Richard. *London.* 1662. Watch case maker. (prob.
Richard Nau, q.v.)
NOEL, Theodore. *Frankfort, Ky, USA.* ca.1830.
NOKE, Penelope. *Brierley Hill.* 1876.
NOLCINI, Joseph. *Cardiff.* 1865-68.
NOLL, ——. *Paris.* ca.1790. W.
NOLSON, John. *London.* a.1689 to Daniel Quare, CC.
1697.
NOMOLAS, ——. *London.* ca.1740. W. (Anagram of
Salomon).
NOON—
Sarah. *Ashby de la Zouche.* 1828-35.
Thomas. *Burton-on-Trent.* 1828-35.
V. *New Radford (Nottinghamshire).* 1849-55.
NOOTT, William. *Haverfordwest.* 1887.

NORCOTT, John. *London.* a. to Samuel Drossatt, CC.
1681-97. W.
NORDBECK, Peter. *Halifax, Nova Scotia.* 1819-d.1861.
NORDMAN—
& HALF. *London.* 1875.
Joseph & Co. *London.* 1869.
Jules. *London.* 1881. (Sucessor to Nordman & Half).
NOREAU, Joseph. *St. Roc, Quebec.* 1848.
NORKETT, I. (?*Sormerset*). Early 18c. T.C.
NORMAN—
Charles. *St. Ives (Huntingdonshire).* 1877.
H. & A. *Wimbledon.* 1878.
Henry. *London.* 1857-81.
James S. *Lincolnton, NC, USA.* 1840.
John. *Castle Combe.* ?Late 18c. C.
John. *Warrington.* mar.1773. W.
Matthew. *Sherborne (Dorset).* ca. 1810-41. C.
R. *Prescot.* 1858. W.
Simon. *Lowestoft.* mar 1782 d. 1837. W. & C.
NORMANDEAU, A. *Montreal.* 1819-25.
NORMINGTON, James. *Halifax.* 1866.
NORRIS—
Alfred. *London.* 1857.
& CAMPBELL. *Liverpool.* 1848-51. Chrons.
Francis. *Liverpool.* 1834.
George. *Woodbridge.* 1858-79. Also Aldborough
1865-79.
George Alfred. *London.* 1863-81.
Henry J. *Coventry.* 1880.
James. *Abingdon.* ca.1710. C. (cf. Joseph).
James. *Whittlesey.* 1875.
John. *London.* 1832-39.
John Alfred. *March (Cambridgeshire).* 1865-75.
Joseph (cf. James). *Abingdon.* Ox Street, 1720.
Joseph. *London.* a.1661 to Edward Norris. CC. 1670.
In *Amsterdam* ca.1675-ca.1693. *London* again
ca.1696. C. & W.
Mary. *Liverpool.* 1848. Chrons.
Patrick. *Philadelphia, USA.* 1844.
Samuel. *Windsor.* 1837.
T. *Weston Jones (Staffordshire).* 1860-68.
Thomas. *Eccleshall.* 1850.
Thomas. *Gnosall.* 1876.
Thomas. *Mortlake.* 1839-55.
Thomas. *Whittlesey.* 1840-65.
W. *Newport (Shropshire).* 1856-79.
William. *USA.* ca.1815. C.
NORTH—
Charles. *London.* 1881.
Ethel. *Torrington, Conn, USA.* 1824-31. C.
Ethel (*brother* of Norris). *Wolcottville, Conn, USA.*
ca.1820. C.
James, jun. (*Great*) *Ilford.* 1839-74.
John. *London.* a.1641 to Ralph Ash, CC. 1650-97.
Jonathan. *York.* ?19c. C.
Lancelot. *York.* b.1601-23. C. (B. then *London* 1639-
64).
Norris. *Torrington, Conn, USA.* 1823-34. C.
Norris. *Wolcottville, Conn, USA.* bro. of Ethel.
ca.1820.
Phineas. *Torrington, Conn, USA.* 1762. mar.1787-
1810. C.
Richard. *Driffield.* (B. ca.1770). 1790-94. W.
Samuel. *Leconfield.* ca.1770-1800. C.
Simeon. *Conn, USA.* 1812. C.
Thomas. *London.* (B. 1817-24) 1828-32.
William. *Leconfield.* 1760-70. C.
William. *London.* CC. 1639-64. W.
William. *York.* 1816-30.
NORTHCOTE—
Samuel, sen. *Plymouth.* 1708-d.1791.
Samuel, jun. *Plymouth.* 1742-d.1813.
NORTHERN—
Edward. *Hull.* 1846. W.
Richard. *Hull.* 1813-40. W.
NORTHEY—
Elijah. *Philadelphia, USA.* 1844-ca.1850.
R. E. *New Haven, Conn, USA.* ca.1820.
NORTHFIELD, Henry John. *London.* 1863-81.
NORTHGRAVES—
Denton. *Hull.* (B. 1806-22) 1830. W.
Frederick. *Picton, Canada.* 1857.

NORTHGRAVES—*continued.*
George. *Brockville, Canada.* 1851.
George. *Perth, Canada.* 1857-71.
George D. *Almonte, Canada.* 1862.
& MADDEN. *Kingston, Canada.* 1853.
William. *Picton, Canada.* 1851.
William. *Quebec.* 1819-26.
William J. *Belleville, Canada.* 1851-63.
NORTHROP (or **NORTHRUP**)—
& SMITH, *Goshen, Conn, USA.* 1831-36. C.
William. *Huddersfield.* 1834-37. C.
William. *Wakefield.* 1743-51. C. Also NAWTHROP.
NORTHWOOD, James. b. ca. 1812 *Caynton (Staffs) Newport (Shropshire).* 1840-75. C.
NORTON—
——. *London.* ca.1820. W.
Benjamin. *London.* 1832-39. Watch cases.
Daniel. *Minty.* Late 18c. or early 19c. C.
Eardley. *London.* b.1728 (in *Lincolnshire*) -d.1792.
Edward. *Sunderland.* 1851.
Elijah. *Utica, NY, USA.* ca.1820.
George. *Ipswich.* Succ. Moores 1791. Insolvent 1775. W.
Henry. *Stamford.* s. of Robert. 1871-76.
John. *Yorktown, Pa, USA.* Late 18c. C.
Nathaniel. *New Haven, Conn, USA.* ca.1840.
Robert. *Stamford.* 1841-d.ca.1870. W. and silversmith.
Samuel. *Hingham, Mass, USA.* 1785.
Samuel. *London.* 1857.
Samuel William. *Fenny Stratford* and *New Wolverton.* 1877.
Thomas. *Philadelphia, USA.* 1789-1811. C.
Thomas & Samuel. *Philadelphia, USA.* ca.1800. C.
V. *Toronto.* 1875.
William. *Leeds (Kent).* 1826-39.
NORWEB, Thomas. *Wetherby* 1766, then *Selby* 1769, *Brigg* 1770-82, *Louth* 1782-6, *Wrawby (Lincs)* ca. 1786 -d. 1809. C. Son apr. *Nottingham* 1779. Name known to be an alias (for Browne?).
NORWOOD, George. *Dover.* Free 1826-33.
NOSEDA, John. *Belfast.* 1774-d.1779. Weather glasses.
NOTERMANN—
Fabian Sebastian. *London.* 1857-63.
Mrs. Fanny. *London.* 1869. (?wid. of Fabian.)

NOTTLE, John. *Okehampton.* mar.1766.
NOUWEN, Francis. *London.* ca.1580-d.1593. W. (also NAUWE).
NOWACKI, Ignatius. *Sheffield.* 1862.
NOWE, Francis. *London.* See Nouwen. (1588 lant. clock.)
NOWLIN, L. *Chicago, USA.* 1840s.
NOXON, Martin. *Edenton, NC, USA.* 1780-1814.
NOYER, Pierre (Peter). *London.* 1851-69.
NOYES—
L. W. *Boston, USA.* 1841.
Leonard W. *Nashua, NH, USA.* ca.1825-ca.1840. C.
NUMA-HINFRAY. *St. Nicolas d'Aliermont.* 1887.
NUNN—
H. *St. Albans.* ca.1800. C.
James. *Stowmarket.* 1879.
NURENBERG, Nathan. *Croydon.* 1878.
NURSE—
R. *Holbeach.* 1861.
Richard. *Kings Lynn.* 1858-75.
Richard. *Snettisham (Norfolk).* 1830-46.
Richard. *Whaplode (Lincolnshire).* 1868-76.
Thomas. *Kings Lynn.* 1830-36.
NUSZ, Frederick. *Frederick, MD, USA.* 1819.
NUTSEY, Isaac. *Alford (Lincolnshire).* 1876.
NUTSFORD—
William. *Whitehaven.* 1848. Same man as below?
Wilson. *Whitehaven.* 1834. Same man as above?
NUTTALL—
C. *Manchester.* 1858. W.
James. *Ormskirk.* 1854.
NUTTER—
E. H. *Dover. NH, USA.* ca.1825. C.
John D. *Mount Vernon, NH, USA.* ca.1825.
Laurence. *St. Neots (Huntingdonshire).* 1731. C. and gunsmith.
NYE—
Edward. *Worthing.* 1839.
Edwin. *Woolwich.* 1874.
John. *Hurstpierpoint.* 1839-51.
Thomas John. *Chatham.* 1847-66.
Thomas John. *Rochester.* 1832-39.

O

OAKES—
Frederick A. *Hartford, Conn, USA.* 1828.
Henry. *Hartford, USA.* ca.1830. W.
Henry & Co. *Hartford, USA.* ca.1830.
John. *Oldham.* 1822-34.
William. *Oldham.* 1834-51.

OAK(E)LY—
James. *Oxford.* mar.1731-d.1749. C. & W.
John. *Oxford.* 1704-05. C. May be from *London* where one such a.1685.
William. *Bilston.* 1876.

OBER, Henry. *Elizabethtown, Pa, USA.* ca.1820.

O'BRIEN—
B. *St. John, New Brunswick.* 1857.
James. *Philadelphia, USA.* 1850.
John. *Philadelphia, USA.* 1844-49.

OCHELTREE, James. *Sunderland.* 1827. May be from *Newcastle-on-Tyne* where one such 1801.

OCHS, G. *New York.* 1872.

O'CLAIR, Narcis. *Albany, NY, USA.* 1819.

O'CLEE—
Alfred. *Harrold* nr *Risley (Bedfordshire).* 1839.
Alfred. *Shefford (Bedfordshire).* 1847-77.
E. A. *Bromley.* 1874.
Edward. *Market Street (Bedfordshire).* 1839.
Frederick A. *Oldham.* 1848-51.
John. *Bexley Heath.* 1832.
John. *Luton.* 1839-54.
John. *Ramsgate.* 1823-27.

O'CONNELL, Maurice. *Boston, USA.* 1842.

O'DANIEL, Perry. *Philadelphia, USA.* 1837-ca.1850.

ODDY, George. *Downham.* 1875.

ODELL—
H. *Barnsley,* 1862.
T. *Yarmouth.* 1846.
Thomas B. *Sherbrooke, Canada.* 1862.
Thomas H. *Littleport (Cambridgeshire).* 1875.

ODERY, William C. *Nottingham.* 1876.

ODGER, Thomas. *Helston.* ca.1800-23. C. & W.

O'DONOHUE—
Jean. *Quebec.* 1890.
Thomas. *Quebec.* 1826-65.

O'DRISCOLL, Jeremiah. *London.* 1857.

ODY, J. *London.* 1840. W.

OEHLER, D. *Ammanford (Wales).* d.1918.

OFFORD/OFFARD, W. H. *Louth.* 1861-68.

OGBURN, J. *Portsmouth.* 1848.

OGDEN—
——. *Halifax.* Prob. Thomas, q.v.
Bernard. *Darlington.* s. of John, b.1707-ca.1740.
G. W. *Middlesbrough.* 1898.
Isaac. *Halifax.* bro. of Samuel, 1693-1700.
James. No town. Same man as at *Halifax.*
James. *Halifax.* mar.1713-d.1716. bro. of Samuel. C.
Jane. *Sunderland.* wid. of John II of *Darlington,* 1741-d.1788.
John I. *Askrigg.* bro. of Samuel. From *Halifax* ca.1680 -post-1707. Later at *Darlington.* Where d.1741. C.
John II. s. of John I. *Darlington* post-1707, then

OGDEN—*continued.*
Alnwick 1732-34, then *Sunderland* 1739. Died pre-1788. Succeeded by widow, Jane.
Samuel. No town. Same man as Samuel I of *Halifax.*
Samuel I. *Halifax.* b.1669-post-1712, then *Benwell* by 1727.
Samuel. *Ripponden.* Same man as Samuel I of *Halifax.*
Samuel II. *Halifax.* b.1689. s. of Samuel I. -1719, then *Alnwick* where d.ca.1760.
Samuel III. *Newcastle-on-Tyne.* s. of Samuel II. mar.1751. To *Halifax* ca.1770-d.1773.
Samuel. *Sowerby Bridge.* Refers to Samuel III of *Halifax.*
Samuel & Thomas. *Ripponden.* Combination of Thomas of *Halifax.* and Samuel II.
T. *Oldham.* 1858. C.
Thomas. *Halifax.* b.1693, s. of Samuel I. d.1769. Fine maker.
Thomas. *London.* a.1651 to Thomas Knifton, CC. 1659.
Thomas. *Ripponden.* Same man as at *Halifax.*
Thomas. *West Chester, Pa, USA.* 1824-30. C.
William. *Bacup.* 1851.
William. *Oldham.* 1848-51.

OGG & McMILLAN. *Aberdeen.* 1846.

OGLE, William. *Philadelphia, USA.* 1828.

OGLETHORPE—
John. *Penrith.* 1869-79.
John. *Kirkby Thore (Westmorland).* 1851.
John. *Temple Sowerby.* 1858.
Joseph. *Kirkby Thore.* 1828.
Samuel. *Kirkby Thore.* 1848-58. W.

O'GORMAN, John. *Toronto.* 1857.

O'HANLON—
Hugh. *Newry.* 1785.
James. *Portadown.* 1858-68.

O'HARA—
Charles. *Philadelphia, USA.* 1899-1900.
James. *Liverpool.* 1824. W.
Robert. *Carrickfergus.* 1865-68.

OHLSCHLAGER, Theodore. *London.* 1863-81.

OHMAN & LINDSTROM. *St. Johns, Newfoundland.* 1885.

OILETTE, A. *Buckingham.* 1823. (?Error for Ortelli.)

OLDER, Peter Upton. *Cranleigh (Surrey).* 1855-78.

OLDFIELD—
——. *Ayton Banks* (nr *Gateshead).* ?19c. C.
E. *Tideswell.* 1855.
Edward. *Place unknown.* ca.1760-ca.80. C.
John. *Watton.* ca.1760. C.
John. *Manchester.* mar.1799-1800. W.
Samuel, sen. *Tideswell.* 1849-76.
Samuel, jun. *Tideswell.* 1876.

OLDHAM—
Charles. *Southam (Warwickshire).* (B. pre-1776) 1842.
John. *Southam (Warwickshire).* 1835-60.
Joseph. *Prescot.* 1834-48.
Richard. *Chorley.* 1848.
Thomas. *Hyde.* 1848.

OLDING—
Henry. *Milbourne Port (Dorset)*. 1824.
J. C. *Landport*. 1848.
John. *Wincanton*. 1801. C.
Nathaniel. *Wincanton*. 1830. C.
OLDS—
C. *Bath*. 1856.
Charles. *Bath*. 1861-83.
OLEWINE—
Abraham. *Pikeland, Pa, USA*. 1834-66. C.
Henry. *Pikeland, Pa, USA*. 1812-67.
OL(L)IFF(E)—
J. *Broughton (Hampshire)*. 1848.
W. *Broughton (Hampshire)*. 1859.
OLIN, H. *Mitchell, Canada*, 1861.
OLIPHANT—
Alexander. *Anstruther*. (B. b.1784-1818) 1837.
Robert. *London*. 1857.
William. *London*. 1844-51.
OL(L)IVE—
Frederick. *Penryn*. 1844-73.
John. *Falmouth*. 1844.
Michael. *Penryn*. 1847, *Falmouth* 1849-73. C., W. and
chrons. Bro. and successor to Thomas.
& READER. *Cranbrook*. 1809. C.
Samuel. *Tonbridge*. ca.1760. C. (B. 1773-94).
Thomas. *Cranbrook*. 1778-1827. C.
Thomas. *Falmouth*. bro. of Michael. d.1849.
Thomas. *Penryn*. From *London* (where one such
a.1756). 1768-1823. W.
OL(L)IVER—
——. *Armagh*. ?From *Enniskillen* ca.1800-d.1822. C.
——. *Dungiven, Co Derry*. ca.1800. C.
Charles. *Enniskillen*. ca.1770-75. C.
Charles. *London*. 1857-63.
Charles. *Spalding*. 1828-35 (?&-1856). Clock case
maker.
D. *Plainfield, NJ, USA*. 1800s.
Daniel. *Beaminster*. 1875. Clock cleaner.
Daniel. *Wellington*. 1861-83.
David. *Reading*. 1830.
Griffith. *Philadelphia, USA*. 1785.
John. *Kirkburton*. pre-1755. W.
John. *Leeds*. mid-18c. C.
John. *Manchester*. (b.ca.1683), believed retired 1749,
d.1766. C.
John. *New York, USA*. 1854.
John S. *Reading, Pa, USA*. ca.1830.
Matthew. *London*. 1832.
R. K. *Toronto, Canada*. 1847-51.
Samuel. *Manchester*. mar.1804. C.
Thomas. *March (Cambridgeshire)*. 1875.
Welden. *Bristol, USA*. ca.1820-30. C.
William. *London*. 1857.
OLLENDORFF, M. *Montreal*. 1847-63.
OLLIER, Adolphe. *Paris*. 1877-ca.1900. Carriage clocks.
OLLIVANT, Thomas and John. *Manchester*. 1828-51.
OLLIVIER, T. *Jersey*. 1837-75. C.
OLMSTEAD—
Gideon. *Charlotte, NC, USA*. 1832.
Nathaniel. *New Haven, Conn, USA*. ca.1820.
Norman. *Brooklyn, NY, USA*. ca.1829. C.
OLORENSHAW—
J. *Coleshill*. 1854.
Joseph & Co. *Coventry* and *Clerkenwell*. 1842-54.
Joseph & Co. *London*. 1857.
OMOND, G. *Tisbury (Wiltshire)*. 1848.
O'NEIL(L)—
Charles. *Ballymena*. ca.1775-ca.1800. Clock case maker
who left his trade labels in cases.
Charles. *New Haven, Conn, USA*. ca.1820.
Daniel. *Belfast*. 1843-54.
James. *Ballymena*. 1843-90. C.
Thomas. *Durham*. ca.1830-40. C.
William. *Ballymena*. 1858-1900. W.
ONIONS, Peter. *Brosley*. ca.1760. C.
OOSTERHOU(D)T—
Dirk. *Yonkers, NY, USA*. ca.1840.
Peter E. *Kingston, NY, USA*. ca.1800.
OOSTERWIJCK—
Jacobus. *Rotterdam*. ca.1700. C.
Severijn. *Amsterdam*. 1658-88.

ONTARIO. 1914 and later, product of Pequegnat Clock
Co.
OPPENHEIM—
Gabriel & Co. *London*. 1863.
William. *London*. 1857-81.
OPPERMAN, Carl. *London*. 1869-81.
ORAM(S)—
Alexander Frederick. *London*. 1875-81.
George Jarvis (& Son). *London*. 1844 (1851-57).
George John. *London*. 1863-69.
George & Son. *London*. 1875-81.
James. *London*. 1869-75.
Jeremiah. *Daventry*. 1864-77.
Uriah. *Castle Cary*. 1861-75.
ORCHARD—
Henry. *Fowey*. 1844-47.
M. *St. Austell*. 1856.
Paul. *Bristol*. 1766. Serviced ch. clocks.
ORD—
Matthew. *Hexham*. 1848-68. C. & W.
Robert. *Kinross (Scotland)*. 1860.
ORDOYNO—
G. S. *Birmingham*. 1854.
George. *Nottingham*. 1811. C. and math. instr. maker.
May be same as next man.
George, sen. *Nottingham*. 1835.
George Sambrook, jun. *Nottingham*. 1864-76.
ORFORD—
John. *Prescot*. 1819-25. W.
Robert. *London*. 1811. W.
William. *Prescot*. 1819. W.
ORILLA. 1914 and later, product of Pequegnat Clock
Co.
ORKNEY, James. *Quebec*. 1780-1826.
ORME(S)—
Thomas. *Manchester/Oldham*. mar.1794-1828. C.
William. *Glossop*. 1855-76.
ORMOND, J. R. *Peterboro, Canada*. 1826-76.
ORMSBY—
Henry. *Philadelphia, USA*. 1839-ca.1850.
James. *Baltimore, USA*. 1771.
ORNE, R. S. *Boston, USA*. ca.1840.
OROIDE Watches. *Newark, NJ, USA*. ca.1870.
ORPWOOD—
George. *London*. (B. 1810) 1832-51.
R. *Cambridge*. 1846.
Richard. *London*. (B. a.1797-1824) 1828-32.
ORR—
Archibald. *Dumfries*. 1860. C.
James. *Newtownards*. ca.1810-46.
Thomas. *Philadelphia, USA*. 1809-17, to *Louisville,
Ky, USA*, ca.1825.
William. *Bangor*. Late 18c.-early 19c.
William. *Eaglesham (Scotland)*. 1860.
William. *Saltcoats (Scotland)*. 1850-60.
ORRELL, John. *Preston*. 1822-48. W.
ORSON, R. *Melton Mowbray*. 1855.
ORTELLI—
A. (& D.). *Oxford*. ca.1790-1846. C. & W.
Andrew Matthew. *Godalming*. 1839-66.
& PIZZI. *Buckingham*. 1830.
ORTON—
Charles. *Kings Lynn*. 1858-75.
Edward. *London*. b.ca.1665, a.1680 to Samuel Clyatt,
CC. 1687. mar.1691. C.
G. *Epsom*. 1851-55.
John. *Manchester*. 1824.
Joshua. *Sheffield*. 1871.
PRESTON & Co. *Farmington, Conn, USA*. ca.1820-
ca.1830.
William. *London*. 1832.
OSBORN(E)—
Ann & Co. *Birmingham*. wid and successor to Thomas.
1793. Dial makers.
A(nn) & J. *Birmingham*. Successor to Thomas. 1800-
1803. Dial makers.
F. *Ixworth*. 1875-79.
Francis. *Bedford*. 1877.
George. *Bedford*. 1839-69.
George. *Belfast*. (b.1813) d.1831 at *Dundonald*. W.
H. *London*. 1863.
Henry. *Colchester*. b.1812-1865.
Henry. *London*. 1857.

OSBORN(E)—*continued.*
James. *Birmingham.* 1808-13. Dialmaker.
John. *Bedford.* 1877.
John. *Bognor.* 1851-55.
John. *Fenny Stratford* (also *Water Eaton*). b.1753-1807. W.
John. *Lynn, Mass, USA.* 1827. W.
John. *Sheffield.* 1871.
Joseph. *Bridgnorth.* 1875-1900. also *Much Wenlock. 1895-1900. W.*
Joseph. *Lane End.* 1835.
Joseph. *Longton.* 1842-50.
Joseph. *Newtown (Monmouth).* 1874.
Mrs. *London.* 1832.
Robert. *Hamilton, Canada.* 1851-71.
Robert. *Rochester, NY, USA.* 1847.
& Son. *Ixworth.* 1865.
Thomas. *Barton Mills nr Soham (Suffolk).* 1875-79.
Thomas. *Bloxham.* 1830-75.
Thomas. *Bolton.* d.1665. C.
Thomas/Anne. *Birmingham.* 1778-98. Dialmakers.
(Thomas Hadley) & WILSON (James). *Birmingham.* 1772-77. Makers of japanned dials.
W. G. *Leamington.* 1860.
W. R. *Watertown, NY, USA.* 1850.
William. *Warwick.* b. 1814 d. 1873. W.

OSDER, P. *Petworth.* 1851.

OSGOOD(E)—
Charles. *Dunham, Quebec.* 1871.
John. *Andover, Mass, USA.* 1790-95; *Haverhill* 1795-1840. C.
John, jun. *Boston, USA.* 1825-42. C.
Orlando F. *Haverhill, Mass, USA.* mar.1843.

OSMAN, F. *Southampton.* 1848.

OSMOND—
——. *Barnstaple.* ?18c. C.
George Frederick. *Hindon (Wiltshire).* 1875.
T. *Wardour nr Tisbury (Wiltshire).* 1848-67.

OTIS—
F. S. *New York.* ca.1860.
Frederick S. *Forestville, Conn, USA.* Also *Bristol, USA.* 1853-56. C.

OTLEY—
Jonathan. *Keswick.* 1828-34. Also author of 'Guide to the English Lakes'.
John. *London.* 1783. W.

OTT, William. *Edinburgh.* 1850-60.

OTTAWA. 1914 and later, product of Pequegnat Clock Co.

OTTICK, M. *Baltimore, USA.* ca.1790.

OTTO—
A. F. *Chicago, USA.* 1855.
Edward. *London.* 1844-81.

OTTY, Adam. *Halifax.* 1866-71.

OUDIN, Joseph. *Philadelphia, USA.* 1814.

OUDIN-CHARPENTIER. *Paris, London, New York, Moscow, St. Petersburg, Madrid* and *Geneva.* 1862 and later.

OUR(R)Y—
(Auguste) & HEIDLAUF. *St. Imier (Switzerland).* 1881.
Louis. *Paris.* ca.1680.

OUSMAR, O. *West Malling.* 1838.

OUTHWAITE, Thomas & Co. *Liverpool.* 1814.

OVENDEN, E. *Greenwich.* 1855.

OVER—
J. *Dearham (Cumberland).* 1873.
John. *Maryport.* 1879.
Thomas. *Liverpool.* 1848-51.

OVERALL—
& BRADSHAW. *Dover.* 1802.
Francis. *Wellingborough.* 1777. W.
Henry. *Daventry.* 1780.
John. *Wellingborough.* 1765.
Joseph. *Wellingborough.* 1777-84. W.

OVERLAND MAIL. Used by Otay Watch Co. California, USA. 1889.

OVERTON—
Henry. *Coventry.* 1880.
Walter. *Docking.* 1875.
William. *Manchester.* 1848. Imported clocks.

OVES, George. *Lebanon, Pa, USA.* mar.1805.

OVINGHAM—
James. *London.* ca.1780. W.
Samuel. *London.* (B. pre-1776)-1779. W.

OWEN—
——. *Pembroke.* ca.1800. C.
Charles F. & Co. *New York.* ca.1850.
& CLARK. *New York.* 1857.
Cornelius. *Newtown (Monmouth).* 1874-86.
Edward. *Birmingham.* 1803-20. Dialmaker.
Evan. *Dolgellau.* 1828-35.
G. B. & Co. *New York.* 1854.
George. *Winsted, Conn, USA.* 1870s-1880s.
George B. *New York.* 1854-64.
Griffith. *Llanrwst.* 1818-1902. W. & C.
Griffith. *Philadelphia. USA.* 1790-1814. C.
Griffith Ellis. *Barmouth.* 1885-d.1930.
H. *Beaumaris.* Prob. same man as at *Bangor.*
Harry. *Portsmouth.* 1878.
Hugh. *Amlwch.* Prob. same man as at *Bangor.*
Hugh. *Bangor.* from *Llanidloes.* 1874-1933. W. & C.
Hugh. *Llangefni.* Prob. same man as at *Bangor.*
Hugh. *Llanidloes.* s. of Griffith of *Llanrwst.* 1845-74. Later at *Bangor.*
& HULK. *Coventry.* 1854.
Humphrey. *Caernarvon.* 1856-90. C. & W.
Humphrey. *Llanrwst.* 1844. C.
James L. *London.* 1881.
John. *Amlwch.* 1844-74.
John. *Bangor.* May be same man as at *Amlwch.* C.
John. *Conway.* d.1790.
John. *Denbigh.* 1679-1715. Rep'd. ch. clock.
John. *Llanrwst.* ca.1760.
John. *Llanrwst.* ca.1875.
John. *Pwllheli.* 1835-44. C.
John P. *London.* 1869.
M. T. *Abbeville, SC, USA.* ca.1848-60.
(Thomas) & MEYLER. *Haverfordwest.* 1830.
Owen. *Amlwch.* 1835-56.
Owen. *Llangefni.* Prob. same man as at *Amlwch.*
R. *Portsmouth.* 1848-59.
& PRICE. *Birmingham.* 1800. Dialmakers.
(S.) & READ. *Cincinnati, O, USA.* 1840s.
Richard. *Holyhead.* 1856.
Richard. *Llannerch-y-Medd.* 1869.
Robert. *Caernarvon.* 1895.
Shadrach. *Amlwch.* 1868. Later *Ffestiniog* 1886-90.
Shadrach. *Llannerch-y-Medd.* 1866. C.
& SILE. *Chester, Pa, USA.* 1800s.
Thomas. *Ffestiniog.* 1874.
Thomas. *Haverfordwest.* ca.1810-60. C. & W.
Thomas. *Weymouth.* 1824.
W. *Coventry.* 1860-80.
Watkin. *Llanrwst.* 1702.
William. *Edern (Wales).* Believed b.1751-d.1824. C.
William. *Egham.* 1878.
William. *Holyhead.* 1840-44.
William. *Llanrwst.* 1810-35. W.
William. b. ca. 1811 *Llanwchaiarn (Montgomery).* to *Oswestry* by 1841-67. Succ. by sons William & Thomas qv.
William. *Trawsfynydd.* 1832.
William B. *Coventry.* 1860-80.
William John. *Dolgellau.* 1866-90.
William & Thomas. *Oswestry.* 1870.

OWENS—
——. *Enniskillen.* Early 19c. C.
Daniel. *Caernarvon.* 1844-56. C. & W.
Daniel. *Cricieth.* 1868-74. W.
Edwin. *Wrexham.* 1887-90.
John. *London.* 1863-75.
Lewis. *Holyhead.* 1874.
Owen. *Liverpool.* 1851.
Owen. *Llannerch-y-Medd.* 1835-44. C.
Owen. *Nebo (Wales).* Prob. same man as at *Llannerch.*
Owen. *Pwllheli.* Prob. same man as at *Llannerch.*
Richard. *Llannerch-y-medd.* 1844-74.
William. *Doagh, co. Antrim.* Retired 1789. W.
William. *Harlech.* 1840.
William. *Utica, NY, USA.* 1839-50.

OWNS, Edward. *Wymeswold (Leicestershire).* ca.1750-ca. 1800. d. 1816 aged 83. Also signed 'E. O. Wymeswold'.

OWSTON—
John. *Keighley*. 1853-71.
Michael. *Scarborough*. 1823-34, then *Bridlington* 1840-51.
OWTTRIM, Thomas. *London*. 1857-75.
OXBROW, William. *Braintree*. 1851-74.

OXFORD. 1914 and later, product of Pequegnat Clock Co.
OXLEY—
Joseph. *Fakenham*. Late 18c.
William. *Worksop*. 1788. C. & W.
OYSTER, Daniel. *Reading, Pa, USA*. 1764-1845. C.

P

P, I, IP, is supposed monogram of John Powley of *Asby* (*Westmorland*).
PACE—
C. *San Francisco, Cal, USA.* ca.1850.
Charles. *London.* 1839-51 (successor to Henry).
Edmund. *London.* 1839-44.
H. *Loughton* (*Essex*). 1855.
Henry. *Brampton, Canada.* 1862.
Henry. *London.* 1832-63. C.
Henry. *Ottawa.* 1875.
Hugh. *Newry.* 1865-92. W.
John. *Bury St. Edmunds.* 1823-55. C. & W.
Thomas. *Chelmsford.* ca.1800.
Thomas. *London.* 1828.
PACHL, Andreas. *Innsbruck.* 1581. Bells for clocks.
PACKARD—
(Jonathan) & BROWN. *Albany, NY, USA.* 1815.
Isaac. *Brockton, Mass, USA.* Early 19c.
Jonathan. *Albany, NY, USA.* 1815. Later *Rochester, NY.* 1818.
(Jonathan) & SCHOFIELD. *Rochester, NY, USA.* 1818.
PACKEM, Benjamin. *Steyning.* 1723. C.
PACKER—
John. *Newbury.* 1837-77.
John. *Tingewick.* 1776-1825. May be same man as at *Buckingham.*
Richard. *Tingewick.* 1847.
W. *Tingewick.* 1854.
William. *Buckingham.* (?1793) 1823-69.
William. *London.* 1839.
William. *Tingewick.* 1846.
PACKMAN, William. *London.* 1839.
PACKWOOD, David. *Coventry.* 1842-54.
PADBURY—
A. *Hemel Hempstead.* 1851.
Andrew. *Epsom.* 1828-39.
C. *Lymington.* 1859-67.
J. *Bishops Waltham.* 1848-67.
J. *Gloucester.* 1856.
J. *Landport.* 1859.
J. *Minchinhampton.* 1840.
James. *Bishops Waltham.* 1878.
John. *Bishops Waltham.* 1830-39.
John. *Cirencester.* 1830.
Matthias. *Burford.* b.1751-85. C.
Mrs. C. *Lymington.* 1878.
Theo. F. *Landport.* 1867-78.
Thomas. *Chertsey.* 1828-66.
PADDON—
——. *Modbury* (*Devon*). ca.1790.
George. *Kingsbridge* (*Devon*). a.1737-d.1771. C.
PADGETT, Joseph Henry. *London.* 1881.
PADGHAM—
David. *Northam* (*Sussex*). 1862-78.
John. *Bethersden* (*Kent*). 1855.
PAGE—
Brothers (Richard & Clement). *St. John, New Brunswick.* 1850-58.
Charles. *Leighton Buzzard.* 1830-47.

PAGE—*continued.*
Charles. *London.* 1857.
David. From *Haverhill, Mass, USA* to *Truro, Nova Scotia.* 1794-1840.
Edward. *Oxford.* a.1684. W.
George. *Donnington* (*Leicestershire*). 1876.
George. *Grantham.* 1849-76.
George. *London.* 1755. W.
Henry. *Margate.* 1839.
J. *Moulton* (*Northamptonshire*). 1854.
James. *Dartford.* 1839-47.
James. *London.* 1869.
John. *Brighton.* 1839.
John, sen. *Market Rasen.* 1835-49.
John, jun. *Market Rasen.* 1849-76.
John Austin. *Caistor.* 1876.
John George. *London.* 1881.
KEEN & PAGE. *Plymouth.* ca.1860. W.
Mrs. Charlotte. *London.* 1875 (?wid. of James).
Richard & Co. *London.* 1881.
Richard R. *St. John, New Brunswick.* 1858-64.
Robert. *Harwich.* d.1759. W.
Robert. *Southwold.* 1830.
Robert. *Wivenhoe.* ca.1720. C.
Robert. *Yarmouth.* (B. mar.1817) 1836.
S. *South Heigham.* 1875.
Thomas. *Bishop's Stortford.* ca.1750. C.
Thomas. *Dartford.* 1838-47.
Thomas. *Manningtree.* 1874.
W. *Clare.* 1853.
William. *London.* 1875-81.
William Robert (& Co.). *London.* 1857-63(-69).
PAG(G)ET(T)—
A. *Cheltenham.* 1879.
James. *Reading.* 1830.
John. *Halifax, Nova Scotia.* 1770-1783.
Thomas. *Cheltenham.* 1870.
W. *Coventry.* 1854.
PAIGE, Walter. *Tunbridge Wells.* 1874.
PAILLARD, VAUCHER FILS (& Co.). *London* and *Switzerland.* 1875(-81).
PAILLE—
Charles. *St. Hyacinthe, Canada.* 1857.
Eli. *Galt, Canada.* 1853.
Eli. *St. Hyacinthe, Canada.* 1862.
Paul. *Galt, Canada.* 1857.
PAILLOT, Leon. *Baltimore, USA.* 1842.
PAILTHORP—
——. *Kirkby Lonsdale.* ?mid-19c. C.
Charles. *Louth.* 1861-76.
Daniel. *Grimsby.* 1861-76.
PAIN(E)—
Benjamin. *Littlethorpe* (*Leicestershire*). ca.1720. C. (One such a.1672 in *London.* May be same man.)
Charles. *London.* 1869-81.
G. *Dover.* 1855.
George, jun. *Stow-on-the-Wold.* 1830-50.
& GRANT. *Windsor.* 1837.
& HEROY. *Albany, NY, USA.* 1813.

178

PAIN(E)—*continued.*
J. P. *London.* 1828.
John. *Banbury.* mar.1824-55.
John P. *London.* 1832-44. Also T.C.
Joshua. *Shoreham.* 1828.
Mark. *Ash (Kent).* 1826-47.
Mark. *Sandwich.* 1845-62.
Ralph. *London.* 1869-75.
Samuel. *Birmingham.* 1868.
Stephen. *Sandwich.* 1858.
Thomas. *London.* 1869-81.
Thomas Richard. *London.* 1881.
PAINTER—
John. *Tingewick.* b.1776—mar.1801. W.
W. *Landport.* 1867.
PALETHORP(E) (see also **PAILTHORPE)**—
William. *Grantham.* 1835.
William. *Liverpool.* 1834.
PALFREY, John. *London.* a.1646 to Richard Scrivener,
    CC. 1654-d.1670. C.
PALIN—
J. *Nantwich.* 1857-78.
William. *Nantwich.* 1828-34.
PALLANT—
Charles. *London.* 1869-75.
Charles Nathan. *London.* 1881.
PALLANTYNE, William. *London.* 1828.
PALLETT, J. *Coventry.* 1860.
PALLISER—
John. *Hull.* 1838-51.
John. *Thirsk.* ?1798-1823. C.
Robert. *Thirsk.* 1834-44.
PALLWEBER, Josef. *Salzburg.* 1885-91.
PALMEN, F. *Great Malvern.* 1872.
PALMER—
Adam. *Haddenham (Cambridgeshire).* 1875.
Bailey. *Madeley.* 1870.
Benjamin. *Bromyard.* 1870-79.
Charles. *Bristol.* a.1810-42.
Charles. *Newport (Monmouth).* 1852.
(Beriah) & CLAPP. *New York.* 1831.
& Co. *Boston, USA.* Early 19c.
D. D. *North Bridgewater, NY, USA* 1838-58; *West
    Winfield, NY,* 1858-64; *Waltham, Mass,* 1864-75.
Edwin. *Banwell (Somerset).* 1883.
H. *London.* 1869.
H. W. *Margate.* pre-1813. W. (Not pre-1713 as in B.)
Harriet. *Haverhill.* 1839.
Henry. *Birmingham.* 1860-68. Also clock cases.
Henry. *Birmingham.* 1880. Importers.
Henry (Harry). *Walton (Felixstowe).* mar. 1816,
    then *Haverhill* 1823-44.
Henry Phillips. *Leominster.* 1870-79.
Horatio. *Monmouth.* 1850-52.
J. *Ludlow.* 1856.
John. *Birmingham.* 1868-80.
John. *Bristol.* ca.1830. C.
John. *Lincoln.* 1876.
John. *London.* (B. a.1796) 1828-32.
John. *London.* 1828. Watch cases.
John C. *Oxford* and *Haywood* and *Salisbury* and
    Raleigh, NC, USA. (1806)-1830-89(-93).
John Henry. *London.* 1881.
Joseph. *Collingwood, Canada.* 1857-63.
Joseph Gunn. *London.* (B. a.1801-CC. 1809-25) 1828.
& PRATT. *Worcester.* 1850.
R. *Axbridge.* 1861.
R. B. *Bath.* 1856.
(John C.) & RAMSAY. *Raleigh, NC, USA.* 1847-55
    (later Palmer alone).
Robert. *Cambridge.* 1875.
Robert. *Cleator Moor (Cumberland).* 1869-73.
Robert (jun.?). *London.* 1828-51.
Robert. *London.* 1869.
Robert. *Ramsgate.* 1874.
Robert. *Yatton (Somerset).* 1875.
S. *Grimsby.* 1868.
Thomas. *Coventry.* 1850-54.
Thomas. *Croydon.* 1866-78.
Thomas. *Nottingham.* 1876.
Thomas. *Shefford (Bedfordshire).* ca.1690. C.
Thomas. *Woolwich.* 1845-47.
W. *Clare.* 1846.

PALMER—*continued.*
W. *Willenhall.* 1868.
W. *Wolverhampton.* 1876.
W. & Co. *Boston* and *Roxbury, Mass, USA.* 1857.
W. E. *Worcester.* 1860-76.
(D. D.) Watch Co. *Waltham, Mass, USA.* ca.1864-75.
William. *Clare.* 1830-39. Also silversmith.
William. *Coventry.* 1880.
William. *Droitwich.* 1842.
William. *Longtown.* 1869-73.
William. *Loughborough.* ca.1815. C.
William. *Milford Haven.* 1871-75.
William. *Newbury.* 1837.
William. *New York.* ca.1802-1818. Clock case maker.
William. *Westerham.* 1826.
William. *Worcester.* 1828.
William Ball. *Bristol.* 1850-79.
William Edward. *Droitwich.* 1828.
William Edward. *Worcester.* 1828-42.
PANCHEN, E. *Yarmouth.* 1865.
PANKHURST—
A. *Maidstone.* 1866.
& MUNN. *Maidstone.* 1851.
PANNELL—
Hugh. *Stokesley.* b.1721-ca.1750, then *Northallerton*
    where d.1788. C.
Joshua. *Northallerton.* s. of Hugh. b.1757-d.1803. C.
PANTER, William. *Choseley.* 18c. C.
PANTHENON. 1914 and later, product of Pequegnat
    Clock Co.
PANTHEON & PANTHEON 'B'. 1914 and later,
    product of Pequegnat Clock Co.
PANTON—
James & Co. *Glasgow.* 1860. Clock importers.
Nicholas. *London.* a.1651 to *Humfrey Downing.* At
    Oxford 1663.
PAPE—
John. *Blyth (Northumberland).* 1827-58. C.
Mrs. Elizabeth. *London.* 1857-81. Clock case maker
    (?wid. of William).
Thomas. *Rothbury.* (B. 1808) 1827-34.
William. *London.* 1828-51. Clock case maker.
PAPEN, L. Place unknown. 1574. W.
PARDELLIAN, J. B. *Montreal.* 1853.
PARE—
& DURAND. *Montreal.* 1875.
Thomas. *London.* ca.1700. C. (May be Parre?)
PARIS(?), Thomas. *Warwick.* ca.1740. C.
PARK(E)—
Brothers. *Cornwall, Canada.* 1862.
George. *Fraserburgh.* 1836.
Halsey. *Brantford, Canada.* 1892.
Henry. *Barnsley.* 1871.
James. *Kilmalcolm.* ca.1760. C. (B. 1802.)
James. *Preston.* 1822-58. Also silversmith.
John. *Inverurie.* 1837.
Joseph. *Liverpool.* 1734-d.1766. C.
Robert. *Greenock.* 1860.
W. J. *Cornwall, Canada.* 1851.
William. *Treorchy.* 1875.
PARKER—
Benjamin. *Bury St. Edmunds.* 1834-51. C.
E. *Knutsford.* 1878.
Eustace. *Altrincham.* 1878.
Francis. *Knaresborough.* ca.1750-retired 1773. C.
George. *Daventry.* 1841-54.
George. *Ithaca, NY, USA.* 1832. Later at *Utica.*
George. *Leeds.* 1853-71.
George. *Tadcaster.* 1822-26.
George. b.1800 *Dalton (Lancashire).* *Ulverston.*
    ca.1824-58. W. Also jeweller and engraver.
Henry. *London.* 1881.
Isaac. *Deerfield, Mass, USA.* 1780.
Isaac. *Philadelphia, USA.* 1818-50.
James. *Heptonstall.* 1838.
James. *Hull.* 1838. W.
James D. *Retford.* 1828-64.
John. *Lindfield (Sussex).* (B. 1813-) 1839-78.
John. *Liverpool.* 1763-d.1795. W.
John. *London.* 1857-69.
John. *Maryland, USA.* 1774. Watch movements.
John. *Prescot.* mar.1781. W.

PARKER—*continued.*
John Robert. *New Walsingham (Norfolk).* 1846-75.
Joseph. *Princeton, NJ, USA.* 1785.
Joseph. *Pudsey* nr *Leeds.* 1853-66. C.
M. *Coventry.* 1860-68.
Mary. *Canterbury.* 1782.
Richard. *Bailieborough, co. Cavan.* 1868-80.
Richard. *Hereford.* 1879.
Robert. *Fakenham.* 1865-75.
Samuel Timothy. *Altrincham.* 1865.
T. H. *Philadelphia, USA.* 1833.
Thomas. *Hull.* b.1814-pre-1842, then *Bradford* 1842-1844, then *Otley* 1844-71. C.
Thomas. *Leeds.* 1871.
Thomas. *London.* 1863-75.
Thomas. *Philadelphia, USA.* 1761-1833.
Thomas S. *London.* 1863-81.
W. *Stow-on-the-Wold.* 1863.
William. *Coventry.* 1828.
William, jun. *Philadelphia, USA.* 1835.
PARKHURST, Alexander. *Maidstone.* 1858.
PARKINS—
Bartholomew. *Manchester.* mar.1792. C.
& GOTTO. *London.* 1881.
Joseph. *Philadelphia, USA.* 1837.
PARKINSON—
& BOUTS. *London.* 1851-81.
& FRODSHAM. *Liverpool.* 1834. Also chrons.
& FRODSHAM. *London.* 1828-present day. ('Est'd. 1801?)
Henry. *London.* 1832-44.
J. *Bacup.* 1858. W.
James. *London.* (B. 1820) 1844.
John. *Settle.* 1834-53.
Joseph. *West Derby (Lancashire).* 1818. W.
Nathaniel. *Kendal.* ca.1825-pre-1837.
Nathaniel. *Preston.* 1848-51.
Robert. *Lancaster.* Free 1732, mar.1739-d.1760. C.
Roger. *Richmond (Yorkshire).* 1736. C. (One such recorded at *Edinburgh* 1745-61.)
Thomas. *Bury.* 1822-34.
PARK(E)S—
G. D. *Cincinnati, O, USA.* 1850s. C.
Henry James. *Birmingham.* 1880.
Hiram. *St. Catharines, Canada.* 1877.
Hugh. *New York.* ca.1840.
J. *Wolverhampton.* 1876.
John. *Prescot.* 1819-24. W.
Jonas. *Bennington, Vt, USA.* ca.1770. C.
T. H. *Rotherfield (Sussex).* 1862.
PARKYN, John. *Bristol.* 1879.
PARLIN, A. S. *Norwich, Conn, USA.* ca.1860.
PARMELE—
Abel. *Branford, Conn, USA.* 1703-ca.1766. C.
Ebenezer. *Guildford, Conn, USA.* 1690-1777. Probably the first Connecticut clockmaker.
John Peter. *Philadelphia, USA.* 1793.
PARNELL—
Frederick Charles. *London.* 1869-81. (Successor to Thomas & Henry.)
Thomas. *Canterbury.* 1784-1802.
Thomas. *London.* (B. 1815-24) 1832-51.
Thomas & Henry (Brothers). *London.* 1857 (-1863).
PARR(E)—
Benjamin. *Grantham.* 1828.
Benjamin. *Halifax.* 1786 -d. 1805. C.
David. *London.* 1662. (Prob. Parry, q.v.)
Edward (& Son). *London.* 1857(-63-69).
Edward (?jun.). *London.* 1875-81.
Henry. *Cavan.* 1790.
Henry. *Leeds.* 1817. W.
Henry. *Prescot.* 1823-25. W.
Henry. *Spilsby.* 1828.
John. *Farnworth.* mar.1828. W.
John. *Liverpool.* (B. 1805-) 1814-34.
John. *Prescot.* mar.1768.
Peter. *Liverpool.* 1834 (B. 1816).
R. S. *Nottingham.* 1849-55.
Robert. *Prescot.* 1795. C.
& Son. *London.* ca.1820. *Cavendish Street.*
Thomas. *Cavan.* 1785.
Thomas. *Liverpool.* ca.1760. C.

PARR(E)—*continued.*
Thomas. *London.* 1662.
W. *Manchester.* mar.1798. C.
William. *Farnworth.* mar.1795-98. W.
William. *Liverpool.* 1848.
William. *Prescot.* mar.1779. W.
William. *St. Helens.* mar.1784-85. C.
PARROT(T)—
Frederick (F. W.). *Philadelphia, USA.* 1847.
Henry. *London.* 1832.
Joseph. *Philadelphia, USA.* 1835-43.
Samuel. *Grassrigg* nr *Killington (Westmorland).* ?18c. C.
William. *London.* 1839.
PARRY—
Daniel Henry. *Nottingham.* 1876.
David. *Caernarvon.* 1895.
David. *London.* CC. 1646-62.
David. *Raglan (Monmouth.).* b.1765-d.1847. C.
Edmund. *Llandeilo.* 1848-87. W. & C.
Edward. *Tregaron.* 1831.
Evan. *Llandeilo.* 1844. (?Error for Edmund.)
& GRIFFITHS. *Pwllheli.* 1844.
Henry. *Pwllheli.* mid-19c. C.
John. *Lampeter.* 1831-55. C. & W.
John. *Pencarreg.* 1831-55. C. & W.
John. *Ruthin.* 1835-44. C. & W.
John. *Tremadoc.* 1835-40. C. & W.
John. *Trenton, NJ, USA.* 1788-1814.
John. *Wrexham.* 1856-68.
John F. *Philadelphia, USA.* 1824.
John J. *Philadelphia, USA.* 1793-1835.
Owen. *Barmouth.* 1886-90.
Owen. *Pwllheli.* 1886-90.
Robert. *Pwllheli.* 1856-90. C.
T. *Stoke-on-Trent.* 1868-76.
Thomas. *Llandeilo.* 1835-44.
W. *Pontypool.* 1849-52.
William. *Aberdare.* 1848.
William. *Holywell.* 1856.
William. *Newport (Monmouth).* 1858-80.
William. *Pontypool.* 1835-58.
William J. *Liverpool.* 1848-51.
PARSONS—
A. A. *Harwich.* 1851-74.
Allen. From *Cuckfield* to *London,* a.1716 to John Wright of *St. James Westminster.* W.
Frederick. *St. Catharines, Canada.* 1851-62.
Garsholm. *London.* ca.1735. C.
George. *Bristol.* a.1829-38. C.
George F. *Bristol.* 1868-79. (May be same as above.)
George Frederick. *London.* 1857-81.
Henry Edmund. *Eastbourne.* 1870.
Henry R. *Philadelphia, USA.* 1840-49.
John. *Gosport.* 1878.
John. *London.* 1869.
Joseph. *Chipping Norton.* 1853.
Joseph Edgecumbe. *Bristol.* 1867.
Robert. *Lisburn.* 1819.
Thomas. *London.* 18c. C.
W. T. *Ottawa.* 1875.
William T. *Toronto.* 1874-76.
& WILLIAMS. *Bristol.* 1842.
PARTIN, William. *Sunderland.* 1827.
PARTINGTON, William. *London.* (B. 1815-24) 1828-44.
PARTRIDGE—
Christopher. *London.* 1832-57.
Horace. *Bristol* and *Boston, USA.* 1868.
Thomas. *Kibworth (Leicestershire).* 1795. C.
William (Captain). *London.* CC. 1640-60. Captain in Royal Life Guard for whom he raised a troop at his own cost. Was denied post of Royal Clockmaker in 1660 in favour of Edward East.
PASCAL, Claude. *The Hague.* ca.1654-d.ca.1670. W.
PASCOE—
Charles. *Camelford.* 1847.
Charles. *Liskeard.* 1856.
Hodgson. *Helston.* a.1829. Later *Penzance* 1835-73.
J. *Truro.* 1815.
PASS—
Clement. *Oldbury.* 1860-76.

PASS—*continued.*
Clement. *Smethwick.* 1850.
& STOW. *Philadelphia, USA.* Late 18c. bells and clock parts.
William H. *Smethwick.* 1876.
William Henry. *Birmingham.* 1880.
PATAY, ——. *Paris.* 1860-62.
PATCHIN, T. A. *Syracuse, NY, USA.* 1846.
PATCHING, H. *Shoreham.* 1855.
'PATEK'. *London.* 1875-81 (Holloway & Co., agents).
PATEK, Phillipe & Co. *London.* 1857-81 and later. (James Xavier Lutiger, agent.)
PATERSON (see also **PATTERSON**)—
George (& Co.) *Cotton (Aberdeenshire).* 1836(-1860).
James. *Perth.* (?ca.1820)-1841-d.1852. Japanner and dial maker.
John. *Airdrie.* 1860.
John. *Coatbridge.* 1860.
John. *London.* 1851.
R. *Lindfield.* 1851.
Walter. ?*Edinburgh.* pre-1744. Sundial.
William. *Largs.* 1860.
William H. *St. John, New Brunswick.* 1857-77.
PATES, Thomas. *Northallerton.* 1834. C.
PATIENT, William (& Co.) *London.* 1857-75.
PATON, Alexander. *Kirriemuir.* 1860.
PATRICK—
George L. *Strood.* 1866-74.
John. *Maidstone.* 1838-47.
John. *Strood.* 1851-55.
Mary Ann. *Greenwich.* 1847.
Mrs. L. *Strood.* 1866.
William. *Greenwich.* 1827-39. W.
PATTE, Jaques. *London.* From *Geneva* 1655.
PATTEN—
(Richard) & FERRIS. *New York.* ca.1820.
James. *St. Helens.* mar.1765-80. W.
Richard. *New York.* ca.1820.
PATTERSON—
A. *Omagh.* ca.1830-40. C.
George. *London.* 1828-39.
James. *Banff.* Early 19c. C.
James. *Liverpool.* 1848.
James. *Omagh.* 1865-68.
James. *Perth.* ca.1820-d.1852. Dial maker.
M. *Comber (co. Down).* 1816. W.
Mrs. *Omagh.* 1854-58. (Also 1880-90).
Robert. *Liverpool.* 1851
Thomas. *Newtownards.* b.1819-d.1840.
Thomas. *Toronto.* 1859.
(Robert) & WHITTLE (James). *Belfast.* 1789.
William. *London.* 1844. Clock case maker.
PATTINSON—
John. *Burnley.* Error for John Battinson.
Thomas. *Winton (Westmorland).* Later 18c.-early 19c. C.
PATTISON—
Alexander. *Coldstream.* 1860.
Alexander. *Stockton.* 1834.
John. *Halifax.* ca.1760-ca.1800. C. (One-time partner with Burton, q.v.)
PATTON—
——. *Castlewellan, co. Down.* Early 19c. C.
Abraham. *Philadelphia, USA.* 1799-1819.
David. *Philadelphia, USA.* 1799.
James. *Clones (co. Monaghan).* ca.1800-1824. C.
James. *Newtownards.* 1854-83. W.
(Abraham) & JONES (Samuel G.). *Philadelphia* and *Baltimore, USA.* 1804-14. (May be white dial makers).
& PARSONS. *Quebec.* 1871.
PATTRICK, T. M. & Son. *Wisbech.* 1865.
PAUL—
——. *London.* ca.1710. C.
Alexander. *Trinaltinagh (co. Londonderry).* ca.1830.
James. *Gravesend.* 1847.
James. *Strood.* 1847.
Mary. *London.* 1832. (wid. of Philip?).
Nowell. *London.* 1668. Alien.
Philip. *London.* (B. 1802-25). 1828.
Thomas. *Dunoon.* 1860.
Thomas. *Glasgow.* 1837-60.

PAUL—*continued.*
Thomas. *London.* a.1663 to John Fromanteel. CC. 1670. C.
PAULET, Luc Seraphin. *Levis* nr *Quebec.* 1896.
PAULUSS, Hans. *Altstadt in Prague.* 1587. C.
PAVELY, James Duval. *London.* 1863-81.
PAVEY, John. *Philadelphia, USA.* 1803-20.
PAVIN, Jean. *Paris.* 1673.
PAVIOUR, W. *Peterborough.* 1864.
PAWSON, John. *North Collingham (Nottinghamshire).* 1876.
PAXTON—
John. 1808. a. of Davidson of *Dunse.* At *Kelso.* 1837.
John. *St. Neots.* (B. 1784-95). 1830-47. W.
Joseph. *St. Neots.* 1839.
William. *St. Neots.* 1847-64. C.
PAY, James. *Hythe.* W.
PAYNARD, James. *Strood.* 1808. W.
PAYNE—
Ann. *Hadleigh.* Wife of William, whose business she cont'd. 1814-44.
C. (& R.). *Abingdon.* 1837-64.
Charles. *Oakham.* 1864-76.
Frederick. *London.* 1857-63.
George. *Ludlow.* b. ca. 1800 -d. 1868. C. & W.
George. *Hadleigh.* 1846-90. W.
George. *Abingdon.* 1877.
J. *Grantham.* 1868.
James. *London.* (B. 1794-1825). 1828-32.
John. *Hadleigh.* s. of William of *Hadleigh.* a.1795-1820.
John. *Maldon.* a.1789-1820.
John. *Melbourne.* 1876.
John. *Wallingford.* 1830.
John & Co. *Abingdon.* 1830.
Jonathan. *Bristol.* s. of William of *Hadleigh.* a.1805-1812. W.
Lawrence. *New York.* 1732-55.
Richard. *Jersey.* 19c. C.
Robert P. *Hull.* 1834-58. C.
& Son. *Bath.* 1800. W.
& Son. *Oxford.* 1880 and later.
& Son. *Wallingford.* 1837.
T. E. *Tunbridge Wells.* 1874.
Theophilus C. *Leicester.* 1876.
Thomas. *Baldock.* 1828.
Thomas. *Kenilworth.* 1868-80.
Thomas. *London.* ca.1750. C. Prob. same as Thomas Paine. a.1709.
W. *Wallingford.* 1864-77.
William. *Colchester.* b. 1749. a. 1763 to Nathaniel Hedge, free 1770, then *Hadleigh* where ret'd due to illness 1814. C.
William (& Co.). *London.* 1828-51. (1857-81). Also pedometers.
William. *Ludlow.* 1828-50.
William. *Wymondham.* 1849-76.
William Petty. *Banbury.* b.1818-d.1897.
Zacharias. *Colchester.* b.1799-d.1839. W.
PEABODY—
Asa. *Wilmington, ND, USA.* 1821.
John. *Fayetteville, NC, USA.* 1823-25.
John. *Wigston Magna (Leicestershire).* 1876. C.
M. M. *Kingston, Canada.* 1851-63.
PEACH—
H. *Bridport.* 1848-55.
Henry. *Beaminster.* 1824-48.
PEACOCK—
George. *Bristol.* 1879.
George. *London.* 1863-81.
H. *Montreal.* 1847-65.
Henry Arthur. *Swansea.* 1887-99. C.
J. *Huntingdon.* 1854-64.
J. T. *Montreal.* 1857-63.
John. *Huntingdon.* 1830-39.
John. *Lincoln.* (B. 1791-95)-ca.1800. C. & W.
John. *Montreal.* 1842-90.
John (& STEVENS, Henry). Place unknown (*Lancashire*?). ca.1680. Lant. clock.
Joshua. *Macclesfield.* 1828-34.
Robert. *Spalding.* 1828.
Samuel. *Kimbolton.* 1830-39.
Thomas. *Highworth.* 1830.
W. *Kimbolton.* 1854-64.

PEACOCK—*continued.*
W., jun. *Middlesbrough.* 1898.
William. *Banbury.* mar.1788-1823.
William. *Chatteris.* 1830.
PEAFF, C. *Brighton.* 1855.
PEAK(E)—
Thomas. *London.* 1857-81.
William. *Workington.* 1869.
William Henry. *Codnor.* 1876.
PEALE, James. *Philadelphia, USA.* 1814-17.
PEAR—
R. S. *Nottingham.* 1849.
W. *Spalding.* 1868.
PEARCE—
Adam. *London.* a.1657 to Edward East. CC. 1664.
Benjamin. *Machynlleth.* s. of George. 1887-90.
Edward. *Cardiff.* 1887-99.
Edward. *Launceston.* 1844-73.
Edwin Roper Johnstone. *London.* 1881.
F. *London, Newington Causway.* Later 18c. C.
George. *Buckley (Wales).* 1868. Later *Machynlleth.*
1874.
George. *Liverpool.* 1851.
George. *Nottingham.* 1876.
Henry. *Grantham.* 1850-76.
Henry. *Huddersfield.* 1871. C.
James. *Redditch.* 1828-35.
John. *Chard.* ca.1790.
John. *Crewkerne.* 1861-75.
John. *London.* 1839-44.
John. *St. Agnes.* 1844. W.
John. *St. Austell.* mar.1850-56.
John. *Stratford-on-Avon.* 1828-54.
Mrs. Maria. *London.* 1875. (?wid. of William).
Thomas. *Bourne (Lincolnshire).* 1850-76.
W. *Evesham* and *Stratford-on-Avon.* 1868.
W. *Derby.* 1855.
W. *Nottingham.* 1855.
William. *Charleston, SC, USA.* post-1780.
William. *London.* 1851-69.
William. *Plymouth.* mar.1781. (B. 1795).
William. *Stratford-on-Avon.* 1860.
William A. *Ellesmere.* 1879.
William Callaway. *Bristol.* 1870-79.
William George. *London.* 1881.
William Philip. *Saltash.* 1873.
PEARCEY, S. *London.* Late 18c. C.
PEARCY, Richard. *London.* 1869-81.
PEARDON, Richard. *London.* 1857.
PEARE, Charles. *Uxbridge, Canada.* 1871.
PEARL, Howard. *Sherbrooke, Canada.* 1862.
PEARMAN—
E. *Hemel Hempstead.* 1851-74.
William. *Richmond, Va, USA.* ca.1834. (One such *London,* a.1823).
PEARS—
George. *Coventry.* 1880.
H. *Leamington.* 1880.
PEARSALL—
& EMBREE. *New York.* ca.1786. C.
Joseph. *New York.* 1773-75. Imported English clocks.
Joseph & James. *New York.* ca.1770-73.
Joseph & Thomas. *New York.* 1770-dissolved 1773.
Thomas. *New York.* 1773.
PEARSE—
A. *Bruton (Somerset).* ca.1800. C.
George. *Dunster.* 1883.
Henry. *Wiveliscombe (Somerset).* 1875.
Isaac T. *Enfield, Conn, USA.* ca.1830. C.
John. *Littlehampton.* 1878.
Thomas. *Cirencester.* 1840-56.
W. *Wellington.* 1861.
PEARSON—
——. *Dore.* ca.1740. C.
Ananiah. *Bedale.* 1834-40.
Charles. *Towcester.* 1841-54. C.
Edmund. *Buckingham.* 1830-42.
Edmund. *Stony Stratford.* 1830.
Edward. *London.* 1857-63.
& GREY. *Georgetown, SC, USA.* 1768.
Hawtin. *Oxford.* s. of Richard. 1825-53.
Henry. *Brackley.* 1830.
Henry. *Oxford.* s. of Richard. mar.1827-62. W.

PEARSON—*continued.*
Henry Andrew. *London.* 1875-81.
(Isaac) & HOLLINSHEAD (Joseph). *Burlington, NJ, USA.* 1750-60.
Isaac. *NJ, USA.* ca.1685, mar.1710-49. C.
James. *Worcester.* 1828.
James Molesworth. *Bridgnorth.* 1835-56. Also gilder and dentist.
John. *Buckingham.* 1823.
John. *Louth.* 1828-76.
John. *Towcester.* 1830.
John Leigh. *London.* 1881.
Ralph. *Scarborough.* 1851.
Richard. *Oxford.* a.1785-1823.
Samuel. *Halifax.* 1790-1822. C.
T. *Ballyclare.* 1832. W.
T. *Earlsdon.* 1860.
Thomas. *Berwick-on-Tweed.* (B. 1795) 1820. W.
Thomas, jun. *Coventry.* 1880.
Thomas. *Lowestoft.* 1875-79.
Thomas & Son. *Earlsdon.* 1880.
W., jun. *Nottingham.* 1864.
William. *Berwick-on-Tweed.* 1843-60. C.
William. *Blackburn.* 1822-34.
William. *Maidstone.* 1826-38.
PEASE—
Isaac T. *Enfield.* 1833-37. C.
John William. *St. John, New Brunswick.* 1857.
PEASELEY, Robert. *Boston, USA.* 1735.
PEAT(T)—
David. *Crieff.* 1844. W.
John. *Darlington.* ca.1800. C.
PEATLING—
John. *Bourne.* 1828.
William. *Market Deeping.* 1828.
PEATY—
Alfred. *Cerne Abbas.* 1875.
D. *Frome.* 1861.
D. *Swanage.* 1867.
T. *Cerne Abbas.* 1867.
PEBERBY, John. *Leicester.* ca.1805-ca.1813. C. & W.
PEBODY, J. *Welford.* 1869.
PECHELOCHE, V. *Paris.* 1851. C.
PECK—
Ambrose. *Bristol, USA.* 1864.
Benjamin. *Providence, RI, USA.* 1824.
& Co. (Julius). *Litchfield, Conn, USA.* ca.1830.
Edson. *Derby, Conn, USA.* 1827.
Epaphroditus. *South Carolina, USA.* 1811-57. To *England* with first shipment of Jerome clocks. Died *London.*
HAYDON & CO. *St. Louis, Mo, USA.* ca.1840. C.
& HOLCOMB. *Sanbornton, NH, USA.* ca.1820.
& JEROME. *Liverpool (England).* 1843. C.
John. *Wellingborough.* (B. 1795). 1830-47.
John (II?). *Wellingborough.* 1864-77.
Moses. *Boston, USA.* 1753-89. Imported London watches.
Timothy. *Middletown, Conn, USA.* 1765-90. *Litchfield.* 1790-1818.
W. *Rushden.* 1864.
William. *Sharnbrook.* 1839-54.
PECKETT, John. *London.* a.1683 to Henry Merryman. CC. 1691. (B. -1712).
PECKHAM—
John. *Lewes.* a.1733 to Thomas Barratt. C.
& KNOWER. *Albany, NY, USA.* 1814. C.
PEDAN, Samuel. *Newtownards.* 1890-99. W.
PEDDUCK, Henry. *Hanley.* See Pidduck.
PEDLER, Joseph. *Withiel (Cornwall).* b.ca.1739-ca.1770. C.
PEDRONE, Louis. *Liverpool.* 1848-51.
PEEBLES, James. *Selkirk.* 1836-60.
PEEL(E)—
J. B. *Salem, Mass.* 1818.
James. *Bradford.* 1871.
John. *Dublin.* 1788. W.
Robert. *Lincoln.* 1876.
PEERLESS Watch Co. *San Francisco, Cal, USA.* 1892.
PEERS, George. *Chester.* 1834-65.
PEET—
John. *Wigton.* 1869-79.

PEET—*continued.*
Joseph. *Wigton.* 1848-58.
Thomas. *Stirling.* 1820.
PEGDEN—
George Robert. *Deal.* 1866-74.
Thomas. *Sandwich.* 1802-45. C.
Vincent. *Deal.* 1851-58.
Vincent. *Herne Street (Kent).* 1826-28.
PEGG—
J. *Ashby-de-la-Zouch.* 1849.
Robert. *Ashby-de-la-Zouch.* 1795-1828.
William Hackett. *Ashby-de-la-Zouch.* 1835.
PEGLER—
Alfred. *Southampton.* 1859-78.
Brothers. *Norwich.* 1875.
Daniel. *Southampton.* (B. 1795-)1830.
Edwin S. R. *Halifax.* 1866.
George. *Southampton.* 1839.
Samuel L. *Blandford.* 1824-30.
& Son. *Blandford.* 1848.
PEIRAE/PEIRAS, Pasquier. *London.* 1648. CC.
PEIRCE, J. *St. Johns, Canada.* 1842.
PEIRSON—
Charles. *London, East Smithfield.* 1781. W.
John. *London.* 1839.
John & William. *London.* 1835.
PELFRYMAN & PYKE. *Birkenhead.* 1878.
PELLAI, Antonio. *London.* 1869-81.
PELLING, James. *Lesmahagow (Scotland).* 1860.
PELTON, George. *Chesham.* 1847.
PEMBROKE—
J. *Ampthill.* 1847.
J. E. *Bedford.* 1847-64.
James. *Bradford.* 1866-71.
PENDLEBURY & Son. *Todmorden.* 1840.
PENDLETON—
Peter. *Prescot.* mar.1803-23. W.
Samuel. *Prescot.* mar.1768-d.1778. W.
PENDRAY, Digory. *Launceston.* mar.1734-d.1741.
PENFIELD—
Josiah. *Savannah, Ga, USA.* ca.1820.
Sylvester. *New York.* 1847.
PENFOLD/PENFORD, Joshua. *London.* a.1684 to
Francis Stamper. CC. 1695.
PENISTAN, John. *Horncastle.* 1828-50.
PENKETHMAN, Thomas. *London.* a.1682 to Thomas
Tennant. CC. 1692.
PENKETT, Thomas. *Liverpool.* 1832.
PENMAN, Peter. *Dunfermline.* ca.1820-ca.1840. C.
PENN—
John. *Peckham.* 1878.
John & Son. *London.* 1869-81.
& KNIGHT. *London.* 1881.
Thomas. *London.* Believed ca.1660. W.
Thomas. *London.* b.ca.1654. mar.1689. W.
Walter E. *Brixton.* 1878.
PEN(N)ELL—
John. *London.* 1700. W.
Mrs. M. *Lichfield.* 1860.
PENNIMAN, John R. *Boston, USA.* 1783-1828. (Dial
painter for S. Willard).
PENNINGTON & PENLINGTON—
& BATTY. ? *London.* ca.1835. C.
Bernard. *Bodmin.* 1662. Rep'd ch. clock.
Christopher. *Kendal.* ca.1790-believed d.ca.1840. C. &
W.
James. *London.* 1844.
James. *Prescot.* 1851. Also chrons.
John. *London.* 1857.
Joseph. *Liverpool.* 1824-51. Also chrons.
Robert (& Co.). *London.* (B. 1780-1824) 1839-44(-51).
Samuel & Thomas. *Liverpool.* 1851. Also chrons.
T. M. *Birkenhead.* 1865.
& TYPKE. *London.* 1863-81. (Successor to GROHE.)
PENNOCK, John. *York.* pre-1638. Later *London.* Lant.
clocks.
PENNWOOD Co. *Pittsburgh, Pa, USA.* ca.1940. Electric
clocks.
PENNY—
Alexander. *Dover.* 1845.
Charles. *Bristol.* a.1772-1801. C. Son of James of *Bath.*
Charles. *Wells.* 1699-1720.
James. *Bath.* 1772. C.
James. *Wells.* ca.1750. C.

PENROSE, Richard. *Redruth.* 1791-1810.
PENSOTTI, J. *Gravesend.* 1855.
PENTECOST—
Charles. *London.* 1857.
Charles Frederick. *London.* 1863-81.
William. *St. Austell.* 1799-1844. W.
PENTNEY, George William. *Wells.* 1875.
PENTON—
Charles. *London.* 1764. C.
Isaac. *London.* ?late 17c.-ca.1720. C.
PENTY, William. *York.* b.1807. Free 1831. d.1833. C.
PENZINK, Martin. *Birmingham.* 1835.
PEORIA Watch Co. *Peoria, Ill, USA.* Successor to
Independent Watch Co. of *Fredonia, NY.* 1885.
PEPLOW—
F. *Shiffnal.* 1863.
Francis. b. ca. 1830 s. of William Peplow of
Wellington. watchmaker. To *Ironbridge* ca. 1870.
d. *Birmingham* 1887. W.
Samuel Kirk. b. 1834 at *Wellington* s. of William
Peplow. watchmaker. Mar. 1856 *Ironbridge. Madeley*
1863.
Samuel Kirk. *Ashby-de-la-Zouche.* 1876.
William I. b.1794, a.1804. In *Shrewsbury, Watling
Street.* 1818. Then *Wellington.* 1828-56. Then *Shiff-
nal.* 1856-85. Died 1895.
William II. *Stourbridge.* s. of William I. b.1815-d.1885.
William III. *Stourbridge.* s. of William II. b.1864-
d.1963.
PEPPER—
Charles. *Potton.* 1854-77. C.
Edward. *Royston.* 1874.
Edwin. *Hinckley.* 1876.
G. *Biggleswade.* 1847-69.
H. J. *Philadelphia, USA.* ca.1850.
H. S. *Philadelphia, USA.* 1837.
Henry. *Maldon.* 1828-51.
Henry J. & Son. *Philadelphia, USA.* 1846-ca.1850.
John. *Biggleswade.* 1830-47.
John. *Newcastle-under-Lyme.* 1850-76.
W. G. *Stoke-on-Trent.* 1876.
PEPPIETT(E)—
Augustus William & Son. *London.* 1869-75.
C. *London.* ca.1780. C.
PEPYS, Richard. *London.* a.1667 to John Harris. CC.
1674-88.
PEQUEGNAT, Arthur V. *Kitchener, Ontario, Canada.*
From *Switzerland.* 1874. Set up first successful
pendulum clock factory in *Canada.* Died 1927.
Arthur Pequegnat Clock Co. ran till 1941.
PERCE, John. *Dartford.* 1874.
PERCIVAL—
James. *Woolwich.* 1839.
Thomas James. *Woolwich.* (B. 1817-24) 1826-55.
PERCY—
Charles. *Mendlesham (Suffolk).* b. 1832. To
Stradbrook 1855. then *Framlingham* 1864-85.
Frederick. *Mendlesham.* 1875-79.
J. *Eye.* 1865.
James. *Stradbrook (Suffolk).* b. 1788-1853.
John. *Debenham (Suffolk).* s. of John Percy.
b. ca. 1838-91.
John Frederick. *Mendelsham* b. ca. 1829
-91. Also at *Eye* and *Stonham* later.
John Hotspur. *Debenham.* 1879.
William. *Liverpool.* mar.1834. W.
Zephaniah. *Mendlesham.* 1806-64. C.
PERDUNE, Robert. *Margate.* 1826-28.
PEREE, (?Jac.). *Stratford.* ca.1800. C.
PEREN, Thomas. *Smarden.* ca.1760-ca.1780. C. (See
PERRIN).
PERFITT—
E. J. *Wymondham.* 1846-75.
J. *New Buckenham (Norfolk).* 1875.
PERKIN(S)—
C. *Aldborough (Norfolk).* 1846-75.
Francis Joseph. *Frome.* 1875-83.
John. *Altofts (Normanton).* 1723-ca.1750. C.
John. *Fitzwilliam, NH, USA.* 1821. C.
Joseph. *Birmingham.* 1868-80.
Moses. *Leeds.* 1850-53.
Robinson. *Jaffray, NH, USA.* ca.1825.
Thomas. *Olney.* 1798. W.
Thomas. *Peterborough.* 1854-77.
Thomas. *Philadelphia, USA.* 1783-99. *Pittsburgh, Pa,
USA.* 1820s-ca.1840.

PERKIN(S)—*continued*
W. *Coventry.* 1880.
PERKS—
S. J. *Dudley.* 1876.
Samuel Joseph. *Wednesbury.* 1860-68.
PERO(O)NE—
Isaac. *London.* 1622-32.
Richard. *London.* 1662.
PERRETT—
James. *Clevedon.* 1883.
Philip H. *Cincinnati, O, USA.* From *Switzerland.* 1820s.
William Edward. *Weston-super-Mare.* 1875-83.
PERRIER—
M. A. *London.* 1881.
Peter. *London.* a.1660-c.1680. W. and goldsmith.
Ulysse. *London.* 1844-51.
PERRIGO—
James. *Wrentham, Mass, USA.* 1737-1808.
James, jun. *Wrentham, Mass, USA.* ca.1800.
PERRIN—
Brothers & Co. *London* and *Geneva.* 1851.
Thomas. *Smarden.* 1705-34. Kept ch. clock. (See Peren).
PERRINE, W. D. *Lyons, NY, USA.* 1850.
PERRING—
Henry. *London.* 1832.
Henry, sen. *London.* 1851-57.
PERRINS, J. W. *Chester.* 1865.
PERROTT, James. *Bristol.* a.1834-50.
PERRY—
Adolphus. *London.* 1881. Clock cases.
Albert. *Salem, Mass, USA.* 1828-61. Later a gunsmith.
Benjamin William. *Stockton.* 1851.
& Co. Ltd. *London.* 1881.
Elias. *Philadelphia, USA.* 1804.
Horace. *Port Hope, Canada.* 1851-63.
J. *Taunton.* 1861.
John James. *Nottingham.* 1876.
John (& Son). *London.* 1839-57. (1863-81).
Marvin. *New York.* From *London.* 1768-76.
R. D. Used by Otay Watch Co. 1889.
Richard. *Winchester.* 1878.
Thomas & Marvin. *New York.* 1767. W.
William. *Bristol.* a.1822-30.
PERRYMAN—
James. *Bristol.* a.1754-74.
John. *Barnstaple.* 1805. W.
PERSCHKY, John. *London.* 1881.
PERSE—
Frederick. *Chatham.* 1874.
J. *Winchester.* 1859.
John. Place unknown. ca.1800.
PERSTON, J. & D. *Glasgow.* 1860. Clock dials.
PERTON, George. *Birmingham.* 1835.
PETCH—
Alfred. *Conway.* 1887-90.
F. *Ulceby (Lincolnshire).* 1868.
J. *Ulceby (Lincolnshire).* 1861.
John Turner. *Huddersfield.* 1834.
PETER—
——. *Niagara, Canada.* 1853.
John. *Camelford.* pre-1832. When went to *USA.* W.
T. *St. Austell.* 1856.
PETERBORO. 1914 and later, product of Pequegnat Clock Co.
PETERKIN, J. *Edinburgh.* 1825. Clock dial maker.
PETERS—
A. R. *Marietta, Pa, USA.* ca.1850.
Bernard. *Hastings.* 1862-70.
David. *Arbroath.* 1837-60.
Edward. *New York.* 1846.
Edward. *Sheffield.* 1787-1822. W.
J. C. *Cambridge.* 1858-65.
James. *Cambridge.* 1830-40.
James. *Philadelphia, USA.* 1821-50. C.
John J. & Co. *Bristol.* 1870. (Successor to Chas. Taylor & Son.)
Peter. *Chatham, Canada.* 1861.
& Son. *Cambridge.* 1846.
PETHERICK—
John Cater. *Stratton (Cornwall).* 1844-47.

PETHERICK—*continued.*
William. *St. Austell.* 1781-d.1784. W.
PETIT, Edward. *London.* 1869.
PETITCLAIR, Joseph. *Quebec.* 1791.
PETITPIERRE & Co. *London.* 1881.
PETRALI, Angelo. *Cardiff.* 1865-75.
PETRIE—
John & William. *New Deer (Aberdeenshire).* 1846.
Mary. *Kirkcaldy.* 1860.
PETTAR, John. *Battle.* a. to Samuel Hammond. Also one of this name at *Rye* C.
PETERMAND, Charles Auguste. *London.* 1832-39. Watch cases.
PETTIBONE—
Lyndes. *Brooklyn, NY, USA.* ca.1840.
& PETERS. *New York.* 1848.
PETTICREW, ——. *Belfast.* ?late 18c. T.C.
PETTIGREW—
John. *Ardmillan* and *Ballynahinch.* 1750-d.1760. C.
Richard. *Port Hope, Canada.* 1857.
William. *London.* 1881.
PETTINGER, T. *Crowle.* 1861.
PETTIT—
Elizabeth. *London.* 1839.
George. *London.* 1875-81.
Isaac. *London.* 1832.
John. *Chipping Ongar.* 1781. W.
John. *London.* (B. a.1785-1824). 1828-51.
William. *London.* 1622-62. Watch case maker, *Westminster.*
William. *London.* 1851-69.
PETTS—
J. *Chelmsford.* 1851-66.
James. *Chipping Ongar.* 1828.
W. N. *Hertford.* 1866.
William. *Hertford.* 1839-59.
PETTY, Henry. *Philadelphia, USA.* 1829-33.
PETVIN, John Finner. *Street (Somerset).* 1883.
PEYTON, ——. *Newtownards.* Late 18c.-early 19c. C.
PFAFF—
Andrew. *Merthyr Tydfil.* 1840-48.
Augustus P. *Philadelphia, USA.* 1830-50. C.
Charles. *Brighton.* 1839.
(Anthony) & HAAS (Michael). *Vaughan, Canada.* 1844.
Joseph. *Truro.* ca.1800. W.
Magnus. *Louth.* 1828. German clocks.
Romanus. *Merthyr Tydfil.* 1848-68.
PFAHLONER, Albert. *London.* 1881.
PFALTZ, J. William. *Baltimore, USA.* ca.1800-12. Partner with Philip Stadtler, later alone.
PFLEGER, Abraham. *Augsburg.* 1623.
PFLUEFFER, Herman. *Philadelphia, USA.* 1849.
PFRANGLEY, Vincent. *Wimborne.* 1855-75.
PHARR, Benjamin Y. *Atlanta, Ga, USA.* 1874.
PHELPS—
& BARTHOLOMEW. *Ansonia, Conn, USA.* ca.1880.
Silas. *Lebanon, Conn, USA.* ca.1720-1785.
(E. S.) & WHITE (G. W.). *Northampton, Mass, USA.* 1828-30.
PHEYSEY, Mrs. J. *West Bromwich.* 1876.
PHILADELPHIA—
Watch Case Co. *Riverside, NJ, USA.* 1927 and later.
Watch Co. *Philadelphia, USA.* ca.1868.
PHILCOX—
George. *Canterbury.* 1823.
George. *Chatham.* 1826-28.
George. *London.* 1839-75.
William. *Wandsworth.* 1878.
PHILIP—
John. *New York.* ca.1717. C. and bells.
Leslie. *Vankleekhill, Canada.* 1851-53.
PHILIPS—
Edwin. *Ottawa.* 1875.
Ephraim. *Hull.* 1851-58. W.
PHILIPSON—
Henry I. *Winster.* b.1754. s. of John. To *Ulverston.* ca.1800-d.1834. C.
Henry John. s. of Henry I. b.1793. Succeeded father. 1834-ca.1865. C. *Ulverston.*
John. *Winster.* b.1726, d.1788. C. Journeyman to the Barbers.

PHILLIP—
Myer. *Rochester.* pre-1767. W.
Peter. *London.* 1633. CC.
PHILLIPPO—
G. *Docking.* 1858-65.
J. *Docking.* 1858.
James. *East Rudham.* 1836.
Z. J. *Burnham Market.* 1865-75.
Mrs. Eliza. *Docking.* 1875.
PHILLIPP, Charles. *Carshalton.* 1878.
PHILLIPPE & LE GRAS. *Baltimore, USA.* 1796.
PHILLIPS—
——. *Ashbourn.* ca.1770. C.
——. *Loscoe.* ca.1750. C.
Abraham. *London.* 1832-57.
Barnard. *London.* 1839.
Benjamin. *Birmingham.* 1850.
Brothers (& Son). *London.* 1839-69 (1875-81).
Charles. *Banbury.* 1832.
Charles. *London.* 1844-75.
David. *Narberth.* 1868.
David & Isaac. *Cardiff.* 1852.
David & Lewis. *London.* 1839. Watch cases.
Edward. *Coventry.* 1880.
Edwin. *Truro.* From *Wadebridge.* 1855.
Francis. *Aldershot.* 1867-78.
George Anderson. *Llanelly.* 1875-87.
Hood. *Haverfordwest.* s. of Phillip. 19c.
Isaac. *London.* 1869.
J. *Great Ringstead.* 1846.
J. *London.* 1851-57.
J. *Southam (Warwickshire).* 1868.
James. *Charleston, SC, USA.* ca.1790.
James. *London.* 1844-81.
James. *Salisbury.* 1830.
James. *Southam (Warwickshire).* 1880.
James & Charles. *London.* 1839.
Joel. *London.* (B. 1820) 1851-81.
John. *Hayle (Cornwall).* ca.1820. C.
John. *Llantrisant.* 18c. C.
John. *Milford Haven.* 1835-68.
John. *Pen-Tyrch.* 18c. C.
John W. *Brampton (Cumberland).* 1869-79.
Joseph. *Brampton (Cumberland).* 1834-58.
Joseph. *Bristol.* ca.1745-ca.1770. C.
Joseph. *London.* (B. 1817-24) 1828-32.
Joseph. *New York.* 1713-35.
Joseph Isaac. *London.* 1832.
Louis. *London.* 1863-69.
Michael. *Banbury.* 1853. C.
Moses. *Aldershot.* 1867-78.
P. *London.* 1857.
Peter. *London.* 1633.
Philip. *London.* 1832-44.
Philip. *Sunderland.* 1851-56.
& PHILLIPS. *London.* 1869-81.
Robert James. *London.* 1869-75.
S. *Norwich.* 1846.
S. M. *Norwich.* 1858-65.
Samuel. *Chelmsford.* ca.1680.
Samuel. *London.* 1839-44.
Saunders. *London.* (B. 1802-24) 1828-32.
Solomon. *Canterbury.* 1832.
T. *Kidderminster.* 1868.
T. *Southam.* 1860.
Thomas. *Haverfordwest.* 1835-50.
Thomas. *Ludlow.* (B. 1791-95) 1828-50.
W. H. *Newport (Mon.).* 1875-85.
William. *London.* 1828.
William. *Tredegar.* 1839. C.
PHILLIPSON. W. *High Bentham (Yorkshire).* 1871.
(See also Philipson.)
PHILP—
James. *Callington (Cornwall).* 1844-47. W.
Robert. *?London.* ca.1740-1781. C.
Robert James. *Brighton.* 1828.
PHILPOT—
John. *Bardfield (Essex).* 1839.
William. *London.* 1851-75.
PHINN, Thomas. *Edinburgh.* 1761. Engraver of clock dials.
PHINNEY, Walker & Co. *New York.* 20c.

PHIPPS, William. *Evesham.* ca.1785. C.
PHITHIAN (also Phythian and Ffithian)—
John. *Whiston (nr Prescot).* mar. 1718-23. W.
John. *Prescot.* 1754-71. W.
John. *Prescot.* mar.1760-?d.1767. W.
John. *Prescot.* 1795. W.
PIAGET—
Henry A. *London and Geneva.* 1851-63.
Victor. *London.* 1857-69.
PICARD—
Antoine Olivier. *St. Sulpice, Quebec.* 1695.
Henry. *?Paris.* ca.1850. Carriage clocks.
J. C. *Cincinnati, O, USA.* ca.1830.
Joseph. *London.* 1851-57.
Luke. *St. Anne de la Pocatiere, Canada.* 1857.
Miss S. A. *London.* 1857-61.
Salomon. *London.* 1857-63.
PICKARD, Joseph and Miss Sarah Ann. *London and Geneva.* 1869-81.
PICKAVANCE, Samuel. *St. Helens.* mar.1822-28. W.
PICKEN—
A. *Northampton.* 1864.
Charles. *Edinburgh.* a.1784-d.1800. W.
John. *Leith.* 1822-30.
PICKERING—
Bernard. *Tarporley.* 1848.
G. *Barnet.* 1866.
George. *Cincinnati, O, USA.* 1867. C.
George. *Hull.* 1858.
J. *Eccleshall.* 1850.
John. *Cincinnati, O, USA.* ca.1830.
John Hunter. *Stone.* 1851.
Joseph. *Lutterworth.* Early 18c. C.
Joseph. *Philadelphia, USA.* 1816-46.
Thomas. *Lutterworth.* ?mid.18c. C.
PICKETT—
George. *Abingdon.* 1877.
George. *London.* 1832.
Joseph. *Hexham.* 1783. C.
Joseph. *Lurgan.* ca.1750-65. C.
Richard, jun. *Newburyport, Mass, USA.* 1815-40.
PICKFORD—
& DIAPER. *Ipswich.* 1875.
Festus. *Wells.* 1866.
John. *Liverpool.* (B. 1814-29) 1828-51.
John. *Richmond (Surrey).* 1828.
Joseph. *Stockport.* (b.1770) d.1844. C.
Richard. *Liverpool.* 1848-51.
W. *Newton Abbot.* ca.1800. C.
William. *Stalybridge.* 1848-57.
PICKINNG(?), John Hunter. *Stone.* 1851.
PICKLES, A. *Bradford.* 1866.
PICKMAN, William. *London.* (B. 1811-24) 1828-32. W.
PICKRELL, J. L. *Greenville, SC, USA.* 1851.
PICNOT or **PIENOT,** Andrew. *London.* 1828-32.
PICKWORTH, John. *Stratford-on-Avon.* 1868-80.
PICTON, The. 1914 and later, product of. Pequegnat Clock Co.
PIDDUCK, Henry (& Sons). *Hanley.* 1842-68(-76).
PIDGEN, John William. *New Bradwell (Buckinghamshire).* 1869-77.
PIENOT or **PICNOT,** Andrew. *London.* 1828-32.
PIERCE—
Daniel D. *Wrexham.* 1876-95. W.
Daniel D. *Ruabon.* Same man as at *Wrexham.* W.
George. *Eastbourn.* From *Westham,* a.1726 to Richard Weller, C.
George. *Westham.* 1718.
Griffith. *Caernarvon.* 1779. Serviced town clock.
Thomas. *Ash (Kent).* 1866-74.
Thomas. *Berkeley (Glos).* 1588, d. 1665. C. & W. & TC.
Thomas. *New Romney.* Late 18c. C.
William S. *Philadelphia, USA.* 1841.
PIERCY, William. *Malton.* ca.1780. C.
PIERCEY, William. *London* 1839.
PIERPOINT, John. *Liverpool.* (B. ca.1820) 1822-24. W.
PIERPONT & Co. *Unionville, Conn, USA.* 1840s.
PIERRET—
Henry S. *Portland, USA.* 1834.
Mathey. *Philadelphia, USA.* 1795. French. W.
PIERS, Charles. *Liverpool.* 1848. Chrons.

PIERSON—
Thomas. *Stokesley.* d.1791.
William. *Kirkby Moorside.* 1790-95.
PIGGIN—
& DYBALL. *Norwich.* 1858.
John. *Norwich.* 1830-46.
PIGGOTT—
H. *Stockport.* 1878.
Henry. *London.* a.1672 to *Sampson Crooke,* CC. 1687 (B.-1692).
Samuel. *New York.* ca.1830. C.
PIGOT, John. *Tewkesbury.* 1879.
PIKE—
H. *London.* 1851. C.
H. C. *Peckham.* 1866. Prob. same as next man.
Henry Clark. *London.* 1857-63. Prob. same as previous man.
James. *Eltham.* (B. 1805-08) 1832. C.
James, jun. *Newport (Isle of Wight)* 1839-78.
James, sen. *Newport (Isle of Wight).* 1830-48.
James, jun. *Newton Abbot.* ca.1765-92. C.
PIKE—
John. *Eltham.* 1820-42.
Lawrence J. *Eltham.* 1839.
Robert. *Wells.* Late 17c. C.
Ruth. *Eltham.* 1839-55.
Mrs. Sarah. *Eltham.* 1845.
William. *Barnstaple.* ca.1720-ca.1790. C.
William. *Cirencester.* 1830.
William. *Newton-Abbot.* ca.1795. C.
William. *Totnes.* 1776. C.
William. *USA.* Late 18c. C.
PILCH—
F. *North Walsham.* ca.1770. C.
Robert. *North Walsham.* 1830-36.
T. *North Walsham.* ca.1780. C.
PILE—
Francis. *Honiton.* mar.1755-ca.1760. C. & W.
William. *London.* 1851-75.
PILFORD, B. *Brighton.* 1878.
PILGH, Thomas. *North Walsham.* ca.1800. C. (cf. PILCH).
PILGRIM, Henry. *London.* 1881.
PILKINGTON—
George. *Tynan (co. Armagh).* ca.1750-70. C.
Hugh Joseph. *Chorley.* 1834-58. W.
J. *Woolwich.* 1826-39.
James. *St. Helens.* mar.1771-83. W.
John. *Darlington.* 1866.
PILSH, Robert. *Bradford.* 1866.
PIMBLET/PIMBLOTT, John. *Macclesfield.* 1865-78.
PINCHBECK, Edward. *Edinburgh.* 1745. From *London,* short stay.
PINCHIN, Samuel Broadbent. *Brighouse.* b.1812-d.1872.
PINDAR, James Joseph. *London.* 1851-63.
PINDER, Robert. *Driffield.* 1840.
PINE, David. *Strasburg, Pa, USA.* 1772.
PINFOLD—
George. *Stony Stratford.* 1869-77.
Thomas. *Middleton Cheney* (?1751) 1762-68, then *Banbury* ca.1768-d.1789. C. & W.
PINGEON, John Louis. *London.* 1851.
PINGLE, Yoh. C. *Montreal.* 1875.
PINGO, John. *London.* b. ca.1668, a.1684-mar. 1690. W.
PINHEY, Richard. *Plymouth Dock.* ca.1795?
PINK, Osborne & Conrad. *Philadelphia, USA.* ca.1840.
PINKARD, Jonathan. *Philadelphia, USA.* 1773. W.
PINKERTON—
———. *Dungannon.* ca.1800? C.
Robert C. *London.* 1863-69.
PINKNEY—
Richard. *Skipton.* b.1682-d.1749. C.
Robert. *Thirsk.* 1866. W.
& SCOTT. *Newcastle-on-Tyne.* 1784.
PINN, Richard. *Sidmouth.* mar.1763-ca.1800. C. & W.
PINNEY—
Francis. *Stamford.* 1844-68.
Francis & Son. *Uppingham & Stamford.* 1876.
Richard Matthew. *Stamford.* 1830-57. W.
PINNINGTON—
James. *Prescot.* mar.1779. W.

PINNINGTON—*continued.*
Thomas. *Liverpool.* (b.1810-29) 1814-48. W.
PINTOU, Paul. *Enniskillen.* ca.1750. W.
PINZING, Martin. *West Bromwich.* 1835. German clocks.
PIPE, Isaac. *Manchester.* 1834.
PIPER—
James. *Chesterton, Md, USA.* 1749-1802.
John Edward. *London.* 1869.
PIPPIN, R., jun. *Bristol.* 1856.
PIQUET, Abraham. *London.* 1839. Watch cases.
PIRIE, Andrew. *Portsea.* 1878.
PIRKIS—
G. *Strood.* 1851-55. Prob. same as next man.
George. *Strood.* 1838-47.
George. *Gravesend.* 1847.
PIRRIE—
James. *Cullen (Scotland).* 1830-d.ca.1870. T.C.
John. *Perth.* 1820-d.1857.
PITCAIRN & ROBERTSON. *Paisley.* 1836.
PITCHER—
E. & Co. *London.* 1880-1900 and later.
John. *London.* b.ca.1659. mar. 1693-97. C.
PITCHFORD, J. *Newcastle-under-Lyme.* 1860-76.
PITKIN—
& Co. *New York.* 1841.
George. *London.* 1869-81.
Levi. *Hartford, Conn, USA.* 1774-1800; *Rochester NY,* 1800-54; also patented a beer pump.
PITMAN—
Arthur. *Wellington (Shropshire).* 1870.
Henry Card. *Castle Cary.* 1883.
James. *London.* 1881.
John. *Birmingham.* 1842.
John. *Philadelphia, USA.* 1818.
Saunders. *Providence, RI, USA.* 1780.
Thomas. *Bristol.* 1842.
William R. *New Bedford, Mass, USA.* 1836.
PITT—
———. *Wareham (Dorset).* Late 18c. C.
Ann. *Chipping Barnet.* 1828.
Caleb. *London.* (B. 1795-1825) 1832. W.
Calvin. *Columbus, O, USA.* ca.1840-70.
Charles. *London.* 1832-51.
John. *London.* 1839-44.
John. *Tetbury.* 1830-56.
Thomas S. *Abingdon.* 1837-64.
Thomas Serjeant. *Newport (Monmouth).* 1848-52.
W. *New York.* 1853.
William. *London.* 1857.
William Golledge. *London.* 1828-44.
William Hilson. *London.* 1881.
William Whetstone. *Chipping Barnet.* 1839.
PITTMAN—
& DORRANCE. *Providence, RI, USA.* ca.1800.
John. *Wilmington, Del, USA.* 1813-25.
John. *Chester Co., Pa, USA.* Early 19c. Travelling clock repairer.
PITTS—
J. *Ramsey (Huntingdonshire).* 1864.
Robert. *Epworth.* 1828-50. C.
PIZZI, ———. *Buckingham.* Partner with Ortelli 1830. Later Pizzi & Cetti, ca.1830-42. W.
PLACE—
James. *Preston.* 1787-1828. W.
Thomas. *Bedale.* 1840. C.
PLAFF, Anthony & Co. *London.* 1844.
PLAMONDON, P. *Quebec.* 1854.
PLANT—
Edward. *London.* a.1655 to William Godbed, CC. 1664.
Robert. *Bakewell.* 1828.
Robert. *Tideswell.* 1835.
Robert. *Stockport.* 1848.
Thomas. *Manchester.* mar.1802. C.
Thomas. *Tideswell.* 1828-49.
William. *Manchester.* 1848-51.
PLASKETT—
Charles. *Bristol.* 1879.
Charles. *London.* 1857. Successor to Reuben.
J. *London.* 1863-81.
Reuben. *London.* 1832-51 (B. has *Bath* 1812-26).

PLASKETT—*continued.*
William. *London.* 1857-69.
William Reuben. *London.* 1881.
PLATE, John. *London.* ?Early 18c. Lant. clock.
PLATER & COLLIER. *London.* 1869-81.
PLATIEL, Daniel. *London.* 1774. W.
PLATNAUER—
Brothers. *Bristol.* 1879-80.
Joseph. *London.* 1857.
Joseph & Samuel (Brothers). *Bristol.* 1850-70 (?-80).
Louis. *Birmingham.* 1880.
PLATT—
A. S. & Co. *Bristol, USA.* 1849-56. C.
Alanson S. *Bristol, USA.* 1849-50. C.
Augustus. *Columbus, O, USA.* 1844. Math. instrs.
Benjamin. 1757. mar.1776. *New Milford, Conn, USA.*
1787-1800; *Lanesboro, Mass,* 1800-1817; *Columbus,*
O, 1817 and later.
& BLOOD. *Bristol, USA.* 1840s.
Ebenezer Smith. *New York.* 1774. C.
Edward. *Leicester.* ca.1800.
G. W. & N. C. *New York.* 1816-20.
J. *St. Helens.* 1858.
James. *Prescot.* 1795. W.
John. *Philadelphia, USA.* 1843.
John. *St. Helens.* 1848-51.
Joseph. *Hyde.* 1878. •
Thomas. *London.* CC. 1637-62. *Cannon Street.*
Thomas. *Prescot.* 1825. W.
William A., jun. *Columbus, O, USA.* ca.1840-ca.1870.
William A., sen. *Columbus, O, USA.* ca.1830-ca.1850.
Grandson of Benjamin.
William Augustus. *Columbus, O. USA.* 1809-44. W.
PLATTS—
J. *Chesterfield.* 1864.
John. *Mansfield.* 1828-35.
PLAYER—
H. H. *Wokingham.* 1877.
Horatio. *London.* 1869.
Horatio James. *London.* (B. 1820-24) 1828-57.
James. *Coventry.* 1880.
John. *Reading.* 1830-77.
Joseph. *Coventry.* 1868-80.
Robert (the elder). *London.* CC. 1678-97.
PLAYSTEAD, William. *Newham.* 1840-42.
PLEASANCE, Frederick. *London.* 1875-81.
PLEASS, William. *Taunton.* 1800-48. C. & W. May be
from *London.*
PLEDGER, ——. *Stanstead.* ca.1810. C.
PLEISTER, Henry. *London.* 1869-75.
PLELPA, Carl. *New York.* 20c.
PLEW, R. *Tewkesbury.* ca.1790-ca.1800. C.
PLIMMER—
Abraham. b. 1757 *Elerdine (Shropshire)* To
Wellington by 1792-1822. C.
William. *Uttoxeter.* 1860-76.
PLOERS, Johann Gregario. *Vienna.* ?ca.1660.
PLOWMAN—
Anthony. *Selby.* 1826.
George. *Chichester.* 1878.
Joseph. *Chichester.* 1828-51.
Mrs. Elizabeth. *Chichester.* 1855-70.
Thomas. *Lancaster.* 1824-51. W. and jeweller.
PLUETT, Anthony. *London.* a.1688 to Jonathan Lown-
des, CC. 1697.
PLUM—
E. *Maidenhead.* 1854.
Mrs. E. *Maidenhead.* 1864-77.
PLUMBLY—
James. *London.* (B. 1820) 1832. W.
& PARR. *London.* 1839-51.
& TUPMAN. *London.* (B. 1825) 1828. W.
PLUMLEIGH, Thomas. *Exeter.* Bankrupt 1725. (May
be from *London,* where one such a.1687 to Thomas
Taylor.)
PLUMLEY—
Charles. *London.* (B. a.1806, CC. 1819-25). 1832-39.
Charles. *Cowansville, Canada.* 1851-58.
Charles & John. *London.* (B. 1820-24) 1828.
PLUMMER—
Charles W. *Norwich.* 1875.
Elijah T. *Atlanta, Ga, USA.* 1871.
J. *North Walsham.* 1858-65.

PLUMMER—*continued.*
John. *East Ruston (Norfolk).* 1836.
Robert. *North Walsham.* 1875.
S. *Jersey.* 1832. C.
William L. St. John, *New Brunswick.* 1850.
PLYMSELL, Henry Vowler. *London.* 1863-81.
PODMORE, Thomas. *Kings Lynn.* 1858-75.
POETON, Josiah. *Coventry.* 1880.
POIGNAND—
Fils. *Jersey.* ca.1807. C.
Louis. *Jersey.* 1781. C.
Pierre. *Jersey.* 1807. C.
POINDEXTER, William. *Lexington, Ky, USA.* 1818-48
POINTER, T. *Bracknell.* 1864.
POINTON—
Aaron. *Hadfield (Derbyshire).* 1876.
J. *Congleton.* 1857.
POLACK, Francis C. *York, Pa, USA.* ca.1850.
POLGLAZE, Henry. *Helston.* 1734. C.
POLKINGHORNE, Philip. St. *Austell.* mar.1764;
Truro 1764-d.1801. C.
POLL, Philip. *Bungay.* s. of Robert II. b. 1775 d. 1846.
POLLARD—
B. *Newark.* 1849-55.
E. *Sherborne (Dorset).* 1848-55.
George. *Crediton.* 1770.
J. *Canterbury.* 1855.
J. *Wye (Kent).* 1866.
James. *Eastry.* 1874.
James. *Lenham.* 1838.
John. *Crediton.* mar.1756.
Peter. *Crediton.* ca.1760. C.
Samuel. *Canterbury.* 1840-51.
Samuel. *Whitstable.* 1838.
& Son. *Camborne.* 1873. W.
T. C. *Gravesend.* 1851.
W. *Gravesend.* 1855.
William. *Camborne.* 1844-56. W.
William R. *Cleator.* 1879. Also fancy goods.
POLLEY, William. *Montreal.* 1808.
POLLHANS—
Adams. *St. Louis, Mo, USA.* 1879-90.
Henry. *St. Louis, Mo, USA.* ca.1860.
Philip. *St. Louis, Mo, USA.* ca.1870. T.C.
POLLI, Angelo. *Leeds.* 1866-71.
POMEROY—
——. *Hartford, Conn, USA.* 1886-1900. C.
Chauncey. *Bristol, USA.* ca.1835.
Eltweed. *Bolton, Mass, USA.* ca.1640. Gunsmith.
(Noah) & HILL (Edward). *Bristol, USA.* ca.1850.
Hunt. *Elmira, NY, USA.* 1832.
John & Co. *Bristol, USA.* 1845.
Noah. *Bristol, USA.* 1847-78. C.
Noah & Co. *Bristol, USA.* 1849-51.
(Noah) & PARKER (George H.). *Bristol, USA.* 1852-
57. C.
& ROBBINS. *Bristol, USA.* 1847-49.
POMFRET, Horatio. *Manchester.* 1848-51. Also im-
ported clocks.
POMINVILLE, Joseph. *Montreal.* 1819-42.
POMPHREY, William. *Kilbride.* 1860.
PONCEAU, Francis. *London.* 1857-63.
PONCY, François. *London.* 1857.
POND—
& BARNES. *Boston, USA.* 1848. C.
C. H. *New Haven, Conn, USA.* 1888. Electric clocks.
L. A. *Boston and Chelsea, Mass, USA.* ca.1845.
Philip. *Bristol, USA.* 1840.
Robert. *London.* 1828.
W. *Abingdon.* 1854-64.
William. *Abingdon.* 1830.
William. *Boston, USA.* 1842.
William & Co. *Boston, USA.* 1830.
William Samuel. *London.* 1881.
PONEY, Abraham. *London.* 1839-51.
PONS (De-Paul), Honore. St. *Nicolas D'Aliermont* and
Paris. 1806-ca.1847. C.
POOL—
Charles. *Barnsley.* 1788-89. C.
David L. *Philadelphia, USA.* 1810-61. Engraver and C.
James M. *Washington, NC, USA.* 1846.
Thomas. *Cincinnati. O. USA.* ca.1830.

POOL—*continued.*
W. *Derby.* 1855-64.
POOLE—
Clement. *London.* ca.1770. C.
Clock Co. *Westport, Conn, USA.* 20c.
George. *London.* 1839. Watch cases.
Henry. *Nantwich.* 1878.
James. *Liverpool.* 1828. Also jeweller.
James (& Co.). *London.* 1851-69(1875-81).
John. *London.* 1832-81.
Joseph. *Hornsea (Yorkshire).* 1823.
T. *Coventry.* 1880.
T. G. *Middlesbrough.* 1898.
Taylor. *Dublin.* 1786-98.
W. *Burnham (Somerset).* 1861. Repairer.
William. *Birmingham.* 1880.
William. *Portsea.* 1830.
POOLEY—
William. *Fenny Stratford.* 1847.
William. *Toddington.* 1854.
POPE—
J. *Cheltenham.* 1863.
William. *London.* 1839. Watch cases.
POPLEWELL, William. Place unknown. ca.1690.
C.
POPPLEWELL—
Benjamin. *Bridlington.* 1823-40. C.
John. *Bridlington.* b.1747-d.1818. C.
PORKER—
William. *Chertsey.* 1839.
William. *London.* 1828.
PORTA, Peter. *Cardiff.* 1868.
PORTE, William James. *Picton, Canada.* 1854-99.
PORTEOUS, Mrs. Ann. *Worcester.* 1876.
PORTER—
Abel & Co. *Waterbury, Conn, USA.* ca.1802-08.
Alfred. *Hartley Wintney.* 1848-78.
C. *Mildenhall.* 1853.
Charles. *Dartford.* 1823.
Charles. *London.* 1828-69.
Daniel. *Williamstown, Mass, USA.* ca.1790-99. C.
& GARDEN. *London.* 1839.
George. *Boston, USA.* ca.1835.
George E. *Utica, NY, USA.* 1834.
(Eleazar), HORACE & Co. *Boston, USA.* 1830.
James. *Farnworth.* mar.1809. W.
L. A. *St. Catharines, Ontario.* 1877.
Mark. *Oakingham.* (B. pre-1775-95) 1830.
Peter. *Prescot.* mar-1808-24. W.
Philip. *Ilminster.* 1883.
Philip. *Martock.* 1875.
Rev. Edward. *Waterbury, Conn, USA.* 1765-1812;
Farmington. 1812-28. C.
Robert. *Galston (Scotland).* 1850.
Robert. *Prescot.* 1794. W.
Rufus. *Billerica, Mass, USA.* 1832. C.
S. & J. *Oakingham.* ca.1770-ca.1810. C.
Samuel. *Bracknell.* 1864. Also at Wokingham.
Samuel. *London.* 1857-81.
Samuel. *Wokingham (Oakingham).* 1837-64.
Samuel & John. *Oakingham.* 1830.
Thomas. *London.* 1863-81.
Thomas. *Prescot.* mar.1773. W.
Thomas. *Prescot.* mar.1774. W.
Thomas. *Childwall (Lancashire).* mar.1813. W.
W. *Bury St. Edmunds.* 1846.
William. *Great Marlom* 1842-54.
William. *Prescot.* mar.1783-31. (May be two such.)
PORTHOUSE—
& FRENCH. *London.* 1863. (Successor to Thomas
Porthouse.)
William. *Barnard Castle.* b.ca.1738-1774; *Darlington*
1784-d.1815. C.
William. *Penrith.* Several of this name. One b.1706-
d.1790. Another working 1828.
PORTLAND, ——. *Liverpool.* ca.1805. W.
PORTLOCK, Henry Arthur. *London.* 1869-81.
POST—
Jordan. *York, Canada.* 1833-53.
Samuel, jun. *New London, Pa, USA.* 1760-85;
Philadelphia 1785-94.
William. *London.* (B. pre-1727-66) 1776. W.

POSTON—
Robert. *Liverpool.* 1824-34. Watch and clock dial
makers.
& WOLDNOUGH. *Liverpool.* 1828. Dial enamellers.
POTCHETT—
John. *Hedon.* 1817. W.
Thomas. *Hull.* mar.1820-40.
POTONIE—
D. & Co. *London.* 1851.
Leon. *Paris.* 1851.
POTTENGER, John. *Belfast.* From *London* 1788. C. and
engraver.
POTTER—
Albert H. *New York* and *Chicago* and *Albany* and *Cuba*
and *Minneapolis.* bro. of William C. Potter, 1836-
1908.
Baldwin. *Beltwibet (co. Cavan).* ca.1790. C.
Brothers. *Chicago, Ill, USA.* 1872-75. After which
William C. Potter alone.
George. *Carlisle.* 1869-79.
H. J. *Bristol, USA.* 1849.
Henry. *London.* 1789. W.
J. O. & J. R. *Providence, RI, USA.* 1849.
John. *Birmingham.* 1835. Dialmaker.
John. *Farmington, Conn, USA.* ca.1780.
John. *London.* 1832. Watch cases.
John. *St. Helens.* 1851.
Joseph. *London.* 1839. Watch cases.
Richard. *Manchester.* 1848-51.
Thomas. *Gainsborough.* 1828-35.
Thomas Wilson. *London.* 1881.
William. *Bristol.* 1754-81.
William. *Farnworth.* mar.1825. W.
William. *St. Helens.* 1848-51.
William. *Wootton-under-Edge.* 1830-50.
William Templer. *Bristol.* 1850.
POTTLE, Henry. *Winchester.* 1830-67.
POTTS—
Edward. *Newcastle-on-Tyne.* 1848-58.
Henry. *Bishopwearmouth.* 1851-56.
John. *Howden.* 1802. W.
John. *Hull.* 1813. W.
John. *Patrington.* 1774-95. W. (not *Partington*).
John. *Redcar.* 1823.
John S. *York.* 1858.
Joshua. *York.* a.1802-51. C. & W.
William. *Eastry.* 1788.
William. *Pudsey.* 1834-53; *Leeds* 1866-77 (to present
day). T.C.
POLIGNOT, ——. *Montecheroux.* 1834-60.
POULIN—
P. E. *Quebec.* 1864.
Peter & Son. *Quebec.* 1861-64.
Pierre. *Quebec.* 1852-55.
POULSON, Edward. *Newtown (Monmouth).* 1835-56.
W.
POULTNEY, John. *Philadelphia, USA.* mar.1781.
Clock casemaker.
POULTON—
E. *Woolwich.* 1874.
R. *London.* 1844.
Robert. *Derby.* 1858-88.
POUND, John. *South Carolina, USA.* 1746.
POVEY—
G. *West Bromwich.* 1868. Clock cleaner.
George. *West Bromwich.* 1876. Cleaner and repairer.
William. *Oldbury.* 1850. Cleaner.
POWELL—
Bartholomew. *London.* a.1650 to John Freeman, CC.
1666. (B. -1673).
Charles. *Coventry.* 1842-50.
David. *Brighton.* 1878.
Francis. *Leeds.* 1850-53.
Frederick. *Birmingham.* 1880. Clock cases.
Henry H. *Milford Haven.* 1887.
James. *London.* 1828.
James. *Witney.* 1852.
James. *Worcester.* ca.1790-1835. C.
John. *Annapolis, Md, USA.* 1745. W. *London* born.
John. *Baltimore.* From *England.* ca.1745.
John. *Liverpool.* 1851. Imported clocks.
John. *London.* 1869-81.

POWELL—*continued.*
Joseph. *Sheffield.* 1806. W.
& PENDRY. *Merthyr Tydfil.* Early 19c. C.
Thomas. *Fenny Stratford.* b.1754 mar.1783-d.1830.. C.
Thomas. *Ollerton.* 1828.
Thomas. *Ravensthorpe (Yorkshire).* mid-18c. C.
Thomas. *Southampton.* 1878.
William. b. ca. 1832 *Nailsworth. Ludlow* 1861-85.
and *Hull.* a. 1756-93 C.
William. *Crickhowell.* 1887.
POWER—
Thomas. *London.* ca.1650. C.
Thomas. *Wellingborough.* ca.1665. C. (See next entry.)
Thomas. *Wellingborough.* 1697-d.1709. C.
POWERS, Hiram. b. *Woodstock, Vt, USA.* 1805; ca.1819
to *Cincinnati,* mar.1832, 1834 to *Washington.* later to
Italy.
POWIS, Richard. *London.* 1811-28. W.
POWLEN, Thomas. *Thame.* 1488-1500. Rep'd ch.
CLOCK.
POWLEY, William. *Asby (Westmorland).* ca.1750.
(Believed b.1681-d.1768). C.
POWLIN, Daniel. *Colchester.* b.1714, a.1728.
POWLTON—
John. *Bishop Auckland.* 1851.
William. *Kendal.* b.1791-1820. Then *Lancaster* work-
ing as journeyman for Russells. Alone 1832-51. C.
POWNE, Thomas. *Looe.* 1804-56. C. & W.
POWTON, R. *Stockton.* 1898.
POYNARD, Noah. *Rochester.* 1847.
POYNTER—
William. *Bracknell.* 1877.
William. *Farnworth.* mar.1828. W.
POYNTON, Joel. *Congleton.* 1848.
POYSER, Edward S. *Nottingham.* 1876.
POXON, R. J. *Alford (Lincolnshire).* 1868.
PRADBURY, John. *Gloucester.* 1850.
PRAEFELT, John. *Philadelphia, USA.* 1897.
PRATT—
——. *London.* ca.1810. C. (Prob. Charles q.v.)
A. Sons. *Haverfordwest.* ca.1870. C.
Azariah. *Marietta, O, USA.* Early 19c.
Charles. *Epping.* 1839-51.
Charles. *Leyburn.* 1840. Also *London* and *Darlington.*
Charles. *London.* (B. 1817-24) 1828-44.
Charles. *Ripon.* 1866-71. W.
& Co. *Yeovil.* 1883.
Daniel, jun. *Reading, Mass, USA.* 1797-1846; *Boston*
1846-71.
(Daniel, jun.) & FROST (Jonathan). *Reading, Mass,
USA.* 1832- dissolved 1835. C.
Henry, jun. *Brighton.* 1862-78.
J. *Huntingdon.* 1854.
James. *Askrigg.* mar.1787-d.1850. C.
James (& Co.). *London.* 1857-81.
John. *Askrigg.* b.1799. s. of James. d.1861. C.
John. *Bishop's Stortford.* ca.1820. C.
John. *Dover.* Free 1793.
John. *Epping.* 1828.
John. *Helmdon (Northamptonshire).* 18c.
John. *London.* 1832.
John. *London.* 1881.
John. *Nottingham.* 1835-64.
John. *Preston.* 1858.
John. *Saffron Walden.* 1751. W.
John. *Waltham Abbey.* 1828-39.
John W. *Bishop's Stortford.* 1828-74.
M. H. *East Dereham.* ca.1790. C.
Matthew. *Swaffham.* 1830-65.
Phineas. *Saybrook, Conn, USA.* 1747-1813. Join
inventor of earliest American submarine.
Richard B. *Haverfordwest.* 1844-50. C.
T. *Sawbridgeworth.* 1866.
Thomas. *Canterbury.* ca.1800-55. C.
Thomas. *Sawbridgeworth.* 1828-39.
W. *Hereford.* 1856.
W. *Norwich.* 1865.
W. G. *Long Sutton (Lincolnshire).* 1861.
William. *Askrigg.* b.1788-d.1857. s. of James.
William. *Bishop's Stortford.* 1828.
William. *Boston, USA.* 1842.
William. *Harlow.* 1828-74.

PRATT—*continued.*
William. *Masham (Yorkshire).* 1840-66.
William. *Sawbridgeworth.* 1839-59.
William & Brother. *Boston, USA.* 1847.
William T. *Washington, NC, USA.* 1834.
PRATTEN, Joseph T. *London.* 1869.
PREBBLE—
Charles. *Margate.* 1823-51.
Edward. *Folkestone.* 1858.
James. *Margate.* 1826-74.
PRECEY, Edmund. *Fordingbridge.* 1839-59.
PREDDY—
C. *Taunton.* 1861-66.
William. *Taunton.* Late 18c. W.
PREDY, Thomas. *London.* 1635.
PREECE—
Redolphus. *Hay.* 1753-92. W.
Rodolphus. *Hereford.* (B. 1753-92) 1830-44. ?Two
such.
William. *Bristol.* 1830-56.
William. *Hereford.* From *Bristol.* 1826. W.
William & Co. *London.* 1863.
PREFONTAINE—
L. M. *Montreal.* 1853.
Louis. *Montreal.* 1862.
PREISACHER, Ge. *Closterneuburg (Austria).* Later 18c.
PREISS, Edward. *London.* 1857.
PREMIER, The. 1914 and later, product of Pequegnat
Clock Co.
PRENTISS—
C. M. C. *USA.* 1875.
Calendar & Time Co. *New York.* 1892.
John. *London.* 1811 (B. 1815-24). W.
John H. *Utica, NY, USA.* 1832.
PRESCOT—
Mrs. E. *Chester.* 1865.
W. *Chester.* 1865.
PRESCOTT—
G. *Chester.* 1878.
George. *Chester.* 1834-48.
John. *St. Helens.* mar.1827-34. W.
Jonathan. *Kensington, NH, USA.* ca.1765.
Thomas. *Prescot.* 1818-23. W.
Thomas. *Chester.* 1878.
PRESLAND, Frederick. *London.* 1869.
PRESS, J. *Rochester.* 1874.
PREST—
George. *Norwich.* ca.1770. C.
John. *Bradford.* 1866-71.
Samuel Hinchliffe. *Shipley (Yorkshire).* 1871.
Thomas. *Bradford.* 1866-71.
Thomas. *Chigwell.* (B. 1820) 1866-74.
William. *Bradford.* 1853.
PRESTIDGE—
Daniel. *Towcester.* 1830-41.
Walter. *Towcester.* 1777. C.
William. *Eydon/Eyton (Northamptonshire).* ca.1740. C.
PRESTON—
, The. 1914 and later, product of Pequegnat Clock
Co.
——. *London.* 1770. W.
Henry. *Bristol.* 1863.
Henry. *York.* 1841-66. W.
J. *Spalding.* 1849.
J. W. *Castleton, Vt, USA.* ca.1830. W.
James. *Londonderry.* 1788.
Job. *Burtonwood.* 1770-88. W.
Job. *Prescot.* 1818-26. W.
John. *Prescot.* 1799-1826. W.
John. *Skipton.* 1749-d.1762. C.
John. *Spalding.* 1850-76.
Josias. *London.* a.1640 to Christopher Vernon.
Michael. *London.* ca.1730. C.
Paul. *Buckingham, Pa, USA.* ca.1860.
R. *Long Clawson (Leicestershire).* 1849.
William. *Farnworth.* mar.1828. W.
William. *Lancaster.* 1818. *Ulverstone* 1820-34. C.
PRESTRIDGE, William. *Eydon.* ca.1770. C. (See
Prestidge.)
PRETTY, S. *Coventry.* 1868.
PREVEAR (Edward) & HARRINGTON (Samuel, jun.).
Amherst, Mass, USA. 1841.

PREVOST—
Jean Adolphus. *London.* 1832-44.
William. *London.* Early 18c. C.
William. *Newcastle-on-Tyne.* b.1663, mar.1690. C.
PREW—
Richard. b.1745. From *Tewksbury* to *Evesham,* where d.1827.
Thomas. *Tewksbury.* s. of Richard, mar.1823-50. C.
PREY, H. W. *Newport, RI, USA.* 'Est'd. 1830'.
PREYLE, J. *Charleston, SC, USA.* ca.1780.
PRICE—
Charles. *London.* 1851-81. (Successor to Robert.)
Charles Edwyn. *London.* 1881.
Daniel. *Wye (Kent).* 1832-66.
David Lloyd. *Beaufort (Wales).* 1844.
David Lloyd. *Brecon.* 1844.
E. D. *Kingston, NY, USA.* 1850s.
Evan. *Llanfyllin.* b.1715, still alive in 1815 aged 100. C. & W.
Frederick J. *London.* 1844-51.
George. *Hereford.* 1856-70.
H. *Birmingham.* Early 19c. C.
Isaac. *Philadelphia, USA.* 1768-98.
J. & Sons. *Weston-super-Mare.* 1861.
James. *Ashford.* 1855-74.
James. *Kingston.* Late 18c. C.
James. *London.* 1869.
James. *Rhayader.* 1830-87.
John. *Ashford.* 1852-74.
John. *Chichester.* a.1751 to Josiah Whitman, C. (B. 1784. W.).
John. *Hanley.* 1850.
John. *Kennington.* 1845-58.
John. *Olney.* d.1713. C.
Joseph. *Chatham.* 1823.
Joseph. *Dover.* a.1814.
Lewis William. *Brecon.* 1835-71.
Nathaniel. *Wolverhampton.* 1860-76.
Paul. *Chester.* 1848-78.
Philip. *Chester Co., Pa, USA.* mar.1791.
Philip, jun. *Philadelphia, USA.* 1813-25. C.
Philip P. From *Philadelphia* to *Cincinnati* 1815, then to *Lebanon, O,* 1823.
Richard. *Bristol.* 1830.
Richard. *Hythe.* 1858-74.
Robert. *London.* 1832-44.
Robert. *New York.* ca.1830.
Samuel. *London.* 1869.
T. *Wye (Kent).* 1855.
Thomas. *Bromyard.* 1844-50.
Thomas. *Clun (Shropshire).* 1879.
Thomas. *Farnworth.* mar.1818. C.
Thomas. *Kington (Hereford).* 1830.
Thomas. *Morriston.* 1875.
Thomas. *Newbridge-on-Wye.* 1840-75.
Thomas R. *Maidstone.* 1845-74.
W. & L. *Pembroke.* ca.1870. C.
William. *Knighton.* 1835-44.
William. *Wolverhampton.* 1876.
William H. *West Chester, Pa, USA.* 1822-79.
PRICHARD, W. *Ewell.* 1862.
PRICKETT, Horatio. *Staplehurst.* 1874.
PRIDDEN, Joseph. *Keighley.* 1871.
PRIDE, William. *Salisbury.* ca.1750. C. (B. pre-1762. W.).
PRIDGEON, W. R. *Kings Lynn.* 1846-75.
PRIDGIN, William. *York.* a.1756-93. Later *Hull.* C.
PRIDHAM, Lewis. *Sandford* nr *Crediton.* mar.1730. C.
PRIEST—
Elizabeth. *Newark.* 1828.
George. *Norwich.* (B. 1796-) 1830-65.
James & John. *Newark.* 1835-76.
John. *Liverpool.* 1834.
Jonas. *Birstwith (Yorkshire).* ca.1850. C.
Jonas. *Harrogate.* 1866-71. W.
Jonas, jun. *Harrogate.* 1871.
Jonas. *Knaresborough.* 1851-66. W.
Jonathan. *Newark.* 1864.
Joseph. *Bristol, USA.* 1775.
William. *Boston.* 1870.
William. *Bristol.* a.1751-93. C.
William. *Newark.* 1835.

PRIEST—*continued.*
William. *Southwell (Nottinghamshire).* 1828.
PRIESTLY, J. B. *Newark.* Late 18c or early 19c. C.
PRIESTLEY—
Sarah. *Wakefield.* 1821. W.
William. *Sowerby Bridge.* ca.1840. C.
PRIM, William W. *West Chester, Pa, USA.* ca.1823.
PRIMAVESI—
A. C. *Haywards Heath.* 1870.
A. C. *Shirley (Hampshire).* 1878.
Brothers. *Bournemouth.* Prob. mid-19c.
Brothers & Co. *Poole.* From *Geneva.* 1867.
Brothers & Co. *Wareham.* From *Geneva.* 1867-75.
F. *Cardiff.* 1887.
F. *Swansea.* 1887.
J. *Poole.* 1855.
J. & Co. *Warminster.* 1842-48.
PRIME—
Abraham. *London.* s. of Andrew. b.1648, a.1665 to father, CC. 1672-(B. -1686). C.
Andrew. *London* (b.1619 *Norwich*), a. in B. Co. 1632, CC. 1646-82 and later. Died *Norwich* 1710. C.
H. *Buxton.* 1864-76.
William. *Alfreton.* 1876.
PRIMROSE, J. *Pensford (Somerset).* 1861-66. Cleaner.
PRINCE—
E. *Landport.* 1859-67.
Edward. *Southsea.* 1878.
George W. *Dover, NH, USA.* ca.1825.
Henry. *Birmingham.* 1880. Dial maker.
Isaac. *Wantage.* 1854-77.
James. *Leeds.* b.1780-d.1837. T.C.
John. *London.* 1844.
John. *Manchester.* 1821. W.
John. *York.* b.1763-d.1835. (Partner in Hampston, Prince & Cattle.)
& LEVY. *Hamilton, Canada.* 1857.
Margaret. *Portsmouth.* 1830.
Richard. *London.* a.1659 to John Clarkson, CC. 1680.
Robert. *Bristol.* a.1770-80.
PRINGLE—
Alexander H. *Cobourg, Canada.* 1851-71.
G. & Son. *St. Thomas, Canada.* 1861.
Henry. *London.* 1844.
Henry A. *London.* 1863-69.
John. *Earlstoun (Scotland).* 1860.
Robert. *Niagara, Canada.* 1851.
Thomas. *Dalkeith.* 1830-36. Succeeded by widow.
Thomas. *Sunderland.* 1827-47.
William. *Edinburgh.* 1825.
William. *Sunderland.* 1834-51.
PRINSKY—
A. *Hartlepool.* 1898.
M. & Son. *West Hartlepool.* 1898.
PRINT, Richard. *London.* a.1681 to John Martin. CC. 1698.
PRIOLLAUD, E. *New Orleans, LA, USA.* ca.1850-60.
PRIOR—
C. *Portsea.* 1859-67.
Daniel. *New Haven, Conn, USA.* ca.1840.
Edward. *London.* 1794-1875. W. (?Two such).
George. *Skipton.* b.1782. s. of John -1807, later *Leeds* 1807-53, also *London.*
Henry. *Leeds.* b.1783. s. of John -1820. C. & W.
John. *Hastings.* 1870-78.
John. *Leeds.* 1834.
John. *Skipton.* b.1747. s. of William. Worked 1782-86, later *Nessfield* 1788-1807-d.pre-1822. W.
Nathaniel. b.1766. *Hertford* pre-1793, then *Leeds* 1793-1808, then *Newington, Surrey,* where d.1821. W.
William I. *Nessfield (Ilkey).* b.1701-d.1794. C. & W.
William II. *Skipton.* b.1793. s. of John -1822. C.
PRITCHARD—
Bartholomew William. *Ewell.* 1839-55.
Buel. *Dayton, O, USA.* ca.1830.
David. *Caernarvon.* 1886-90.
George Thompson. *London.* 1875-81.
Griffith. *Liverpool.* mar.1834. W.
Griffith. *Llanbedrog.* ca.1770-d.1794. C.
& HOLDEN. *Dayton, O, USA.* 1840s.
John. b. ca. 1800 *Stanton-on-Hineheath (Shropshire).* To *Wem* 1849-75.

PRITCHARD—*continued.*
(Eben) & MUNSON (Lemuel H.). *Bristol, USA.* 1844.
C.
Richard. *Caernarvon.* 1895.
Robert. *?Wales.* 1790.
& SPINING. *Dayton, O, USA.* 1830-33.
Thomas. *Runcorn.* 1848-78.
William. *Birmingham.* 1835-42.
PRITCHETT—
F. *Croydon.* 1862-66.
Frederick. *Birmingham.* 1850.
PROBASCO—
Jacob. *Philadelphia, USA.* 1822.
John. *Trenton, NJ, USA.* 1800-23; *Lebanon, O, USA.* 1823.
PROBERT, James. *Wigan.* 1816.
PROBYN, Thomas. *London.* (B. 1820) 1832. Also T.C.
PROCTER, Alexander. *Tarland (Scotland).* 1837-60.
PROCTOR—
B. *Sutton-in-Ashfield.* 1849-64.
E. *Sutton-in-Ashfield.* 1876.
G. K. *Beverly, Mass, USA.* 1860-ca.1880.
Gooday. *Colchester.* b.1706-d.1740. C.
James. *Launceston.* 1801. C.
Joseph. Place unknown. ca.1670. Lant. clock.
Joseph. *Louth.* 1876.
Matthew. *Shap.* 1757-63. C.
Thomas. *Ross.* 1850-56.
W. M. *New York.* 1737-60.
PROFAZE, George. *London.* 1851-63.
PROFAZI, Constantine. *London.* 1869-81.
PRONTAUT, Anthony. *New York.* ca.1840.
PROSSER—
George Henry. *Bristol.* 1870-79.
John. *London.* 1832.
L. & Co. *Birmingham.* 1868. Importers.
Timothy. *Monmouth.* 1830-44.
PROWETT—
James F. *Bannockburn.* 1860. Also ironmonger. (?From *Colchester*).
James Fryers. *Colchester.* b.1824-47. W.
PROWSE, Henry. *Andover.* 1830-39.
PRUE, E. *Stratford-on-Avon.* 1854.
PRUJEAN, John. *Oxford.* 1676-89. Sundials and astrolabe and math. insts.
PRYOR—
E. J. *Dover.* 1866.
Joseph. *Liverpool.* 1676-d.1720. W.
Mrs. E. *London.* 1851-63 (wid. of Robert?).
Robert. *London.* 1828-44.
Thomas. *London.* 1832.
PUCKERIDGE—
Alfred J. *London.* 1839-51.
John. *London.* 1828-39.
PUDNEY, G. *New York.* 1824. W.
PUGH—
Charles. *Rotherham.* 1871.
E. *Newcastle-on-Tyne.* ca.1820. C.
R. *Newtown (Monmouth).* Prob. same man as at *Llanbister.* ca. 1810-20. C.
Richard. *Llanbister.* ca.1820. C.
Richard Dyke. *Liverpool.* 1848-51.
PUGSLEY, William. *Bristol.* 1842-52.
PULCIFER, Ebenezer. *Nottingham.* 1876.
PULLAN—
Benjamin I (Benoni). *Bradford.* b.1697-d.1750. C.
Benjamin II. *Leeds.* b.1737. s. of Benjamin I -d.1787. W.
PULLEN—
H. *Bromley (Kent).* 1874.
James. *London.* a.1669 to William Elmes. Math. inst. mr.
Jonathan. *London.* a.1676 to John Miller, CC. 1683 (B.-1706). W. & C.
PULMAN, P. *Axminster.* ca.1820. C.
PULSIFER, F. L. *Boston, USA.* 1854.
PUNG, T. *Kings Lynn.* 1846-65.
PUNNET(T)—
John. *Cranbrook.* 1669. Kept ch. clock.

PUNNET(T)—*continued.*
Mrs. Jane. *Lynstead (Teynham) (Kent).* 1866-74.
Thomas. *Battle.* Made ch. clock. 1656; then *Cranbrook* 1657-63; then *Rye.* pre-1674-d.ca.1713.
William. *Tenham (Kent).* 1847.
PUNSHON—
Anthony. *Bradford.* 1834-37.
Ralph. *South Shields.* 1856.
PURCELL, Charles. *Richmond, Va, USA.* ca.1819.
PURCHASE, William. *St. John, New Brunswick.* 1864-77.
PURDY, Elisha. *York, Canada.* 1800.
PURINGTON—
Elisha, jun. *Kensington, NH, USA.* 1736-58. C. Also gunsmith.
James. *Kensington* or *Hampton, NH, USA.* 1663-d.1718. C.
James. *Kensington, NH, USA.* s. of Jonathan, 1759-ca.1816, then to *Marietta, O, USA.*
Jonathan B. *Kensington, NH, USA.* s. of Elisha, 1732-1816. C.
PURKIS, Isaac. *Haverhill.* 1830.
PURNELL, John. *London.* 1839-63.
PURRATT, Richard. *Newport Pagnell.* b.1722, mar.1751-d.1792. C.
PURSALL—
Henry. *Birmingham.* 1880.
Thomas. *Birmingham.* 1854-68.
PURSE—
Hans. *Hillsborough.* ca.1800. C.
Hans. *Newtownards.* Early 19c. C.
John. *Philadelphia, USA.* 1803.
Thomas. *Baltimore, USA.* 1796-1812.
W. *Charleston, SC, USA.* ca.1810.
PURSER—
Joseph. *London.* 1857-63.
Richard Stephen. *London.* 1863-81.
PURSEY, William. *Sheerness.* 1847.
PURVES, James. *Middleton-in-Teesdale.* 1856. C.
PURVIS—
Alexander. *London.* 1828-44.
& BISHOP. *London.* 1851-81.
PUTLEY—
Francis. *Dover.* 1874.
Francis. *London.* (B. a.1819) 1828-63.
PUTNAM—
James. *Chesham.* mar.1763. C.
Jonathan. *New York.* 1848.
Norris. *Walthamstow.* 1839-74.
PYATT, Richard. *Leeds.* 1827-53. W. & C.
PYBUS—
John. *Caistor.* 1861-76.
John. *London.* 1832.
William. *Caistor.* (b.1776-d.1861) 1810-61. C. & W.
William. *Wisbech.* 1875.
PYE—
James. *Farnworth.* mar.1799. W.
John. *Manchester.* mar.1791. W.
John. *St. Columb.* d.1740. C.
John. *St. Helens.* 1785. W.
Robert. *Manchester.* mar.1792. C.
PYKE—
J. *Chatham.* 19c. C.
John. *Wivelscombe.* ca.1750. C.
T. *Bridgwater.* ca.1820-25. Clock dial maker.
PYLE—
A. *Tynemouth.* 1850.
Andrew. *North Shields.* 1848-58.
Benjamin. *Washington, NC, USA.* d.1812.
Benjamin II. *Fayetteville, NC, USA.* 1838-41.
John Howard. *London.* 1869.
PYNE, Nathaniel. *London.* a.1667 to Thomas Wheeler, CC. 1677 (B. -1706).
PYNOCK, William. *Ely.* 1830.
PYOTT—
D. *Lewes.* 1862.
David. *Eastbourne.* 1862.
James. *London.* 1869.
PYPER, John. *Belfast.* b. *Enagh, Dromara.* 1865-94. W.

Q

QUAIFE—
Thomas. *Mountfield (Sussex)*. b.1806-ca.1833. Then *Battle* 1833-59. W. & C.
Thomas. *Hawkhurst.* 1866-74.
Thomas Samuel. *Canterbury.* 1874.
Walter. *Newhaven.* s. of Thomas of *Battle.* 1878.
QUANDALE, Lewis. *Philadelphia, USA.* 1813-45. W.
QUANTOCK, Edward. *Sheffield.* 1871.
QUARTERMAIN(E)—
George. *London.* 1875.
Joseph. *Aylesbury.* 1781-1804. C.
QUARTERMAN—
G. *Bournemouth.* 1867.
William. *Chalgrove (Oxfordshire).* 1725. W.
William. *Poole (Dorset).* 1875.
QUASH, Joseph. *London.* a.1637 to George Smith, CC. 1646-64. *Blackfriars.*
QUEBEC, The. 1914 and later, product of Pequegnat Clock Co.
QUELCH—
John. *Oxford.* s. of Richard. a.1652-94. W.
Joseph. *Oxford.* 1684. W.
Martin. *Oxford.* a.1650-53. W.
Richard I. *Oxford.* a.1608-d.1652. W.
Richard II. *Oxford.* a.1652-67. W.
QUESNEL—
Charles William. *Jersey.* 1832-37.
Jacques. *Jersey.* 1787. C.

QUESNEL—*continued.*
Michael. *Jersey.* 1771. C.
QUEST—
Henry. *Marietta, Pa, USA.* bro. of Samuel. ca.1810-1830.
Samuel. *Maytown, Pa, USA.* bro. of Henry, ca.1813. C.
QUIBEL, Eugene. *St. Nicolas d'Aliermont.* 1889.
QUICK—
Francis. *Bristol.* 1840-50.
James. *Penzance.* 1873.
QUIGLEY—
Patrick. *Kilgolough (co. Cavan).* 1813.
Robert John. *Toronto.* 1882-88.
QUILLIAM, Samuel. *Liverpool.* (B. 1825) 1848-51.
QUIMBY—
Henry. *Portland, Me, USA.* ca.1830.
Phineas. *Belfast. Me, USA.* 1830-50.
William. *Belfast, Me, USA.* 1821-50.
QUINGNET, Egidus. *Antwerp.* 1557.
QUIN(N)—
James. *Belfast.* (b.1840) 1865-d.1874.
Thomas. *London.* 1839.
Thomas. *London.* 1875-81.
Thomas D. *London.* 1844.
Thomas. *New York.* 1775. *Philadelphia.* 1776.
QUINTON, George. *Downton.* ca.1770. C.
QUY, ——. *Bath.* 1798.

R

R, H, HR, monogram on Northern English longcase clocks ca.1740, maker unknown.
R, J, JR, monogram on Northern English longcase clock dated 1772.
R, V, VR, trademark of V. Reclus, q.v.
R, W, WR, monogram on longcase clock of ca.1775. Maker unknown.
RAAY, Franz de. *London.* 1857-69.
RABAN—
  John. *Baldock.* 1839.
  W. R. *Luton.* 1847-69.
  William. *Harrold (Bedfordshire).* 1830.
RABOURS, ——. *Paris.* ca.1740. C.
RABY, Louis. *Paris.* 1855.
RACINE—
  David. *Baltimore, USA.* ca.1800.
  David. *Quebec.* 1890.
RACKSTRAW, W. W. *Leek.* 1860. W.
RADCLIFF(E)—
  ——. *Barkisland (Yorkshire WR).* Same man as at *Elland.*
  C. *Birkenhead.* 1857.
  Charles. *Liverpool.* (B. 1677) 1693-d. 1700. W.
  Charles. *Liverpool.* 1848-51. Chrons.
  J. N. *Birchrunville, Pa, USA.* ca.1874.
  Nathaniel. *Elland.* 1800-34. C.
RADCLYFFE, Brothers. *London.* 1875-81.
RADERMACHER—
  A. *Beccles.* 1858.
  Alfred. *London.* 1863-69.
RADFORD—
  Edward. *Hanley.* 1835-50.
  Henry. *Hanley.* 1828.
  James. *Ottawa.* 1875.
  T. & Sons. *Leeds.* Same as Thomas Radford I, q.v.
  Thomas I. *Leeds.* b.1731-d.1801. C.
  Thomas II. *Leeds.* b.1761. s. of Thomas I. d.1793 C.
  Thomas. *Stoke-on-Trent.* 1828-50.
  William. *Leeds.* b.1764. s. of Thomas I. d.1826. C.
RADGATE, John. *London.* ca.1760. C.
RADGES—
  Joseph & Co. *Coventry.* 1868-80.
  Joseph & Co. *London.* 1881.
  Mendel. *Sheffield.* 1862. W.
RAE, Alexander. *Dumfries.* ca.1760. C.
RAFARD, Charles. *London.* 1863.
RAFFAELLI, Giacomo. Place unknown. 1804. C.
RAFTERY—
  Thomas. *London.* 1857.
  William. *London.* 1881.
RAGGETT—
  J. *Hemel Hempstead.* 1851.
  James. *Ramsgate.* 1823-27.
  James. *Thorpe Walton* and *Great Oakley (Essex).* 1839
  William. *London.* 1869.
RAHMER, G. *New York.* ca.1840.
RAIGHT, George. *Eastry (Kent).* 1826-38.
RAILROAD Watch Co. *Albany, NY, USA.* 1896.

RAINBOW—
  J. E. *Brierley Hill (Staffordshire).* 1868.
  W. O. *Weedon.* 1869.
RAINE—
  J. *Hartlepool.* 1834.
  Joseph. *Durham.* (B. 1804-20) 1827-34. C.
  Joseph. *Wolsingham.* 1827.
  Nathan. *Philadelphia, USA.* 1773. (One such a *Durham* pre-1756. W.).
RAINER, John. *Rayleigh (Essex).* 1866-74.
RAINES, William. *London* a.1653 to William Almond, CC. 1660-68. (One such d. *York* 1694.) C.
RAINEY—
  John. *Belfast.* 1846-49.
  Thomas. *Belfast.* (b.1849) ca.1880-d.1911. W.
RAINFORTH, John Tonge. *Bridgwater.* 1875.
RAINGO, Freres. *Paris.* 1834-ca.1860.
RAINTON, George. *Leeds.* 1834-53.
RAIT—
  D. C. *Paris.* Mid-19c. C.
  David Crichton. *Glasgow.* 1841-60. W.
RALFS, William. *Norwich.* 1865.
RALPH—
  Alexander. *Ottawa.* 1875.
  George. *Sandwich.* 1832.
  William. *Sandwich.* 1826.
RAMAGE, J. D. & J. *Kingston, Canada.* 1857.
RAMBLES, Robert. *Leeds.* 1871.
RAMPTON—
  C. *Alton.* 1859.
  Charles. *Farnham (Surrey).* 1878.
RAMSAY—
  Andrew. *Londonderry.* 1865-80.
  David. From *Scotland* via *France* to *London,* where appointed 1613 Clockmaker to James I. In *Tutle Street* 1622. First Master of CC. 1632. Near the *Wounded Hart in Holborn* 1653. Died 1660. C. & W.
  Mark. *Edinburgh.* 1795. Watch cases.
  Robert. *Dumfries.* 1785-1823.
  Walter J. *Raleigh, NC, USA.* 1831-56.
RAMSBOTTOM—
  ——. Hall Green (*Yorkshire WR*) same as John of *Wakefield.* 18c. C.
  ——. *Newmiller Dam (Yorkshire WR)* same man as at Wakefield 18c.
  John. *Horbury (Yorkshire WR).* 1838.
  John. *Wakefield.* ca.1760-ca.1790. C.
  John. *Wakefield.* 1851-62.
RAMSDELL & WHITCOMB. *USA.* 1870.
RAMSDEN—
  Charles William. *Wakefield.* 1853-71. W.
  Wright. *Brooklyn, NY, USA.* ca.1830.
RAMSEY—
  George. *Billericay.* 1866-74.
  William. *Letterkenny.* 1824-28.
  William. *Liverpool.* 1780. W.
RAND—
  Daniel. *Boston, USA.* (B. 1825) 1830. C.
  William. *London.* ca.1730. C.

RANDALL—
A. *Greenwich.* 1866-74.
Christopher. *Taunton.* d.1666. W.
H. *Landport.* 1859-67.
Henry Pearce. *Witney.* 1852.
John. *Holt.* 1830-46.
Nathan. *Ixworth.* 1865-75(-79) see Randell.
Richard. *London.* 1857.
S. *Coventry.* 1880.
Thomas. *Warminster.* 1842.
Walter. *London.* 1875.
William. *Holt.* 1858-75.
William. *London.* 1869-81.
William. *Wareham.* (B. 1795) 1824-30.
RANDEL(L)—
Henry. Name used by US Watch Co. ca.1869-70.
Nathan. *Ixworth.* (1865) 1875-79. See Randall.
Thomas. *Cromer.* 1830-36.
RANDLE, J. *Atherstone.* 1868-80.
RANKIN—
Alexander. *Greenock.* 1836-60.
Alexander. *Philadelphia, USA.* 1829-33.
Brothers. *Belfast.* 1880-94.
Duncan. *Oban.* 1860.
John A. *Elkton, Md, USA.* 1896.
RANKLYN, Thomas. *Oxford.* 1604-d.1658. T.C.
RANLET—
Noah. *Gilmanton, NH, USA.* 1777-mar.1800. C.
Samuel. *Monmouth, Me, USA.* ca.1800. C.
RANNIE, Alexander. *Turriff.* 1836.
RANSINGER, M. *Elizabethtown, Pa, USA.* ca.1850.
RANSOME—
Henry. *Norwich.* 1830-46.
J. *Norwich.* 1858-65.
RANT—
John. *London.* a.1676 to William Cattell, CC. 1687 (B-1708).
Jonathan. *London.* a.1680 to Francis Munden, CC. 1687-97 (B. -1725).
RANYELL, George Joseph. *Boston.* 1868-76.
RANZ, C. *Walsall.* 1868.
RAPER—
Sydney. *Thirsk.* 1840-51. C.
Thomas. *Barnard Castle.* 1844-56.
Thomas. *Bedale.* 1823-40.
RAPHAEL, I. *Birmingham.* 1860.
RAPP—
Augustus. *London.* 1851.
William D. *Philadelphia, USA.* 1828-ca.1850.
RATAGE, John. *Kingston, Canada.* 1851-69.
RATCLIFF(E)—
John. *Manchester.* 1834.
Joseph. *London.* 1828. Watch cases.
Joseph. *Wrexham.* mid-18c. C.
RATHBONE, Thomas. *Congleton.* 1865-78.
RATHMER, I. *Winterswijk (Holland).* 1800. C.
RATLEY—
Thomas (William). *Clapham (Surrey).* 1828(-39).
William. *Battersea.* 1878.
RATLIFF, Joseph. *London.* 1832.
RATTENBURY, Frederick. *Stratton (Cornwall)* 1847-73.
RATZERHOF, Josef. *Vienna.* 1838-58. C.
RAVEN, Mary. *Sheffield.* 1862-71. W.
RAVENSCROFT, G. O. *Dover.* 1874.
RAW—
Edwin. *Hull.* 1858.
Henry. *Whitby.* 1858-66. C. & W.
Peter. *Vienna.* ca.1810. C.
William. *Hull.* 1846-51.
William. *Whitby.* 1814-58. C. & W.
RAWE—
George Dunn. *Wakefield.* 1851. W.
James. *Fowey.* 1778-d.1781. W. (also ROWE).
Nathaniel. *Durham.* pre-1756. W.
RAWFINGER, John. *London. Deptford.* 1622. Alien.
RAWLINGS, Robert. *Wisbech.* 1876.
RAWLIN—
William. *Newark.* 1876.
William. *Sheffield.* 1862. W.
RAWLINS, John. *Stone.* ca.1750-ca.60. C.
RAWLINSON, George. *Sleaford.* 1876.

RAWORTH—
Charles. *Plymouth.* ca.1740.
Henry. *Plymouth.* s. of Charles. mar.1761-95.
RAWSON—
George. *Boston.* 1828-35.
Jason R. *Holden, Mass, USA.* 1839. W.
John. *Morpeth.* (B. 1808-20) 1827.
John. *Penrith.* (B. ca.1770-1820) 1828. Also jeweller.
& ROYCE. *Wigan.* 1858.
Smith, E. G. *Saratoga Springs, NY, USA.* 1867.
RAWSTHORNE—
John. *Clitheroe.* 1848-58.
John. *Ormskirk.* 1830.
RAY—
Benjamin W. *Elgin, Ill, USA.* 1801-83.
D. *Brandon (Norfolk) and Bury St. Edmunds* 1844-46.
Daniel. *Battle.* ca.1790-d.1809. C.
Daniel (1). *Sudbury.* b. ca. 1666, d. 1723.
Henry. *London.* 1839.
(Benjamin) & INGRAHAM (Elias, Andrew). *Bristol, USA.* 1841-43.
James. *Leeds.* 1826.
Lydia. *Battle.* 1839.
Miss S. *Battle.* 1851-62.
W. *Brighton.* 1851.
William. *Battle.* b.1800. s. of Daniel. 1830s.
William. *Bury St. Edmunds.* 1830-79.
William Stevens. *Brighton.* b.1826. s. of William of *Battle,* -1870. W.
RAYE, John. *Oxford.* 1617-48. T.C.
RAYHAM & JONES. *London.* 1863. Clock cases.
RAYMENT—
John. *Stamford.* (B. 1795) 1798-1800. C. Later master of Ladies Boarding School, Huntingdon, where d.1803.
Richard. *Bury St. Edmunds.* b. ca. 1686, d. 1754, C. & W.
Thomas. *Stamford.* (B. 1760) 1775-d.ca.1791. C. & W.
William. *Stowmarket.* Mar. 1706, d. 1760. C.
RAYMOND—
Alfred. *London.* 1851.
Edmund Alexander. *Hungerford.* 1837.
Edne. *Devizes.* 1830.
Richard. *Bury St. Edmunds.* 1700-43. cf. Rayment.
RAYMONT, John. *London.* 1622. Alien.
RAYMORE, James F. & Co. *Montreal.* 1862.
RAYNELL, G. *Boston.* 1861.
RAYNER—
James, *Sandwich.* b.1839-61. W.
James. *Sheffield.* 1784. C.
RAYN(E)S—
Joseph. *Lowell, Mass, USA.* 1834-47.
William. *York.* d.1694. C. (See Raines.)
RAYSON, William. *Bingham (Nottinghamshire)* 1855-76.
REA—
Archelaus, jun. *Salem, Mass, USA.* 1750, mar.1777-d.1792. C. & W.
George. *Flemington and Pittsburgh, USA.* 1795-1838. C.
READ—
A. E. J. *Westbury (Wiltshire).* 1867.
Abner, Ezra and Isaiah. *Ohio.* Brothers, ca.1809.
C. *Coventry.* 1860.
C. *Wolverhampton.* 1876.
Charles. *Coventry.* 1850. Watch case maker.
Charles. *Helston.* mar.1788-1803. C. & W.
Charles. *Coventry.* 1828-80.
Charles, jun. (& Son). *Coventry and Clerkenwell.* 1854-80(-81). Wholesaler.
Daniel. *Ipswich.* 1814-58.
D. & A. *Ipswich.* 1865.
D. F. & A. *Ipswich.* 1858-65.
Daniel. *Ipswich.* 1830-39.
Edwin. *Sevenoaks.* 1874.
Frederick. *Ipswich.* s. of Daniel b. ca. 1831-64.
G. A. *New Jersey.* 1874. C.
George. *Chelmsford.* 1832-59.
Godfrey. *Baldock.* 1828.
Henry. *London.* 1863-69.
James. *Stevenage.* 1828.
James. *Weymouth.* 1824-48.

READ—*continued.*
John. *Southport, Portsea (Hants).* 1830.
John Mabyn. *Helston.* ca.1815-44. C. & W.
John Mabyn. *Pontypool.* 1849.
Jonathan. *Beccles.* 1865-79.
Matthew. *Aylsham.* b.1756-d.1826.
Richard. *Buntingford (Hertfordshire).* 1828-39.
Richard. *Chelmsford.* Late 18c. C.
S. W. *Coventry.* 1880.
Samuel. *Coventry.* 1880.
Silas G. *New Brunswick, NJ, USA.* 1812-14.
& Son. *Coventry.* 1880.
Thomas. *York.* b.1689-mar.1710. Engraver.
(Abner, Ezra, Isiah) & WATSON (Thomas). *Cincinnati, O, USA.* ca.1809. C.
William. *Buntingford.* 1839-74.
William. *Coldharbour Lane, London.* 1878.
William. *Coventry.* 1850-80.
William H. J. *Philadelphia, USA.* 1831-ca.1850, and later.
READER—
Nathaniel & William. *Hull.* ca.1760. C.
Richard. *Hull.* mid-18c. C.
Thomas Oliver. *Cranbrook.* 1839-45.
READING, George. *Coventry.* 1850-60.
READMAN, J. *Stockton.* 1898.
REASNORS, John. *Rochester, NY, USA.* 1841.
REAVES, Hamilton. *London.* 1869.
RECAS, Thomas. *London.* 1632. CC.
RECHARD, Karl Louis. *Sheffield.* 1861-71. W.
RECHT, Joseph. *Brantford, Canada.* 1851.
RECKUMAB, ——. *London.* ca.1780. W.
RECLUS, V. *Paris.* 1867-78.
RECORD (?RECO), John. *London.* 1632-62.
REDAELLI, John. *London.* 1863.
REDDALL—
Mrs. M. *Froxfield (nr Woburn).* 1864-69.
Robert. *Woburn.* 1864-77.
T. *Froxfield (nr Woburn).* 1877.
W., T. & R. *Froxfield (nr Woburn).* 1854.
REDDICK, Thomas. *Bytown, Canada.* 1853.
REDFE(A)RN—
Charles. *London.* 1839.
Charles Edward. *London.* 1857-69.
Thomas. *London.* 1863-81.
REDGATE, Edwin. *Sheffield.* 1862-71.
REDGRAVE—
& CHURCH. *London.* 1863.
Thomas. *London.* 1851-57.
REDPATH, ——. *Kelso.* 1835.
REDRUP, Eli. *Chesham.* 1780.
REDSHAW, John. *Newcastle-on-Tyne.* ca.1740. C.
REDWOOD, William. *London.* 1844-75.
REED—
Alfred. *Cambridge.* 1875.
Benjamin. *Bristol, USA.* ca.1775.
Daniel I. *Philadelphia, USA.* 1798.
Edward T. *Hastings.* 1862-70.
Ezekiel. *Brockton of Bridgwater, Mass, USA.* pre-1800. C.
Frederick. *Philadelphia, USA.* 1814-23.
George. *Moulsham (Essex).* 1839-55.
G. *Washington. Philadelphia, USA.* 1839-50.
Isaac. *Philadelphia, USA.* 1820-46.
Isaac. *Shelburne, Nova Scotia.* 1786.
Isaac. *Stamford, Conn, USA.* (1746) to *Stamford* 1790-ca.1808. C.
Isaac & Son. *Philadelphia, USA.* 1830-50. C. & W.
James. *Dunnville, Canada.* 1861.
James. *Liverpool.* 1851. W.
James. *Rochester.* 1849.
James R. *Pittsburgh, Pa, USA.* ca.1840.
John W. *Philadelphia, USA.* 1846.
Jonathan. *Brampton.* 1858-79. Also silversmith.
& LEHMAN. *Swansea.* ?Later 19c.
Levi. *Fishguard.* 1840-75.
Osmon. *Philadelphia, USA.* 1831-41.
Richard, jun. *Chelmsford.* s. of Richard sen., 1765-1816. W.
Simeon. *Cummington, Mass, USA.* ca.1770.
Thomas. *Cambridge.* 1846-65.
William. *Chelmsford.* s. of Richard. sen., 1770-d.1809.

REED—*continued.*
Zelotus. *Goshen* and *Cummington, Mass, USA.* s. of Simeon, ca.1796.
REEDER—
Abner. *Philadelphia* (1766) 1793-98, *Trenton, NJ,* 1798-1841.
Westoby. *Hull.* 1851-58.
REEKIE, William. *Ballymacarrett.* 1880.
REEPE, Joseph. *Plymouth.* ca.1650-ca.1670. Lantern clocks.
REES—
Alexander Hugh. *London.* 1851-75.
Daniel. *Narberth.* 1830.
Daniel M. *Pembroke Dock.* 1844-68.
David. *Llanarth.* 1868-75. C.
John. *Carmarthen.* 1829-40.
John. *Mydroilyn.* b.1813, then *Machynlleth* where d.1895.
John, jun. s. of John, sen., *Machynlleth* 1868, then *New Quay* 1868-75.
Phillip. *Basaleg (Wales).* mid-19c. C.
Robert. *Towyn.* 1868.
T. *Cyffig (Wales).* mid-18c. C.
REESE—
David. *Aberaeron.* s. of John of *Machynlleth,* 1868. May be same man as at *Llanarth.*
James J. *Bala.* Same man as at *Portmadoc.* 1885. W.
James J. *Portmadoc.* 1874-95. s. of John of *Machynlleth.*
REET, George P. *Melrose, Mass, USA.* 1828-87.
REEVE—
Benjamin. *Philadelphia,* and *Greenwich, NJ, USA.* ca.1760-1801. C.
& Co. *New York.* 1854.
G. H. *Lindsay, Canada.* 1862.
George. *Philadelphia, USA.* 1804.
George, sen. *Zanesville, O, USA.* 1800. C.
H. W. *Martham (Norfolk).* 1858-65.
Henry. *London.* a.1658 to William Comford, CC. 1682-97.
J. *Lowestoft.* 1858.
J. *Wymondham.* 1865.
James Webster. *Martham.* 1875.
John. *Holt.* 1875.
John. *Leicester.* (B. 1791)-ca.1830.
Joseph. *Brooklyn, NY, USA.* ca.1840-50.
Joseph. *Philadelphia, USA.* Early 18c. C.
Michael. *Huddersfield.* 1866-71. W.
Richard. *Philadelphia, USA.* 1804.
Richard. *Zanesville, O. USA.* s. of George R., sen. ca.1815.
Richard & George. *Philadelphia, USA.* 1804.
Roger. *London.* a.1650 to father, Thomas.
Samuel. *Diss (Norfolk).* 1722. C.
Samuel. *Stonham (Suffolk).* b. 1684 d. 1718. C.
Thomas. *New York.* ca.1840-50.
William. *Rye.* 1720-25. W.
REEVES—
Alfred. *London.* 1832. Watch cases.
Benjamin. *Berkeley.* 1830-56.
Caleb. *Crawley.* 1862-70.
Charles. *Frampton (Gloucestershire).* 1840.
Charles. *Hereford.* 1844-70.
Daniel. *Dorking.* 1878.
David S. *Philadelphia, USA.* 1830-35.
E. *Boston.* 1868.
E. *Lewes.* 1851-62.
George. *Leeds.* 1866. W.
J. *Dorking.* 1862-66.
James. *Capel (Surrey).* 1839.
James. *Chatham, Canada.* 1851-62.
James. *Quebec.* 1815-26.
James William. *Bawtry.* 1871.
& KING. *London.* 1832.
Mrs. Harriet. *Hereford.* 1879.
Robert. *York.* b.1636-72. W.
Stephen. *Bridgeton, NJ, USA.* 1776.
W. *Cranley (Surrey).* 1862.
W. *Newent.* 1879.
William. *Birmingham.* 1828.
William. *Hereford.* 1879.
William (& Son). *London.* 1839-51(57-63).
William. *Newent.* 1850-56.

REEVES—*continued.*
William. *Rye.* 1725.
Y. *Philadelphia, USA.* 1808.
REGALLY, M. *Boston, USA.* 1842-46.
REGAMEZ & Co. *London.* 1881.
REGARD, Raymond. *London.* (B. CC. 1677)-1697. W.
REGENSBURG, Moses A. *West Chester, Pa, USA.* 1835-43. W.
REGINA, The. 1914 and later, product of Pequegnat Clock Co.
REGULATOR No. 1, The. 1914 and later, product of Pequegnat Clock Co.
REHFIELD, Charles J. *London.* 1869-81.
REIBLEY, Joseph. *Philadelphia, USA.* 1845.
REICHLE, Anton. *London.* 1844.
REID—
Adam. *Toronto.* 1857.
Adam. *Woolwich.* 1847.
Alexander. *Greenwich.* 1839-55.
Andrew. *Biggar.* (b.1767-) d.1860.
George. *London.* 1828.
John. *Sanquhar (Scotland).* 1837.
& Sons. *Newcastle-on-Tyne.* 1827-58. C., W. and chrons.
Thomas. *Auchtermuchty.* 1837.
Thomas. *Edinburgh.* 1788. Watch cases.
William. *London.* (B. CC. 1824) 1828. W.
William Otto. *Biggar.* (b.1820 at *Sanquhar*) -d.1849.
REIDER—
& Co. *London.* ca.1820. C.
Joseph. *London.* 1875-81.
REILLY—
& Co., J. C. *Louisville, Ky, USA.* ca.1816.
& GRAHAM. *Belfast.* 1800-03. W.
REIMERS, Henry Frank. *London.* 1881.
REINER—
Andrew. *London.* 1875-81.
Brothers. *Landport.* 1878.
Casper & Co. *Dudley.* 1850.
REINHOLD, H. G. *Strasbourg.* 1643-d.1653. W.
REIRTSALPEL. Reverse signature of Leplastrier on watch.
REISBYE, Anthony. *London.* See Risby.
REISER—
Albert. *London.* 1875.
Augustin. *Philadelphia.* 1772.
REISLE, John. *Clapham (Surrey).* 1839.
RELF, Edward. *London.* 1832.
RELPH, George Brown. *London.* 1875-81.
REMELLARD, David. *Quebec.* 1864-65.
REM(M)INGTON—
E. *Sheffield.* 1871.
O. H. *Akron, O, USA.* 1860s.
RENARD, Nicholas. *La Rochelle.* ca.1680. Sundial.
RENDA, ——. *Paris.* 1830. C.
RENK, A. & C. *Salford.* ca.1840. C. & W.
RENNARD, Thomas. *London.* a.1785-1815. C.
RENNIE—
Alexander. *Insch (Scotland).* 1860.
Alexander. *Turriff.* ca.1835.
Alexander David. *Arbroath.* 1837.
Charles. *Auchterarder.* 1860.
James. *Carlisle.* 1828-48.
Joseph. *Sheffield.* 1834.
RENNISON—
Charles. *Dumfries.* 1860.
George H. *Bishopwearmouth.* 1851-56.
Michael. *North Shields.* 1867.
W. & M. *North Shields.* 1858.
William. *Newcastle-on-Tyne.* 1827-48.
William (& Son). *North Shields.* 1827 (1847-55).
RENNOLES, T. *London.* 1851-57.
RENON, Jacob. *London.* Foreigner. Temporary journeyman 1655 to Jeremy Gregory.
RENOUF, Edouard. *Jersey.* 1748-ca.1770. C.
RENSHAW—
James. *Prescot.* 1818. W.
James. *Farnworth.* 1795-1800. W.
W. *Northampton.* 1864-69.
RENTON—
James. *Bradford.* 1853.
James. *Ilkley.* 1871.

RENTON—*continued.*
James. *Toronto.* 1874-77.
James. *York.* 1840. W.
RENTZHEIMER, Henry. *Salisbury, Pa, USA.* 1785-88. C.
RENTZSCH—
Frederick. *London.* 1851.
Sigismund. *London.* (B. 1813-40)-44.
RENVOISE/RENVOUZE, George. *London.* 1828-44.
RENWICK—
George. *Halifax, Nova Scotia.* 1863-74.
James R. *Halifax, Nova Scotia.* 1873.
James W. *Bala.* 19c. W.
RERBORN, Christofle. *Naples.* ca.1730. C.
REQUIER, Charles L. M. *Paris.* 1878-89.
RESENTHALL, Archy. *Leeds.* 1871.
REST, John Peter. *London.* 1863-81.
RESTELL—
Richard. *Croydon.* 1839-78.
Thomas. *Tooting.* 1828-39.
RETALLICK, Richard. *Liskeard.* 1810-23. C.
RETHROE, T. *Letterston (Wales).* 19c.
RETTICH, Joh. *Wien (Vienna).* 1826-71. C.
REUBEN, Jacob. *Dover.* 1823-32.
REVELL—
E. *Eltham.* 1866.
Edward. *Eton.* 1854.
Edward. *Slough.* 1847-54.
REVERE—
Clock Co. *Cincinnati, O, USA.* ca.1930.
Paul. *Boston, USA.* 1792-1818.
Paul (& Son). *Boston, USA.* 1804-d.1818. (Son Joseph continued till 1828).
REVILL, William. *Tickhill (nr Doncaster).* 1834-37. C. & W.
REVIS, Thomas. *Cambridge.* 1840.
REY—
Claude. *Montreal.* 1862-76.
Gustave J. *London.* 1869.
REYMAOND, C. D. *Ottawa.* 1862.
REYMOND, Brothers. *London.* 1881.
REYNOLDS—
(E. J.) & BENTON (J.). *Rochester, NY, USA.* ca.1850.
George. *Eye.* 1879.
George. *York.* 1641-d.1680. W.
Henry. *Launceston.* 1873. W.
Henry. *London.* 1857-69.
Henry. *Potton.* 1839.
Henry A. *Rochester, NY, USA.* 1847.
J. *Fordingbridge.* 1867.
J. *Sittingbourne.* 1855.
J. *St. Columb Minor (Cornwall).* 1851-56. W.
J. *Wadebridge.* 1856.
James. *Bodmin.* 1844.
James. *Wadebridge.* 1844. (See J. above.)
John. *Egloshayle (Cornwall).* ca.1780-d.1786? C.
John. *Hagerstown, Md, USA.* ca.1790-1832.
John. *Newquay.* 1851-59.
John. *Oxford.* 1702-33.
John. *Padstow.* 1820-73. C.
John. *Potton.* 1830.
John. *Wadebridge.* ca.1780. C.
John C. *Cobourg, Canada.* 1871.
John & Co. *Stourbridge.* 1850.
Joseph. *London.* a.1683 to Richard Whitehead. CC. 1691.
Morgan. *Usk.* 1844-80.
O. *Sittingbourne.* 1866-74.
R. *Southam (Warwickshire).* 1868.
Richard. *Dover.* ca.1710. C.
Thomas. *London.* (B. 1817-24) 1828-44.
Thomas. *Oxford.* s. of John. 1745-d.1799. C. & W.
Thomas. *New York.* 1891.
Thomas. *Wadebridge.* 1786. C.
Thomas & Ernest. *London.* 1869.
William. *Launceston.* ca.1820-56. C.
William jun. *Launceston.* 1856.
William. *Neath.* 1822-44.
William. *Padstow.* 1844-56.
REYNOLDSON—
John. *Ulverston.* 1786. W.
Thomas. *Hull.* 1846-58.

REZ, C. *Quebec.* 1861.
RHEAD, William. *Rugeley.* 1835.
RHIND—
Thomas. *Manchester.* 1834-51. Foreign clocks.
Thomas. *Paisley.* 1836.
RHODES—
——. *Bradford.* Same as Manoah Rhodes, q.v.
Edmund. *Kendal.* b.1837-61.
John. *Manchester.* 1848-51.
Joseph. *Hull.* 1813. C.
Manoah (& Son). *Bradford.* 1837-66 (and later).
Thomas. *Kendal.* b.1830-71. W.
Thomas & Edmund. *Kendal.* 1869. C.
William. *Leeds.* 1826. C.
William. *Stoke.* 1850.
RICAUT, Isaac. *London.* 1635.
RICE—
Charles. *Lewiston, Minn, USA.* post-1850.
George F. *Quebec.* 1871.
George Sance. Name used by NY Watch Co., 1872-75.
Gideon. *New York.* ca.1840.
H. P. *Saratoga Springs, NY, USA.* 1827-30. W.
(Charles W.) & HARRINGTON (John E.). *USA.* 1859.
Hugh. *Newry.* 1789.
J. T. *Coventry.* 1880.
Joseph. *Baltimore, USA.* 1784-1801. C.
Joseph T. *Albany, NY, USA.* 1813-50.
Luther G. *Lowell, Mass, USA.* 1835.
Nicholas. *Hayle.* 1844.
Nicholas. *Penryn.* 1813.
Nicholas. *Redruth.* 1823. W.
Phineas. *Boston* and *Charleston, Mass, USA.* 1830.
William (W. C.). *Philadelphia, USA.* 1835-50.
RICEE (or **RICCE**), John. *Ulverston.* 1786-87. C.
RICH—
Abraham. *Bridgwater.* ca.1810-27. W.
Alexander. *Charleston, SC, USA.* ca.1790.
Andrew. *Worcester.* 1872-76.
Charles. *London.* 1828-32. Clock case maker.
Gideon. *New York.* ca.1840.
James. *Bridgwater.* 1861-83.
John. *Bristol, USA.* 1763-1812. C.
Thomas. *Chatham.* 1792.
Thomas. *Worcester.* 1860-76.
& WILLARD (Benjamin Franklin). *Boston, USA.* ca.1842-44.
William. *London.* (B. 1727-77). Clock *case* maker.
William. *Salem, Mass, USA.* pre-1857. W.
RICHARD—
C. A. *Columbus, O, USA.* 1835.
C. A. & Cie. *Paris.* ca.1848-1880s.
& Co. *London.* From *Paris.* Opened 1857-81.
Henry. *Llandeilo.* 18c.
John. *Exeter.* Aged in 1803. W.
John. *Llansamlet.* Later 18c. C.
R. *London.* 1814. W.
RICHARDS—
——. *St. Ives.* 1787. C.
A. *London.* 1857.
Alanson. *Bristol, USA.* ca.1825-ca.1840.
B. (Bryan) & A. (Alanson). *Bristol, USA.* 1828-35.
Bryan. *Bristol, USA.* 1823-33.
(Gilbert) & Co. *Chester, Conn, USA.* ca.1830.
David. *Pontllanfraith.* mid-19c.
David. *Pontypridd.* 1851. C.
George. *London.* 1832. Watch cases.
Henry. *London.* a.1688 to Joseph Armiger. CC. 1699 (B. -1703).
Henry. *London.* 1851-57.
Isiah. *Llansamlet.* ?mid-18c. C.
J. *Edinburgh.* ca.1850. W.
J. *Lancaster.* 1798. W.
J. T. *Llanengan.* 18c. C.
James. *Birmingham.* 1816.
John. *Bala.* 1856-76.
John. *Bilston.* 1828-50.
John. *Llanuwychllyn.* 1795.
John. *London.* a.1654 to William Pettit. CC. 1662.
John. *Tiverton.* ca.1710. C.
L. *Towyn.* Later 19c. W.
Lewis. *Bromyard.* 1870-79.
Llewellyn. *Aberkenfig.* 1887.

RICHARDS—*continued*
Luke. *London.* 1633. CC. 1646-62. Whitefriars.
& MORRELL. *New York.* 1809-32.
Owen. *Bala.* 1844-56. C.
Owen. *Bangor.* 1868-76.
R. *London.* 1876. W.
R. *Willenhall.* 1860.
Robert. *London.* a.1691 to Isaac Lowndes.
Robert. *Uttoxeter.* 1835-42. C.
S. R. *Philadelphia, USA.* 1805.
Samuel. *Paris, Me, USA.* ca.1860.
Seth. *Bristol, USA.* ca.1815.
(Seth) & Son. *Bristol, USA.* 1815.
Thomas, sen. *London.* 1828. Watch cases.
Thomas. *Westerham.* ca.1730. C.
Thomas James. *London.* 1828.
William. *Aberavon.* See Townsend & Richards.
William. *Bala.* s. of Owen. Later 19c.
William. *Poole (Dorset).* 1824-30.
William Henry. *Bristol.* 1850-70. Chrons.
William Joseph. *London.* 1875-81.
William, jun. *Philadelphia, USA.* 1813.
William R. *Petersfield.* 1878.
William Rush. *Bristol, USA.* 1816-85.
RICHARDSON—
Ann. *Easingwold.* 1823. Widow of James, sen.
C. *Oakham.* 1864.
Charles. *Earl Shilton.* 1876.
& Co. *Darlington.* 1898.
David. *London.* 1869-75.
E. *Coventry.* 1860.
F. *Stockton.* 1898.
Francis. *Philadelphia, USA.* 1681-1729.
Henry. *Southport.* 1848.
Holford. *Northwich.* 1834. W.
Holford. *Tarporley.* 1828.
Hugh. *London.* a.1691 to Henry Bradley.
I. *Burton-on-Trent.* 1876.
J. & N. *Boston, USA.* 1803.
James. *Coventry.* 1854-80.
James, sen. *Easingwold.* mar.1802-d.pre-1823. W.
James, jun. *Easingwold.* b.1807. s. of James, sen. -1866. W. & C.
James. *Helmsley.* b.1735-d.1812. C.
James. *Stony Stratford.* 1760. C. & W.
John. *Belfast.* 1692. C.
John. *Boston, USA.* 1805.
John. *Bridlington.* 1690. From *South Cave.* C.
John. *Bubwith (East Yorks).* ca.1780-d.1805. C.
John. *Carlisle.* 1858-69.
John, sen. *Howden.* 1823-46.
John, jun. *Howden.* 1844-58.
John. *Malton.* Late 18c. C.
John. *Manchester.* 1828-51.
John. *Seaham Harbour.* 1856.
John. *Selby.* 1846-71.
John. *St. Helens.* 1770. W.
Joseph. *Hexham.* 1762-d.1777. C.
Joseph. *Northwich.* 1857-78.
Martin. *Little Falls, NY, USA.* 1837-39.
Moses. *Belfast.* 1678. -Free 1680. C. & sundials.
Mrs. Mary Ann. *Carlisle.* 1873-79.
Philip. *Goole.* 1844-71. W.
Philip. *Selby.* 1846.
Richard. *Lancaster.* ca.1840. Clock case maker.
Richard. *London.* a.1667 to Richard Ames, CC. 1675.
Richard. *London.* a.1677 to John Bellard.
Robert. *Middlesbrough.* 1866-98. W. & C.
& RODHOUSE. *Coventry.* 1880.
Thomas. *Brampton.* (B. 1815-)1828. W.
Thomas. *Howden.* 1851-58.
Thomas. *Manchester.* (B. 1804-)1822-51. W.
Thomas. *Richmond Hill, Canada.* 1853.
Thomas. *Yorkville, Canada.* 1861-63.
Thomas Edwin. *Goole.* 1851-71. W.
William. *Balfron (Scotland).* 1828. W.
William. *Bishop Auckland.* 1834-56.
William. *Brampton.* 1828-58.
William (& Son). *Coventry.* 1842-50(54-80).
William. *Easingwold.* b.1803. s. of James, sen. -1844.
William. *Edinburgh.* 1793. (Succeeded later by widow.)
William. *Goole.* 1846-51. W.

RICHARDSON—*continued.*
William. *Helmsley.* 1793-94. C.
William. *Huddersfield.* 1850.
William. *Norfolk, Va, USA.* ca.1796. C.
William. *Selby.* 1823-44.
RICHES—
G. *Halesworth.* 1846.
G. *Lowestoft.* 1853.
John. *Norwich.* ca.1740. C.
RICHMAN—
Henry. *London.* 1863-75.
Isaac. *Philadelphia, USA.* 1850.
RICHMOND—
——. *Chelmsford.* 1808. W.
——. *Springfield (Essex).* 1808. W.
A. *Providence, RI, USA.* ca.1810.
Franklin. *Providence, RI, USA.* 1824-49.
G. *Providence, RI, USA.* ca.1810.
James. *York.* 1823-51.
John. *Chelmsford.* ca.1813-24.
Joseph. *Gatehouse.* ca.1805. C.
Joseph. *York.* b.1757-d.1818. C.
Thomas. *Birmingham.* 1850.
RICHMORE, ——. *London.* ca.1740. W.
RICHSON, J. *London.* 1754. W. (B. pre-1778).
RICHTER—
Frederick. *Carlisle.* 1869.
& STORBECK. *Carlisle.* 1873.
RICKARD—
——. *London.* 19c. C.
Edmund. *Ryde (Isle of Wight).* 1867-78.
Hercules. Probably *Penzance.* mar.1812. C. & W.
John. *Exeter.* Free 1776-1803. Locksmith, clock repairer & W.
John. *Dursley.* 1840-70.
Joseph. *Wootton-under-Edge.* 1850-56.
Thomas. *Wootton-under-Edge.* 1840-42.
RICKETT, William. *London.* 1828-32.
RICKETTS—
Edward. *London.* 1869.
John. *Pershore.* 1835-60.
W. *Leamington.* 1868.
RICKSECKER—
Israel. *Dover, O, USA.* From *Bethlehem, Pa.* 1834-72. C.
J. *St. Clairsville, O, USA.* 1839. C.
RIDDEL—
Donaldson & James. *Aberdeen.* 1846-60.
James. *Aberdeen.* 1853.
John & Co. *Belfast.* ca.1843-80. Clock wholesaler.
Robert. *Belfast.* ca.1825-45. C.
RIDDLE—
David. *Larkhall* nr *Hamilton.* 1860.
James. *Fermanagh, Pa, USA.* ca.1780.
James. *Strabane.* 1813.
Robert. *Old Meldrum.* 1870.
RIDER—
Arthur. *Baltimore, USA.* 1822-24.
Job. (b. ca.1749). *London, Dublin* then *Hillsborough.* ca.1791, then *Belfast* ca.1791-d.1833. C. & W. and math. instrs.
John. *London.* 1832-44.
Richard. *Welshpool.* 1822-40.
Samuel. *Bristol.* a.1844-59.
Tryall. *Manchester.* ca.1750-60. C.
William. *Liverpool.* mar.1790 (B. 1800-03).
RIDGWAY—
John & Son. *Boston, USA.* 1842.
Joseph. *Dukinfield.* 1878.
RIDING, John. *Birmingham.* 1816. Dialmaker.
RIDLEY—
Henry. *London.* 1857-81.
Henry & Son. *London.* 1863-81. (Successor to T. & Son.)
Josiah. *London.* a.1677 to John Clowes. CC. 1685-97. C.
Moses. *London.* 1828.
T., jun. *London.* 1857.
T. & Son. *London.* 1857.
Thomas, sen. *London.* 1832-51.
Thomas. *Woodford.* 1828-51.
William. *Thirsk.* 1795. W.

RIDSDALE, Thomas. *Harrogate.* 1866. W.
RIEDEL, Ludwig. *Chatham.* 1866-74.
RIEDER—
Gehr & Co. *Wimbledon.* 1878.
John. *London.* 1869.
John & Lorenz (& Co). *London.* 1851(57-63).
RIEGO, Miguel del. *London.* 1875-81.
RIEL, George. *Philadelphia, USA.* 1805.
RIEPLE, George. *Swansea.* 1868-71.
RIESLE—
Egedens. *London.* 1839. (See next entry.)
Egidius. *New York.* ca.1840. (See previous entry.)
John. *London.* 1881.
RIESS, Albert. *London.* 1863-81.
RIGBY—
J. *Ashton-in-Makerfield.* 1858.
James. *Liverpool.* (B. 1813-29) 1822-34.
John. *Coventry.* 1828-42.
John. *London.* 1839.
Joseph. *Prescot.* 1797. W.
Nicholas. *Ormskirk.* Died 1754. W.
Richard. *Howden.* 1858.
Thomas. *London.* (B. 1791-1815) 1828.
Thomas. *Prescot.* 1834.
W. *Leigh.* 1858. W.
William. *Coventry.* 1835-54.
William. *Newton (Lancashire).* 1848. W.
William. *Prescot.* 1818-26. W.
William D. *Bala.* 1887.
William D. *Ruabon.* 1887.
RIGG—
John. *Guisborough.* ca. 1740-90. C.
Thomas. *Guisborough.* mid-18c. C.
RIGGINS, Edward. *Oxford.* 1725. Turret clock dial maker.
RIGGS—
Brothers. *Philadelphia, USA.* 1872-79.
(Daniel) & Brother (Robert). *Philadelphia, USA.* 1866-70.
(Daniel) & Co. *Philadelphia, USA.* 1864.
(William H. C.) & Son (Daniel). *Philadelphia, USA.* 1863.
William H. C. *Philadelphia, USA.* 1819-61. C.
RIGHTON, Thomas. *London. Grays Inn Lane.* ?ca.1685. Lant. clock.
RIHILL, William. *Moreton (Hampstead).* Early 19c.
RIHL, Albert (A. M.). *Philadelphia, USA.* 1849.
RILEY—
——. *Elland.* Same man as at *Todmorden.*
——. *Ripponden.* Same man as at *Todmorden.*
David Robert. *Coventry.* 1854.
George. *Halifax.* 1810-14. C.
Gillingham. *Todmorden.* a.1773-1822.
John. *Philadelphia, USA.* 1773-1818. W.
& READ. *Coventry.* 1880.
RIMELL, George. *Gloucester.* 1850-56.
RIMMER—
Henry. *Cronton (Lancashire).* mar.1794. W.
Robert. *Liverpool.* (B. 1825) 1828.
William. *Liverpool.* 1771. W.
RIMMINGTON—
Eli & Son. *Lubenham.* 1876.
(G.) & Son. *Lubenham.* 1849-64.
RIMONDI, Charles. *Halifax.* 1837-66.
RINECKER, Cajetan A. *London.* 1863-81.
RING, John. *London.* a.1686 to Thomas Martin, CC 1693.
RINGLAND—
John. *Dublin.* 1753. W.
William. *Belfast.* ca.1715-50. W. and goldsmith.
William II. *Belfast.* ?From *Dublin.* ca.1757-66.
RIORDAN, Matthew. *Ulverston.* 1869. C., W. and jeweller.
RIOU—
& BOELL. *New York.* ca.1840. French clocks.
E. *New York.* 1840.
RIPLEY—
William. *Grimsby.* 1849-61.
William. *Louth.* 1835.
RIPPIN—
James. *Holbeach.* 1849-61.
James (sen.?). *Spalding.* 1828-35.

RIPPIN—*continued.*
James. *Spalding.* 1868-76.
Mrs. A. *Holbeach.* 1868-76.
Mrs. M. *Holbeach.* 1861.
William. *Holbeach.* 1849-50.
RIPPON—
Charles. *Chelmsford.* 1874.
Ebenezer W. *Brentwood.* 1874.
Edwin. *Sheffield.* 1871.
J. *Chelmsford.* 1866.
Ralph. *Bishop Auckland.* ca.1740. C.
William. *South Shields.* 1834-56.
RISBRIDGE, William. *Dorking.* b.ca.1649, mar.1680. C.
RISBY, Anthony. *London.* 1617-22. W.
RISCH, H. *Ipswich.* 1865.
RISLE, Alois. *London.* 1881.
RISLEY—
John. *Haverfordwest.* 1844.
John. *Londonderry.* 1865.
RITCHIE—
——. *Dundee.* 1831.
Andrew. *Edinburgh.* a.1822.
Benjamin. *Maryland, USA.* 1774 (from *Scotland*).
George. *Arbroath.* 1837.
George. *Philadelphia, USA.* 1785-1811.
James. *Hull.* 1813. W.
James. *Muthill (Scotland).* 1836.
John. *Coupar Angus.* 1847.
Samuel. *Forfar.* 1800-37. (Not *Cupar, Fife* as in B.)
Thomas. *Cupar, Fife.* 1833.
RITTENHOUSE—
Benjamin. *Norristown, Pa, USA.* bro. of David. 1740-ca.1819.
& POTTS. *Morristown* and *Philadelphia, USA.* ca.1770.
RITTERBAND, Henry. *New York.* ca.1830. (One such *Bristol, England.* 1825-30).
RITTSON, John. *Terrill (sometimes Tirril).* (Prob. *North Yorks near Reeth).* Mid to late 18c. C. & barometers.
RIVERS—
Henry. *Nottingham.* 1876.
William M. *Hanley.* 1868-76.
RIVETT, Thomas. *Winslow.* 1847-69.
RIVIERE, John. *Cheltenham.* 1830.
RIVOLTA—
& DEL VECCHIO. *Wellington (Shropshire).* 1863.
Francis. *Wellington (Shropshire).* 1870-79.
RIX—
Isaac. *London.* 1851. C.
Thomas Julians. *London.* 1869-81.
W. P. *Brighton.* 1851.
Wilton Parker. *Brighton.* 1870.
RIXON, John. *Oxford.* 1660-76. Turret clock dial painter.
ROACH, Patrick. *Manchester.* 1848. Imported clocks.
ROATH, R. W. *Norwich, Conn, USA.* ca.1830. W.
ROBATS, Alexander. *Charing (Kent).* 1826.
ROBB—
David. *Alyth (Scotland).* 1860.
John. *Quebec.* 1820-26.
ROBBINS—
& APPLETON. *New York.* ca.1871. W.
G. *Huntingdon.* 1864.
George. *Philadelphia, USA.* 1833-51.
J. *London.* 1844.
Royal E. *Philadelphia, USA.* To 1902.
Thomas. *Buckland (Kent).* 1839-66.
Thomas. *Chatham.* ca.1740-ca.1780. C.
William. *Canterbury.* 1826-32. W.
ROBERDO, Isaac. *London.* ca.1710. (B: has Roberdeau, 1712).
ROBERT—
H. *USA.* 1874.
Jacob. *Stourbridge.* 1850.
Stauffer Son & Co. *London.* 1839-51.
ROBERTO, William. *Oswestry.* 1879.
ROBERTON, George. *London.* 1857.
ROBERTS—
——. *Llanelly.* ca.1820-40.
——. *Otley.* Same man as William Roberts.
Alexander. *Charing (Kent).* 1826-55.
C. *Llanrwst.* 1826.

ROBERTS—*continued*
Cadwaladr. *Tremadoc.* 1840-56.
Candace. *Bristol, USA.* dtr. of Gideon. Dial painter.
Charles. *Leeds.* b.1797-d.1832. C.
D. *Horncastle.* 1861-68.
David. *Prion (Denbigh).* a.1858, d.1906.
E. *Derby.* 1855.
Edward. *Llangefni.* 1820-1904. C.
Elias. *Bristol, USA.* 1727-d.1778. C. Killed by Indians.
Elias & Co. *Bristol, USA.* s. of Gideon. 1808.
Enoch. *Philadelphia, USA.* 1816.
F. *Philadelphia, USA.* 1828.
Frederick. *London.* 1869.
Gabriel. *Ruthin.* Later 19c.
George. *Birkenhead.* 1848.
George. *Liverpool.* 1848.
George. *London.* (B. 1820) 1828-32.
Healey. *London.* 1881.
Hugh. *Gaerwen.* Prob. same man as at *Holyhead.*
Hugh. *Holyhead.* 1868. C.
Hugh. *Llangefni.* ca.1819-42. C.
Hugh II. *Llangefni.* s. of Hugh I. Later 19c. and early 20c. C.
Hugh. *Llanerch-y-medd.* Same as Hugh II of *Llangefni.*
Isaac. *Ruabon.* Late 18c. W.
J. *Darlaston.* 1868.
J. *Downham Market.* 1846-58.
J. H. *Birkenhead.* 1865.
James. *Bath.* ca.1830. C.
James. *Tipton (Staffordshire).* 1876.
Jesse. *Ruthin.* 1887-90. W.
Joah. *Brighouse (Yorks, WR).* 1871.
John. *Aberystwyth.* 1835.
John. *Battle.* 1828.
John. *Burnley.* 1822-34.
John. *Cardiff.* 1858-68.
John. *Cranbrook.* 1826-28.
John. *Criccieth.* 1868-76.
John. *Dudley.* 1850-60.
John. *Folkestone.* 1847-55.
John. *Grimsby.* 1876.
John. *Holyhead.* 1776-91. C.
John. *Llanrwst.* 1887-90.
John. *London.* (B. CC. 1805-20) 1828.
John. *London.* 1869.
John. *Long Sutton.* 1868-76.
John. *Ruabon.* pre-1759-64, then *Wrexham* 1764-post-1771. C.
John. *Talsarnau (Criccieth).* 1874. W.
John E. *London.* 1863-69.
John Richard. *Dudley.* 1868-76.
Jonathan. *Fewston nr Otley.* b.1744. s. of Joseph. d.1785. C.
Joseph. *Fewston nr Otley.* b.ca.1720. s. of William. -1783. C.
Joseph. *Newburyport, Mass, USA.* 1808-52.
& LEE. *Boston, USA.* 1772.
Mrs. Mary Ann. *Selby.* 1871.
Michael. *Bury.* 1848-58. W.
N. H. *Philadelphia, USA.* 1848.
Oliver. *Lancaster, Pa, USA.* ca.1800. Partner with Montandon 1802-08. Later at *Eaton, O, USA.* C.
Owen. *Amlwch.* Prob. same man as at *Caernarvon.* C.
Owen. *Caernarvon.* 1868-90. C. & W.
Owen. *Gaerwen.* 1868-77. C. & W. Prob. same man as at *Caernarvon.*
Owen. *Holyhead.* Prob. same man as at *Caernarvon.* C.
Owen. *Llangefni.* 1835.
Owen. *Llangefni.* *Amlwch.* Same man as at *Llangefni.* 1886-90.
Owen Hugh. *Llangefni.* 1886-90. C.
Peter. *Rye.* 1718. W.
Piercy. *London.* (B. CC. 1816) 1832.
R. M. *Downham.* 1865.
Rees Pughe. *Dolgellau.* 1886-90.
Richard. *Llanymynech.* 1787-1864. Inventor. Made some clocks.
Richard Henry. *Bangor.* 1868.
Richard William. *Birmingham.* 1880.
Robert. *Bangor.* 1835-56. C. & W.
Robert. *Holyhead.* Prob. same man as at *Bangor.* C.
Robert. *Llangefni.* 1827. Prob. same man as at *Bangor.*

ROBERTS—continued.
Robert. Llanrwst. Early 19c. C.
Robert. Over (Cheshire). 1878.
Robert. Ruabon. 1874.
Robert E. London. 1863.
Robert Edward. Canterbury. 1874.
Robert Hugh. Bala. Late 19c.-early 20c.
Roger. Hempnall (Norfolk). 1836.
Roger. London. 1851-81.
S. E. Toronto. 1877.
Silas. Trenton, NJ, USA. 1790-1820.
Thomas. Easton, Pa, USA. 1815-35.
Thomas. London. 1811. W.
Thomas. Ruabon. Mid-18c. C.
Titus Merriman (T. M.). Bristol, USA. s. of Gideon, 1793. mar.1817-56.
W. Bath. 1856-66.
W. Coleshill. 1854.
W. R. Stratford, Canada. 1861-63.
W. R. Toronto. 1875.
Walter James. Mold. 1887.
William. Annapolis, Md, USA. 1745.
William. Bangor. 1856.
William. Fewston nr Otley. b.ca.1700-d.1773. C.
(alias Johnson), William. Flixton (Lancashire). b.1675-d.1755. C.
William. Llangefni. 1835-56.
William. London. 1636-46. CC. Lantern clock.
William. New Hamburg, Canada. 1857.
William. Philadelphia, USA. 1821.
William. Rhyl. 1874.
Wolston. Derby. 1818-49. C.
Wyllys. Bristol, USA. s. of Gideon. 1795-1841. C.
ROBERTSHAW—
J. Pontefract. 1826.
John. Manchester. 1848-51. Imported clocks.
ROBERTSON—
A. Southampton. 1859.
Daniel (& Co.) Glasgow. 1836(-60).
Daniel. Perth. 1860.
David. Kilconquhar. 1860.
David. Perth. 1837.
Duncan. Blairgowrie. 1837.
Ebenezer. Ipswich. 1830-55.
James. St. John, New Brunswick. 1818.
John. Newcastle-on-Tyne. ca.1790-1800. W.
John. Liverpool. 1689-97. W.
Joseph. Kendal. b.1813-61. From Scotland.
Joseph. St. John, New Brunswick. 1842.
Matthew. Biggar. 1860.
Matthew. Mauchline. 1837.
Peter. London. 1851-81.
Thomas. Glasgow. 1837.
Thomas. Rothesay. 1837-60. W.
Thomas William. London. 1875.
William. Dunbar. 1803. W. (Not Dundee as in B.)
William. Falkland (Fife). 1830-60.
William. Perth. 1860.
William. Rutherglen. 1860.
ROBESON—
Isaac. Philadelphia, USA. 1843-46.
William. Newcastle-on-Tyne. (b.1778) 1834-d.1838.
ROBIE—
J. C. Binghamton, NY, USA. 1835.
John. Plattsburg, NY, USA. ca.1817.
ROBINS—
Charles. Bristol. ca.1780. C.
E. W. Gravesend. 1874.
E. W. Maldon. 1866.
Edward W. London. 1857.
J. Brentwood. 1866.
J. Ramsgate. 1866.
John. London. 1828.
Thomas. Barham (Kent). 1874.
ROBINSON—
Abel. Bradford. 1871.
Alexander. Ballyclare. ca.1770-ca.1800, later Larne. Numbered his clocks.
Anthony. Trenton, NJ, USA. 1788-96; Philadelphia 1796-1802. (Partner with Dickman pre-1798.)
Benjamin. Manchester. 1800-51. W.
Brothers. Shrewsbury and Ludlow. 1879.

ROBINSON—continued.
Burnham & Co. Danvers, Mass, USA. 1831.
Charles. London. 1869-81.
Charles. Warrington. mar.1775. W.
Charles & Co. Toronto. 1861-77.
D. Knossington (Rutland). 1864.
Dalton. Guisborough. b.1821, s. of William of Leyburn, -1866.
E. Crewe. 1857-65.
Edward. Rathfriland. ca.1770-80. C.
Edward. Welshpool. 1887-95.
Edward & Co. Shrewsbury. 1879.
& FLINT. Uppingham and Hallaton. 1855.
Francis. Northampton. 1704 (one such at London, a.1685. CC. 1707).
George. Leeds. 1866-71. W.
George. Liverpool. 1828.
H. T. Chesterfield. 1876.
Henry. Halifax. 1866-71.
Henry. Liverpool. 1848-51.
Henry. Shrewsbury. 1870.
Isaac. Philadelphia, USA. 1829-35.
Isiah. Leeds. 1837.
Isiah. Rochdale. 1851.
J. Armagh. ca.1820-ca.1840. C.
J. Ipswich. 1846.
Jacob F. Wilmington, Del, USA. ca.1840.
James. Bala. 1887.
James. Corwen. 1887.
James. Leyburn. 1837. W.
James. London. ca.1760, Well Close Square.
James. Middleham. 1834.
James. Warrington. 1824-34. C. & T.C.
James Greenleaf. Chichester. 1870-78.
Jeremiah A. Lowell, Mass, USA. 1835-37.
John. Barrow-in-Furness. 1866-69.
John. Blackburn. 1893.
John. Chesterfield. 1835-76.
John. Dewsbury. 1834-53.
John. Doncaster. 1725. C.
John. Dungannon. ca.1770-85. C.
John. Haverhill, Mass, USA. 1640; Exeter, NH, USA. 1653.
John. Horsforth nr Leeds. 1871.
John. Winterton (Lincolnshire). 1828-50.
Joseph I. Clones (co. Monaghan). 1824-ca.1855. W.
Joseph II. Clones. 1854-92.
Joseph. Dewsbury. b.1766-d.1834. C.
Joseph. Northampton. 1864-97.
Obed. Attleboro, Mass, USA. ca.1790.
Philip. Bishop Auckland. 1851-56. C.
R. Bradford, Canada. 1862.
Richard. Crosby Garrett. 1849-58.
Richard. Lloydtown, Canada. 1851.
Richard. Sheffield. 1862-71. W.
Robert. London. a.1645 to John Pennock. CC. 1652-62, Lothbury. Lantern clock.
Samuel. Cookstown (co. Tyrone). ca.1800? C.
Samuel. Pittsburgh, Pa, USA. ca.1830.
T. Birmingham. 1860-65. Dial makers.
& TATE. Brigg. 1861.
Thomas. Carlisle. 1828-34. Also jeweller.
Thomas. Kendal. b.1794-1858.
Thomas. London. a.1691 to Jonathan Rant. CC. 1703(-B. 1716).
Thomas. Newcastle-on-Tyne. 1848.
Thomas. Sheffield. 1822-62. W. & C.
W. B. Stockton. 1898.
W. H. Lisnaskea (co. Fermanagh). 1830-45. C.
William. Birmingham. 1842.
William. Birmingham. 1880.
William. Brigg. 1828-50.
William. Chillicothe, O. USA. From Scotland. 1809.
William. Darlington. 1827.
William. Dudley. 1835.
William. Leyburn. 1821-40. C.
William. London. a.1655 to Henry Fetters. CC. 1667. d.1691. W.
William. Middleham. 1866.
William. Rugby. 1850-80.
William. Tadcaster. 1833. Clock repairer.
William & Co. Liverpool. 1848.

ROBINSON—*continued.*
William F. *Philadelphia, USA.* 1835.
William K. *Brownville, NY, USA.* 1828.
William Spence. *Richmond (Yorks).* 1866.
ROBITAILLE, Brothers. *Ottawa.* 1875.
ROBJOHN, Thomas. *New York.* ca.1840.
ROBLIN, L. J. H. & Sons. *London.* 1863-81.
ROBOLD, Zeprian. *Greenock.* 1836 (German).
ROBSON—
Allatson. *Whitby.* mar.1749-51. C. & W.
Brothers. *London.* 1875.
Elizabeth. *Chester-le-Street.* 1827-34.
G. *Brierley Hill.* 1860-76.
J. S. *Newcastle-on-Tyne.* 1858.
James. *Newcastle-on-Tyne.* 1834-36. C.
Joseph. *Bishop Auckland.* 1827.
Joseph. *Coundon.* 1827.
Joseph. *Houghton-le-Spring.* 1827-34.
Mathew. *Belford.* 1848-58. C.
Samuel. *North Shields.* pre-1784-ca.1790. C. & W.
W. *Wooler.* ?18c. C.
William. *Banff.* 1860.
William. *North Shields.* 1827. C.
William. *Wolsingham.* 1851.
William Edward. *Newcastle-on-Tyne.* 1848-58. C.
ROBYN, Walter. *Launceston.* 1460. Set ch. clock.
ROCE, J. H. *Ramsgate.* 1855. Chrons.
ROCHAT—
Henry. *London.* 1851.
Jules. *London.* 1839-69.
ROCHE—
Richard. *Knighton.* 1840-71.
Richard. *Liverpool.* 1848.
ROCHFORD, (W?). *Sunderland.* ca.1830. C.
ROCK—
Conrad & Joseph. *London.* 1851-75.
David. *Leeds.* 1871.
John. *Lichfield.* ca.1750-ca.1760. C.
Thomas. *Derby.* 1849-80.
ROCKE, George. *Kidderminster.* 1860-72.
ROCKFORD—
Watch Co. *Rockford, Ill, USA.* 1873-1901.
Watch Co. Ltd. *Rockford, Ill, USA.* 1901-1915.
ROCKISLAND Watch Co. *Rock Island, Ill. USA.* 1871-74.
ROCKWELL—
Henry. *New York.* ca.1840.
Samuel. *Providence, RI, USA* and *Middletown, Conn.* 1722-73.
RODDEFORD, J. *St. Albans.* 1859.
RODE, William. *Philadelphia, USA.* 1795.
RODELEY, S. S. *Oakham.* 1855.
RODGER—
Alexander. *Campbeltown.* 1837.
Arthur. *Larne.* 1880.
RODGERS—
George. *Larne.* 1865-68.
George. b. ca. 1819 *Whitchurch (Shropshire).*
Market Drayton. 1840-71.
William. *Market Drayton.* 1879.
William. *Philadelphia, USA.* 1824.
RODHOUSE, Frank. *Coventry.* 1868.
RODLEY, Stephen. *Oakham.* 1863.
RODMAN, John. *Burlington, NJ, USA.* 1760-68.
RODMANS, William. *Rhode Island.* ca.1780. W.
RODWELL, William. *Leicester.* 1876.
ROE—
E. *Coventry.* 1854-60.
George Hartwell. *Cambridge.* 1840-75.
Joseph Adolphus. *Ipswich.* 1844-68.
& JACOB. *Liverpool,* 1851.
James. *Swansea.* 1830-35.
Richard. *Epperstone (Notts).* 1680 -d. 1720. C.
Richard. *Midhurst.* (B. 1791-95) 1828. W.
William. *Midhurst.* (B. pre-1821) 1828-62.
ROFFEY—
Edward. *London.* 1815. W.
Edward John. *Birmingham.* 1880.
ROGAN—
Edward. *Liverpool.* 1851.
Thomas. *Belfast.* 1854.
ROGER—
William, jun. *Liskeard.* 1650. Rep'd. town clock.

ROGER—*continued.*
William. *Stonehaven.* (B. 1820)-1846.
ROGERS—
Abner. *Berwick, Me, USA.* 1777. (Prob. later at *Portland, Me,* 1799.)
Caleb. *Newton, Mass, USA.* 1765-1839. W.
Charles, sen. *London.* a.1649 to William Almond. CC. 1657 (B. -1678). *Blackfriars.*
Charles, jun. *London.* s. of Charles, sen., a.1677 to his father.
Christopher. *London.* 1828.
David. *Saffron Waldon.* 1828.
Hannah. *Dudley.* 1828.
Henry. *London.* 1881.
Henry Thwaites. *Ore (Hastings).* 1878.
Isaac. *Marshfield, Mass, USA.* 1800-28.
J. *Halifax, Nova Scotia.* 1818.
J. *Lyndhurst.* 1867.
J. H. *London.* 1875.
James. *Amersham.* 1798-1823.
James. *Birmingham.* 1868.
James. *Dudley.* 1835-50.
James. *New York.* 1872. W. & C.
James M. *Troy, NY, USA.* 1836-40.
John. *Billerica, Mass, USA* to 1770, *Newton, Mass,* 1770-1800. (?*Boston* 1765.)
John. *Daventry.* 1693.
John. *Devizes.* ca.1770. C.
John. *London.* 1851-81.
John. *Redbridge (Hants).* 1839-59.
John. *Stevenage.* 1839.
Joseph. *Cambridge.* 1865-75.
Joseph. *Ripon.* 1851.
Mark. *Romsey.* 1830.
Misses M. A. & B. *Stevenage.* 1851.
Nathaniel. *Windham, Me, USA.* ca.1800.
Paul. *Berwick, Me, USA.* Late 18c.
Peter. *New York.* ca.1830.
Richard. *Dudley.* 1835-72.
Samuel. *Brighton.* 1870.
Samuel. *Coleraine.* 1858-90.
Samuel. *Plymouth, Mass, USA.* 1766-1839.
Thomas. *Bilston.* 1828.
Thomas. *New York.* ca.1820.
William. *Boston, USA.* 1860.
William. *Dudley.* 1828.
William. *Hartford, Conn, USA.* 1837. W.
William. *London.* CC. 1640-d.ca.1665, *Chancery Lane.*
William. *London.* 1844.
William H. *Brooklyn, NY, USA.* ca.1850.
ROGG—
John. *Salisbury.* 1859-75.
& PFRANGLEY. *Salisbury.* 1848.
ROI, Henry. *Hamburg, Pa, USA.* Early 19c.
ROLAND—
Henry. *Albany, NY, USA.* 'est'd. 1832.' C.
& SMYTHMAN. Place unknwon. 18c. C.
ROLEWRIGHT, Thomas. *Oxford.* a.1584-88. Rep'd. T.C.
ROLFE—
Robert. *London.* 1832-81.
Robert. *Streatham.* 1878.
ROLLAND, J. B. *Montreal.* 1857.
ROLLESTON (or **ROULSTON**), James E. *Belfast.* 1865. W.
ROLLIN, ——. *Paris.* pre-1850 -55. C.
ROLLING, William. *St. Austell.* d.1839. W.
ROLLINGS—
W. *Poole (Dorset).* 1848.
Walter. *Lyme Regis.* 1830.
ROLLINS, John G. & Co. *London.* 1881. American clocks.
ROLLISON—
Dollif I. *Halton nr Leeds.* 1721-d.1752. C.
Dollif II. *Halton nr Leeds.* b.1752, after death of father, Dollif I. There till 1773 then *Sheffield* pre-1779-91.
'ROLLO', Edwin. Used by US Watch Co. ca.1868.
ROLLS, Alling. *London.* 1832.
ROMBACH—
Augustus. *London.* 1857-81.
Brothers. *London.* 1869-81.
G. *Aberdare.* 1871-75.

ROMBACH—*continued.*
Joseph. *London.* 1832-57.
Joseph. *Neath.* 1865-71.
Matthias. *London.* 1875-81.
Peter & Co. *London.* 1851.

ROME—
Joseph. *London.* 1863-81.
William. *London.* 1875.

ROMET(T), William. *London.* a.1650 to William Partridge.

ROMLEY, Robert. *London.* ca.1760. C.

ROMNEY—
George. *Sandwich.* 1891.
John. *London.* 1697.

RONCATI, P. *Southampton.* 1848.

RONCHETTI—
Joseph. *Scarborough.* 1858-66. W.
Joseph. *York.* 1837-66. W.

RONCORONI, P. *Basingstoke.* 1839.

RONSON, Peter. *Philadelphia, USA.* ca.1796.

RONTREE, R. *Ebor.* See Rowntree, Ralph. *York.*

ROOE, Richard. *Eperstone.* Lantern clocks. Early 18c.

ROOK, William Henry. *Bradford.* 1871.

ROOKE—
George F. *Cirencester.* 1879.
John. *Cirencester.* 1856-79.

ROOKER, William. *London.* ca.1725. C.

ROOKES, Barlow. *London.* a.1656. CC. 1665. d.1680. *Fore Street.* C.

ROOKS, Edwin. *Taunton.* 1875-83.

ROOME, James H. *New York.* 1854.

ROORBACK, M. *New York.* ca.1770.

ROOT—
George F. Used by Cornell Watch Co., ca.1871-74.
Joel C. *Bristol, USA.* s. of Samuel. ca.1852-67.
Lafayette. *New Haven, Conn, USA.* ca.1850.
Samuel Emerson. *Bristol, USA.* 1820-96. C.
Sylvester S. *Bristol, USA.* 1842-44. *New York* 1848.

ROPER—
James. *Shepton Mallet.* b. 1740. s. of Samuel of *Crewkerne* -ca. 1790. C.
Martin. *Penrith.* (B. 1820-29) 1828-34. W.
Nicholas. *Oakhill (Somerset).* ca.1800-1863. C.
Samuel. *Crewkerne.* b.1707-d.1759. C.
T. *Nottingham.* 1849-55.
William. *Lewes.* a.1743 to John Feron. C.

ROQUES, P. *Villereal.* 18c. C.

ROSE—
Daniel. *London.* CC. 1685-97.
Daniel. *Reading, Pa, USA.* 1749-1827.
Daniel, jun. *Reading, Pa, USA.* ca.1820-37.
& DICKINSON. *Bath.* 1883.
Henry. *Wigan.* (b.1759)-d.1825. W.
Herbert Wesley. *Bath.* 1883.
I. *Castleton (Derbyshire).* 1849.
Jacob. *London.* 1869-81.
James. *Bristol.* 1850-70.
James H. *Ramsgate.* 1866-74.
John. *Caledonia, Canada.* 1861.
John. *Farnworth.* mar.1792. W.
John. *London.* 1832.
Joseph. *Birmingham.* 1854-80.
Joseph. *Bromsgrove.* 1842.
Joseph. *Liverpool.* 1822. W.
Michael. *London.* a.1663 to Thomas Claxton. CC. 1676. C.
T. *Oxford.* d.1772. W.
Thomas, sen. *Wigan.* 1815-24.
Thomas, jun. *Wigan.* 1824-34.
William. *Liverpool.* 1790. W.

ROSEN, Israel. *London.* 1857.

ROSENBAND—
& DREY. *Hamilton, Canada.* 1857.
Leopold. *Hamilton, Canada.* 1862.

ROSENBERG—
Anshell. *Hull.* 1858.
Henry. *Belleville, Canada.* 1862.
Leon. *Cardiff.* 1887-99.
Mayer. *Wolverhampton.* 1850.

ROSENBLATT, Benjamin. *Birmingham.* 1880.

ROSENBLOOM, Michael. *Aberaron.* 1868-75.

ROSENBOHM, Ernest. *London.* 1857.

ROSENTHAL—
B. & Co. *Montreal.* 1875.
H. *Swansea.* Later 19c. C.

ROSEVEAR, John. *St. Austell.* 1788-d.1798. C.

ROSIGNE, John Henry. *London.* 1869-81.

ROSKELL—
John. *Liverpool.* 1848-51. Chrons.
Joseph. *St. Helens.* 1818. W.
Robert & Son. *Liverpool.* 1834. Chrons (B. 1825).

ROSKILLY, William. *Liskeard.* 1856-73.

ROSS—
——. *Winchester.* 19c. carriage clock.
Alexander. *Hull.* 1846. W. Successor to John Ross.
Alexander Coffman. *Zanesville, O, USA.* From *Brownsville, Pa, USA.* 1804-1883.
Charles. *Broughty Ferry (Scotland).* 1828-60.
David. *Dysart.* 1836-60.
George. *Inverary.* 1835.
Henry. *London.* 1881.
Henry. *Montreal.* 1875.
J. *Halifax, Nova Scotia.* 1818.
J. *Stalybridge.* 1865.
J. H. *Battersea.* 1866.
James. *Wrexham.* 1856-74. W.
John. *Hull.* 1834-40. W.
M. *Carlisle.* 1858.
M. S. *Pictou, Nova Scotia.* 1879.
& PECKHAM. *London.* 1811.
Richard. *Ballibay.* 1846.
Richard. *Banbridge.* mar.1820-30. C.
Robert. *New York.* 1854.
Thomas. *Crowle (Lincolnshire).* ca.1770. C.
Thomas. *Hull.* ca.1790-d.1812. C.
Thomas. *Tain.* 1836.
W. G. (& Co.). *Halifax, Nova Scotia.* 1876-78(-81).
William. *Dingwall.* 1849-60.
William. *Dover.* a.1827.
William. *Huntly.* 1836-60.
William. *Stonehaven.* 1846.
William, jun. *Pictou, Nova Scotia.* 1879.
William F. & Co. *Toronto.* 1875-77.

ROSSE—
Samuel. *London.* a.1664 to Thomas Taylor. CC. 1672(B. -1689).
William. *London.* a.1683 to Thomas Wheeler. CC. 1691.

ROSSELOT, P. A. *New York.* 1849.

ROSSENBURG, Samuel. *Toronto.* 1861-65.

ROSSET—
——. *Grenoble.* ca.1600. C.
T. E. & MULFORD. *Elizabethtown, Pa, USA.* 1860-73. Clock case makers.

ROSSI—
G. *Norwich.* 1846.
Theodore. *Norwich.* 1875.

ROSSIN—
M. Brothers. *Toronto.* 1847-61.
Michael & Co. *Montreal.* 1848-50.

ROSSITER—
Ebenezer. *Clevedon.* 1875.
George. *Bridgwater.* 1861-66. C.
George. *London.* 1832.
George. *Weston-super-Mare.* 1875-83.
John. *Weston-super-Mare.* -1861-83.
Thomas. *Penzance.* 1854-73. C. & W.
Walter. *Truro.* 1823-56. W.

ROSTANCE, James. *Manchester.* mar.1791. C.

ROSWELL, William Thomas. *London.* 1857-69.

ROTH—
——. *London.* ca.1720-ca.1740. C.
J. & Co. *London.* 1881.
N. (Nelson). *Utica, NY, USA.* 1837.

ROTHCHILD—
Isaac. *Bristol.* 1840-42.
Joseph. *Bristol.* 1850-56.

ROTHERHAM—
John. *Colchester.* b.1694-1710.
Richard Kevitt & Sons. *Coventry.* 1842.
& Sons. *Coventry.* 1850-80.

ROTHERY, Joseph. *Halifax.* 1866-71.

ROTHROCK, Joseph. *York, Pa, USA.* 1783-90. C.

ROTHSCHILD, Moss Joseph. *Birmingham.* 1868-80. Wholesale.
ROTHWELL, John. *Ormskirk.* 1824-34. C.
ROTTACKER, Matthew. *London.* 1881.
ROUBEL, John. *Bristol.* a.1758-74, later *Bath.* 1774-81.
ROUCKLEIFFE, John. *Totnes.* 1745. W. ?Later at *Bridgwater.*
ROUGH—
Robert. *Bristol.* ca.1725-ca.1755. C. & W. (Also ROACH.)
James. *Kirkcaldy.* 1836-60. C.
ROUGHLEY, Henry. *Burtonwood.* mar.1784-93. W.
ROUGHT, John. *Swineshead (Lincolnshire).* 1850.
ROULSTONE, John. *Boston, USA.* 1786-1803.
ROUMIEU—
Adam I. *London.* a.1657. CC. 1687-97. W.
Adam II. *London.* s. of Adam I. 1689-1707.
James. *London.* b.ca.1668 (B. CC. 1692-1707).
ROUND—
G. *Carshalton.* 1862-66.
Isiah. *Kingswinford.* 1876.
ROUNTON, Percival. *Bridlington.* 1645. Rep'd. ch. clock.
ROUSAILLE, Denis Jacques. *Paris.* 1889.
ROUSCELIN, ——. *Cherbourg.* ca.1790. C.
ROUSE—
Emanuel. *Philadalphia, USA.* 1747-68.
Frederick. *Cheltenham.* 1850-70.
William. *Cheltenham.* 1840-42.
William. *Westbury (Somerset).* 1883.
William Madison. *Charleston, SC, USA.* 1812-88.
ROUTHIER, Alfred C. *Quebec.* 1890.
ROUTIER, Charles. *Quebec.* 1844-63.
ROUTLEDGE—
Adam, jun. *Carlisle.* 1828-58.
Charles Henry. *London.* 1875.
James N. *Carlisle.* 1869-79.
Joseph. *London.* 1832.
R. *Carlisle.* 1873.
William. *Brampton.* 1879. Also silversmiths.
ROUTLEIGH (also **ROUGHLEIGH**), George. *Launceston.* b.1745-d.1802. C.
ROUTLEY, Edwin. *Bath.* 1861-83.
ROUX-BORDIER-ROMAN. *Geneva.* ca.1790. W.
ROUX & CIE. *Montbeliard.* 1867.
ROW—
F. *Liskeard.* 1850.
James. *Farnham.* 1828-39.
Mrs. Ann. *Alton.* 1839-48.
& Son. *Alton.* 1878.
W. B. *Bournemouth.* 1867.
W. T. *Alton.* 1859-67.
William. *Alton.* 1830.
ROWBOTHAM, William Stathern. *Melton Mowbray.* 1855-76.
ROWDEN—
Charles. *Stapleford (Wiltshire).* 1859-75.
Charles John. *Chipping Ongar.* 1874.
John. *London.* a.1683 to Francis Dinnis. CC. 1691-97.
ROWE—
Benjamin Chalwen & J. *Falmouth.* 1823-47.
Frederick. *Great Marlow.* 1869-77.
J. *Beaminster.* 1867.
James. *St. Austell.* 1823. W.
John Smith. *St. John, New Brunswick.* 1855.
Thomas. *London.* a.1691 to George Mertins. CC. 1699.
Thomas Squire. *Abergavenny.* 1858-80.
ROWELL—
George. *Oxford.* b.1770-d.1834. C. & W.
Philip George. *Brighton.* 1851-78.
Richard Rouse. *Oxford.* s. of George. 1834-65. W. & C.
Thomas. *Brighton.* (B. 1822) 1828-55.
ROWLAND—
David. *Aberystwyth.* (B. 1779-) 1816-30. C.
David W. *St. Thomas, Canada.* 1851-62.
Henry. *Liverpool.* 1824.
John. *Berwick-on-Tweed.* 1837.
Joseph. *Burton-on-Trent.* 1850.
Joseph. *Leicester.* 1855-76.
Theophilus. *Dover.* a.1829-55.
Walter. *Yetholm (Scotland).* 1833.
William & Thomas. *London.* 1839.

ROWLANDS—
Christopher. *London.* 1832-39.
David. *Harlech.* 1887.
David. *Pwllheli.* 1844.
Evan. *Meiford and Llanfyllin.* 1890-95. W.
James. *London.* 1881.
John. *Caernarvon.* Prob. same man as at *Pwllheli.* C.
John. *Pwllheli.* 1844.
Robert. *London.* 1832-39. Gold watch cases.
Thomas I. *Bwlch-y-Garreg (Montgomery).* Early 19c.-1841. C.
Thomas II. *Caersws.* s. of Thomas I. b.1841-95.
Thomas. *Machynlleth.* 18c. C.
William. *Aberdaron.* Prob. same man as at *Pwllheli.*
William. *London.* (B. CC. 1820-24) 1828-39. Watch cases.
William. *Pwllheli.* 1821-56. C.
ROWLETT—
Robert. *King's Lynn.* 1865-75.
Robert. *Loughborough.* 1876.
Samuel. *King's Lynn.* 1875.
ROWLEY—
Alfred Francis & Henry George. *London.* 1869.
Arthur & Henry. *London.* 1881. ('Est'd. 1808').
Henry. *Liverpool.* 1834.
Henry. (b. ca. 1768) *Shrewsbury.* 1796-1842. W.
Joseph. *Tarporley.* 1878.
Miss Elizabeth. *Birmingham.* 1868-80.
William. *Liverpool.* 1848.
ROWNTREE—
Ralph. *York.* b.1673-1696. C.
Robert. *York.* 1822-34.
ROWORTH, William. *Bath.* 1875-83.
ROWSE, Thomas. *St. Agnes.* mar.1852-73.
ROWSON—
Christopher (& Sons). *Liverpool.* 1848-51(-1880) Wholesalers.
Christopher & Son. *London.* 1881.
ROY—
David. *London.* CC. 1682-97.
Gabriel. *Helensburgh.* 1860.
John. *Carlisle.* 1873.
ROYAL, The. 1914 and later, product of Pequegnat Clock Co.
ROYALL, Joseph H. *Fortuneswell (Dorset).* 1867-75.
ROYCE, Harvey. *Morrisville, NY, USA.* 1834.
ROYCROFT, Thomas. *London.* a.1681 to Benjamin Harris. CC. 1699-(B. -1717).
ROYDOR, Francis. *Boston, USA.* 1813.
ROYLE—
John. *Bolton.* ca.1760. C. (B. 1738).
& RAWSON. *Wigan.* 1858. W. Successor to George Esplin.
RUBENSTONE, Harris. *London.* 1863-81.
RUBIE, Francis. Name used by National Watch Co. ca.1870-78.
RUBINS, Richard. *Grantham.* a.1761 to John Wood.
RUCK, John Henry. *Chatham.* 1847.
RUDD—
Joshua. *Bradford-(?on-Avon).* ca.1775. C.
Mrs. M. A. *Dudley.* 1872-76.
Thomas. *Dudley.* 1842-68.
RUDGE—
E. *West Bromwich.* 1868.
Emanuel *Berkeley (Gloucestershire).* 1879.
RUDISILL, George. *Manheim, Pa, USA.* ca.1820.
RUDKIN—
Mrs. Henrietta Mary. *Leicester.* 1876.
Thomas. *London.* a.1673 to John Wright, jun. CC. 1683.
RUDLING—
Joseph. *Harleston (Norfolk).* 1875.
W. *Wymondham (Norfolk).* 1846-58.
RUDOLPH, Samuel. *Philadelphia, USA.* 1803.
RUDRUPP, John. *Amersham.* ca.1710.
RUE, Henry. *Philadelphia, USA.* 1835.
RUEMPOL, Hendrik. *Laeren.* 1753. C.
RUFF—
Anton & Godfrey. *London.* 1857-69.
George Henry. *London.* 1875-81.
John. *London.* 1875-81.
Joseph. *Glasgow.* 1860.

RUFF—*continued.*
Loafler & Co. *London.* 1875.
Pius. *London.* 1875-81.
& TRITCHLER. *Bolton.* 1848.
William E. *Washington* and *Halifax, NC, USA.* 1822-29.
RUFFELL, William. *Tunbridge Wells.* 1874. (?Russell.)
RUGENHEIMER, Moses. *New York.* 1854.
RUGLESS—
Samuel. *London.* (B. 1820) 1828.
T. *Bury St. Edmunds.* 1865-79.
T. *London.* 1844-51.
RUH—
J. G. *Camberwell.* 1866.
John George. *London.* 1857.
RUICKBIE, William. *Innerleithen (Scotland).* 1860.
RULE—
James. *Dundee.* 1837.
John. *Kelso.* (B. 1791-1836)-1860. W. May be two such.
John. *St. John, New Brunswick.* 1798.
RUM, John. *Hemel Hempstead.* 1828. (May be Ruth).
RUMBALL, Bryan. *Newbury.* ca. 1645 -d. 1685. C.
RUMBOLD, John. *Salisbury.* 1859-75.
RUMETT, William. *London.* a.1651.
RUMSEY, Charles. *Salem, NJ, USA.* 1820-41.
RUNCORN, Richard & Robert. *Manchester.* 1782-95.
RUNDELL—
Edward. *Norton St. Philip nr Bath.* ca.1710. C.
W. H. *Hamilton, Ontario.* 1875.
RUNDLE, John. *Bratton.* ca.1770-ca.90. C.
RUNGE, Heinrich. *Birmingham.* 1860-80.
RUNGEL, August. *Augsburg.* ca.1660.
RUSH, Edward. *Botesdale (Suffolk).* 1879.
RUSHBRIDGE, William. *Dorking.* 1725. Clocksmith.
RUSHBY, G. *Grimsby.* 1876.
RUSHEN, George. *Highbridge (Somerset).* 1861-83. Repairer.
RUSHTON—
James. *Colne.* 1848.
Joseph. *Oakengates.* 1879.
T. *Ham.* 1862.
RUSSEL(L)—
A. G. *Hamilton, Ontario.* 1895.
A. W. *Toronto.* 1871.
Alexander & Co. *Kirkcaldy.* 1834. Sundial.
Benjamin. *Norwich.* 1830-75. C.
Benjamin. *Thirsk.* 1823.
Benjamin Edward. *London.* 1881.
Charles. *London.* 1811-28. Clock case maker at 18. Barbican.
Charles H. *Lynn, Mass, USA.* ca.1848.
& CLARK. *Woodstock, Vt, USA.* ca.1830.
Cornelius. *London.* b.1608-33. W. (Also Rousel).
D. *Leith.* 1833. W.
Edward. *Cawston (Norfolk).* 1875. Same man as at *Foulsham.*
Edward. *Foulsham.* 1836-75.
Edward. *South Erpingham (Norfolk).* 1836.
Frederick. *Maidstone.* 1874.
Frederick George. *Great Malvern.* 1868-76.
George. *Philadelphia, USA.* 1832.
George. *Walsall.* 1835-42.
Henry. *Hackford (Norfolk).* 1865-75.
Henry. *St. Leonards.* 1870.
Henry George. *London.* 1881.
Hugh. *Moffat.* 1837.
Isaac. *London.* 1828. Watch cases.
James. *Belfast.* 1804-09. W.
James. *Buxton (Norfolk).* 1858-75.
John. *Cawston (Norfolk).* 1830-65.
John. *Falkirk.* 1850.
John. *London.* b.1633. s. of Cornelius. (Prob. in *The Hague.*) In *London* as journeyman to John Nicasius. 1655-56. Then *The Hague* again 1657.
John. *London.* 1828.
John. *Sheffield.* 1822-37. W.
John E. *London.* 1844-51.
John Edward. *London.* 1828-32.
Jonathan. *Geneva, NY, USA.* 1807.
& JONES Clock Co. *Pittsfield, Mass, USA.* 1884-88.
Ltd. *Liverpool.* 1903. C. & W.

RUSSEL(L)—*continued.*
Major John. *Deerfield, Mass, USA.* 1765.
Mark. *Birmingham.* 1880.
Nicasius. b.1642. In *The Hague.* s. of Cornelius. *London.* a.1656 to William Rogers. CC. 1663(-B. -1701).
Nicholas. *St. Neots (Cornwall).* 1609-16. Rep'd ch. clock.
R. *Hamilton, Ontario.* 1862-95.
R. *Upper Sydenham.* 1866-74.
Richard. *Hamilton, Ontario.* 1861-88.
Richard. *Leicester.* 1855-76.
Robert. *Ballymena.* ca.1790-1832. C. & W.
Robert. *Oshawa, Canada.* 1852.
Samuel. *Middletown, Mass, USA.* 1846.
Samuel. *Selkirk.* 1837.
T. *Brandon (Suffolk).* 1858-65.
Thomas. *Charleston, SC, USA.* 1855. *Columbia, SC, USA.* 1856.
Thomas. *Halifax.* 1840.
Thomas. *Ingleton.* mar.1808. From *Lancaster.*
Thomas. *London.* 1844.
Thomas & Son. *Liverpool.* 1848-1902. W.
Thomas & Son. *London.* 1869-75.
Thomas & Son. *Toronto.* 1874-77.
W. *Wells.* 1846.
William. *Wootton (Bedfordshire).* s. of Thomas. b. 1710, d. 1770. C.
William. *Glasgow.* 1827.
William. *Liverpool.* 1834-51.
William. *Wells.* ca.1790. C. (See W. above).
RUSSELLS. *London.* 1881. ('Est'd. 1797').
RUSSWURM, Michael. *Wien.* ca.1710.
RUST—
Benjamin. *Stony Stratford.* 1842.
Jeremiah. *London.* 1844.
William. *Hull.* 1790-1823. W.
RUSTEIN, August. *London.* 1869.
RUSTIN, Joseph. *Oxford.* a.1686. W.
RUTHERFORD—
Alexander. *London.* 1875-81.
George. *Huddersfield.* 1837.
J. *Bradford.* 1866.
John. *Fisherrow, Musselburgh.* 1860.
Walter. *Jedburgh.* 1836.
William. *Hawick.* 1837-60.
William. *Ipswich.* 1865-79.
RUTHVEN & WATSON. *Hamilton, Canada.* 1853.
RUTTER—
Henry. *London.* 1881.
John E. *Ripon.* 1871.
Samuel. *London.* 1851-81.
RUXTON, John. *Carrickmacross (co. Monaghan).* ca.1780. C.
RYALL, Thomas. *Manchester.* 1848-51.
RYALLS, Frederick. *Barnsley.* 1871.
RYAN, Charles Thomas. *Stamford.* 1876.
RYANN—
James. *Armagh.* ?late 18c.-early 19c.
William. *Armagh.* ?late 18c.-early 19c.
RYCKMAN, George R. *Hamilton, Ontario.* 1875-78.
RYCROFT, Timothy. *Liverpool.* d.1704. W.
RYDER—
John. *Camberwell.* 1828.
Peter. *Northwich.* 1865-78.
RYDORE, Francis. *Halifax, Nova Scotia.* 1802.
RYE, G. *Ipswich.* 1865-79.
RYLANCE, William. *Coventry.* 1880.
RYLAND—
James. *Ormskirk.* 1786. (B. 1795-d.1803). W.
John. *Ormskirk.* 1822-29.
RYLEY—
C. *Louth.* 1849.
Elizabeth. *Coventry.* 1828.
Frederick. *Birmingham.* 1880.
John & Co. *Coventry.* 1868.
John Taylor & Son. *Coventry.* 1835-50.
& Son. *Coventry.* 1833-68. W.
William. *Otley.* 1866-71.
William Henry. *Coventry.* 1854-60.
William Taylor. *Coventry.* 1880.
RYNEN, H. *South Holland.* 1804. C.

S

S, J, JS, trademark of Soldano. q.v.
SABER—
George. *Reading, Pa, USA.* Early 19c.
Wolff & Lewis. *Liverpool.* 1848-51.
SACH—
Edmund. *London.* 1857-81.
Edward. *Chelmsford.* ca.1848.
William. *Bocking.* 1828-39.
SACK, Charles. *London.* 1869-81.
SACKERSON, William. *Manchester.* mar.1802. W.
SADD, Harvey. 1776. To *New Hartford* (*USA*). 1798.
mar 1801. To *Austinburgh.* 1829-d.1840.
SADDLETON, James P. *King's Lynn.* 1830-36. C.
SADLEIR, Samuel. *London.* a.1687 to Samuel Vernon.
CC. 1694-(B. 1723). C.
SADLER—
Charles. *Leeds.* 1866. W.
J. G. *Hamilton, Canada.* 1865.
James G. *Montreal.* 1862.
Thomas. *Norwich.* 1725-97. W.
SAER, Joseph. *London.* CC. 1687-97. C.
SAFELY—
Isabella. *Carluke.* 1860. (wid. of John?).
John. *Carluke.* (b.1803), d.1857.
SAFFORD—
(Charles B.) & KAIL (Emil). *Kingston, NY, USA.*
ca.1850. C.
Thomas. *Cambridge.* 1811-30.
SAGAR—
E. *Kirkby Stephen.* Early 18c. C.
Edmund. *Askrigg.* b.1746. mar.1767. Later *Middle-
ham.* 1772-ca.1778. Then *Skipton.* 1793-d.1805. C.
James. *Blackburn.* 1838-51.
John. *Blackburn.* 1834-51. Also jeweller.
Matthew. *Skipton.* b.1778. s. of Edmund. d.1804. W.
Robert Holgate. *Blackburn.* 1822-51. Also jeweller and
silversmith.
SAILOR, Washington. *Pa, USA.* 1825-33. C.
SAINSBURY—
Henry. *Bath.* 1835.
Henry. *London.* 1875-81.
Richard. *London.* 1844-51.
Robert. *Bristol.* a.1675-82.
William. *London.* 1881.
SAINT—
Brothers & Co. *London.* 1869.
Martin. *Paris.* ca.1710. W.
SALIK, Morris. *Kendal.* From *Poland.* b.1829-51.
SALISBURY—
E. *Coventry.* 1854-60. Watch cases.
W. *Ashby-de-la-Zouche.* 1834.
SALKIND, Solomon. *Norwich.* 1858-75.
SALMON—
Alfred. *Cincinnati, O, USA.* ca.1820.
Alfred. *London.* 1844-57.
H. S. *Chatham.* 1855.
John. *London.* a.1648 to Nicholas Tomlins. CC. 1654.
Salmon. *London.* 1851.
Thomas. *Chelmsford.* 1702.

SALMON—*continued.*
William. *London.* (B. 1817-24). 1828.
William H. *Morrisville, NY, USA.* 1830. Also
Cazenovia, NY. 1830-36.
SALMONS, H. *Leyton* (*Essex*). 1866.
SALSBURY—
R. *Upton* (*Worcestershire*). 1860-68.
Robert. *Guildford.* 1878.
W. *Pershore.* 1860.
William. *Evesham.* 1850.
William, jun. *Pershore.* 1868-72.
SALTBY—
John. *Grantham.* 1849-61.
Thomas. *Grantham.* 1828-35.
SALTER—
William. *Manchester.* 1828.
William T. *Southampton.* 1878.
SALTMARSH, Samuel. *London.* 1839. Watch cases.
SALUSBURY, William. *Ormskirk.* 1834.
SALYBACKER, I. *Columbia, SC, USA.* ca.1850.
SALZBERG, Sampson. *London.* 1875.
SAMBROOK, Timothy. *Cardigan.* 1887.
SAMM, Edward. *Linton nr Cambridge.* ca.1770. C.
SAMMON, William. *London.* 1668. CC.
SAMPER, M. & Co. *London.* 1881.
SAMPLE, William. *Boston, USA.* ca.1850.
SAMPSON—
——. *London.* 1622.
Alexander. *Hagerstown, Md, USA.* 1799-1805.
David. *Toronto.* 1862.
E. *London.* 1851.
Henry. *Penzance.* ca.1780. C.
John. *Penzance.* 1752-ca.1775. C.
John. *Truro.* ca.1750. C.
Richard. *Liskeard.* b.1734-d.1814. W. & C.
Richard. *St. Columb.* 1734-1814. C.
Thomas. *Liverpool.* 1851.
Thomas. *Toronto.* 1861.
William. *Philadelphia.* 1803.
SAMSON—
——. *London.* 1787. W.
Lazarus. *London.* 1851-57.
SAMUEL—
A. *Louth.* 1849.
Abraham (& Son). *London.* 1828-39(44-81).
David. *Jersey.* 1832. (One such a. *York.* 1809. Free
1820.).
Edwin L. *Liverpool.* 1848. Wholesaler.
Eliza. *Liverpool.* 1834.
Ephraim. *London.* 1857.
F. & C. *Liverpool.* 1822-24. W.
Flora & Co. *Liverpool.* 1828-34.
George & Israel. *Liverpool.* 1848-51.
Henry. *Liverpool.* 1848-51.
Henry. *London.* 1863.
Hyman. *Charleston, SC, USA.* 1806-09.
J. *Tenterden.* 1811. Bankrupt.
John. *London.* 1662-65. C.
John. *London.* 1857-75.

205

SAMUEL—*continued.*
Lazarus. *Neath.* 1848.
Lewis. *Liverpool.* 1814-48. Also jeweller.
Lewis Henry & Co. *Liverpool.* 1848-51.
Mrs. M. *Richmond* (*Surrey*). 1851.
Moss. *London.* 1851.
Nathan. *Liverpool.* 1828. Also jeweller.
Samuel. *Chichester.* 1828.
Samuel. *London.* 1851-63.
Samuel J. *Liverpool.* 1834.
Samuel Morris. *London.* 1851-81.
Saul. *Liverpool.* 1851.
Sylvester L. *Liverpool.* 1848-51.
Simpson & Co. *Liverpool.* 1822-24.
Thomas. *Richmond* (*Surrey*). 1839.
SAMUELS—
Alexander. *London.* 1869-81.
& DUNN. *New York.* ca.1844. C.
Isaac. *London.* 1828-32.
Lazarus. *Newport* (*Monmouth*). 1844.
Moses. *Cardiff.* 1858-75.
Samuel. *London.* 1857-75.
Samuel. *Louth.* 1828-35.
SANBROOK, John. *London.* a.1663 to William Elmes, math. inst. maker. CC. 1680.
SANCTON, John. *Bridgetown, Nova Scotia.* 1866.
SANDELL, Edward. *Baltimore, USA.* 1817.
SANDERS—
A. *Newbury.* 1854-64.
Charles. *London.* 1839.
G. *Atherstone.* 1854.
George. *Atherstone.* 1828-50.
George. *Featherstone.* ca.1850. C. (?error for *Atherstone?*).
George. *Hinckley.* 1835.
J. *Leicester.* 1849.
J. *Longton.* 1860-76.
James. *Birmingham.* 1880.
James. *Eastover* (*Somerset*). 1861-83.
L. S. *Barrie* (*Canada*). 1862.
Mrs. M. *Atherstone.* 1860.
Mrs. Rachel. *London.* 1875-81. (Successor to Samuel & Co.).
Nathaniel. *Manchester.* (b.1742). mar.1768-ca.1800. In 1825 at *Altrincham,* d. 1828 *Bowden, Cheshire.* C.
Samuel & Co. *London.* 1869.
Thomas. *London.* 1811. Wholesaler.
William. *Stanstead.* 1839-74.
SANDERSON—
George. Place unknown. 18c. C.
John. *Huddersfield.* 1837.
John. *Wigtown* (*Scotland*). 1715.
Robert. *Hull.* 1858.
Samuel. *London.* 1839.
William. *Doncaster.* b.1756-d.1805. W.
SANDFORD, William Charles. *London.* 1828-32.
SANDIFORD, James. *Salford, Manchester.* (b.1725), mar.1762-d.1775. C.
SANDILAND, Charles H. *Welshpool.* 1887.
SANDLES—
John. *Alcester.* 1828.
John. *Feckenham.* 1842-60.
SANDOW, Thomas. *St. Ives* (*Cornwall*). 1844-56.
SANDOZ—
Charles H. *Philadelphia, USA.* 1800-02.
Frederick. *Charleston, SC, USA.* ca.1800.
Gustave. *Paris.* 1867-1900. C.
Louis. *Philadelphia, USA.* 1845.
SANDRIN, Daniel. *London.* CC. 1692-97. (B. French engraver).
SANDS—
J. N. *Macclesfield.* 1857.
James Naylor. *London.* 1828.
SANDY, James. *Alyth* (*Scotland*). 1780-d.1819.
SANFORD—
Abel. *Hamilton, NY, USA.* (1798) 1834-43.
Eaton. *Plymouth, Conn, USA.* ca.1810. C.
Judson. *Hamilton, NY, USA.* 1843-44.
Ransom. *Plymouth, Conn, USA.* 1840.
Samuel. *Plymouth, Conn, USA.* 1845-77.
SANGAMO ELECTRIC Co. *Springfield, Ill, USA.* 1926-28.

SANGSTER, Alexander. *Peterhead.* 1837.
SAN JOSE WATCH Co. *San Jose* and *Alviso, Cal, USA.* 1891.
SANKEY, James. *Warrington.* 1767. W.
SANSOM—
Charles. *London.* 1863-81.
John. *Hemel Hempstead.* mar.1737. C.
William. *Thrapstone.* 1824.
SANSON, John. *New York.* d.1847. Casemaker.
SANT—
R. *Windsor.* 1847.
Robert. *Hull.* 1834.
Samuel. *Norwich.* 1830-36.
SANTON, ——. *London.* Late 17c. Lant. clock.
SANTUCCI, Marcho. *Naples.* Late 17c. C.
SAPHIN—
Peter. *London.* 1828-75.
Thomas. *London.* 1828-63.
SARE, Henry Frank. *Bath.* 1883.
SARGE(A)NT—
Benjamin. *Hastings.* 1862-78.
Henry. *London.* 1828-32.
Jacob. To *Mansfield.* 1784. mar.1785. To *Springfield, Mass.* ca.1787. To *Hartford.* 1795-d.1843. C.
John. *London.* 1881.
John. *Wilmington, NC, USA.* 1821-22.
Joseph. *London.* (B. 1794-1820). 1828.
Joseph. *Springfield, Mass, USA.* 1820 and later.
Thomas. *London.* 1875-81.
SARJE(A)NT, Joseph. *London.* 1832-57.
SARL, John & Sons. *London.* 1844-69.
SARL(E)S—
F. *Fareham.* 1867.
Henry. *Hastings.* 1839.
W. *Peckham.* 1862.
SARNIA, The. 1914 and later, product of Pequegnat Clock Co.
SARRAT, ——. Place unknown. ca.1635. W.
SARTIN, G. *Lewisham.* 1866.
SASSETER, William. *Arundel.* ca.1770. C.
SASSEVILLE—
Francis. *Quebec.* b.1797-d.1864.
& LESPERANCE. *Quebec.* 1854-56.
SATER, Joseph & Co. *Oldham.* 1851.
SATTEL(L)E—
A. *Lincoln.* 1861-68.
George. *Lincoln.* 1876.
SATTELY, Z. *Sherston Magna* (*Wiltshire*). 1859-75.
SAUBER, Herman August. *London.* 1869-81.
SAUDE, Pierre. *Paris.* ca.1660-65.
SAUER, Frederick. *West Chester, Pa, USA.* 1857.
SAUL, Matthias. *Lancaster.* (b.1787-d.1860). C.
SAULET, ——. *Paris.* 1820-50.
SAUM—
John & Co. *London.* 1857.
Joseph. *Mold.* 1887-90.
Matthew. *Perth.* 1860.
SAUNDERS—
Albert. *Birmingham.* 1868-80.
Charles. *Banbury.* mar.1794-1823. W.
Daniel. *London.* (B. 1820-24). 1832-44.
David. *Dumbarton.* 1860.
G. *Atherstone.* ca.1820. C.
George. *London.* 1869.
H. & A. *Montreal.* 1862-76.
& JACOB. *Dorchester.* 1875.
John. *Brackley.* 1824.
John Knight. *Warminster.* 1842.
Joseph. *London.* 1857.
Richard. *Oxford.* a.1674. W.
Thomas. *Banbury.* mar.1795-1810. C. & W.
Thomas. *Dorchester.* 1830-67.
Thomas. *London.* (B. a.1792-1824) 1828-32.
W. *Storrington* (*Sussex*). 1878.
William. *Beaconsfield.* 1869-77.
William. *London.* 1875.
William. *Wallingford.* 1830.
SAUSSE, Richard. *Philadelphia, USA.* pre-1778. Then to *New York.*
SAUTER, Richard. *Hanover, Pa, USA.* ca.1770.
SAUTEUR, freres. *St. Nicolas d'Aliermont.* 1870.
SAUZE, Eugene. *London.* 1881.

SAVAGE—
D. *Montreal.* 1847.
David. *Guelph, Canada.* 1851-63.
George. *Elland.* mar.1787.
George. *Huddersfield.* Late 18c.-1808. Later to Canada.
George. *London.* 1828.
George. *Montreal.* 1818-55.
George. *Toronto.* 1843-53.
George & Son. *Montreal.* 1842-49.
James. *London.* 1881.
John. *Exeter.* Free 1608-d.1627. C.
John. *Exeter.* s. of Peter. Free 1658-71. C.
John. *Clifton (Westmorland).* ca.1730-ca.1740. C.
John. *Shrewsbury.* 1812 -ret'd. 1834, d. 1848.
John Y. *Raleigh, NC, USA.* 1808. (With Stedman 1819-20.)
Joseph. *Montreal.* 1842.
Joseph. *Ormskirk.* ca.1750. C.
(John Y.) & KUNSMAN (Henry). *Salisbury, NC, USA.* 1823.
& LYMAN. *Montreal.* 1851-67.
LYMAN & CO. *Montreal.* 1871-76.
Peter. *Exeter.* s. of John. Free 1628-d.1657. C.
T. *Oxford.* ca.1750. C.
Thomas. *London.* a.1659 to Joseph Quash.
Thomas. *London.* (B. CC. 1825). 1828-32.
Thomas James. *London.* 1839-51.
& VINCENT. *London.* 1811. Wholesale watches.
W. M. *Columbus, O, USA.* 1838-41.
William. *Glasgow, Ky, USA.* 1805-20. Later with Eubanks.
SAVARD—
Albert. *Quebec.* 1870.
Joseph. *Quebec.* 1862.
SAVERY—
——. *Taunton.* ca.1790. C.
Andrew. *London.* a.1669. To Thomas Parker. CC. 1676-97.
John. *Kingsclere.* 1878. Clock cleaner.
SAVIL, William. *Dublin.* ca.1760. C.
SAVILLE, John II. *London.* s. of John I. a.1671 to his father. CC. 1678-(B. 1690). C.
SAVIN—
D. W. *Norwich.* 1858.
David William. *Cromer.* 1865-75.
GAVOI, Gio. *Florence.* 1765. Sundial.
SAVORY—
A. B. & Sons. *London.* 1839-63.
Thomas Cox (& Co.). *London.* 1839 (1851-57). C.
SAW, William. *New Haven, Conn, USA.* ca.1840.
SAWER, Joseph Pobjoy. *Maidstone.* 1839. Clock case maker.
SAWIN—
(John) & DYER (John Wild). *Boston, USA.* 1822-28.
John. *Boston, USA.* 1801-d.1863.
Silas. *New York.* ca.1820.
SAWTELL—
Edwin. *Bristol.* 1838-50.
S. *Bristol.* 1863.
SAWTER, Thomas. *Doncaster.* b.1792-d.1813. W.
SAWYER—
Francis. *Widcombe* nr *Bath.* 1883.
H. *Herne Hill.* 1878.
Henry James. *Coningsby (Lincolnshire).* 1868-76.
& LEVERTT. *Salem, Mass, USA.* (1804). ca.1830-39.
Samuel. *Huntingdon.* (?B. late 18c.). 1830.
& Son. *Peterborough.* 1869-77.
W. H. *Peterborough.* 1854.
SAXTON—
John. *New Market, Canada.* 1857-63.
John. *Toronto.* 1851.
Joseph. *Philadelphia, USA.* 1823.
& LUKENS. *Philadelphia, USA.* ca.1830-40.
SAY, Nehemiah. *London.* a.1648 to Henry Child. CC. 1656.
SAYER(S)—
Andrew. *London.* 1857-63.
J. (?or T.). *Lindfield.* 1696. C.
James. *Bromyard (Hereford).* 1830.
Matthew. *Exeter.* mar.1734-63. C. & W.
Will. *London.* 1758. W.

SAYFRITZ, H. *Wolverhampton.* 1860-68.
SAYLE, Daniel. *Ramsey, IoM.* 1860.
SAYLES, F. A. *Sheffield.* 1822. W.
SAYLOR, George H. *Scarborough.* 1851.
SAYRE—
Elias. *Elizabethtown, NJ, USA.* Late 18c. C.
& FORCE. *New York.* ca.1810.
John. *New York.* 1771-1852. Partner with Thomas Richards.
(John) & RICHARDS (Thomas). *New York.* 1802-1811.
SCALES—
——. *Rydal.* ?mid-19c. C.
Arthur. *Kendal.* 1913.
Edward. *Manchester.* 1848-51.
Henry Fawcett. *Kendal.* s. of William I. b.1829-71.
Isaac. *Kendal.* 1828.
John. *Stockbridge (nr Winchester).* 1878.
Thomas. *Kendal.* b.ca.1825-79.
Thomas & Henry F. *Bowness.* 1858. Same man as at Kendal.
William I. *Kendal.* b.1792-1851.
William II. *Kendal.* b.1852-71. W. s. of Thomas.
SCALPIN, Augustus. *St. Johns, Newfoundland.* 1885.
SCAMBLE, Peter. *Liverpool.* 1851.
SCANLAN, David. *London.* 1851-57.
SCANLON, Thomas. *Armagh.* 1843-46.
SCARFE, Francis. *Colchester.* b.ca.1700-d.1767. C.
SCARISBRICK—
Anthony. *Prescot.* mar.1795-97. W.
James. *Liverpool.* mar.1663. C.
SCARLES, W. H. *Nottingham.* 1876.
SCARRON, James. *Wigan.* 1797.
SCAWTHORPE, William. *Wainfleet (Lincolnshire).* 1828-35.
SCHAAF, John. *Thorold, Canada.* 1861-74.
SCHAAP, Henry. *London.* 1851.
SCHAEFFER, Herman. *Halifax, Nova Scotia.* 1867-81.
SCHAFFER—
John. *London.* 1832-39. Dial maker.
T. C. *Portsmouth, NH, USA.* ca.1840. Invented a fan driven by a clock movement.
SCHAFLE, Matthias. *Colchester.* 1874.
SCHALLER, Andrew Francis. *London.* 1869.
SCHAMEL, ——. *Prague.* Early 19c. C.
SCHARF, J. *Selinsgrove, Pa, USA.* (ca.1778) from *Switzerland.* ca.1810-1859. C.
SCHEBBLE, Joseph & Paul. *Dartford.* 1855.
SCHEID, Daniel. *Sumneytown, Pa, USA.* (1782). mar.1809-1876. C.
SCHELBLE, Joseph. *Dartford.* 1866-74.
SCHELL, Samuel F. *Philadelphia, USA.* 1829-35.
SCHEM—
& FALCONNET. *Charleston, SC, USA.* ca.1800.
J. F. *Charleston, SC, USA.* ca.1785-90.
SCHENK—
F. *Brighton.* 1851-70.
& HEAPY. *Brighton.* 1839.
SCHERER—
Charles S. *London.* 1869-81.
George Frederick. *London.* 1832-51.
SCHERR, Lewis. *Philadelphia, USA.* 1843 and later.
SCHERZINGER—
A. & Co. *London.* 1839.
C. *London.* 1844.
Matthew. *London.* 1869.
P. & M. & Co. *Newcastle-on-Tyne.* 1836. German clocks.
SCHIFER, Andreas. *Graz.* 1573-93.
SCHILSKY, Joseph Henry. *London.* 1844.
SCHINKLE, John. *Philadelphia, USA.* 1810.
SCHLANGE—
Jacob. *Copenhagen.* 1642. C.
Joseph. *Helsingor.* 1647. C.
SCHLEER, Michael. *London.* 1881.
SCHLEICHER, Nicholas. *Brighton.* 1862-70.
SCHLENKER, John M. *Guelph, Canada.* 1861-63.
SCHLESINGER, Casper Wolff. *Liverpool.* 1834.
SCHLIENTZ, Auguste. *Waterloo, Canada.* 1857.
SCHLITZ, Lorenz. *Vienna.* ca.1820. C.
SCHMALZE, F. *York County, Pa, USA.* ca.1850.
SCHMID, John G. *Philadelphia, USA.* 1850.

SCHMIDT—
Aine. *Trier*. Same as Jean I, q.v.
Jean I. *Trier* (also signed Johann). b.ca.1730-d.1805.
  C. & W.
Jean II. *Trier*. s. of Jean I. b.1780-1805.
KETTERER & Co. *Leeds*. 1850. C.
Thomas. *London*. ca.1810. C.
SCHMIT, George. *Neustat*. 1843. W.
SCHMITNAGEL, G. J. *Brighton*. 1851-62.
SCHMITZ, Henry. *Swindon*. 1875.
SCHNABEL, Louis. *London*. 1881.
SCHNEEBERGER, Hans. *Augsburg*. Early 17c.
SCHNEIDER, Peter. *Philadelphia, USA*. ca.1840. C.
SCHNELL, M. *Berwick-on-Tweed*. 1858.
SCHOEMAKER, Peter. *New York*. ca.1810.
SCHOENER—
Christoff. *Augsburg*. ca.1720. W.
Daniel. *Vienna*. pre-1660. W.
SCHOLEFIELD (and variants).
Edmund. *Rochdale*. b.1730-d.1792. C. s. of Major, sen.
Francis. *Dewsbury*. 1834-71. W.
James. *Barnsley*. b.1738-d.1775. C.
John. Place not stated. Prob. the Barnsley maker.
John. *Barnsley*. b.1721-d.1788. C.
Josiah. *Manchester*. mar.1803. C.
Major, sen. *Salford* and *Rochdale*. b.1707-d.1783. C.
Major, jun. *Salford* and *Manchester*. b.ca.1749-d.1813.
  C.
Robert. *Rochdale*. d.1736. C.
Robert. *Rochdale*. d.1759. C.
Robert III. *Rochdale*. 1752-94. C.
William. *Putney*. 1839.
SCHOLLABERGER, John. *Philadelphia* or *Maryland,
  USA*. ca.1800.
SCHOMO, Thomas. *Philadelphia, USA*. ca.1820.
SCHONHARDT—
Andrew. *Dowlais*. 1875-87.
Frank. *Dowlais*. 1871-87.
Karl. *Blackheath*. 1874.
SCHOOF, William George. *London*. b.1830-d.1901. W.
SCHOOLTINK, A. *Steenderedn*. 1796. C.
SCHREIBLMAYR, Johann. *Vienna*. ca.1775. C.
SCHREINER—
Charles W. *Philadelphia, USA*. 1813-33.
Henry M. *Lancaster, Pa, USA*. s. of Martin II.
  b.ca.1855-1939.
Martin I. *Lancaster, Pa, USA*. b.1767-d.1866.
Martin (II) & Philip. *Lancaster*. Sons of Martin I.
  1830-38.
P. & Son. *Columbia, Pa, USA*. ca.1850.
SCHRIBER, Hieronimus. *Brierly Hill*. 1850-68.
SCHRIEBNER, Nicholas. *Cardiff*. 1829-35.
SCHUESSLER, Louis. *London*. 1863-81.
SCHUEZ, George. *Graz*. 1532.
SCHULEN—
& BOBY. *Ipswich*. 1865-79.
Charles (& Co.). *Ipswich*. 1839-53(-58).
SCHULER—
J. *Philadelphia, USA*. 1845.
J. & Co. *London*. 1857.
Joachim & Co. *London*. 1863-81.
M. *Bromley*. 1851.
M. & J. *London*. 1844-51.
Matthew. *London*. 1839.
Philip. *London*. 1863.
SCHULTHEISS—
J. & Son. *Germany*. ca.1870. C.
J. T. *Chatham*. 1874.
SCHULTIS, Sebastian. *London*. 1875-81.
SCHULTZ—
George. *Koningsberg*. 17c. C.
Gottlieb. *Philadelphia, USA*. 1821-44.
Gustav. *London*. 1869-81.
Jacob. *Lancaster, Pa, USA*. ca.1850.
SCHUNDLER, Maurice. *Sheffield*. 1871.
SCHWAB—
B. & Son. *London*. 1869. (Successor to Schwab &
  Marx.)
John George. *London*. 1869.
joseph. *London*. 1851-63.
Lawrence. *Northampton*. 1841. German clocks.
&·MARX. *London*. 1863.

SCHWANFELDER, James & John. *Leeds*. Dial pain-
  ters. ca.1773-1837.
SCHWAR—
Andrew & Co. *London*. 1857-63.
Benitz & Co. *London*. 1875-81.
George. *London*. 1869-75.
& SAUM (& Co.). *London*. 1844(-51).
SCHWARCK, D. *Leamington*. 1860-68.
SCHWARER—
Batz. *London*. 1863-75.
John. *Sheffield*. 1862-71.
SCHWAR(T)Z—
Anthony. *Flint*. 1887-90.
Anthony. *Holywell*. 1868-90.
Franciscus. *Brussels*. ca.1630-pre-1660. W.
George. *York, Pa, USA*. 1775.
Lorenz. *Chipping Barnet*. 1851-74.
Peter. *York, Pa, USA*. 1758-90. C.
SCHWEHR, C. *Wolverhampton*. 1860-68.
SCHWEIG, Conrad. *London*. 1839.
SCHWER—
Alois & Co. *Bristol*. 1850-70.
Ashley & Co. *Bristol*. 1850.
C. *Southampton*. 1848.
C. & Co. *Bristol*. 1856.
Cosman. *Bristol*. 1879.
Joseph. *Bryn-mawr*. 1875.
Xaver. *Salzburg*. Early 18c.
SCHWERER (and Swearer, etc.)—
Charles. *Aberdare*. 1871-87.
Conrad. *London*. 1869-81.
Edward. *London*. 1881.
George. *Bradford*. 1834-37.
George. *Colchester*. ca.1846-55 (also Swearer).
& HOFFMEYER. *London*. 1863-69.
I. *Chelmsford*. 1866. C.
J. *Truro*. 1856.
Jacob. *Falmouth*. 1844.
John. *Sheffield*. 1862.
Joseph. *London*. 1857-63.
Joseph & Charles. *Northampton*. 1864-77.
Karle. *London*. 1857-69.
& KLEISER. *York*. 1866. W.
Lawrence (& Son). *Sheffield*. 1822(-37).
Matthew. *Hull*. 1790-1834. C.
Matthew. *York*. 1839-58. C. & W.
Nicholas. *Colchester*. ca.1840-50.
Peter. *York*. 1827-40. Brother of Philip.
Philip. *York*. 1834-46. C. Brother of Peter.
SCHWERSENSKY—
Isaac. *Chatham*. 1832.
Isaac. *Liverpool*. 1848-51.
SCHWETZER, George. *Fredericktown, Md, USA*. Late
  17c. C.
SCHWING, John G. *Louisville, Ky, USA*. (1783)
  1803-68.
SCHWOB, Brothers. *Montreal*. 1871-76.
SCHWORER, Joseph. *London*. 1881 (prob. Schwerer).
SCMOLTZ, William. *San Francisco*. ca.1850.
SCOBELL, John. *Bodmin*. 1645. Rep'd ch. clock.
SCORAH, John B. *Bradford*. 1866-71.
SCORELL, ——. *Newbury*. ca.1770. C.
SCORPION, ——. *Southwell*. ca.1775. C.
SCOTCHER, Nicholas S. *Wrexham*. 1868-76.
SCOTCHMER—
Charles & Son. *London*. 1857-69.
John Davis. *London*. 1875-81 (successor to Charles).
SCOTT—
——. *Dromore*. Late 18c.-early 19c. C.
A. H. & J. *Montreal*. 1845-63.
(David) & ANDERSON. *Greensboro, NC, USA*. 1829.
Andrew. *Dingwall*. 1794.
Andrew. *London*. 1832.
Anne. *Harrisburg, Pa, USA*. ca.1850. C.
Austin. *London*. 1857-75.
Charles. *Aylesbury*. 1792-1831.
Charles & William & Co. *London*. 1857-63.
& Co. *Glasgow*. 1837. Clock dial makers.
& Co. *Quebec*. 1898.
David. *Dundee*. 1850.
David. *Greensboro, NC, USA*. 1797-d.1875 (of Scott &
  Anderson).

SCOTT—*continued.*
Edward. *Patrington.* 1876.
Edward. *Tadcaster.* 1844.
Frederick. *Dundee.* 1837.
George. *London.* 1832. Watch cases.
Henry. *London.* 1863.
Henry. Place not stated. ca.1760. Prob. Northern English.
J. *Heanor (Derbyshire).* 1855.
J. A. *St. Johns, Newfoundland.* 1877-85.
James. *Armagh.* 1787.
James I. *Ardmillan.* 1750. C.
James II. *Ballynahinch.* ca.1792-1846. C.
James. *Lancaster.* b.1803-32, then *Kendal* 1832-71.
James. *London.* 1832-39.
James. *Rochester.* 1823.
James. *Selkirk.* 1837.
James Wilson. *Kendal.* s. of James. b.1832-79.
John I. *Lisleen (co. Down).* ca.1800. C.
John II. *Ballynahinch.* s. of James. ca.1846-58.
John. *Charleston, SC, USA.* 1803-d.1809.
John. *Jedburgh.* 1860.
John. *Penzance.* 1864.
John. *Peters Township, Pa, USA.* Late 18c.
John. *Sandgate.* 1776-95.
John (Hargreaves). *Burnley.* 1848-51.
Joseph. *Dorchester.* 1848-75.
Joseph. *Leeds.* 1834-66.
Joseph. *Whitehaven.* 1858-73.
Joshua. *London.* a.1674 to Cornelius Harbert.
& KITTS. *Louisville, USA.* 1843-44.
M. *Derby.* 1855.
Morris. *Birmingham.* 1880.
Nicholas. *Maidstone.* 1780. W. (B. 1791-95).
Robert. *London.* (B. 1805-20) 1828-44.
S. *Newtownards.* 1866. W.
Samuel. *Aylesbury.* 1798.
Samuel. *Concord, NC, USA.* 1825.
& Son. *Kendal.* 1858-79 (James and James Wilson).
Thomas. *Downington, Pa, USA.* 1834-50. W.
Thomas. *Gainsborough.* (B. 1795) 1828-50.
Thomas. *London.* 1881.
Thomas. *Sheffield.* 1779. C.
Thomas. *Workington.* 1879.
W. *Christchurch (Hampshire).* 1846. C.
W. D. & Co. *Louisville, Ky, USA.* 1848.
Walter. *Longtown.* 1848-58.
William I. *Ballygowan.* (b.1760.) ca.1782-d.1834.
William II. *Ballynahinch.* (b.1807.) ca.1844-d.1849.
William III. *Ballynahinch.* 1865.
William IV. *Saintfield (co. Down).* Late 18c.-early 19c. C.
William. *Beith.* ca.1790. C.
William. *Christchurch.* 1830-39.
William. *Ferryport-on-Craig (Fife).* 1837.
William. *London.* (B. a.1816-)1832.
William. *London.* 1863-75.
William. *Thornhill (Scotland).* 1855-60.
William & Co. *London.* 1851.
William D. (& Co.). *Louisville, Ky, USA.* 1841, with John Kitts 1843-44, alone 1845-47 (& Co. 1848-49).
William Henry. *Cobourg, Canada.* 1886.
Wing & Co. *London.* 1844-51.
SCOVILLE Manufacturing Co. *New York.* 1872. C.
SCOWN, John. *Launceston.* 1847.
SCRAFTON—
William. *Helmsley.* 1816 from *Rivaulx.* W.
Wilson. *Wilmington, Del, USA.* 1868. C.
SCRIVENER—
——. *Diss.* ca.1790. C. (Prob. Edward, q.v.)
E. *Halesworth.* 1846.
E. *Fressingfield* nr *Halesworth.* 1853.
E. *Soham.* 1846.
Edward. *Diss.* 1830-58.
Edward. *London.* 1863.
J. *Diss.* 1846.
John R. *Kimbolton.* 1854-77.
Philip. *Reading.* 1877.
Philip. *Stowmarket.* 1839-79.
Richard. *London.* 1639- (B. 1668). *Fleet Street.* Engraver.
William Camden. *London.* 1857-81.

SCROCK, John M. *Millerburg, O, USA.* 1846 (with John G. Fischer).
SCRYMGEOUR, James. *Aberdeen.* 1846 (perhaps from *Glasgow* where one such 1816-37).
SCUDDER, Captain John. *Westfield, NJ, USA.* 1767-1848. Clock case makers.
SCULLEN, Benjamin. *Yarmouth.* ca.1770. C.
SCULLY, Michael. *Armagh.* 1784.
SCURR, Richard William. *Thirsk.* 1834-66.
SEABOURNE—
Richard. *London.* 1642.
William. *London.* a.1651 to Samuel Horne. CC. 1659. d.1668. Watch case maker in *Chancery Lane.*
SEABROOK, S. *Hornchurch.* 1855-66.
SEAGER—
John. *Liverpool.* 1822-34.
Robert. *London.* 1857.
SEAGRAVE, Thomas. *Bulwell (Nottinghamshire).* 1876.
SEAGULL, A. *Hull.* 1861. W.
SEALEY—
John. *Egremont.* 1869-79.
W. & J. *Egremont.* 1858.
William. *Egremont.* 1828-34.
William. *Kendal.* ?1828.
William. *Preston.* 1848.
SEALY—
Albert Robert. *London.* 1857.
R. *Peckham.* 1866.
T. *Corston (Wiltshire).* 1859.
Thomas. *Tetbury.* 1840.
SEAMAN—
Edward. *Shifnal.* 1879.
Samuel. *Lowestoft.* 1830.
Thomas. *Edenton, NC, USA.* 1790-1802. Maybe later *New York.*
SEAMARK, Thomas. *Harborough.* ca.1780-ca.1790.
SEAMER—
Abel I. *York.* 1649-d.1682. W.
Abel II. *York.* 1677-1713. W. s. of Abel I.
William. *York.* 1627-49. W.
SEARLE—
George. *London.* 1832-44.
J. J. B. *Great Berkhampsted.* 1874.
SEARLES, H. *Hastings.* 1851-55.
SEARS, John. *Chillicothe, O, USA.* ca.1810.
SEARSALL, Thomas. *New York.* ca.1780.
SEARSON—
John. *New York.* 1757.
Tobias. b. 1683 *Ipswich.* a. 1698 to T. Holborough of *Colchester.* Then *Ipswich* again where mar. 1711, ret'd. 1734, d. 1767. C.
SEARUM, Robert, jun. *Yarmouth.* 1772. W.
SEASON, T. J. *Ottawa.* 1875.
SEBBER, Ann. *Dover.* 1823.
SEBIRE, A. P. *Jersey.* 1832-34.
SECKEL, Myer. *London.* 1844.
SECKER, Richard. *London.* 1839-44.
SECOURABLE, Isidor. *London.* 1875.
SEDDEN, Henry. *Farnworth.* mar.1806. W.
SEDDINGER, Margaret. *Philadelphia, USA.* 1846.
SEDDON—
Alfred. *Forest Hill (Kent).* 1874.
Alfred. *Sydenham.* 1851-74.
Charles. *London.* 1875.
Edward. *Nether Hoyland (Yorkshire).* 1871.
James. *Farnworth.* mar.1800. W.
James. *London.* a.1655 to Isaac Ploveire. CC. 1662.
John. *Wigan.* 19c. W.
Josiah. *Manchester.* mar.1764. W.
Peter. *London.* 1839.
William. *Manchester.* 1834-51. Imported clocks (also at Hulme).
SEDGWICK—
& BISHOP. *Waterbury, Conn, USA.* '1820'.
& BOTSFORD. *Watertown, Conn, USA.* ca.1820.
Samuel. *Bristol.* 1812.
SEDLEY, John. *London.* a.1686 to Cuthbert Lee. CC. 1701 (B. -1732. W.).
SEDMAN, John F. *Scarborough.* 1858-66. W.
SEDWELL, Edward. *London.* a.1656 to Thomas Loomes. CC. 1664. Believed died of Plague 1665.
SEED—
George. *Hamilton, Canada.* 1861.
John. *Liverpool.* 1828.

SEELEY—
George Henry. *Llangollen.* 1887-90.
T. *Coventry.* 1860-68. Cleaner.
SEELY & FREEMAN. *Ogdensburg, NY, USA.* 1830.
W.
SEFTON—
——. *Selby.* ca.1800. C.
Edward. *Tadcaster.* 1775-80. C.
SEGAR, Dornick. *Cleveland, O, USA.* ca.1830.
SEGER, James S. *New York.* 1832. C.
SEGSWORTH—
J. & Co. *Toronto.* 1877.
John C. *Toronto.* 1861-87.
SEIF(F)ERT—
Charles. *Dumfries.* 1887.
Christopher. *Regensburg.* ca.1800. W.
Gustavis & Sons. *Quebec.* 1857-63.
SEIGNIOR, Robert. *London.* a. to John Nicasius to
whom he was related. CC. 1667-d.1687. C. In 1674
appointed Royal Clockmaker *without* fee (East being
the Clockmaker with fee).
SEIGWART, K. *Ipswich.* 1875-79.
SELBY, Peter. *Wareham.* (B. 1760-95) 1824. C. Also
gunmaker.
SELF(E)—
F. *Chatham.* 1858.
F. *Sheerness.* 1851.
F. & H. *Bristol.* 1840.
Frank. *Worcester.* 1868-76.
Francis William. *Erith.* 1874.
Henry. *Bristol.* 1842.
Henry. *Melcombe Regis.* 1875.
Henry. *Poole.* 1867-75.
William. *Brighton.* 1878.
William. *Croydon.* 1878.
William. *Gravesend.* 1866-74.
William. *Sheerness.* 1855.
SELF-WINDING—
Clock Co. *New York.* ca.1888 -present day. C.
Watch Co. *New York.* 1887.
SELIG—
Benjamin. *Penzance.* 1844-47. C.
Edwin & Co. *London.* 1881.
Kaspar. *London.* 1881.
SELINE—
Isaac. *Swansea.* 1875-87.
Moses. *Swansea.* 1868-75.
SELKIRK, Samuel. *Kalamazoo, Mich, USA.* ca.1850.
SELLAR—
Alexander. *Reading.* 1854-77.
John. *Elgin.* (B. 1820-37)-60.
SELLEN, George. *Moulsham (Chelmsford).* 1609-39.
SELLER, Thomas. *Toronto.* 1861.
SELLICK, James. *Marazion (Cornwall).* ca.1740. C.
SELLMAN, Aaron. *St. Leonards.* 1851-78.
SELWOOD (or SELLWOOD)
John. *London.* CC. 1641-d.1651. C. Bro. of William.
William. *London.* Bro. of John. CC. 1633-d.1653. C.
The Mermaid in Lothbury.
William. *Wootton Bassett.* 1867-75.
SEMMENS, Herman. *Truro.* 1834-44. W.
SENDRE, Lawrence. *London.* 1662.
SENECAL, Louis & Co. *London.* 1875-81.
SENG & HESS. *New York.* 1854.
SENNERT, F. L. *Lititz, Pa, USA.* ca.1850. C.
SEPHTON—
Edward. *Prescot.* 1826. W.
John. *St. Helens.* mar.1768. W.
Kendrick. *St. Helens.* 1826-34. W.
Kenwright. *Prescot.* 1795-98. W.
Luke. *Prescot.* 1796. W.
Peter. *Prescot.* 1796-97. W.
SERGEANT, Edward. *London.* 1869.
SERJEANT, Joseph. *London.* 1863.
SERMON, Joseph. *London.* a.1675 to Morgan Cave.
SERVOSS—
Charles. *Philadelphia, USA.* 1849.
Joseph S. *Philadelphia, USA.* 1850.
SESSFORD—
Joseph. *Newcastle-on-Tyne.* (B. 1816-24) 1827-34.
C.
Robert. *Newcastle-on-Tyne.* 1836.

SESSIONS—
Calvin (& Sons). *Burlington, Conn, USA.* 1845.
Clock Co. *Bristol-Forestville, Conn, USA.* 1903 and
later.
William E. *Bristol, Conn, USA.* Grandson of Calvin.
1857-1920.
SETH, Thomas. Clock Co. *London.* 1875-81.
SETTER, Robert. *London.* 1760.
SETTLE—
Brothers. *Coventry.* 1868-80.
& Co. *London.* 1881.
Hannah (& Sons). *Coventry.* 1842-50 (1854-60).
SEVIER, George Albert. *Gloucester.* 1863-79.
SEWEL(L)—
Charles. *Birmingham.* 1880.
Charles. *Toronto.* 1837-47.
Christopher. *Bradford.* 1814-52. C.
Henry. *Bradford.* 1866.
J. *Buntingford (Hertfordshire).* 1851-59.
John. *Louth.* ca.1770-ca.1790. C.
John William. *Grantham.* 1876.
Thomas. *Newcastle-on-Tyne.* 1848-58.
W. *Hadleigh.* 1858-65.
SEWILL—
John. *London.* 1875-81.
Joseph. *Liverpool.* 1848-51. Chrons.
Joseph. *London.* 1881. Chrons. 'Maker to the Queen of
Spain'.
SEXTON, W. *Holt.* 1846-65.
SEXTY—
Brothers. *Grantham.* 1868-76.
R. W. *Grantham.* 1861.
Richard. *Spilsby (Lincolnshire).* 1849-50.
SEYER—
Andrew. *London.* 1857-81.
John. *London.* 1869-81.
SEYMOUR(E)—
BALL & Co. *Unionville, Conn, USA.* ca.1830. C.
(H. A.) & CHURCHILL (John). *Bristol, USA.* 1846-52.
Henry Albert. *Bristol, USA.* (b.1818. mar.1844) 1846.
John. *Wantage.* 1712. Lant. clock. (B. d.1760.)
& MOORE. *London.* 1839.
Sylvester. *Pittsburgh, Pa, USA.* ca.1840.
Williams & Porter. *Farmington, Conn, USA.* 1832-37.
C.
SHACKELL, Joseph. *London.* 1869-81.
SHACKLETON—
John. *Heptonstall.* mar.1694-ca.1717, then *Foulridge
(Lancashire).* d.1725. C.
Percival. *Sowerby Bridge.* 1871.
SHACKLOCK—
——. *Stanfree (nr Bolsover).* ca.1810. C.
Francis. *Bolsover.* 1835-64.
Francis. *Stanfree.* ca.1750. C.
Godfrey. *Stanfree & Bolsover.* 1795-1849. C. & W.
SHADE, John V. *Philadelphia, USA.* 1845-47.
SHADFORTH, Whitacker. *Richmond, Va, USA.*
ca.1796.
SHAEFFER, Benjamin. *Elizabethtown, Pa, USA.*
ca.1850.
SHAFFER, Philip. *Lancaster, Pa, USA.* 1788-1802.
SHAKEL, Robert George. *London.* 1881.
SHAKESHAFT, Lawrence. *Preston.* 1782-1848.
SHAKESPEARE, Thomas. *Burton-on-Trent.* 1828-35.
SHALLENBERGER, John. *Philadelphia, USA.*
ca.1840.
SHALLER, Nicholas. *London.* a.1672 to James Grimes,
engraver.
SHALLON, George John. *London.* 1828.
SHANKS, Robert. *Blyth.* 1848.
SHANNAN, Samuel. *Castleblaney (co. Monaghan).*
1824-46.
SHAPPENE, Solomon. *Walsall.* 1850.
SHAPPS, Obadiah. *Amherst, Nova Scotia.* 1866.
SHARF, John. *Mifflinburg, Pa, USA.* 1820-26. C.
SHARKEY, Richard. *Birmingham.* 1828-50.
SHARLAND—
Frederick George. *Bath.* 1875.
W. *Speenhamland (Berkshire).* 1864.
SHARMAN—
George. *Lowestoft.* 1830-65.
John. *Melton Mowbray.* ca.1800-1835. C.

SHARMAN—*continued.*
& Son. *Melton Mowbray.* 1849.
SHARP(E)—
Francis. *Dumfries.* 1837-49.
G. & H. *London.* 1844.
George. *London.* 1863. (Successor to Hadarezer.)
George & Hadarezer. *London.* 1851.
George Holderness. *Sheffield.* 1861-71. W.
Hadarezer. *London.* 1857.
J. *Stamfordham.* 1858. C.
James. *Northampton.* (B. from 1812). 1830.
John. *Doncaster.* 1717. C.
John. *Newcastle-on-Tyne.* 1834-36.
John. *North Shields.* 1848.
John. *Sheffield.* 1861-71. W.
John. *York.* 1866. W.
John Bunyea. *Faversham.* 1823-32.
John & Son. *London.* 1828-39.
L. L. *St. John, New Brunswick.* 1871-77.
Robert. *Coldstream.* 1825.
Robert. *Jedburgh.* 1815-25.
Samuel. *East Retford.* 1864-76.
Thomas H. *Thornley Chapelry.* 1856.
Thomas H. *Mexborough.* 1871.
Walter. *Howden.* ?later 18c. C.
William. *Bourne.* 1835-50.
William. *Coventry.* 1880.
William. *London.* a.1672 to John Ebsworth. CC. 1681
   (B. -1699). W. & C.
William. *Retford.* 1828-55.
William. *Snaith.* 1834-44. W.
William Henry. *London.* 1869-81.
SHARPEY, John. *Canterbury.* 1714. W.
SHARPLES, James. *Liverpool.* mar.1785-1815. (B.
   -1829). W.
SHARPLEY, Rice. *Montreal.* 1857-76.
SHARPS, Thomas. *Birkenhead.* 1878.
SHARRATT, James. *Stone.* 1828-50.
SHARROCK(S)—
George. *Downpatrick.* 1774-1800. W.
Henry. *Wigan.* 1822. C.
SHAVE, Richard. *London.* 1863.
SHAW—
B. E. *Newport, Vt, USA.* ca.1860.
C. *Coventry.* 1880.
C. S. *Ottawa.* 1875.
Caleb. *Kensington, NH, USA.* 1717-91. C.
David. *Leicester.* 1828-49.
David. *Plainfield, Mass, USA.* 1792-1882.
E. *Aberavon.* mid-19c. C.
E. *Billingborough.* ca.1800. C.
E. S. *Chesterfield.* 1864.
Edward S. *Worksworth.* 1828-55.
G. & A. *Providence, RI, USA.* 1810.
George. *London.* 1869.
George. *New York.* ca.1840.
H. C. *Middlesbrough.* 1898.
Harry. *Lymington.* 1839.
Hunter. *Downpatrick.* 1840-54. C.
J. *Nottingham.* 1855.
Jacob. *Whitehouse (co. Antrim).* 1849-80.
James. *Halifax.* 1820-50. C.
James. *Kilrea (co. Derry).* 1846. C.
James. *Leicester.* 1864-98.
James. *Toronto.* 1843.
James F. *Halifax, Nova Scotia.* 1790-1830.
John. *Lancaster.* Free 1801. (s. of Thomas of
   Lancaster).
John Henry. *London.* 1881.
Joseph. b. 1827 *Malpas, Chester* 1849,
   Wellington 1854-75.
Joseph. *Whitchurch.* 1879.
Joseph K. *Philadelphia, USA.* ca.1770.
Mrs. & Co. *Leicester.* 1855.
P. *Olney, Ill, USA.* ca.1860. C. & W.
Richard. *St. Helens.* 1781. W.
Robert. *Banbridge.* 1825-66. C.
Robert. *Belfast.* Successor to Henry Montgomery.
   1818-49. W. & C. One time partner of Cochran &
   Shaw.
Robert. *Halifax.* a.1762. C.
S. B. *Boston.* 1849.
Samuel. *Spilsby (Lincolnshire).* 1861-76.

SHAW—*continued.*
Samuel B. *Falkingham (Lincolnshire).* 1861-76.
Seth W. *Providence, RI, USA.* 1856.
T. *Spalding.* 1849-68.
Thomas. *Donnington.* 1850.
Thomas. *London.* 1828.
Thomas. *Lymington.* 1830.
Thomas. *Toronto.* 1837.
Thomas M. *London.* 1832.
William. *Liverpool.* 1790. W.
William. *London.* 1863-81.
William James. *Lincoln.* 1868-76.
William John. *Belfast.* 1846-68.
SHEAM, Francis. *New York.* ca.1810.
SHEARD, J. J. *Dukinfield.* 1878.
SHEARER—
James. *Bromley.* 1839.
James. *London.* (B. 1825-)1832-51.
Marvin. *Akron, O, USA.* 1933. 'Electric wonder clock'.
Mrs. Charles. *Glasgow.* 1836.
& WALKER. *Glasgow.* 1836.
SHEARING, Henry. *Sheffield.* 1861-71. W.
SHEARMAN, Martin. *Hingham, Mass, USA.* 1821.
SHEARSMITH, Robert. *London.* (B. 1817-24). 1832-51.
SHEATHER, Daniel. *Cranbrook.* 1713. Rep'd ch. clock.
SHEEN—
John Samuel. *London.* 1832.
William. *Hope nr Shrewsbury.* 1863-70.
SHELDON, James. *Birmingham.* 1842-60.
SHELDRICK, R. *Whittlesford (Cambridgeshire).*
   1858-65.
SHELICHER & RABOLD. *Birmingham.* 1860.
SHELLEY, Francis. *Brighton.* 1862-78.
SHELTON—
John George. *Botesdale (Suffolk).* 1875-79.
Samson. *London.* BC. 1623. MCC. 1634-D.1648.
Thomas. *London.* CC. 1635. (Clerk).
William. *Alfreton.* 1876.
SHENFIELD—
John. *Manchester.* 1848-51. Imported clocks.
John. *Tenbury.* 1868-76.
SHENTON, ——. *Leicester.* ca.1840. C.
SHEPHERD/SHEPHARD/SHEPPARD and variants.
   (See also SHIPPARD.)
——. *Sheffield.* ca.1750-60. C.
Benjamin. *Edenbridge.* 1874.
& BOYD. *Albany, NY, USA.* 1810-29.
C. *Harleston (Norfolk),* 1846.
Charles. *Lewes.* 1870.
Charles (& Son). *London.* 1832-63. (1869-81).
Charles. *Walsall.* 1860-76.
Edward. *Bristol.* 1830-79.
Eli. *St. Neots.* 1877.
Frederick. *London.* 1863.
George. *Cleobury Mortimer.* 1870.
J. H. *Hanley.* 1860.
Nathaniel. *New Bedford, Mass, USA.* ca.1810. C. & W.
& PORTER. *Waterbury, Conn, USA.* ca.1800-07. C.
& REED. *London.* 1881.
Thomas. *Helston.* Watch dates unknown.
Thomas. *Liverpool.* 1790-95. W.
Timothy B. *Utica, NY, USA.* 1834.
William. *Liverpool.* 1848-51. Chrons.
William Robert. *Pontefract.* 1822-66.
SHEPHERDSON, Richard Porter. *Doncaster.* 1858-71.
SHEPLEY, Edward. *Manchester.* mar.1788. C. (B.
   1790-1800).
SHEPPERLEY—
Anthony. *Nottingham.* (B. 1814-18). 1828.
G. *Nottingham.* 1864.
& PEARCE. *Nottingham.* 1835.
SHERBIRD, Jer. *London.* 1828. (B. J. 1820).
SHERBROOKE, The. 1914 and later, product of
   Pequegnat Clock Co.
SHERBURD, Jonn. *London.* 1857-69.
SHERGOLD—
R. *Epping.* 1855-66.
Richard. *Chipping Ongar.* 1874.
SHERIDAN, Frederick. *Pocklington.* 1830. C.
SHERLOCK, George. *Dunadry (co. Antrim).* 1785-
   1800.

SHERMAN—
C. R. & Co. *New Bedford, Mass, USA.* ca.1860.
Robert. *Philadelphia, USA.* 1799.
Thomas. *Bristol.* a.1783-1830.
William. *Philadelphia, USA.* Early 19c.
SHERRARD, H., L. J. & E. *Belfast.* 1858.
SHERRATT—
Enoch. *Burslem.* 1868-76.
J. *Congleton.* 1857-65.
J. *Wigan.* 1858. W.
R. *Tilley (Gloucester).* ca.1725-30. C. (One such (Ralph) a.*London.* 1700. May be same).
SHERRING—
John. *Alford (Lincolnshire).* 1876.
John K. & Co. *Manchester.* 1848-51.
SHERRY & BYRAM. *Sag Harbour, NY, USA.* 1851. T.C.
SHERWIN—
Joseph. *Burton-on-Trent.* 1850-76.
William. *Buckland, Mass, USA.* ca.1830. C.
SHERWOOD—
James. *Yarm.* b.1800. s. of Thomas I. -1834. C.
John. *Faversham.* 1845-55.
John. *Hythe.* 1839.
R. *San Francisco.* ca.1850.
Thomas. *Doncaster.* b.1748-d.1792. W.
Thomas. *Prescot.* 1796-98. W.
Thomas. *Prescot.* 1826. W.
Thomas I. *Yarm.* b.1744. To *Yarm.* 1771-d.1821. C.
Thomas II. *Yarm.* s. of Thomas I. b.1776-d.1836. C.
Thomas III. *Yarm.* s. of Thomas II. b.1811-1840. C.
Thomas. *York.* mar.1802. Later at *Leeds.* W.
W. *Sittingbourne.* 1855.
SHERZINGER—
Robert. *London.* 1881.
T. *Farnborough.* 1851-55.
T. *Frimley (Surrey).* (1851). 1855-66.
SHETHAR, Samuel. *Litchfield, Conn, USA.* 1795-98.
SHEWELL. John. *Oxford.* b.1632-1676. T.C. (Also Showells.)
SHEY, Joseph. *Scarborough.* 1846-66. W. & C.
SHIDET, V. *Shreveport, LA, USA.* ca.1850.
SHIELDS, Thomas. *Philadelphia, USA.* 1769-94.
SHIELS, John. *London.* 1828-39.
SHIER, Thomas. *Banff.* 1837.
SHIERS, Martin R. *Barnsley.* 1871.
SHIERWATER & LLOYD. *Liverpool.* 1912. T.C.
SHILDRAKE, William. *Norwich.* 1830-58.
SHILL—
& PLATNAUER. *Pontypool.* 19c. W.
Samuel. *Abersychan.* 1830. C.
Samuel. *Bristol.* a.1800-1812.
SHILLING—
E. *Milton (Kent).* 1866-74.
Edward. *Milton (Kent).* ca.1800-38. C.
Edward. *Sittingbourne.* 1839.
James. *Milton (Kent).* 1823-28.
John. *Ashford.* 1847.
Mrs. Harriet. *Milton (Kent).* 1847-55.
SHILTON, George. *Cockermouth.* 1879.
SHIMELS, T. *Market Rasen.* 1868-76.
SHIMER, John. *Philadelphia, USA.* 1811.
SHIN, Thomas. *Mathern (Wales).* mid-18c. C.
SHINDLER—
Thomas I. *Canterbury.* ca.1690. W.
Thomas II. *Canterbury.* s. of Thomas I. Free 1707. W.
Thomas III. *Canterbury.* s. of Thomas II. Free 1747.
Thomas. *Romney.* pre-1746. W.
SHINKLE, John P. *Philadelphia, USA.* 1824.
SHINWELL, Henry. *Bakewell.* 1849-76.
SHIPHAM, John. *Hull.* 1834-40. W.
SHIPHERD, Arthur. *New York.* 1764.
SHIPLEY—
James. *Derby.* (B. 1809). 1828-35. W.
Mrs. A. *Derby.* 1849.
SHIPMAN, Nathaniel. *Norwich, Conc, USA.* b.1764-ca.1800. C. Later a farmer. Died 1853.
SHIPP—
& COLLINS. *Cincinnati, O, USA.* ca.1830.
S. A. M. *Cincinnati, O, USA.* From *Va.* 1820.
SHIRE, Isaac. *Hamilton, Canada.* 1862.
SHIRT, William. *London.* (B. a.1784-)1828-44.

SHIRTON, John Gaffee. *Richmond.* 1843. C.
SHOEMAKER—
Abraham. *Philadelphia, USA.* 1846.
Benjamin. *Philadelphia, USA.* ca.1800.
David. *Mount Holly, NJ, USA.* 1802-10. C.
SHOLE—
Simeon (?or Samuel). *Deptford.* 1839-49.
Simeon. *London.* 1851-57.
SHOLL (or SHOLE), Robert. *Truro.* 1765-1815. C.
SHON, Shenkyn. *Pont-nedd-Fechan.* 1795-1883. C. (Not earlier, as one clock dated 1714 suggests).
SHONBECK, Frederick. *Hamilton, Ontario.* 1921.
SHORE—
A. *Canterbury.* 1851-55.
Alfred. *Birmingham.* 1850.
William. *London.* 1863.
SHOREY—
W. *Melcombe Regis.* 1867.
William. *Swanwich (Swanage).* 1830-55.
SHORT—
John. *Halifax, NC, USA.* From *London.* 1792-1819.
Joshua. *London.* a.1656. To Lancelot Meredith. CC. 1665(-B. 1682).
William. *Lymington.* 1839.
SHORTILL, William J. *Markdale, Ontario.* 1913.
SHORTLAND—
Samuel. *Hanslope.* b.1722-49. Then *Ampthill.* 1749-57.
Thomas. *Stony Stratford.* bro. of Samuel. ca.1778. C.
William. *Stony Stratford.* bro. of Samuel. b.1716-mid century.
SHORTMAN—
J. S. *Newnham.* ca.1820. C.
Josiah William. *Newnham.* 1842.
S. *Brentwood.* 1855.
Samuel. *Newnham.* 1830-50. Also cabinet maker.
Samuel. *Penzance.* 1823-64.
Samuel Tompson. *Bristol.* 1812.
Samuel & William. *Newnham.* 1856-70.
William. *Newnham.* 1879.
SHORTO—
David. *London.* 1857.
Edward. *Yarmouth.* 1836-46.
William Charles. *London.* 1863-69.
William Charles. *Midhurst.* 1878.
SHORTSINGER, Engelbart & Co. *Belfast.* 1868-94. C.
SHOTTLANDER, J. W. & Co. *Stratford-on-Avon.* 1868.
SHOURDS (SHROUDS), Samuel. *Bordentown, NJ, USA.* 1718-58.
SHOWELLS, John. See Shewell.
SHREVE—
BROWN & CO. *Boston, USA.* 1856.
CRUMP & LOW. *Boston, USA.* 1869.
G. C. & Co. *San Francisco.* ca.1860.
STANWOOD & CO. *Boston, USA.* 1860.
SHREEVE, William. *Halifax.* b.1774-d.1817. Clock dial painter.
SHREIBER, W. *Maes-teg.* 1875.
SHREWSBURY CLOCK & INSTRUMENT Co. *New York.* 1935-47.
SHRIVELL—
Mrs. R. *Brighton.* 1878.
R. *Brighton.* 1851-62.
Richard. *Brighton.* (B. 1822). 1828-39.
Robert. *Brighton.* 1870.
SHROETER, Charles. *Baltimore, USA.* 1807-17.
SHUFFLEBOTHAM—
Charles. *Coventry.* 1880.
Joshua. *Coventry.* 1850-68.
Samuel & Joshua. *Coventry.* 1835.
William. *London.* 1863-69.
SHUKER—
Aaron. *Birmingham.* 1854-80.
Aaron. *Birmingham.* 1868. Clock cases (evidently same man as above).
SHULER—
David C. *Trappe, Philadelphia, USA.* 1846-1901. Then to *Norristown, Pa.*
John. *Philadelphia, USA.* 1849.
SHUMAN, John. *Easton, Pa, USA.* ca.1790.
SHURLEY, John. *Albany, NY, USA.* 1839.
SHUTE, George. Place unknown. ca.1700. C.
SHUTT, Caspar. *London.* CC. 1647.

SHUTTLEWORTH—
Francis. *Sarum.* ca.1750-ca.1760. C.
Henry. *London.* a.1662 to Edmund Gilpin. CC. 1669.
Henry. *London.* 1875.

SHUTZ—
Gustavus. *Philadelphia, USA.* 1825-33.
Peter. *York, Pa, USA.* 1758-90. C. (Swiss).

SIBBALD, William. *London.* (B. 1817-)1828-39.

SIBLEY—
Asa. *Woodstock, Mass, USA.* 1764. mar.1787. To *Walpole, NH.* 1790. To *Rochester, NY.* 1808-d.1829. C.
Charles E. *Portsmouth.* 1878.
Clark. *New Haven, Conn, USA.* 1810-17.
Gibbs. *Sutton, Mass, USA.* b.1765, mar.1788. Later to *Canandaigua, NY.*
J. *Hampstead Norris.* ca.1830. C.
James. *Rochester, NY.* 1847.
& MARBLE. *New Haven, Conn, USA.* 1801-07. C. & W. and swords.
Richard S. *Boston, USA.* 1842.
Stephen. *Great Barrington, Mass, USA.* ca.1790-1816. C.
William S. *Pontesbury.* 1879.

SICHEL—
Leon. *Birmingham.* 1880.
Leon. *Chaux de Fonds (Switzerland).* 1881. W.

SIDEBOTHAM/SIDEBOTTOM—
George & Sons. *Rothwell* nr *Leeds.* 1871.
S. *Stockport.* 1878.

SIDERY—
C. *Kingsclere.* 1848.
Charles. *Kingsclere.* 1878.

SIDLE—
& BARTBERGER. *Pittsburgh, Pa, USA.* ca.1840.
Matthias & Nicholas. *Pittsburgh, Pa, USA.* 1850.

SIDNEY, Arthur. *London.* 1881.

SIDWELL—
Charles. *Yeovil.* 1883.
Robert. *Nuneaton.* 1800-mar.1813. C.

SIEAR, Joseph. *Huddersfield.* 1853. C.
SIEBE, August. *London.* 1857.
SIEBER, Albin. *Windsor.* 1877.

SIEDLE—
A. *Greenwich.* 1855.
E. *Merthyr Tydfil.* Later 19c. C.
L. *Woolwich.* 1851-66.

SIENGER, Ferdinand. *London.* 1863-75.
SIGNIO/SYNEO, Richard. *Bodmin.* 1658-d.1660. Rep'd. ch. clock.
SIGOURNEY, Charles, jun. *Hartford, Conn.* ca.1800. Clock parts.
SILBER & FLEMING. *London.* 1875-81.
SILBERRAD, S. I. B. *London.* 1805. Sundial.
SILCOX, G. E. *Bath.* 1866.

SILK(E)—
John. *Elmstead (Kent).* ca.1670-pre-1765. C. & W.
Robert. *London.* ca.1675. (Also gunmaker.)

SILL, M. & F. *New York.* ca.1840.
SILVANI, Vincent. b. ca. 1818 at *Brighton. Leamington* 1850.

SILVER—
& JAMES & CO. *London.* 1828-32.
& WAY. *Bristol, USA.* 1864-66.

SILVERMAN—
Lyon. *Montreal.* 1875.
Mrs. *Montreal.* 1862.
Soloman. *Montreal.* 1853-63.
Solomon & Son. *London.* 1881.

SILVERSIDES, Henry William. *London.* 1881.

SILVERSTON(E)—
Jacob. *Brighton.* 1828.
Simeon. *London.* 1857-81.

SILVERTHAW & SON. *New Haven, USA.* 'Est'd. 1846'. W.

SILVESTER—
J. *Grimsby.* 1868.
J. *London.* 1851.
John. *London.* a.1686 to Henry Jones. CC. 1693-98.
Joseph. *London.* 1857-75.
T. *Grimsby.* 1868.

SILVESTON, F. (& Co.). *Coventry.* 1854-60 (1868-80).
SILWOOD. William. *Abingdon.* 18c.

SIMCO, Charles. *London.* 1832-57.

SIMCOCK—
Henry. *Daintree (= Daventry).* ca.1690. C. (B. 1714).

SIMCOCK (& SYMCOCK)—
John. *Daventry.* 1719-27.
Joseph. *Nantwich.* 1834.
Thomas, sen. *Warrington.* mar.1809-34.
Thomas, jun. *Warrington.* 1834-51.
William. *Warrington.* 1848-58. W.

SIMCOE, The. 1914 and later, product of Pequegnat Clock Co.

SIMCOX—
John. *Billing (Northamptonshire).* 1709.
William. *London.* a.1674. to James Markwick. CC. 1682-97.

SIMES, Isaac. *London.* 1622. Petitioner. (Also SYMMS).

SIM(M)—
Charles. *Farnworth.* mar.1799. W.
George. *Farnworth.* mar.1799. W.
George. *Prescot.* 1795. W.
John. *Longside (Scotland).* 1837-60. Also hardware.
Joseph. *Farnworth.* mar.1828. W.
Joshua. *Prescot.* mar.1815-26. W.
Thomas. *Prescot.* mar.1809-23. W.

SIMMONDS— .
——. *Longton.* 1876.
Herbert. *Aberavon.* 1865.
Herbert. *Aberdare.* 1868.
Herbert. *London.* 1863.
J. J. *Hereford.* 1879.
Samuel. *Sheffield.* 1862. W.

SIMMONS—
Abel. *Buffalo, NY, USA.* 1836.
Alexander. b. ca. 1810. from *Henley in Arden* to *Warwick* 1835-80.
Charles Sadler. *Redditch.* 1842-76.
& Co. *London.* ca.1825-30. C.
Ebenezer Leadbeter. *London.* (B. 1815-25) 1832-69.
Edward & Son. *London.* 1875.
George Henry. *Builth.* 1887.
H. *Balsall (Birmingham).* 1880.
Herbert. *London.* 1863.
Isaac. *Manchester.* 1834-51.
J. *Streatham.* 1866.
J. O. *Fareham.* 1848.
John Saddler. *Henley-in-Arden.* 1828-60.
Joseph. *London.* 1857-63.
Morrice. *London.* 1844.
O. *Fareham.* 1839.
Owen. *Fareham.* 1867-78.
Sarah. *Fareham.* 1830.

SIMMS—
Benjamin. *Witney.* Early 19c. C.
Charles P. *Chipping Norton.* s. of Samuel. b.1820-1910. C.
Frederick. *Chipping Norton.* 1816-94. C.
George. *Canterbury.* ca.1710-20.
John I. *Chipping Norton.* 1772-79. W.
John II. *Chipping Norton.* s. of John I. 1757-1823.
Samuel. *Chipping Norton.* s. of John II. 1790-1869. C. & W.
William. *Chipping Norton.* 1785-1844. C.

SIMNET, John. *Albany, NY, USA.* 1783.

SIMON—
D. *Crowbridge.* ca.1800.
Henri. *Paris.* 1880.
Thomas. *Perth.* 1860.

SIMONDS, Thomas C. *London.* 1869.

SIMONS—
David. *Hull.* 1851.
Elijah. *Mass, USA.* ca.1800.
George. *London.* 1839-44.
George & Moses. *London.* 1851-57.

SIMONTON, Gilbert. *New York.* 1820.
SIMPKINS, William. *Mansfield.* 1828-76.

SIMPLEX—
Co. *Gardner, Mass, USA.* Late 19c. Time recorders.
Time Recorder Co. *Gardner, Mass, USA.* 20c.

SIMPSON—
——. *Londonderry.* ca.1800. C.
——. *Winton (Winchester).* 1738. W.
Alexander. *Hagerstown, Md, USA.* 1799-1801. Then Johnson & Simpson till 1804.

SIMPSON—*continued.*
Benjamin. *Halifax.* 1761. C. (B. 1775).
Caleb. *Hadleigh.* 1830-53.
& Co. *Harrogate.* 1826. C. Also *London.* ('To the Royal
Family').
Daniel. *Workington.* 1828-34.
David. *Pictou, Nova Scotia.* 1866-80.
David. *Portree (Isle of Skye).* 1837.
Edmund. *Gloucester.* 1850-79.
Edmund. *Preston.* s. of Stephen of *Greta Bridge.*
b.1794-d.1820.
Edmund. *Preston.* s. of Isaac. ca.1840-57. W. Later
gold thread maker.
Frederick W. *London.* 1857-81.
G. *Southampton.* 1859.
G. H. *Windsor.* 1864.
George. *Hertford.* 1839.
George. *Maidstone.* 1839.
Isaac. s. of Stephen. b.1800. *Greta Bridge.* mar.1820.
To *Chorley.* Then ca.1831. To *Preston* where d.1859.
C.
Isidor. *Gosport.* 1859-78.
J. *London.* 1863.
J. *Wigton.* ca.1790. C. (See also John).
J. (& Son) *Wolverhampton.* 1860 (1868-76).
James. *Iver.* 1798. *Colnbrook* by 1830.
James. *Iver.* b.1807. *Aylesbury* by 1842-87.
James. *Lincoln.* 1828.
John. *Bedale.* 1834-40.
John. *Garstang.* 1822-28.
John. *Girvan.* 1850-60.
John. *Hull.* 1858.
John. *Irvinestown (co. Fermanagh).* 1846.
John. *Lancaster.* bro. of Stephen of *Greta Bridge.*
b.ca.1744-76. Later became a soldier.
John. *Liverpool.* ca.1830.
John. *Spalding.* 1849-50.
John. *Wigton.* 1828-34.
Jonathan. *Ky, USA.* ca.1820. C.
Jonathan Wortley. *Preston.* mar.1820-58. C.
Joseph. *Preston.* 1828.
Mary. *Carlisle.* 1834.
Mary. *Cockermouth.* 1828. Also toy dealer.
R. *Lancaster.* ?mid-19c. C.
Robert. *Garstang.* 1834.
Robert. *London.* 1808 -d. 1839. *(Strand
then Great Castle St. then 15 New St., Marylebone).*
Robert. *Poulton-le-Fylde.* 1822-48.
Robert William. *Croydon.* 1878.
Samuel. *Colne.* b.1774-1833. Mason by trade, but
dressed clocks.
Samuel. *Liverpool.* 1834.
Stephen. b.1752, mar.1788 at *Lancaster.* To *Caton.*
1801. To *Preston.* 1804. At sign of Tup's Clock. Died
1821. C.
Stephen. *London.* 1832.
Stephen. *Oakham.* 1828-55.
Stephen, jun. *Preston.* b.1791. s. of Stephen, sen. To
Mansfield. 1823, where became gas engineer. Died
1840.
Thomas. *Armagh.* 1784.
Thomas. *Durham.* ca.1810-20. C.
Thomas. *London.* 1828-44.
Thomas Archer & Co. *London.* 1863.
W. *Rickmansworth.* 1859-74.
W. *Watford.* 1866.
William. *Philadelphia, USA.* 1801.
William. s. of Stephen. b.1781. *Preston.* 1804. Later
Bingley. Where d.1846.
William Frederick. *Bognor.* 1878.
William M. *Yarmouth.* 1830-65.
SIMS—
Charles. *Watford.* 1851-74.
Edward. *King's Lynn.* 1830-46.
Frederick Coltonham. *London.* 1881.
Henry. *London.* 1839.
Henry. *Portsea.* 1859-78. Chrons.
Js. *London.* 1828.
James. *Stockport.* 1828-34.
Richard. *Amersham.* 1830-77.
Richard. *Watford.* 1828-39. (B. has *Walford.* 1795?).
Thomas. *Brighton.* 1878.
Vincent W. *Southampton.* 1878.

SIMS—*continued.*
W. *Wandsworth.* 1862.
William. *Peckham.* 1878.
SIMSON—
G. (exors of). *Hertford.* 1859.
& GROOMBRIDGE. *Hertford.* 1866-74.
Mrs. M. *Hertford.* 1851.
Stephen. *Southampton.* 1830-39.
Thomas. *Hertford.* 1828.
SINAVER & LEVEY. *Ottawa.* 1862.
SINCKE, Jah Jacob. *Copenhagen.* ca.1700. C.
SINCLAIR—
James. *Alloa.* 1835.
James. *Wick.* 1860.
Peter. *Glasgow.* 1837.
Walter. *London.* 1881.
William. *Philadelphia, USA.* 1837.
SINDERBY, Mrs. E. W. *London.* 1844-51 (?wid. of
Francis H.).
SINDRY—
James. *Oxford.* 1788-ca.1790. C.
Lawrence. *London.* a.1649 to Henry Ireland, CC. 1661
(B.-1705). (Also Sendry).
SINGER—
George. *Baltimore, USA.* 1842.
J. W. *Frome.* 1800-61.
Jacob. *Toronto.* 1877.
SINGLETON—
H. *Leamington.* 1880.
John Arthur. *Manchester.* mar.1798-1811. W.
Robert. *Greensboro, NC, USA.* 1839.
T. *London.* 1851.
SINISTER, ——. *Birmingham.* ca.1790. C.
SINNET, John. See Simnet.
SINNOTT, Patrick. *Philadelphia.* Till 1760 then *Balti-
more* from 1761.
SINWELL, Richard. *Pittsburgh, Pa, USA.* 1837.
SISWICK, John. *Barnsley.* b.ca.1730-88. C.
SITHERWOOD—
John. *Lisburn.* mar.1829-46. C.
John H. *Lurgan.* By 1813, then *Portadown* 1820-46.
William. *Richhill.* ca.1800. *Tanderagee* 1824. C.
SIVEL—
Matthew. *Belfast.* 1839.
Matthew. *Londonderry.* 1839.
SIZELAND, J. H. *London.* 1857-69.
SIZER, John. *Grimsby.* 1876.
SKARGEL, Samuel. *Sheffield.* 1714. mar. 1718 d. 1732. W.
SKARRATT/SKARRETT—
——. *Kington.* 1834.
& Co. *Worcester.* 1872-76.
Elizabeth. *Kington.* 1844-50.
Henry. *Hay.* 1835.
Henry. *Kington.* 1863-70.
John Martin (& Co.). *Worcester.* 1828-60(-68).
Mrs. E. *Kington.* 1863.
Thomas Carleton. *Kington.* 1830-35.
SKEATES, William. *Bristol.* 1863-79.
SKEENE, John. *Liverpool.* 1834.
SKEGGS, William. *London.* (B.1795-1824) 1832-39. C.
SKELLHORN, Richard. *New York.* 1775. C. & W. &
guns.
SKELLORN, John. *Liverpool.* 1851.
SKELTON—
Coultas. *Malton.* 1799-1858. C. s. of Robert.
Edward. *London.* 1857.
George. *London.* 1851-69.
Robert I. *Malton.* (B.1752) ca.1760-1785. C. & W.
Robert II. *Malton.* b.1764-94. W.
Samuel. *London.* a.1662 to Thomas Mills. (Prob.
Shelton, s. of Sampson).
Thomas Alfred. *Lower Tooting.* 1851-78.
William. *London.* 1875-81.
SKEOCH, James, sen. *Stewarton.* 1837-60.
SKERMAN—
J. W. *Hertford.* 1866.
William & James. *Hertford.* 1839. T.C.
SKERRETT, Frederick. *Newcastle-under-Lyme.*
1868-76.
SKERRITT, Abraham. *Heanor.* 1864-76.
SKERRY—
Brothers. *London.* 1869-81.

SKERRY—*continued.*
William. *London.* 1832-63.
SKIDMORE—
John Timothy. *Arkrigg.* 1857-d.1914. C.
& Son. *Coventry.* 1850.
Thomas. *Lancaster, Pa, USA.* 1767.
SKILLINGTON, T. *Leicester.* 1864.
SKINNER—
Alvah. *Boston, USA.* 1830-42.
Carlos. *Waterloo, Canada.* 1871.
Charles. *London.* 1839-57.
Charles. *Peckham.* 1878.
Charles. *Sheffield.* 1871.
Edward. *Lewes.* 1870-78.
George. *Eye.* 1853-75.
James. *Exeter.* b.1783-1803.
John senior. *Exeter.* b. ca. 1747, mar 1774 d.
1818. C.
John. *Exeter.* b.1777-1803. W. (B. d.1818).
John. *Hamilton, Ontario.* 1871-95.
John. *Sheffield.* 1837. C.
John. *Toronto.* 1875.
Joseph. *Winterbourne.* ca.1790-1820. C.
Mark. *Eye.* b. 1767-1846. Also at *Scole (Norfolk)*
1839.
Richard. *Exeter.* 1774-1818. W.
Thomas. *Birmingham.* 1880.
W. H. *Leamington.* 1880.
William. *Aberdare.* 1887.
William. *Harwich.* 1839.
SKIPTON, James Frederick. *Cirencester.* (B. 1820)
1830-50.
SKIRROW—
James. *Wigan.* (B.1783-1814) 1822-34. C.
Robert. *Halifax.* 1810-53. C.
SKYRME, James. *Truro.* 1847-56.
SLACK—
Joseph. *Ipstones (Staffordshire).* ca.1800-1818, W. and
farmer.
S. *Chesterfield.* 1864-76.
Uriah. *Chesterfield.* 1876.
SLADE—
Edward. *Mevagissey.* 1844-47. C.
George. *Monksilver (Somerset).* 1710. Made ch. clock.
John. *Penryn.* 1844-56.
William. *Penryn.* 1844-47. C.
SLADER, William. *Brading (Isle of Wight).* 1830.
SLARK, William. *London.* 1857-63.
SLATER—
G. *Kidsgrove.* 1860.
G. *Tunstall (Staffordshire).* 1860.
George. *Burslem.* 1842-60. W. & C.
George. *Hyde.* 1857-78.
George. *Liverpool.* 1848.
George. *Uttoxeter.* 1828-35.
J. *Barking.* 1866.
James. *Ormskirk.* 1848-58.
James. *Uttoxeter.* 1828-50.
John. *Cheadle.* 1850.
John. *Mayfield.* 1862-70.
John. *Steyning.* 1828.
John. *Toronto.* 1874-76.
John. *Uttoxeter.* 1842.
John M. Q. *Cheadle.* b.1820 in *East Indies,* worked at
Uttoxeter, then moved to *Cheadle* ca.1845-51. C.
Joseph. *Uttoxeter.* (b.1773) d.1822. C.
Mrs. B. *Rugby.* 1860.
T. *Shoreham.* 1851.
Thomas. *Lichfield.* 1835.
W. *Brighton.* 1851.
W. *Uttoxeter.* 1860.
Walter. *London.* 1869-81.
William. *Barking.* 1874.
William. *Lichfield.* 1828.
William. *London.* 1832-69.
William. *Shoreham.* 1839.
William. *Yarmouth.* 1875.
SLEATH, John. *Sandwich.* a.1747.
SLEEMAN, Henry. *Penzance.* 1833. C.
SLEEP, John. *Liskeard.* 1844-47. C. & W.
SLEIGH—
Edward. *London.* 1828. Dialmaker.
Joseph Pedley. *Cheadle.* 1876.
William. *Stockton.* (B. mar.1818) 1827-34. W.

SLEIGHTHOLM, J. *London.* Late 18c. W.
SLIBBY, R. *Port Hope, Canada.* 1862.
SLICER, William. *Annapolis, Md, USA.* 1769. C. and
clock cases.
SLIGH, Samuel. *West Cahn, Pa. USA.* 1796. (Later a
plasterer-1799.)
SLIGHT, John. *Sheffield.* 1871.
SLIMAN, Archibald. *Cumnock.* 1837-50.
SLIMEN, William. *Ayr.* 1836.
SLOAN—
(George), McMurray & Co. *York, Canada.* 1833.
William. *Ballymena.* ca.1851-99. W.
SLOSS, James. *Middleton in Teesdale.* 1851.
SLOUGH, William. (Also Sloogh and Slouch). *London.*
a.1676 to Benjamin Bell, CC. 1687-97.
SLOWE, William. *Manchester.* 1848.
SLUCE, William. *London.* 1844-81.
SLY—
John. *Weymouth.* 1824.
Richard. *London.* ca.1760. C.
Robert. *Oxford.* 1823-53.
Samuel. *Norwich.* (B. 1795) 1830-36. C. & W.
Thomas. *Salisbury.* 1859-75.
SLYTH, John. *Audlem.* 1834-78.
SMAILES, Richard. *Rochdale.* 1848-51.
SMALLBONES, John Thomas. *Chatteris.* 1858-75.
SMALLEY—
J. *London.* 1851.
John. *Blackburn.* 1721-d.1725. C.
John. *Lancaster.* ca.1720. C. (Prob. same as above).
Joseph Smoult. *London.* 1869-75.
Peter Marsh. *London.* 1857-81.
SMALLPAGE—
——. *Halifax.* 1721-23. Rep'd ch. clock. May be from
Wakefield or *Leeds?*
Daniel. *Leeds.* b.1668 -mar.1691.
Daniel. *Wakefield.* 1729-36. W.
Edwin. *Leeds.* 1871.
James. *Leeds.* 1866. W. Successor to John S.
John. *Leeds.* 1822-53.
John. *Norton nr Malton.* 1840-46.
SMALLWOOD—
George. *Birmingham.* 1880.
John. *Chelford.* ca.1710. C. (Prob. same as one who
d.1715 at *Macclesfield*).
T. *Coventry.* 1860. Cleaner.
SMART—
Alexander. *London.* 1828-51.
George. *Lexington, Ky, USA.* From *Britain,* 1794-
1810. C.
John. *Coventry.* 1842-68.
John. *London.* a.1674 to Jeremiah Johnson, CC. 1682.
(B. -1696).
John. *Philadelphia, USA.* 1839-50, and later.
Thomas. *London.* (B. 1815-20) 1828.
Thomas. *New York.* 1773. C.
William. *Sileby (Leicestershire).* 1876.
SMEATON—
John. *York.* 1646-d.1686. W.
John. *London.* (b.1724. *York*) 1742-d.1792.
SMEDLEY, Lindsey. *Matlock Bath.* 1876.
SMEED, George. *London.* 1839.
SMEETON—
——. *Kibworth (Leicestershire).* Pre-1835. C.
John. *Leicester.* 1864-76.
SMETHURST—
H. *Stafford.* 1868-76.
Henry. *Farnworth.* mar.1786. W.
SMILLIE, David. *Montreal.* 1842-63.
SMITH—
A. *Boston, USA.* 1854.
A. B. *New York.* 1835-40.
A. D. *Cincinnati, O, USA.* ca.1860. C.
A. P. *Dundee.* 1850.
Aaron. *Ipswich, Mass, USA.* 1775-85.
Abraham. *London.* a.1771.
Abraham. *London.* 1857-63.
Alexander. *Banff.* 1845.
Alexander. *Keithhall nr Inverurie.* 1846-60.
Alexander. *Tranent (Scotland).* 1837.
Alfred. *Birmingham.* 1880.
Alfred. *Herne Bay.* 1866-74.
Alfred. *Huddersfield.* b.1828. s. of William II of

SMITH—*continued.*
  Huddersfield. -1871.
Alfred. *King's Lynn.* 1875.
Amable. *Sorel, Canada.* 1851-58.
Andrew. *Prestonpans.* 1830.
Andrew. *Tain.* 1860.
Anthony. *Wellington.* 1841-47.
Arthur. *Wrexham.* 1662-70. Rep'd. ch. clock.
B. *New York.* Early 19c. C.
B. *Philadelphia, USA.* 1835.
Benjamin. *Guernsey.* C. (date unknown). Prob. 19c.
Benjamin. *Leeds.* 1761. C. & W.
& BLAKESLEY. *Bristol, USA.* ca.1830.
BLAKESLEY & Co. *Bristol, USA.* ca.1840. (Later
  Smith alone.)
& Brother. *Philadelphia, USA.* 1843.
& Brothers. *New York.* 1841.
Brothers. *Walton-on-Trent.* 1864.
C. *Leeds.* Early 19c. C.
C. *Tonbridge.* 1866.
C. C. *Fayetteville, NC, USA.* From *London* 1841-43.
  Succeeded by Edwin Glover.
Charles. *Cranbrook.* 1874.
Charles. *Aberdeen.* 1846.
Charles. *Falmouth.* 1813-23.
Charles. *Liskeard.* 1803-16.
Charles. *Reading, Pa. USA.* Early 19c.
Charles. *Saxmundham.* 1810-30. C. & W.
Charles A. *Brattleboro, Vt, USA.* ca.1890-ca.1940.
Charles James. *London.* 1869.
Charles N. *Philadelphia, USA.* 1835-50. and later. C.
Charles Roe. *Bangor.* 1835.
Charles William. *Surbiton.* 1878.
Charlotte. *Long Sutton (Lincolnshire).* 1835.
Co., The. *Philadelphia, USA.* pre-1850-present day.
& Co. *Canton, O, USA.* 1831. C. & math. instrs.
& Co. *Nottingham.* 1876.
D. *Reading.* 1854.
Daniel. *Chelmsford.* ca.1823.
Daniel. *Dorking.* Early 18c. C.
Daniel Treadway. *Salem, Mass, USA.* (1824) 1846-64.
David. *London.* a.1654 to Andrew Prime. CC. 1661.
David. *Pittenweem.* 1827-34.
David. *St. Andrews.* 1835-73.
E. S. *Sittingbourne.* 1851-55.
E. A. (Edward) & D. T. (Daniel T.) *Salem, Mass,*
  USA. ca.1850.
Ebenezer. *Brookfield, Conn, USA.* Late 18c.
E(dmund). *Ashby-de-la-Zouch.* 1849-64.
Edmund. *New Haven, Conn, USA.* ca.1837.
Edward. *Barton (Staffordshire).* 1842-50.
Edward. *Louth.* 1828-50.
Edward. *Newbury.* 1830.
Edward. *Newcastle-on-Tyne.* 1827-36.
Edward Augustus, jun. *Salem, Mass, USA.* 1850-64.
(Capt.) Elisha. *Sanbornton, NH, USA.* Early 19c.
Elisha, jun. *Sanbornton, NH, USA.* Early 19c.
Elizabeth. *Scarborough.* 1858.
Ernest. *Philadelphia, USA.* 1830-33.
F. *Blandford St. Mary (Dorset).* 1848-67.
F. C. *Philadelphia, USA.* 1844.
& FENN. *Baltimore, USA.* 1842.
Frederick. *London.* 1857.
Frederick Sinnott. *Epping.* 1874.
G. *Brighton.* 1851-62.
G. *Leamington.* 1860-68.
G. *Mortlake.* 1862.
G. J. *Earlsdon nr Coventry.* 1860-68.
George. *Derby.* 1871-1932.
George. *Forres.* 1837-60.
George. *Haverfordwest.* ca.1855. C.
George. *Huntly.* 1837-46.
George. *Liverpool.* 1784-90. W.
George. *London.* Alien 1622, CC. 1632-38.
George. *London.* 1832.
George. *London.* 1881.
George. *Manchester.* mar.1803. W.
George. *March (Cambridgeshire).* 1830.
George. *Peckham.* 1828-45.
George. *Ripon.* 1834-51. C.
George. *Whelpington.* 1827.
George A. *Belfast.* 1880.

SMITH—*continued.*
George Edward. *Lower Norwood (Surrey).* 1862-78.
George Thompson. *London.* 1869.
George W. G. *Wymondham.* 1875.
& GOODRICH. *Bristol, USA.* 1847-52. C.
& GOODRICH. *Philadelphia, USA.* 1850.
Guy. *Burton-in-Lonsdale* and *Thornton-in-Lonsdale.*
  b.ca.1774-1808. C.
H. *Wednesbury.* 1868.
H. C. *New York.* 1840.
Harry. *Leicester.* 1876.
& HARTHILL. *Birmingham.* 1850.
Henry. *London.* a.1650 to Nicholas Coxeter, CC.
  1658-65.
Henry. *Petersfield.* ca.1730. C.
Henry. *Portslade.* 1878.
Henry (& Son). *Reading.* 1830-37.
Henry A. *Rochester, NY, USA.* ca.1830.
Henry C. *Waterbury, Conn, USA.* ca.1814, then
  Plymouth Hollow. ca.1829-40. C.
Henry S. *Leeds.* 1871.
Hezekiah. *Philadelphia, USA.* 1845.
& HIND. *Hartlepool.* 1851-56.
& HIND. *Stockton.* 1851.
Horatio. *London.* 1844-63.
Horatio. *York.* a.1815-51. C.
Isaac. *Philadelphia, USA.* 1840-43.
Isaac Cragg. *London.* 1832.
Isiah. *Cefn, Ruabon.* 1874.
J. *Billericay.* 1855-66.
J. *Birmingham.* 1860.
J. *Leamington.* 1860.
J. *Leicester.* 1849-55.
J. *Maldon.* 1866.
J. *Newark.* 1864.
J. *Wednesbury.* 1876.
J. A. *Chipping Sodbury.* 1863.
J. B. *Brill (Buckinghamshire).* Early 19c. C.
J. B. *Shepton Mallet.* 1861-66.
James. *Bryn Mawr.* 1887.
James. *Carlisle.* 1869-73.
James. *Grantown (Inverness).* 1837-60.
James. *Grimsby.* 1868-76.
James. *Irvine.* 1837-60.
James. *Kendal.* b. 1853-79. W.
James. *King's Lynn.* 1774 W. & silversmith in High
  Street.
James. *London.* 1875.
James. *Monaghan.* 1846.
James. *Ottawa.* 1862.
James A. *Croydon.* 1878.
James Arthur. *Devizes.* 1867-75.
James B. *Cumnock (Scotland).* 1860.
James B. *Kington (Hereford).* 1850-79.
James R. *Lower Mitcham (Surrey).* 1878.
James S. *Philadelphia, USA.* 1827-50. Wholesale. C.
Jasper. *London.* 1648.
Jesse, jun. *Salem. Mass, USA.* 1756-1844. (Partner of
  Benjamin Balch 1807-32.)
John. *Amersham.* 1798.
John. *Ashton-under-Lyne.* 1828-34.
John. *Bathgate (Scotland).* 1860.
John. *Berwick-on-Tweed.* b.1750-d.1785. C.
John, jun. *Cranbrook.* 1859-74.
John. *Derby.* 1855-76.
John. *Keighley.* 1853-66.
John. *Lancaster, Pa, USA.* ca.1800. C.
John. *Leeds, Canada.* 1857.
John. *Leek.* 1830.
John. *Leicester.* 1846.
John. *London.* a.1647 to Edward Daniell, CC. 1654.
John. *London.* a.1648 to Thomas Wolverston. CC.
  1656-57. W.
John. *London.* 1839-57.
John. *London.* 1869-81.
John. *Manchester.* 1834-51.
John. *March (Cambridgeshire).* 1830-46.
John. *Newcastle-on-Tyne.* 1827-36.
John. *New York.* 1841-48. W.
John. *Nottingham.* 1864-76.
John. *Oldbury.* 1842-50.
John. *Peterhead.* Late 18c. C.

SMITH—*continued.*

John. *Philadelphia, USA.* 1835.
John. *Scarborough.* 1834-51.
John. *St. John, New Brunswick.* 1861.
John. *Skipton.* 1772. Made ch. clock dial.
John. *Thirsk.* 1790-1807. W.
John, jun. *Wrexham.* s. of John, sen. b.1803. Succeeded father in 1830, d.1868.
John I. *York.* 1750-64. C.
John II. *York.* a.1758. S. of John I.
John C. *London.* 1857.
John P. *Broadstairs.* 1826-28.
John P. (& Sons). *Sittingbourne.* 1839-45(-47).
John R. *Arichat, Nova Scotia.* 1838.
John S. *Manchester.* 1848.
John & Sons. *London.* 1844-81. Clock case makers.
John & Sons. *London.* 1844-81 and later.
Joseph. *Bristol.* 1725-d.1778. W.
Joseph. *Brooklyn.* NY, USA. ca.1830.
Joseph. *Coleford.* 1863-79.
Joseph. *Gainsborough.* 1828-35.
Joseph. *London.* 1844.
Joseph. *New York.* 1841. C.
Joseph. *Sheffield.* 1862. W.
Joseph. *Skipton.* b.1717. d.1795. W.
Joseph. *Workington.* 1848.
Josiah. *Reading, Pa, USA.* 1778-1860.
Lawrence. *Keighley.* 1834-37. C.
Levi. *Bristol, USA.* ca.1820. (Of Smith & Blakesley. ca.1840-42). C.
Luther. *Keene, NH, USA.* ca.1785-1840. C. & T.C.
Lyman. *Stratford, Conn, USA.* 1802. C. & W., also silver and jewellery. Successor to Nathan Wade.
Morris. *London.* a.1693 to Robert Webster. CC.1702.
Morton. *Bramley (Surrey).* 1878.
Mrs. Martha. *Huddersfield.* 1850. W.
N. *Southampton.* 1839.
Nathaniel. *Columbus, O, USA.* ca.1814. W.
Nathan. *London.* a.1680 to Isaac Carver, math. inst. mr., CC. 1689 (-B. 1719).
P. R. *Salisbury.* 1867.
Peter. *Deal.* 1838-51.
Peter. *Mevagissey.* 1789-d.1844.
Philip. *London.* 1667. CC., math. inst. mr.
Philip L. *Marcellus, NY, USA.* ca.1830. C.
R. & Co. *Watertown, Conn, USA.* ca.1830. C.
Ransom. New York. 1848. C. Wholesaler.
Richard. *Birmingham.* 1777. Dialmaker.
Richard. *Birmingham.* 1868.
Richard. *Brill (Buckinghamshire).* (?b.1765)-late 18c.
Richard. *Bristol.* mar.1674.
Richard. *Newport.* mid-18c. C.
Richard. *Outwell (Norfolk).* 1875.
Richard. *Scarborough.* 1866. W.
Robert. *Farnworth.* 1818. W.
Robert. *Inverness.* 1840-60.
Robert. *N. Berwick (Scotland).* 1835-60.
Robert. *Philadelphia, USA.* 1818-22.
Robert. *Quebec.* 1854-57.
Robert. *Thame.* 1442-55. Rep'd. T.C.
Ross. *Lerwick.* 1860.
S. *Lyons, NY, USA.* ca.1830. W.
S. *Wolverhampton.* 1860-68.
S. B. & Co. *New York.* 1854-67.
S. W. & Bros. *Napanee, Canada.* 1889.
Samuel. *Birmingham.* 1868-80.
Samuel. *Coventry.* 1814. W.
Samuel. *London.* 1857-81.
Samuel. *New York.* Late 18c. C.
Samuel. *Philadelphia, USA.* 1845.
Samuel. *Sandwich.* a.1752.
Samuel. *Tamworth.* 1850. C.
Samuel. *Toronto.* 1877.
Samuel. *Walton-on-Trent (Derbyshire).* 1835-49.
Samuel G. *Coatesville, Pa, USA.* 1865.
Samuel Henry. *Leicester.* 1835-55.
Samuel Howard. *Norwich.* 1865.
Samuel M. *Listowell, Canada.* 1871.
& SEARS. *Rochester. NY, USA.* ca.1830.
& SILL. *Waterbury, Conn, USA.* 1831.
Solomon. *London.* 1828.

SMITH—*continued.*

& Sons. *Barton (Staffordshire).* 1860. .
& Sons. *Birmingham.* 1860.
& Son. *London.* 1881.
& Sons. *Walton-on-Trent (Derbyshire).* 1855.
Stephen. *Great Dalby (Leicestershire).* 1835. W. & C.
Stephen. *London.* a.1651 to Ahasuerus Fromanteel, CC. 1658.
Stephen. *Melton Mowbray.* 1849-76.
T. *Birkenhead.* 1857.
T. *Lancaster.* ca.1760. C.
& TAYLOR. *New York.* ca.1840.
Theophilus. *Wisbech.* 1840-75.
Thomas. *?London.* Lant. clock dated 1663.
Thomas. *Ashbourne.* 1828-49.
Thomas. *Bath.* 1883.
Thomas. *Bilston.* 1828-35.
Thomas. *Birmingham.* 1835-42.
Thomas. *Chatteris.* 1830-65.
Thomas. *Gateshead.* 1827.
Thomas. *Leighton.* ca.1760. C.
Thomas. *Lichfield.* 1850.
Thomas. *Lincoln.* 1861-76.
Thomas. *Liverpool.* From *London.* 1724. W.
Thomas. *London.* 1828-32.
Thomas. *London.* 1863-69.
Thomas. *Norwich.* pre-1769.
Thomas. *Oxford.* a.1669. C.
Thomas. *Sheffield.* 1777-ca.1790. C.
Thomas. *Strathaven.* 1860.
Thomas & Son. *London.* 1875.
Thomas William. *London.* 1828.
Vincent. *Poundon (Buckinghamshire).* 1711-mar.1718. T.C.
W. *Bingham (Nottinghamshire).* 1855-64.
W. *Greenburn.* ca.1800. C.
W. *Hereford.* 1863.
W. *Kingston, Canada.* 1851-63.
W. *Shaftesbury (Dorset).* 1867.
W. *Shipston-on-Stour (Warwickshire).* 1860.
W. J. *Bath.* 1861.
Walter C. *Louisville, Ky, USA.* 1831-50s.
Walter Fletcher. *Lincoln.* 1868-76.
William. *Askrigg.* b.1748-74. C.
William. *Bristol.* mar.1619.
William. *Broadstairs.* 1838-74.
William. *Cambridge.* 1875.
William. *Canterbury.* 1845-55.
William. *Dudley.* 1842.
William. *Elgin.* 1860.
William. *Fort William.* 1837.
William. *Herne Bay.* 1845-55.
William, sen. *Huddersfield.* b.ca.1755-1795. C.
William, jun. *Huddersfield.* b.1795-1837. C.
William. *Llandeilo.* 1830.
William. *Loanhead.* 1836.
William. *London, Fleet Street.* 1646.
William. *London.* ca.1710. C.
William. *London.* 1844.
William. *London, Canada.* 1851.
William. *Londonderry.* 1846.
William. *Market Deeping.* 1876.
William. *March.* 1840-65.
William. *Measham (Derbyshire).* 1828.
William. *Middle Tysoe (Warwickshire).* 1880.
William. *Musselburgh.* 1860.
William. *Napanee, Canada.* 1843.
William. *New York.* 1810. Cabinet maker.
William. *New York.* 1831. W.
William. *Quebec.* 1857.
William. *Wingham.* 1832-39.
William, jun. *Crowland.* 1850.
William, sen. *Crowland.* 1828-50.
William & Co. *Derby.* 1876.
William Henry. *London.* 1875.
William Henry & Sons. *Bury St. Edmunds.* 1875. American clocks.
William John. *London.* 1844.
William Sellers. *Bristol.* a.1819-26.
William W. *Brighton, Canada.* 1857-63.
Zebulon. *Bangor, Me, USA.* ca.1830. W.

SMITHERS—
Benjamin. *Chichester.* 1870.
Benjamin. *Southsea.* 1878.
Charles E. *Portsea.* 1830-78.
Henry. *Coventry.* 1880.
J. *Portsmouth.* 1848-59.
SMITHIES—
George. *Masham.* 1732. Rep'd. ch. clock.
John. *Otley.* 1732-d.1746. C.
SMITHS Clock Manufactory. *New York.* 1841.
SMITHYES, John. *London.* ca.1790? C.
SMITTEN, R. T. *Philadelphia, USA.* 1844-47.
SMITTON, Peter. *London.* (B. 1820-) 1828-32.
SMORTHWAIT, John. From *Westmorland* to *Colchester*
1706-d.1739. C. & W.
SMYTH(E)—
Charles. ? *Saxmundham.* 1830. Also gunsmith.
Charles. *Tonbridge.* 1838-74.
Israel. *Woodbridge.* 1839.
James. *Garvagh (co. Derry).* 1846.
Luke. *Yoxford (Suffolk).* 1830-58. Also silversmith and
ironmonger.
SNAITH, George. *Wigton.* 1869-79.
SNAKENBURGER, George. *Birmingham.* 1868.
SNARY, Robert. *Malton.* mar.1666. C.
SNASHALL—
George. *Brighton.* 1870.
William. *Marden (Kent).* 1847.
SNATT—
John. *Ashford.* 1774-1807. W.
John. *Ashford, Mass, USA.* ca.1700. (?) See previous
entry.
John. *Lewes.* From *Warnham.* a.1736 to Thomas
Barrett. C.
SNEAD, J. *Lancaster.* 1858. W.
SNELL—
Albion. *London.* 1857-81.
Christopher. *Knottingley.* 1866-71. W.
Christopher. *Selby.* 1846.
George. *London.* a.1679 to Thomas East. CC. 1688 (B.
-1700).
W. *Honiton,* ca.1820. C.
SNELLEN, W. ? *Dordrecht.* ca.1770. Chron.
SNELLING—
Burgess & Edward. *London.* 1881.
Henry. *Philadelphia, USA.* From *London.* 1776. W.
J. *Thetford.* 1846.
James. *Alton (Hampshire).* 1793-1830. Also silver-
smith.
John. *Ashford.* 1826-55.
John. *Romney.* 1823.
Robert. *London.* 1869-75.
Thomas. *London.* a.1672 to Cuthbert Lee, CC. 1680
(B. -1682).
SNOOK—
George. *Landport.* 1878.
J. *Bristol.* 1879.
John. *Tai-Bach (Glamorgan).* ?18c. C.
John. *London.* 1632.
Robert P. *Southampton.* 1867-78.
William. *Blaenavon (Monmouthshire).* 1868-99.
SNOS(S)WELL—
C. *London.* 1863.
William. *London.* (B. a.1803-20) 1828-57. Musical
clocks.
SNOW—
Charles James. *Worksop.* 1876.
F(rancis?). *Otley.* ca.1830. C. Prob. son of William
II.).
John. *Frome.* ca.1750-ca.1760. C.
John, jun. *Frome.* ca.1780. C.
John. *London.* 1731. *Queen's Arms Alley in Shoe Lane,*
parish of St. Brides. W. (B. a.1716).
Jonathan. *Frome.* ca.1760. C.
Joseph. *Bradford.* 1853.
Richard. *Pateley Bridge.* b.1769, s. of William I. -1837.
C. & W.
Richard R. *Ripon.* 1866-71. W.
Thomas. *Birstwith.* From *Otley* -1822. C.
Thomas. *Bradford.* 1811-22. C. & W.
Thomas. *Knaresborough.* 1834-44. C. Prob. from
Bristwith.

SNOW—*continued.*
Thomas. *Otley.* b.1774. s. of William I. -ca.1810. Later
at *Birstwith.* C.
Will. No town. See Will of *Padside.*
William. *Marlborough.* ca.1650-late 17c. W. & C.
William. *Otley.* ca.1790. C. Prob. some of clocks of
William II of *Padside* were signed thus.
William I. *Padside* nr *Pateley Bridge.* b.1736-d.1795. C.
and farmer.
William II. *Padside.* b.1766. s. of William I. d.1807. C.
SNOWBALL, George. *London.* 1857-69.
SNOWDEN/SNOWDON—
James. *North Shields.* 1848-58.
Joseph. *London.* 1863-69.
SNOWSILL, John. *London.* 1832.
SNYDER—
George. *Philadelphia, USA.* 1801.
Peter, jun. *Exeter, Pa, USA.* 1779 and later.
SOADY, William Henry. *Torpoint (Cornwall).* 1873. W.
SOAR, James. *London.* (B. a.1823) 1844-69.
SOEUFEN, Isaac. *London.* Alias Swale, Isaac. q.v.
SOFTL(E)Y—
Thomas. *Barnard Castle.* 1856.
Thomas. *Louth.* 1868-76.
SOLARI, Giovanni. *Cardiff.* 1887-99.
SOLDANO—
——. *Paris* and *Geneva.* 1855-78.
Joseph. *London.* 1851.
Joseph Fils. *London* and *Geneva.* 1863-69.
Louis & Son. *London.* 1857. (Successor to Joseph.)
Vve de J. Fils. *London* and *Geneva.* 1875-81.
SOLLER, Iohn Henry. *London.* 1863-69.
SOLLIDAY (and variants)—
Benjamin. *Rock Hill, Pa, USA.* s. of Frederick. Late
1700s. C.
Calvin. *New Hope, Pa, USA.* ca.1900-1915. Later
Lambertville, NJ. C.
Charles. *Doylestown, Pa, USA.* To 1860. C. & W.
Daniel H. *Sumneytown, Pa, USA.* Post-1800. Then
Philadelphia. 1829-50(-1873). W.
Eli. *New Hope, Pa, USA.* s. of Samuel. Retired 1887.
C.
Frederick. 1717. To *USA.* 1740-92.
Henry. *Towamensin, Pa, USA.* 1814. C.
Jacob. bro. of Benjamin. *Reading, Pa, USA.* Also
Bedminster and *Northampton, USA.* 1748-1815. C.
John. *Richland, Pa, USA.* 1782.
John N. *Reading, Pa, USA.* 1794-1816. Then *Bucks*
County. 1816-81. C.
Peter. s. of Jacob. *Bedminster, Pa, USA.* 1795-1807.
Samuel. s. of Benjamin. *Doylestown, Pa, USA.* (1805).
1828-34. Then *Hope Town, Pa.* 1834-80. C. Also
jeweller and coal merchant.
& Sons. *Dublin, Pa, USA.* 1830.
SOLO, Henry. *Quebec.* 1730.
SOLOMAN—
Henry. *Boston, USA.* 1820-25. W.
Morris. *King's Lynn.* 1836.
SOLOMON—
Abraham. *London.* 1828-39.
& Co. *Birmingham.* 1815. Dial maker.
& Co. *Coventry.* 1868.
Edward. *Margate.* 1769. W.
Emanuel. *Canterbury.* 1785-1826. W.
Henry & Co. *London.* 1844-69.
J. *Coventry.* 1860-68.
J. *Hanley.* 1868-76.
Jacob Henry. *London.* 1844.
Jonas. *London.* 1875.
L. T. *Quebec.* 1830.
Leon & Co. *London.* 1857.
Mark. *London.* 1875.
Miriam & Bella. *Canterbury.* 1838.
Moses. *London.* (1817-25) 1828.
Phineas. *London.* 1832-44.
Samuel. *London.* 1839-70.
Samuel & Lazarus. *Rochester.* 1839.
Simon. *St. Johns, Newfoundland.* 1792.
Solomon. *London.* 1851.
Thomas. *Penzance.* d.1793.
SOLOMONS—
A. *Yarmouth.* 1846-65.

SOLOMONS—*continued.*
S. & L. *Chatham.* 1839.
SOMER, Johannes. Place unknown. ca.1640. Table clocks.
SOMERS—
Albertus. *Woodbury, NJ, USA.* ca.1820. Successor to George Hollinshead. To *Woodstown.* To succeed John Whitehead. 1821-33.
& CROWLEY. *Philadelphia, USA.* 1828-33.
Lawrence. *London.* 1828.
SOMERVELL, James. *Great Boughton.* d.1746. C.
SOMERVILLE, James. *Montreal.* 1862. (?see Sommerville.)
SOMES—
J. *Margate.* 1874.
William. *Margate.* 1838-74.
SOMMERVILLE, James G. *Abergavenny.* 1858. (?see Somerville.)
SON, Henry. *Worcester.* 1876.
SONES, Benjamin Elisha. *Lowestoft.* 1875-79.
SOO, The. 1914 and later, product of Pequegnat Clock Co.
SOPHIN, Thomas. *London.* 1839.
SORET et JAY. *Switzerland.* ca.1625. W.
SORG—
Charles. *London.* 1844-57.
F. X. *Norwich.* 1875.
Joseph. *Neustadt.* ca.1800.
SORRELL—
Edward. *Brighton.* 1870-78.
George. *London.* 1857-81.
SOSAR, J. *Port Dover, Canada.* 1857.
SOULBY, H. *Sleaford.* 1849.
SOURDERIAT, ——. *Paris.* ca.1715. W.
SOUTH BEND. Used by South Bend Watch Co. ca.1902-1930.
SOUTH BEND WATCH Co. *South Bend, Ind, USA.* 1903-33. Formerly American National Watch Co.
SOUTH—
Henry. *Rotherham.* (B. 1710). 1731-d.1750. C.
James. *Charlestown, Mass, USA.* ca.1810.
SOUTHCOTE, William. *Bristol.* a.1820-30. (Also Southcott).
SOUTHEE, H. S. E. *Canterbury.* 1866.
SOUTHERN—
Calendar Clock Co. *St. Louis, Mo, USA.* 1875-89. C.
Daniel. *St. Helens.* 1822-28. W.
SOUTHEY, J. *Leicester.* 1849.
SOUTHGATE, William. *Maidstone.* a.1750.
SOUTHWOOD, James. *Havant.* 1878.
SOUTHWORTH—
E. *Caernarvon.* ca.1820. C.
Elijah. *New York.* 1793-1830.
John. *London.* a.1668 to John Matchett. CC. 1689. (B. -1701).
Peter. *London.* a.1656 to Jeremy East. CC. 1664-97.
SOWER—
Christopher. *Germantown, Pa, USA.* ca.1724-40. C. Also doctor, author, farmer, printer, etc.
Daniel. *Phoenixville, Pa, USA.* ca.1845-49. W.
SOWTER—
John. *London.* a.1671 to Solomon Bouquet. CC. 1683. (B. -1694).
John. *London.* 1810-18. Later *Oxford.* 1818-53. C.
SPAHN—
F. *Leighton Buzzard.* 1854-69.
Frederick. *Ampthill.* 1877.
SPALDING—
Anthony. *Somersham (Huntingdonshire).* 1830-39.
William. *Liverpool.* 1784. W.
SPANGENBERG, G. *Kingston, Canada.* 1851-69.
SPANGLER—
Jacob. *Hagerstown, Md, USA* and *York, Pa.* s. of Rudolph. 1790-1810. C.
Rudi. Same as Rudolph.
Rudolph. *Hagerstown, Md* and *York, Pa, USA.* ca.1764-1805. C.
SPARK(E)—
J. *Shipdham.* 1846.
James. *Mildenhall.* 1830-39.
James Brook. *Colchester.* b.1801-d.1866.
John. *Chatham. Canada.* 1861.

SPARK(E)—*continued.*
Reuben. *London.* 1869.
SPARKES—
Angel. *Plymouth.* ca.1795-ca.1830. C.
J. *Wolverhampton.* 1868.
J(ames?). *Uppingham.* 1855-64.
Mrs. James. *Uppingham.* 1876.
Nicholas. *London.* 1646.
SPARKS—
J. J. *Ipswich.* 1853-58.
J. J. *Lymington.* 1848.
SPARROW—
Albert James. *Frome.* 1883.
John. *London.* (B. a.1779, CC. 1787-1824) 1828.
Thomas. *St. Neots (Cornwall).* 1722. C.
SPASSHAT(T)—
J. *London.* 1851-57.
Joseph. *Dartford.* 1845-47.
SPATT, Johann Antoin. *Aichach.* ca.1775. C.
SPAUEN, C. J. *Middlesbrough.* 1898.
SPAUL, Alexander. *Battersea.* 1878.
SPAULDING—
Abraham. *Brooklyn, USA.* ca.1840.
& Co. *Providence, RI, USA.* ca.1800. C.
Edward. *Providence, RI, USA.* (1732). ca.1770-85. C.
Edward, jun. *Providence, RI, USA.* ca.1800.
SPEAKMAN—
Edward. *London.* s. of William. a.1686 to his father. CC. 1691. (B. -1712).
T. *Warrington.* 1858. W.
Thomas. *London.* s. of William. a.1675 to father. CC. 1685. (B. -1714). C.
William. *London.* a.1654 to Andrew Prime. CC. 1661. (B. -1701). C.
William II. *London.* s. of William I. a.1688 to father. -1716?.
SPEAR—
& Co. *Charleston, SC, USA.* ca.1850.
Jacob. *London.* 1832-51. Musical clocks.
James E. *Charleston, SC, USA.* 1846-71. W.
SPEARS, Frederick. *Liverpool.* 1834.
SPEDDING, Richard. *Blackburn.* 1834.
SPEER(S)—
Charles. *Liverpool.* 1828-34. (B. 1825).
Samuel. *Newtown Stewart.* ca.1811-46. C.
SPEIGELHALTER, J. N. *Bradford.* 1866. See Spiegelhalter.
SPEIGHT, R. *Lincoln.* 1876.
SPIERS, William. *Hamilton.* ca.1830. C.
SPELLIER, August. *Philadelphia, USA.* ca.1870.
SPENCE—
Gavin. *New York.* ca.1810.
Hugh. *Alford (Scotland).* 1860.
Hugh. *Huntly (Scotland).* 1837.
John. *Boston, USA.* 1823-30.
John. *Stromness.* 1836-60.
SPENCER—
——. *Bilston.* Late 18c.
& COOK. *London.* 1875-81.
David. *Coventry.* 1860.
Edward. *Chelmsford.* ca.1848.
Eli. *Bolton.* 1824-34.
Frederick Augustus. *Needham Market.* 1879.
Henry. *Chester.* 1848-57.
HOTCHKISS & CO. *Salem Bridge, Conn, USA* ca.1830.
J. *Chelmsford.* 1859-70.
J. *Stalybridge.* 1857-65.
James. *Stalybridge.* 1848. (See previous entry.)
John. *Chelmsford.* 1874.
John I. *Colne.* b.1711, mar.1738, d.1794. C.
John II. *Colne.* b.1739. s. of John I. mar.1785, d.1809. C.
John III. *Colne.* b.1791. s. of John II. -1833. C.
John. *London.* 1869.
Julius. *Utica, NY, USA.* ca.1820.
Mrs. Emma. *Chester.* 1865-78.
Noble. *Wallingford, Conn, USA.* From *London.* 1796. C. & W.
Robert. *Southminster.* 1839.
& Son. *Chelmsford.* 1839.
T. *Chelmsford.* 1851.

SPENCER—*continued.*
Thomas. *Birmingham.* 1880.
Thomas. *Chelmsford.* ca.1823-32.
Thomas. *Cheltenham.* 1856-70.
Thomas. *London.* a.1671 to Solomon Bouquet. CC. 1685.
Thomas. *Manchester.* 1824.
W. G. *North Brixton (Surrey).* 1878.
Wooster & Co. *Salem Bridge, Conn, USA.* 1828-37. C.
SPENDLOVE—
Henry. *Metfield (Suffolk).* 1721 -d. 1750. C.
John, ?jun. *Thetford.* 1830-46.
SPERRY—
& BRYANT. *Williamsburg, LI, USA.* ca.1850. C.
& BUNKER. *New York.* 1848.
C. S. *New York.* 1848.
Elijah. *New York.* 1848.
F. S. *New York.* 1854.
& GAYLORD. *New York.* 1854.
Henry. *New York.* 1848.
Henry & Co. *New York.* 1854. C. Also looking glasses.
J. T. *New York.* ca.1830.
& SHAW. *New York.* 1846-48. C.
Silas. *New Haven, Conn.* ca.1840.
William. *Baltimore, USA.* 1842.
William. *Philadelphia, USA.* 1843-49.
William S. *New York.* 1840. C.
SPETH, Andrew. *Liverpool.* 1848. Importer of German clocks.
SPICE, William. *Hanover, Pa, USA.* Early 19c.
SPIEGELHALDER—
Andrew. *London.* 1863.
U. *London.* 1863.
SPIEGELHALTER—
& Co. *Southsea.* 1839.
& FEHRENBACH. *London.* 1875.
George (& Co.). *London.* 1844 (1851-81).
George & W. *London.* 1869-81.
J. N. *Bradford.* 1871.
Leo. *London.* 1857-75.
Leo. *Sutton (Surrey).* 1878.
Matthew & John. *Malton.* 1851-66. C. Claimed 'est'd. 1741' as succ. to Newby, who succ. Bartliffe, who succ. Kidd, all in the same shop.
Otto & Co. *London.* 1881.
Savory. *St. Austell (Cornwall).* 1873.
SPIERS—
James. *London.* 1863.
Joseph. *Leighton Buzzard.* 1864-77.
Maurice. *Birmingham.* 1842-60.
Saul Cohen. *Birmingham.* 1880.
Solomon Cohen. *Birmingham.* 1850-54.
SPIESS, William. *London.* 1869-81.
SPIKE—
Edward Lloyd. *Halifax, Nova Scotia.* 1837-1900s.
Thomas Daniel. *Halifax, Nova Scotia.* 1840-1926.
SPILLER—
Francis John. *Taunton.* 1883.
John. *New York.* ca.1820.
SPILLING, James. *London.* 1863.
SPINK, Marshall & Son. *London.* 1832. (B. -later 18c.?).
SPIRIDION—
Wladyslaw. From *Poland. London.* 1838. Then *Fareham.* 1838-44. Then *Cardiff.* 1844. Succeeded Henry Grant 1855. d.1891.
& Son. *Cardiff.* 1891-1920.
SPITTAL(L)—
James. *Whitehaven.* 1828-48. Also engraver.
John. *Whitehaven.* 1858-79.
SPITTLE—
D. *Coventry.* 1854-60.
Henry. *Hathern.* 1795-1835.
Henry. *Loughborough.* 1828-35.
John. *Wymeswold (Suffolk).* b. 1787, s. of Henry Spittle qv. -1840s, when believed moved to *Loughborough.*
Richard. *London.* a.1691 to William Davison. CC. 1699. (B. -1701).
SPITTLEHOUSE, J. *Loughborough.* 1849.
SPIVEY, Edward. *London.* 1828. (B. late 18c.?).
SPOONER—
T. *Liverpool.* 1847.
William. *Sheffield.* 1871. Clock cleaner.
SPORN, Martin. *Prague.* d.1580.

SPOTSWOOD & Co. *Baltimore, USA.* 1785. W.
SPRAIN, William. *London.* 1828.
SPRATLEY—
J. *Chesham.* 1869.
Joseph. *Luton.* 1877.
SPRATT—
Brothers. *Saintfield.* 1880-91.
Charles. *Saintfield.* ca.1825-80.
Gregory. *London.* Alien 1622.
John I. *Ardmillan.* ca.1750. C.
John II. *Saintfield.* s. of William II. ca.1820-d.1856.
John III. *Saintfield.* d.1872.
Samuel I. *Ardmillan.* 1750-d.1821. C.
Samuel II. *Ardmillan.* ca.1790-d.1821.
Samuel I. *Elkton, Md, USA.* Pre-1831 when sold to Thomas Howard, jun.
& Son. *Saintfield.* ca.1825-47.
William I. *Ardmillan.* ca.1750.
William II. *Ardmillan.* Then *Saintfield.* s. of Samuel I. ca.1786-d.1848.
SPREAT & Co. *Manchester.* 1848-51.
SPRENGER, George. *London.* 1857-75.
SPRING—
Robert. *London.* 1851-63
Solomon C. & Co. *Bristol, USA.* 1864-68. Successor to Birge, Peck & Co. C.
Solomon Crosby. *Bristol, USA.* 1826-1906. (Working 1858-95).
SPRINGALL—
——. *Norwich.* ca.1790-ca.1800.
Isaac. *South Erpingham.* 1836-46.
SPRINGFIED, Thomas Osborne. *Norwich.* b. ca. 1781 -d. 1858. C. Sheriff & mayor. Known as T. O.
SPROGELL, John. *Philadelphia, USA.* 1764-99. C.
SPROSTON, W. *Trellegg (nr Monmouth).* ca.1730. C.
SPRUNT, David. *Perth.* 1848.
SPRY—
Jacob. *Liskeard.* 1634. Rep'd ch. clock.
Nathanael. *Launceston.* 1623.
William. *Hereford.* (ca.1803). 1830-35. C.
SPUR, George. *Aylesbury.* pre-1719-d.1770.
SPURGE—
J. *Faversham.* 1866.
J. *Sheerness.* 1874.
James. *Great Dunmow.* 1839.
William. *Woolwich (Kent).* 1826-55.
SPURGEON, Robert. *Norwich.* 1875.
SPURGIN—
Jane. *Colchester.* wid. of Jeremy. 1699-ca.1706. C.
Jeremy. *Colchester.* b.1666-d.1699. C.
SPURGING, Thomas W. *Waltham Abbey.* 1866-74.
SPURR—
James. *Liverpool.* 1834. Dial enamellers.
Thomas R. *Annapolis, Nova Scotia.* 1838.
SPURRIER—
John. *London.* a.1677 to John Harris of *Oxford.* CC. 1684-86.
John. *Upton.* 1842.
John (& Son). *Tewkesbury.* 1830-50(-56).
John. *Wimborne.* 1824-30.
William Henry. *Tewkesbury.* 1879.
SPURWAY, R. *London.* 1851.
SPYCHER, Peter, jun. *Tulpehocken, Pa, USA.* 1784-1800.
SPYER, Samuel. *London.* 1828.
SPYERS, Moses. *Philadelphia, USA.* 1830.
SQUIER—
James. *Darlington.* 1827.
John. *Chichester.* a.1711 to Francis Mitten. C.
SQUIRE—
& Brothers. *New York.* ca.1860.
Horatio N. & Son (George H.). *New York.* 1838. W.
James. *Penrith.* 1848-58.
John. *Ambleside.* 1834.
& LANE. *New York.* 1857.
Thomas. *Gloucester.* ca.1775. C.
SQUIRES, James. *Penrith.* 1847-58.
SQUIRREL(L)—
Arthur H. *London.* 1881.
C. *Sutton (Surrey).* 1866.
S. *Croydon.* 1862-66.
S. *Ipswich.* 1846.
S. *Woodbridge.* 1853.
Samuel. *London.* 1875.

SQUIRREL(L)—*continued.*
W. *London.* 1869.
William. *Bildeston.* 1730. W.
William Knibb. *London.* 1875.
STACE—
John. *Folkestone.* 1847-66.
Mrs. M. A. *Folkestone.* 1874.
STACEY—
Edwin. *Sheffield.* 1862. W.
George. *Worksop.* 1828-55.
John. *London.* a.1675·to Thomas Wheeler. CC. 1683.
John Henry. *Wivelscombe.* 1883.
T. B. *Brockville, Canada.* 1862-71.
STADDON—
James Summerson. *Eastwood (Nottinghamshire).* 1876.
James Summerson. *Ripley (Derbyshire).* 1864-76. C.
STADEN & ROE. *London.* 1881.
STADLINGER, John R. *Buffalo, NY, USA.* 1891.
STADTER, P. B. & Co. *Baltimore, USA.* 1871.
STAFFORD—
George. *Liverpool.* 1851.
Napoleon. *Hereford.* 1870.
Richard T. *London.* 1857-69.
Robert, jun. *Leicester.* 1876.
T. *London.* 1851.
W. *London.* ca.1760. C.
STAGG—
John. *London.* 1851.
Leonard. *Bridport.* 1867-75.
STAHLBERG—
A. *Waterloo, Canada.* 1862.
August N. *Hamburg, Canada.* 1857.
STAINARD, Philip. *Morriston (Wales).* Mid-19c. C.
STAINTON, James. *London.* 1828.
STAIPLES, ——. *Winton.* Late 18c. W.
STALKER—
& MITCHESON. *Newcastle-on-Tyne.* 1784.
William. *Kirriemuir.* 1860.
STAMFORD, Richard. *London.* a.1652 to Peter Willierme.
STAMP, George. *Grimsby.* 1850.
STAMPER, Francis. *London.* a.1675 to Samuel Davis, jun. CC. 1682-d.1696. At the Golden Ball in *Lombard Street.* Quaker. C. & W.
STANBURY—
E. *Canterbury.* 1866-74.
W. *Bath.* 1800-35. W.
William. *London.* 1875-81.
STANCLIFFE—
Eli. *Honley* and *Clifton (nr Huddersfield).* 1757-58. C.
John I. *Askrigg.* b.1770-d.1850. C.
John II. *Askrigg.* s. of John I. b.1807-d.1850. C.
John. No place. Same as *Halifax* man.
John. *Burnley.* 1824-35. C. & W.
John. *Halifax.* b.1706-d.1780. C.
John. *Halifax.* 1850.
Joseph. No town. Same as Joseph of *Halifax.*
Joseph. *Barkisland (nr Halifax).* b.1740. s. of John. d.1812. C.
Joshua. *Halifax.* b.1708. bro. of John. -1770. C. & W.
STANDARD—
Electric Time Co. *New Haven, Conn, USA.* ca.1888. C.
Watch Co. *Minneapolis, Minn, USA.* 1893.
Watch Co. *Syracuse, NY, USA.* 1888-95. Later silversmith.
STANDFAST, W. *Taunton.* 1866.
STANDFORD, John. *Esher.* 1828.
STANDISH—
John. *Prescot.* d.1795. W.
John. *St. Helens.* 1824-27. W.
John. *Warrington.* mar.1769. W.
William. *London.* a.1660 to Jeffrey Baily. CC. 1668. (B. -1687).
STANDRING—
J. H. *Bradford.* 1866.
Jeremiah. *Bolton.* b.1712 (at *Rochdale).* mar.1742-d.1782. C.
STANDVEN—
Thomas. *Broadstairs.* 1839-40.
Thomas. *Ramsgate.* 1840-51.
Thomas Emery. *Ramsgate.* 1832-47.

STANFORD—
J. *Middlesbrough.* 1898.
William. *South Walsham.* ca.1780.
STANGER—
Hugh. *London.* 1832.
Hugh Peter. *London.* 1844-63.
STANHOPE—
J. *Guiseley* nr *Leeds.* ca.1870. C.
Thomas. *Preston.* ca.1730-ca.40. C.
STANILAND—
John. *Malton.* 1846-66. W.
Robert. *Malton.* 1840.
STANLEY—
Charles. *Liverpool.* ca.1780-90. W.
C. *Bilston.* 1860.
G. *Shrewsbury.* 1879.
George. *Malvern Link.* 1872.
George. *Southwell.* 1853-76.
J. *Chillicothe* and *Zanesville, O, USA.* 1807.
James Hill. *London.* 1863-81.
James Hill. *Upper Lewisham Road, London.* 1874.
John. *Alton (Hants).* 1878.
John. *Manchester.* 1851.
John. *Prescot.* mar.1822-25. W.
Phineas. *Lowell, Mass, USA.* 1837.
Richard. *London.* 1857.
Salmon. *Cazenovia, NY, USA.* 1831.
Samuel. *London.* 1881.
Samuel Ferdinando. *London.* 1857-81.
T. *London.* 1851.
Thomas. *Prescot.* 1819-25. W.
W. *Boston.* 1861-68.
STANNARD, Noel. *Ipswich.* 1879.
STANSELL, Walter D. *London.* 1869.
STANTON—
Alfred. *Market Deeping.* 1849-68.
Alfred. *Spalding.* 1876.
Baxter. *Donnington (Lincolnshire).* Clock case maker. 1828-35.
& Bro. *Rochester, NY, USA.* 1845-66.
Edward. *London.* a.1655 to Nathan Allen. CC. 1662-99. (B. -1707). C. & W.
George S. *Providence, RI, USA.* 1824.
John. *London.* a.1684 to Richard Farmer. CC. 1692. (B. -1725). C.
Originall. *London.* 1662. Almost certainly error for Reginald. q.v.
Reginald. *London.* 1662-left CC. pre-1669. C.
Robert. *Shrewsbury.* To *Oswestry* by 1836 -d. 1861.
Samuel. *London.* a.1692 to Benjamin Tebbatt. CC. 1719. (B. -1733).
W. *Fenny Stratford (Buckinghamshire).* 1847-54.
W. *Providence, RI, USA.* ca.1810.
William. *Ampthill.* 1854-77.
William. *Bloxham (Oxfordshire).* b.1804-ca.1830. Then *Buckingham.* 1830-47. Then *Fenny Stratford.* 1851-54.
William P. & Henry. *Rochester, NY, USA.* 1826-1845.
STANWAY, H. *Newcastle-under-Lyme.* 1860.
STANWOOD, Henry B. & Co. *Boston, USA.* ca.1850.
STAPLES—
& DOBBS. *New York.* ca.1780.
George P. *St. John, New Brunswick.* 1859.
John. *Boston.* 1835.
John. *Leicester.* 1805-ca.1815. C. & W.
John L., jun. *New York.* 1793.
STAPLETON—
James. *Bradford.* 1866.
Thomas. *London.* a.1686 to Richard Watts. CC. 1693. (B. -1723). W.
Thomas H. *Battersea.* 1878.
William. *Thirsk.* 1724. C.
STAPLYTON, Myles. *Armagh.* 1622. C.
STAREE, Anthony. *Birmingham.* 1850.
STARES, W. J. *St. Johns, Newfoundland.* 1864-85.
STAREY, Robert. *Ledbury.* 1835. Repairer.
STARK—
J. *Croydon.* 1866.
Jabez. *Wellington (Somerset).* 1883.
W. T. *Xenia, O, USA.* ca.1830.
STARKIE, Edward. *Liverpool.* 1683-89. W.

STARLEY—
Francis. *London.* 1641. Banned from clockmaking.
William Henry. *London.* 1863-81.
STARR—
Frederick. *Rochester, NY, USA.* 1834.
William. *London.* 1844-51.
William James. *London.* 1857-75.
STARRETT, James. *East Nantmeal, Pa, USA.* 1796-
1804. C. & W.
STARTURGE, ——. *Modbury.* mar.1740.
STASINON, P. Joseph. *Tournai.* ca.1710. W.
STATON, John. *Chatham.* 1845-47.
STATTON, R. J. *Toronto.* 1856.
STATZELL, P. M. *Philadelphia, USA.* 1845 and later.
STAUFFER—
Samuel C. *Manheim, Pa, USA.* (1757). 1785-1825. C.
& Son & Co. *London.* 1857-81.
STAVNER, John. Place unknown. Late 17c. Lant.
clocks.
STAYNES, Thomas. *London.* CC. 1658.
STEAD—
George. *Leeds.* 1866-71. W.
James. *London.* 1875.
STEANE—
I. *Coventry.* 1854-60.
J. *Abingdon.* 1854.
Mrs. S. *Abingdon.* 1864.
Thomas. *London.* 1869-75.
STEAT, Frank. *London.* 1881.
STEBBINS—
& Co. *New York.* 1832. W. Also jeweller and silver-
smith.
& HOWE. *New York.* ca.1830.
Lewis. *Waterbury, Conn, USA.* 1811. Dialmaker for
Chauncey Jerome.
Thomas. *New York.* 1830.
STEBER, David. *Dover.* 1826-32.
STECKELL, Valentine. *Frederick, Md, USA.* 1793.
STECKMAN, H. *Middletown, Pa, USA.* ca.1850. C.
STEDMAN—
Charles. *Steyning.* 1862-70.
George. *Petworth.* 1828-55.
Henry. *Bristol.* 1794. Watch case maker.
J. S. *Godalming.* 1866.
James. *Storrington (Sussex).* 1839-70.
James Walter. *Petworth.* 1870-78.
John C. *Raleigh, NC, USA.* (Of Savage & Stedman
1819-20.) Alone 1822-d.1833. W. and silversmith.
Mrs. A. & Son. *Petworth.* 1862.
Richard. *Godalming.* (B. pre-1773-)1828-39. W.
Sone. *Midhurst (Sussex).* 1862-78.
William H. *Colchester.* 1874.
STEEDMAN, William. *Milnthorpe.* ca.1820. C.
STEEL—
G. *Ipswich.* 1858.
George. *Horton, Nova Scota.* 1831.
George Edward. *Chester-le-Street.* 1851-56.
John. Town not stated. ca.1750-60. Northern English.
(Prob. the *Killamarsh* man.)
John. *Killamarsh (nr Rotherham).* b. 1711 d. 1792. C.
& LAWSON. *London.* 1869-75.
R. F. *Adams, NY, USA.* ca.1850.
Samuel. *New Haven, Conn, USA.* ca.1840.
Samuel. *London.* 1857.
Thomas. *Kirkintilloch.* 1837.
W. *Reading.* 1877.
W. F. *Brixton.* 1862.
William. *Albion, Mich, USA.* ca.1860.
William. *London.* 1875.
William F. *London.* 1869-81.
William Francis. *London.* 1839.
STEELE—
James. *London.* (B. a.1793, CC. 1801-40) 1844.
James Smith. *London.* 1869-75.
Joseph. *London.* 1851-81.
Joseph. *Oxford.* Successor to Samuel Denton. 1831-46.
W.
& Sons Ltd. (David and Robert). *Belfast.* 1880-94.
Clock importers.
William Francis. *London.* 1851-63.
STEER—
Henry. *Derby.* 1864-76.

STEER—*continued.*
James. *Guildford.* 1839-51.
James. *Ripley (Surrey).* 1828.
John. *Burton-on-Trent.* 1849-60.
John. *Derby.* 1835-49.
William. *Bristol.* 1863-70.
William. *Newport (Monmouth).* 1880.
STEERS, William. *Keighley.* 1837. C. & W. dealer.
STEIERT—
& KREIZ. *Camberwell.* 1862.
M. *London.* 1869.
Peter. *London.* 1857.
STEIGHTHOLM, John. *Sleaford.* ca.1770. C.
STEIKLEADER, John. *Hagerstown, Md, USA.*
1791-93.
STEIN—
Albert H. *Norristown, Pa, USA.* 1837.
Daniel. *Norristown, Pa, USA.* ca.1830.
George. *Allentown* or *Northampton, Pa, USA.* s. of
Jacob, ca.1820. C.
Jacob. *Allentown, Pa, USA.* 1771-1842. C.
Jacob & Son (George). *Allentown, Pa, USA.* ca.1840.
C.
STEINBACH—
Charles. *London.* 1851-57.
Mrs. Mary. *London.* 1863-81.
STEINER, Leon. *London.* 1875-81.
STEINMANN—
Daniel. *London.* 1832-57.
& Son. *London.* 1863.
STEINMETSSEL, Hans. *Prague.* 1550. C.
STEINSEIFFER, John. *Hagerstown, Md, USA.* 1743-
ca.1777. Perhaps later at *Williamsport, Pa.*
STEIRT, Martin. *London.* 1863-75.
STELERT & DOTTER. *Bromley (Kent).* 1874.
STELJES, James. *London.* 1869.
STELL, G. *Todmorden.* 1858. Cleaner.
STELLWAGEN, Charles K. *Philadelphia, USA.*
1840-48.
STENHOPE, Thomas. Place not stated. See Thomas
Stanhope.
STENING, Cornelius. *Hove.* 1862-70.
STENNETT—
Robert. *Bath.* ca.1775-95. C.
William. *Toronto.* 1833-47.
STEPHEN—
Charles. *Old Meldrum.* 1860.
David. *Thurso.* 1860.
James. *Old Meldrum.* 1837-60.
STEPHENS—
——. *Penzance.* ca.1805. C.
& DAVIES. *Neath.* ca.1840. C.
George. *Comrie (Scotland).* 1860.
Henry. *London.* 1851-57.
Henry. *London.* 1875-81.
T. C. & D. *Utica, NY, USA.* 1840.
Thomas. *Forest Hill (Kent).* 1874.
William. *Albany, NY, USA.* 1840-42.
William. *Godalming.* ca.1750 (B. late 18c.).
William. *Stratford-on-Avon.* 1850.
STEPHENSON—
Daniel. *London.* 1828.
George. *Pickering.* 1866. W.
George. *Warminster.* 1830.
(L. S.), HOWARD (E.) & DAVIS (D. P.) *Boston* or
Roxbury, Mass, USA. ca.1849. W.
James. *Leeds.* 1853.
John. *Leeds.* 1826-50.
Robert. *Pickering.* 1823-66. W.
Thomas. *Carlisle.* 1879.
Thomas. *Dromore.* 1846. C.
W. *Stratford-on-Avon.* 1854-60.
William. *Dorchester.* 1830.
STEPTO, William. *London (St. Giles).* a.1695. CC.
1703-ca.1710. C.
STERL, Franz. *Mauer.* 1820-55. C.
STERLING—
David. *Dromore.* 1854.
Richard. *South Woodstock, Vt, USA.* 1811. C. and
chairs.
Robert. *Dromore.* 1865. Later cabinetmaker at *Ban-
bridge.*

STERN—
John Charles. *Grimsby.* 1868-76.
Samuel. *Macclesfield.* 1848.
Samuel. *Toronto.* 1871-77.
William. *Philadelphia, USA.* 1820-22.
STERNBERG, John & Bros. *Manchester.* 1851.
STEUERT, Lambert. *London.* 1857.
STEVENS—
Abraham. *London.* 1875.
Charles G. *New York.* 1840.
Daniel. *London.* a.1653 to Jeffrey Baily, CC. 1661 (B. -d.1703).
Daniel J. *London.* 1857.
Edward. *London.* 1857-69.
Edwin. *London.* 1875-81.
Ezekiel. *London.* 1839-44.
George M. *Boston, USA.* 1880s, T.C.
& HEATH. *Chillicothe, O, USA.* ca.1815.
Henry. See Peacock, John.
HODGES & Co. *New York.* 1872.
I. T. *London.* 1851.
Isaac Thomas. *London* (?1851)1863.
J. *Market Deeping.* 1868.
J. *Weston-super-Mare.* 1866.
J. P. *Utica, NY, USA.* 1837.
Jacob. *London.* 1839.
John. *Bangor, Me, USA.* ca.1840.
John. *Bourne.* 1849-61.
John. *Colchester.* From *London.* 1691-ca.1720. Lant. clocks.
John. *Great Milton* (*Oxfordshire*). mar.1784-1819. C.
John. *St. Ives* (*Cornwall*). 1823. W. & C.
& LAKEMAN. *Salem. Mass, USA.* 1819-30. (Later Stevens alone.)
M. *Chillicothe, O, USA.* 1815.
Mark. *London.* 1851-69.
Robert. *London.* 1869-75.
S. *Lowell, Mass, USA.* 1853.
Samuel. *London.* a.1672 to Edward Stanton, CC. 1680 (B. -1706).
T. *Henfield* (*Sussex*). 1862-78.
Thomas. *Exeter.* 1795. (Said to be a. in *London* 1785).
Thomas. *Huntingdon.* ca.1760-ca.1770. C.
Thomas. *London.* a.1692 to Samuel Marshall, CC. 1700 (B. -1724).
Thomas Henry. *London.* 1857-75.
(J. P.) Watch Co. *Atlanta, Ga, USA.* 1877-87. Later sold to Ezra Bowman.
William, jun. *London.* 1869-81.
William, sen. *London.* 1828-63(-69).
William G. *London.* 1844-57.
STEVENSON—
John. *Hanley.* 1860-76.
John. *London.* 1828-39.
Mrs. Mary. *London.* 1875-81 (wid. of Richard James?).
Richard James. *London.* 1851-69. (Successor to Robert).
Robert. *London.* 1844.
W. M. *Leicester.* 1855-64.
William. *Nottingham.* 1828-35.
William. *Strabane.* 1827.
STEVENTON, ——. *Drayton.* ca.1770. C.
STEVER—
(J.) & BRYANT. *Burlington, Conn, USA.* ca.1845. C.
(J.) & HILL (William S). *Bristol, USA.* 1852-56.
(J.) & PRINDLE. *Burlington, Conn, USA.* Believed 1850s. C.
(J.) & WAY (John A.). *Bristol, USA.* 1864-66.
STEWARD—
Aaron. *Philadelphia, USA.* 1843.
Charles. *Newport Pagnell.* pre-1775-mar.1785. W.
James Henry. *London.* 1857-75.
John. *Henley-in-Arden.* 1717-47. Lant. clocks and T.C.
Thomas. *Newport Pagnell.* ca.1740. C.
Thomas B. *Hamilton, Ontario.* 1867-73.
William. *London.* 1857-81.
STEWART—
Alexander (& Co.). *London.* 1832-75(-81).
Alexander. *Richmond, Canada.* 1862.
Alexander Baikie. *Kirkwall.* 1835.
Allan & Robert. *Glasgow.* 1835.
Arthur. *New York.* ca.1820.

STEWART—*continued.*
Charles. *Blairgowrie.* 1836.
Charles. *Builth.* 1887.
& Co. *Philadelphia, USA.* 1824.
Daniel. *Salem, Dorchester* and *Fitchburg, Mass, USA.* 1732-d.1802. C.
Francis. *Brechin.* 1837.
G. *Belfast.* ca.1790-1820. Clock dial painter and engraver and watch case maker.
George. *Philadelphia, USA.* 1837.
Hannah Jane. *Londonderry.* wid. of Thomas. 1839.
James. *Baltimore, USA.* 1792.
James. *Dunkeld.* 18c. C.
John. *Auchterarder.* 1837.
John. *Belfast.* 1785.
John. *Belfast.* 1827.
John. *Dunbar.* (B. 1792). ca.1810. C.
John. *Tobermory.* 1860.
Joseph. *Liverpool.* 1848-51.
& McFERRAN. *Manchester.* 1848. Chrons.
Richard. *Inverness.* 1860.
Robert. *Glasgow.* 1841-60.
Robert. *Blairgowrie.* 1836.
Robert. *Helmsley.* 1814-23. W.
Robert. *Newport* (*Wales*). 1848.
Robert John. *Bristol.* 1812-38.
T. *Gravesend.* 1855.
Thomas. *Londonderry.* mar.1833-d.1836.
William P. *Funkstown, Md, USA.* ca.1820.
STICHLER, John. *Marietta, Pa, USA.* ca.1850.
STICKLAND, William. *London.* 1851.
STICKLER—
Henry Charles. *Frome.* 1883.
Thomas. *Botney.* 1749 (or Stockler).
STICKNEY, Moses. *Boston, USA.* 1823.
STIER—
& DILGER. *Bath.* 1856-66.
Frederick. *Mere* (*Wiltshire*). 1859-75.
Michael (& Son). *Bath.* 1875(-83). Successor to Stier & Dilger.
STIERT, L. *Bromley.* 1866.
STIFFELL & CARTIER. *London.* 1869. (George Frederick Jepson, agent.)
STILES—
(Samuel) & BALDWIN (Jedidiah). *Northampton, Mass, USA.* 1791.
Samuel. *Northampton, Mass, USA.* 1785. Poss. to *Windsor, Conn.* ca.1795.
William. *London.* 1832.
STILL—
Francis. *London.* a.1691 to Robert Halstead, CC.1698 (B-1710). W.
William. *Aberdeen.* 1846.
STILLAS, John. *Philadelphia, USA.* 1783. To *Baltimore.* 1790-93.
STILLINGER, ——. *London.* 1641. Not in CC.
STILLMAN—
Barton. *Burlington, Conn, USA.* 1790-95, then *Westerly, RI.* ca.1810 and later.
Deacon William. *Burlington, Conn, USA.* (1767) ca.1786-91; *Westerly, RI,* 1793-1809. C.
Ira. *Newport, RI, USA.* ca.1830.
Paul. *Westerly, RI, USA.* (1782). ca.1803-1810.
STILLSON, David. *Rochester, NY, USA.* ca.1830.
STIRSKEY, Franz. *Halifax, Nova Scotia.* 1871-81.
STOCK—
Isaac. *London.* Late 17c. C.
Jabez. *London.* ca.1700-ca.1740. C.
STOCKALL—
James John. *London.* 1863-81.
Thomas. *Cleobury Mortimer* (*Shropshire*). 1856-79.
Thomas & Henry. *Cleobury Mortimer* (*Shropshire*). 1870.
STOCKDALE, Thomas. *York.* Post-1614.
STOCKELL & STUART. *Newcastle-on-Tyne.* (B. 1795) 1799-1801. C.
STOCKER—
Henry. *London.* 1881.
Joseph. *London.* 1863-69.
William. *London.* 1881.
STOCKFORD—
Joseph. *Thame.* 1770-74. T.C. & C.

STOCKFORD—*continued*
Thomas. *Great Haseley (Oxfordshire).* Successor to Thomas Holloway. 1764. C.
Thomas. *Thame.* 1778-85. Prob. same man as above.
STOCKIN, Charles. *Northampton.* 1658.
STOCKLER (or Stickler), Thomas. *Bolney (Sussex).* 1735. C.
STOCKMAN—
   J. *Emsworth (Hampshire).* 1848-59.
   James. *Portsmouth.* 1830.
   James. *Winchester.* 1839.
STOCKS, John. *Bocking.* 1839.
STOCKTON—
   ——. *London.* 1722. W.
   ——. *Stokesley.* ca.1780-85. C.
   Francis. *Yarm.* b.1751-d.1829. C.
   Francis. *York.* mar.1819. W.
   George, jun. *South Shields.* 1827-44. C.
   Mark. *Whitby.* b.1764-1800. W.
   Samuel W. *Philadelphia, USA.* 1823-31, then perhaps to *New York* ca.1840.
   Thomas. *Yarm.* 1761-d.1794. C.
STOCKWELL—
   Joseph. *Manchester.* 1851.
   Thomas. *Cleobury Mortimer.* 1849-85.
STODDARD—
   ——. *Litchfield, Conn, USA.* 1820s. C.
   & KENNEDY. *New York.* 1794.
STODDART—
   ——. *Quebec.* 1884.
   Benjamin. *South Cave (East Yorkshire).* 1801-41. C. & W.
   J. *Birmingham.* 1860.
   J. *Huddersfield.* 1871.
   James. *London.* 1844-63.
   John. *London.* 1839-44.
   John. *Holmfirth.* 1866. W.
   Robert. *London.* 1828.
   Robert. *London.* 1851.
   Robert & James. *London.* 1832-39.
STOESSIGER—
   Alexander. *London.* 1881.
   Frederick. *London.* 1857-81.
   Frederick & Alexander. *London.* 1863.
STOFF, M. Place unknown. ca.1770. C.
STOFFEL, Ferdinand A. *Paris.* 1889.
STOKEBERRY, George. *Philadelphia, USA.* 1837.
STOKEL, John. *New York.* 1820-43.
STOKES—
   E. *Coventry.* 1854.
   George. *Birmingham.* 1880.
   I. *Knutsford.* 1857.
   J. *Bocking.* 1851-74.
   J. *Macclesfield.* 1857-65.
   James. *Stourbridge.* 1868-76.
   John. *Congleton.* 1842.
   John. *Knutsford.* 1834-48.
   John. *St. Ives.* ca.1720-ca.30. C.
   William. *Dudley.* 1842.
   William. *London.* 1863-69.
   William. *Walton (Surrey).* 1878.
STOLL—
   Anthony. *Birmingham.* 1850.
   George. *Lebanon, Pa, USA.* ca.1850.
   H. *Northwich.* 1865.
   Henry. *Runcorn.* 1878.
STOLLENWERCK—
   & Brothers. *New York.* ca.1820.
   P. M. *Philadelphia, USA.* 1813-14; *New York.* 1815.
STOLZ—
   Columban & Co. *Birmingham.* 1868-80.
   Matthew. *London.* 1869-81.
STONE—
   ——. *Newry.* Late 18c.-early 19c.
   Alfred. *Cheltenham.* 1879.
   & ALLEN. *Shrewsbury.* 1828. W.
   Charles. *Liverpool.* 1797. W.
   Edwin. *London.* 1881.
   Ezra. *Boston, USA.* 1810.
   Frederick. *Bristol.* 1842.
   George. *Brighton.* 1828.
   George. *Weston-super-Mare.* 1875.

STONE—*continued.*
   H. *Maidstone.* 1855.
   Isaac. *Lurgan.* ca.1750-ca.1760. C.
   J. *Newnham (Gloucester).* 1856-70.
   James. *Aylesham.* ca.1800. C.
   John I. From *London* to *Thame,* 1764. Then *Aylesbury,* mar.1764-89. W. & C. and silversmith.
   John, jun. *Aylesbury.* s. of John, sen. b.1768. mar.1789-1842.
   John. *Cheltenham.* 1850-79.
   John. *Henley-on-Thames.* ca.1795. W.
   John. *Reading.* 1830.
   John James. *London.* 1881.
   Richard. *Thame.* 1761-83. C. & W. & T.C.
   Robert. *Bristol.* 1870-79.
   Robert. *Manchester.* 1848-51.
   Thomas. *Thame.* 1801-09. W.
   William. *London.* a.1692 to Thomas Fletcher, CC. 1700 (B. -1703).
   William G. *Somers, NY, USA.* 1809.
   William Henry. *London.* 1832-39.
STONEHOUSE—
   Ann. *Leeds.* 1820. Widow of Jonathan.
   C. *Newcastle-on-Tyne.* 1858. C.
   John. *Leeds.* b.1789, s. of Jonathan, -1834. C.
   Jonathan. *Addingham.* mar. 1788-89. Later *Leeds.* 1798-1817. C.
   Robert. *Leeds.* 1826-53. C.
STONER (sometimes **STEINER**), Rudy. *Lancaster, Penn, USA.* b.1728-d.1769. C.
STONES—
   George. *Blackburn.* 1822. C.
   Thomas. (alias SCONES). *London.* a.1684 to John Westoby, CC. 1692 (B. -1747). C.
STOPP, Max Hermann. *London.* 1851.
STORCH—
   Johann. *Bayreuth.* ca.1750.
   William. *London. St. Clements, Middlesex.* 1765.
STORER—
   J. & R. *London.* 1811. Wholesale W.
   James. *London.* 1828.
   James. *London.* 1832. Watch cases.
   John. *Greenock.* 1860.
   W. *Taunton.* 1861.
STOREY (includes **STORY**)—
   Benjamin. *Tuxford.* 1835-55.
   Cole. *London.* 1797. C.
   Edward. *Ulverston.* 1842-51; *Cartmel Town* 1849-69.
   F. *Tuxford.* 1864.
   George. *Ipswich.* 1853-75.
   J. *King's Lynn.* 1865.
   James. *Bury St. Edmunds.* 1879.
   John. *Liverpool.* 1678-d.1721. W.
   Joseph. *Liverpool.* 1703-15. W.
   & SEDMAN. *Dewsbury.* 1866-71. W.
   William. *Barrow-in-Furness.* b.1838-d.1912. (Believed also *Halifax* and *Dewsbury.*)
   William. *Manchester.* mar.1803. C.
STORKEY, J. *Ramsgate.* 1866.
STORM, Edward. *Ipswich.* 1875.
STORR—
   Batty. *York.* b. 1710-d.1793. W. & C. Bro. of Marmaduke.
   & GIBBS. *London.* 1741 (B. 1752). W. & C.
   Isaac. *York.* b.1750, s. of Batty. d.1773. C. & W.
   Jonathan. *York.* b.1739. s. of Marmaduke. d.1805. C. & W.
   Marmaduke. *York.* b.1702-pre-1728, later *London* d.1750. C.
   William. *York.* b.1742. s. of Batty. To *London* pre-1778-d.1812. W.
STORRS—
   (Nathan) & COOK (Benjamin F.). *Northampton, Mass, USA.* 1827-33. (Also at *Amherst* in 1829).
   Nathan. *Northampton, Mass, USA.* 1791. With Baldwin 1792-94, later with Cook, q.v.
   Thomas. *Boston, USA.* 1803.
STORZ, J. C. *Taunton,* 1861.
STOTT—
   Eli. *Wakefield.* 1773-ca.1810. C.
   John. *Ashby-de-la-Zouch.* 1876.
   Ormerod. *Newchurch.* 1851. Clock dealer.

STOUGHTON, Dulcina. *Shelburne, Nova Scotia.* 1787.
STOUR—
    James. *London.* 1828.
    Robert & Son. *London.* 1828.
STOUT, Samuel. *Princeton, NJ, USA.* (1756) ca.1779-95.
STOW—
    D. F. *New York.* ca.1830.
    Edward. *Camberwell.* 1862-78.
    Edward. *London.* 1857.
    P. M. *Philadelphia, USA.* ca.1810.
    Solomon. *New York* and *Southington, Conn, USA.*
        1823-34. (Later with PECK-1837.)
STOWEE, Frederick. *Philadelphia, USA.* 1802.
STOWELL—
    A. *Charleston* or *Boston, Mass, USA.* ca.1850.
    Abel. *Worcester, Mass, USA.* ca.1800-ca.1820. C. &
        T.C.
    Abel, jun. *Boston, USA.* s. of Abel I. ca.1820-56. T.C.
    John. *Boston* and *Medford, Mass, USA.* 1815-25. C.
    John. *Charlestown, Mass, USA.* 1825-36.
    John J. *Boston* and *Charlestown, Mass, USA.* ca.1831.
    & Son. *Charlestown, Mass, USA.* ca.1845. T.C.·
    W. H. *Birmingham.* 1860.
STOY—
    George. *Colchester.* 1855-80. C.
    Gustavus. *Lancaster, Pa, USA.* 1806. C. & W.
STRACHAN—
    James. *Cumbernauld.* 1860.
    William. *Barton-on-Humber.* 1828-35.
STRADLING, Alfred. *Newbury.* succ. to John Joyce.
    1866-91.
STRAEDE, Charles. *Lititz, Pa, USA.* ca.1850. C.
STAFFORD, Thomas. *Dewsbury.* 1851. W.
STRAHN, Samuel. *London.* 1832. Watch cases.
STRAITON, David. *Montrose.* (B. 1820) 1837-60.
STRAM, Alfred. *London.* 1871. Gold watch case maker.
    Successor to Louis Comtesse.
STRANG—
    David. *Falkirk.* 1860.
    James. *Glasgow.* 1834.
    Robert. *Alloa.* 1842-89
STRANGE—
    Charles. *High Wycombe.* s. of Thomas, 1829-76.
    F. *Croydon.* 1855.
    Mrs. C. *Croydon.* 1862-66.
    & Son. *Kingston (Surrey).* 1862.
    T. F. *Littlehampton.* 1851.
    T. F. *Kingston (Surrey).* 1855-66.
    Thomas. *Banbury.* Successor to C. W. Drury. 1823-66.
        T.C.
    Thomas (& Son). *High Wycombe.* Successor to John
        Lee 1825-29 (1830-39).
    William. *Kingston (Surrey).* 1828-51.
STRATFORD—
    The. 1914 and later, product of Pequegnat Clock Co.
    James. *London.* 1857.
    Thomas. *High Wycombe.* 1877.
    William.   *Cheltenham.* 1840-42.
STRATH, Benjamin & Brothers. *London.* 1851-69.
STRATHON, John Thomas. *Saltash.* 1873. W.
STRATTEN, Henry. *Birkenhead.* 1878.
STRATTON—
    Charles. *Worcester* and *Holden, Mass, USA.* ca.1830-
        ca.1840. C.
    George John. *London.* 1828-32.
    John. *Devizes.* 1830.
    Joseph. *London.* 1839-44.
    Nelson Pitkin. *London.* 1875 (agent).
STRAUB—
    C. *Norwich.* 1846.
    Ferdinand. *Camberwell.* 1878.
    & HEBTING. *London.* 1844-81.
    Michael. *London.* 1844.
STRAUS, Morris. *Farringdon.* 1837.
STREET—
    John. *Montreal.* 1866.
    & PIKE. *Bridgwater.* ca.1760. C.
    Richard. *Bridgnorth.* 1828-35.
    Richard. *Charlbury.* mar.1795.
    Robert. *Hitchin.* 1851-74.
STREETER—
    Edwin W. (F.R.G.S.). *London.* 1875-81.

STREETER—*continued.*
    Gilbert L. *Salem, Mass, USA.* 1846-50, W. Later a
        publisher.
    John. *Wisborough Green (Sussex).* 1862-78.
STREETIN, Richard. *London.* From *Hunston (Sussex).*
    a.1738 to Richard Conyers, C.
STRENT, Frank. *London.* 1881.
STRETCH—
    Benjamin. *Bristol.* 1723-64. W.
    Isaac. *Philadelphia, USA.* 1732-52. C. & W.
    James. *Bristol.* s. of Samuel of *Bristol.* 1733.
    John. *Bristol.* 1704. C.
    Peter. b.1670, worked *Leek.* ca.1700, to *Philadelphia,
        USA.* ca.1702-46. C.
    Samuel. *Bristol.* s. of James of *Birmingham.* a.1753.
        Later at *Birmingham.*
    Samuel. From *England* to *Philadelphia, USA.* 1711-32.
        C. See next entry?
    Samuel. *Wolverhampton.* Early 18c. C.
    Thomas. s. of Peter, b. *England.* To *Philadelphia,
        USA.* 1746-82. C.
    William. s. of Peter. To *Philadelphia, USA.* ca.1720-
        55. C.
STRETTON—
    Mrs. C. *London.* 1869. (?wid. of William).
    William. *London.* 1832-63.
STREVETT, George. *London.* 1839.
STRIBBLEHILL, Samuel. *London.* 1832.
STRIBLING, Benjamin. *Stowmarket.* b. ca. 1663 d.
    1720 C.
STRICKLAND—
    Benjamin John. *Bristol.* 1840.
    Charles. *London.* 1881.
    George. *Aberdare.* 1875.
    George. From *Bath* to *Bristol.* 1770-ca.1780. C.
    George & WELCHMAN, Mordecai. *Bristol.* Late 18c. C.
    J. P. *Gravesend.* 1866.
    James. *Ulverston.* 1849-51. (One such at *Manchester*
        1813).
    Sarah. *Bristol.* ca.1790. wid. of George.
    William. *London.* 1828.
    William. *London.* 1863.
    William. *Tenterden.* 1802. C.
STRIEB, C. H. *Wooster, O. USA.* 1822-47.
STRIEBY, Michael. *Greensburg, Pa, USA.* ca.1790-
    1830. C.
STRINGER—
    Edwin. *Stourbridge.* 1860-76. C.
    F. J. *Willenhall.* 1868.
    Frederick. *Bilston.* 1842-50.
    James George. *Coventry.* 1880.
    Thomas. *?London.* ca.1725. C.
STRINGFELLOW—
    John. From *Rochdale.* 1694 to *Halifax*-d.1718. C. (One
        such a. *London* 1681 to Edward Stanton, CC. 1691.
        May be same man.)
    Mrs. Thomas. *Taunton.* 1875.
    T(homas?). *Taunton.* 1861-66.
STRIPE, John. *Chichester.* 1742. C.
STRIPLING—
    Francis. *Barwell.* s. of Thomas. ca.1775.
    Joseph. Place unknown. ca.1690-ca.1700. C.
    Thomas. *Barwell.* b.1707-d.ca.1777. C.
    Thomas (?II). *Lichfield.* ca.1820. C. (One such 1761-
        d.1775 evidently not the same man.)
    Thomas & William. *Lichfield.* 1828-42. C.
    William (?& Thomas). *Lichfield.* 1835.
STRIPP, William. *East Grinstead.* 1878.
STROBEL, Kaspar & Melchior. *Prague.* 1585. C.
STROH, Augustus. *London.* 1857.
STROND, George. *London.* 1857.
STRONG—
    Charles. *London.* 1857-69.
    Ebenezer George. *Cardiff.* 1887.
    John. *London.* 1839.
    Peter. *Fayetteville, NC, USA.* (1764) 1788-97. C. & W.
STROUB, Constantine. *Treorchy.* 1875-87. W.
STROUD—
    Abraham. *Seacon (co. Antrim).* ca.1750. W.
    Elizabeth. *London.* 1839 (?wid. of Joseph). See next
        entry.
    Mrs. E. *London.* 1844. See previous entry.

STROUD—*continued.*
G. *Brixton.* 1862-66.
J. *London.* 1787. W.
Joseph. *London.* (B. 1817-24) 1828-32. W.
STRUGNELL, Charles. *London.* 1832. Clock case maker.
STRUTT, Francis. Place unknown. 18c.
STUART—
Alexander. *Kirkwall.* 1836.
George, sen. *Newcastle-on-Tyne.* (B. 1790-1820) 1827-36. C. & W.
George, jun. *Newcastle-on-Tyne.* 1827-58.
H. *Coventry.* 1854.
H. & Co. *Coventry and Liverpool.* 1868-80.
Henry. *Liverpool.* 1834-51. (B.-1825?).
James. *London.* 1863-69.
James. *Philadelphia, USA.* 1837-50.
Robert. *Hunmanby.* ca.1830. C.
Thomas. *Philadelphia, USA.* 1839.
STUB(B)S—
Frederick. *Loughborough.* 1876.
Gabriel. *London.* a.1670 to Robert Seignior, CC. 1675 (B.-1677).
George. *Leamington.* 1868-80.
John. *Kegworth (Leicestershire).* 1828-35.
Peter. *Warrington.* Famous toolmaker and supplier. ca.1800-ca.1830.
Thomas. *Kegworth.* 1804. W.
STUBER, C. *Bedford.* 1854.
STUBINGTON—
Henry. *Gosport.* 1859-78.
Henry. *London.* 1851-57.
Henry Compton. *Gosport.* 1830-48.
Walter C. *Winchester.* 1878.
STUBLER, Joseph. *Wigan.* 1799. C. (One such *Manchester* 1808).
STUBLEY—
Benjamin. *Liverpool.* 1834-51.
John (& Abraham). *Liverpool.* 1834-51 (1848).
STUDEBAKER—
Name used by South Bend Watch Co. 1903-30.
Of *Canada.* Name used by South Bend Watch Co. 1902.
STUDER, Mathias (Mathew). *London.* 1839 (1844).
STUDLEY—
David. *Hanover, Mass, USA.* ca.1806. *N. Bridgewater* and *Brockton, Mass, USA.* 1834.
Luther. *N. Bridgewater* or *Brockton, Mass, USA.* bro. of David, 1840s.
STUMBELS—
William. *Aveton Gifford (Devon).* 1728. Same man as at *Totnes* later.
William. *Totnes.* mar.1723-d.1769. C.
STUMP—
Joseph. *Reading, Pa, USA.* Early 19c.
Richard. *Oldham.* 1824. Also jeweller.
STUNT—
Charles. *London.* 1844-63.
Robert. *London.* 1844-51.
STURGE, H. *Bristol.* 1856.
STURGEON, Samuel. *Shippensburg, Pa, USA.* ca.1810.
STURGES, William. *Bristol.* a.1702.
STURGIS, Joseph. *Philadelphia, USA.* 1813-17.
STURROCH, James. *Montrose.* 1860.
STURROCK, William. *Edinburgh.* 1855.
STUTSON, James. *Rochester, NY, USA.* 1838.
STYLE, T. *Chesterton (Cambridgeshire).* 1865.
STYLES, William. *Belfast.* 1622. C.
SUDBURY, John. *London.* a.1675 to Christopher Maynard. CC. 1686.
SUDLOW, Benjamin. Mar. 1745 *Beccles* then *Yarmouth* 1763 -d. 1787.
SUFFOLK—
Samuel. *London.* 1851-81.
Watch Co. *New York.* 1901.
SUFFRIN, Mathew Mawhinney. *St. John's, Newfoundland.* 1850.
SUGDEN, Samuel. *Selby.* ca.1700. C.
SUGGATE, George senior. *Halesworth.* b. 1720 d. 1807. W. Also silversmith and jeweller.
SUGGETT—
John. *Little Walsingham (Norfolk).* 1836.

SUGGET—*continued.*
Thomas. *Walsingham (Norfolk).* 1830.
SUGGIT, John. *Buttercrambe (East Yorkshire).* 1840. W.
SULEAU, Louis. *London.* 1869.
SULEY, John. *Baltimore, USA.* 1810-12.
SULLEY see SULLY.
SULLIVAN—
C. D. & Co. *St. Louis, Mo, USA.* ca.1850.
George. *Greenwich.* 1866-74.
George J. *Greenwich.* 1847.
J. T. *St. Louis, Mo, USA.* ca.1850.
Richard. *Deptford.* 1839-47.
Richard. *London.* 1857.
William. *Anerley (Surrey).* 1866-78.
SULLY—
Henry W. *Newport (Monmouth).* 1875-80.
John. *Langport.* b.ca.1824-51. W.
Richard. *Nottingham.* 1835-64.
William. *Langport.* b.ca.1805, mar.ca.1825-41. C., W. and later a silversmith.
SUMMERHAY(E)S—
John. bro. of Robert & b. 1775. *New York* ca. 1800-20
R. *Milverton.* 1861-66.
Robert. *Buckland St. Mary.* b. 1784, also at *Ilminster & Taunton* where d. 1857.
W. *Taunton.* 1861.
SUMMERS—
Edward. *Coventry.* 1880.
Frederick. *Landport.* 1878.
James. *Hamilton.* 1860.
SUMMERSGILL, Robert. *Preston.* b.ca.1800, a.1813-81. W.
SUMMERTON, George. *Oxhill* nr *Kineton.* 1854-80.
SUMNER—
James. *Liverpool.* 1800. (B. has 1700, which must be an error).
Richard. *London.* 1792-1811. W.
William. *Boston, USA.* 1684. Blacksmith. Kept town clock. (Poss. from *London?*).
William. *London.* a.1654 to Robert Robinson. CC. 1661. (May be to *USA?*).
SUNDERLAND—
Thomas. *Birchencliffe (nr Huddersfield).* b.1846-d.1930. C.
Thomas. *Leeds.* 1770. Dial engraver and painter.
SUNNER, Hans. *Nuremberg.* 1570.
SUNTER, Robert. *Bradford.* 1866-71.
SUPREME, The. 1914 and later, product of Pequegnat Clock Co.
SURMAN & KALTENBACK. *Manchester.* 1834.· Imported clocks.
SURMOIRE, John. *London.* (B. 1640) 1648-poor member of CC.
SUTCLIFFE—
H. R. *Horsforth (nr Leeds).* ca.1830-40. C.
J. *Manchester.* 1858. W.
James. *Oldham.* 1834-51.
John. *London.* 1857.
Joseph Thomas. *Sheffield.* 1871.
Stephen. Town not stated. Late 18c. C. (Prob. *Yorkshire*).
T. *Oxenhope (nr Keighley, Yorkshire).* ?Late 18c. C.
Thomas. *Elland.* 1871.
SUTHERLAND—
G. S. *Ottawa.* 1862-76.
George. *Stonehaven.* 1830.
James. *Forres.* 1837.
James. *Helmsdale (Scotland).* 1860.
John. *Aberdeen.* 1836.
William. *Pulteney (Wick).* 1837.
SUTLOW, Thomas. *Manchester.* mar.1803. C.
SUTTON—
Charlotte. *Tonbridge.* 1823.
Enoch. *Boston, USA.* 1825-42.
F. W. *Littlehampton.* 1855-62.
G. *Norwichville, Canada.* 1857.
Henry. *London.* (B. 1659 planisphere). 1664. Made calculating machine with Samuel Knibb.
Isaac. *London.* Error for ISAAC. Sutton, q.v.
John. *Bristol.* 1839.
John. *Burton-on-Trent.* 1850-60.
John. *London.* 1869-81.

SUTTON—*continued.*
John Windram. *Maidstone.* 1826-39.
Julius. *Barton-on-Humber.* 1849-50.
Robert. *Barton-on-Humber.* 1828-35.
Robert. *New Haven, Conn, USA.* 1825. C. & W.
Robert. *Whitehaven.* 1858-79. Also optician and chrons.
Thomas. *Maidstone.* 1790-1823.
William. *Liverpool.* 1795-1828.
SWABY, Israel. *Dover.* 1775-85. W. Strand Street. Also silversmith.
SWADLING—
Henry. *Sydenham.* 1847.
John. *Stourbridge.* 1842.
SWAIM, James. *Philadelphia, USA.* 1833.
SWAIN(E)—
Alfred. *Cleckheaton.* 1866-71. W.
E. *Coventry.* 1854-60.
John. *Bradford.* 1871.
John. *Leeds.* 1834-53.
Llewellyn George. *Yarmouth, Nova Scotia.* 1864.
Reuben. *Kensington, NH, USA.* 1778.
Richard. *Stratford-on-Avon.* 1828-42. Also broker.
Thomas. *Macclesfield.* 1878.
SWAINSON, George Robert. *London.* 1832-39.
SWAITES, Isaac. *Settle.* 1866. W.
SWALE, Isaac. *London* (alias Soeufew). 1668. Alien.
SWAN(N)—
The. 1914 and later, product of Pequegnat Clock Co.
Benjamin. *Haverhill, Mass, USA.* (1798). ca.1810, later *Augusta, Me,* to 1867.
Justus. *Richmond, Va, USA.* ca.1819.
Moses M. *Augusta, Me, USA.* s. of Benjamin. ca.1840.
Robert. *Bridlington.* mar.1760-ca.1780. C.
Samuel. *Hyde.* 1878.
Thomas. *Dumfries.* 1893-96.
Thomas. *Randalstown (co. Antrim).* 1854.
Timothy. *Suffield, Conn, USA.* 1787-95.
William James. *London.* 1881.
SWANSON, George. *Thurso.* 1860.
SWARTZ, Peter. *York, Pa, USA.* ca.1790.
SWAYNE, E. J. *New York.* ca.1830. C.
SWEARER (see also SCHWERER).
G. *Chelmsford.* ca.1859.
J. *Colchester.* 1874.
John. *London.* 1828-44. C.
John Diamond. *Colchester.* ca.1865-ca.1870.
Solomon. *Colchester.* ca.1860-ca.1870.
SWEBY, John. *London.* a.1662 to William Speakman, CC. 1671 (B. -1698).
SWEEPER, Richard. *Romsey.* 1830.
SWEET—
James S. *Boston, USA.* 1842.
Timothy. *Southsea.* 1867-78.
SWENEY, Thomas. *Philadelphia, USA.* ca.1850. C.
SWERER, George. *Ipswich.* 1839. (See also SWEARER, etc.)
SWETE, Mayne. *Modbury (Devon).* ca.1700-1705. C.
SWIFT—
Anthony. *Cardigan.* Late 18c. W.
John D. *Cazenovia, NY, USA.* 1816-21.
Richard. *Cowes.* 1830.
Samuel. *Wakefield.* 1796.
William. *Walsall.* 1818. W.
SWINBURN (and variants)—
Charles. *Whitehaven.* 1879.
George. *London.* 18c. C.
John. *Durham.* 1851.
John, jun.? *Hexham.* 1762-95. C.
John. *Hexham.* 1827-34.

SWINBURN—*continued.*
John & George. *Durham.* 1856.
John Robert. *Workington.* 1869-79.
T. *Workington.* 1873.
Thomas. *Whitehaven.* 1858-79.
William. *Hexham.* d.1782. C.
William, jun. *Hexham.* 1799. C.
William. *Hexham.* 1827-48.
SWINDEL(L)S—
John. *Macclesfield.* 1828-34.
John. *Stockport.* 1865-78.
Thomas. *London.* 1828. Watch cases.
William. *Stockport.* 1878.
SWINDEN—
Francis Charles. *Birmingham.* 1828 (and 1835-50. Dial maker).
& Sons. *London.* 1875.
& Sons. *Birmingham.* 1858-80. Wholesalers and importers. Also dial makers.
SWINGLER, ——. *Grantham.* 1828.
SWINYARD, George. *Maidstone.* a.1750.
SWIRE, George. *Colchester.* 1830-40. Dutch clocks.
SYBERBERG, Christ. *New York* 1756, *Charleston, SC,* 1757-68. W.
SYDERMAN, Philip. *Philadelphia, USA.* 1785-94.
SYDNEY, The. 1914 and later, product of Pequegnat Clock Co.
SYKES—
Benjamin. *Lancaster.* 1869.
George. *Malton.* 1823. C.
James. *Huddersfield.* 1850-71.
John. *Gainsborough.* 1861-76.
John. *Manchester.* mar.1793. C.
Samuel. *Huddersfield.* 1871.
Saville. *Stalybridge.* 1865-78.
Thomas. *Leeds.* 1807-17.
Thomas Henry. *Huddersfield.* 1866-71. W.
William. *Holbeck (Leeds).* b.1754-d.1835. C.
William Henry. *Heckmondwike.* 1850-71.
SYLVESTER, Joseph. *Alford (Lincolnshire).* 1876.
SYMES—
Mrs. R. *London.* 1844.
Richard. *London.* 1828. Dial maker.
SYMINGTON, Andrew. *Kettle (Fife).* 1834-60. W.
SYMM also **SYMME(E)S**—
Cleadon. *Newton, NJ, USA.* bro. of Daniel. 1789.
Daniel. *Walpack, NJ, USA.* s. of Timothy. (1772) 1792, then to Ohio to become a lawyer in 1802.
Luke. *Prescot.* mar.1780. W.
Timothy. *Walpack, NJ, USA.* 1744. mar.1765-80, then to *Newton,* later to *Ohio.*
SYMON(D)S—
Benjamin. *London.* 1851-81.
John. *Hackford (Norfolk).* 1793.
John. *Launceston.* 1847-73.
John. *Reepham (Norfolk).* 1734-1815.
Joseph & Co. *Liverpool.* 1795. W.
Iulia. *Hull.* 1834-58. W.
Moses. *Hull.* 1823. Also jeweller.
S. *Yarmouth.* 1846.
Samuel. *Sheffield.* 1834-37. W.
Thomas. *London.* a.1655 to Hugh Cooper, CC. 1661. C. (also Simonds).
W. B. *Helston.* 1856. C. & W.
SYNCHRONOME CLOCK Co. *Birmingham.* 1880.
SYRETT, George. *London.* 1857-81.
SYXFORTHE, John. *Masham (Yorkshire).* 1542. Kept ch. clock.
SZAPIRA, Jacob. *Brighton.* 1870-78.

T

TABER—
Elnathan. *Roxbury, Mass, USA.* 1784, mar.1797-1854. C.
H. *Boston, USA.* ca.1850.
J. *Saco, Me, USA.* ca.1820. C.
Stephen N. *New Bedford, Mass, USA.* 1777-1862. C.
TABOR, L. A. *Holyoke, Mass, USA.* ca.1860.
TABRAHAM, Arthur. *Cambridge.* 1875.
TAF, John James. *Philadelphia, USA.* 1794.
TAFFINDER—
Abraham. *Rotherham.* 1841-71. W. & C.
Alfred. *Mexborough.* 1871.
TAGNON, Adolphus. *London.* 1875-81. Marble and alabaster clock cases.
TAILLE, Eli. *Christieville, Canada.* 1851.
TAILOR, J. *Costessey (Norwich).* 1875. (See Taylor).
TAINSH—
Brothers. *Cardiff.* 1887.
Mrs. Susannah. *Weston-super-Mare.* 1875.
TAIT—
Archibald. *Edinburgh.* a.1821.
Charles. *Peebles.* 1836-60.
George Burgess. *Wainfleet.* 1850.
John, *Brampton.* 1879.
Thomas. *Alnwick.* 1827.
W. *Nottingham.* 1849-55.
TALBOT—
Sylvester. *Dedham, Mass, USA.* ca.1815-17.
W. *Cambridge.* ca.1790-mid-19c. C.
TALLACK, John. *St. Austell.* ca.1800. C.
TALLER, Jacob. *Caschau.* Table clock (?early 17c.).
TALLIBART, Louis. *London.* 1844-51.
TALLMADGE, Elliott C. *Torrington, Conn, USA.* 1831-40. C.
TALLON, T. *Luton.* 1864.
TAMPLIN, R. *Bristol.* 1856-63.
TANDY—
C. *Birmingham.* 1854. Clock case maker.
William. *Birmingham.* 1868-80. Clock cases.
TANNENBERG, Simon. *Leeds.* 1866.
TANNEHILL, Thomas Tatlow. *Bassano, Canada.* 1918.
TANNER—
Edward. *Llancarfan (Glamorgan).* a.1771 (not 1791).
Edwin S. *Reading.* 1864-99.
John & George. *Bristol.* 1801.
Joseph. *Cirencester.* 1840-56.
Joseph. *London.* a. to Isaac Day, CC. 1682 (-B. 1698).
Joseph Seymour. *Cirencester.* 1870-79.
Lloyd. *Low Norwood.* 1878.
S. *Lewes.* 1878.
Thomas. *Lewes.* 1839.
William. *Hailsham (Sussex).* 1828-39.
William (& Son). *Lewes.* 1839-55(-1862).
William. *London.* 1844-51.
TANSLEY—
Sarah. *Birmingham.* 1842.
Thomas, sen. *Birmingham.* (B. 1808) 1828-35. C.
Thomas, jun. *Birmingham.* 1828-42.
TAPPEN, John. *Flemington, NI, USA.* 1839-43.

TAPSON, ——. *St. Germans (Cornwall).* 1829. W.
TARBELL, Edmund. *Boston, USA.* 1842.
TARBOCK, John. *Manchester.* d.1732. W.
TARBOX, H. & D. *New York.* ca.1830.
TARBUCK—
John. *Farnworth.* mar.1808. W.
John. *Prescot.* mar.1765.
Joseph. *Prescot.* 1823. W.
TARLETON—
Richard. *Liverpool.* 1795-1815. W.
Thomas. *Liverpool.* 1848-51.
TARNER, Herman Frederick. *London.* 1857.
TARONI—
F. *Haverfordwest.* Early 19c.
George. *Hull.* 1840-51.
TARR—
G. *Bath.* 1861.
George. *London.* 1875.
Joseph. *Brixton.* 1839.
Thomas. *Manchester.* 1755. C.
William Ferris. *Bristol.* 1879.
TARRY—
William. *London.* ca.1720. C.
William. *Woburn.* Lant. clock. Early 18c.
TASKER—
George. *London.* 1832-63.
William. *Banbury.* 1813-53. C. & W.
William. *Bristol.* 1840-56.
TATE—
James. *Liverpool.* 1828.
James. *Maidstone.* 1847.
John I. *Downpatrick.* ca.1780-90.
John II. *Castlewellan (co. Down).* 1846.
John. *Stanhope.* 1851.
Richard Henry. *Horncastle.* 1868-76.
Thomas. Godfrey. *Winterton (Lincolnshire).* 1876.
TATHAM—
Christopher. *Appleby.* 1858-73. W.
Christopher. *Carlisle.* 1834.
TATLOCK, James. mar.1727-40. C. & W.
TATTERSALL, Robert William. *London.* 1857.
TAUTE—
Herman. *Kingston (Surrey).* 1878.
Herman. *London.* 1869-75.
TAVENDER, John. *London.* 1857.
TAWNEY, Robert. *Oxford.* a.1755-pre-1771. Later *London.* ca.1771-76. W.
TAWS, Charles. *Philadelphia, USA.* 1802. Math. instrs.
TAYLOR—
——. *Bakewell.* ca.1830. C.
——. *Ilminster.* ca.1800. C.
Alfred. *Croydon.* 1878.
Alfred. *London.* 1857.
Alfred. *London.* 1857.
Alfred Henry. *London.* 1863.
Andrew. *New York.* 1854.
Arthur. *London.* 1881.
(John) & BALDWIN (Isaac). *Newark, NJ, USA.* ca.1825.

228

TAYLOR—*continued.*
& BETTS. *Coventry.* 1868. Also chrons.
Charles. *Bristol.* 1830.
Charles. *Cardiff.* 1858.
Charles. *Kinross.* 1836.
Charles. *London.* 1839.
Charles. *London.* 1875.
Charles. *Sedalia, Mo, USA.* ca.1870.
Charles. *Uttoxeter.* 1868-76.
Charles & Co. *Bristol.* 1879.
Charles & Son. *Bristol.* 1842-63.
Charles & Son. *London.* 1851-81.
Charles (& WILKINS, William). *Coventry.* 1880.
D. *Manchester.* 1858.
David. *London.* 1851-69.
Dennis. *Diss.* 1846-75.
Edward. *Prescot.* 1819. W.
Edward. *Prescot.* 1823-25. W.
& ELLIOTT. *London.* ca.1800-20.
F. *Haverhill.* 1858-65.
Francis. *Oshawa, Canada.* 1857-63.
George. *London.* 1875.
George. *Manchester.* mar.1830-57. C.
George. *Preston.* d.1738. C.
George. *Stoke-on-Trent.* 1876.
George. *Wolverhampton.* 1850.
George & Co. *Providence, RI, USA.* ca.1875. C.
George Rodney. *Sunderland.* 1827-56.
H. Henry. *Yarmouth.* 1875.
Henry. *Liverpool.* 1828.
Henry. *London.* 1863-69.
Henry. *Manchester.* 1760.
Henry. *Philadelphia, USA.* 1760. C.
(John) & HINSDALE (Horace Seymour). *Newark, NJ,
  USA.* ca.1810.
J. *Sandgate.* 1851-74.
J. *Woodbridge.* 1846.
J. & Co. *Hamilton, Ontario.* 1895.
J. T. *Dumfries.* 1855.
James. *Doune.* 1837.
James. *Mormond (Scotland).* 1847. W.
James. *Ripon.* 1871.
James. *St. Helens.* 1786. W.
James. *Sandgate.* 1874. (see J.).
James. *Sunderland.* 1851.
James. *Tillicoultry (Scotland).* 1837.
James Barton. *Sandgate.* 1845-57. W.
John. *Ashton-under-Lyne.* 1713-d.1744. C.
John. *Birmingham.* 1828.
John. *Brighton.* 1878.
John. *Costersey (Norfolk).* See Tailor, J.
John. *Dover.* 1858.
John. *Dukinfield.* ca.1700. C.
John. *Dunning (Scotland).* 1860.
John. *Farnworth.* mar.1824. W.
John. *Grantham.* 1849-61.
John. *Gravesend.* 1838.
John. *Leeds.* 1866. W.
John. *Liverpool.* pre-1755. W.
John. *London.* (B. a.1813) 1828-44.
John. *London.* 1881.
John. *Manchester.* 1745-81. C.
John. b. 1803. *Stannington. To Morpeth* by 1841
  -d. 1858. W. & C. Succ. by widow. Mary. 1858-71.
John. *Mt. Holly, NJ, USA.* ca.1773-ca.1816.
John. *Northampton.* 1698.
John. *Norwich.* 1830. C.
John. *Prescot.* mar.1804-23. W.
John. *Soham.* 1830.
John. *Strood.* 1826-47.
John. *Sunderland.* 1827-34.
John. *Swansea.* Successor to B. R. Hennessy in 1875.
John. *Woodbridge.* 1830-39.
John & Daniel. *Liverpool.* 1834-51.
John & Edmund. *Rochdale.* 1848-58. W.
Jonathan. *Ashton-under-Lyne.* (b.1742)-d.1808. C.
Joseph. *Hamilton, Ontario.* 1868-76.
Joseph. *Pontefract.* 1851-71.
Joseph. *Strichen (Scotland).* 1837.
Joseph. *Wakefield.* 1851-71.
Joseph. *Wolverhampton.* 1828-50. Also vertical jack
  maker.
Joseph. *York, Pa, USA.* ca.1785.

TAYLOR—*continued.*
Joseph T. *Bala.* 1874-77. C. & W.
Michael. *Newcastle-on-Tyne.* 1852.
Mrs. S. *London.* 1863. (?wid. of David.)
Noah C. *Salisbury, NC, USA.* 1844.
P. L. & Co. *Brooklyn, NY, USA.* ca.1830.
Peter. *Prescot.* 1795, W.
Philip. *Oshawa, Canada.* 1857-73.
Philip. *Philadelphia, USA.* d.1709. C.
Randell John. *London.* 1863-69.
Richard. *Boston, USA.* 1657. Rep'd. ch. clocks.
Richard. *London.* a.1648 to Abraham Guyott, CC.
  1655-66 (B: -1719). Watch case maker.
Richard. *York, Pa.* ca.1785.
Richard. *Foots Cray (Kent).* 1866-74.
Richard. *Wainfleet.* 1828-61.
Richard. *Yeovil.* ca.1790. C.
Robert. *Colchester.* b.ca.1786-1824.
Robert. *Leeds.* 1866. W.
Robert. *London.* a.1693 to James Hatchman, CC.
  1703-d. pre-1716.
Robert. *London.* 1832-69.
Robert Roger. *London.* 1875-81.
Rothes. *Newtownards.* ca.1780-90. C.
S. *Liverpool.* ca.1785-90. W.
S. *Loughborough.* 1864.
Samuel. *Chipping Campden.* 1850-56.
Samuel. *Debenham.* 1823.
Samuel. *Halesworth.* 1839.
Samuel. *Liverpool.* 1834.
Samuel. *Middleton nr Manchester.* d.1743. C.
Samuel. *Rochdale.* 1834-58. C.
Samuel. *Worcester, Mass, USA.* 1780-1864. Partner
  with J. Barker to 1807, alone to 1856.
Samuel E. *Bristol, USA.* 1849-58. Partner of Birge,
  Peck & Co.
Sarah. *Birmingham.* 1835.
Son & WARD. *Coventry.* 1854-60.
T. *Ilkeston.* 1855.
T. W. *Sittingbourne.* 1874.
Thomas. *Bradford.* 1866-71.
Thomas. *Canning, Nova Scotia.* 1866.
Thomas. *Ellesmere.* mar.1792-1835.
Thomas. *Horton in Ribblesdale (Yorkshire).* 1729.
  Clocksmith.
Thomas. *Kinross.* 1836.
Thomas, sen. *London.* a.1638. CC. 1646. d.1684.
Thomas, jun. *London.* CC. 1659-d.1690.
Thomas. *London.* 1851.
Thomas. *Manchester.* 1824-34. C.
Thomas Lee. *Pontefract.* 1826-37. W.
Thomas Longmore. *Birmingham.* 1860.
W. *Caistor.* 1868.
W. (& C.). *Coventry.* 1860.
W. & J. *Hamilton, Canada.* 1862-66.
W. S. *Troy, NY, USA.* 1847-49.
William. *Berwick.* 1834.
William. *Bold.* d.1768. W.
William. *Buffalo, NY, USA.* 1835.
William. *Dumfries.* 1787-1823.
William. *Farnworth.* mar.1816. W.
William. *Hamilton, Canada.* 1851-65.
William. *Leeds.* b.1660, mar.1684. W.
William. *Liverpool.* 1822-51.
William. *London.* 1857.
William. *Nailsworth.* 1840-56.
William. *Oxford.* b.1795-d.1854. T.C.
William. *Scunthorpe.* 1876.
William. *Wick.* 1860.
William. *Wimbledon.* 1878.
William. *Wolverhampton.* 1860-76.
William & Co. *Coventry.* 1856-60(-80?).
William Edward. *Liverpool.* 1848-51.
William Gayler. *London.* 1869-81.
William & Son. *Hamilton, Canada.* 1861-63.
William Thursfield. *Liverpool.* 1828.
TAYSPILL, John. *London.* 1768. C.
TAYTON, James. *Coventry.* 1880.
TAZEWELL, S. S. *Bridgeton, NJ, USA.* ca.1864-68. C.
TEAGE, Thomas. *St. Columb (Cornwall).* 1585. Kept
  ch. clock.
TEASDALE, James. *Airdrie.* 1860.

TEDDEMAN, ——. *Canterbury.* ca.1785-ca.1795. C.
TEISSENRIEDER, Wolfgang. *?Vienna.* 1572.
TELFER—
Alexander. *Antigua.* (From *Aberdeen*). d.1805.
Peter. *Montreal.* b.1806-d.1834.
TELFORD—
Edward. *Carlisle.* 1848-58.
John. *Carlisle.* 1848.
John. *Keswick.* 1828.
John. *Maryport.* 1858-79.
John. *Wigton.* (B. 1820) 1828-58. Also hardwareman.
Thomas. *London.* 1869-75.
W. *Whitehaven.* 1858.
W. N. *Chester.* 1865.
William H. *Whitehaven.* 1869-79.
TEMPLE—
Joseph. *Sidestroud* (= *?Sidestrand, Norfolk*). ?Early 18c.
Robert. *Devizes.* ?19c. C.
TEMPLER—
Charles. *London.* a.1665 to Charles Rogers, CC. 1673.
Richard. *Exeter.* Free 1776. W.
Richard. *Plymouth.* 1759-64.
Thomas. *Barnstaple.* Free 1780-1802. W.
Thomas. *Chard.* Free 1802. W. (s. of Thomas T. of *Exeter.* W.).
TEMPLETON—
David. *Maybole.* 1850-60.
John. *Ayr.* 1836-60.
Matthew. *Beith.* 1850-60.
Robert. *Ayr.* 1850-60.
Robert. *Newport, RI, USA.* ca.1785.
Robert. *Newry.* 1792. C.
Samuel. *Dalmellington (Scotland).* 1860.
Thomas. *Dalmellington (Scotland).* 1837-50.
TENNANT—
Thomas. *London.* a.1660 to Carr Coventry, CC. 1668 (B. -1692).
Thomas. *London.* 1857-63. Clock cases.
Thomas Edward. *London.* 1875-81. Clock cases.
TENNENT, Thomas. San Francisco, Cal, USA. ca.1850.
TENNEY, William. *Washington, NY, USA.* ca.1790. C.
TENNISON, Reuben. *Chesterfield.* 1828-49.
TERHUNE—
& BOTSFORD. *New York.* mid-19c. C.
& EDWARDS. *New York.* 1860-72. C.
H. *New York.* 1853.
TERNBACH, M. *St. Louis, Mo, USA.* ca.1830.
TERRIER—
James. *London.* a.1685 to Isaac Day, CC. 1694 (B. -1706).
Thomas. *London.* a.1689 to William Laughton, CC.1694 (B. -1722). W.
TERRY—
Alfred. *London.* 1875-81.
Edward. *Sunderland.* 1834-56.
Eli, jun. & Co. *Plymouth, USA.* 1829-41. C.
Eli & Samuel. *Plymouth, USA.* 1824-27.
Eli & Sons. *Plymouth, Conn, USA.* 1823-33. C.
Francis. *Masham.* ca.1790-ca.1800. C.
Henry. *Leeds.* 1837-53.
Isaac (& Son). *London.* 1839-57 (1863-69).
James. *Newcastle-on-Tyne.* 1848-56.
John. *Banbury.* 1700-d.1736. C.
John. *Oakville.* 1851.
John I. *York.* b. pre-1673-d.1757. C.
John William. *Brighton.* 1870.
Thomas. *Manchester.* mar.1822-51. W.
Thomas. *York.* s. of John I. b.ca.1705-33.
Reuben. *York.* s. of John I. 1713-24. W.
Thomas. *London.* 1875. (Successor to William.)
William. *Bedale.* ca.1770-d.1820. C.
William. *Hull.* 1823-46.
William. *London.* 1869.
William. *Masham.* ca.1770-1840.
William. s. of William of *Bedale.* To *Richmond (Yorks)* by 1807-1840.
William. *Thoralby* (nr *Aysgarth, Yorkshire*). ca.1760-ca.1770. C.

TERT, John. *Allendale (Northumberland).* 1848.
TESSEYMAN, William. *Northallerton.* b.1842, s. of George. d.1912.
TESTER—
H. *Hastings.* 1878.
T. *Wellingborough.* 1864-77.
TETEBLANCHE, I. *Paris.* ca.1760. C.
TETERGER, Hippolyte. *London.* 1863-75.
TETLOW—
George. *Marple.* 1878.
J. *Eccles.* ca.1775. C.
TEUTONIA Clock Manufactory. *Germany.* ca.1880. C.
TEW (see also **TUE** and **CHEW**)—
G. *Bury (?St. Edmunds).* Early 18c. Lant. clock.
George. *Carmarthen.* 1887-99.
George. *Hereford.* 1870.
J. *Hereford.* 1856-63.
Thomas. *London.* a.1674 to William Watmore. Engraver.
William. *Yarm.* 1829. C.
TEWILL, Henry. *Birmingham.* 1880.
THACHE, Robert. *London.* a.1681 to John Benson, Engraver, CC. 1681 (B. -1706).
THACKE, Philip. *London.* a.1676 to Thomas Wheeler, CC. 1685. (B. -1701). C.
THACK(H)ALL—
John. *Oundle.* 1830-41.
W. & Son. *Oundle.* 1824.
THACKRAY—
James. *York.* 1858. W.
Mrs. R. *Middlesbrough.* 1898.
Thomas. *Otley.* 1826. C.
THACKWELL—
Charles. *Cardiff.* 1848-65.
Charlotte. *Gloucester.* 1840.
G. *Cheltenham.* 1856.
George. *Gloucester.* 1842.
George. *Newent.* 1840.
John. *Cardiff.* 1740-1830 (two makers?). C.
Joseph. *Gloucester.* 1830.
Thomas. *Bristol.* 1830.
Thomas. *Monmouth.* 1822-44.
THATCHER—
Cornelius Octavius. *Cardiff.* 1875.
Henry J. *Cardiff.* 1875-87.
J.(ames?). *Lambourn.* 1854. C.
James. *Wantage.* 1854-77.
THEAKER, Thomas. *Nottingham.* 1876.
THEARSBY, Peter. *York.* See *Thoresby.*
THELWALL, Charles John. *Manchester.* 1851.
THELWALL, Richard. *Manchester.* 1822-34.
THEOBALD, Thomas. *Burneside (Westmorland).* Late 18c. C.
THEODOR & WEHRLEY. *Bishop's Stortford.* 1851.
THEVENAZ, Freres (Brothers). *London.* 1863-81.
THICK—
Frank. *Frome.* 1883.
William. *Frome.* 1875-83.
THICKBROOM—
Alfred & James. *London.* 1839. Watch cases.
George. *London.* 1832. Watch cases.
THICTHENER (see also Thitchener)—
Joseph. *London.* 1881.
W., jun. *London.* 1851.
William. *Barnsley.* 1871.
THIELE, Edward & Frederick. *London.* 1869.
THIRKELL—
George. *Hartlepool.* 1834-51.
James Lawrence. *Middlesbrough.* 1866. W.
THIRLWALL & BIBBY. *Rotherham.* 1871.
THITCHENER (also Thicthener, q.v.)—
Henry. *London.* 1869.
John. *London.* 1851.
Thomas. *Chelmsford.* 1855-74.
Thomas C. *London.* (?1832) 1839-75.
William. *London.* 1832-44.
THODEN—
Ferdinand. *London.* 1875-81.
Julius. *London.* 1869-75.
THOERMER, Adolph Ferdinand Erdmann. *London.* 1881.

THOM—
Charles. *Liverpool.* Alone post-1799, formerly partner with Crump, see below.
& CRUMP. *Liverpool.* Partnership 1790-99.
THOMA—
Conrad. *London.* 1857-81.
Frederick. *London.* 1875.
Joseph. *Hanley.* 1850-68.
William & Co. *London.* 1875-81.
THOMAS—
Benjamin. *Fishguard.* 1875-87.
Benjamin. *Pembroke Dock.* 1875-87.
Benjamin & D. *Carmarthen.* 1847-52. W.
Celestin & Co. *London.* 1857.
Charles. *Ross.* 1830-44.
D. *Carmarthen.* 1844.
D. *Fernlake (Wales).* 1887.
D. *Llandovery.* 1848. W.
Daniel. *London.* a.1675 to John Browne, math, inst. mr., CC. 1682-(B. -1711).
David. *Aberdare.* 1868-80.
David. *Aberystwyth.* 1864-87. W.
David. *Cellan (Wales).* b.1792. d.1846.
David. *Kidwelly.* Later 18c. C.
David. *Swansea.* 1875-99.
F. *Halesowen.* ca.1810. C.
George. *Pembroke.* 1868.
George. *Swansea.* 1887.
H. *Coventry.* 1868.
Henry. *London.* 1851-57.
Hugh. *Carmarthen.* 1830-35.
J. *Newport.* 1817. (Made a year clock.)
J. & J. *Carmarthen.* 1868. W.
J. J. *Portmadoc.* ?From *Carmarthen.* W.
James. *Bridgend.* s. of Samuel. 1852-1917.
James. *Merthyr Tydfil.* ca.1840-ca.1850. C.
James. *Tredegar.* 1850.
John. *Builth.* 1868-71.
John. *Carmarthen.* 1871-87. C. & W.
John. *Ebenezer (Caernarvonshire).* Late 19c. W.
John. *Halifax.* 1866.
John. *Llandovery.* 1875-87.
John. *London.* 1832-51.
John. *Newcastle Emlyn.* 1868-75.
John. *Pen-y-Groes.* 1868-90. C.
John. *Pontypridd.* 1887.
John. *Treherbert.* 1875.
John B. *Cardiff.* 1858.
John B. *Southampton.* 1867-78.
John & Henry. *London.* 1857.
John and James. *Llandovery.* 1868.
John Marshall. *Redruth.* 1844-73. Also at *Penryn* in 1873. C.
John Thomas. *London.* 1857.
Maria & Son. *London.* 1857.
Margaret. *Tremadoc.* 1868.
Matthew. *Bilston.* 1850. C.
Matthew. *Shelton (Staffordshire).* 1842.
Matthew Henry. *Perranuthnoe (Cornwall).* d.1844.
Moses. *Conway.* 1874.
Moses. *Pen-y-Groes* 1875-80, then *Cricieth* 1880-d.1936. C.
Owen. *Northampton.* 1752. (B-one such *Birmingham.* 1753.)
Philip. *Doncaster.* 1834.
Philip. *Sheffield.* 1862-71. W.
Richard I. *Helston.* 1781. W.
Richard II. *Helston.* mar.1804-ca.1815.
Richard. *London.* 1844-51.
Richard. *Norwood (Surrey).* 1839.
Richard. *Oswestry.* 1835.
Richard. *St. Day (Cornwall).* ca.1810. C.
Richard Bonner. *Portmadoc.* 1868-95. C. & C.
Samuel. *Bridgend.* 1819-1900.
Thomas. *Cardigan.* 1868-93, C. & W.
Thomas & Co. *London.* 1851.
William. *Lincoln.* 1828-50.
William. *Liverpool.* mar.1834. W.
William. *Llangollen.* 1874.
William. *London.* 1828-32.
William. *Trenton, NJ, USA.* 1780. Also gold and silversmith.

THOMAS—*continued.*
William James. *London.* 1863. (Successor to Maria & Son.)
THOMEQUEX, Peter. *Northampton, Mass, USA.* 1802. W.
THOMLINSON, J. *Thame.* ca.1800. C.
THOMPSON—
——. *Wigton.* ca.1770. C.
Andrew. *Campbeltown.* 1836.
Andrew. *Downpatrick.* mar.1828.
Avery J. *Cherry Valley, NY, USA.* 1852-1917. s. of Lyman. W.
Avery J., jun. *Cherry Valley, NY, USA.* s. of Avery J., sen. 1883.
C. & W. *Bath.* ca.1800.
Charles. *Birmingham.* 1868.
Christopher. *Belford.* 1834-58.
David. *Hamilton, Ontario.* 1875.
Ebenezer. *London.* 1839-63.
Ebenezer Henry. *London.* 1869-75.
Edward. *London.* 1844-81.
Edward J. *London.* 1844-69.
& FLOYD. *Glens Falls, NY, USA.* ca.1830.
George. *Alston.* 1869.
George. *Hull.* 1858.
George. *London.* 1844-57.
George B. *London.* 1863-69.
H. C. Clock Co. *Bristol, USA.* 1903-07 and later.
Henry. *Philadelphia, USA.* 1847.
Hiram C. *Bristol, USA.* 1830. Successor to Noah Pomeroy, 1878.
Isaac. *Litchfield, Conn, USA.* 1796.
Isaac (II). *London.* s. of Isaac I, a.1689 to his father, CC. 1699.
J. *Alnwick.* 1827.
J. *Bedford.* 1869.
J. *Blackburn.* 1858. W.
J. *Brampton.* 1858.
J. *Dean. Louth.* 1868-76.
James. *Baltimore, USA.* ca.1800; *Pittsburgh, Pa,* ca.1820. Later made steam engines.
James. *Batley.* 1871.
James. *Cannock.* 1876.
James. *Darlington.* ca.1770-d.1825. C.
James. *Dewsbury.* 1866. W.
James. *Huddersfield.* 1850-71. W.
James. *Liverpool.* 1722-34. W.
James. *Selby.* 1834-44. C.
James. *Streetsville, Canada.* 1857.
James T. *Belfast.* d.1860. W.
John, sen. *Belfast.* mar.1834-ca.1850. W. & C.
John, jun. *Belfast.* ca.1850-68.
John. *Chesterfield.* 1828-49.
John. *Coventry.* 1880.
John. *Doncaster.* 1709-24. C.
John. *Kirkby Stephen.* ?mid-19c.
John. *London.* a.1655 to Jeremy Gregory. CC. 1662.
John. *London.* 1857.
John. *New York.* 1780. Successor to Isaac Heron.
John. *Pickering.* 1866. W.
John. *Saltaire (Yorkshire).* 1871.
John. *Truro.* 1698. Kept ch. clock.
John I. *York.* b.1614-d.1692. W.
John II. *York.* b.1639. s. of John I. d.1714. W.
Joseph. *Cirencester.* ca.1720-ca.1740. C.
Joshua. *Fleetwood.* 1848.
Joshua. *Leeds.* 1834. W.
Joshua. *Staindrop.* 1827.
Joseph. *Atherstone.* 1828.
Joseph. *Cockermouth.* 1848-58. Also jeweller.
Joseph. *Whitehaven.* 1828-34.
Lyman W. *Cherry Valley, NY, USA.* 1825-1910. W.
Mark Graystone. *Tonbridge.* 1847-74.
Mrs. M. *Chesterfield.* 1855-76.
Philip. *Woodbridge.* 1830.
R. *Keswick.* 1858.
R. *Newcastle-on-Tyne.* 1856-58.
& RANGER. *Brattleborough, Vt, USA.* ca.1810.
Richard. *St. John, New Brunswick.* 1842-76.
Robert. *London.* a.1666 to Edward Eyston, CC. 1681 (B-1684).
Rowland. *Wigton.* 1848-73.

S. N. *Roxbury, Mass, USA.* 1853.
Samuel. *Darlington.* s. of James, q.v. 1825-34.
Silas G. *New Haven, Conn, USA.* ca.1850.
Thomas. *Liverpool* and *Lancaster.* Free 1747-67. W.
Thomas. *Laugharne.* 1847-50.
Thomas. *London.* 1844.
& VINE. *London.* 1869-81.
W. *Bedford.* 1864.
W. *Halifax, Nova Scotia.* 1863.
William. *Baltimore, USA.* 1762-1800. Repairer.
William. *Carlisle, Pa, USA.* 1780.
William. *Coventry.* 1854-60.
William. *Coventry.* 1880. (May be same man as above.)
William. *Dalkeith.* ca.1850. C.
William. *Gateshead.* 1851-56.
William. *Hull.* 1858.
William. *Pembroke.* 1844.
William. *Wilmington, NC, USA.* 1834. C. & W.
William James. *Ashford.* 1874.
William John. *London.* 1839.
William George. *London.* 1839.
& WILLIAMS. *Ludlow.* 1870.
THOMS, William. *Grampound* (*Cornwall*). 1663. Kept ch. clock.
THOMSON—
——. *Ashford.* ?Late 19c.
Adam. *London.* 1839-57.
Alexander. *Berwick.* 1819.
Alexander. *Kirkwall.* 1836.
Andrew. *Glasgow.* 1827-41.
George. *Glasgow.* 1833-60.
James. *Dumfries.* 1713-20.
James. *Dumfries.* 1887.
James. *Leslie* (*Scotland*). 1825.
John. *Stirling.* 1836.
Matthew. *London.* 1851-69.
& PROFAZE. *London.* 1863-81. (Successor to Adam.)
Richard. *Edinburgh.* ca.1810. C.
Robert. *Greenock.* 1860.
William. *Dalkeith.* 1840.
William. *Edinburgh.* 1849.
William. *Wigtown.* 1860.
William & Co. *Dalkeith.* 1860.
THOREAU, ——. *Jersey.* 1740. W.
THORESBY, Peter. *York.* b.1646-1680. W.
THORN, James, sen. *Colchester.* b.1707-d.1762. C.
THORNBOROUGH, ——. *Richhill* (*co. Armagh*). ca.1745? C.
THORNDELL, Richard. *Bampton.* mar.1762-68.
THORNE—
J. C. *Witham* (*Essex*). 1851-66.
James. *London.* (B. 1820-)1839.
John. *London.* 1811.
Simon, *Tiverton.* (B. 1720). mar.1740. C.
Thomas. *Tiverton.* mar.1742.
THORNELOE—
Charles. *Lichfield.* 1835-76. C. Also bellfounder and brassfounder.
Henry. *Birmingham.* 1854-68.
John T. *Coventry.* 1880.
Richard. *Coventry.* 1868-80.
THORNHAM, George. *Hedon* 1806. Later (B. 1806-34) *Hull.* C.
THORNHILL—
Henry Dyer. *London.* 1857.
John & Son. *London.* 1857.
T. *Yeovil.* 1861-66.
THORNLEY—
Charles. *Altrincham.* 1878.
George. *Altrincham.* 1857-65.
THORNTON—
Andrew. *Philadelphia, USA.* 1811.
G. *London.* 1804. W.
G. H. *Bradford.* 1866.
H. *Birmingham.* 1854. Clock case maker.
Henry. *London.* a.1692 to Samuel Steevens, CC. 1699 (B. -1732). C.
J. *Birmingham.* 1860.
J. *Lydney.* 1863.
J. *Oldbury.* 1860.
J. G. *Scarborough.* 1866. W.

THORNTON—continued.
John. *Bradford.* 1717-d.1725. C.
John. *Grantham.* 1849-50.
John. *London.* 1832.
Joseph. *Leeds.* 1871.
T. *Tipton.* 1860.
Thomas. *Birmingham.* 1880. Clock cases.
Thomas. *Much Wenlock.* 1870.
W. *Leamington.* 1880.
William. *York.* a.1741-1760. C.
Yates. *Romford.* ca.1745. C.
THOROWGOOD—
Edward. *London.* CC.1668-70. Math. inst. mr.
John. *London.* a.1652 to Richard Scrivener. CC. 1660-62. Engraver, *Fleet Street.*
Thomas. *London.* 1662.
William. *London.* a.1652 to Thomas Belson, CC. 1660 (B. -1687).
THORP—
Benjamin. *Oxford.* a.1726. C.
H. W. *Beaver Dam, Wis, USA.* ca.1870.
Richard. *Market Weighton.* 1823-46. W.
Thomas. *Colchester.* b.1717-d.1804. C. & W.
W. *Bath.* 1800-35.
THORPE—
E. & Co. *Upper Alton, Ill, USA.* ca.1860.
Edward. *Bethnall Green, London.* 1785. From *Colchester* where he returned after only one year. W.
Edward. *Colchester.* s. of Thomas. b.1752-d.1831. W.
Edward. *Stockwell* (*Surrey*). 1828.
F. *Tunbridge Wells.* 1874.
John. *London.* a.1641 to John Harris, CC. 1657-70 (when blind and poor).
John Hughes. *London.* 1875-81.
Joseph. *Ashby-de-la-Zouch.* 1835. C. and bell-hanger.
Thomas. *Pateley Bridge* (*N. Yorkshire*). 1871. T.C.
THORY, John. *Holbeach.* 1835.
THOWNSEND—
Charles. *Philadelphia, USA.* 1799-1828.
Charles, jun. *Philadelphia, USA.* Successor to father, 1829-50.
THRALL, J. H. *Almonte, Canada.* 1871.
THRASHER, Charles. *Birmingham.* 1868-80.
THREADCRAFT, B. *Charleston, SC, USA.* pre-1800.
THRESHER, Edwin. *Coventry.* 1880.
THRISCHER, John. *London.* 1869.
THRISTLE—
——. *Dunster.* 1866.
Francis. *Melcombe Regis* (*Dorset*). 1875.
Francis. *Williton.* 1883.
J. *Nether Stowey* (*Somerset*). 1861-75.
J. *Stogursey* (*Somerset*). 1861-66.
James (& Nephew). *Williton.* 1861-66(-75). C. (Perhaps earlier too.)
William. *Warminster.* 1842.
THUM, Charles. *Philadelphia, USA.* Early 19c. C.
THURLOW, Edward. *Ryde* (*Isle of Wight*). 1848-78.
THURMOTT, Edward. *Colchester.* Successor to James Brown in 1863-69. C.
THURROWGOOD, ——. *Hertford.* ca.1720. C. (May be one of the *London* Thorowgoods, q.v.)
THURSFIELD, P. *Congleton.* 1857-65.
THWAITE(S)—
John. *Penrith.* 1869.
John Thomas. *London.* 1875.
& REED. *London.* 1828-present day. ('Est'd. 1740'.)
Robert. *Barnard Castle.* b.ca.1775-d.1845. W.
Robert. *Barnard Castle.* 1856. (May be a second of this name.)
S. Place unknown. ca.1790. C.
THWING, James. *London.* a.1679 to Daniel Rolfe, CC. 1688 (B. -1712). W.
TIBBETTS, Edwin. *Birmingham.* 1880.
TICE & ROBERTS. *Brooklyn, NY, USA.* ca.1840.
TICEHURST, Henry. *Rye.* 1839.
TICHENOR, David (Isaac). *Newark, NJ, USA.* ca.1820.
TICKLE (& TICKELL)—
——. *Kingsbridge* (*Devon*). ca.1800. C.
Peter. *Farnworth.* mar.1819. W.
Richard. *Sutton* (*Lancashire*). 1809-23. W.
Robert. *Farnworth.* 1806-mar 1809. W.

TICKLE (& TICKELL)—*continued.*
William, sen. *Newcastle-on-Tyne.* 1763-1790. C.
William, jun. *Newcastle-on-Tyne.* b.ca.1760-d.1801. C.
William. *Prescot.* mar.1826. W.
TIDMARSH—
Charles. *London.* 1851.
Charles Parsons. *London.* 1875.
Charles & Samuel. *London.* 1857-69.
Joseph. *London.* 1875-81.
TIDMAS & FULLER. *London.* 1851.
TIEBOUT, Alexander. *New York.* 1798.
TIERNEY, John. *Philadelphia, USA.* 1820-24.
TIFFANY—
Charles Lewis. *New York.* 1837-1888.
Edward. *Leeds.* 1834. W.
George S. *New York.* 1904.
John. *Leeds.* 1850. W.
Never Wind Clock Corp. *Buffalo, NY, USA.* 1904.
TIFFT, Horace. N. *Attleboro, Mass, USA.* ca.1800-1840.
TIGG, George F. *Southampton.* 1878.
TIGHT, George. *Reading.* 1830.
TILDEN—
John. *Cranbrook.* 1713. Made church clock dial.
Thurber. *Providence, RI, USA.* 'Est'd. 1856'.
TILLER, Alfred. *Richmond (Surrey).* 1878.
TILLEY—
M. H. *Dorchester.* 1867-75.
W. *Nottingham.* 1849.
TILLINGHAM, James. *St. Helens.* 1820. W.
TILLINGHAST, Stephen. *Liverpool.* 1742-90. W.
TIMBY—
Globe Timepiece. *Saratoga, NY, USA.* 1863. C.
Theodore Ruggles. *Saratoga Springs* and *Baldwinsville, NY, USA.* 1863-65.
TIMINGS, George Henry. *Dudley.* 1828-60.
TIMM, ——. *Derby.* Late 18c. C.
TIMME, M. *Brooklyn, NY, USA.* ca.1840.
TIMMIS—
J. W. *Burslem.* 1860.
John W. *Tunstall (Staffordshire).* 1850.
John Wesley. *Newcastle-under-Lyme.* 1868-76.
TIM(M)S—
Richard. *Moreton-in-the-Marsh.* 1879.
Thomas. *Birmingham.* 1854-68.
Thomas. *Aspley Guise* nr *Woburn.* 1839-47.
Thomas. *Dunstable.* 1830.
Thomas. *Smethwick.* 1850-68.
TIMSON, William W. *Newburyport, Mass, USA.* 1854-60.
TINDAL(L)—
George Henry. *Woolwich.* 1845-47.
James Frederick. *London.* 1839. Watch cases.
TINGES, Charles. *Baltimore. USA.* 1796-1815.
TINKER—
& EDMONDSON. *Leeds.* ?ca.1800. C.
Joseph. *Dewsbury.* 1871.
TINKLER—
Nicholas Watson. *Newcastle-on-Tyne.* 1834-58. C.
Strachan. *Gateshead.* (B. 1811-) 1828-ca.1830. C
TINSLEY, Mrs. Joseph. *Altrincham.* 1878.
TIPLING, William. *Leeds.* mar.1692-d.1712. C.
TIPPEN—
J. *Lenham (Kent).* 1851-55. (See TIPPING).
James. *Ashford.* 1874.
James. *Headcorn.* 1866-74.
William. *Charing.* 1866-74.
TIPPING—
George. *London.* a.1664 to Francis Bicknell. CC. 1674. (B.-1680).
James. *Lenham (Kent)* 1826-47 (see TIPPEN).
William. *Charing.* 1858.
TISDALE—
Col. Nathan. *New Bern, NC, USA.* 1795.
E. D. *Taunton, Mass, USA.* ca.1870.
William II. *Washington* and *New Bern, NC, USA.* 1816-50. C. Also silversmith.
TISSEMAN—
& Son. *Bath.* 1875.
William. *Bath.* 1883.
TISSOT—
Alexander. *New York.* 1805.
& BONETTO. *London.* 1863-81.

TITCOMB—
Albert. *Bangor, Me, USA.* ca.1850.
Enoch J. *Boston, USA.* 1834. C.
TITHERTON, John. *London.* Late 17c. C.
TITLEY, William. *Newcastle-under-Lyme.* 1876.
TITTERINGTON—
Alexander. *Belfast.* 1868.
Alexander. *Lisburn.* 1840-58. (?Later *Belfast.*)
TITUS, James. *Philadelphia, USA.* 1833.
TOBIAS—
Evan. *Llandeilo.* 1810-40.
S. I. & Co. *New York, USA.* 1829. (Cf. Samuel Isaac at *Liverpool* 1811-13.)
TOBIN, John. *Beaver River, Nova Scotia.*1847.
TODD—
Daniel. *Glasgow.* 1848-60. W.
David. *Kinross.* 1860.
J. *Brigg.* 1861.
J. *North Meols (Lancashire).* 1858. W.
John. *Dumfries.* 1828-49.
M. L. *Beaver, Pa, USA.* ca.1830.
R. I. *Liverpool.* 1857. W.
Richard. *Strasbourg, Pa, USA.* ca.1769.
Richard J. *New York, USA.* ca.1830.
Tracy. *Lexington, Ky, USA.* 1840s.
William. *Carlisle.* 1848.
William. *Easingwold.* ca.1740. C.
William. *Glasgow.* 1838-48.
William. *Scarborough.* d.1752. C.
TODE—
G. P. *London.* 1858.
George. *Canterbury.* 1858.
TODMAN, Richard & Son. *London.* 1857.
TOFT, Thomas. *Hanley.* 1860-76.
TOKIO, The. 1914 and later, product of Pequegnat Clock Co.
TOLE—
Henry. *Stony Stratford.* 1842-77.
John. *Newport Pagnell.* 1854-77.
Thomas. *Newport Pagnell.* mar.1820-69. Later Tole & Son.
TOLEMAN—
John. *Caernarvon.* s. of William. 1844-56.
John. *Llangefni.* Same man as at *Caernarvon.*
John. *Pwllheli.* s. of John of *Caernarvon.* 1837-88. C.
Richard H. *Caernarvon.* s. of John of *Caernarvon,* 1868-90.
William. *Caernarvon.* 1791-1844. C. & W.
TOLFORD, Joshua. *Kennebunk, Me, USA.* 1815. Then to *Portland, Me.*
TOLHURST, Henry. *Ticehurst.* 1851-78.
TOLLADY—
Alfred Edward. *Bury St. Edmunds.* 1865-75.
D. *Bury St. Edmunds.* 1853-65.
TOLLES, Nathan. *Plymouth, Conn, USA.* Sold out 1836.
TOLLEY, Charles. *London.* a.1676 to John Brewer, CC. 1683-(B.-1730). C.
TOLMAN, Jeremiah. *Boston, USA.* 1810.
TOLPUTT—
James. *Dover Castle.* ca.1714. C.
John. *Portsea Common.* ca.1765. C.
TOMBS, Daniel. *London.* (?*Westminster*). ca.1690. C.
TOMEY, Joshua. *Dublin.* (B. 1774-80). 1786. W.
TOMKIES—
R. B. *Atherstone.* 1868.
Richard Brown. *Rhyl.* 1887.
TOMKINS—
John. Place unknown. Late 18c. C.
Thomas. *Leominster.* 1844-79.
& Son. *Leominster.* 1830-44.
TOMKINSON, Joseph. *Nantwich.* 1834.
TOMLIN(E)—
Charles. *Dover.* 1874.
George. *Lymm (Cheshire).* 1878.
TOMLINS, Nicholas. *London.* a.1639 to Peter Closon, CC. 1646-d.1658.
TOM(B)LINSON—
E. *Basingstoke.* 1859.
Edward. *Basingstoke.* 1830-39.
Edward. *Newcastle-under-Lyme.* 1828-50.
Elias. *Bakewell.* 1876.
George. *Newcastle-under-Lyme.* 1850.

TOM(B)LINSON—*continued.*
Job. From *Bicester* to *Thame*, 1819-65. T.C., C. & W.
John. *Bicester.* 1843. C.
John. *Horncastle.* 1835-61.
John. *Leeds.* 1850.
John. *Thame.* 1772-1819. T.C.
John. *York.* 1851. W.
Nicholas. *London.* ca.1667. CC.
Thomas. *Bicester.* 1823. C.
Thomas. *London.* CC. 1646-64. *Chancery Lane.*
Thomas. *London.* ca.1780. W.
Thomas. *London.* 1828.
Timothy. *St. Helens.* 1763-65, d.1782 at *Flixton.*
William. *Colchester.* b.ca.1654-d.1709. W.
TOMMISSON, John. *St. Helens.* 1768-84. W.
TOMPION, T. *London.* 1826.
TOMPKINS, James. *Olney.* 1798.
TOMPSON—
Joseph. *London.* 1844-57.
Joseph William. 1857.
TOMS—
Thomas. *London.* 1828.
William, jun. *Puddletown (Dorset).* 1867-75.
TONCHON, Henry. *London.* 1857.
TONGE, George. *Oxford.* b.1730-d.1803. C.
TONKIN—
James. *Chacewater (Cornwall).* Believed error for James Hocking.
Thomas. Place unknown. Early 18c. May be man below.
Thomas. *Penzance.* pre-1768. W.
TOOGOOD, James Thomas. *Uckfield.* 1878.
TOOKE—
Samuel. *King's Lynn.* 1830-36.
W. H. *Liverpool.* ca.1850. (Late F. L. Hausberg).
TOOLEY—
A. *Woburn.* 1854.
George (& Son). *Aylesbury.* 'Est'd. 1836'-1854 (-1869).
Stephen. *Aylesbury.* 1877.
TOOMER, John. *Glastonbury.* 1861-83.
TOON—
James. *Kidderminster.* 1850.
Thomas. *Kidderminster.* 1842.
TOOTELL—
———. *St. Helens.* Late 18c. C.
John. *Eccles (Lancs)* ca. 1785-1828. C.
William. *Chorley.* (B. 1790-1824) 1822-28.
TOPHAM—
J. *Nantwich.* 1857.
James. *Nantwich.* 1834.
John. *Bradford.* 1735-43. C.
John. *Southowram (Halifax).* d.1778. C.
W. I. *Newark.* 1855.
William. *Ipswich.* 1879.
TOPLEY, Charles. *London.* 1875-81.
TOPLIS, John & Sons. *London.* 1875.
TOPPING—
John. *London.* a.1691 to William Grimes. (B. d.1747).
Thomas. *Newcastle-on-Tyne.* 1787-1801. Watch glass makers.
TORBOCH, Gottfried. *Munich.* ca.1705. W. (B. ca.1710).
TORCHER, Alexander. *Greenlaw (Scotland).* 1837.
TORK, Clock Co. Inc. *Mt. Vernon, NY, USA.* 1918-present day.
TORKINGTON, William Henry. *Manchester.* 1834.
TORLEY—
J. *Bilston.* 1860-68.
John. *Hull.* ca.1790. C.
TORONTO, The. 1914 and later, product of Pequegnat Clock Co.
TORR—
Edwin. *London.* 1869.
Edwin Elijah. *London.* 1881.
Eli. *London.* 1875-81.
TORREY, Benjamin B. *Hanover, Mass, USA.* Early 19c. C.
TORRINI, Emanuel. *London.* 1875-81.
TORRY—
Alexander. *Banchory Ternan.* 1846-60.
John. *Birkenhead.* 1878.
TOSEN, John James. *London.* 1863-81.

TOTTEM, George. *London.* 1844.
TOTTINGTON, Joshua. *Barnsley.* d.1776. C.
TOUGH—
David. *Forfar.* 1860.
Rev. George. *Ayton (Berwickshire).* 1837.
TOURANGEAU, Joseph. *Quebec.* 1890.
TOUTAIN, Louis D. *Jersey.* 1853.
TOWER, Reuben. *Plymouth. Hingham* and *Hanover, Mass, USA.* ca.1810-30.
TOWERS—
———. *Wincanton.* 1822. C.
Alfred. *London.* 1875.
TOWILL, Ar. *Taunton.* 1745. W.
TOWN(E)—
Ira S. *Montpelier, Vt, USA.* ca.1860. W.
Thomas G. *Port Colbourne, Canada.* 1923. Patented a self-winding watch.
TOWNL(E)Y—
———. *Temple (?Temple Cloud, Somerset).* ca.1725. C.
John. *Liverpool.* (B. 1825) 1828-34.
& QUILLIAM. *Liverpool.* 1834.
Thomas (& Son). *Liverpool.* 1822-48.
TOWNSEND—
A. *Coventry.* 1880.
Charles. *Philadelphia, USA.* 1799.
Charles, jun. *Philadelphia, USA.* ca.1824-50.
Christopher. *Newport, RI, USA.* 1773-ca.1810. Casemaker.
Elisha. *Philadelphia, USA.* 1828.
H. *Conway, Mass, USA.* ca.1870.
James. *Helmsdon (Northamptonshire).* 1720.
John. *London.* 1632. Expelled from CC.
John. *Philadelphia, USA.* 1813-33.
John, jun. *Philadelphia, USA.* 1849.
Joseph. *Baltimore, USA.* 1792.
R. *Woolwich.* 1874.
& RICHARDS (William). *Aberavon.* 1887.
Thomas A. *Bath.* ca.1790. W.
William A. *Montreal.* 1842-67.
TOWNSHEND, William. *London.* 1839.
TOWNSON—
J. *Ulverston.* ca. 1704
John. *Barrow-in-Furness.* 1866-64. C.
Samuel. *?London.* ca.1760. C.
Thomas. *Pennington* nr *Ulverston.* mid-18c. C.
TRAC(E)Y—
———. *Magherafelt.* Early 19c. C.
(E.) & BAKER. *Philadelphia, USA* and *Waltham, Mass, USA.* 1857. Watch case maker.
C. & E. *Philadelphia, USA.* 1847. Watch case makers.
Charles. *Philadelphia, USA.* 1842-50. Watch case maker.
Erastus. *Norwich* and *New London, Conn, USA.* (1768) 1790-96. s. of Isaac.
Gurdon. *New London, Conn, USA.* (1767) 1787-92. C. Also silversmith.
James. *Ottawa.* 1862-67.
James. *Pembroke.* 1835-87.
William H. *Bytown, Canada.* 1851-76.
William H. *Smiths Falls, Canada.* 1857-63.
TRAEXINGER, Franz. *Aschach.* ca.1750.
TRAHN, Peter & Co. *Philadelphia, USA.* 1843-49. W.
TRAIES, William. *Dartmouth.* Free at *Exeter.* 1817. W.
TRAIL—
Edwin. *London.* 1839-51.
William. *London.* (B. 1802-24) 1828-57.
TRAINER, Richard. *Keady.* 1868. Also gunsmith.
TRAIRS, W. & Son. *Leek.* 1868.
TRAMPLEPLEASURE, James. *Jersey City, NJ, USA.* 1850-54.
TRANMER—
Richard. *Pickering.* 1866. W.
William. *Wakefield.* 1851-71.
TRANTER, Nicholas. *London.* ?mid-18c. W.
TRAUNWIESER, F. *Halifax, Nova Scotia.* 1864.
TRAVEL, John. *Dublin.* 1845.
TRAVERS—
Henry. *Liverpool.* d.1723. W.
Matthew. *London.* 1811. Watch case maker.
TRAVIL—
R. *Cambridge.* 1858.
T. *Cambridge.* 1858.

TRAVIS—
——. *Sheffield.* pre-1710. W. (Prob. Joshua, q.v.)
George. *Rotherham.* mar.1774-95.
James. *Rotherham.* mar.1805.
John. *New York.* 1860.
John. *Sheffield.* 1862-71. W.
Joshua. *Sheffield.* d.1735. W.
R. W. *Kirton-in-Lindsey.* 1868-76.
Samuel. *Leek.* 1868-76.
Samuel. *Macclesfield.* 1878.
T. *Doncaster.* ca.1745-ca.1760. C. (May be same man as at *Thorne*).
Thomas. *Manchester.* 1878.
Thomas. *Thorne.* 1746-57. C.
William. *Leek.* 1828-60.
William Robert. *Kirton-in-Lindsey.* 1849-61.
TREADWAY, Amos. *Middletown, Conn, USA.* 1787. W.
TREADWELL, Oren B. *Philadelphia, USA.* 1847-49.
TREAT—
(Sherman) & BISHOP (Daniel F.). *Bristol, USA.* 1830s. C.
George. *Newark, NJ, USA.* 1850-58.
Orrin. *New Haven, Conn, USA.* ca.1840-ca.1850.
Sherman. *Bristol, USA.* 1828-35. C.
TREE—
Alfred. *London.* 1869.
Charles. *London.* 1881.
James. *London.* 1851-81.
TREGOE, Timothy. *Great Marlow.* 1769-98. W.
TREGONING, ——. *Breage (Cornwall).* 1770. C.
TREGOSSE, Stephen. *Grampound (Cornwall).* 1686. Kept ch. clock.
TREHANE, Sampson. *Exeter.* 1803. W.
TREINEN, Michael. *London.* 1881.
TRELEAVEN—
James. *Bodmin.* 1847.
Thomas. *Bodmin.* 1844-73.
TRELFORD & Co. *Belfast.* ca.1870-77.
TREMBLAY, Thelesphor. *Montreal.* 1884. Shoemaker who patented a watchman's clock.
TREMLETT, Alfred. *Bristol.* 1830.
TREMONT, Watch Co. *Boston, USA.* 1864. To *Melrose, Mass, USA.* 1866-68. W.
TRENCKNER, William Charles. *London.* 1881.
TRENDELL—
& BRACHER. *Reading.* 1854.
George. *Maidenhead.* 1837-77.
James. *Reading.* (B. 1819-) 1830-37.
TRENERRY—
John. *Penryn (Cornwall).* 1652. Rep'd. ch. clock.
John. *Redruth.* mar.1806-ca.1835. C.
TRENHAM/TRENHAN—
Daniel. *Helmsley.* 1832-pre-1851. W.
Elizabeth. *Helmsley.* 1851. Widow of Daniel.
John Richard. *Helmsley.* b.1837. s. of Daniel. -1866. W.
TRENTON Watch Co. *Chambersburg, NJ, USA.* (Formerly New Haven Clock Co. 1883-86) 1907-sold out to Ingersoll 1908.
TRESHER, Fesser & Co. *Wolverhampton.* 1850.
TREVENA—
William I. *Redruth.* 1706-1786. C.
William II. *Redruth.* 1799-1873. W. & C.
William. *St. Agnes.* 1844-46.
TREVERTON, John. *St. Breoke (Cornwall).* mar.1728. C.
TREWEEK, John. *Callington (Cornwall).* 1823. W.
TREWIN, John. *St. Blazey (Cornwall).* 1823-56. W.
TREWINNARD—
Alfred. *London.* 1869-81.
Arthur. *London.* 1875-81.
Edward. *London.* (B. 1820-25) 1828.
George. *London.* 1832-44.
Joseph. *London.* 1828-63.
Joshua. *London.* (B. 1805-1810) 1832.
Joshua George. *London.* 1863.
& Son. *London.* 1869-75.
William. *London.* 1828-32.
TREZIES, James. *London.* 1839-44.
TRIBE—
Daniel. *Thakenham. (Sussex).* b.1717 (a.1730 to John Tribe of *Petworth*) -ca.1740, then *Petersfield.*

TRIBE—*continued.*
George I. s. of John. b.ca.1714-ca.1750. *Thakenham.*
George II. s. of George I. *Thakenham.* b.1746-74.
Gilbert. *Newark, NJ, USA.* 1850-52.
J. *London.* ca.1770. W.
John. *Thakenham.* ca.1702-d.1747.
John. s. of John of *Thakenham.* b.1706. *Petworth* till d.1777. C.
John Carr. *London.* 1863-81.
T. *Southsea.* 1848.
Thomas. *Petworth.* mar.1784-ca.1790.
TRICKETT, W. T. *Coventry.* 1880.
TRIEBLER, Johan. *Friedburg (Germany).* ca.1700. W.
TRIGG—
Edward. *London.* 1869.
Thomas. *London.* a.1692 to John Westoby, CC. 1701 (-B. 1718). C.
TRIGGS—
& BUSBY. *Guildford.* 1839.
R. W. *Blackheath.* 1866-74.
W. *Guildford.* 1838-62. W.
TRIMBLE—
I. *Newry.* 1827.
& KEANE. *Newry.* 1821-24.
TRIMNELL, William Henry. *Canterbury.* 1847-55.
TRINDER, Albert. *Swindon.* 1867-75.
TRINGHAM, George. *London.* (B. 1799-1824) 1828-44.
TRIPPETT—
John. *Kingston-on-Thames.* CC. 1668-d.1700. T.C.
Robert. *London.* a.1688 to James Hatchman, CC. 1700. (B.-1723). C.
William. *Hull.* 1670-71. Repaired Beverley Minster clock.
TRIPPLIN, Julien. *London and Besançon.* 1881.
TRIST, James. *Torquay.* 1850.
TRISTON, William. *Bedford.* Next to St. Paul's churchyard in 1774. W.
TRITCHLER, John. *Bolton.* 1851-58. C.
TRITSCHLER—
Ferdinand H. *Carlisle.* 1869-79.
John & Co. *London.* 1844-81.
Joseph. *London.* 1851.
Joseph. *London.* 1881.
Joseph & Co. *London.* 1839-44.
Lawrence. *Whitehaven.* 1848. C.
& MILLER & Co. *London.* 1851-63.
William & Co. *Carlisle.* 1848-69. Also dealer in writing desks.
& WILLMAN. *London.* 1851-57.
TROBRIDGE, William James. *Bath.* 1875-83.
TROIN, James Bray. *Cambridge.* 1875.
TROLL—
George. *London.* 1851-75.
Matthew. *London.* 1869.
TRONE, Peter. *Philadelphia, USA.* 1844.
TROTH—
James. *Pittsburgh, Pa, USA.* ca.1820.
Thomas. *Pittsburgh, Pa, USA.* ca.1820.
TROTMAN, Benjamin. *London.* 1881.
TROTT—
Andrew C. *Boston, USA.* 1779-1812.
(John Proctor) & CLEVELAND (William). *New London, Conn, USA.* 1792. Bought stock of Gurdon Tracy.
G. *Halifax, Nova Scotia.* 1866.
John Proctor. *New London, Conn, USA.* 1769-1852.
Jonathan, jun. *New London, Conn, USA.* (1771) 1800-13. C.
Peter & Andrew C. *Boston, USA.* 1805.
TROTTER—
Charles. *Abingdon.* 1837.
Charles. *Lincoln.* 1861.
J. *Alnwick.* 1827-58.
Jeremiah. *New York.* ca.1820-ca.1830.
Joseph. *Newcastle-on-Tyne.* (B. 1820-) 1827-36. C.
Joseph. *Quebec.* 1822.
Joshua. *Stokesley.* ca.1790. C.
Thomas. *Whitby.* 1834-40.
William. *Blackwatertown.* (b.1794) 1824-d.1838.
TROUGHTON, John. *Helston.* 1707-ca.1750. C.
TROUNSON—
Thomas. *St. Just in Penwith (Cornwall).* 1873. W.

TROUNSON—*continued.*
William Henry. *Penzance.* 1856-73.
TROUP—
Alexander, sen. *Halifax, Nova Scotia.* b.1776-d.1856.
Alexander, jun. *Halifax, Nova Scotia.* (b.1806) 1838-73.
James. *London.* (B. CC. 1825) 1828-32.
John. *London.* 1839-69.
& Sons. *London.* 1875-81.
Thomas. *Halifax, Nova Scotia.* (b.1819) 1869-74.
TROUTBECK, William. *Leeds.* mar.1709-d.1738. C.
TROW, Ephraim. *Haverhill, Mass, USA.* 1832. C. & W.
TROWBRIDGE, John. *Monmouth.* 1868.
TROWER, William. *Prittlewell (Essex).* 1828-39.
TROY, John. *London.* 1832.
TROYON, J. *Poitiers.* ca.1630. W.
TRUAX, De Witt. *Utica, NY, USA.* 1842.
TRUDEL(E), Michel. *Quebec.* 1844-49.
TRUDGILL, Jacob. *Ipswich.* 1879.
TRUEN, J. *Pocklington.* ca.1790. C.
TRUMAN, Jeffrey. *Waynesville, O, USA.* 1822. W.
TRUMBALL & HASKELL. *Lowell, Mass, USA* ca.1860.
TRUMBLE, J. *East Dereham (Norfolk).* 1858.
TRUNDLE, Charles. *Harwich.* 1839.
TRUSCOTT—
Alfred. *Tenby.* 1868-87.
George. *Cardiff.* 1844.
George. *Haverfordwest.* 1835.
George. *Milford Haven.* 1868.
George. *Newport (Monmouth).* 1868.
George. *Pembroke.* 1844-87.
George. *London.* 1878.
Henry. *Solva (Wales).* 1875.
James. *Aberystwyth.* b.1822-d.1874.
James. *Milford Haven.* 1875.
James. *St. Columb (Cornwall).* ca.1830-73. C.
James. *Tenby.* 1844-87.
John F. *Narberth (Wales).* 1868-75.
Joseph. *St. Austell.* 1842-47.
Lewis. *Mevagissey.* ca.1800-1812. C.
Lewis. *Pembroke.* 1844.
Lewis. *Truro.* ca.1820-33. C. & W.
Mrs. Matilda Jane. *Swansea.* 1887.
William H. *Welshpool.* 1874.
William Howell. *Aberystwyth.* 1887.
TUCK—
J. *Portsea.* 1859-67.
John. *Margate.* 1874.
John. *Romsey (Hampshire).* 1859-78.
TUCKER—
David. *Carmarthen.* 1868-75.
Elisha. *London.* 1851-81.
Frederick Squire. *Bristol.* 1870.
George. *Southampton.* 1867-78.
J. *Corsham.* 1859.
J. W. *San Francisco, Cal, USA.* ca.1850-ca.1860.
John. *Exeter.* b.1760-1803. W.
John. *St. Ives (Huntingdonshire).* 1877.
John O. *London.* 1869.
Rees. *Carmarthen.* Successor to John Howell, 1830-40 W.
W. J. *St. Johns, Newfoundland.* 1885.
William. *Gosport.* 1830-67.
TUCKFIELD—
Charles. *London.* 1869.
Joseph. *London.* 1851-63.
TUDMAN, John. *Langport.* 1875.
TUDRIE—
Edward Thomas. *Eastover (Somerset).* 1883.
Edward Thomas. *New Alresford (Hampshire).* 1859-78.
TUE (?also Tew), Thomas. *King's Lynn.* Lantern clocks dated 1663 and 1697. (Signs them 'de L'inn'.)
TUERLING, James. *New York.* 1857.
TUGBY, W. *Whitwick (Leicestershire).* 1855-64.
TULLER, William. *New York.* 1831.
TULLES, PALLISTER & McDONALD. *Halifax, Nova Scotia.* 1810-12.
TULLOCH, James. *Campbeltown (Scotland).* 1860.
TULLOH, John. *Nairn.* 1836.
TUMMON, W. *Sheffield.* 1871.

TUNKS, D. *Accrington.* 1858. W.
TUNNELL, John. *Deptford.* 1847-49. (Prob. from *London.*)
TUNNICLIFF—
George. *Derby.* 1835.
J. *Coventry.* 1854-60.
TUNSTALL, John from *Prescot.* mar.1775 at *Warrington.* W.
TUNWELL, W. *Bishopwearmouth.* 1856.
TUPMAN—
George. *London.* (B. 1794-1820) 1828. W.
George (& Co.). *London.* 1875 (-1881).
George & Henry. *London.* 1844-69.
James. *London.* 1828-39. Also foreign clocks.
TURCOT, Narcisse. *Quebec.* 1844-82.
TURELL, Andrew C. *Boston, USA.* 1800-1810.
TURFORD, James. *Ludlow.* 1870-79.
TURNBULL—
Charles. *Oxford.* 1573-81. Sundials.
Edwin. *Scarborough.* 1851.
John. *Baltimore, USA.* 1798.
John. *Dunfermline.* (B. 1780)-1784. W.
John. *Hawick.* 1827. W.
John. *Stokesley.* 1834-40.
Richard. *Wooler.* 1848-58.
Robert. *Coldstream.* 1860.
Thomas I. b. 1749 *Darlington.* s. of William I mar. *Whitby* 1784. d. 1796. C. & shipowner.
William. *Whitby.* 1823-66. W. & C.
TURNER—
Albert. *Welshpool.* 1887.
Alfred. *Cheddar.* 1875-83.
Allison. *Ashtabula, O, USA.* ca.1830. C.
Arthur. *Tetbury.* 1879.
Charles. *Finchingfield (Essex).* 1839.
Charles & Alfred. *London.* 1881.
Cornelius. *Taunton.* 1883.
F. P. *Surbiton.* 1878.
Franklin. *Cheran, SC, USA.* ca.1814-23.
Frederick. *London.* 1844. Dial silverer.
George. *London.* 1869-81.
Henry (& Son). *Colchester.* s. of Thomas. b.1792-1860. (& Son 1860-73, when died).
Henry Clow. *Colchester.* s. of Henry above. With father 1860-73; alone 1873-93; with Bonner 1893-1913.
Herbert. *Kingston-on-Thames.* 1878.
Isaac. *Birmingham.* 1842-60.
Isaac. *Penrith.* 1858-79. Also jeweller. Also at *Kendal* in 1873.
James. *Larne (co. Antrim).* 1824-68. W.
John. *Chipping Ongar.* 1839-66.
John. *Ilminster* 1861; *Langport* 1861-83.
John. *King's Lynn.* 1830-65.
John. *Leytonstone.* 1874.
John. *Longton.* 1876.
John. *New York.* 1848.
John. *Southwell.* 1828-35.
John. *Spaldwick (Huntingdonshire).* 1854-77.
John. *Yearasley-cum-Whaley (Cheshire).* 1865-78.
John Appleby. *Grimsby.* 1849-50.
Jonathan. *Doncaster.* 1858.
L. *Kingston-on-Thames.* 1851-55.
Mrs. Lydia Maria. *Kingston-on-Thames.* 1862-66.
Richard. *Lewes.* 1715-27. C.
Theophilus. *Middleton (Lancashire)* 1785; *Chadderton (Lancashire)* 1799. C. & W.
Thomas. *Colchester.* 1792. W.
Thomas. *London.* 1654. Journeyman of Peter de Landre.
Thomas. *Tregony.* ca.1790. C.
William. *Birmingham.* 1880.
William. *Boston (Lincolnshire).* 1861-76.
William. *Haverfordwest.* 1835.
William. *London.* 1828-39.
William. *London.* 1881.
William. *Nayland (Suffolk).* 1879.
William Jamison. *Larne (co. Antrim).* mar.1841. W.
TURNLY, James. *Belfast.* Partnership with John Knox III dissolved 1815.
TURPIN—
Benjamin. *London.* (B. 1817-24) 1832-39.
Brothers. *London.* 1857-81. (Successor to George.)

TURPIN—*continued.*
George. *London.* 1851. (Successor to Mrs. Susannah.)
Mrs. Susannah. *London.* 1844. (?wid. of Benjamin.)
TURRELL, Samuel. *Boston, USA.* 1789-1820.
TURRETT & MARINE CLOCK Co. *Boston, USA.*
1860. T.C.
TURTON—
George. *Killamarsh (Yorkshire).* ca.1790. C.
M. *Leeds.* 1871.
Nathaniel. *Manchester.* (B. 1804-14) 1824.
William. *Sheffield.* (B. 1817) 1822.
TURVEY, John F. *Belfast.* 1858.
TUSHINGHAM (or TUSSINGHAM), John. *London*
b.1664, a.1682 to Richard Prince, mar.1687. W.
TUSTEN, Hiram S. *Abbeville, SC, USA.* ca.1842-85.
Later to *Monroe, La, USA.*
TUSTIN—
Charles. *Leicester.* 1864-76.
Septimus. *Baltimore, USA.* 1814.
TUTHILL—
Daniel M. *Saxtons River, Vt, USA.* ca.1842. Shelf clock
cases.
R. J. *Axbridge (Somerset).* 1866.
TUTIN—
John Allan. *Leeds.* 1871.
S. & J. *Nottingham.* 1855.
TUTTELL, Thomas. *London.* a.1688 to Henry Wynne.
CC. 1695. (B. -1699).
TUTTLE, Eliada. *Owego, NY, USA.* 1833. Shelf clocks.
TUTTON, Edward. *Richmond (Surrey).* 1839-55.
TUXFORD—
Thomas Haynes. *Boston.* 1849-76.
Weston Ingram. *Boston.* 1828-35.
TWANBY, George. *Birmingham.* 1880.
TWEEDY, William. *Newcastle-on-Tyne.* 1834-52.
TWEMLOW, J. *Boughton (Cheshire).* 1857.
TWICHELL, George Hearn. *Maidstone.* 1839.
TWINN, James George. *London.* 1881.
TWISS—
Austin, Benjamin, Joseph & Ira. *Montreal* and *Ligorin*
and *Longueuil, Canada.* 1821-51.
B. & H. *Meriden, Conn, USA.* 1828-37. C.
Hiram. *Meriden, Conn, USA.* 1834. (Of B. & H.
Twiss.)
TWIST, Joseph. *Ormskirk.* 1848-58.
TWITCHELL—
Marcus. *Utica, NY, USA.* 1829.
Samuel. *Maidstone.* 1847.
TWOMBLY & CLEAVES. *Biddeford, Me, USA.*
ca.1870.
TWYFORD—
Josiah. *Manchester.* 1792-1822. W.
William. *London.* (B. 1805-11). 1832-39.
TYACKE, George. *Breage (Cornwall).* ca.1820-d.1853.
C.
TYAS—
John. *London.* 1832-39. Watch cases.
John. *Manchester.* mar.1824. C.

TYAS—*continued.*
William Thomas. *London.* (B. CC. 1820-25) 1828-32.
TYDEMAN—
G. *Stowmarket.* 1865.
George. *Needham Market.* 1865-75.
H. *Rayleigh.* 1866.
James. *Bildeston (Suffolk).* 1830-79.
TYE—
Alfred James. *Birmingham.* 1868-80.
Robert. *Spalding.* 1861-76.
TYLER—
A. *Nottingham.* 1849-55.
Charles. *London.* 1851-57.
Charles. *London.* 1875. (May be same as above.)
E. A. *New Orleans, La, USA.* ca.1840-ca.1850.
Edward. *Bristol.* a.1780-pre-1828.
Edward. *London.* 1851-63.
Edward, jun. *London.* 1881.
George. *London.* a.1692 to Robert Dingley. CC. 1699.
(B. -1723). C.
George. *London.* 1839-69.
John. *Ashby-de-la-Zouch.* 1835.
John. *Leicestershire.* 1828.
John William. *London.* 1863-81.
Mrs. Ann *London.* 1869.
Richard. *Adderbury.* b.1733-d.1800. C.
Robert. *Melton Mowbray.* 1828-49.
W. *Hinckley.* ca.1790-ca.1800. C.
W. *Nottingham.* 1849.
William. *Cuckfield.* 1870.
TYMMS—
Edward. *Cheltenham.* 1879.
John. *London.* a.1656 to Nicholas Tomlins, then to
Richard Beck in 1658 when former died.
William Robert. *Cheltenham.* 1840-70.
TYRELL—
Henry. *Hastings.* 1839-78.
Walter. *Hastings.* 1828.
TYRER—
Edward. *Manchester.* 1644. (B. has one such at *Chester.*
1638. C.). W.
H. *London.* 1811.
Henry J. *Sheffield.* 1871.
James. *Prescot.* a.1797. W.
James Henry. *London.* (B. a.1801-24) 1828-51.
TYSON—
Henry. *Egremont (Cumberland).* 1834.
Henry. *Whitehaven.* 1828-34.
Leech. *Philadelphia, USA.* 1823-31.
Leonard. *Hawkshead (Cumberland).* 1741. Ch. clock
dial.
Matthew. *Gosforth (Cumberland).* 1834-58. W.
T(homas). *Bolton.* 1848-58. W.
TYTE—
Cornelius. *Swansea.* 1844-52. C.
Cornelius. *Wells.* 1861-75.
Samuel. *Warminster.* 1830. Also silversmith and
jeweller.

U

UFEN, Wrt. *Norden (Holland)*. Early 19c. C.
UGLOW—
    Abel. *Camelford*. ca.1770. C. (May be too early an
        estimate?).
    Abel. *Launceston*. 1823.
    George. *Stratton (Cornwall)*. 1791-pre-1841. C.
    W(m.?). *Collumpton*. ca.1800. C.
    William. *Truro*. ca.1820-47. C.
UHL, J. *Prague*. mid-18c.
ULLMAN & ENGELMANN (Fuerth). *London*. 1881.
ULPH, Robert. *Stalham (Norfolk)*. 1836-46.
ULRICH, John. *London*. 1832.
UMBRECHT, John. B. *Chicago, USA*. 1878.
    Casemaker.
UNDERHILL—
    Cave. *London*. a.1647 to Isaac Daniel. CC. 1655. (B.
        -1669).
    Daniel. *New York*. ca.1810.
    Samuel. *Wolverhampton*. 1828-35. Also bottle jack maker.
    William. *Newport (Shropshire)*. (b. 1782 s. of
        Thomas of *Albrighton*). 1828-36. C.
UNDERWOOD—
    Caesar. *London*. (B. 1795-1820) 1828-32.
    David. *Goshen, Pa, USA*. 1796-1801. W.
    James. *Rochford (Essex)*. 1874.
    Richard. *London*. 1863-75.
    Robert. *London*. (B. mar.1769-1808) 1811. W.
    Roger. *Rugby*. 1850-80.
    William. *Bristol*. a.1660-67. Maybe later to *London*
        where one such ca.1705.
UNDY, Richard. *Retford*. 1800-35. Successor to William
    Chumbley. C.
UNION—
    Clock Co. *Bristol, USA*. 1843-45.
    Manufacturing Co. *Bristol, USA*. 1843-45. C.
UNITE, Matthias. *Oxford*. a.1681. Lant. and br. clocks.
UNITED—
    States Clock Co. *New York*. 1872.
    States Clock & Brass Co. *Chicago, USA*. ca.1860-68. C.
    States Clock Case Co. *Cincinnati, O, USA*. ca.1870-
        ca.1880. Shelf clock cases.
    States Clock Manufacturing Co. *New Haven, Conn,
        USA*. To *Austin, Ill.* 1866.
    States Time Corp. *New York, USA* and *Waterbury,
        Conn*. Successor to Waterbury Clock Co. in 1944.
    States Time & Weather Service Co. *New York*. 1897.
UNSWORTH—
    Edward. *Farnworth*. 1818. W.
    James. *Chorley (Lancashire)*. 1834-51.
    John. *Farnworth*. mar.1768(-B. d.1810. W.).
    Peter. *Liverpool*. 1795. W.
    Thomas. *Liverpool*. 1715-24. Spring maker.

UNSWORTH—*continued*.
    Thomas. *Prescot*. 1797. W.
    Thomas. *Prescot*. 1826. W.
UNTHANK—
    George. *Stokesley*. 1840. C.
    Robert. *Stokesley*. ca.1835. C.
    & Son. *Stokesley*. 1832. C.
    Thomas. *Stokesley*. 1866. W.
UPCROFT, W. *Feltwell (Norfolk)*. 1845-65.
UPJOHN(S)—
    & BRIGHT. *London*. 1839-57.
    BRIGHT & WOOD. *London*. 1863-81.
    James. *Exeter*. 1803. (Apprentice. W.)
    John. *Exeter*. 1803-mar.1806. W.
    John & Thomas. *London*. 1828-32. 'To his Majesty'.
    Mrs. E. *Maldon*. 1855.
    Nathaniel. *Plymouth*. ?later 18c. W. & C.
    Peter. *London*. (B. 1791-1825) 1828-32.
    Richard. *London*. 1869-81.
    Thomas. *Launceston*. 1765.
UPSALL, Samuel. *Yeovil*. 1883.
UPSON—
    Lucas. *New York*. 1846.
    MERRIMAN & CO. *Bristol, USA*. 1831-38. C.
UPTON, George. *Liverpool*. 1851.
URCH, William. *Wedmore* nr *Weston-super-Mare*.
    1875-83.
URE, William. *Cumbernauld*. 1836.
URLETIG, Valentine. *Reading, Pa, USA*. ca.1755-1783.
    C.
URQU(H)ART—
    —. *Elgin*. post-1820.
    John. *London*. ca.1760. C.
URSEAU (and variant spellings), Nicholas. *London*.
    Believed French. Supposed working by 1531. *Char-
    ing Cross*. Clockmaker to Queen Elizabeth. d.1590.
USHAR, William. *Patrington*. 1851.
USHER—
    —. *Town Malling*. ca.1820. C. Cf. Uslar.
    Charles & Co. *Leicester*. 1876.
    & COLE. *London*. 1869-81. (Successor to Joseph).
    James (& Son). *Lincoln*. 1835-50(-76). Also clock case
        maker.
    Joseph. *London*. 1863.
USLAR, G. *West Malling*. 1838. Cf. Usher.
USMAN, C. *West Malling*. 1838. Cf. above.
USMAR, O. *Wrotham (Kent)*. 1851-55.
USMER—
    Oliver. *West Malling*. 1832-51.
    William. *West Malling*. 1845-74.
UTELEY, H. *Niagara Falls, Canada*. 1833.

V

VACHERON—
Cesar & Co. *London.* 1869. (Charles Lamy, agent).
& CONSTANTIN. *London.* 1863. (Charles Lamy, agent).
VAHA, Kelly. *Straw, Draperstown (co. Derry).* Early 19c. C.
VAIL—
Edward. *Laporte, Ind, USA.* ca.1840-ca.1870.
Elijah M. *Albany, NY, USA.* 1840.
VALE—
George Frederick. *London.* 1863.
John. *Bury St. Edmunds.* 1839-65.
Joseph. *Lutterworth.* 1864-76.
& RICHARDSON. *Bury St. Edmunds.* 1875-79.
& ROTHERHAM. *Coventry.* 1828-35.
W. & Sons. *Lichfield.* 1860.
William. *Lichfield.* 1850.
William. *London.* (B. 1805-24) 1828-57.
William Edward. *Lichfield.* 1828-42.
VALENTINE, William B. *Boston, USA.* 1821.
VALIANT/VALIENT—
David. *Yoxford (Suffolk).* 1879.
George. *Eye.* 1875-79.
John. *Harleston.* 1875.
S. *Brockdish (Norfolk).* 1858-75.
Samuel. *Dickleburgh (Norfolk).* 1846 -d. 1890.
Samuel. *Leiston (Suffolk).* 1879.
VALIN, François. *Quebec.* 1750.
VALLANCE, Thomas. *London.* (B. one such *Liverpool* 1816). 1828. Watch cases.
VALLANT, T. *Hoxne (Norfolk).* 1846.
VALLE, D. *Three Rivers, Canada.* 1853.
VALLIN—
John. b.ca.1535 in *Flanders.* To *London* ca.1590 where d.1603. C.
Nicholas. s. of John. b.ca.1565. From *Brussels.* to *London.* pre-1590. d.1603. C.
VALOGNE, Charles. *London.* 1844-81.
VANACKER, Abraham. *London.* 1655. Journeyman to William Rogers.
VAN ALLYN, ——. *Conn, USA.* ca.1630.
VAN BARKER, Daniel. *Holland.* ca.1740. C.
VAN BRUSSEL, J. *Amsterdam.* ca.1710. C.
VAN BUREN, William. To *Newark, NJ, USA.* 1792. *New York.* 1794.
VAN BURGH, Jacob. *West Bromwich.* 1835-42. Also silversmith.
VANCE—
James. *Toronto.* 1837-47.
Joseph. *Toronto.* 1843.
VANCHER, Fritz. *London.* 1839.
VAN COTT, A. B. *Racine, Wis, USA.* ca.1850.
VANCOUVER, The. 1914 and later, product of Pequegnat Clock Co.
VAN DE HAGUE, Henry. *Croydon.* b.ca.1664-mar.1687. C.
VAN DER MARCKT, Cornelis M. ?*Wessaanen.* 1729. C.
VAN DER NATEN EN ZOON. *Amsterdam.* 1789. C.
VAN DER SLICE, John. *Wormelsdorf, Pa, USA.* Early 19c. C.

VAN DER VEER, Joseph. *Somerville, NJ, USA.* Early 19c.
VAN DUSEN, R. *Tweed, Canada.* 1857.
VAN EPS, George K. *New York.* 1830s-40s. C. Importer.
VAN GOOR, Colman. *London.* 1851-57.
VAN GUNTON—
Charles. *St. Marys, Canada.* 1862.
John. *Hamilton, Canada.* 1861-65.
VAN HOUTEN, J. *Amsterdam.* 18c. W.
VAN KUICK, G. *St. Antonis (Holland).* 1768. C.
VAN LEEUWEN, Symon. *Amsterdam.* ca.1725. C. (B. has Simeon. 1742. W.).
VAN LONE, James. *Philadelphia, USA.* 1775.
VANN, J. *Birmingham.* 1865. Dial maker.
VANNORNAM, C. H. & Co. *Hamilton, Canada.* 1857-66.
VAN RAALTE (VAN RAELTE), Joseph. *London.* 1857-63.
VAN SCOLINA, Richard. *London.* 1839-44.
VAN STEENBERGH, P. *Kingston, NY, USA.* ca.1780.
VAN STRIPE, Nicholas. *London.* ca.1680-1700. Forged watches by 'C. Gretton'.
VAN STRYP, J. Bernard. *Antwerp.* ca.1660. C.
VANTASSEL, W. H. *Brockville, Canada.* pre-1867.
VANTINE, John L. *Philadelphia, USA.* 1829-47.
VAN VALKENBURGH, Charles. *New York.* 1854.
VAN VLEIT, B. C. *Poughkeepsie, NY, USA.* 1830-47. C., W. and silver dealer.
VAN VOORHIS, Daniel. (1751). 1780. *Philadelphia, USA.* Then *Princeton, NJ.* Till retired 1819.
VAN WAGENER, John. *Oxford, NY, USA.* 1843.
VARCO, Robert, jun. *Fowey.* 1873. W.
VARIER, Vaughan. *Swansea.* mid-19c. C.
VARIGION/VERIGION, Daniel. *London.* ca.1815. C. (?see VAUGINON below).
VARLEY—
John. No town. Same man as J. V. of *Huddersfield.*
John. *Huddersfield.* b.1733-d.1785. C.
John. *Slaithwaite.* Same as J. V. of *Huddersfield.* q.v.
Joseph. *Dukinfield (Cheshire).* 1848.
Thomas. *Fakenham.* 1836.
William. *Huddersfield.* s. of John. post-1785. C.
William. *Leeds.* 1834. C.
VARNDELL, Mrs. S. *Odiham.* 1848.
VARNEY, George. *Hemel Hempstead.* 1828-39.
VARNISH, John. *Rochdale.* Error for John Barnish.
VARSOVIE, ——. *Leningrad.* 1914.
VASQUETO, ——. *Racunis.* ca.1790.
VASSALI, Jerome. *Scarborough.* 1840-58. C.
VAUGHAN—
Alfred C. *Bristol.* 1879.
Charles. *London.* 1832-86. W. (perhaps two of this name?)
David. *Philadelphia, USA.* From *London.* 1695-1702. C. & W.
E. Kelleher. *Bristol.* 1879.
Edwin. *Bristol.* 1850-79.
Frederick. *Bristol.* 1870.
George. *London.* (B. 1802-20) 1828. W.
J. *Staplehurst.* 1855.

239

VAUGHAN—*continued.*
James. *London.* 1832.
John. *Rye.* 1828.
Richard. *Watlington (Oxfordshire).* a.1732. C.
Rowland. *Pontypool.* mid-18c. C.
William. *Dyffryn (Merionethshire).* Late 19c.
William. *Newport (Mon).* 1844-80.
VAUGINON, Daniel. *London. St. James, Westminster.* 1756 took Joseph Jacobs as apprentice. (See VARIGION above.)
VAUSE, George James. *Leeds.* 1850-53. W.
VAUTROLER, ——. *London.* Alien 1622. (Not same as James Vautrolier, q.v.)
VAUTROLIER, James. *London.* 1622. Petitioner for CC. One of first assistants of CC. in 1632. (Not same as Vautroler, q.v.)
VEAL(E)(S)—
——. *Bodmin.* Late 16c. C.
Charles. *Bristol.* 1850-79.
(John) & GLAZE (William). *Columbia, SC, USA.* 1838-41.
Henry. *Bristol.* 1850.
John. *Bristol.* s. of Thomas. 1680-1713.
John. *St. Day (Cornwall).* 1844-73. C.
Joseph. *Mevagissey.* ca.1790. C.
Thomas. *Bristol.* 1652.
Thomas. *Southampton.* 1878.
William. *Bath.* 1883.
VEAZIE, Joseph. *Providence, RI, USA.* 1805.
VEEN, Dirk. *Nederlands.* 1729. C.
VEIJNS, ——. *Paris.* 'A la vielle cite'. 1740. W.
VEILLARD, ——. *Geneva.* ca.1800. W.
VEIT—
——. *Vienna.* 1500. C.
Johann. *Schrujvlethall.* 18c. C.
VEITCH, Andrew. *Haddington (Scotland).* ca.1775. C.
VELLAM, Charles. *Crowland (Lincolnshire).* 1828-35.
VELLENOWETH—
Mrs. E. *London.* 1875-81. (?wid. of William Henry).
William Henry. *London.* 1857-69.
VENABLES—
Charles. *High Wycombe.* 1830-42.
William. *Wareham.* 1824-30.
VENN, James. *Toronto.* 1874-77.
VENNALL/VENNELL—
Charles. *Deal.* 1866-74.
James. *Deal.* 1845-55.
James. *Sandwich.* 1832-40. W. & C.
Thomas. *Deal.* 1826-29. W.
VENNING—
Henry John. *Birmingham.* 1880.
J. H. *Balham (Surrey).* 1878.
John. *St. John (New Brunswick).* 1825.
William Morris. *St. John (New Brunswick).* 1816.
VENT, John. *Cheltenham.* 1850.
VERBACK, William. *London.* a.1681 to Dorcas Bouquett.
VERDIN—
Jacob. *Cincinnati, O, USA.* ca.1850. T.C.
Michael. *Cincinnati, O, USA.* ca.1850. T.C.
VER(I)LEY, Daniel or John Daniel. *Liverpool.* 1834-48. Also mus. boxes.
VERITY—
Henry. *Lancaster.* 1869. Also jeweller.
William. *Hunslet (Leeds).* 1866.
William. *Rothwell (nr Leeds).* ca.1850. C.
VERMONT, Clock Co. *Fairhaven, Vt, USA.* 1910-21.
VERNEUIL, ——. *Paris.* ca.1780-ca.1820.
VERNEY—
& ELLICOTT. *Hemel Hempstead.* ca.1800. C.
P. *Hemel Hempstead.* ca.1800. C.
VERNON—
The. 1914 and later, product of Pequegnat Clock Co.
Christopher. *London.* 1639. CC.
Henry. *Alnwick.* 1848-58. C.
James George. *London.* 1863.
Joseph. *Watlington (Oxfordshire).* 1852.
Norton (or Morton). *Toronto.* 1861-77.
Samuel II. *London.* s. of Samuel I, to whom a.1677. CC. 1685-97.
Thomas. *London.* ca.1740-ca.50. W.
Thomas. *Nottingham.* 1864-76.

VERNON—*continued.*
W. *Haddenham.* 1854.
W. *Watlington.* 1847.
W. J. *Crewe.* 1865.
VEROW, John. *Barton (Leicestershire).* ca.1785-ca.1790.
VERRIER, James. *North Curry (Somerset).* ca.1750. C.
VESEY, Robert. *Oxford.* a.1682-94. W.
VESPER—
Thomas. *London.* 1828-51.
Thomas & William. *London.* 1839.
William. *London.* 1844-57.
VESSEY—
Charles. *Ramsey.* 1830-77.
Francis. *Nottingham.* 1876.
VETTEE, John. *Hull.* 1838. W.
VEY, John. *Wimborne.* 1830-55.
VIAN, William. *Liskeard.* 1729-ca.1740. C. & W.
VIBBER, Russell. *Westtown, Pa, USA.* 1810-15.
VIBERT—
Charles. *Penzance.* 1750-d.1809. C.
John Pope. *Penzance.* (B. ca.1780). 1790-1865. C.
Henry Pope. *Gloucester.* 1870.
Henry Pope. *Penzance.* 1864. W.
VICARINO, Adolphe. *London.* 1857-63.
VICAR(E)Y—
John. *London.* 1857.
Robert. *London.* 1869-75.
VICK—
J. K. *Stroud.* 1856.
James. *London.* ca.1735-ca.1745. W.
John L. (& Son). *Stroud.* 1830-42(-50).
Richard. *London.* a.1692 to Francis Asselin. CC. 1702. (B. -d.1750). C.
VICKERS—
George. *Fergus, Canada.* 1851.
George. Henry. b. ca. 1834 *Wrockwardine,* to *Oakengates* by 1856-75, also *Wellington* 1861.
Isaac. b.ca.1820. *Irton (Cumberland).* To *Lancaster.* pre-1848-d.1905. C.
Joseph. *Coventry.* 1880.
VICKERY, Joseph. *Bristol.* 1850-79.
VICTORIA, The. 1914 and later, product of Pequegnat Clock Co.
VIDLER, William. *Greenwich.* 1874.
VIEL(L)—
Charles. *London.* a.1678 to Richard Jarret. CC. 1686-97.
George Hypolite. *London.* 1839-63.
VIENER, Moritz Seigfried. *London.* 1832.
VIET, Henry. *London.* ca.1705. C.
VIEYRES—
Antonio. *London.* 1832-44.
& REPINGTON. *London.* 1857-81.
VIGN & LAUTIER. *Bath.* 1809. Chrons.
VIGOR, C. *Tunbridge Wells.* 1874.
VILE, Jacob E. *Deal.* 1809-47. W.
VILES—
Henry. *Aston.* 1880.
Henry. *Birmingham.* 1880.
VILLEBRAND, Johan. *Augsburg.* ca.1690. Sundial.
VILLENEUVE & DELONDE (& Co.). *London.* 1875(-81).
VILLENGER—
Ferdinand. *London.* 1875-81.
William. *London.* 1869-81.
VILLON, Albert. *St. Nicolas d'Aliermont.* 1867-89.
VIMPANI/VIMPANY—
A. E. *Shefford (Bedfordshire).* 1877.
Edmund. *London.* 1869.
H. D. *Fareham.* 1867.
Harry Daniel. *Rickmansworth.* 1874.
John Dean. *London.* 1851-57.
Richard. *Cheltenham.* 1850-70.
William. *Fareham.* 1878.
VINCE, Mrs. Ann. *Hingham (Norfolk).* 1858-75.
VINCENT—
A. *Bath.* 1819-35.
F. H. *Taunton.* 1861.
J. B. *Putney.* 1862.
James. *Gravesend.* 1858.
John, jun. *Melcombe Regis.* 1855-75.
John. *Weymouth.* 1830.
Mrs. *Bath.* 1856.

VINCENT—*continued.*
W. A. *Haywards Heath.* 1878.
William. *London.* 1869.
William. *York.* Free 1764-1797. W.
William Sharp. *London.* 1857-75.
VINER—
& Co. *London.* ca.1845. C.
Charles Edward (& Co.). *London.* (B. a.1802-CC. 1813-40) 1828-69.
& HOSKINS. *London.* 1832.
Joseph. *Martock.* 1866-1904.
Mrs. S. *Martock.* 1861.
VINING, L. S. *Cincinnati, O, USA.* ca.1830-ca.1840.
VINTON, David. *Providence, RI, USA.* 1792.
VIRGO(E)—
James. *Sevenoaks.* 1874.
Thomas. *London.* a.1674 to Samuel Davis. CC. 1682-d.1685. C.
VITTY, George. *Granby, Canada.* 1857.
VITU, Samuel. *London.* Later 18c. C.
VIVIAN, James. *Fowey.* 1873. W.
VIVIER, Octave. *London.* 1863-81.
VOAK, James. *London.* 1851-69.
VOGEL—
Frederick. *Middleburg* and *Schoharie, 'NY, USA.* 1820s-30s. Imported English clocks.
Herman. *Halifax, Nova Scotia.* 1869-81.
VOGT—
A. F. *Northampton.* 1877.
Anthony. *London.* 1839-81.
Charles. *London.* 1839.
Charles & Frederick. *London.* 1844.
Fidelius. *Thrapston (Northamptonshire).* 1854-77.
Frederick & Co. *London.* 1851-57.
Henry. *Dunnville, Canada.* 1851.
Ignatius Christian. *New York.* 1764.
John. *London.* 1839.
William. *Glasgow.* 1860.
VOIGHT—
Henry. *Reading, Pa, USA.* 1780. To *Philadelphia.* 1814.
Sebastian. *Philadelphia, USA.* 1793-1800.
Thomas H. *Reading, Pa, USA.* s. of Henry. 1811-35. C.

VOKES—
Edward. *London.* 1857.
Edwin James. *Bath.* 1875-83.
Frederick. *Newport.* 1875.
William. *Bath.* 1856-66.
VOLANT, Elias (or Ely). *London.* Alien 1622-34. Worked 'at a baker's house in Holborn'.
VOLK—
& FLOESSEL. *Liskeard.* 1873.
M. *Brighton.* 1851-62.
Plazi. *London.* 1839-51.
VOLOGNE, ——. *Paris.* 1860-65.
VON DIECK, Henry. *London.* 1857.
VON GUNTAN, John. *Preston, Canada.* 1851.
VON GUNTEN, C. S. *Hamilton, Canada.* 1862-78.
VON QUINTER, John. *Hamilton, Canada.* 1856.
VOOGHT, William Francis. *London.* 1863.
VORHEES & VAN WICKLE. *New Brunswick, NJ, USA.* ca.1840.
VORLEY/VORLER—
H. *Market Harborough.* 1855.
Henry. *Thrapston (Northamptonshire).* 1841.
John. *Corby.* 1841.
Samuel John. *Crowland.* 1850.
Thomas. *Market Harborough.* 1835.
VOSE, John. *St. Helens.* 1818-34. W.
VOTIER, Lewis. *London.* Alien 1622. Worked with Abel Monpas. (Prob. Louis Vautier from *Blois,* who b.1581-d.1638.
VOUTE, Lewis C. *Brisgeton, Pa, USA.* 1826-35.
VOYCE, George. *Dean.* 1717. C. (B. 1791).
VOYSEY, J. *Taunton.* 1866.
VRIJTHOFF, Jan B. *The Hague.* ca.1775. C.
VUILLE, Alexander. *Baltimore, USA.* 1766. (Cf. FRANCIS & VUILLE.)
VUILLEUMIER—
& AMEZ-DROZ. *London.* 1863-69.
Charles. *London.* 1875-81.
VURLEY—
Peter. *Wisbech.* 1830-58.
Thomas. *Fakenham.* 1830.
Thomas. *Wisbech.* (B. 1791) 1840-46.

W

W, W, Monogram of William Whittaker of *Halifax*, q.v.
WAAGE & NORTON. *Philadelphia, USA.* 1798.
WADDELL—
George. *Aughnacloy (co. Tyrone).* 1846.
John. *Tenterden.* 1866-74.
William. *Kendal.* From *Scotland.* b.1809-51. W.
WADDINGTON, Robert. *Coventry.* 1880.
WADDLE, ——. *Marden (Kent).* 1858.
WADDLETON, John. *Birmingham.* 1842-50.
WADE—
Benjamin. *London.* 1881.
Charles. *Boston, USA.* 1830.
Charles Rice. *Aylsham.* 1830-75.
Edward. *Quebec.* 1826-45.
J. *Wootton-under-Edge.* 1856.
J. S. *Staythorpe (Nottinghamshire).* 1849-55.
James. *Farnham (Surrey).* 1828.
James. *Wootton-under-Edge.* 1830.
John. *Northampton.* 1777. W.
Nathaniel. a. *Norwich, Conn, USA.* 1793, *Newfield*
1793-98, *Stratford* 1798-1802. Sold to Lyman Smith.
R. *Aylsham.* ca.1800. C.
Robert. *Kineton.* 1860-80.
WADESON, Richard. *Burton* and *Holme (Lancashire).*
1879. W.
WADEY—
Frederick. *Henfield.* 1878.
James F. *Storrington (Sussex).* 1878.
WADGE—
Francis. *Callington.* b.1794-d.1856.
Josiah. *Callington.* 1754. C.
Josiah. *Callington.* 1823. W.
Josiah & Francis. *Callington.* 1844-56.
WADHAM(S)—
George. *Bath.* 1856-75.
George D. From *Cornwall, England* to *Wolcottville,*
Conn, USA. 1825-37. C.
WADSWORTH(S)—
Arthur. *Birmingham.* 1850-54.
J. C. (James) & A. (Amos). *Litchfield, Conn, USA.*
1830s. C.
Jeremiah. *Georgetown, SC, USA.* 1820s.
LOUNSBURY & TURNERS. *Lichfield, Conn, USA.*
ca.1830. C.
T. *Cheshunt.* 1859-74.
The Arthur. Name used by Newark Watch Co.
ca.1868.
& TURNERS. *Litchfield, Conn, USA.* ca.1820-30. C.
William. *Hornsea.* ca.1750-d.1811. C. and blacksmith.
(Also *Wadforth*).
WADY, James. *Newport, RI, USA.* ca.1750-55.
WAGGITT—
Charles. *York.* s. of Michael II, b.1790-d.1857. C.
George William. *York.* b.1820. s. of Charles. -d.1872.
C.
John W. *Tadcaster.* b.1824. s. of Charles from *York*
1850. C.
Michael I. *Richmond (Yorkshire).* mar.1752-post-1777.
C.

WAGGITT—*continued.*
Michael II. *York.* s. of Michael I, b.1770. From
Richmond 1822. C.
WAGHORN(E)—
C. *New Brompton (Kent).* 1866-74.
James. *Maidstone.* 1826-47.
WAGLAND, William. *Windsor.* 1877.
WAGNER—
Anthony. *Hull.* 1851-58. C.
& GERSTLEY. *London.* 1881.
Joseph. *London.* 1881.
WAGSTAFF—
Alfred. *London.* 1869-81.
Edward D. *London.* 1869-75.
George James. *London.* 1857-81.
James. *London.* 1839-57.
Joseph. *London.* 1844.
Thomas. *Banbury.* b.1724-ca.1756. C. & W. Later
London.
William. *London.* 1875-81.
WAIGHT—
John. *Birmingham.* 1828-35.
William. *Birmingham.* (B. 1790-1808)-1835.
William, sen. *London.* 1875-81.
William, jun. *London.* 1875.
William, sen. (& Son). *London.* 1851-57 (1863-69).
WAIN, William. *Burslem.* 1828-42. C.
WAINMAN—
Samuel Truman. *Leeds.* 1866-71. W.
William. *Hessle.* 1834-51.
William. *Hull.* 1840.
WAINSCOTT, Charles. *Hereford.* 1863-79.
WAINWRIGHT—
F. *Liverpool.* 1841. W.
George. *Manchester.* mar.1795. C.
John. *London.* a.1671 to William Thorowgood. CC
1679.
John (& Son). *Nottingham.* Est'd 1779-1820. C.
John. *Ormskirk.* 1834-58.
Nathan. *Liverpool.* d.1790. W.
Thomas. *Prescot.* mar.1779. W.
W. *Leicester.* 1864-76.
William. *Nottingham.* s. of John, 1814-28. C.
WAIT(E)—
DEWEY & Co. *Ravenna, O, USA.* ca.1855.
F. & Son. *Cheltenham.* 1879.
George W. Name used by Cornell Watch Co. ca.1871.
John. *Bradford.* ca.1850-ca.1860. C.
L. D. *Skaneateles, NY, USA.* 1838-47.
Thomas. *Cheltenham.* 1830-70.
W. *Grimsby.* 1869.
William. *Cheltenham.* 1842.
William. *London.* 1869-81.
WAITES—
John. *Whitehaven.* 1828-34. C.
M. *Ross.* 1863.
WAITHMAN—
Anthony. *Leeds.* 1825-34. W.
E. *Coventry.* 1854. Watch case maker.

242

WAITHMAN—*continued.*
Mary Ann. *Leeds.* 1837. Widow of Anthony. W.
WAITLEY, ——. *Worthington, O, USA.* ca.1893.
WAIZENE(C)KER—
Julius. *London.* 1863.
Julius. *Maidenhead.* 1877.
WAKEFIELD—
——. *London.* Early 19c. W. Prob. Todd, q.v.
Edwin. *Gateshead.* 1850-56.
John. *Ayton Banks.* (B. 1804) 1827. (?d.1846).
John. *Gateshead.* 1836.
John. *London.* 1839-57.
John Samuel. *Hartlepool.* 1856.
Thomas. *Redcar.* Late 19c. W.
Todd. *London.* 1832. Mus. clocks.
Timothy. *Lancaster.* s. of William. Free 1811-51. W.
William. *Lancaster.* Free 1782-d.1847. W. and silver-
smith.
WAKELIN—
David. *Aston (Warwickshire).* 1880.
David. *Birmingham.* 1880.
William. *Coventry.* 1880.
William. *London.* 1857-63.
WAKEMAN, C. J. *London.* 1869-81.
WALBANK, John. *Keighley.* 1866.
WALDEN, Samuel. *Wokingham.* 1877.
WALDER, Thomas. *Arundel.* 1828-51. C.
WALDIE—
T. *Morpeth.* 1858.
Thomas. *Blyth.* 1827-48.
William. *Dunse.* 1831.
WALDFOGEL—
BEHA & Co. *King's Lynn.* 1836.
G. & Co. *King's Lynn.* 1846.
J. *Peterborough.* 1847.
Kreutz & Co. *Peterborough.* 1841(-47). German clocks.
(G) SIEDLE (& Co.). *Lynn.* 1846. C.
WALDRETT, C. *Lewes.* 1833. C. See Hooker, William.
WALDRON—
John. *Tiverton.* mar.1723-32 (B. 1737). C.
John (?II). *Tiverton.* 1800.
WALDVOGEL—
Anthony. *London.* 1832-81.
Matthew. *Bury.* 1848-58.
Paul. *London.* 1881.
WALE—
Francis. *Reading.* 1830-37.
Frederick W. *Leicester.* 1876.
WALES—
James. *Leeds.* 1837-53. W.
& McCULLOCH. *London.* 1863-81.
Samuel H. *Providence, RI, USA.* 1849-56. W.
WALFORD—
Charles. *Kidderminster.* 1828-42.
Henry. *Banbury.* 1778-1861. C. (?Two such).
John A. *Leek.* 1850-66.
John George. *Banbury.* Est'd. 1814-53. C. & W.
Thomas. *London.* a. to Elisha Dodd. CC. 1690
(B.-1716).
William. *Brackley.* 1841-54.
William. *Buckingham.* 1854-77.
WALING, N. C. *Vaudreuil, Canada.* 1851.
WALIS, Barent. *Amsterdam.* ca.1750. C.
WALK, Mark. *London (Islington).* ca.1740. C.
WALKDEN—
James. *Blackburn.* 1834-58.
John. *Blackburn.* 1851.
Thomas. *London.* a.1682 to Solomon Bouquet, CC.
1694 (B.-1719).
WALKER—
——. *Stokesley.* C. Dates unknown.
A. *Brockport, NY, USA.* 1831.
A. *Silsden.* 1880-1902. Clock repairer.
Abraham. *Ilford.* 1828.
Benjamin. *Liverpool.* 1828-34.
& BLUNDELL. *London.* 1832. Also T.C.
& CARD. *London.* 1844.
& CARPENTER. *Boston, USA.* 1807.
Charles. *Coventry.* 1828.
D. *Kineton.* 1860-68.
Edward. *London.* 1832. Watch cases.
& FINNEMORE. *Birmingham.* 1808-11. Dial makers.

WALKER—*continued.*
Francis. *Maryport.* 1828-48.
Frederick. *London.* 1851-75.
George. Name used by New York Watch Co. ca.
1871.
George. *Fletton* nr *Peterborough.* 1877.
George. *London.* a.1676 to Thomas Creed. CC. 1684.
George. *Oxford.* 1689. Lant. clocks.
Henry. *Barnoldswick.* 1871.
Henry. *Nottingham.* 1864-76.
& HUGHES. *Birmingham.* 1815-35. Clock dial makers.
Isaac. *Long Plain, Mass, USA.* ca.1800.
J. *Chapel-en-le-Frith.* 1849.
J. *Coventry.* 1854-60. Watch case maker.
J. *Mildenhall.* 1875-79.
J. H. *Grimsby.* 1868.
J. W. *Birmingham.* 1860.
James. *Dunmow.* (B. 1785-91) 1793. C. & W.
James. *Ellon (Aberdeenshire).* 1860.
James. *Leeds.* 1817-22.
James. *Newport (Mon.).* 1848-71.
James. *Penicuik (Scotland).* 1836-50.
James. *Wigan.* 1844.
James Thomas. *London.* 1832.
John. *Bradford.* 1871.
John, sen. *Colchester.* b.1707-80. C.
John, jun. *Colchester.* b.1758. s. of John, sen. -1795,
then *Woodbridge* -1818.
John. *Darlington.* 1834.
John. *London.* 1811-81 (?two such).
John. *London.* ca.1890. T.C. (May be above man.)
John. *Long Sutton (Lincolnshire).* 1850.
John. *Maryport.* 1858-79.
John. *Middlesbrough.* 1840-47. W.
John. *Pocklington.* ca.1770. C.
Jonadab. *London.* a.1678 to Michael Cornish. CC. 1687
(B.-1729). W.
Joseph. *Workington.* 1828-69. Also jeweller.
Joseph. *London.* 1875-81.
Joseph. *Prescot.* 1826. W.
Julius. *Buffalo, NY, USA.* 1840-48.
Lewis. *Stamford.* Free 1720. W.
Matthew. *Aughnacloy (co. Tyrone).* 1846.
Matthew. *Litcham (Norfolk).* 1836.
Michael. *Bolton.* 1822.
Mrs. S. *Litcham (Norfolk).* 1846.
Peter. *London.* a.1656 to Samuel Betts. CC. 1663.
Richard. *Eastbourne.* 1862-70.
Robert. *Beverley.* 1834-40. C.
Robert. *Glasgow.* 1836.
Robert. *Hartlepool.* 1856.
Robert. *Maghera* 1824; *Dungiven* 1846. C.
Robert. *York.* 1820-66. W.
Robert. *York.* b.1828 from *Glasgow*-d.1913.
Thomas. *Derby.* 1876.
Thomas. *Fredericksburg, Va, USA.* 1760-75. C.
Thomas. *Howden.* 1834-40.
Thomas. *Hull.* 1846.
Thomas. *Leeds.* 1850-53.
Thomas. *London.* 1828-44
Thomas. *Oxford.* 1665-1715. Rep'd. ch. clock.
Thomas. *Retford.* 1835.
Thomas. *Sculcoates.* b.1743-d.1815.
& THORNTON. *Gloucester.* 1863.
William. *Bradford.* 1853.
William. *Keighley.* 1866-71.
William. *London.* 1839.
William. *Loughborough.* 1828. Clock pinion maker.
William. b. ca. 1812 *Hull. Shrewsbury* 1841-70. C.
William S. *Montreal.* 1862-76.
WALKEY, John. *St. Columb.* ca.1800-1817. W.
WALKMAN, William. *Dursley.* 1830.
WALL—
& ALMY. *New Bedford, Mass, USA.* 1820-23.
Benjamin. *Richmond (Surrey).* Late 18c. C.
John. Place unknown. 1857.
John. *Chatham.* 1812. W.
John Golding. *Ross.* 1879.
William. *Coventry.* 1828.
William. *Wandsworth.* 1828-39.
William A. *New Bedford* and *Hanover, Mass, USA.*
Early 19c.

WALLACE—
A. H. Name used by U.S. Watch Co. ca.1870.
Alexander. *Goderich, Canada.* 1851.
Hugh. *Portadown.* 1865.
James. *Kircubbin (Portaferry).* 1846-54.
James. *Sheffield.* d.1727. C.
(Colonel) James. *Whitby, Canada.* Prop. of Canada Clock Co.
John. *Belfast.* (B. 1810-24) 1819-66. W. (Partner of Edward Gribben till early 1820s.)
John. *Coleraine.* mar.1812.
John. *Dumfries.* 1882.
John. *Musselburgh.* 1836.
John. *Pittsburgh, Pa, USA.* ca.1830. C.
Joseph. *Dumfries.* 1882.
Robert. *Croydon.* 1878.
Robert. *Halifax, Nova Scotia.* 1875-81.
Robert. *Newtownards.* 1781-1824. W.
Robert. *Philadelphia, USA.* ca.1860. C.
Samuel. *Larne.* ca.1770-ca.1785. C.
William. *Goderich, Canada.* 1857.
William. *Newtownards.* 1840-46. W.

WALLEN—
Thomas. *Coventry.* 1860-80.
W. *Odiham.* 1848.
William. *Coventry.* 1860-80.
William. *Henley-on-Thames.* ca.1725. W.

WALLER—
James. *London.* From *Horsham.* a.1735 to Charles Cabrier. C.
P. *Southend-on-Sea.* 19c. C.
Robert. *London.* 1828-32.
William. *Briston (Norfolk).* 1875.
William John. *Woolwich.* 1845-66.

WALLEY & WHALLEY—
Joseph. *Liverpool.* 1781. In 1788 dissolved partnership with Robert Jones.
Richard. *Liverpool.* mar.1712-d.1743. W.
Robert. *Bolton.* d.1675. C.
Samuel I. *Manchester.* mar.1733-d.1744. C.
Samuel (?II). *Manchester.* Believed alive 1750. C.
Thomas. *Manchester.* mar.1759. C.

WALLIN—
Nicholas. *Ashby Parva (Leicestershire).* 1826. W.
Robert. *Philadelphia, USA.* 1845.

WALLING, Richard C. *Portsmouth.* 1878.

WALLINGTON, W. *Devizes.* 1848.

WALLIS—
E. J. *Birmingham.* 1860.
Ebenezer. *London.* 1863-81.
George. *Gillingham (Dorset).* 1830.
H. I. *Peckham.* 1855.
J. *London.* ca.1830.
James William. *Camberwell.* 1862-78.
Richard. *London.* 1869.
Richard. *St. Austell.* 1748. W.
Richard. *Truro.* 1735-d.1770. C.
William. *Leamington.* 1880.
William. *St. Austell.* 1741-d.1750. C. & W.
William J. *London.* 1857.

WALLIT, Richard. *London.* a.1686 to Thomas Fletcher. CC. 1693 (B.-1709).

WALLS, John. *Canterbury.* ?Late 18c. C.

WALLWOOD, William. *Richhill (co. Armagh).* ca.1760-75. C.

WALLWORTH, H. *Crewe.* 1878.

WALM(E)SLEY—
Edward. *Manchester.* 1848.
Mr. *Ormskirk.* 1711. W.
Robert. *Sale.* 1878.
William. *Maidstone.* a.1800.

WALPOLE, Thomas. *London.* ca.1750.

WALSALL (?Walshaw), William Robert. *Idle (Yorks, West Riding).* 1838. Later at. *Knaresborough.* See *Walshaw).*

WALSH—
——. *Forestville, Conn, USA.* ca.1825.
——. *Londonderry.* ca.1800. C.
——. *Reading.* ca.1850. C.
Arthur. *London.* 1857.
Arthur Paul. *London.* 1869-75.
Edward. *Abergavenny.* 1868.

WALSH—continued.
Henry. apr. 1827 to Robert Howard Fish of *London. Newbury* 1835-39. C.
Henry. *Reading.* 1837.
James. *Montreal.* 1864.
& OLIPHANT. *London.* 1851.
Ralph. *Newry.* 1778.

WALSHAW (see also Walsall).
M. *Bradford.* 1866.
William Robert. *Knaresborough.* From *Idle.* 1851-71. T.C.

WALSHAY (may be corruption of Walshaw), I. Town unknown, prob. *West Riding.* 1789. C.

WALTER—
Albert. *Treorchy.* 1887.
Augustus. *Wells.* 1875-83.
Edmund. *Burford.* 1853. W.
Edmund. *Faringdon.* 1853.
Edmund. *Witney.* 1853. W.
Edward. *Caterham.* 1878.
Edward. *Godstone.* 1839.
Edwin S. *London.* 1869.
Edwin S. *Wimbledon.* 1878.
Jacob S. *Baltimore, USA.* 1782-1865.
James William. *Godstone.* 1828.
M. F. *Hartford, Conn, USA.* ca.1850.
William. *Stourport.* 1850.
William John. *Woolwich.* 1847. W.

WALTERS—
Charles D. *Harrisburg, Pa, USA.* ca.1850-ca.1860.
David. *Llandeilo.* 1814-44.
E. *Stafford.* 1868.
George. *Pontypool.* 1868.
John. *London.* a.1638 to Thomas Howse, CC. 1645 (B.-1697).
John. *London.* 1857.
Nicholas. *London.* 1622. Petitioner for CC. (B.-1630). W.
Richard. *Llandeilo.* 1868.
W. *Ashbourne.* 1855-64.
William. *Clun.* 1842-56.
William S. *London.* 1851.
William Shaw. *Colnbrook (Buckinghamshire).* 1869-77.

WALTHAM Clock Co. *Waltham, Mass, USA.* Sold 1913 to Waltham Watch Co.

WALTHER, Julius. *Hamilton, Canada.* 1856.

WALTON—
Christopher. *London.* 1839.
E. *Stafford.* 1860.
H. R. *Bristol.* Early 19c. C.
Henry Mabson. *Brighton.* 1862-70.
Hiram. *Cincinnati, O, USA.* ca.1820.
James. *Lancaster.* b.1832-51.
James. *Barford* 1797. *Warwick* 1826-51.
John. *Alston.* 1834-48.
Percy. *Brighton.* 1878.
Phillip. *Cowbridge.* ca.1710. C. (One such a.*London* 1707).
Richard. *Waddesdon (Buckinghamshire).* 1869-77.
S. B. *Livermore Falls, Me, USA.* ca.1850. Year clock.
Thomas. *Birkenhead.* 1878.
William. *Swineshead (Lincolnshire).* 1828.

WANGLER—
Joseph. *Birmingham.* 1880.
Joseph. *Worcester.* 1868-72.
Luke. *Oxford.* 1852. C.
M. *Banbury.* 1842.
Richard. *London.* 1869-81.
W. *Yarmouth.* 1858.

WANLESS, John. *Toronto.* 1867-77.

WANSEY, ——. *London.* 1646.

WARBURTON—
Ellis. *Stockport.* 1878.
John. *Liverpool.* 1824-51.
Joseph, jun. *Bollington (Cheshire).* 1878.
Thomas & Son. *Bollington (Cheshire).* 1857-78.
William. *London.* a.1685 to Thomas Besely. CC. 1693 (B.-1701).

WARD—
The. 1914 and later, product of Pequegnat Clock Co.
Abraham. *Grimsby.* 1835.
Anthony. *Truro.* 1705. Later *USA.* C.
Benjamin. *Snaith.* 1858.

WARD—*continued.*
Benjamin. *Swinefleet*, nr *Goole.* 1871.
Benjamin. *Toronto.* 1856-58.
C. *Worcester.* 1860.
Edward. *Grimsby.* 1828.
Edward. *London.* CC. 1638-54.
Edward. *Maidstone.* 1874.
Edward H. *Philadelphia, USA.* 1839-42.
G. *Toronto.* 1877.
George. *Helmsley.* mar.1784-ca.1810. C.
Henry. *Evesham.* 1850-60.
Henry. *York.* 1830-66. W.
Isaac. *Philadelphia, USA.* 1811-18.
J. *Birkenhead.* 1865.
J. *Sutton-in-Ashfield.* 1849-64.
J. & Co. *Philadelphia, USA.* 1843.
J(John) & W(William). *Philadelphia, USA.* 1839-42.
J. & W. L. & Co. *Philadelphia, USA.* 1839.
James. *Hartford, Conn, USA.* ca.1800.
James. *Stratford-on-Avon.* 1828-35. Also optician.
John. *Boroughbridge.* 1826. W.
John. *Liverpool.* (B. 1825) 1828. W.
John. *Patrington (East Yorks.).* 1846.
John. *Philadelphia, USA.* 1803-48.
John Henry. *London.* 1875.
Joseph. *Grimsby.* 1868-76.
Joseph. *New York.* 1735-60.
Lauren. *Salem Bridge, Conn, USA.* ca.1830.
Lewis. *Salem Bridge, Conn, USA.* ca.1830.
Macock. *Wallingford, Conn, USA.* 1702-83. C.
Nathan. *Fryerburg, Me, USA.* ca.1800.
Richard. *Dublin.* 1786-88. W. & C.
Richard. *London.* (B. 1817-40) 1828-63.
Richard. *Prescot.* 1794. W.
Richard. *Salem Bridge, Conn, USA.* 1829.
Thomas. *Baltimore, USA.* 1755-77.
Thomas. *Ipswich.* 1865-79.
Thomas. *Lisburn.* 1784.
Thomas. *Spilsby (Lincolnshire).* 1828.
Thomas. *Warrington.* mar.1773. W.
Thomas. *Wrexham.* ?18c. C.
Thomas Henry. *Cardiff.* 1868-99.
W. W. *Winnsboro, SC, USA.* 1841.
William, sen. *Grimsby.* 1850-61.
William, jun. *Grimsby.* 1861-76.
William. *Litchfield, Conn, USA.* ca.1830. C. and silversmith.
William (Henry). *Liverpool.* 1824-28. (B. 1796-1829.)
William. *London.* 1857-81.
William J. *Toronto.* 1875-77.
William L. *Philadelphia, USA.* 1831 and later.
WARDE, William. *Skipwith.* b.1582-mar.1640. C.
WARDEN—
Hugh. *Ballycastle* and *Ballygraney (co. Down).* ca.1797-1810. C.
J. *Waltham Abbey.* 1855.
Samuel. *London.* 1828.
WARDY, Jean. *London.* ca.1730. C. (Prob. WADY?)
WARDLAW, Henry. *Liverpool.* 1814-34. (B. 1805-29). C.
WARE—
Beacon. *Salem* and *Greenwich, NJ, USA.* 1789-1820.
D. T. & Co. *London, Canada.* 1851-67.
George. *Camden, NJ, USA.* 1820-22.
George. *London.* 1863.
P. T. & Co. *Hamilton.* (b.1821 *Port Hope*). 1851-58.
Robert. *London.* a.1693 to Samuel Steevens. CC. 1701 (B.-1712).
William. *Milbourne Port (Dorset).* 1824.
WAREHAM—
Charles & William. *Coventry.* 1880.
Ethelbert. *Wimborne.* 1875.
John. *London.* (B. 1805-25) 1828.
WARFIELD—
Alexander I. *London.* ca.1679-87. C. Blacksmiths Co.
Alexander II. *London.* s. of Alexander I to whom a.1683. CC. 1692 (-B. 1719).
J. H. *Baltimore, USA.* ca.1827.
John. *London.* (B. a.1629-1662). *Fleet Street* 1646.
WARIN, William. *Thirsk.* ca.1760-ca.1770. C.
WARING—
Charles John. *Liverpool.* 1848.

WARING—*continued.*
George. *Doncaster.* 1834. C.
George. *Liverpool.* d.1783. W.
George. *New York.* 1848.
Henry. *Warrington.* mar.1776. W.
James. *Stockport.* 1828-34.
Thomas. *Waringstown (co. Armagh).* Supposedly 1686-1733. C.
WARK, William. *Philadelphia, USA.* 1848.
WARLAND, Edmund. *Handsworth.* 1876.
WARLEY, T. *Fakenham.* 1858.
WARLOCK—
Daniel O. L. *St. John, New Brunswick.* 1857-64.
James Aaron. *St. John, New Brunswick.* 1840.
WARMINGHAM, Andrew. *Manchester.* mar.1747. C.
WARMISHAM, William. *Manchester.* 1828-34.
WARMUNDE, Charles. *Halifax, Nova Scotia.* 1873-78.
WARNE—
A. *West Cowes.* 1848.
Alfred. *Newport (Isle of Wight).* 1848-78.
Edwin. *London.* 1851.
Isaac. *Southsea.* 1878.
WARNER—
Albert. *Bristol, USA.* 1857-88. C.
Catherine. *Evesham.* 1842.
Cuthbert. *Baltimore, USA.* 1799-1807, then *Philadelphia, USA.* 1837.
Elizabeth. *Evesham.* 1828-35.
Elijah. *Lexington, Ky, USA.* 1818-29. C.
George. *Campden (Gloucestershire).* 1842. Also beer retailer.
Isaac. *Campden (Gloucestershire).* 1830-40.
J. *Bridport.* 1855.
Joel S. *Cornwall, Canada.* 1857-63.
John. Place unknown. ca.1750. C. English provincial.
John senior. *Evesham.* Believed b. 1757, d. ca. 1820.
Succ. by widow. Elizabeth 1822-35. *(High Street).* C.
John. *Kington.* 1835.
John. *London.* b.ca.1660, a.1675 to Henry Wynn. CC. 1682. C.
John. *London.* s. of William. a. 1689 to father. CC. 1696. (B.-1716).
John. *New York.* 1790-1802.
John. *Shrewsbury.* (b. ca. 1805) 1840 -d. 1844. W. (succ. to John Savage).
John Smith. *Moreton-in-the-Marsh.* 1840-79.
(George) & REED. *New York.* 1802.
Richard. *Burford.* 1790. C.
Robert. *Pershore.* b. ca. 1798 -d. 1853. C.
Robert. b. ca. 1840 *Pershore*, s. of Robert senior.
Apr. to W. Counsell of *Faringdon.* To *Mumbles (Wales)* post 1861 -d. 1891. C. & Baptist minister.
(George J.) & SCHUYLER. *New York.* 1798.
Thomas. *Chipping Campden.* 1850-79.
Thomas, jun. *Chipping Campden.* 1879.
Thomas. *Cincinnati, O, USA.* ca.1820.
Warren. *Cincinnati, O, USA.* ca.1830.
William. *Leicester.* 1876.
William. *London.* 1689-96.
WINTHROP & Co. *Bristol, USA.* 1853.
WARNICA, Leonard B. *Stayner, Canada.* 1871.
WARNOCK—
Samuel. *Saintfield.* ca.1825-43; *Comber* 1846. C.
T. N. *Barrie, Canada.* 1851-58.
WARR—
Jabez. *Luton.* 1839-47.
William. *Market Street (Bedfordshire).* 1839.
WARREN—
A. *Birmingham.* 1854.
C. *Canterbury.* ca.1770. C.
E. *Swindon.* 1848-59.
Edward De. *Sandwich.* From *Poole (Dorset).* b.1808-61.
Frederick. *Attleborough (Norfolk).* 1875.
G. *Woolwich.* 1866.
George. *Plumstead (Kent).* 1851-66.
George. *Worcester.* 1872.
George Frederick. *Birmingham.* 1868.
Henry. *Stamford.* 1857-1909. W. and optician.
James. *Birmingham.* 1880.
James. *Canterbury.* 1778-1832. W.
James. *Dublin.* 1786-88. (B. 1795).
John. *Attleborough.* 1836-65. C.
John. *Canterbury.* 1874.
John. *London.* 1875-81.
John. *Syston (Leicestershire).* ca.1790. C.

WARREN—*continued.*
Joseph. *Ixworth (Suffolk).* 1830-58.
R. J. *Woolwich.* 1866-74.
Richard. *London.* a.1659 to Richard Bowen. CC. 1668.
  (B. -d.1702). W.
Robert. *Little Houghton (Northamptonshire).* Also
  Daventry. 1727.
Robert. *Pershore.* 1828.
& Son. *Canterbury.* 1838-45.
William. *Coventry.* 1880.
William. *London.* 1863-81.
William. *Olney.* b.1790-1851. W.
WARRINGTON—
John. *Philadelphia, USA.* 1811-33.
John & Co. *Philadelphia, USA.* 1828-31.
S. (Samuel) R. *Philadelphia, USA.* 1828 and later.
S. R. & Co. *Philadelphia, USA.* 1841.
WARRY—
John, sen. *Bristol.* ca.1775-1830. C.
John, jun. *Bristol.* s. of John, sen. 1830-42(-56?).
John & Michael Bevan. *Bristol.* 1850-79.
W. B. *Bristol.* 1856-63.
WARSCHAUER, Sally. *Birmingham.* 1880.
WARSFIELD, John. *Lewes.* From *Horsham.* a.1720 to
  Thomas Barrett. C.
WARSLEN, Samuel. *London.* 1828.
WARTER, Robert. *York.* 1515. Made *Minster* clock.
WARWICK—
Samuel. *London.* 1851-81.
Stephen. *London.* 1869-75. (Successor to Samuel).
Thomas. *Birmingham.* 1828-54.
WASBROUGH—
HALE & CO. *Bristol.* 1830-42.
William. *Bristol.* pre-1755-67.
WAS(H)BOURNE—
Daniel. *Bristol.* s. of John of *Gloucester.* a.1702-09.
John. *Gloucester.* ca.1687-1710. C.
Samuel. *Great Marlow.* 1838-42.
WASHINGTON—
Lady. Name used by American Watch Co. ca.
  1880.
Watch Co. *Washington, DC, USA.* 1872-74.
WASON, William. *Belfast.* 1693-1723. T.C.
WASS, John. *Hull.* pre-1753-ca.1765. C. & W.
WASSALL, Joseph. *London.* 1832.
WASSELL—
Charles Frederick. *London.* 1869.
& HALFORD. *London.* 1875.
WASSON—
David. *Bristol.* s. of Solomon. 1659-1670.
Robert. *Ballymena.* 1843-46.
Solomon. *Bristol.* Free 1642-59.
WASTE, George. *Waltham Abbey.* ca.1730. C.
WATERBURY—
Clock Co. *Waterbury, Conn, USA.* 1857-1944.
Clock Co. *Winsted, Conn, USA.* ca.1850.
Watch Co. *Waterbury, Conn, USA.* 1878-98. Later
  became New England Watch Co. ca.1912.
WATERFALL—
Richard. *Wimborne.* 1867-75.
William. *Coventry.* 1828-35.
WATERHOUSE—
A. *Ticehurst.* 1878.
Frederick. *London.* 1869.
G. *London.* 1844-51.
George. *Birmingham.* 1818-20. Dial makers.
William. *Coventry.* 1850. Watch case maker.
WATERLOW, William Henry. *London.* 1881.
WATERS—
——. *Belfast.* ca.1770-85. C.
Henry David. *London.* 1857.
John. *London.* a.1674 to Nicholas Beck. CC. 1682.
  (B. -1705). W.
Thomas. *Frederick, Md, USA.* Late 18c. C.
William C. *Milnathort (Kinrosshire).* 1834-60.
WATHEN, Thomas. *Bristol.* a.1767-80.
WATHEW—
& BUTT. *Chester.* 1857.
Joel. *Stroud.* 1879.
John. *Wednesbury.* 1876.
John F. *Chester.* 1848.
WATKIN. Isaac. *Mountsorrel (Leicestershire).* 1795.

WATKINS—
A. *Middlesbrough.* 1898.
Alexander. *London.* 1844-81.
Daniel James. *Barking.* 1839.
Edwin. *Lichfield.* 1876.
George. *London.* 1869-75.
Henry. *Birmingham.* 1850-80.
Henry, jun. *Birmingham.* 1880.
J. *Barking.* 1851.
James. *Barking.* 1828.
James. *London.* 1863-69.
John. *London.* 1839. Watch cases.
John. *Manningtree.* 1828.
John & Co. *Northampton.* 1877.
John Stickley. *Birmingham.* 1828-60.
Martha. *Manningtree.* 1839.
Rachel. *Abergavenny.* 1822-35.
Theophilus Henry. *Birmingham.* 1842.
Thomas. *Birmingham.* 1868.
Thomas. *Stratford (Essex).* (B. 1809-24) 1828.
Thomas Henry. *Birmingham.* 1835.
W. *Coventry.* 1854-60.
William. *Abergavenny.* From *Cwm-Du.* ca.1750-
  d.1765. C.
William. *Abergavenny.* 1830-50.
William. *Blaina.* 1868-75.
William. *Hereford.* 1844-50.
WATKINSON, H. *St. Helens.* 1858. W.
WATLEY, Samuel. *Neath.* 1875.
WATMORE, William. *London.* Engraver. 1671. (B.
  1672. CC.).
WATMOUGH, William. *Wigan.* 1851.
WATSON—
——. *Lowestoft.* ca.1780. C.
——. *Monaghan.* ?early 19c. C.
——. *Norwich.* ca.1770. C.
——. *Paris Hill, Me, USA.* ca.1840.
——. *Truro.* 1681. Rep'd. ch. clock.
Alexander. *Glasgow.* 1835.
& BELL. *York.* 1834-40.
C. *West Hartlepool.* 1898.
Charles. *Cambridge.* 1830.
Charles. *London.* 1844.
Christopher. *Kirkby Moorside.* 1773. (B. 1775).
Christopher. *York.* 1819-23. Later Watson & Bell, q.v.
  W.
& DICKSON. *Quebec.* 1893.
Edward. *Lincoln.* 1828-35.
Edward. *London.* (B. CC. 1815-40) 1828-63.
Edward. *London.* 1881.
Edward & S. *London.* 1869.
Francis. *Beverley.* 1823-76.
G. *Cincinnati, O, USA.* 1820s.
George B. *Airdrie.* 1860.
Henry. Place believed *Blackburn.* ca. 1755. C.
Henry. *London.* 1857.
Inglis. *London.* 1869-81.
J. *Chelsea and Boston, Mass, USA.* ca.1840.
J. *King's Lynn.* 1846.
James. *Aberdeen.* 1840-60.
James. *Haslingden.* 1848-58.
James. *New London, Conn, USA.* 1769-1806. C.
James. *Philadelphia, USA.* 1820. And later of Hil-
  deburn & Watson.
John. *Accrington.* 1834. W.
John. *Blackburn.* 1822-34. C.
John. *Boston, USA.* 1842.
John. *Bradford.* 1871.
John. *East Retford.* 1876.
John. *Grasmere.* 1822. Clock cleaner.
John. *Kirriemuir.* 1837-60.
John. *Manchester.* 1848-51.
John Forrest. *London.* 1863-69.
Jonathan. *Hornsea.* 1866.
Jonathan. *Hull.* 1858.
Joseph. *Cambridge.* 1830-40. Succeeded by T. Reed.
Joseph. *Hornsea.* 1848. W.
Joseph. *Market Rasen.* 1828-35.
Joseph. *Newcastle-on-Tyne.* 1836.
Luman. *Cincinnati, O, USA.* s. of Thomas. (1790).
  1825-34(-41).

WATSON—*continued.*
Luman & Son. *Cincinnati, O, USA.* post-1830.
Michael. *Newcastle-on-Tyne.* 1801-27. W.
R. *Peckham.* 1862.
& REED. *St. Albans, O, USA.* 1828. C.
Robert. *Alyth (Scotland).* 1836.
Robert. *Newcastle-on-Tyne.* 1834-58.
Robert. *Cardiff.* 1887.
Stephen. *Embsay nr Skipton.* 1871.
Thomas. *Cincinnati, O, USA.* Of Read & Watson. 1809.
Thomas. *Haverfordwest.* 1850-75.
Thomas. *London.* a.1662 to John Henderson (prob. = John Hilderson).
Thomas. *Newcastle-on-Tyne.* 1834-36.
Thomas. *Wakefield.* 1866-71. W.
Thomas. *Yarmouth.* ca.1760. C.
William. *Elland.* 1834.
William. *London.* a.1684 to Sarah, wid. of Humfrey Peirce. CC. 1691-97.
William. *London.* 1828-39. Watch cases.
William. *London.* 1832-63.
William. *Patrington.* 1855.
William. *South Cave.* 1840-51. W.
WATT—
Charles. *Dunkeld.* 1860.
John. *Irving.* ca. 1780-1850. C.
Robert. *London.* 1851.
WATTERS—
John. *London.* 1869.
William. *Inverness.* 1837.
WATTIS, George & Frederick. *Birmingham.* 1880.
WATTLES, W. W. (& Sons). *Pittsburgh, Pa, USA.* ca.1850.
WATTS—
Benjamin. *Cardiff.* 1887.
Benjamin. *Cheltenham.* 1850-63.
Charles. *Bourn.* ca.1770. C.
Charles. *Oxford.* 1823-29.
Charles. *Rochester, NY, USA.* 1844 and later.
E. *Christchurch (Hampshire).* 1848-78.
Edward. *Bournemouth.* 1878.
Edward. *Christchurch (Hampshire).* 1830.
Gilford. *London.* 1881.
Harry. *Poole (Dorset).* 1824-30.
Henry. *Christchurch.* 1839.
Henry Robert. *Westbury (Wiltshire).* 1875.
J. *Peterborough.* 1864.
James. *Lurgan.* Late 18c.-early 19c.
John. *Canterbury.* 1836. W.
John. *Northampton.* 1673.
John. *Stamford.* 1690-d.1719.
Robert. *Stamford.* 1757-59. W.
Stuart. *Boston, USA.* ca.1740. C.
(T.?). *London.* mid-18c. C.
W. G. *Wandsworth.* 1862-66.
Walter. *London.* a.1688 to Charles Halstead. CC. 1695 (B. -1698).
William. *Gravesend.* 1845.
William. *Tenbury (Worcestershire).* 1835.
William F. *Gloucester.* 1879.
WAUGH—
Alexander. *Armagh.* Free 1760-1780. C.
James. *Armagh.* ca.1785-1805. Later *Dublin.* C.
John. *Schenectady, NY, USA.* 1803.
William. *Belfast.* 1835-44.
William. *Liverpool.* 1824.
WAWNE—
John. Town not stated. Same man as at *Kirkby Moorside.*
John. *Kirkby Moorside.* Later 18c. C.
WAY—
John. *Waggontown, Pa, USA.* 1766-96. C.
John. *Wincanton (Somerset).* 1805-37.
Thomas. *Wincanton.* 1837.
Thomas. *Newport (Mon.).* 1850.
WAYCOTT, Peter. *Ashburton.* 1790-1800.
WAYLAND—
Henry (J. & J.). *Stratford (Essex).* 1828-74.
John. *Colchester.* b.ca.1680-d.1718. C.
WAYLETT, Richard S. *London.* 1839.
WAYMAN, J. *Haddenham.* 1846-58.
WAYMONT, James. *London.* 1869.

WEAKLEY, George. *?USA.* ca.1725. C.
WEARE—
Charles. *Chichester.* 1878.
John. *Chichester.* 1870.
Joseph. *Wincanton.* 1813-d.1886.
Josiah. *Wincanton.* 1861-83. (s. of Joseph).
Robert. *Birkenhead.* 1848.
Robert. *Welshpool.* 1840-44.
S. *Birkenhead.* 1865.
W. *Birkenhead.* 1865.
W. S. *Birkenhead.* 1878.
William. *Wincanton.* 1861-83.
WEARN—
Roger. *Hayle (Cornwall).* ca.1785. Prob. worked at *St. Erth.* q.v.
Roger I. *St. Erth.* 1752. C.
Roger II. *St. Erth.* 1748-d.1820. C. & W.
WEATHERALL, Thomas. *Hexham.* (B. 1796-)1827-34.
WEATHERBURN, Robert. *Berwick.* (B. 1820) 1827-48.
WEATHERBY, Thomas. b. ca. 1842 *Drayton in Hales.* *Market Drayton* 1861-1900. W.
WEATHERELL, Charles. *Tipton.* 1835-42.
WEATHERHEAD—
J. *Kirkby Lonsdale.* ca.1730-40. C.
Leonard. *Kirkby Lonsdale.* 1766. C.
Miles. *Kirkby Lonsdale.* 1732-35. C.
R(ichard). *Kirkby Lonsdale.* ca.1750. C.
WEATHERILT, Samuel. *Liverpool.* 1795-1834.
WEATHERL(E)Y—
David. *Philadelphia, USA.* 1805-50 and later. C.
Philip. *Colchester.* b.1751-ca.1798. Then *Ipswich* where d.1819.
WEATHERS, Michael. *Shelburne, Nova Scotia.* 1786.
WEAVER—
George. *Droitwich.* 1850-60. C.
Nicholas. *Rochester, NY, USA.* (1791) 1815-46(-53).
Peter. *Beamsville, Canada.* 1857.
WEBB—
——. *Bampton (Oxfordshire).* 1776.
——. *Thaxted.* ca.1795. C.
——. *Wellington.* ca.1775. C.
& ABRAHAM. *Liskeard.* 1823.
Alfred Martin. *Cardigan.* 1875-87.
C. *Bishop's Stortford.* 1859-66.
Charles. *Exeter.* mar.1728.
Charles. *Hornton (Oxfordshire).* 1850-53. C. & W.
Charles Clark. *London.* 1839-81.
Charles John. *London.* 1875.
E. S. *Southsea.* 1867.
Edward. *Bourton.* ca.1715. C.
Edward. *Bristol.* a.1719-d.1761. C.
Edward. *Chewstoke (Church Stoke).* 1688-93. Lant. clocks.
Edwin. *Birmingham.* 1868-80.
Francis. *Watlington.* ca.1710-32. C.
G. *Newbury.* 1854.
George Alexander. *Cheadle.* 1868-76.
H. *Portsmouth.* 1848.
Henry. *Frome.* 1861-83.
Henry. *Gloucester.* 1870.
& HUDSON. *London.* 1881.
Isaac. *Boston, USA.* 1708. C.
Isaac. *London.* 1668-97.
James. *Bristol.* ca.1680-1711. Lant. clocks.
James. *Frome.* 1760-1814. W. & C.
James. *London.* 1875-81.
John. *Birmingham.* 1868-80.
John. *Wells.* 1684. C.
John. *Haywards Heath.* 1878.
John. *Liskeard.* 1811-15. W.
John. *London.* 1875-81. Clock cases.
Joseph. *London.* a.1650 to wid. of Richard Masterson. CC. 1662-1668.
Joseph. b. ca. 1834 *Ludlow. Shrewsbury* 1851-75. W.
Lawrence. *Thaxted.* 1828-39.
Matthew. *Chewstoke.* 1688. Lant. clocks.
R. *Taunton.* 1861-66.
Richard. *Brecon.* 1868-99. C.
Richard. *Chatham.* 1874.
Robert. *London.* 1828.
& Sons. *Frome.* ca.1800. C.
T. *Walthamstow.* 1866.

WEBB—*continued.*
Thomas. *Cardigan.* 1850-87. C.
Thomas. *Hook Norton (Oxfordshire).* 1788-1819. C.
Thomas. *New Quay.* Same man as at *Cardigan.*
Thomas. *Newcastle Emlyn (Wales).* Same man as at *Cardigan.*
W. H. *Liskeard.* 1856.
William. *Brecon.* Successor to Richard. 1887.
William. *London.* 1828-32. Watch cases.
William. *London.* 1844-69.
William. *Thaxted.* 1851-74.
William. *Warminster.* d.1776. W.
William & Son. *Newbury.* 1830-37.

WEBBER—
Henry. *London.* 1863.
J. *St. Columb (Cornwall).* 1856.
James. *St. Columb (Cornwall).* 1847-73. W.
John. *Falmouth.* 1844-73.
John I. *Falmouth.* 1791-d.1804. C.
John II. *St. Columb.* 1809-11. W.
John. *Woolwich.* 1832-47.
John (?II). *Woolwich.* 1866-74.
John & Son. *London.* 1851-63.
Joseph A. *Cardiff.* 1868.
Lewis. *London.* 1832. Tortoiseshell watch cases.
Lewis J. *London.* 1869-81.
R. W. *Grimsby.* 1868.
Thomas. *Bristol.* 1850.
William. *Falmouth.* 1845. W.
William. *Lavenham.* 1839.
William. *Norwalk, O, USA.* ca.1830. C.
William. *St. Columb (Cornwall).* 1847.
William. *Woolwich.* 1826-55.

WEBER—
Galun. *Prague.* 1581.
Joseph Valentine. *London.* 1857-81.
Louis. *London.* 1881.

WEBLEY—
F. *Feckenham (Worcestershire).* 1868-72.
G. *Bromsgrove.* 1860-76.

WEBSTER—
Charles. *London.* 1844-57.
Ernest. *London.* 1881.
George. *Nottingham.* 1814-40.
Henry (& Sons). *London.* 1857-69(75-81).
J. *Horbury.* 19c. C.
James. *Bradford.* 1853-71.
James. *Edinburgh.* 1840-68. s. of John of *Edinburgh.*
John. *Edinburgh.* From *Peterhead.* 1826-69.
John. *Leslie (Scotland).* 1860.
John. *Prescot.* 1795-97. W.
John. *Shrewsbury.* (B. has 1772, may be not same man). 1828-35.
John. *Whitby.* 1823.
Richard. *London.* 1832. Watch cases.
Richard, jun. & Co. *London.* 1851.
Richard, F.R.A.S. *London.* 1857-81. ('Est'd. 1711'.)
Richard, Son & Co. *London.* 1844.
Son & Co. *London.* 1839.
William. *Prescot* 1824-26. W.

WEEDEN—
William. *Great Chesterford (Essex).* 1851-74.
William Farnham. *London.* 1881.

WEEDON, William. *London.* a.1686 to Nathan Barrow. CC. 1695-97.

WEEK(E)S—
Frank. *Newport (Isle of Wight).* 1867-78.
G. *Havant.* 1859.
G. *Warblington (Havant).* 1867.
J. *Haslemere.* 1855-66.
Jason. *Bangor, Me, USA.* 1840-60.
John. *London.* 1828. Clock case maker.
John. *London.* 1811. Clock case maker. (Prob. same as above.)
Johnson. *London.* a.1671 to Robert Cooke. Math. inst. mr. CC. 1683-97.
T. *Derby.* 1849.
Thomas. *Wednesbury.* 1835-42.
William. *Petworth.* 1855-78.
William. *Trotton (Sussex).* 1862-70.

WEGG—
Konrad. *Innsbruck.* 1542.

WEGG—*continued.*
N. *Deptford.* 1849.
Nathaniel. *London.* 1851-81.

WEHL(E)N—
Gustavus. *London.* 1863-69.
Gustav Paul J. & Co. *London.* 1869-75.
Gustav Paul & Co. *London.* 1881.

WEHRLE/WEHRLEY and variants. Including **WHERLY**—
A. *London.* 1851.
Andrew. *Cambridge.* 1858-75.
Brothers. *London.* 1875-81.
Carl. *Penarth.* 1887.
& Co. *Bishopwearmouth.* 1856.
F. *North Shields.* 1858.
Henry. *London.* 1875-81.
Henry & Co. *London.* 1851.
Matthew. *London.* 1857.
Mathias. *London.* 1875-81.
P. *Ipswich.* 1865.
P. *Leeds.* Late 19c. C.
Paul. *Cambridge.* 1865-75.
Paul. *London.* 1869-75.
Philip & Co. *Leeds.* 1837-53. C.
(M. & C.) & SCHWAR & CO. *London.* 1839.
Sebastian (& Co.). *London.* 1875-81.
Silvester. *Bishop Auckland.* 1851. C.
Silvester. *Crook.* 1856. C.
Silvester. *South Shields.* 1830-38.
Stephen. *London.* 1857-69.
Sylvester. *Newport (Mon.).* 1858.
Theodore. *Leeds.* 1871.

WEICHERT, William. *Cardiff.* 1865-87.

WEIDA, Solomon. *Rochester, NY, USA.* 1847.

WEIDEMEYER, John M. *Fredericksburg, Va, USA.* 1790-1820.

WEIGEL, Henry. *York, Pa, USA.* ca.1820.

WEIGHT—
Edmund. *Aberdare.* 1868. *Merthyr Tydfil.* 1875.
Henry. *Gloucester.* 1840.
Henry. *Malmesbury.* 1842-67. Also T.C.
W. *Sherston Magna (Wiltshire).* 1848.

WEIGHTMAN—
Edwin. *Croydon.* 1878.
Edwin. *London.* 1869-75.

WEILL & HARBURG. *London.* 1869-81.

WEIR—
Alexander. *Coleraine.* ca.1825-40. C.
John. *Newcastle-on-Tyne.* 1759. W.
Robert. *Coleraine.* 1839-54. Also optician.
William. *Dumfries.* 1710-13.
William. *Newcastle-on-Tyne.* 1852.

WEISBARTH, Charles William. *Truro.* 1873. W.

WEISS—
Jedediah. *Philadelphia, USA.* 1777.
Jedediah. b. *Bethlehem, Philadelphia, USA.* 1796-d.1873. C.
Joseph. *Allentown, Pa, USA.* ca.1840. C.

WEISSENBORN, Charles A. *Southampton.* 1878.

WEISSER—
Joseph. *Brighton.* 1839-62. (German clocks).
Joseph. *Gravesend.* 1845.
Martin. *Philadelphia, USA.* ca.1830. C.

WEITTENBORG, Auguste. *London.* 1863.

WEL(L)BO(U)RN(E)—
Edward. *Middlesbrough.* 1866.
Frederick. *Thirsk.* 1866. W.
R. *Worksop.* 1864.

WEL(L)BY—
J. *Northampton.* 1864.
J. *Whitstable.* 1851.
John. *Canterbury.* 1839.
John. *Whitstable.* 1845-51.
O. *Northampton.* 1869.
& Son. *Northampton.* 1877.

WELCH—
Arthur. *Toronto.* 1877.
D. *Queensborough (Kent).* 1847.
E. N. (Manufacturing Co.). *London.* 1875(-81).
Elisha N. b. *Chatham, Conn, USA.* 1809-d. *Forestville.* 1887.
F. *Kingston (Surrey).* 1878.

WELCH—*continued.*
F. *Surbiton.* 1866.
George W. *New York.* 1851-72.
J. *Basford (Nottinghamshire).* 1864.
J. *Hanley.* 1868.
James. *Houghton-le-Spring.* 1834-56. C.
James B. *Burslem.* 1868-76.
John. *Bangor.* 1874-87. W.
John. *Coxhoe nr Sedgefield.* 1827-56. W.
John Thomas. *Port Dinorwic (Wales).* 1887.
Joseph. *Redditch.* 1868-76.
(E. N.) Manufacturing Co. *Bristol, USA.* 1864-1903. Later became Sessions Clock Co.
SPRING & CO. *Bristol, USA.* 1868-84. Partnership of Elisha N. Welch & Solomon C. Spring.
W. T. *Kingston (Surrey).* 1862.
William, sen. *Kingston (Surrey).* 1828-51.
William, jun. *Kingston (Surrey).* 1855.
William. *New York.* 1805.
WELCHMAN, Mordecai. *Bristol.* Late 18c. See Strickland, George.
WELDON—
Oliver. *Bristol, USA.* 1841.
W. *London.* 1764. W.
WELLARD, R. W. *Sandwich.* 1867.
WELLER—
& Co. *Croydon.* 1878.
Richard. *Croydon.* (B. 1808-11) 1828.
Richard. *Eastbourne.* b.1693, a.1718. mar.1726-d.1765. C.
Thomas. *Croydon.* (B. 1802-24) 1828-66.
William. *East Grinstead.* 1870-78.
William Holloway. *Littlehampton.* 1855-78.
WELLES, Gelston & Co. *Boston, USA.* ca.1825. Also silversmith.
WELLINGS, W. *Ellerdine (Shropshire).* 1832-56.
WELLMAN, James. *Charmouth (Dorset).* 1855-75.
WELLMINN, Lawrence. *London.* 1863.
WELLS—
C. H. *Colebrook, Conn, USA.* ca.1840.
Calvin. *Watervleit and New Lebanon, NY, USA.* 1817. C.
D. *Ogdensburg, NY, USA.* 1813.
Edward. *London.* 1875-81. Clock cases.
George. *Shipston-on-Stour.* 1835-42.
(Joseph) & HENDRICK (Ebenezer N.) & Co. *Bristol, USA.* 1845.
J. *Deptford.* 1849.
J. *Frome.* 1861.
J. & F. *Chipping Barnet.* 1859.
J. H. *Brixton.* 1862-66.
James. *London.* 1857-81.
James. *Solihull.* 1842.
John. *Banbury.* mar.1774.
John. *Chipping Norton.* ca.1823. Same man as at *Shipston.*
John. *London.* a.1672 to John White, CC. 1682. (B. -1688.) C.
John. *London.* 1839. Watch cases.
John. *St. Albans.* 1851-74.
John. *Sibford Gower (Oxfordshire).* mar.1785. Then *Shipston.* ca.1790-d.1809. C. & W.
John II. *Shipston-on-Stour.* s. of John I. Succeeded father. b.1787-d.1847.
Joseph A. *Bristol, USA.* 1832-47. C.
Joseph Hadley (& Son). *Chipping Barnet.* 1828(-51).
Matthew. *Bradford.* 1853.
Matthew. *Lurgan.* 1820-46. C.
& MCKENZIE. *Quebec.* 1822.
Neddy. *Shepley nr Huddersfield.* ca.1770. C. (Not *Shipley).*
Nugent. *Newport (Mon.).* 1858-99. Also optician.
Richard. *Solihull.* 1850-80.
Robert. *Ballynahinch.* 1775-83. W. & C.
Stephen. *London.* ca.1740. C.
T. *Ballynahinch.* ca.1800. C.
T. *Cambridge.* 1846-65.
Thomas. *Banbury.* 1832-53. C.
Thomas. *Chipping Barnet.* 1839.
Thomas. *Donaghadee (co. Down).* 1820-30. C.
Thomas. *Shipston-on-Stour.* s. of John. b.1786-d.1855. Worked with bro., John, q.v.

WELLS—*continued.*
Thomas. *Solihull.* 1850.
W. *McIvern Link.* 1876.
Walter James. *Barnet.* 1866-74.
& WILCOX. *Lynn, Mass, USA.* ca.1860.
William. *Gravesend.* 1847-66.
William. *London.* CC. 1689-97. C. *Southwark.*
William. *Solihull.* 1828-42.
William and Charles. *Liverpool.* 1848-51.
WELLSTEED, G. J. *Landport.* 1859-67.
WESBY—
Jonathan. *Prescot.* 1851.
Peter. *Farnworth.* mar.1834. W.
WELSH—
Alexander. *Baltimore, USA.* 1800.
Arthur Paul. *London.* 1863.
Bela. *Northampton, Mass, USA.* 1808. W. and jeweller.
David. *Lincolnton, NC, USA.* 1849. C. & W.
James. *Houghton le Spring.* 1827-51.
John. *Chesham.* Said to be ca.1700. C.
P. F. *Montreal.* 1862.
Robert. *Ballymena.* ca.1775-85. C.
WELSTEAD, Andrew. *London.* 1772. W.
WELTON—
H. & H. *Plymouth, Conn, USA.* Same as H. W. & Co.
Hiram. *Terryville or Plymouth, Conn, USA.* 1841-45. Successor to Eli Terry, jun.
Hiram (Herman) & Co. *Terryville, Conn, USA.* 1841-45. C.
John. *London.* ca.1750-55. W. (?error for Melton).
M. *New York.* 1840-44. C., later lawyer.
WENDELL, ——. *Albany, NY, USA.* 1839-42.
WENDWIN, Aloys. *Vienna.* ca.1800. W.
WENHAM/WINHAM—
David. *East Dereham.* 1830-46.
George. *Watton (Norfolk).* 1836-58.
J. *East Dereham.* 1865.
J. *Swaffham.* 1846-58.
John. *Dereham.* (B. mid-18c.-1795). d.1823.
Mrs. E. *Watton (Norfolk).* 1865.
William. *Newmarket.* 1840.
William. *Wymondham.* 1836.
WENSCH, Laurenz Christian. *Germany.* ca.1677. C.
WENTERSPOCHER, ——. *Cremster.* ca.1755. C.
WENTHAM, John. *Norwich.* 1763. W.
WENTWELL, H. *Charleston, SC, USA.* 1795.
WENTWORTH—
George. s. of Thomas of *Salisbury. Oxford.* b.1692-d.1746. C.
Joshua L. *Lowell, Mass, USA.* 1834-37, with Joseph Raynes.
Robert. *Sarum (Salisbury).* mid-18c. W.
Thomas. *Sarum (Salisbury).* 1692 -d. 1740. C. & W. Succ. by son, Thomas II, who d. 1769.
WENZEL—
Co. *Washington, DC, USA,* also *Baltimore* and *New York.* 1880s.
Herman J. b.*Germany* 1830, to *USA (San Francisco).* 1851-1906.
WERNER, Hans. *Germany.* Early 16c. C.
WERNET, August. *Newport (Wales).* 1875-99.
WESLAKE. See also Westlake—
James. *London.* 1875-81.
John. *London.* (B. 1820-40) 1828-69.
WESONCRAFT, Joseph. *Dublin.* 1695.
WEST—
David. *Quebec.* 1844-48.
E. *Gravesend.* 1851.
& EASTMAN. *Montreal.* 1809.
Edward. *Stafford co. Va, USA.* 1757-88; *Lexington, Ky,* 1788-1827. C. & W.
Edwin. *London.* 1851-81.
George Yaxley. *Docking (Norfolk).* 1830-65.
H. *Southbridge, Mass, USA.* 1830. C., W. and silversmith.
James. *Bolton.* 1824-51.
James. *Sudbury.* ca.1750-ca.1760. C.
James L. *Philadelphia, USA.* 1829-33.
John. *Northwich.* 1878.
John. From *Riccarton (Linlithgow)* 1850 to *Canada,* then to *Oregon, USA,* ca.1860. C.
John Joseph. *Putney.* 1866-78.
Joseph. *Holbeach.* 1876.

WEST—*continued.*
Josiah. *Philadelphia, USA.* 1798-1808.
Richard. *Bexley Heath.* 1839.
Robert. *Leek.* 1868-76.
Samuel Kent, *Swansea.* 1887.
Samuel Thomas. *London.* 1875.
Thomas. *London.* a.1687 to Thomas White, CC. 1695. (B. -d.1723).
Timothy. *London.* 1832. Watch cases.
William. *Bolton.* 1828.
William. *Bexley.* 1847.
William. *Helston.* 1751-1831. Also at *St. Ives* and *Hayle.*
William. *Penzance.* 1832. W.
William. *St. Ives.* 1751-1831. C. & W.
William. *Twyford.* 1877.
WESTAWAY—
H. *Portsea.* 1859.
Henry. *Woolwich.* 1845-55.
John. *London.* 1839.
WESTCOTT—
John. *London.* a.1691 to William Mason, CC. 1703.
(John?). *London.* 1767 C. (B. has one a.1726).
John. *Oxford.* ca.1780. C.
WESTERMAN, Richard. *Leeds.* 1834-53. W.
WESTERN—
Clock Co. Ltd. *Peterborough, Canada.* Incorporated in 1919.
John. *Wragby (Lincolnshire).* 1849-76.
WESTGATE, T. *Eastbourne.* 1878.
WESTHORN—
J. *Northampton.* 1847.
John. *Derby.* 1855-64.
WESTLAKE (see also *Weslake*)—
——. *St. Dominic.* ca.1840. C.
B. *London.* 1863-69.
William. *Chipping Barnet.* 1839-66.
WESTLEY, J. *Soham (Cambridgeshire).* 1858-65.
WESTMORE—
James H. *Ventnor.* 1878.
Robert I. *Fazakerly (Lancashire).* Free 1761-1801. W.
Robert. *Preston.* 1794-1825. W.
Robert II. *West Derby.* s. of Robert of *Fazakerly.* Free 1785. W.
WESTMOR(E)LAND—
John. *Hartlepool.* 1851-98.
W. *West Hartlepool.* 1898.
WESTOBY, John. *London.* a.1669 to Thomas Wheeler, CC. 1677-97.
WESTON—
Benjamin. *Shelton (Staffordshire).* 1835.
H. *Bristol.* 1856.
J. *Norwich.* 1846.
J. *Ragby (?Rugby).* ca.1800?
James. *Ipswich.* 1839. C.
James. *Lewes.* 1759. C.
John. *Coventry.* 1828.
John. *Edzell (nr Brechin).* 1860.
John. *London.* (B. 1805-08?) 1844.
Joseph. *Hyde.* 1834.
Matthew. *Eyemouth.* 1860.
Mrs. Charlotte. *Hastings.* 1862-70.
Mrs. Mary M. *Bristol.* 1870.
Ralph. *Wolsingham (Co. Durham).* 1827-34 (d.1838?). C.
Robert. *Hastings.* 1828-78.
Samuel. *Stratford (Essex). Ham Lane.* 1724-30. C. & W. (One such a. *London* 1698.)
Thomas. *London.* 1857-63.
W. *Coventry.* 1854-60.
W. *Northampton.* 1877.
William. *Coventry.* 1850. Watch case maker.
Wiltiam. *Hinckley.* 1864-76.
William. *London.* 1869.
William. *Stratford (Essex).* s. of Richard of *Limehouse,* a.1730 to Samuel Weston.
William & Son. *Cartmel.* 1848. W.
WESTRAP, H. *Coventry.* 1854-80.
WESTWOOD—
Richard. *London.* a.1684 to Thomas Birch, CC. 1691 (B. -1710). C. & W.
Robert. *London.* (B. 1820-29) 1828-32. W. & C.
Samuel. *Birmingham.* 1880. Dialmaker.

WETHERALL, ——. *Parr.* ca.1740. C.
WETHERELL—
——. *Dudley Port.* ca.1840. C.
John & Co. *London.* 1875.
WETHERILL, J. *Stockton.* 1898.
WEYLANDT, H. *Amsterdam.* ca.1725. C.
WEYMAN/WEYMSS, James. *Appleby.* 1828-34.
WEYSEL, J. M. *London.* ca.1750. C.
WHALE, Roger. *Rugby.* 1880.
WHALEY—
John. *London.* 1839-44.
John Burn. *Houghton le Spring.* 1834.
William. *Hartlepool.* 1851-56.
Wiltiam. *Maidstone.* a.1779.
WHALING, J. *Mossley (Manchester).* 1878.
WHALLEY, Henry. *Wrington (Somerset).* 1883.
WHAM—
G. *Ely.* 1865.
T. *Great Marlow.* 1847.
William. *London.* (B. 1820) 1832.
William Thomas. *Beaconsfield.* 1842. Then at *Great Marlow.* 1847.
WHARIN, William. *Toronto.* 1853-77.
WHARTON—
J. *Grimsby.* 1861.
Mrs. Sarah. *Grimsby.* 1868-76.
W. G. *East Dereham.* 1875.
WHATLEY, Robert. *London.* (B. a.1814, CC. 1821) 1828.
WHEATLEY—
James Atkinson. *Carlisle.* b. 1834, s. of Thomas, 1860-79, and business continues today.
James A. *Penrith.* 1869. Also goldsmith and jeweller.
John. *London.* a.1657 to Jeffrey Bailey. CC. 1668.
Thomas. *Carlisle.* b. 1797, ret'd 1860 d. 1864.
W., jun. *Falmouth.* 1856.
William. *London.* a.1690 to John Steevens, CC. 1698. (B. -1700).
WHEBLE, Thomas. *London.* 1869-81.
WHEELDON, John A. *Barrow.* 1866.
WHEELER—
Cornelius. *Bridgnorth.* b. 1770 d. 1842. W.
D. C. *Dundas, Canada.* 1861.
George. b. 1792 at *Bridgnorth,* s. of Cornelius Wheeler, watchmaker. To *Much Wenlock* by 1828 -50. C. and publican.
J. *Much Wenlock.* 1856.
James. *Colchester.* b.1604-35. C. & W.
John H. *Kettering.* 1854-77.
Jonathan. *Knaresborough.* 1866. W.
Philip. *Worksop.* 1864-76.
Samuel. *Dunnville, Canada.* 1851-65.
Thomas, jun. *Marlborough.* 1830-42.
Thomas. *Preston.* ca.1860. C.
Thomas. *Toronto.* 1846-63.
William. *Lee (Kent).* 1866-74.
WHEELHOUSE, Francis. *Sheffield.* ca.1750-75. W. & C.
WHERLE. See *Wehrle.*
WHELAN—
Charles B. *Pembroke, Canada.* 1866-71.
Christopher J. *London.* 1869-81.
WHELLER, Frank. *Bridgwater.* 1883.
WHETTER, Frederick. *St. Austell.* 1873. W.
WHICHCORD—
John. *Ingatestone (Essex).* 1839. W.
William. *Ingatestone (Essex).* (B. 1787)-1828. C. & W.
WHICHER, Richard. *Mortimer (Berkshire).* ca.1730. C.
WHIFFIN—
Samuel. *London.* 1863.
Swann. *London.* 1869.
WHILLOCK, W. *Bilsworth (Northamptonshire).* 1877.
WHINCUP, William. *Monk Fryston.* 1838. (*West Yorkshire*).
WHIPP—
Thomas. *Bacup.* 1851.
Thomas. *Rochdale.* b.1738-d.1780. C.
Thomas. *Rochdale.* 1834-58. W.
WHISKARD, John. *London.* 1863.
WHISTON—
George. *Stafford.* 1828-50.
Joseph. b. ca. 1797 *Patshull (Staffs). Newport (Shropshire)* 1828-63.
Thomas. b. ca. 1821 *Newport.* -1891. W.

WHITAKER see Whittaker.
WHIT(E)BREAD, John. *Ampthill.* 1830-39.
WHITBY, Robert. *Chester.* 1828.
WHITCHER, John. *London.* 1857-63.
WHITE—
Charles. *London.* 1857-75.
Charles John I. *Bristol.* ca.1800.
Charles John II. *Bristol.* s. of Charles John I, a.1839-46.
Charles Steward. *London.* 1869-81.
& Co. *London.* 1875.
Edward. *Birmingham.* 1816-35.
Edward. *London.* 1863-81.
Edward. *Louth.* 1876.
Frederick. *Leamington.* (b. ca. 1827) 1854-68.
Prob. bro. of Henry.
G. *Bourne (Lincolnshire).* 1861.
George. *Birmingham.* 1860-68.
George. *Blofield (Norfolk).* 1858-75.
George. *Bristol.* 1774-pre-1814. C.
George. *Leicester.* 1876.
George. *London.* 1857.
Henry. (b. ca. 1815) *Leamington* 1849-51.
J. *Coventry.* 1860-80.
J. *Southwold.* 1853.
J. *Warminster.* 1848.
J. A. R. *London.* 1869.
J. S. *Montreal.* 1854-64.
James. *Oxford.* a.1804-12. C.
James. *Wickham Market.* 1830-79.
John. *Caledonia, Canada.* 1861.
John. *Kirton (Lincolnshire).* 1828-35.
John. *Upton (Worcestershire).* 1850-60.
Joseph. *London.* 1832.
Joseph. *Newry.* 1785-87.
& MILLER. *London.* 1869.
Mrs. Emma. *London.* 1881.
Mrs. H. L. N. *London.* 1869.
Philip. *Pembroke Dock.* 1868.
S. *Acle (Norfolk).* 1858.
& Son. *Toronto.* 1877.
T. *Montreal.* 1862.
Thomas. *Witney.* 1741-43. Wooden clocks.
Thomas John. *Haverfordwest.* ca.1840-87. C. & W.
W. H. *Wolfville, Nova Scotia.* 1866.
William. *Cranfield.* 1864-77.
William. *Grimsby.* 1876.
William. *Hull.* 1851-58.
William. *Newry.* 1785-87.
William. *Tewkesbury.* 1842-70. C.
WHITEFIELD, George. *Southsea.* 1878.
WHITEFORD—
———. *Hamilton, Canada.* 1875.
James. *Armagh.* 1843.
James A. *St. Johns, Newfoundland.* 1864-85.
John. *Three Rivers, Canada.* 1851-58.
William A. J. *Three Rivers, Canada.* 1871.
WHITEHALL—
Edwin. *Newport (Mon.).* 1848-75. C.
Robert J. *Newport (Mon.).* 1880.
Thomas Batkin. *Nottingham.* (B. 1814-18) 1835. W.
WHITEHEAD—
Benjamin. *London.* 1869-75.
Benjamin. *Wandsworth.* 1878.
Christopher. *Ripon.* 1837.
Edmund. *Wetherby.* ca.1800-d.1818. C.
Edward. *Knaresborough.* 1826-34.
Edward (& Son). *Sevenoaks.* 1823 (1847-66).
G. *London.* 1839.
G. *Market Harborough.* 1864.
Henry. *Farningham.* 1851-74.
J. *Market Harborough.* 1876.
J. *St. Mary Cray (Kent).* 1839.
James. *London.* 1881.
James. *Sevenoaks.* 1823-38.
Mark. *Chichester.* 1878.
Richard. *London.* a.1663 to Henry Wynne, CC. 1671. (B. -1693).
Richard. *Newry.* 1827.
Thomas. *Warley* nr *Halifax.* 1800 (B. mar.1802)-1830.
Thomas. *Wetherby.* b.1740-d.1792. C.
Thomas. *Wetherby.* 1822-44.
W. *Coventry.* 1880.

WHITEHEAD—*continued.*
William. *Bedale.* 1866. W.
William. *Brackley (Northants).* 1794.
William. *Manchester.* 1828-34.
William. *Stalybridge.* 1834.
WHITEHORN, T. *London.* 1863.
WHITEHOUSE—
Alfred William. *Great Malvern.* 1876.
Edmund. *London.* 1863-69.
George. *Birmingham.* 1868-80.
Samuel. *Lichfield.* 1835-50.
WHITEHURST—
J. *Derby.* 1855.
& Son. *Derby.* 1828-35.
William. *Derby.* 1781-1835. C.
WHITEING, George Henry. *Grimsby.* 1876.
WHITELEY—
John. *Ripponden.* ca.1790-ca.1820. From Bury.
Joseph. *Barnsley.* 1862-71. C.
Thomas I. *Ripponden.* 1829-71. Known as 'Tom Clock'.
Thomas II. *Ripponden.* s. of Thomas I, whom he succeeded. d.ca.1930.
William. *London.* 1881.
WHITERN, William. *Henley-on-Thames,* a.1806.
WHITEROW, John. *London.* 1839-44.
WHITESIDE—
James. *Ormskirk.* 1824-28.
Joseph. *London.* 1857.
WHITEWAY, Foliot. *Ulverstone.* 1864-66.
WHITFIELD—
Edward. *London.* a.1655 to John North, CC. 1663-97 (B. -1700).
George. *Brecon.* 1875-99. W.
James. *Liverpool.* d.1674. W.
James. *Liverpool.* 1734-d.1756. C.
John. *Clifton (Westmorland).* 1759-63. C.
John. *Liverpool.* 1677. W.
John. *Wavertree.* Same as John of *Liverpool.* q.v.
Robert. *Liverpool.* 1678-d.1726. W.
Samuel. *London.* b.1652, mar.1682. W.
Thomas. *Farnworth.* 1818. W.
WHITFORD—
George. *London.* (B. CC. 1810-40) 1828-57. W.
Thomas. *Haverfordwest.* 1771. W.
WHITHAM, Samuel. *Mirfield (West Riding Yorkshire).* 1871.
WHITING, William. *London.* 1881.
WHITLEY, George. *Market Harborough.* ca.1740. C.
WHITLOCK, William. *Blisworth (Northamptonshire).* 1889.
WHITMAN, Josiah. *Chichester.* 1751. C.
WHITMARSH, E. *Trowbridge.* 1875.
WHITMORE—
G. *Leicester.* 1855-64.
George. *Bristol.* 1728. C.
J. *Bewdley.* 1868.
John & Son. *Northampton.* 1864.
Samuel. *Daventry.* 1771-77. W.
& Son. *Northampton.* 1877.
William (& Son). *Northampton.* 1830 (41-54).
William. *Stockport.* 1848.
WHITNALL/WHITNELL—
James. *Newcastle-on-Tyne.* 1834-48.
Thomas. *Newcastle-on-Tyne.* 1848-56. C.
WHITNEY—
A. *Brockville, Canada.* 1857.
Henry. *Brockville, Canada.* 1862.
Marcus. *Quebec.* 1819.
WHITROD—
Benjamin. *Diss (Norfolk).* s. of John of *E. Harling,* b.1820-d.1891. C. & W.
J. *Stoke Ferry (Norfolk).* 1865.
John. *East Harling (Norfolk).* b.1797-d.1873. C. & W.
WHIT(T)AKER—
George. *Providence, RI, USA.* 1805.
George. *Rawtenstall.* 1848-58. W.
George. From *Finningley* to *Thorne.* b.1748-d.1780. C.
Gen. Josiah. *Providence, RI, USA.* ca.1800-05. Partner of Nehemiah Dodge.
Isaac. *Stalybridge.* 1848.
J. *Birkenhead.* 1857.

J. *Bury.* 1858. Clock cleaner.
James. *Bawtry.* ca.1770. C.
James. *Middleton* (nr *Manchester*). d.1720. C.
John. *Camberwell.* 1828.
John. *Warley* nr *Halifax.* d.1714. C.
Last. *Aylsham* (*Norfolk*). 1875.
Richard. *London.* 1863.
Sarah. *Leeds.* 1834-37. Clock dial painter.
& SHREEVE. *Halifax.* ca.1810-17. Dial makers.
Thomas. *Providence, RI, USA.* ca.1820.
Thomas. *Thorne.* mar.1780-92. C.
William. *Camberwell.* 1878.
William. *Halifax.* ca.1820-ca.1840. Dial maker.
William. *Halifax.* 1866-71. C.
William. *London.* 1832-69.
William. *New York.* 1731-55.
WHITTEMORE, J. *Boston, USA.* 1856.
WHITTEN, C. *Coventry.* 1868.
WHITTINGHAM—
  Edward. *London.* 1863.
  William. *London.* 1869.
WHITTINGTON—
  Henry. *Bawtry.* 1834-37.
  Henry. *Blyth.* 1828.
  Henry. *Manchester.* 1848-51.
  Richard. *Bristol.* 1797. Clock engraver.
WHITTLE—
  Edward. *Prescot.* mar.1826. W.
  John. *Liverpool.* 1861. W.
  Peter. *Tyldesley* 1848; *Shaw* nr *Oldham* 1851-58. C.
  Thomas. *London.* a.1671 to Henry Harper, CC. 1683.
    (B. -1691).
WHITTON. Francis. *Norwell (Notts).* ca. 1740. C. Some-
    times Witton.
WHITWELL, Robert. *London.* a.1642 to Robert Grin-
    kin, CC. 1649, (B. -1673), W. at *Whitefriars.*
WHITWORTH, Henry. *Rochdale.* 1834. C.
WHYATT, C. D. *Hadleigh.* 1875-79.
WHYMAN, William. *Wymeswold* (*Leicestershire*). 1876.
WHYTE, Alexander. *Montrose.* 1860.
WICHITA Watch Co, The. *Wichita, Kansas, USA.* 1887.
WICKENS—
  ——. *Rye.* 1828.
  George C. *Southampton.* 1859.
  Obed. *New York.* 1849.
  W. *Southampton.* 1848.
WICKES—
  Alfred Nelson. *London.* (B. a.1821-CC. 1833) 1832-51.
  George. *London.* 1828. Watch cases.
  Henry. *London.* 1857. (Successor to Alfred Nelson,
    W.).
  John. *London.* 1828-32.
  Valentine. *Merton* (*Surrey*). 1878.
  Valentine. *London.* 1869-75.
  William. *London.* 1839.
  William Gibson. *London.* (B. CC. 1814-25) 1828.
WICKLIFF(E), William. *Liverpool.* 1724-34. W.
WICKMAN, E. R. *Borga* (*Porvoo*). Early 19c. C.
WICKS—
  Thomas. *London.* (B. 1805-20) 1828-39.
  William. *London.* 1832.
WIDDIFIELD—
  & GAW. *Philadelphia, USA.* 1820-22.
  William. *Philadelphia, USA.* 1817.
  William, jun. *Philadelphia.* 1820-23.
WIDDOP, W. *Burnley.* 1858. W.
WIDEMAN—
  Harvey. *Markham, Canada.* ca.1902. W.
  Jacob. *Augsburg.* 1650. C.
WIDENHAM—
  & ADAMS. *London.* 1839-44.
  Richard. *London.* (B. 1824) 1832. (W. to the Hon.
    Board of Admiralty).
  Thomas. *London.* 1851-57.
WIDLUND, Martin. *Kokkola* (*?Sweden*). 1782. C.
WIDMER, James. *London.* 1881.
WIECKING, George. *London.* 1844.
WIEDEMEYER, J. M. *Baltimore, USA.* 1800.
WIELAND—
  Charles. *London.* 1832-69.
  Charles. *Swansea.* 1868.

WIELAND—*continued.*
  Frederick. *London.* 1832-75.
  Frederick. *Philadelphia, USA.* 1848.
  Robert & William. *London.* 1828-39.
  & Son. *London.* 1828.
  W. T. & R. G. *London.* 1844.
WIESSNER, Leopold. *London.* 1869-81.
WIGG—
  Nathaniel. *Deptford.* 1847.
  William. *Halesworth.* 1839-79.
  William Smith. *Yarmouth.* 1875.
WIGGALL, George. *London.* 1857.
WIGGIN/WIGGAN—
  Henry. *Newfields, NH, USA.* Early 19c. Clock
    cases.
  Mrs. Rebecca. *Colne.* wid. of Robert. 1848-58.
  Robert. From *Colne* to *Skipton* 1806, then *Colne* again
    1807-pre-1848. C.
  Thomas W. *Birmingham.* 1880.
  William. *Wigan.* 1718. Sundial.
WIGGINS—
  & Co. *Philadelphia, USA.* 1831.
  T. & Co. *Philadelphia, USA.* 1831-46.
  Thomas. *Leytonstoke* (*Essex*). 1839.
  Thomas. *London.* 1832.
WIGGINTON, William. *London.* (B. a.1783-1824) 1828.
WIGGLESWORTH, William. *Leeds.* 1871.
WIGHT—
  Andrew. *Ayr.* 1848.
  James. *London.* (B. 1815-24) 1828-69.
  Richard. *London.* ca.1790. W.
WIGHTMAN—
  Alexander. *Moffat.* 1837-60.
  Allen S. *New York.* 1840.
  David. *Belfast.* ca.1875-94. W.
  James. *London.* a.1663 to Thomas Fenn. CC. 1670
    (B. -1684), *Lombard Street.*
  Thomas. *London.* a.1692 to Henry Hester, jun, CC.
    1701 (B. 1745). W.
  William. *London.* a.1686 to Joseph Windmills, CC.
    1696 (B. -1744). C.
WIGLESWORTH—
  Thomas, sen. *Caistor.* 1828-35.
  Thomas, jun. *Caistor.* 1835-50.
WIGLEY, Henry John. *Dursley.* 1840-42.
WIGNALL—
  George. *Ormskirk.* 1834.
  Henry. *Burnley.* 1851-58. W.
  John. *Ormskirk.* 1798-1848.
  Mary & Sons. *Ormskirk.* Late 18c. C.
WIGNEY, S. *Brighton.* 1878.
WIGSTON—
  David. *Carlisle.* 1879.
  T. *Coventry.* 1854.
  William. *Derby.* 1828. C.
WILBUR—
  Charles. *New York.* 1846.
  Job B. *Newport, RI, USA.* 1815-49.
WILCOCK, Stephen. *Toronto.* 1877-95.
WILCOCKSON, Henry. *Liverpool.* 1834-48. (B. 1821-
    29).
WILCOX—
  Cyprian. *New Haven, Conn, USA.* ca.1820.
  George. *Bradford* (*Wiltshire*). 1875.
  J. *Dyke* (*Lincolnshire*). 1856. Clock case maker.
  John. *Bourne* (*Lincolnshire*). 1850. Clock case maker.
  John. *Woolwich.* 1838-55.
  Sidney. *Birmingham.* 1880.
  Wesley W. *Chicago, USA.* 1881. W.
  William. *Liverpool.* 1848.
  William. *London.* 1851.
  William. *Wolverhampton.* 1860-76.
WILD(E)—
  Abraham. Place unknown. 1694. C.
  Francis. *Beverley.* 1846-58.
  Francis Joseph & Co. *Dundee.* 1844-60.
  John. *Monmouth.* 1852-99.
  Michael. *Ryhill* nr *Wakefield.* Grandson of John Day,
    q.v. ca.1790-1838. C.
  W. *Brighton.* 1855.
WILDBAHN, Thomas. *Reading, Pa, USA.* 1763-1805.
  C.

WILDER—
Ezra. *Hingham, Mass, USA.* s. of John. 1825. mar. 1841-50. C.
Joshua. *Hingham, Mass, USA.* b.1786. mar.1812-1860. C.
Joshua & Ezra. *Hingham, Mass, USA.* 1845. C. & W.
L. H. & Co. *Philadelphia, USA.* 1845. Dealers.
WILDMAN—
Richard. *Lancaster.* Free 1757. W.
William. *Caton (Lancaster).* mar.1796 at *Thornton in Lonsdale.* C.
WILDMORE—
C. D. *Swaffham.* 1836.
C. D., sen. *Sleaford.* 1849-76.
WILES—
Edwin James. *London.* 1881.
& JOHNSON. *London.* 1875.
S. *Marlborough.* 1848.
Stephen. *Chippenham.* 1842-67.
Stephen. *Malmesbury.* 1830.
WILEY, Alexander. *Lisburn.* ca.1785-1825. C.
WILKES—
& BAKER. *Birmingham.* 1815-20. Dial makers.
George Henry. *London.* 1869-81.
Henry. *Bristol.* a.1793.
John (& Son). *Birmingham.* 1803-15 (1822-30). Dial makers.
John. *Birmingham.* 1880. Clock cases.
John. *London.* 1755. W. (-1772).
Robert. *Montreal.* 1866-76.
Robert. *Toronto (from England).* 1847-80.
Samuel. *Birmingham.* 1835-50. Dial maker.
W. T. *Walthamstow.* 1874.
WILKIE—
John. *Cupar, Fife.* 1830-60.
Robert. *Leven (Fife).* (b.1805) 1825-d.1875.
WILKIN, James. *Armagh.* 1854-90.
WILKINS—
Asa. *Wiscasset, Me or RI, USA.* ca.1810.
Charles. *Epsom.* 1878.
E. *Newport (Isle of Wight).* 1859.
Edward. *Newport (Isle of Wight).* 1830-39.
George. *London.* 1768-86, then *Oxford* 1786-98. C.
Jane. *London.* 1828.
John. *Launceston.* mar.1735. C.
John. *Loughborough.* Later 18c. C.
John. *Oakham.* mid-18c. C.
Matthew. *London.* 1844.
Richard Samuel. *Great Malvern.* 1868-76.
Robert. *London.* a.1660 to Thomas Fenn, CC. 1670 (B. -d.1706). W. *Ludgate Street.*
Samuel. *London.* 1832-81.
Samuel Thompson. *Birmingham.* 1880.
Thomas. *Moreton-in-the-Marsh.* 1870-79.
William. *Coventry.* See under Taylor, Charles.
WILKINSON—
——. *Spalding.* ca.1790. C.
Abel. *Halifax.* 1769.
Benjamin. *Birmingham.* 1880.
Charles. *Canton, NY, USA.* 1820s. C.
& Co. *Toronto.* 1871-77.
Henry. *Ribchester (Lancashire).* 1822.
Isaac. *Leicester.* s. of Joseph, whom he succeeded 1846-76. C. (With son, John, 1884-98).
J. W. *Brantford, Canada.* 1861.
James. *London.* 1839.
John. *Alston.* 1848-58. Also gunsmith.
John. *Cartmel.* 1823-38.
John. *Leeds.* 1826-37. W.
John. *Leicester.* 1801. C. (B. 1815-26).
John. *Stoke.* 1842.
John. *Whitby, Canada.* 1853-63.
John & Thomas. *Leeds.* 1866-71.
Joseph. *Carlisle* and *Annan.* 1843.
Joseph. *Coventry.* 1828-50.
Joseph. *Dewsbury.* 1866. W.
Joseph. *Eamont Bridge.* 1829.
Joseph. *Leicester.* 1828-35.
Joseph. *Penrith.* s. of William of *Penrith.* b.1793-early 19c. W. (B. 1820).
Joseph. *Sheffield.* 1862-71.
M. *Nantwich.* 1857.

WILKINSON—continued.
Samuel. *Stamford.* a.1770 to Thomas Rayment.
(Thomas) & SMITH (Henry). *Leeds.* 1850-53. W.
Thomas. *Hull.* 1840. W.
Thomas. *London.* 1832.
W. S. & J. B. *Chicago, USA.* ca.1880.
William. *Leeds.* ca.1775-1807. C. (B. 1809).
William. *Penrith.* (B. ca.1790)-d.1812. C.
William. *Prescot.* mar.1818. W. (B. 1825).
William. *St. Helens.* 1824.
WILKS—
John. *?London.* Early 18c. Lant. clocks.
Joseph. *Liverpool.* 1834.
Richard M. *Wrexham.* 1887.
WILL, Thomas. *Huntly (Scotland).* 1860.
WILLACY, Edward. *Liverpool.* 1828. Gold dial maker.
WILLANS, James. *North Ossett* nr *Dewsbury.* 1866. W.
WILLARD—
Aaron, jun. *Boston, USA.* b.1783. s. of Aaron, sen., succeeded father 1823-64. C.
Alexander Tarbell. *Ashburnham, Mass, USA.* (1774) 1796-1800, then *Ashby, Mass.* 1800-1830(-1850). C.
B. *New York.* 1818.
Benjamin Franklin. *Boston, USA.* s. of Simon, 1803-47. C.
Henry. *Boston.* s. of Aaron, 1802-87. Clock cases.
John. *Boston, USA.* 1803-42.
John Mears. *Boston, USA.* s. of Simon, 1800.
Philander Jacob. *Ashburnham, USA.* 1772-1825, then *Ashby* 1825-40. C.
Sylvestre. *Bristol, Conn, USA.* 1835.
Zabadiel. *Boston, USA.* s. of Simon, jun. a. to father 1841-retired 1870. C. Also lecturer and chemist.
WILLBANK, John. *Philadelphia, USA.* 1839-41. C. Also bellcaster.
WILLBRAHAM, John. *Alfreton.* 1835. C.
WILLBY, John. *Canterbury.* 1839.
WILLCOCK, John. *Sutton (Lancashire).* 1752. W.
WILLCOX, Alvan. *Norwich, Conn, USA.* 1783-1819; *Fayetteville, NC,* 1819, then to *New Haven, Conn,* 1823.
WILLDAY—
Alfred. *Birmingham.* 1880.
Charles James. *Coventry.* 1880.
WILLER, Henry. *Newcastle-on-Tyne.* 1834. German clocks.
WILLERTON, Richard. *Southwell (Nottinghamshire).* 1876.
WILLEY—
Joseph. *Leeds.* 1834. C.
Joseph. *Mosborough* (nr *Rotherham*). 1871. Clock dealer.
WILLIAMS, Watkin. *Trecastle (Wales).* Early 19c. C.
WILLIAMS—
Adrian. *London.* 1881.
Alfred. *London.* 1875.
Arthur. *Falmouth.* 1814-56. W.
Augustus. *London.* 1875-81.
Benjamin. *Elizabeth, NJ, USA.* 1788-94.
Charles. *Coleford.* 1850.
Charles White. *Bristol.* 1846.
& Co. *Coventry.* 1854.
Daniel. *Bangor (Wales).* 1886-90.
Daniel. *Builth.* 1830-44.
Daniel. *Crickhowell.* 1868-75.
Daniel. *Neath.* 1770-82.
David. *Builth.* 1840-44.
David. *East Cahn, Pa, USA.* 1795. C.
David. *Newport* and *Providence, RI, USA.* Early 19c. C.
Edward. *Bilston.* 1876.
Edward. *Caerphilly.* Same man as at *Llancarfan.*
Edward. *Llancarfan.* b.1708-d.1763. W.
Edward. *Maesteg.* 1875.
Evan. *Crickhowell.* 1835-68.
Evan. *Llan-y-crwys* nr *Lampeter.* mid-19c. C.
Evan. *Merthyr Tydfil.* 1822.
Evan. *Trecastle.* 1840. C. & W.
Evan. *Newport (Mon.).* 1830-35.
Evan Griffith. *Bryn-mawr.* 1844.
Evan Griffith. *Newport (Mon.).* 1835.
F. *Landport.* 1867.

WILLIAMS—*continued.*
F. J. *Horn Dean (Hampshire).* 1859.
Frederick. *Carmarthen.* 1868-75.
G. *St. Davids.* 1832. W.
George. *Axminster.* 1790-1800.
George. *Bristol.* 1830-42.
George. *Halifax, Nova Scotia.* 1866.
George. *Montgomery.* 1832-35.
George. *Swansea.* 1848.
George Robert. *Charleston, SC, USA.* 1786.
Griffith. *Brecon.* 1830-44.
Griffith. *Newport (Mon.).* 1760-95. C.
H. *Wimbledon.* 1862-66.
& HATCH. *N. Attleboro, Mass, USA.* 1850s.
Henry. *Harrogate.* 1866. W.
Henry. *Llancarfan.* (b.1727 *Gloucestershire*) 1753-
d.1790. C. & W.
Henry. *Wandsworth.* 1878.
Henry J. *Carmarthen.* 1875-99.
Hinds P. *Boston, USA.* 1860. W.
Humphrey (& Co.). *Caernarvon.* 1856-74(-87). W.
Ichabod. *Elizabeth, NJ, USA.* 1836-42.
Isaac. *Coed-Poeth (Wales).* 1868-74. W.
J. *Bideford.* Late 18c. C.
J. *Coleford.* 1856.
J. *Leeds.* ca.1850-ca.1860. C.
J. *Newcastle-on-Tyne.* ca.1840-50. C.
J. *Salisbury.* 1842-67.
J. Hood. *Haverfordwest.* 1872-87. W.
James. *Bristol.* 1840-50.
James. *Exeter.* mar.1766.
James David. *Merthyr Tydfil.* 1868-87. W.
Jasper. *St. Ives (Cornwall).* 1844-47.
John. *Aberdare.* 1843-1935.
John. *Abergele.* 1835-44.
John. *Aberystwyth.* 1856-58.
John. *Brackley.* 1864-77.
John. *Caernarvon.* Late 19c.-d.1917. C.
John. *Carmarthen.* 1859-99.
John. *Denbigh.* 1822-44.
John, jun. *Denbigh.* 1822.
John. *Haverfordwest.* 1835. C.
John. *Holyhead.* Prob. same man as at *Llangefni.*
John. *Llangefni.* 1856.
John. *London.* 1828-39. Watch cases.
John. *Neath.* 1822-40.
John. *Newcastle-under-Lyme.* 1842-76.
John. *Philadelphia, USA.* 1818.
John. *Tregony.* 1792. W. & C.
John. *Upper Solva (Pembrokeshire).* 1830-68.
John. *Yeovil.* 1875.
John G. *Chicago, USA.* 1882. W.
John G. *Denbigh.* 1841-56. W.
John Shortman. *Bristol.* a.1833-70.
John Thomas. *Callington (Cornwall).* 1856-73.
Joseph. *Adderbury.* b.1762-d.1835. C.
Joseph. *Denbigh.* 1856-74.
Joseph. *Liskeard.* 1873.
Joseph. *Londonderry.* mar.1834-90. C.
& LESLIE (William J.). *New Brunswick, NJ, USA.*
1780-91, then *Trenton, NJ*, 1791-1806. (Also Leslie
& Williams).
Lewis. *Cerrig-y-Drudion.* Later 19c. C.
Matthew. *Penmachno (Wales).* 1874. C.
Nicholas. *Libertytown, Md, USA.* 1792.
ORTON & PRESTON. *Farmington, Conn, USA.*
ca.1830. C.
Owen. *Tremadoc.* 1844.
Rees. *Brecon.* 1868-99.
Richard. *Dolgellau.* 1844.
Richard. *Liskeard.* 1774. C.
Robert. *Bangor.* Prob. same man as at *Llanrwst.*
Robert. *Bristol.* 1830.
Robert. *Llanrwst.* 1844-74. W.
Robert. *London.* 1832.
Robert. *Wrexham.* 1822-44.
Sara. *Harrogate.* 1866. W.
Stephen. *Providence, RI, USA* with Nehemiah Dodge
1799. Later alone.
Stephen. *Wolverhampton.* 1860-68.
T. *Earlstown (Lancashire).* 1858. C.
Thomas. *Axminster.* 1790-1800.

WILLIAMS—*continued.*
Thomas. *Brackley.* 1841-54.
Thomas. To *Flemington, NJ, USA.* 1792-1808.
C.
Thomas. *King's Sutton (Northamptonshire).* ca.1752.
Thomas. *Liverpool.* 1834.
Thomas. *Llancarfan.* d.1728. C.
Thomas. *Manchaster.* 1834.
Thomas. *Pembroke Dock.* 1868.
Thomas. *Pontardulais.* 1868. Later *Aberdare*, where
d.1884.
Thomas. *Preston.* ca.1750-1771. C.
Thomas Henry. *Castleford (Yorkshire, West Riding).*
1871.
Thomas John. *Cardiff.* 1875-87.
& VICTOR. *Lynchburg, Va, USA.* ca.1810.
Walter. *London.* 1875.
Walter M. *Bangor (Wales).* 1887.
William. *Adderbury.* s. of Joseph, b.1793-d.1862. C. &
W.
William. *Aberystwyth.* 1851-55.
William. *Brecon.* 1830-44.
William. *Bethesda.* 1887.
William. *Caernarvon.* b.1726-d.1762.
William. *Cheltenham.* 1840-63.
William. *Cross (Carmarthen).* 1814. W.
William. *Dolgellau.* 1868-87.
William. *Frodsham.* 1828-34.
William. *Lampeter.* 1835-44.
William. *Liverpool.* 1848-51.
William. *Llangefni.* 1813.
William. *Llanidloes.* 1874-87.
William. *London.* Alias of William Hanslap, q.v.
William. *London.* 1804. W.
William (& Co.). *London.* 1844(-63)-75.
William. *Merthyr Tydfil.* 1822-37. C.
William. *Newport (Mon.).* 1858. Also optician and
math. inst. mr.
William. *Pontypool.* 1844.
William. *Portsea.* 1830.
William. *Swansea.* 1887-99.
William F. *Merthyr Tydfil.* 1848.
& PRITCHARD. *Caernarvon.* 1895.
William Pritchard. *Holyhead.* 1835-44.
William Ruffe. *Newport (Mon.).* 1848-52.
WILLIAMSON—
——. *Ulverstone.* mar.1763. W. (B. 1762).
Andrew. *Downpatrick.* 1790.
Charles. *London.* 1851-63.
Christopher. *London.* (B. CC. 1821) 1839-44. W.
& Co. *Toronto.* 1871-77.
Henry. *Baltimore, USA.* 1808.
Henry. *London.* 1844-81.
J. *London.* 1844.
J. *Rochdale.* 1858. W.
James. *Berwick-on-Tweed.* 1827.
James & Co. *London.* 1863.
John. *Armagh.* Free 1742.
John. *Armagh.* 1796-98.
John. *Bold.* 1757. W. (B. 1765).
John. *Downpatrick.* Late 18c.-early 19c. C.
John. *Leeds.* From *London* 1683-d.1748. Maker of high
repute. C.
John. *Liverpool.* 1710-d.1716. W.
John. *London.* 1857.
John. *Stamford.* 1795.
John. *Warrington.* 1824-34.
Joseph. *Rochdale.* 1848-51.
Richard. *Liverpool.* mar.1707-32. W.
Robert. *London.* a.1658 to John Harris. CC. 1666. (B.
MCC. 1698-1714). W. & C.
Robert. *Wilmslow.* 1857-78.
Samuel. *Wigan.* Free 1684-d.1726 at *Cronton.* W.
T. *Stalybridge.* 1857.
T. *Stamford.* 1855.
Thomas. *London.* a.1661 to John Harris. CC. 1668
(B.-84). W.
Thomas Henry. *Castleford.* 1866. W.
William. *Coventry.* 1880.
William. *London.* a.1655 to Thomas Weekes, CC. 1663-
(B. d.pre-1679).
WILLIANN, Theodore. *London.* 1863.

WILLIARME, Pierre (Peter). *London.* (B. from *Geneva*). CC. 1633-54.
WILLIBALD, John. *Abertillery.* 1880.
WILLIN(S), William. *London.* (B. 1800, CC. 1807) 1828-39. Watch cases.
WILLIS—
Charles. *Liverpool.* 1834. Watch dial enamellers.
Ephraim. *Reepham (Norfolk).* 1836-46.
G. *Northwich.* 1857-65.
(J?). *Harthill nr Sheffield.* ca.1740. C.
John. *Burlington, NJ, USA.* 1745-53. C.
John. *London.* 1828-32. Dial makers.
Richard. *Liverpool.* 1834.
Richard. *London.* 1869-81.
Thomas. *Dunmow.* 1855-74.
Thomas. *Leeds.* 1871.
Thomas. *London.* 1839. Dial maker.
Thomas. *Prescot.* mar.1803-23. W.
WILLMAN—
G. & M. *Chester.* 1857.
& IMBERY. *Wolverhampton.* 1860.
J. *Woolwich.* 1874.
John & Philip. *London.* 1863-69.
Joseph. *Bangor.* 1856-87.
Philip. *London.* 1875-81.
R. & Co. *London.* 1857.
Sebastian. *Bangor.* 1856-74.
T. *Upper Sydenham.* 1874.
WILLMER—
J. *Northampton.* 1869.
William. *Olney.* mar.1787-98. W.
WILLMOTT (see also WILMOT), Benjamin. *Easton, Md, USA.* 1797-1816.
WILLOCK, John. *Pittsburgh, Pa, USA.* 1830s.
WILLOCKS, John. *Brechin.* 1860.
WILLOUGHBY—
Benjamin. a.*London* 1676 to Robert Dingley, to *Bristol* where free 1691-ca.1700.
John. *London.* a.1679 to Thomas How. CC. 1686 (B.-1710).
John Edward. *Sandy (Bedfordshire).* 1877.
William. *Windsor.* 1877.
WILLOW (sometimes WELLOW), John. *London.* a.1609 to Robert Grinkin, BC. 1617-d.pre-1655. W.
WILLOX, J. *Wolverhampton.* 1868.
WILLS—
Joseph. *Philadelphia, USA.* (b.1700) 1725-59. C.
Richard. *Truro.* 1752-ca.1805. W. & C.
T. *Cambridge.* 1858.
Thomas. *St. Austell (Cornwall).* 1710-d.1739. C.
William. *Truro.* s. of Richard, 1756-1819. C. & W.
WILLSON (see also Wilson).
Claude Vair. *London.* 1839.
George. *Lincoln.* 1861.
George. *Saxelby (Lincolnshire).* 1861.
George V. *London.* 1828-32.
Joseph. *Lurgan.* ca.1750-60. C.
Mrs. A. *London.* 1844.
Mrs. L. *Lincoln.* 1861.
R. *Sleaford.* 1861.
Richard. *Lincoln.* 1835-50.
& WOODRUFF. *Somersham (Huntingdonshire).* 1864-77.
WILMIRT, John J. *New York.* 1793-98.
WILMOT (see also WILLMOT).
George. *London.* a.1651 to John White, CC. 1670.
Joseph. *Northampton.* 1806.
& RICHMOND. *Savannah, Ga, USA.* 1850s.
Samuel. *New Haven, Conn, USA.* 1808; *Georgetown* and *Charleston, SC,* 1825.
Stephen. *London.* a.1667 to Edward Staunton, CC. 1674 (B.-1713).
Thomas. *London.* a.1653 to Thomas Loomes, by 1668 was in the *Hague* with the elder Fromanteel.
WILMSHURST—
——. *Sandwich.* ca.1695. W.
James. *Brighton.* 1814. W.
Joseph. *Brighton.* (B. 1822) 1828-39.
Ninyon. *Brighton.* ca.1760. C.
Samuel. *Burwash.* mar.1768.
Stephen. *Basingstoke.* ca.1695-1700. (B. has 1791. May be two such.) C.

WILMSHURST—*continued.*
Stephen. *Oldham.* 1752. C. (May be error for *Odiham,* where one such known pre-1755).
T. *Chichester.* 1855.
Thomas. *London* (from *Mayfield*). a.1713 to Mrs. Finch, widow. C.
WILSDON—
T. W. *Canterbury.* 1865-74.
William. *London.* 1875-81.
WILSHAK, Joseph. *Yarmouth.* 1858-75.
WILSON—
——. *Richmond (Yorks).* ?Later 18c. C.
A. A. *Brighton.* 1878.
Alexander. *Cardiff.* 1761-80.
Alexander. *Gibraltar.* ca.1785. C.
Alfred. *London.* 1844. Clock case maker.
Andrew D. *Providence, RI, USA.* ca.1890-ca.1920. C.
Becket. *Manchester.* 1848.
Charles. *Birmingham.* 1842-68.
David. *Beith.* 1837-60.
David. *London.* 1828. Clock case maker.
& DUNN. *New York.* 1844.
Edmund, jun. *Sheffield.* 1871.
Edmund Skidmore. *Sheffield.* (B. 1825) 1834-62. W.
Edward. *Huntingdon.* 1830.
Edward. *London.* a.1663 to Robert Smith. CC. 1670 (B.-82).
& FAIRBANK. *Bradford.* 1850-53. Formerly Allott & Wilson.
G. *Keswick.* 1858.
G. *Lincoln.* 1868.
& GANDAR. *London.* 1851-75.
George. *Appleby.* 1829-58.
George. *London.* a.1681 to Zachary Mountfort. CC. 1692 (B.-1700). C.
George. *London.* 1863.
George. *Penrith.* 1869-79. Also jeweller.
George. *Peterborough.* 1830-54.
George. *Prescot.* 1851.
George. *Sculcoates.* 1810. C.
& GRAVENER. *Coventry.* 1880.
& HAYNES. *Stamford.* ca.1840. C.
Hosea. *Baltimore, USA.* 1817.
Hugh. *Rothesay.* 1860.
J. *Castleton.* 1855.
J. *Chatham.* 1851.
J. *Chichester.* 1851. W.
J. *Gainsborough.* 1868.
J. T. *Stamford.* 1861-68.
James. *Askrigg (North Yorkshire).* ca.1746-d.1781. C.
James. *Belfast.* b.1720s, wkg ca.1749-d.1789. Numbered clocks known 4-549. Claimed to have imported first white dials into *Belfast* in 1782. C.
James. *Belfast.* 1854-68.
James. *Birmingham.* 1778-d.1809. Clock dial maker/japanner.
James. *Lurgan.* 1865.
James. *Stamford.* Uncle of Joseph. 1786-99. Then partner with Thomas Haynes to 1803, when he retired. C.
James. *Trenton, NJ, USA.* b.ca.1745. Free 1769-73.
James Lamport. *Woodstock, Canada.* 1875.
Jeremiah. *Gateshead.* 1836-56.
John. *Bourne.* 1828-35.
John. *Bradford.* 1871.
John. *Broughton-in-Furness.* mar.1783-88. C. & W.
John. *Carlisle.* 1858-79.
John. *Chatham.* 1845-51.
John. *Durham.* 1770.
John. *Kilmarnock.* ca.1760-80.
John. *Lincoln.* 1850.
John. *Liverpool.* 1848.
John. *London.* 1637.
John. *London.* 1844-57.
John. *Londonderry.* 1757-73.
John. *Moira (co. Donegal).* ca.1780. C.
John. *Nuneaton.* 1828-54.
John. *Oban.* 1837.
John. *Peterborough.* 1757-95.
John. *Peterborough.* 1830-41.
John. *Sheffield.* 1834-37. W.
John. *Sheffield.* 1871.

WILSON—*continued.*
John. *Tobermory.* 1860.
John. *Wakefield.* Early 19c. C.
John. *Winster (Derbyshire).* 1835.
Joseph. b.1844 in *Isle of Man.* Died *Bradford* 1895.
Joseph. *Chichester.* 1828-62.
Joseph T. *Stamford.* Nephew of James, 1818-55. Also silversmith.
Mrs. Eliza. *London.* 1863.
Mrs. Emma. *Towcester.* 1877.
P. *Keith.* 1846.
P. G. *London.* 1875.
Peter. *Dufftown (Scotland).* 1860.
R. *Lincoln.* 1868.
R. T. *Ballyclare.* 1894. C.
Ralph. *Stamford.* (b.1729) 1795-1801. (Died at *Peterborough* in 1829 aged 100).
Richard. *Keswick.* 1869-73.
Richard. *York.* Free 1586.
Robert. *Belturbet (co. Cavan).* 1824-46.
Robert. *Manchester.* d.1638. w.
Robert. *Nuneaton.* 1860-80.
Robert. *Philadelphia, USA.* 1835.
Robert. *Williamsport, Pa, USA.* 1832.
Robert Emilius. *London.* 1875-81.
Samuel. *Chesterfield.* 1876.
T. *Birkenhead.* 1865.
T. *Burton (Staffordshire).* 1860.
T. *Cambridge.* 1830-58.
T. *Henley-in-Arden.* 1868.
T. *Whitehaven.* 1858.
T. H. *Edinburgh.* 1850.
T. & J. *Philadelphia* or *Nottingham, Pa, USA.* 1796. C.
Thomas. *Barnard Castle.* 1847-51.
Thomas. *Birmingham.* 1842-68.
Thomas. *Bourne.* 1828-35.
Thomas. *Burton-on-Trent.* 1835-42.
Thomas. *Cambridge.* 1823-58. C.
Thomas. *Guisborough.* 1790-1834. C.
Thomas. *London.* a.1651 to Simon Hackett, CC. 1659 (B.-94). W.
Thomas I. *Londonderry.* ca.1760-d.1799. W.
Thomas II. *Londonderry.* 1813.
Thomas. *Manchester.* 1828.
Thomas. *Mauchline.* 1860.
Thomas. *Peterborough.* 1830.
Thomas. *Richmond Hill, Canada.* 1862.
Thomas. *Somersham (Huntingdonshire).* 1839-54.
Thomas. *Spalding.* 1828-50.
Thomas. *Stewarton.* 1837.
Thomas & John. *Guisborough.* 1866.
Titus, jun. *Kendal.* s. of Titus, sen. b.1803-71. W.
W. *Chesterfield.* 1864.
William. *Apperley Bridge* nr *Bradford.* ca.1750. C.
William. *Auchterless (Scotland).* ca.1830.
William. *Cardiff.* 1787-98. C.
William. *Hull.* ?Late 18c. C.
William. *London.* a.1686 to William Ames, CC. 1693 (B.-1712).
William. *London.* 1828-32. Clock case maker.
William. *London.* 1828-32.
William. *Maryport.* 1869-79.
William. *Sheffield.* 1871.
William. *York.* Free 1607. s. of Richard. C.
William Henry. *London.* 1875-81.
William John. *Leeds.* 1866. W.
William & Son. *London.* 1839-44.
William Walls. *Newcastle-on-Tyne.* 1848-56.
WILTBERGER, Charles H. *Washington, DC, USA.* ca.1825.
WILTON—
Walter. *Gloucester.* 1870-79.
William. *St. Day (Cornwall).* ca.1830-47. C.
WILTSHIRE—
Donald. *Wootton-under-Edge.* 1879.
George. *London.* 1869-81.
WIMBLE—
John. *Ashford (Kent).* 1716.
Thomas. *Ashford.* b.1694, a.1716 to John Wimble. C.
WIMPEN, David. *Penryn.* 1736. C.
WINCH—
——. *Maidenhead.* ca.1790. W.

WINCH—*continued.*
Amos. *London.* a.1670 to Robert Halsted, CC. 1677 (B.-1690).
Charles. *Burnham (Buckinghamshire).* 1842.
James. *Maidenhead.* 1830-37.
Joseph. *Uxbridge.* ca.1760. C.
Richard. *Reading.* (B. from 1817) 1830-37.
William. *Reading.* 1854-77.
WINCHESTER, J. *Dover.* 1813. W.
WINDELER, Henry. *London.* 1851.
WINDER—
Edward. *Eastbourne.* 1870-78.
J. *Kendal.* 1875.
John Christopher. *Canterbury.* 1847-74.
Thomas. *Kirkby Lonsdale.* 1839-79. W.
Thomas. *Lancaster.* 1823-34. W.
Thomas, jun. *Lancaster.* Free 1825. W. s. of Thomas, sen.
William. *Lancaster.* s. of Thomas, sen. Free 1830. W.
William Prior. *Reading.* 1864-77.
WINDLE—
Edward. *Prescot.* 1798. W.
William. *Stockton.* 1851-56. C.
WINDRIDGE, James. *Coventry.* 1880.
WINDSOR—
(The). 1914 and later, product of Pequegnat Clock Co.
James. *London.* (B. a.1778, CC. 1787-96) 1828-44.
WINE, A. *Walsall.* 1868.
WING—
Jeremiah. *Braintree.* 1828-39.
Mark. *London.* (B. CC. 1811-24) 1828-51.
Mark. *Stratford (Essex).* 1839. Prob. the *London* man.
Miss S. *Braintree.* 1855.
Moses. *Windsor, Conn, USA.* 1760-1809. C. and silversmith.
WINGATE—
Frederick. bro. of Paine. b.*Haverhill, Mass, USA.* 1782. To *Augusta, Me, USA.* 1806-64. C.
George. *Baltimore, USA.* 1816.
Paine. b.1767. To *Boston. USA.* 1789. To *Newburyport, Mass,* 1803. To *Augusta, Me,* ca.1811. To *Haverhill.* 1817-33. C.
WINKELMAN, James. *London.* 1765. W.
WINKELS, James Orlando. *Whitby.* 1866. W.
WINKLE—
Frederick. *Walsall.* 1876.
J. *West Bromwich.* 1860-68.
James. *Oldbury.* 1850-54.
WINKLER, P. Nth. *Great Malvern.* 1876.
WINN, Robert. *Birmingham.* 1815-39. Dial maker.
WINNIPEG, The. *Canada.* Model of Pequegnac Clock Co.
WINNOCK—
Daniel. *London.* s. of Joshua. a.1695, CC. 1707, d.1726. C.
Joshua. *London.* b.1649, a.1664 to stepfather, Ahasuerus Fromanteel the elder, whose daughter he later married, CC. 1672-d.1718. C.
Joshua. *London.* Grandson of Joshua the elder. a.1716-alive 1725.
WINSER—
Albert. *East Grinstead.* 1870-78.
Albert (& Son). *Brighton.* 1851-70(-78).
F. *Worthing.* 1878.
WINSHIP—
David. *Litchfield, Conn, USA.* ca.1830..Clock cases.
Robert. *Barton-on-Humber.* 1835-68.
WINSLADE, William. *Bridgewater.* 1883.
WINSLOW—
Ezra. *Westborough, Mass, USA.* 1860. C.
Jonathan. b.*Harwich, Mass, USA.* 1765-1847. C.
WINSTANLEY—
Charles. *Holywell.* 1840-56.
Edward. *Mold.* 1835.
Edward. *Preston.* 1822.
Edward. *Wigan.* 1806-d.ca.1828. W.
Ellen. *Holywell.* 1856-68.
Henry. *Brooklyn, NY, USA.* ca.1840.
James. *Liverpool.* 1675(-ca.1715?). W.
Jeremiah. *Holywell.* a.1687-ca.1750.
John. *Holywell.* 1791-1840.
John. *Manchester.* 1834.

WINSTANLEY—*continued.*
John. *Ruabon.* 1835.
Joseph. *Liverpool.* 1848-51.
Michael. *Liverpool.* 1848-51.
Robert. *Holywell.* 1844-56.
Robert. *Ormskirk.* 1822-28.
William. *Liverpool.* 1675-80. W. & watch spring maker.
William. *Neston.* 1828.
WINSTON, A. L. W. *Bristol, USA.* ca.1840. C.
WINSTONE—
Harry. *Cardiff.* 1887-99.
William H. *Cardiff.* 1875.
WINTER—
Abraham. *Hereford.* 1830-44.
Anthony. *Belfast.* 1774.
J. C. *Canterbury.* 1851.
John. *Bristol.* Early 18c. C.
MANTLE & CO. *Tredegar.* 1880.
Richard (& Sons). *Chieveley (Berkshire).* 1854-77.
Richard. *Thatcham (Berkshire).* 1877.
Robert. *Edinburgh.* 1837.
Robert. *London.* (B. 1820) 1832.
Samson. b. ca. 1799 *Prussia, Germany. Wellington (Shropshire).* 1849-63.
Thomas. *Liverpool.* 1828-34.
WINTERBOTTOM, Thomas. *Philadelphia, USA.* 1750.
WINTERHALDER—
——. *Doncaster.* ?19c. C.
——. *Swansea.* 1887.
A. *London.* 1869-81.
Alois. *London.* 1869-81.
Andrew. *London.* 1857.
August. *Pontypool.* 1880.
C. *London.* 1844.
Charles. *Santa Cruz, Cal, USA.* ca.1860.
Cirel. *London.* 1857.
& Co. *London.* 1857.
D. & G. *Maidstone.* 1865-74.
J. *London.* 1844-75.
J. (& Co.). *Leicester.* 1855-76.
John & Joseph. *London.* 1844.
Joseph. *London.* 1875-81.
L. *Kettering.* 1877.
L. & Co. *London.* 1851.
M. *London.* 1857-75.
Matthew. *Falmouth.* 1873. W.
Maximilian. *Pontypool.* 1868-80.
P. *Leeds.* 19c. C.
& QUENNETT. *Sheffield.* 1871.
W. *Belfast.* 1865. Clock cleaner and repairer.
William. *London.* 1881.
WINTERMUTE, O. *Newton, NJ, USA.* ca.1870-ca.1890. W.
WINTERS, Christian. *Easton, Pa, USA.* ca.1800. C.
WIRE—
Samson. *Coggeshall.* 1839.
William. *Colchester.* b.1804-d.1857. W.
WIRRALL, Copley. *London.* a.1637 to Thomas Alcock. CC. 1647.
WISDEN, Edward. *Brighton.* (B. 1822) 1828.
WISE—
A. *Yarmouth.* Early 19c. C.
Doiley. *Bodicote (Oxfordshire).* a.1720-d.1788. W.
Featherstone. *Hull.* 1823-51.
Luke senior. *London.* CC 1694. s. of John Wise senior. b. ca. 1650. mar. *Tilehurst* 1679. To *Reading* by 1691. *Minster Street.* d. 1735. C. & W..
'& Gentleman'.
Matthew. *Banbury.* b.1759-85.Later at *Daventry.* C.
Thomas. *Charlbury.* Early 19c. C.
Thomas. *London.* b.ca.1665. s. of John. a.1678 to father. CC. 1686. (B. -1704). C.
William. *Brooklyn, USA.* ca.1835.
William & Sons. *Brooklyn, USA.* ca.1930.
WISEMAN—
James. *Hamilton.* 1849-60.
Thomas. *London.* 1839-44.
WISH, Henry. *London.* 1875-81.
WISMER, Henry. *Plumstead, Pa, USA.* ca.1798-1828. C.
WISSER, John. *London.* 1875-81.
WISTER, Charles J. *Germantown, Pa, USA.* ca.1820-ca.1860.
WISTON, G. *Stafford.* ca.1790-1800. C.

WITHAM—
Henry. *Abingdon.* 1837.
Mrs. P. *Abingdon.* 1854.
William. *Abingdon.* 1864-77.
William. *Cains Cross (Gloucestershire).* 1840.
William. *Stonehouse.* 1850-70.
WITHER, John. *London.* a. to James Gould. CC. 1699. (B. -1720).
WITHERS—
Ann. *Lewes.* From *Steyning.* Spinster. a.1712 to Isaac Guepin. W.
C. *Bristol.* 1856.
Charles. *Bristol.* 1879.
William. *Bristol.* 1840-70. W.
WITHERSPOON—
Alexander. *Dundas, Canada.* 1861-66.
Alexander. *Edinburgh.* 1831-34. W.
Alexander. *Tranent (Scotland).* 1841.
L. *Dundas, Canada.* 1851-53.
WITHILL, Thomas. *Hotham Sand nr Hull.* 1823-40. C.
WITHINGTON—
Peter. *Mifflinburg, Pa, USA.* ca.1820.
W. B. *Devizes.* 1859.
William. *Taunton.* 1875-83.
WITHNALL, ——. *Holywell.* ca.1780. C.
WITMER, Abel. *Ephrata, Pa, USA.* (1767). 1790-1821. C.
WITSEN, Adolph. *Amsterdam.* ca.1720. C.
WITT—
Dr. Christopher. *Germantown, Pa, USA.* ca.1710-65. C. T.C. Also doctor and scientist.
H. *Southampton.* 1859-67.
I. *Beaminster (Dorset).* 1848.
Isaac. *Rothwell (Northamptonshire).* 1854-77.
WITTE, Samuel. *London.* a.1651 to John Chapman. CC. 1660.
WITTENMYER, Michael. *Doylestown, Pa, USA.* (1772). 1808-46. C.
WITTMAN—
Josiah. *Chichester.* ca.1700. C.
Waltfogel & Co. *Norwich.* 1836.
WITWER, Isaac. *New Holland, Pa, USA.* ca.1850-55. C.
WIXLEY, Alexander. *London.* 1881.
WOFFINDIN, Thomas. *Thorne.* mar.1785-90. C.
WOITOWITSCH, Herman & Gottlieb. *Greenwich.* 1874.
WOLF(F)(E)—
Brothers. *Birmingham.* 1880.
Henry. *Marietta, Pa, USA.* ca.1850. C.
J. *Richmond (Surrey).* 1851.
Mrs. B. *Richmond (Surrey).* 1855.
Thomas D. *Westtown, Pa, USA.* ca.1815.
WOLFLE, Ferdinand. *London.* 1869-75.
WOLHAUPTER, Benjamin. *Fredericton, New Brunswick.* 1873-76.
WOLLARM, J. *Liverpool.* pre-1785. W.
WOLLER—
Charles. *Birmingham.* 1835-68. Also musical boxes.
Charles. *Oldbury.* 1842. *Worcester.* 1850.
Matthew. *Birmingham.* (B. 1801-08) 1816-28.
& STRAUB. *Norwich.* ca.1800? C.
Charles. *London.* 1869.
WOLLMAN, Michael. *Tilsit.* Early 17c. C.
WOLTZ—
George Elie. *Hagerstown, Md, USA.* 1820s.
John. *Shepherdstown, West Va, USA.* Early 19c.
Major John George Adams. *Hagerstown, Md, USA.* (1744) 1770-1813. C.
Samuel. *Hagerstown, Md, USA.* ca.1800.
William. *Oakland, Md, USA.* ca.1825-50.
WOLVERSTONE—
Benjamin. *London.* a.1649 to Richard Rickard. CC. 1656-68. *Cornhill.*
Thomas. *London.* a.1643 to Edward East. CC. 1650-62. (B. -d.pre-1690).
WOMERSLEY—
George. Town not known. Believed *Huddersfield.* ca.1735. C.
George (II?). *Huddersfield.* ca.1790-1814. C.
James. Town not known. Believed *Huddersfield.* ca.1790. C.

WOMERSLEY—*continued.*
Jonathan. *Huddersfield.* Believed s. ot George. 1813-22. C.
William. *Halifax.* pre-1785. Later *Huddersfield.*
WOOD—
——. *Birmingham.* ca.1810-ca.1820. Dial maker.
A. *Ramsbottom* (nr *Manchester*). 1858. W.
Alexander. *Glasgow.* 1836.
Alexander. *Stirling.* 1834.
Alfred. *London.* 1851-81.
Alfred. *Stratford* (*Essex*). 1866-74.
Alfred. *Woodford Green* (*Essex*). 1874.
& ALLAN. *Montreal.* 1866.
Arthur Henry. *London.* 1881.
Benjamin. *Ludlow* b. ca. 1831-68. W.
B. B. *Boston, USA.* 1841. C.
Charles. *Granby, Canada.* 1851.
Charles. *London.* 1875.
Charles. *Montreal.* 1845-48.
David. *Newburyport, Mass, USA.* 1766. mar.1795-ca.1850. C.
Edward. *London.* 1875.
& FOLEY. *Albany, NY, USA.* 1852.
Frank. *Worthing.* 1870.
George. *Armagh.* 1840.
George. *Mortlake.* 1878.
George. *Putney.* 1878.
George Thomas. *Crowland.* 1861-76.
H. *Birmingham.* 1860.
Henry. *Bagshot.* 1878.
Henry. *Brixton.* 1878.
Henry. *Canterbury.* 1826-28.
Henry. *Kidderminster.* 1842-50.
Henry. *Leatherhead.* 1839.
Henry. *London.* 1881.
Henry Samuel. *Canterbury.* 1832.
Henry Samuel. *London.* 1851.
& HUDSON. *Mount Holly, NJ, USA.* ca.1773-ca.1790.
J. *Leicester.* 1864.
J. *Newcastle-on-Tyne.* 1856.
J. *Southampton.* 1859.
J. *Tweedmouth.* 1858.
James. *Great Neston.* 1857-78.
James. *New York.* 1874. Successor to George A. Jones & Co.
James Henry. *Birmingham.* 1880.
James William. *Bristol.* 1879.
James William. *Shrewsbury.* 1870.
John. *Clerkenwell.* b.1793. To *Littlehampton.* Late 1820s. Emigrated to *Canada.* 1832.
John. *Devizes.* 1842-48.
John. *Heckmondwike.* 1866. W.
John. *Liverpool.* 1814-28. (B. 1796-1824). W.
John. *London.* a.1689 to John Nash. CC. 1701. (B. -1743.)
John. *Montreal.* 1807-50. a. in *London, England.* To *Montreal.* 1832.
John. *Mount Holly, NJ, USA.* 1790-1810.
John. *Mountsorrel* (*Leicestershire*). 1876.
John. *Philadelphia, USA.* 1734-38.
John. *Philadelphia, USA.* (1736) 1760-93. C. & W.
John Bolton. *Coventry.* 1860.
John & Son. *Montreal.* 1844-76.
John & Son. *Philadelphia, USA.* 1754-61. C.
Joseph. *Morley* nr *Leeds.* 1871.
Joseph. *Scarborough.* ca.1760-ca.1790. C.
Josiah. *New Bedford, Mass, USA.* Early 19c.
M. *Rockport, Ind, USA.* ca.1840.
N. G. *Boston, USA.* 1856.
P. W. *Quebec.* 1883.
Peter. *Little Hampton* (*Sussex*). 1839.
Peter. *London.* 1828-32.
Richard. *Oxford.* 1676-d.ca.1700. Sundial.
Robert. *Bolton.* 1851-58. W.
Robert. *London.* a.1659 to Jeremy Gregory. CC. 1671-86.
Robert. *London.* (B. 1820) 1828-32.
Robert. *Whitburn* (*Scotland*). 1860.
Robert. *Workington.* (B. 1811-29) 1828-34.
Samuel. *Ashton-under-Lyne.* 1815. C.
Samuel. b. ca. 1786 *Kidderminster.* To *Tenbury* ca. 1823, then *Ludlow* by 1836 -d. 1861. W.
Samuel. *Middlesbrough.* 1847.

WOOD—*continued.*
T. *Clapham* (*Surrey*). 1866.
T. *Knutsford.* 1857. (See Thomas. W.).
Thomas. *Berkhamsted.* 1839-51.
Thomas. *Knutsford.* 1834-48.
Thomas. *London.* a.1682 to Robert Nemes. CC. 1691-97. (B. -1707).
Thomas. *London.* 1869-81.
Thomas. *Reading.* ca.1790.
Thomas. *Southam* (*Warwickshire*). 1835-60.
Thomas James. *London.* 1857-81.
W. *Worthing.* 1878.
W. R. *Newtown.* Later 19c. C.
William. *Bolton.* 1851.
William. *Clerkenwell.* b.1798-1840s. W.
William. *Hartlepool.* ca.1840. C.
William. *Leominster.* 1856-79.
William. *Liverpool.* 1824.
William, jun. *Liverpool.* 1851.
William. *Stanningley* nr *Leeds.* 1871.
William. *Tadcaster.* 1826-34.
William Archibald. *Montreal.* 1898.
William & Sons. *Birmingham.* 1860.
WOODBURN, Samuel. *Southport.* 1851.
WOODCOCK—
Ann. *Colchester.* wid. of William, jun. 1821-29. C.
C. *Salisbury.* 1875.
& Cc. *Baltimore, USA.* 1871.
Frederick. *Shaftesbury* (*Dorset*). 1855-75.
J. *Crowle.* 1861-76.
Joseph Scotter. *Kirton-in-Lindsey.* 1850.
Mrs. Ann. *Huddersfield.* 1871.
Samuel. *London.* 1881.
William. *Baltimore, USA.* 1819-29.
William, sen. *Colchester.* b.1748-d.1802. C. & W.
William Flag. *Colchester.* b.1772-d.1821. C.
William Gonner. *Colchester.* b.1806-33. C.
WOODFORD, Isaac. *New Haven, Conn, USA.* ca.1845.
WOODHAM—
Henry. *Hungerford.* 1864-77.
James. *Hungerford.* 1877.
James William. *Hungerford.* 1830-54.
WOODHAMS—
T. *Sevenoaks.* 1851-55.
Thomas. *Sevenoaks.* 1826-47.
WOODHEAD, Hartwell. *Wibsey* nr *Bradford.* 1871.
WOODHOUSE, Joshua. *St. Hyacinthe, Canada.* 1851.
WOODIN, Riley L. *Decatur, NY, USA.* 1830s. C.
WOODING, Edmund A. *Torrington, USA.* 1831-40. C.
WOODLEY—
Josiah (& Son). *Kineton.* 1828-50. Also *Maltster.*
& WALKER. *Kineton.* 1850.
WOODLOCK, Michael. *Birmingham.* 1868.
WOODMAN—
Maitland W. *Gosport.* 1878.
Mary. *London.* 1839. (wid. of William?).
William. *London.* (B. a.1786-)1832.
WOODMANSEY—
Henry. *Doncaster.* 1858.
Mrs. Sarah. *Doncaster.* 1871.
WOODROFFE, J. *Loughborough.* 1864.
WOODROFE—
Richard. *Liverpool.* 1848.
Richard & Sons. *Birkenhead.* 1865-78. Also *Liverpool.* 1878.
Robert. *Wallasey.* 1848-57.
WOODROW—
R. *Norwich.* 1846.
Robert. *Lowestoft.* 1875-79.
WOODRUFF—
Charles. *Dover.* 1874.
Charles. *Margate.* 1838-66.
Charles & Co. *London.* 1828.
Enos. *Cincinnati, O, USA.* ca.1820-ca.1830.
John II. *New Haven, Conn, USA.* ca.1850.
& Son. *London.* (B. 1822) 1828.
W., C. & C. *Dover.* 1855-66.
& WHITE. *Cincinnati, O, USA.* ca.1840.
William. *Dover.* 1858.
William. *Margate.* 1874.
WOODS—
C. R. *London.* 1844-51.

WOODS—*continued.*
Charles. *Melbourne, Canada.* 1857.
Charles James. *London.* 1857-75.
H. *Chobham.* 1866.
John. *Farnworth.* 1818. W.
John. *Warrington.* mar.1775. W.
Joseph. *King's Lynn.* 1846-75.
M. *Wigan.* 1858. W.
Oliver Ernest. *Caerphilly.* 1887.
Peter. *Liverpool.* 1834.
T. *Chobham.* 1862.
Thomas. *Farnworth.* mar.1765. W.
Thomas George. *London.* 1875.
William. *Farnworth.* mar.1803. W.
William. *St. Helens.* 1787. W.
WOODSTOCK, The. 1914 and later, product of Pequegnat Clock Co.
WOODWARD—
Alfred. *Birmingham.* 1868.
Alfred. *Hanley.* 1876.
Antipas. *Middletown, Bristol, Conn, USA.* (1763) 1791-1812. C. and goldsmith.
Frank. *Derby.* 1855-76.
Henry. *Colchester.* ca.1676-ca.1700.
Horace & Co. *Birmingham.* 1880.
J. *Ashford.* 1857-67.
J. *Derby.* 1849-55.
J. *Kidderminster.* 1872.
J. *Wolverhampton.* 1868.
James. *Philadelphia, USA.* ca.1795. Clock case maker.
John. *Farnworth.* mar.1834. W.
John. *Killamarsh (Yorkshire).* 1876.
John. *London.* (B. a.1803-CC. 1820) 1828-44. C.
John. *London.* 1857-75.
John Richard. *London.* 1875.
Joseph. *Birmingham.* 1880.
Joseph. *Derby.* 1864-76.
Robert. *London.* 1875-81.
T. *Derby.* 1864.
T. *Farnsfield nr Southwell (Nottinghamshire).* 1849-55.
Thomas. *Chichester.* 1870.
Thomas. *Liverpool.* 1851.
Thomas. *London.* 1832.
Thomas. *Worcester.* 1868-76.
Walter John. *Walcott nr Bath.* 1875-1883.
WOODWISS, John. *Birmingham.* 1868.
WOODYEAR, Edward. *Salisbury.* 1830.
WOOG—
Adolf. *London.* 1875-81. (Successor to Jules).
Jules. *London.* 1869.
Moses. *London.* 1863.
Moses & Samuel. *London.* 1857.
Samuel. *London.* 1863-69.
WOOLF—
B. *Charleston, SC, USA.* ca.1800.
E. *Birmingham.* 1854.
Jonas (& Son). *London.* 1857-69(-75).
Lewis. *Liverpool.* 1834-51. Chrons.
Marcus. *London.* 1844-51.
Nathan. *London.* 1857.
WOOLFALL—
Richard, sen? *Liverpool.* 1851.
Richard, jun. *Liverpool.* 1848-51.
WOOLHAUPTER, John. *St. John, New Brunswick.* 1799.
WOOLLARD—
H. *Epping.* 1866.
Thomas. *Cheshunt.* 1839-74.
WOOLLER, Joseph. *Wolverhampton.* 1842-50. German clocks.
WOOLLETT, John. *Maidstone.* 1782. W.
WOOLLEY—
——. *Tenterden.* mar.1789.
C. *West Ham.* 1866.
George. *Bristol.* 1830-63.
George. *Shifnal (Shropshire).* b. 1807, s. of William Woolley, clockmaker. -1836. C.
James. *Codnor (Derbyshire).* b.ca.1700-d.1786. Uncle of John, q.v. Signed clocks Wolley.
John. *Codnor (Derbyshire).* b.1738-d.1795. C. Nephew of James.
John & Co. *Birmingham.* 1860. Importer of Swiss watches.

WOOLLEY—*continued.*
Thomas. *Chester.* 1878.
WOOLIE (also **WOOLLY),** Josias. *Gloucester.* b.1623, mar.1663. W.
WOOLMER, James Shreeve. *Reepham (Norfolk).* 1836-75. C.
WOOLNOUGH—
Frederick. *Liverpool.* 1834. Dial enamellers.
John. *Caernarvon.* 1840-44. C.
Joseph. *London.* 1875-81.
WOOLRICH, John. *Wolverhampton.* 1835-42.
WOOLSEY—
William. *Gainsborough.* 1861-76.
William. *Grimsby.* 1876.
WOOLSON, Thomas, jun. *Amherst, NH, USA.* Early 19c.
WOOLSTON, W. *Yarmouth.* b. 1815 d. 1888. W. & C.
WOOLTERTON, Jerome. *Saxmundham.* 1846-79.
WOOLWORTH—
Chester. *New Haven, Conn, USA.* ca.1840.
R. C. *Philadelphia, USA.* 1816.
WOO(R)SEY, William. *Farnworth and Prescot.* 1824-33.
WOOT(T)ON—
George. *Cardiff.* 1875.
John. *Manchester.* 1828-51. Also imported clocks.
Luke. *London.* 1828-32.
S. *Coventry.* 1854-60.
Thomas. *Birmingham.* 1860-80.
Thomas. *Colchester.* b.ca.1736-d.1797. C.
William. *London.* b.1755, a.1769-76. Then *Chelmsford.* ca.1776-85. Then *Colchester.* ca.1785-d.1818. C. & W.
WORBOYS, Julius. *Baldock.* 1874.
WORCESTER, Martin. *Wednesbury.* 1828-42.
WORDEN, C. M. *Bridgeport, Conn, USA.* ca.1850.
WORDLEY—
Brothers. *London.* 1881.
William. *London.* 1881.
WORGAN, Matthew I. *Bristol, USA.* a.1741-d.1798. C. & W.
WORMALD, John. *Dewsbury.* 1866. W.
WORRALL—
John. *Liverpool.* mar.1708-11. W. (May be same as next man).
John. *Liverpool.* ca.1775. C.
John. *London.* 1839-51.
WORS(T)FOLD, (?John). *Dorking.* 1728. W. (B. has b.1704).
WORTHINGTON—
James. *Lichfield.* 1876.
James. *Liverpool.* 1834.
Thomas. *Burton-on-Trent.* 1860-76.
WORTON—
G. *Coventry.* 1860. T.C.
Robert. *Philadelphia, USA.* 1849.
WOTHERSPON (cf. **WITHERSPOON)**—
James. *Quebec.* 1848.
John. *Glasgow.* 1830.
WRAGG, E. *Ilkeston.* 1855.
WRAIGHT, George. *Eastry (Kent).* 1826-32. C.
WRANGHAM, Francis George. *Gateshead.* 1851.
WRANGLES, Thomas. *Scarborough.* ca.1780-1807. C.
WRAPSON—
Charles. *Chichester.* 1828.
Charles. *Midhurst.* 1839.
Henry. *Havant.* 1830.
James. *Havant.* 1839.
John. *Midhurst.* 1870.
Mrs. M. *Midhurst.* 1851-62.
WRATHALL, James Henry. *London.* 1869-75.
WRATTEN—
C. *Cambridge.* 1858.
Charles. *Chesterfield.* 1876.
Charles. *Thetford (Norfolk).* 1836.
WRAY—
John. *Brigg.* 1828-61.
John. *Ulceby.* 1850-76.
Richard. *Messingham.* 1861-76.
Robert. *Gargrave.* mid-19c. C.
& Son. *Birmingham.* ca.1840-ca.1886. C.
William. *Birmingham.* 1868-80.
William. *Brigg.* 1868.

WRAY—*continued.*
William. *Halifax.* 1682. Rep'd. ch. clock.
WREGHITT (also WRIGHT), John. *Patrington.* ca.1790-1840. C. & W.
WRENCH, J. *London.* ca.1745. W.
WRIGHT—
——. *Linton (Cambridgeshire).* pre-1797. C.
Alfred. *London.* 1875-81.
B. & Co. *Birmingham.* ca.1820. Dial maker.
Benjamin. *London.* a.1678 to Abraham Prime. CC. 1685. (B. -d.1709).
Benjamin. *London.* 1832. Watch cases.
& BENTLEY. *Stafford.* 1835-42. Also gold and silver smiths.
Brothers. *Northampton.* 1877.
Charles Cuching. *New York.* ca.1800-1812. Then *Utica, NY.* 1812-ca.1830.
Christopher. *Birmingham.* 1835-42. Dial maker.
Cook. *North Walsham.* 1865.
Edward. *Birmingham.* 1868.
F. *Yarmouth.* 1875.
Filbert. *Bristol, USA.* 1849-56.
Frederick. *Birmingham.* 1880. Clock cases.
Harvey. *Bristol, USA.* 1831. C.
Henry. *Clapham.* 1839.
Henry or Harry. *London.* 1875.
I. & Son. *Bath.* 1861.
J. & Co. *Bath.* 1856.
John. *Bangor.* 1844-56.
John. *Bristol.* 1740. Sundial.
John. *Doncaster.* Free 1701. W.
John. *Dorking.* (B. 1791-)1828.
John. *Lincoln.* ca.1770. C.
John. *Lincoln.* 1701. W.
John. *Lincoln.* 1828. Clock case maker.
John. *Liverpool.* d.1701. W.
John. *Liverpool.* d.1771. *Island of Nassau, USA.* W.
John. *Liverpool.* (B. 1825) 1834. W.
John. *London.* a.1653 to Thomas Claxton. CC. 1661.
John. *London.* a.1656 to Jeremy Gregory. CC. 1671.
John. *London.* a.1691 to James Markwick. CC. 1700.
John. *London.* a.1693 to Henry Merryman, CC. 1715.
John. *London* at *St. James, Westminster.* 1716. W.
John. *London.* 1844-81.
John. *Manchester.* 1828.
John. *Newcastle-on-Tyne.* 1827.
John. *Southwark.* ca.1770. C.
John Joseph. *Toronto.* 1895.
Joseph. *Dorchester (Oxfordshire).* 1837.
Joseph. *London.* 1828-39. Dial maker.
Julius. *Bristol, USA.* s. of Harvey. 1857.
& MARTIN. *London.* 1881.
Mrs. C. *London.* 1863.
Mrs. M. A. *London.* 1857.
R. *Newcastle-on-Tyne.* 1858.
Richard. *Coventry.* 1880.
Richard. *London.* 1839-75.
Richard. *Manchester.* 1824.
Richard. *Witham.* a.1761-1839. C.
Richard. *Woolwich.* 1845-51.
Robert. *Bury.* 1848-58. W.
Robert. *North Walsham.* 1830-58.
Sampson. *Brockville, Canada.* 1857.
Sampson. *Watford.* 1828-66.
Sampson. *Witham.* s. of Richard of *Witham.* a.1813-20. Later at *Maldon.* 1820-39. C.
Samuel. *Lancaster, NH, USA.* 1808-ca.1830.
& SMITH. *Maldon.* 1855.
Stephen. *Selby.* 1858.
Sydney. *Thirsk.* 1866. W.
T. *Camberwell.* 1866.
T. *Long Sutton (Lincolnshire).* 1868.
T. *Nottingham.* 1855.
T. H. *Lancaster, NH, USA.* Early 19c.
Thomas. *London.* 1828-81.
W. *Cambridge.* 1846.
W. A. *Bolton.* 1858. W.
Walter. *Ecclefechan.* 1837-60.

WRIGHT—*continued.*
William. *Cambridge.* 1830-46.
William. *Chipping Ongar.* 1828.
William. *Dunbar.* 1820-37.
William. *London.* a.1684 to Henry Brigden. *Crown Court, Southwark* in 1731. (B. -d.1758). C.
William. *London.* 1828. Watch cases.
William. *London.* 1828-51.
William. *Maldon.* 1866-74.
William. *Market Weighton.* 1851. W.
William. *Soham.* 1840.
William. *Sutton (Lancashire).* 1779. W.
WRIGLEY, Isaac. *Manchester.* d.1742. C.
WROTH, Samson. *Taunton.* ca.1705. C.
WRUCK, F. A. *Salem, Mass.* 1864.
WUILLEUMIER, Daniel (& Co.). *London.* 1863 (69-75).
WULFF, Henry. *Birmingham.* 1880.
WULFSON, Lewis. *Manchester.* 1851.
WURNEMUNDE, C. *Quebec.* 1862.
WYAND, John. *Philadelphia, USA.* 1847.
WYAND/WYARD, Stephen. *Framlingham.* ca.1730.
WYATT—
——. *Blandford.* 3rd qtr. 18c. C.
Henry. *London.* 1839.
John. *Altrincham.* (B. pre-1799) 1828. W.
Lewis. *Hanley.* 1860-76.
Kewis. *Shelton (Staffordshire).* 1842-50.
Lewis. *Macclesfield.* 1828.
Lewis. *Stockport.* 1834.
& Son. *Bournemouth.* 1878.
Thomas. *Altrincham.* 1848.
W. *Romsey (Hampshire).* 1848.
William Thomas. *Blandford.* 1867-75.
WYCHE, David. *London.* a.1686 to Thomas Taylor, jun. CC. 1694 (B. -1717). W.
WYCOFF, Peter. *London,* Canada. 1857-62.
WYCKOFF, P. & Brother. *London,* Canada. 1853-63.
WYLD(E)—
John I. *Nottingham.* b.ca.1710., mar.1740, d.pre-1786. C.
John II. *Nottingham.* s. of John I. ca.1786. W.
John. *Sheffield.* ca.1770-ca.1780. C.
William. *Darlaston.* 1876.
WYLEYS, ——. *Charleston, SC, USA.* ca.1785.
WYLIE—
——. *Glasgow.* Early 19c. C.
Alexander. *Dumfries.* 1730-55.
James. *Thirsk.* mar.1814-51.
Matthew. *Paisley.* Late 18c. C.
William. *Stromness.* 1836-60.
WYMAN—
J. W. *Stanstead Plain, Canada.* 1851-58.
ROGERS & COX. *Nashua, NH, USA.* ca.1830-37.
WYMARK—
Mark. *London.* (B. 1816-24) 1828-32. W.
Philip. *Brighton.* (B. 1822-) 1828-39.
& Son. *London.* 1844-69.
WYNHALL, Edwin. *Looe.* 1856-73.
WYNN—
Christopher. *Baltimore, USA.* 1842.
John. *Guildford.* 1878.
John Lawrence. *Alresford.* (B. 1795) 1830.
Thomas. *Bristol.* Late 18c. C.
Thomas. *London.* ca.1765-70. C.
W. *Hinstock (Shropshire).* 1856-63.
William. From *Farnham* to *Dean Street, London* ca. 1822. Believed d. ca. 1841.
WYNNE—
A. *Tattenhall (Cheshire).* 1857.
Henry. *London.* a.1654 to Ralph Gretorix, CC. 1662. (B. mCC. 1690-1708). W.
John. *London.* a.1670 to James Graves, CC. 1678-97.
Richard. *London.* 1786.
Robert. *Salisbury, NC, USA.* 1827 of *Huntington* and *Wynne*; alone 1830.
WYSE, F. H. *Quebec.* 1853.

Y

YABSLEY, James C. *Brighton.* 1862-70.
YACKER, C. & Son. *Brixton (Surrey).* 1878.
YALE Clock Co. *New Haven, Conn, USA.* 1881-83.
YANDELL, William & Co. *London.* 1881.
YARDLEY—
Charles. *Croydon.* 1878.
James. *Bishop's Stortford.* (B. pre-1796 W.) 1839.
John. *Coventry.* 1850. Watch case maker.
Peter. *Bishop's Stortford.* 1839-59.
Thomas. *Ware (Hertfordshire).* 1828.
Thomas. *Warwick.* 1835.
William. *Bishop's Stortford.* 1828.
YARNALL, Allen. *Sugartown* and *West Chester, Pa, USA.* 1803-32. C.
YARNDLEY—
Charles. *Birmingham.* 1842.
Charles. *Southampton.* 1859-78.
Thomas. *London.* 1832.
YARRINGTON, John. *Bristol.* a.1693-1700.
YATEMAN (also Yeatman, q.v.), Andrew. *London.* a.1684 to James Wolfreston, CC. 1692 (B. -1703). W.
YATES (and YEATES)—
Andrew. *Brighton.* 1851-78.
Edward J. *Freehold, NJ, USA.* 1805-09.
George. *London.* 1832. Watch cases.
Henry. *Burtonwood.* 1785-89. W.
Henry. *Huyton (Liverpool).* 1755-73.
Henry. *Odiham.* 1878.
& HESS. *Liverpool.* 1824-51.
James. *St. Helens.* 1765-70. W.
John. *Bootle.* (B. 1788. W.) d.1796. Watch engraver.
John. *Culcheth.* d.1729. C.
John. *Hastings.* 1870.
John. *Liverpool.* 1747-73. W.
John. *Pocklington.* 1816-18. C.
John. *Wandsworth.* 1828-55.
John B. *New York.* 1893.
John R. *Appleby.* 1849-51.
John R. *Penrith.* 1869-79.
John Thomas. *New Glasgow, Nova Scotia.* 1871.
Joseph. *Trenton, NJ, USA.* 1789-96, of Yates & Kent. To *Freehold, NJ, USA.* 1803.
(Joseph) & KENT. *Trenton, NJ, USA.* 1796-98.
Mrs. Mary Ann. *Brighton.* 1878.
Peter. *Haslingden.* 1828.
Peter. *Hyde.* 1834.
Robert. *Evesham.* 1876.
Rowland. *Williton.* 1883.
Samuel. *Liverpool.* 1810. W.
Samuel. *London.* 1881.
Simpson. *Penrith.* 1858-79. C.
Thomas, sen. *Penrith.* 1828-49. Also jeweller.
Thomas, jun. *Penrith.* 1849-79.
Thomas. *Prescot.* mar.1756-85. W.
Thomas. *Preston.* 1851-58. W.
William. *London.* CC. 1640.
William. *Yatebank (Lancashire).* 1724. C.
YEAGER, William. *Philadelphia, USA.* 1837.
YEAKLE, Solomon. *Northampton, Pa.* Early 19c. C.

YEALEY, L. *Burton-on-Trent.* 1868.
YEAR Clock Co, The. *New York.* 1841-1903.
YEATES (see YATES)—
YEATMAN, C. N. *Salisbury.* 1875. (See also Yateman).
YEISER, Frederick. *Lexington, Ky, USA.* 1859.
YELAH—
John. *Wrexham.* Haley reversed, q.v. pre-1773.
Thomas. *Wrexham.* Haley reversed, q.v. pre-1767.
YELF, William. *Lymington.* 1878.
YEOMAN(S)—
Charles. *Hull.* 1846-58.
Charles. *Leeds.* 1871.
(James) & COLLINS (John). *New York.* From *London.* 1767. Yeomans alone 1771-73.
Edward. *Manchester.* 1851.
Elijah. *Hadley, Mass, USA.* ca.1771-94. C.
George. *Birmingham.* 1880.
Henry. *Nottingham.* 1828-64.
James. From *Birmingham, England* 1767. Of Yeomans & Collins 1769-d.1773.
James. *Manchester.* mar.1836-51. W.
Joseph. *Cockermouth.* 1869-79.
Samuel. *Coventry.* 1880.
William. *Guisborough.* ca.1730-50. C.
YERKES, William. *Philadelphia, USA.* 1774. C.
YERWORTH, James. *London.* 1778. W.
YEWDALL, Joseph. *Bradford.* 1853-71. W.
YLMER, Andreas. *Augsburg.* 1559. C.
YOELL—
Samuel & Henry. *Cowes.* 1830.
Yoell. *Portsea (Hampshire).* 1830.
YOLLAND, James V. *London.* 1869-81.
YON, John. *Southampton.* 1839-67.
YONGE—
George. *London.* (B. 1776-1815) 1832.
George & Son. *London.* (B. 1820-25) 1828.
George & Walter. *London.* 1839.
Walter. *London.* 1844-57.
YONGUE, Robert A. *Columbia, SC, USA.* 1852-ca.1857. W.
YORK(E)—
The. 1914 and later, product of Pequegnat Clock Co.
John. *Coventry.* 1850-54. Watch case maker.
John. *London.* 1844.
W. *Long Buckby.* 1854-69.
YORKEE, Jacob. *Manheim, Pa, USA.* Early 19c.
YOU, Thomas. *Charleston, SC, USA.* ca.1760-80.
YOUELL, Robert. *Leicester.* ca.1750-70. C. (One such a. *London* 1691).
YOUNG(E)—
Alexander. *London.* 1828.
Alexander. *London.* 1875-81.
Archibald.·*Dundee.* 1828.
B. *Watervleit, NY, USA.* 1800.
Charles. *London.* 1832.
David. *Hopkinton, NH, USA.* 1776-1800. Prob. case maker only.
Edward James. *London.* 1863-69. W.
F. *Sittingbourne.* 1803. W.

YOUNG(E)—*continued.*
Francis. *New York.* 1780.
Francis. *St. John. New Brunswick.* 1785.
George. *Selby.* 1871.
Henry. *Liverpool.* ca.1800-1810. W.
Henry. *London.* 1659 to Thomas Taylor, CC. 1672 (B. -ca.1700). Strand.
Isaac. *Liverpool.* (B. 1766) d. 1768. W.
Isaac. *Prescot.* mar.1811-19. W.
Jacob. *Philadelphia, USA.* ca.1769. C.
James. *Ballymena.* 1880.
James. *Dundee.* 1828.
James. *Knaresborough.* 1851-66.
James. *London.* 1828-32.
John. *Belfast.* 1865-80.
John. *Glasgow.* 1836-60.
John. *London.* (B. 1820) 1832.
John G. *Dundee.* 1850.
Joseph. *Bewdley (Worcestershire).* 1828-50.
Joseph. *Kinfare (Staffordshire).* 1860-76.
Mark. *Newcastle-on-Tyne.* 1834-36. C.
Mary. *Newcastle-on-Tyne.* 1848-56.
Martin. *Dublin.* 1788. W.
Mrs. Sarah. *London.* 1875-81. (Successor to Edward James).
O. *Newcastle-on-Tyne.* 1858.
R. *North Shields.* ca.1780. C.
& RADFORD. *Ottawa.* 1871.
Richard. *Newcastle-on-Tyne.* 1827-ca.1830. C.
Robert. *Ballymena.* ca.1810-20. C. (B. 1824).
Robert. *Belfast.* 1849.
Robert. *Belfast.* 1894.
Robert. *Sittingbourne.* 1874.
S. E. *Laconia, NH, USA.* 1884.
Samuel. *Bunbury* (nr *Chester*). 1749. C. and sundials.
Samuel. *Charles Town, WV, USA.* ca.1800.
Stephen. *New York.* (B. 1805) 1810-16.
Thomas. *Bewdley.* 1868-76.
Thomas. *Dundee.* 1850.

YOUNG(E)—*continued.*
Thomas. *London.* a.1689 to Richárd Medhurst of *Croydon,* CC. 1699. W.
T(homas?). *Stourport.* 1860.
Walter. *Dublin.* a.1773 to Thomas Blundell.
William. *Auchtergaven (Perthshire).* 1836.
William. *Ballymena.* ca.1780. C.
William. *Harrogate.* 1834-d.1876. C.
William. *Huyton (Liverpool).* mar.1760. W.
William. *London.* a.1656 to John White, CC. 1668-97.
William. *London.* a.1674 to George Harris, CC. 1682-97.
William. *London.* (B. 1820) 1828. C.
William. *Ottawa.* 1862.
William. *Oxford.* 1656-95. T.C.
William. *Portglenone (co. Antrim).* mar.1830-68.
William. *St. Helens.* 1773. W.
William Henry. *Boston (Lincolnshire).* 1861-76.
William Henry. *Cambridge.* 1846-75.
William Henry. *Swaffham.* 1830-36.
William J. & Son. *Philadelphia, USA.* ca.1857. Sundials.
YOUNGS—
Benjamin. *New York.* ca.1800.
Benjamin. *Windsor, Conn, USA.* 1761. To *Schenectady, NY,* ca.1766, later to *Watervleit, NY.* C.
Ebenezer. *Hebron, Conn, USA.* 1756-80. C.
Isaac. *New Lebanon, NY, USA.* 1793-1865. C.
Seth. *Hartford, Conn, USA.* 1711. mar.1735-39. To *Windsor, Conn,* 1742. To *Torrington* 1760.
YOUNGSON, ——. *Scampston* nr *Rillington (East Yorkshire).* ca.1830-ca.1840. C.
YSERS, Anta. *Malines.* 18c. C.
YUIL(L)—
Robert. *Glasgow.* 1840-49.
Thomas. *Castle Douglas (Scotland).* 1836.
William G. *Truro, Nova Scotia.* 1866.
YULE, James. *Castle Douglas (Scotland).* 1836-60.

Z

ZACHARIAH—
Brothers. *London.* 1869.
H. *Gosport.* 1848.
Henry I. *Pickering.* 1840-51. C.
S. *Portsea.* 1859.
S. *Southampton.* 1859.
ZACHARY, John. *London.* a.1687 to Daniel Quare, CC. 1694 (B. -1713).
ZAGNANI, Francis. *London.* 1832-57.
ZAHM—
& Co. *Lancaster, Pa, USA.* ca.1850.
G. M. *Lancaster, Pa, USA.* 1843.
H. L. & E. J. *Lancaster, Pa, USA.* 1850s.
& JACKSON. *Lancaster, Pa, USA.* 1850s.
ZAHNE, Ferdinandus. *Hamburg.* 1670.
ZAHRINGER—
John & Co. *Birmingham.* 1850-54.
Michael & Co. *Birmingham.* 1860-68.
Vincent. *London.* 1863.
ZANLICH, Simon. *Prague.* 1570.
ZARINGER & HOFFMAYER. *Birmingham.* 1842.
ZECHBAU(E)R—
M. *Manningtree.* 1866.
Titus. *Bristol.* 1879.
ZEISSLER, G. A. *Philadelphia. USA.* 1848.

ZEPLER, Louis. *Birmingham.* 1868.
ZICHLER, J. *Crewe.* 1857.
ZIELINSKY, Adam. *London.* 1881.
ZIFFEL, C. *Thetford (Norfolk).* 1846 (cf. Zipfel).
ZIMMERMAN—
Anthony. *Reading, Pa, USA.* 1768-88. C.
C. H. *New Orleans, La, USA.* ca.1850-ca.1860.
ZIPFEL (see also Ziffel)—
Anthony & Co. *Oldham.* 1848-51.
& BEHA. *Birmingham.* 1868.
Bernard. *Norwich.* 1836.
C. *Long Stratton (Norfolk).* 1846.
Charles. *Norwich.* 1846-75.
Charles, jun. *Norwich.* 1865-75.
John. b. ca. 1780 *Lenzkirk, Germany. Norwich* 1830 -d. 1862.
Joseph. b. 1784 *Germany. Norwich* 1823 -d. 1850.
Matthew. *Norwich.* 1830-65.
ZOLKI—
Marks. *Sheffield.* 1871.
Simon. *Sheffield.* 1871.
ZOLLING, Ferdinand. *Frankfort.* 1690. (B. 1750). W.
ZOTTI, Romualdo. *London.* 1863-81.
ZUCKER—
Charles. *London.* 1857-75.
Lewis. *London.* 1857.

ADDENDA

ABBREVIATIONS IN ADDENDA

app apprentice or apprenticed
B   Baillie (quoting dates given in Watchmakers & Clockmakers of
       the World, Volume 1)
b   born
C   Clockmaker
c   when after 17, 18, etc, century
c   when preceding a date, circa (approximately)
cf   confer/compare
CC   Clockmakers' Company, free of CC in the year stated
d   died and -d, until died
Jnr junior
marr year married
prob probably
qv   see also
rep'd repaired
ret'd retired (year)
Snr senior
succ successor to, succeeded by
T C turret clockmaker
W   Watchmaker
wkg working
?   denotes uncertainty

A

ABBOT, Francis. Derby 1821-31
William. Knaresborough c1765 W
ABBOTT, William. Knaresborough 1765 W
ABEL, John. Bungay 1864
ABLITT, Frederick. Ipswich 1844
& Kirk. Stowmarket 1810-14 (John Ablitt & William
Kirk). Succ to James Bethel qv
ABRAHAMS, Michael. Nottingham 1853
ACKERS, Thomas. Collingbourn, Wilts c1760 C
ADAMS, Clement. Stamford app 1762 to Samuel
Haslewood (B pre-1777 W)
Edward. Youghal 1628 C
James. Halesowen c1760 C
John. Harlington, Beds, made church clock 1719
for 12 19s 4d
Nathaniel. Stowmarket b 1753 -1790 W & C
William. Liverpool 1845 W
ADAMSON, William. Leeds c1700 C
ADCOCK, William. Maidenhead 1899- 1903
ADDISON, Joseph. London Lost watch advertised
1802
ADKINS, Thomas. London Son of Thomas Adkins of
Ebisham, barber app 1716 to Moses Meigh (B CC
1745 W)
AGAR, C. Selby c1775 C
James. Malton 1840 W
AINSWORTH, G(eorge). Warrington c1810-15 C (B
1818) Also name found cast into bells
ALAND, R. Deal 1865
ALBERTIN, ——. Geneva 1851 W
ALBORN, Joseph. Windsor 1883
ALAND, R. Deal 1865
ALCOCK, Edward. Southwell 1706-07
ALEXANDER, Issac. Nottingham 1745-60 W
William. Tunbridge Wells 1851
& Son. Coventry 1900 W
ALLAN, David. London Lost watch reported in 1790
William. London c1770 C
ALLEN, H. W. Newbury c1840
T. Shefford, Berks 1800
William. Louth c1760 musical C
ALLISON, Robert. Hull 1806
ALLPORT, Thomas. Derby 1858
ALLSOP, H. A. Derby 1895-1900
AMAT, George & Alfred. Sheffield 1860 (see
also Amatt?)
AMBROSE, J. Dedham 1812 TC (may be the Sudbury
man)
AMES, Thomas. Cleobury Mortimer Believe marr
1768 W
ANDERSON, Christopher. Brigg, Lincs 1780
Christopher. Hull 1768-84, at which latter
date lived at Glanford, Lincs C

ANDERSON —— Continued
David Jnr. Aberdeen Made sundial for town 1597
Ebenezer. Auchtermuchty c1800 C
James. Arbroath c1790-1800 C
Thomas Barker. Hornsea, Yorks 1892 W & tob-
acconist
William. Hull 1787
ANDREW(E)S, Edmund. Shrewsbury 1646-50 W
Thomas. Woodbridge Repaired T C at Ips-
wich 1594-97
William. Sutton in Ashfield 1877
ANDOUIN, Peter. Dublin app 1719 to Daniel Pineau
ANN(ES)S, or ANNIS,(William?) Eye, Suffolk
c1785-87 (prob free 1744)
ANSELL, M. Ipswich 1836 W
ANTHONY, Miss Clara Emley. Ilkeston 1900-32
Wesley Henry. Ilkeston 1888-95
APPLETON, ——. Burnley c1900 C
ARCHER, ——. Stow c1740 C (B 1795)
Walter. Place not known c1700-1720 C
ARDEN, Henry Jnr. Norton 1850 (same man as
Malton)
L. Arklus W no date
ARGYLE & PEAKE. Heanor 1884
ARGYLE, Samuel. Derby 1880-84
ARIS, Philip. Shrewsbury 1623 W
ARMSON, John Edward. Somercotes, Derbys 1862,
Riddings 1876
ARMSTRONG, Thomas. Poulton, Lancs c1830 C
(B 1808)
William. Ellesmere (b 1820) 1841 W
ARNOLD, Henry. Maidenhead 1899
John. Child Okeford, Dorset c1760 C
ASHBURN, ——. Place not known c1740 C
prob Midlands
ASHBY, Thomas & Son. Whaley Bridge 1884
John. Ludlow (b c1824 Coventry) 1848-51
ASHDOWN & BARTLETT. Maidstone 1888 W
ASHFORD, Alfred. Derby 1888
ASHMEAD, ——. Dudbridge c1750 br. clock
ASHOVER, Thomas. Derby 1822
ASHTON, Aron. Tideswell 1773 bro of Samuel of
Ashbourne
(J?). Tideswell c1750 C (B Thomas pre-1795 W)
William. Hull 1826
ASPINALL, J. Place not known 1720 W
William. Shrewsbury 1623 W
ASTON, Charles. Derby 1895-1900
Samuel. (b c1795 Frodsham) Ludlow 1851-61 W
ATKIN, Abraham. Sheffield 1860
S. Nottingham 1877
ATKINS, Daniel. London 1772 W
ATKINSON, ——. Bradford c1830-1840 C

ATKINSON —— Continued
Thomas. Bury St Edmunds 1716 C
ATTERBURY, Charles. Derby 1849-50
Francis. Warwick Repaired T C 1592- 1620
AULT, John. Belper 1835 W
Joseph. Belper 1835-46
John. Derby 1839

AUSTALL, Joshua. Kingston c1720 C
AUSTIN, Charles. Canterbury 1784
AXFORD, John. Bristol marr 1687 & again 1691
C & W
AYERS, Sarah. Beccles 1865 (widow of Edward?)
AYLEFF (or AYLESS), John. London c1730 br. clock
(B Ayloffe pre-1782)

B

BACK, Henry. Boughton Blean, Kent marr 1790 C
BADCOCK, William Cluett. Derby 1841
BADDELEY, George. Newport, Shropshire (b 1730
Tong) Marr 1754 & again 1764, d 1785
John. b 1724 Tong, bro of George. Albrigh-
ton c1766 d 1804 C
John. Pattingham, Shropshire 1797 -d 1828
Thomas. Shifnal 1789
BADDSTON, John. Ipswich repaired T C 1625-36 d
1666 Succ by son Edmund
BAGSHAW, Henry. London 1813 W (B 1808)
BAINBRIDGE & FURNACE. Dublin 1766 W
BAKE, ——. Wetherby, Yorks Succ to Fryers
post- 1871
BAKER, Henry. Appleby, Leics early-mid 18c C
James & H. Easingwold, E Yorks 1840 C
John. Newark c1800 C
W. Chesterfield 1900
William. From London to Shrewsbury 1821 d 1857 W
BALE, John Henry. Derby 1884-95
BALLARD, John. Hornchurch, Essex prob late 18c C
BANCE, Luke. Hungerford 1795
Matthew. Kintbury, Nr Hungerford
late 18c C (cf Bunce)
BANCROFT, Gilbert. Derby 1850
& Woodwood. Derby 1831-46
BANDY, James. Edinburgh date uncertain - 19c? C
BANES, William. Chesterfield 1855
BANISTER, James. Wrexham 1765, Ellesmere by 1773
d 1780
BANKES, Richard. Coventry sundial dated 1630
BANKS, J. Nottingham c1701-19
John. Chester free 1682
BANNISTER, Miss Susannah. Derby 1881-84
BARBER, Charles. Derby 1843
Francis. Hermitage, Berks 1710
John. Nottingham 1832
John. Stratford, London 1771 W (B 1799-
1811)
Mary. Bridgnorth widow of Edwin 1870-75
BARCLAY, T. Place not known c1790 C
BARKER, John. Stamford Journeyman to Thomas
Monck left 1798
& Jones. Llanrwst c1800-1805 C
Sarah. Sheffield 1860
Mrs Susan. Framlingham 1874 (widow of
Samuel Keer?)
Thomas. Framlingham & Debenham b c1757
d 1822
William Keer. Son of Thomas, trained Lon-
don, working Debenham 1809-22 then Beccles
1822-46

BARLOW, I. London no date W
James. London no date W
John. Oldham 1774 C
William. Ashton 1789 W (B 1760)
William. Kings Lynn lantern clock c1700
BARNARD, William. Newark b 1707 d 1785 numbered
his clocks
& Savory. Clockwatch 1779
BARNBY, Samuel Bishop. Derby 1852
BARNES, George. Sheffield 1860
Samuel. No town 1770 W, prob the London
man
Thomas. Birmingham c1820 C
BARNETT, Jonathan. Oswestry kept T C 1719
W. Durham no date W
BARNWELL, Henry. Matlock 1884- 1916
BARR, ——. Dover mid-18c C
John. Port Glasgow c1760 C
BARRACLOUGH, Isaac. Derby 1826
BARRAL, Jeanne Pierre. Geneva 1781 W
BARRETT, Charles. Shrewsbury b 1728 d 1807 W
Isreal. Ramsgate 1847
John Snr. Newport, Shropshire 1696 d 1729 W
John Jnr. Newport b 1698 son of John
Snr, d 1739 C
BARRIDON, A. Rieupeyroux Late 18c C
BARRILLET, Aylmer. London c1705 W
BARRINGER, Laurens. Hull 1791
BARRINGTON, Samuel. Dublin 1667 C
BARRON, Thomas. Dublin watch case maker
BARROW, James. Edinburgh ran away from master,
Andrew Brown 1699 (B one such London early 18c)
John. Halifax 1728 W (wife or daughter
died there)
BARRS, Hewbert. Reading 1899
BARTHOLOMEW, Edward. Sherborne b 1700 d 1766
(son of Thomas)
Thomas. Sherborne b 1650 d 1727
BARTHROP, Walter. Ixworth, Suffolk b c1766 marr
1767, insolvent 1769
BARTLE, F. Heckmondwike 1882 W
BARTLETT, Thomas. Whareham c1760 C
BARTON, Frederick. Sheffield 1860
George. London 1773 W
James. South Collingham, nr Ollerton 1844-48
Luke. Arnold, Notts 1832
BASFORD, Daniel. Newport, Shropshire 1782-1836
BASKETT, ——. Moulton (South?), Devon c1795 C
BASLINGTON, Hurliman. Alford, (Lincs?) early 18c C
BASSETT, George Francis. Ipswich marr 1793-95, then
Long Melford 1795-96, then Philadelphia c1797
d 1798 W

BASSETT —— Continued
Thomas. Bath & Wells marr 1675 W
Thomas. Kinsdale, Eire 1787 C & W
BATE, Cleare. Dublin app 1724
William. Dublin app 1724
BATES, G. London 1762 W
George. Buxton 1852
BATTEN, William. Son of Thomas Batten of Brentford
app 1724 to William Risbridges of Dorking
BATTERSBY, George. Manchester marr 1694 C
BATTY, Edward. Halifax marr 1733 C
James. Sheffield 1787-1815 W
BAUER, Edward. Southwell 1864
BAX, Philip. Son of Elizabeth Bax of Dorking, app
to Henry Smith of London
BAXTER, Mary. Shrewsbury widow of John 1835-51
William. Shrewsbury 1787-1828
BAYLEY, Joseph. Nottingham 1814
William. Wakefield no date W
BAYLIS, Henry Jnr. Reading 1899
BAYNOR, James. Sheffield d 1810 C (May be error
for Raynor?)
BEACH, ——. Brighton c1800 C
BEADLE, James. Woolwich 1874
BEARDSLEY, N. Nottingham 1855
BEAUMONT, John J. Hull 1872
BEEFIELD, George. London no date W
BEER, J. R. Plymouth c1770 C
BEETON, John. Bury St Edmunds 1823
BEHRENS, Morris. Nottingham 1848
BEISLY, J.P. Wallingford 1899
BELL, Ephraim. Warwick app to Joseph Wright 1752-
59 later to London?
James. Bawtry 1828-44
John. London son of William Bell of Tarraby,
Cumberland, yeoman, app 1698 to Nathaniel Smith
BELLAMY, William. Place not known, no date W
BENBOW, Thomas. Northwood, Shropshire b 1739
son of John qv, d 1809 C
BENNETT, ——. Cheadle, Staffs c1740 C
——. Cheapside, London 1856 W
Alexander. Tideswell, Derbyshire 1891-1916
Giles. Malmesbury c1770 C
John. Dublin app 1676 C
John. London 1713 took George Dennis app
Mrs Mary. Tideswell, Derbyshire 1922-32
BERFIELD, Henry. North Walsham, Norfolk 1854
BERINGER, David. Nuremberg c1745 pocket sundial
BERRESS, T. London lost watches reported 1799 &
1818
BERRESFORD, A. & S. Bros. Eckington 1876-1922
BERTINI, W. London 1859? C
BETHEL, James. (from London) Stowmarket, nephew
of John, marr 1807-10, also Needham Market in
1809, succ by Ablitt & Kirk
John. (from London) Stowmarket b c1758,
1791, marr 1799 ret'd 1808 d 1821, succ by James qv
BETHLEY, Joseph. Sheffield 1860
BETTERS, Robert. Middlesborough 1872
BEVAN, Thomas. Marlborough C early 18c
BEVIL, Charles Perry. Ipswich 1799, bankrupt 1822
BIBBY, H. Liverpool 1860 W
BICKERSTAFFE, William. Son of Robert of Langton,
Lancs, app 1633 to Thomas Wright of Chester
BICKERTON, George. Ellesmere 1789-1822
& Son. Ellesmere early 19c C
·BIDDLESCOMBE, John. Sturton Caundle, Dorset
1705 C
BIDHAM, John. London c1740 C
BIGGIN, ——. Wisbech c1750 C
BIGGS, E. T. Maidenhead 1899

BILBIE, William. Chewstoke 1757 W
BILSBORROW, John. Huddersfield c1730-40 C
BINCH, Thomas. Mansfield b 1688, marr 1710, took
John Boot app c1724 (B 1737)
BING, Daniel. Sandwich b 1727, marr 1757 & again
1767, Thanet C
Daniel. See also under Byng
BINKS, John. Worksop 1822-36
BIRCHALL, Samuel. Oswestry 1800 C
Thomas. Derby 1829-31
BIRNESS, T. London Lost watches reported 1799 &
1818
BISHOP, William. London late 19c C
BISPHAM, Isaac. Kendal, son of Joseph Bispham of
Yealand Conyers. app to Isaac Hadwen of Kendal.
Note Hadwen visited America & cf Bispham clock-
makers in Philadelphia
BLACK, James. Kirkcaldy 1820-37 C
James. Sligo, Eire 1763 W
BLACKLEY, Joseph. Canterbury b 1782 marr 1806
BLACKMORE, J. Sundial dated 1826 at Otterford
church, Somerset. May be the Sidmouth maker
BLACKSHAW, J. Nottingham 1877
BLACKWELL, John. Kirk Ireton, Derbyshire 1702-03
BLAGDEN, George. Chichester mid-18c c1780 C
BLAIKIE, William. Edinburgh marr 1726 c1740 C
(may be app 1701 in London as Blakey)
BLAKEBOROUGH, Benjamin. Thorne 1923 W
Richard. Bedale 1872 W
BLAKEWAY, Charles. Albrighton, Shropshire b 1729
bro of Thomas of Rushbury, d 1809 C
Charles. Shifnal 1789 prob the Albrighton
man
John. Rushbury b 1768 son of Thomas of Rush-
bury, marr 1796 C
Thomas. Much Wenlock b 1765, son of Thomas
of Rushbury, 1795
Thomas. Rushbury b 1724 d 1805 C & T C's
BLANCHARD, Robert. Hull 1838-51, son of William qv
BLAND, B. Reading 1770-95
BLATCHLEY, Thomas. Bradford (on Avon?) c1790 C
BLIGHT, William. Plymouth c1770 C
BLOCKLEY, Joseph. Canterbury b 1782 -1806
BLOOMER, Samuel. London c1740 W
BLOUNT, Henry James. Wirksworth 1884-95
BLOW, Henry. Lincoln clock case maker c1800
BLOWERS, Edward Snr. Beccles b 1701, son of
Isaac, marr 1728 & again 1745, d 1762, succ by son,
Isaac III qv
Edward, Jnr. Beccles b 1730, son of Ed-
ward Snr. Ret'd 1787 d 1812 C & later
gentleman
Isaac I. Beccles b 1660 marr c1690 d 1719
locksmith & T C, succ by son Isaac II
Isaac II. Beccles b 1693, son of Isaac I, marr
1724 d 1819
Isaac III. Beccles, son of Edward Snr, b 1732, 1762
BLUNDELL, Thomas II. Dublin goldsmith, nephew
of Thomas I to whom app 1768-1824
Joseph II. Dublin 1766-1824 C
BOARINGTON, William. Stourbridge c1790 C
BOBY, William. Ipswich 1859, partner with C Schulen
until 1885, then with Jannings 1891-1913
BOHUN, Stephen. Waterford 1809 W
BOLT, Richard. Teignmouth 1830-52
William. Dawlish 1844-66, business remained in
family until 1920s
BON, H. London? Sundial dated 1689
BOND, Henry Charles. Reading 1883
W. No place stated, sundial dated 1795
BONSALL, ——. Breaston, nr Derby c1760 C

BOOKER, Nugent. Dublin 1770-82 C & W

BOOT, Elizabeth. Sutton in Ashfield b 1726, daughter of John Boot Snr, made clocks herself until marr 1760, when ret'd
John Jnr. Sutton in Ashfield, son of John
Snr b 1730 marr 1759 (B c1775), clocks signed 'Boot Junior'
John Snr. App to Thomas Binch, Mansfield (b 1704), marr 1725 Sarah Warrender, worked Sutton in Ashfield, d 1767 C
William. Sutton in Ashfield, son of John Snr b 1735 marr 1757 d 1792 C

BOOTH, Abraham. Sheffield 1860
Willis J. Derby 1888-1900

BOREHAM, John. Shimpling, Suffolk 1769 -d 1777, prob from London, see Vol I

BORELLI, Gaetano. Reading 1899

BORRETT, George Snr. Stowmarket marr 1753, insolvent 1782, also at Harwich 1773
George Jnr. Stowmarket b 1754, son of George Snr, moved to Harwich c1773 & d 1773

BOTLY & LEWIS. Reading 1899

BOTT, Thomas. Melbourne, Derbyshire c1810 C

BOTTOM, Francis. Belper 1846

BOUME, George. Kirkby (Lincs?) c1780 C

BOUN, John. Matlock 1764-1800

BOURNE, William. Kirkby c1780 C

BOWDEN, James William. Glossop 1895

BOWEN, Arthur Owen. Alfreston 1884

BOWER, Peter. Redlench (Redlinch, Somerset) c1740 C
Robert. Chester no date W

BOWINE, John. Shoreditch 1832, C at Kingsland Road

BOWKER, John. Market Drayton 1850

BOWLES, Michael. London his app, Thomas Bale, free in CC 1704

BOWLEY, W. Shrewsbury 1818

BOWMAN, James. Son of Henry Bowman of Lambeth, app to Daniel Delander 1712
William. Reading 1883-99

BOWN, James. Matlock 1829-46

BOWRING, G. Jersey 1868 W

BOWYER, William. Newport, Shropshire marr 1799-1806 W

BOX, A. W. Canterbury 1882

BOYCE, Peter. Wangford, Suffolk marr 1720, later Beccles, where marr again 1742 & d 1757 C

BRADBERRY, ——. Middleham, Yorks c1785 C
Robert. Richmond 1847-54

BRADBURY, T. Ilkeston 1891

BRADEY, George. Sheffield 1860

BRADFORD, Thomas. Nottingham b 1669 marr 1695
William. Tiverton 1809 W

BRADLEY, Thomas. Ilkeston c1760-1815
Thomas. Mansfield 1783- 95

BRADSHAW, Edmund. Whittington Moor, Derbyshire 1884
Ellis. Shrewsbury marr 1670 d 1707, (a prisoner)
& Ryley. Coventry 1805 W

BRADY, John. Newport, Shropshire b c1780 1841

BRANCH, Charles. Thetford 1845

BRAND, ——. Liverpool 1810 W

BRANDETH, Uriah. Place not known, date not known early 18c? C

BRANSLEY, Edward. Derby 1727

BRAY, Thomas. Nottingham 1825

BRAYLEY, Joseph. Nottingham (from London?) 1805-15 W

BRECKELL, Edmund. London c1760 C

BRENDRETH, Benjamin. See Brandreth

BRENTNALL, Thomas. Derby 1788 W

BREWER, J. (or T.?). Cheadle, Staffs c1750 C

BREWSTER, W. Dover 1865

BRIANT, John. b 1748 Exning, Suffolk then Hertford by 1778 d 1829 C & bellfounder

BRIDGE, Robert. London 1784 W

BRIDGES, Henry. Thames Ditton, Surrey sundial dated 1720 and 1746
William. b c1771 Ipswich, brother of John, app London 1785

BRIERLEY, John. Upperthorne, Sheffield d 1711 W

BRIGGS, Francis. Place not known (N W England?) pre-1825 C
John. Skipton c1770 C, (same man as there later?)

BRIGHT, Jerome. Saxmundham b c1769 marr 1790 believed ret'd 1829 d 1846 C & gentleman
Thomas. Coventry 1799 W

BRIMBLE, John. Westminster, London c1715 C (B has Bristol 1785-1801?)

BRINDLEY, James. Mansfield 1773-95 C

BRISBOURN, Peter. Roddington, Shropshire b 1723 d 1793 C

BRISBURNE, Peter. Same as Brisbourn

BRISTOW, S. Lincoln c1780 C

BROADHURST, Walter. Litchfield c1700 C

BROCKE, Samuel. London c1620 W

BRODERICK & SONS. Spalding c1810-1820 C (numbered their clocks)

BRODHURST, Walter. Ludlow free 1710

BRODSHAWE, Adam Snr. Shrewsbury 1582-1622 blacksmith & T C
Adam Jnr. Shrewsbury son of Adam Snr b 1582 -1639, repaired T C

BROMLEY, Robert. Hull 1838 W

BROOKES, James Frederick. Middleham, N Yorks 1872

BROOKHOUSE, John. Sheffield no date W (perhaps husband of Sarah)
& Tunnicliffe. Derby 1820-29

BROOKS, Edward. Derby 1754
Robert. Ipswich marr 1765 and again in 1766 ret'd 1768

BROWERTON, S. Northfleet 1874

BROWLZER, William. Whittington Moor, Derby 1884

BROWN(E), Charles, George. Thatcham, Newbury 1883-1903 Church. Newbury c1800
David. London c1740 C
Edward. Bristol marr 1689 C
George. Bingham 1848-53
Henry. b c1818 Newtown, Montgomery - Shrewsbury 1841-95, partner of Evans & Brown
John. Derby 1838
John. Bristol 1689 C
Joseph. b c1847 Oakengates, Coventry 1871 W
Richard. Rumford (Romford) c1695 C
Robert. Botesdale, Suffolk c1740 C
Thomas. Ruyton XI Towns, Shropshire 1772, d 1781 C
William. Yoxford, Suffolk, partner with John John Kemp, succ Robert Poll at Harleston, 1771-92 alone

BROWNSWORD, Mrs R. Nottingham 1853-54

BROUGHAM, Henry. London 1861 W

BRUGGER, John. Sheffield 1860 W
& Ketterer. Leeds 1853

BRUMBY, George. Hull 1872

BRUMWELL, John. Pall Mall, London 1711 W
John. London 1711 W

BRUNNER, James. Hedon, Yorks 1872

BRYAN, William. Ludlow 1755 W

BRYON, Thomas. Stamford 1847 W

BRYSON, John. Edinburgh early 19c C
BUCHAN, Alexander. Bridge End, Perth c1790-
1800 (one of this name there in 1833 may
be same man
BUCK, Nathaniel. London no date W
  John William. London app 1837 to John
  Frodsham
BUCKE, George. b c1740 app 1754 to William
  Mayhew of Woodbridge, then to Bungay c1765-84
BUGDEN, Richard. Godbury (Sudbury?) c1780
BUGLESS, Thomas. Bury St Edmunds 1864-65
BULLINGFORD, F. London 1837 W
  R.B. London 1829 W
BULLOCK, Benjamin. Marshfield (Glos?) late 18c C
  Christopher Jnr. Ipswich marr 1775
  Christopher Snr. Botesdale, Suffolk marr 1740
  d 1758 C
  & Davies. Ellesmere b 1719 son of Edmund,
  -d 1797 C
  Edmund. Ellesmere 1708 d 1734
  Edward. Oswestry 1732-33
  Edward. Loddon, Norfolk c1840
  Gilbert. Bishopscastle c1720-30 C prob
  same as Gilbert of Beaucastle
  Jeremiah. Ellesmere b 1723 son of Edmund -d
  1780 C
  Richard. Ellesmere b 1719 son of Edmund -d 1797 C
  Robert. Halesworth 1823 C
  William. Wootton, Oswestry d 1742 C
  William. Yoxford, Suffolk 1844
  Zeph. Box, Wilts c1810 C
BUNCE, Matthew. See Bance
BUNNETT, Jacob. Ipswich 1779 marr 1783 -89
BURCH, ——. London c1740 W
BURDGE, Nicholas. Galway 1817-24 W

BUREN, ——. No place known, no date W
BURGES, William. London, invented name found on
  watches made by Jasper Harmer
BURN, John. London no date W
  & Son. Newbury 1887-1955
BURNETT, C. London c1780 W
  Charles. Ludlow marr 1729-43 C
BURPUTT, John. Tooley Street, London c1715-1730
  (B has Barputt ante-1773 W) C
BURRELL, Benjamin. North Frodingham, E Yorks
  1850-55, then Bridlington 1867-79 W
  George. Sheffield 1860
BURTON, E. Lindale, Lancs c1790 C
  Thomas. Mildenhall d 1725 C
  Thomas. Nottingham 1853
BUSH, Herman. Hull 1872
BUTCHER, William P. Derby 1891-1908
BUTLER, Carey. Diss 1854
  Charles. Derby 1888-95
  James. Bolton b 1757 marr 1783 Alice
  Reynolds d 1809 C
  Jacob. Bolton b c1761 bro of James
  with whom he worked & succ c1809 -d 1818
  C & silversmith, some clocks signed jointly
  James & Jacob. Bolton 1778 C
  John. Reading 1823-40
BUTTERFIELD, John. Son of John B of Stoke, app
  1717 to William Moore of London
BUTTERWORTH, Joseph. Place not known (Northern
  England?) c1740 C
  Samuel. Rochdale c1760 C
BYNG, Daniel. St Lawrence, Thanet & St Peter's,
  Sandwich 1767
BYRD, (alias Bin & Burd), Richard. Wenlock 1642-
  59 repaired T C

C

CASHELON, ——. Crevecoeur 18c C
CADDEY, Edward. London son of a London gun-
maker app 1701 to Samuel Verne
CADE, James. London prob c1880
  John. Southwell 1812 C
CALAME, ——. Geneva c1795 W
CALCOTT, John Jnr. Coton, nr Wem b 1777 son
of John Snr d 1853 C & W
  John. Prees, Shropshire 1824-33
  John. Whitchurch 1840-51
  Richard. Coton. nr Wem l 1719 Edstaston
  marr Wem 1742 d Edstaston 1784 C
CALDREN, Nathaniel. East Dereham, Norfolk
  c1710-18
  Calin, Jo. Pietro. Genoa c1700 C
CALSHOTT, John. Coton error for Calcott qv
CALOW Joseph George. Swadlincote, Derbys
  1846-84
CALVER, George. Hopton, nr Botesdale, Suffolk 1844
  James. Woodbridge 1774 -d 1792 C
  James. Loddon & Diss marr 1780 -d 1809
  John. Woodbridge b c1695 working 1739
  -d 1751 C & W
  William. Eye 1812-23 W

CAMELL, William. Ludlow free 1778
CAMERON, George. Northallerton 1872
CAMPBELL, Henry. b 1816 Hexham, Oswestry
  1871 W
  John. Campbelltown c1780 C
CANNANS, John. Son of John Cannans of Lambeth,
  London app to Thomas Dennett 1716
CANONA, John Manzia. Halesworth son of Peter
  1871 -d 1906 Also at Southwold from 1883
CAPPER, John. Shrewsbury repaired T C 1578,
  hanged for treason 1581
CARE, Thomas. London no date W
CARESWELL, Francis. Shrewsbury 1776 d 1823 W
CARGEL, William D. Chesterfield 1884
CARLEY, Mrs Charlotte. Bungay 1846 (widow
  of Enoch?
CARLSSON, N. M. Danzig 18c C
CARNABY, John. Newcastle on Tyne 1720-1780 W
CARNHILL. see Carnill
CARNILL, Joseph Henry. Ilkeston 1890
CARR, John. Newark 1790
  Paule. Place not known prob Lancashire
  c1750-1760 C

CARSON, William. Tideswell 1884-88
CARTER, George Edwin. Derby 1884-91
James & Theophilus. Coventry 1807 W
Joseph. Coventry Son of Joseph C. grocer
app 1807 to James & Theophilus Carter of
Coventry d 1881
Philip. Huntington believed marr 1716 C
William. Frome c1800 C
CARTWRIGHT, —. c1840-50 C
Thomas. Wem b 1843 -71 W
William. Ellesmere repaired T C 1758 -1770
CARVALHO, Samuel. Nottingham 1848
CARVER, Jesse. Newnham, Sittingbourne 1874
CARWELL. see Cawell
CASSON, John. Perth c1820 C
CASTELL, Robert. London 1810 W
CATLIN, Daniel. Lynn (Kings Lynn), Norfolk c1740 C
CATTANIO, J. & H. York 1835 W
CATTLE, William. Warwick 1819 C
CAVELL,Nathaniel. Ipswich b c1749 app 1763 to
John Page whom he succ d 1789 C & W
CAVIT, Ebenezer app 1759 to John Clay of Welling-
borough. At Bungay, Suffolk 1784 then Bedford
1785-1808 when partnership with Thomas Clare
dissolved
CAWDELL, T. Shrewsbury c1800 C
CAWELL (or Carwell), James. Hitchin, Herts c1700 C
CAWNE, Charles E. Barrow in Furness 1885
Cay, Joseph. Derby 1871
CETTI, J. & Co. London 1826 W
CHADWICK, Benjamin. London 1836 W
CHAMBERLAIN, John. Place not known, working at
Hatfield House 1763- 4
CHAMBERS, G. C. Canterbury 1878-82
Charles. Deene, Northants c1700 -d 1726 C
CHAMBLEY, F. Newcastle under Lyme c1810 C
CHAMBLER, T. Ramsgate c1870
CHAMPNEY, ——. Selby 1776-90 W
CHAMPTON, W. Place not known 1806 W
CHANDLER, Anthony. Whaddon, Bucks made T C
1673
B. Nottingham 1770
CHANTRELL, Roger. Chester repaired T C 1624
CHAPLIN & FULLER. Bury St Edmunds partnership
of Thomas Chaplin and Thomas Fuller c1815
CHAPMAN, ——. Sheerness c1820 C
John. Minster Sheppey, Kent marr 1784 W
Samuel. Ipswich c1840 C
Thomas. Bath c1740 C
CHARLES, John. Portsmouth c1760 C
CHARLTON, Edmund, Lechmere, Ludlow free 1830
W
CHASTEY, Robert. Hatherleigh, Devon 1823- 38 (B
d c1850)
William. Totnes 1823, then Teignmouth 1830-
52. Also Torquay 1830
CHASTY, Charles. Dawlish 1838
Charles Heard. Shaldon, Devon 1852. May be also
at Dawlish
CHATER, E. London 1789 W
CHATON, ——. Royal Exchange, London 1829 W
CHEADLE, James. b c1855 Staffordshire. Newport,
Shropshire 1871, later at Shifnal
Joseph. Ashbourne 1860- 1937
CHERINGTON, Thomas. London late 19c C, in
style of late 18c
CHILDS, William. Ollerton 1842-53,later
Southwell
CHISVOLI J. Bros. Chatham & Strood 1830-40 C

CHRISTIAN, (Ewan?). Isle of Man Sundials dated
1666 & 1681
John. Calthorpe, Norfolk 1777
CHRISTIE & Co. Ilkeston 1895-1900
CHUNE, Thomas. Shifnal 1781-96 C
CHURCHILL, B. Snr. Marston c1760-70 C
Blencone. Pryors Marston, Warwks early 18c? C
CHURTON, Joseph Jnr. Whitchurch b 1802 son of
Joseph Snr. To Prees c1845 where until 1863
was C & innkeeper
CLARE, Thomas. Warrington c1760-90 C
CLARK, James E. Montrose 1906 clock restorer
John. Petworth sundial 1742
Richard Newport 1814 W
William. Stalbridge c1760 C
William R. Chieveley, Newbury 1899
CLARKE, Elijah. Oswestry b 1843 Dorset, 1871 W
James (or Joseph). Tuxford 1787-1812 W
Samuel. Hinckley 1773-97
W. Bury St Edmunds c1810 W
William. Radford, Notts 1844
CLAY, Eliza. Newark 1844
Henry W. Gringley-on-the-Hill, Notts 1844-53
James. Derby 1739 W (relative of Charles Clay
of London qv)
S. Gainsborough early 19c C
Samuel. Gainsborough c1720 C
CLAYTON, John. Marple, Cheshire c1750 C
John Snr. Marple, Cheshire son of George
qv d 1769 C
Ralph. Marple, Cheshire c1750 C
CLEAK, Adam. b 1723 Churchstanton, Devon, marr
at Luppitt 1753, then again 1772, then to Upottery
then c1782 to Bridport, ret'd c1804 to Upottery
where d 1809 C
William. Bro of Samuel b 1813 Taunton
London c1860 d 1890 C
CLEAVER, J. Canterbury 1888
CLEMENT, John. Tring c1770-1780 C
William. St Albans c1720 C (of the William Clement?)
CLENCH, Richard. Ludlow kept T C 1582 -d 1627
CLERKE, ——. Royal Exchange, London mid-19c C
Francis. Nottingham casemaker 1853
CLIFF, Nathaniel. Hull c1770-1785
CLIFTON, Cuthbert. Shrewsbury marr 1635 -42 W
CLOUGH, John W. Chesterfield 1895-1922
CLOUTING, Tobias. Eye 1797 (later a farmer?)
CLYMER, Richard. Bristol 1675 C
COATES, G. Christchurch, New Zealand No date W
Thomas. Paisley late 18c C prob the Ham-
ilton man
COATS, John. London Son of Luke Coats of Kep-
wick, Yorks app 1705 in CC to Henry (illegigble)
COBB, John. Dunham-on-Trent, Notts 1832
COCKER, & Son. Hathersage, Derbys 1828
COCKINGS, R. Deal 1865
COCKNAY, Charles. Hull 1872
CODLING, William. Hutton Rudby, N Yorks 1872
COGHILL, William. 12 Claremont Street, Shrewsbury
1850 W
COHEN, S. Sheffield No date W
Max. Manchester 1894
Simeon. Glossop 1884
Wolf. Sheffield 1860
COLBORNE, A. Canterbury 1888
COLE, Richard. b 1771 app London 1785, then
Ipswich -d 1833 C & W
Richard Wright. b 1840 son of Richard Stinton
Cole of Ipswich app London 1856 d 1919

COLEMAN, Charles. London (B app 1760) W from London but late of Barbados d St Vincent 1779
COLES, Michael. Scarborough c1760-70 C
COLLETT, John. Chester c1800 C
COLLEY, Robert. London 1793 W
COLLIER, David Snr. b c1721 at Gatley Green, Cheshire until c1780 then Etchells c1780 -d 1792
David Jnr. b 1787 son of John Collier of Cheadle qv marr 1819 worked at Gatley d 1840 C
Thomas. Chapel-en-le-Frith 1760
William. Place not known c1720-1750 C
COLLINGWOOD, Matthew. Berwick c1840 C
COLLIS, W. H. & Son. Bury St Edmunds 1891
COLLIWELL, William. Derby 1812
COLMAN, ——. Bildeston c1800 C
COMBERMAN, Thomas. Leamington b c1766 1841 W
COMING (Cumming?), George. Kincardine c1780 C
COMPIE, J. Perth? prob casemaker. Name stamped into case of clock by A. Buchan qv 1790-1800
COMPIGNE, Michael. Artillery Ground, Duke Street, London 1701 W
CONSTANTIN, Philippe Giradel. Spitalfield Market, London 1702-c1730 W & C
COOK, Harry. Newark 1886
John. Southwell, Notts 1851
John. Nottingham 1853
Robert. Hathersage 1828
Thomas. Chard 18c C
Thomas. Deptford 1874
William. Aberdeen kept town clock 1651 with Patrick Wanhagan
COOKE, Bernard. Hull 1863 C
James Christopher. Derby 1843
John Thomas. Hull 1872
COOMBE, T. Brighton 19c chronometer
COOMBES, Joshua. London son of James Coombes late Kingston-on-Thames, tailor app 1712 to Henry Austen
COOPER, ——. Whitchurch 1742
Charles. Whitchurch repaired T C 1770-74
George Benjamin. East Dereham c 1876-1900 also fire insurance agent and for Shakespeare Sewing Machine Co
Charles. Whitchurch repaired local T C's 1765-73
Mrs Martha. Saxmundham 1763-4
Thomas. Newport, Shropshire marr 1783 d 1789 C
William. Derby 1843
William. Derby 1760-1802 W
William. London 1677
W. Hull 1872
William. Filey 1867
COPE, Benjamin. Franche, Worcs 1786-1802 C
George & Francis. Radford, Notts 1845-77
Joseph. Nottingham 1842-48
CORNELIUS, William R. Dawlish 1866-79 believed succ to Henry Strowbridge
CORNFORTH, Reuben. Stockton c1760, C (from Stokesley)
John. Stokesley 1741-47 C
CORNOW, Andrew. Towednack, Cornwall sundial dated 1737
CORNTHWAITE, E. Place not known 1826 W
CORNU(E), Peter Daniel. London (B c1730) app 1719 to John Lewis then 1719 to John Bayley
CORNWALL, Henry. London 1817 W
CORRALL, Edwin. Lutterworth son of Francis, b 1847 (perhaps to Hanley later) -d 1919
Francis. b 1695 London CC 1720-32, then Lutterworth -d c1779
Francis. Lilbourne, Northants c1730 C (must

CORALL —— Continued
be the Lutterworth man)
George John Adams. Lutterworth son of Francis c1830, perhaps Mansfield later?
Powell. Lutterworth b 1773 son of Francis -c 1777
Thomas Teissier. Lutterworth son of Francis b 1727 d 1795 C & W
William. Lutterworth b 1777 son of Thomas Teissier C
CORRIER, Jean. Dover b 1732 1756 W (B Paris 1772-1812)
CORSON, Joseph & James. Maryport c1850
CORTHORN, John. Newark 1844-49
CORVEHILL, William. Much Wenlock maker of clocks, organs, chimes & bellfounder d 1546
COSTER, Charles. Newbury 1867-74 succ to Webb succ by J. Mansfield
Thomas. Newbury 1763-89
COTHER, William. Derby 1763-89
COTTERILL, Francis. Belper 1828
COTTON, Matthew. Great Yarmouth 1854
COTTONBELT, John. London son of Henry Cottonbelt of Southwark app 1718 to Robert Higgs
COUNSELL, John. Ross c1750 C
COWAN, ——. Airdrie 1889 W
COWLEY, John. Ilkeston 1858
COWPER, ——. London 1803 W
COX, ——. Taunton c1810 or perhaps later dial maker
Henry. Nottingham b 1781 -1818 d 1874 London W
Henry. Nottingham b 1795 cousin to Henry Cox above, Nottingham 1816 - d 1860
J. Nottingham 1832
Thomas. Nottingham 1864
William. Place not known, combined watch and telescope c1780
& Adams. Nottingham 1832
COXON, Frederick. Derby 1892
George. Oldbury c1830-36 W
CRACKNELL, George. Redlingfield, Suffolk c1844 C
CRAGG, Thomas. Horsham 1807 (B c1785 CC) C
CRIAG, Robert. Galston, Scotland c1820 C
CRAMPERN, John Gallis. Newark 1767, to Tuxford 1786 d 1792 C
William. Newark d 1770
CRANNAGE, Alfred. Glossop 1888-95
Frederick. Chesterfield 1891-95
Mrs M. Glossop 1895-1900
CRAWSHAW, Thomas. Retford 1810 W
CREED, John. Yarmouth c 1730-40 C
CREMONINI, ——. Bilston c1840 C
CRESSENNER, Robert. London c1715 C
CRESSWICK, H. Hessle, E Yorks 1872
CRIGHTON, John. Dundee 1789-95 W
CRISP, Edward. Wellington, Shropshire b c1819-1841
William. Wrentham, Suffolk b 1736 died 1810 W & C
CROCKFORD, George. Nottingham 1877
CROFT, W. Matlock, Bath 1895-1932
CROLEE, John. London c1730-40 C
CROOK, John. Hackney, London c1795 C
CROSS, John. Wallingford 1820
William. b c1811 Shewsbury marr 1837 -43 at which latter time was in jail there C
CROSSKILL, Joseph. North Walsham c1760
CROUCH, A. D. Reading 1899
CRUMPE, Richard. Ludlow kept T C 1603-17
William. Ludlow kept T C 1603-17 (same man as Richard?)

CRUNDELL, Walter. Ludlow b c1853 -1871 W
CRUSE, John. Newbury 1854
CRUST, Joseph. Place not known, no date W
CUDDON, James. Bungay 1838
CURRIER, William Thomas. Whittington Moor.

CURRIER —— Continued
  Derbys 1880-1900
CURTIS, John. London (B app 1715 -c1760) 1786 W
  Thomas. Bristol 1757 W
CUSSONS, George. Whitby 1872

D

D. H. Monogram for Hans Ducher qv
DADE, Daniel. Bungay c1790-1800
DADSWELL (sometimes Dodswell), Edward.
  Rotherfield Sussex b 1659 son of Robert Dadswell
  marr 1677   d 1736, yeoman & clocksmith
  John. Burwash b 1727 son of Alexander Dads-
  well of Rotherfield marr 1766, believed d 1790 C
  Thomas. Rotherfield b 1688 son of Edward Dads
  well app to Thomas Muddle his uncle, d 1752 C
  Thomas. Rotherfield b 1719 son of Edward Dads-
  well app to uncle Thomas Dadswell marr 1748
DADY, H. Linthorpe, Yorks late 19c C (cf Dandy?)
DAKIN, John. Nottingham app 1791 to John Hudson
DALE, ——. Newbury c1790
DALMAN, Samuel. Newark 1780-1823 W & cabinet-
  maker
  William. Newark 1822
DALSTON, Thomas. Place not known (London?)
  c1690 C
DALTON, Samuel. Rugby late 18c C
DAMMANT, Samuel. b c1768, to Ipswich c1791
  d 1804 C & W
DANDY, Harry. Bridlington b 1884 d 1949 C (cf
  Dady?)
DANIEL, Henry. Farringdon, London 1820
  Ralph. Stockport marr 1698 C
DANIELL, Hugh. Ludlow kept T C 1603 -d 1645
DARBY, Charles. Derby 1884-95
  Frederick James. Derby 1888-1900
DARKEN, ——. Condover, Shropshire 1682 C
DAVENPORT, George. Derby. 1847.
DAVEY, J. Ipswich 1823
  Peter. St Austell c1810 C
  Robert. Norwich d 1811 W
DAVID, William. Wokingham 1899
DAVIDSON, C. London no date W
DAVIES. Edward. Ellesmere b c1754 1781 -d 1798 C
  Edward. Ellesmere. b c1786 1836-41 C
  Henry. Shrewsbury 1842 C
  John. Dolegeirwch c1760 C
  John. Shrewsbury 1789-97 kept T C
  Joseph. Ratcliffe Highway, Shadwell, London,
  c1700-1710 C
  Samuel. Leamington 1850 W & C
  William. (b c1798 Italy) Wellington, Shropshire 1851
  may be same as William Henry qv
  William. Chester d 1774 W (B 1770)
DAVIS. John. Shifnal b c1840 son of John Davis of
  Shifnal -1875 C
  Joel. Nottingham 1844-53
  Robert. Aylsham c1770 see also Davy
  Robert. Buxton 1888
  Thomas. Shifnal b c1816 son of William Davis

DAVIS —— Continued
  of Shifnal -1843
  William. Warwick 1815-17 C
  William. Chester (B 1770) d 1774 W
  William. Shrewsbury 1789-97 W
  William. Shrewsbury 1863
  William Henry. Shifnal b 1798 son of William
  Davis marr 1826 -75 C
DAVY, Robert. Aylsham late 18c (see also Davis)
  Samuel. marr 1758 Leiston, Suffolk d 1797
  Halesworth
DAWBORNE, Hugh. Ampthill 1744 C
DAWELL, Thomas. Town not stated 1792 W (cf
  Badswell)
DAWKINS, John. Framlingham c1750 -d 1789 C
DAWSON, Samuel. Nottingham 1853
DAYWAILE, Florent. Brussels c1770 C
DEACON, W. Great Smeaton, Yorks sundial dated
  1809
DEADMAN, Arthur William. Leamington 1846 W
  (probably the Aylesbury man)
DEANE, Edward. Place not known c1740 C
DEAVES, Richard. Whitchurch, Shropshire b 1702
  -1780 C
DEBENHAM, Robert Strange. b 1739, Ipswich 1763,
  to Long Melford by 1778 where succ George
  Maynard -1795
DEBNAM, John. Newport, Essex c1800 C (B has
  Debram)
  Laurence. Place not known late 17c lantern clock
DE FRETIS, Nicholas. London 1685 not in CC
DEL Vecchio & Cetti. Wellington, Shropshire 1841-56
DEMAINE, Anthony. Woodbridge 1780-1811
DERN, R. B. Greenwich 1874
DENNIS, George. London son of John Dennis of
  Kingston on Thames, app 1713 to John Bennett
DENTON, Mrs Rebecca. Windsor 1899
DEPREE, Raeburn &Young. Exeter 1906, later
  Depree & Young qv
  & Young. Exeter 1909-19, then John Young qv
DE RAAY. John Hubert. Reading 1864
DEVASTON, John. Place not known sundial dated
  1789
DEVERAL, John. Nottingham 1771
DEVIS, Mrs. London (widow of William who d1763?)
  opposite St Dunstans Church, Fleet Street ret'd Jan
  1764
DEXMIER, Peter. London c1730 C
DICKERSON, Daniel. From Diss to Framlingham
  1755 -c1772 then Yarmouth then by 1778 to Eye
  then by 1789 to Ipswich then 1790 to Framlingham
  again 1801 C

DICKERSON ——Continued

Daniel Jnr. b 1762 son of Daniel Snr. To
Harleston, Norfolk 1791-1807
DIXEY, C.W. & Sons. London sundial dated 1886
DOBBIE, John. Calton, Glasgow c1780-1826 C
DOBSON, Francis D. Driffield no date W
DODSWELL. see under Dadswell
DOIGE, John. Tavistock c1770 C
DOLBY, James. London no date W
DOLEMAN, John. Nottingham 1832- 54
DOMINICE, François. Geneva (B 1685) -c1700 W
DONALD, John. Kilmarnock c1820 C
DONKIN, T. Scarborough 1872
DONISTHORPE, Richard. London c1780 C (prob
the Loughborough man)
Richard. Loughborough 1773-1795 C
DOORNHECK, Cornelis. Amsterdam c1725 C
DORER, Bros. Nottingham 1877
FORLING, Bilby. b 1736 marr 1761 -1796 C
DORMAN, William. Lydd -later 18c? C
DORRELL, John. Tooley Street, London c1730 C
DOUGHTY, Samuel. Derby 1715
DOUGLAS, C. & J. Airdrie c1835 C
DOWNS, Benjamin, Mansfield 1790-1810 C & W
DOWNUM, Samuel. b 1723 app to John Smorthwaite
of Colchester where worked until 1751, then

Sudbury. To Colchester again 1753
DRAKE, S. Reading 1899
DRAPER, John. Maldon c1770 C
DRESCHER, Joseph Albert. Cottingham, E Yorks 1867
Joseph Albert. Hedon, E Yorks 1892
DRURY, William, Nottingham 1832
William Edward. Hull 1872
DU CHESNE, Laurens. Amsterdam early 18c? C
DUCHER, Hans. Sometimes Tucher. Nuremberg
sundial dated 1560-1586
DUCK, Richard. b 1711 Ipswich app to 1727 Thomas
Reynolds London. Ipswich again 1731 -d 1762
C & W
DUCKER, Charles. Chester Repaired T C 1604
George. Chester Repaired T C 1599-1602
DUFF, David. Kinross c1810 C
Thomas. Luncarty, Scotland c1780 C
DUGMORE, John. Derby 1753-91
DUMPER, ——. London 1828 W
DUNCAN, Robert. London W no date
DUNLOP, Andrew. Glasgow c1710 C
DUNN, W. Hull 1796-7 W
DURRAN & Son. Reading 1899
DUTTON, Alfred. Liverpool 1865 W
Thomas. West Bromwich c1810-1815 W
DYBALL, William. Bungay 1810- 40
DYXSON, Roger. Bridgnorth 1550 Repaired T C

E

EADES, William. Dublin app 1734 to Thomas
Blundell C
EADEY, Thomas. In Hallgate, Wigan 1794 C
EAST, John. Dublin app 1656 to Daniel Bellingham.
He was son of John East watchmaker of London
EASTERSON, Thomas. Woodbridge Repaired T C
1722- 3
EASTS,George. Gunthorpe, Notts d 1727 C
EATON, Thomas. Chester Rep'd TC 1599-1624
EAYRE, George. Kettering b 1705 son of
Thomas Eayre d 1749 C
Joseph. St Neots and Kettering b 1707 son of
Thomas Eayre d 1772 C & bellfounder
Thomas Jnr. Kettering b 1691 son of
Thomas Snr. d 1758 C & bellfounder
Thomas III, Kettering. son of Thomas II
marr 1748 d c1760 C & bellfounder
EBBS, John. Dublin 1762-74 W
William. Gorey, Ireland 1858
EBORALL, John Jnr. b Coventry, at Warwick
c1820-50, then Leamington 1850- 54
James. Warwick prob brother of John Jnr.
b c1808, working 1838- 51
Edward. Warwick brother of James b c1816 -1851
ECCLESHALL, Charles. Newport, Shropshire
b 1843 -61 W
ECCLESTON, Edward. Dublin 1753-60 W
ECHTER. See Egter
EGTER, Pieter. Dordrecht c1740 C
EDGE, Griffith. Prees, Shropshire Kept TC 1668

EDMOND, E. Liverpool W no date
EDMONDS, ——. B'Castle (Bishopscastle or
Barnard Castle?) c1760
EDWARDS, ——. Devonport c1825 C
——. Oakengates, Shropshire 1868
Benjamin. Bungay Rep'd TC 1690
Edward. Bishopscastle 1771-91
Edward. Ludlow b c1840 -1866
son of Robert Edwards of Ludlow
George. Ludlow 1850 W
Henry. Birmingham 1842 W
John. Cirencester c1810 C
Robert Jnr. Ludlow b c1846 son of
Robert Snr 1900 W
Samuel. Hull 1872
Samuel Jnr. Bridgnorth b c1825 Son of
Samuel Snr d 1856
Thomas. Haverhill, Suffolk c1770 C
William. Dublin 1774 W
William. Ludlow b c1829 son of Robert -1851 W
William. Shrewsbury 1796 -1807 W
EEDS, John. Dublin app 1734
EGAN, William. Cork 1858
EGGLESTON, Richard. Hull 1747 ret'd 1779
EHRHARDT, William. Birmingham b 1832 Germany
At Birmingham by 1858 (empl 11 men and 2 boys)
Son, William Jnr b 1858 d 1897 W
ELDRED, Dodson. Son of John Eldred of
Newton, Suffolk app 1754 in CC
to John Marigot of London, free London 1782

ELEY, Joseph. Derbys 1843
 William. Scropton, Derbys 1748
 (also London?)
ELIAS, Thomas. Haltwhistle c1770 C
ELLABY, Edward. Derby 1846
ELLEBY, John. Derby 1846 (prob the Ashhbourne
 man)
 John Jnr. Ashbourne b 1803 son of John
 Snr. - d1870
ELLES. see Ellis (variant spelling)
ELLIOTT, Edward. Bolsover 1895
 Ernest. New Brampton, Derbys 1895
 Thomas. Dublin free 1698 W
ELLIS, F.H. Dublin 1868
 George. Oakengates, Shropshire 1842-50
 J. Chapeltown (this is the Leeds man)
 James. Canterbury 1674 W
ELLSON, William. Belper 1827-29
ELOC, Richard. Ipswich signature (reversed)
 of Richard Cole qv
EMBERSON, James. Marden, Kent 1826-28
EMERSON, ——. East Dereham c1880 W
 Robert. Boldon, Northumberland, school-
 master 1770-1805 known to have made
 sundials, one dated 1775
ENGLAND, William. North Petherton, Somerset
 b c1801. 1838-42 d 1860 C
ENGLEFIELD, William. Place not known, repairer
 1838
ENGLIS, D. Auchtermuchty, Fife c1810 C
ENRIGHT, Michael. Dublin 1858
ENSOR, James. Ilkeston 1884
 William. Derby 1839
ENWORTH, Thomas. Hessle, E Yorks 1872
ESSEX, H. London no date W
ETCHES, John. Nottingham 1832-48
 Mrs Mary. Nottingham 1848-53
 Thomas. Ashbourne 1850- 85

ETCHES —— Continued
 William J. Hadfield, Derbys 1895
 Williamson, Nottingham 1834
EVANS, Diego. London c1775 C
 Edward Sheen. b c1855 Builth Wells,
 Bishopscastle 1871
 Eliza. Cork 1858
 James. Shrewsbury b c1821 son of James Evans
 qv d 1858 C
 James. b 1709 Oswestry, to Shrewsbury 1732-74 W
 John. Pwllheli c1780 C
 John. Shrewsbury b 1783 son of Richard
 Evans watchmaker -1824 W
 John. Shrewsbury 1796 C
 John. Wirksworth 1860
 Mary. Shrewsbury 1838- d 1862 widow of
 John qv W
 Philip Henry. Shrewsbury b c1823 son of
 William Evans, Watchmaker qv -1862 W
 Price James. Shrewsbury b 1746 son of James
 Evans watchmaker - 1796
 Richard. Shrewsbury b 1754 son of James
 Evans watchmaker -1783 to Oswestry c1789 -95
 W and goldsmith
 Richard. Shrewsbury 1806-14
 Thomas. Cromford, Derbys 1853- 55
 William. Dublin 1802 W
 William. b c1789 Shrewsbury 1812 -d 1847 C
 William & Son Matlock 1895-1928
EVERALL, John. B c1759 Worked Minsterley,
 Shropshire, but d Presteigne, Radnor 1815 C
EVERETT, James. Damerham (Wilts?) c1740 C
 Jonas. Bardwell 1717 -d 1748 Locksmith
EVERINGTON, A. Nottingham 1864-1912
EVERS, Robert. Dublin app 1693 -1716 C
EWER, Thomas. Clare, Suffolk 1791-98 C
EYRE, James. Limavady, Londonderry 1858
EZARD, Thomas F. Driffield 1867-97

F

FAIRCLOTH. See also Fairclough
 Joseph. Carlow, Ireland 1770 W
FAIRCLOUGH, E. Liverpool b 1774 1803-1831 W
 Evans. Dublin 1823 -38 W some-
 times Faircloth
FAIRFIELD, Warren. Dublin 1742 W
FALLER, Clement. Norwich 1845 German clocks
 Joseph. Ballymoney 1810-20
FANNYMORE. See Finnymore
FARLEY, John. Fairford c1820
 Thomas. Dublin 1805-58 W Two such?
 John. Place not known br. clock c1790
 (the Fairford man?)
FARMER, Joseph. Ludlow 1851 W
FARQUHAR, William. Dublin 1819-23
FARR, Harwood. Doncaster early 19c? C
FARRER, Joshua. Brighouse, Yorks c1750 C
FARRER, Thomas. Saxmundham, Suffolk c1763, succ
 to N. Cooper ret'd. 1798 d 1819 C

FAWCETT, John. Dublin 1775-92 W & optician
 William. Dublin app 1712 free 1729 d 1763 W
FAWKES, Isaac. Abingdon 1780
FEHRENBACH, John. Belfast 1858
FELICETTI, Anito. Naples c1780 W
FELMINGHAM, Robert. Denham & Stradbroke,
 Suffolk marr 1781 -1801 C also innkeeper
 Robert. Bungay 1823
FELTHAM, B.D. Swaffham c1810 C prob R.D. qv
 Robert Daniel. Swaffham 1845
FELTON, Richard. app 1711 at Ludlow to
 Thomas Vernon free 1718 marr 1732. Later
 at Bridgnorth C
FENN, Joseph. London in CC 1715 to
 John Andrews
FENNELL, Thomas. Warwick 1822
FENNY, John. Dublin from 1723 -39 C
 Joseph. Dublin free 1723 C
FENTON, Joseph. Mansfield 1732

FENTON —— Continued
Benjamin. Chesterfield 1800-23
John. Bakewell 1822-55
Samuel. Chesterfield 1827- 29
Samuel. Worksop 1828
FERENBACH, M & D. Edinburgh c1840 C
FERGUSON, John R. Ballinrobe, Ireland 1858 W
Matthias M. Ballina, Ireland 1858
P. Liverpool 1774 W
FERNSIDE, Benjamin. Norwich 1854
FESSER, Andrew. b c1816 Germany. Shrewsbury 1841-51 C
FIELD, George. Dublin 1823-4 W
James. Hemstead (Hemel Hempstead?) c1730 C
Samuel. Hemstead (Hemel Hempstead?) c1710 C (of James?)
William. app in CC 1725 to James Lloyd, free 1734. To Bungay by 1750-57 C
FIELDEN, ——. Leominster c1750 C
FIELDING, Benjamin. son of Benjamin of Yarmouth, Woodbridge then by 1771 to Yarmouth W
FINCH, John. Wolverston, Co Durham sundial dated 1723
FINLAW, Samuel. App 1640 to Thomas Wright of Chester
FINLAY, William. Limerick d 1802 W
FINNYMORE (sometimes Finnimore). Christopher. London, member Merchant Taylors' Co, had Thomas Downes as app 1720. Worked Hatton Garden d 1783 W
FISCHER, R. Dover 1865
FISH, Edwin. Knottingley, Yorks 1841, joiner but longcase clock noted
FISHER, Arthur. Derby 1880
Henry. Preston c1775 C & W
Stephen E. Bridgnorth b c1838 -61 W
William. Derby 1895
FISHWICK, James Snr. Boughton, Cheshire d 1768 C
James Jnr. Boughton, Cheshire d 1743
FISKE, Thomas. Stowmarket 1767 -c1786 C (later farmer?)
FITT, Stephen & Co. Cromer 1854
FITZGERALD, Edward. Dublin app 1694
George. Dublin d 1761 W
Thomas Henry. Nottingham app 1794 to John Hudson
FIVEY, George. Dublin 1782- d 1812 W
FLAGGETT, Bartholomew. Newbury c1645-1707 C & T C
Bartholomew. Newbury. c1690-c1730 C
James. Newbury 1722 repaired T C
William Jnr. Newbury 1707 -1720 C
William Snr. Newbury 1680-1707 C
William. Newbury c1715 C
FLEMING, Hugh. Dublin 1786-96 W
William. Dublin c1710 C
FLETCHER, Charles. Shrewsbury kept T C 1724-39
George. Coalbrookdale 1803 C
George. Shrewsbury 1720 C
John. Thorne, W Yorks 1869
Robert. Chester 1782 - d 1841 C (may be two such)
W. Gainsborough late 18c C?
William. Newbury c1715
of GORDON & FLETCHER. Dublin 1805-30
FLETMAN, ——. Little Denham, Suffolk 1764
FLOWER, Robert. Retford c1750 C
Turvey. Dublin 1821-1831 W (known as Old Flower).
FOGARTH, Thomas. Dublin 1815 W & jeweller

FOLEY, Patrick. Dublin c1790-1800
FORBES & DALE. London W no date
FORD, M. Manchester 1792 W
FORDHAM, Joseph. Braintree c1710 -c1720 C
John. Bocking c1830 C
John. Dunmow c1720 C
Thomas. Braintree c1753 C
FOREST. see Forrest
FORREST, John. London 1897 W
Simon, Kirkfieldbank, Fife c1800-37 C
same man as at Lanark qv
William. Chesterfield 1884-91
FORSTER, Charles. Dublin 1868-74 (succ to David)
David. Dublin 1826-68
David. Dublin 1827-30 W C
Thomas & David. Dublin 1823-25 W & engravers
FORTUNE, Ambrose. Wexfod 1858
FOSS, William. Hull d 1792
FOSTER, Stockton. late 19c clock cleaner
Will. Marnham, Notts mid-18c? C
FOULKS, ——. 'At the Three Lions', Salisbury, 1701 W
FOUQUET, Noel. Tours, France c1800 C
FOWLE, Nathaniel Jnr. Boston 1804 W
FOWLE, Thomas. East Grinstead 1789 W
FOWSTON, John. Hull 1792, partnership with James Russell dissolved 1795 W
FOX, John. Grantham c1750 C
John. Great Easton, Leics c1810 C
Jnr. Portsmouth c1760 C
Thomas. Canterbury 1878-88
FOXTON, James. Sutton in Ashfield, Notts b 1764 son of Richard Foxton & wife Elizabeth (Boot, of the clockmaking family), took over the Boots' shop in King Street d 1835
James. Sutton in Ashfield 1790 -d 1829 C
John. Mansfield 1822
Richard. Derby c1760 C
FRANCIS, A. Dublin c1805 W
Edward. Maesbury, Oswestry 1756-66 C
FRANCOMBE, ——. London 1774 W
FRANKLAND, J. North Walsham, Norfolk 1854
FRANKLIN, Robert. London c1750 C
FRANKS, Zachariah. Hull 1872
FRASER, Alex. Inverness clockcase maker 1840
FREBOUL, John. Dublin app 1741
FREEMAN, Thomas. Oakengates, Shropshire 1863 W
FRENCH, Robert. Dublin app 1773
FRENGLEY Bros. Dublin 1868-80
Jacob. Dublin 1874
FRESTON, Albert. Ipswich 1868
FRIEND, ——. Newton (Abbot?) 1839 prob same as W. Friend
FRIERSON, John. Derby 1710
FROST, Robert. Nottingham 1777-83 C
FRY, John. Hexham 1868 W
John. Sutton (in Ashfield) c1760 C
Samuel. Dublin 1820-26 C
FRYDE, C. Sunderland 1863 W
FRYER, ——. Bramley (not known which one) late 18c C
Edward. Hull 1872
John Harry. Hull 1872
FULLER, Richard. Cork 1817-20 W
Robert. Watton, Norfolk c1775-90 C
FULLERTON, Samuel. Dublin 1858
FULTON, J. Dumfries c1740-50 C
Joseph. Aughnacloich, Argyll 1858 W
FURBER, Thomas. Prees, Shropshire repaired T C 1740

FURBER —— Continued
Thomas. Bridgenorth b c1811, 1841 W
FURNACE. George. Dublin 1751-81 W

FURNIVAL. Benjamin. Stockport c1790-1800 C
FUSTON. John. see Fowston

G

G. G. Monogram of George Graydon. Dublin qv
GADSBY, Thomas. Mansfield 1844
GAINSBOROUGH, John. b 1711 Sudbury brother
Humphrey. To Beccles 1743 then Sudbury again
where marr 1745 C & inventor
GAITE, William. Bowlish, Somerset mid 18c C
William. Shepton Mallet c1740-1750 C
GALABIN, John. Greenwich bc1699-1763 W
GALLEMORE, John. Derby 1847
GANNEY, William. Bacchus Walk, Shoreditch 1832 W
GARDENER, John. Oswestry marr 1744 d 1765 C
GARDNER, John. Inverness 1840 Casemaker's label
inside - Alex Fraser cabinetmaker Inverness C
GARDINER, Henry. Dublin 1788-90 W & C
John. Wrentham 1855-64
Thomas. Sudbury b c1696 marr 1751 d 1751 C
GARRARD, Jacob. Long Melford 1855
James. Woolpit, Suffolk 1841-55
GARRETT, George F. Ripley, Derbys 1884
Derby 1888
GARROD, John Frederick. Woodbridge 1850
GARTY, William. Dublin 1802-30 W
GASCOIGNE, Owen (Owin). Newark b 1647 son of
William Gascoigne Snr d 1719 C
William Jnr. Newark son of Owen Gascoigne
c 1700 -d 1740 C
GASGRAVE ——. Middleham, Yorks sundial dated
1788
GASKIN, John I. Dublin app 1725 -32 W
John II. Dublin 1749
John III. Dublin 1761-65
John IV. Dublin 1816 -d 1834
GATES, Charles. Ipswich b c1788 d 1808 W
GATLAND, Edward. Cuckfield, Sussex c1700 C
George. Cuckfield, Sussex mid-18c
GATZ, W. Place not known (repairer). May be
London 1821
GEILLER, ——. Paris c1810 C
GEORG, Johan. Griessbeckh, Ausburg mid-17c
GEORGE, John. Dublin 1822-30 C & W
William. Leamington 1834- 51
GEORGESON, Edmund. Liverpool no date W
GERVAS, Thomas. Epworth c1750-1780 C
GIANNE, Lewis. Shrewsbury 1809 d 1816,
barometers
GIB, Pieter. Rotterdam c1720-1725 W
GIBB, William. Newcastle (which?) c1775 C
GIBBONS, W. Nottingham 1864
GIBBS, Alexander. Kintore, Aberdeen c1720 C
W. Plymouth c1775C
William. London (B app 1736) 1766 W
GIBSON, ——. Alfreton 1800
Frederick William. Newbury 1884-1903
Henry. London 1776 W

GIBSON —— Continued
Joseph. Batley, Yorks b1708, travelling
clockmaker
GILL, Caleb. Wellington, Shropshire b 1844 -68 W
John. Idle, Yorks late 18c C
Peter. Aberdeen c1785 C
William. Winster, Derbys 1795
GILLBERT, H. Coventry late 19c C
GILLESPIE/GILLESPY. Charles. Dublin 1747 -d
1769 (one such app 1734 London) C & W
GILLINGS. James. Brother of Thomas Coventry to
1851 then Liverpool by 1861 then Manchester
by 1871 then to South Africa c1880 W
GILSON. Henry. Place not known 1698 W
Thomas. Pocklington 1851
GITTENS. George. b c1829 Kinnersley. To
Oswestry by 1851 W
GITTINS. W. Shrewsbury 1786-1806 barometers
GLASCOE. Philip I. Dublin 1729 W
Philip II 1740- 67
Thomas. Dublin 1774-85 (B ante 1756) W
GLASE, Edward. Bridgnorth c1770-1807 C
GLAVERT. Edward Albert. Buxton 1888-1908
GLOVER, George. Shrewsbury from London 1818-23
GODDARD, P. H. Darlington c1840 C
GOLDEN. John. London c1780 pedometers
GOLDSBY, ——. Derby 1821
GOLDSMITH, ——. Liverpool 1882 W
GOLLAND, Edward. Retford late 18c? C
GOODALL, ——. Tadcaster sundial dated 1846
prob George II qv
Matthew. Leeds no date W
GOODHUGH, R. & R. Welbeck Street, London
c1830 C
GOODMAN, A. Hull 1871 W
GOODO. Francis. London 1705 W
GOODWIN. F. H. Canterbury 1888
Henry Jnr. Newark 1842 -56 W & C
Joseph. Newark 1844-50
GOODYER. James. Guildford c1740-50 C
GORDON, Alexander. Dublin 1756 -d 1787 C & W
James London (B app 1806) 1825
John. Dublin 1784-85 W
& FLETCHER. Dublin (B 1795) 1805-30 W
GORE, William. Nottingham 1853- 68
GORELY, Harry. Wokingham 1899
GORSUCH, Thomas, (also Gossage). Shrewsbury
b 1683 free 1701 d 1727 W
GOTHARD, ——. Newark repaired T C 1658
GOTOBED, Richard. London c1780 C
GOUGH, Nicholas. Newbury c1690-1700 W
GOWAN, ——. Bonnyrigg, Midlothian 1882 W
GRAHAM, John. Carnwath, nr Lanark c1760 C
GRAINGER, Abner. Condor, Derbys 1877-84

GRANGER, Marc. London c1690 C
GRANT, Daniel. London c1820 C
J. Shrewsbury 1828
GRANTHAM, John. Newbury (believed from London 1770-90
GRAY, Henry. Lesslie, Scotland c1830 C
James. Kinross c1810 C
Joseph. Durham marr 1715 -d 1768 C & W
Townsend. Dublin free 1752 C & W
GRAYDON, George. Dublin app 1753 to Charles Gillespy d 1805 W
GRAYSON, Matthew. Gosforth, Cumberland 1811, clock cleaner
GRAYSTON, Joseph. Ipswich 1844
GREAVES, J. Mansfield 1770
GREEN, G. L. Beverley 1892
George Jnr. Son of George Snr of Bridgnorth b c1837 Liverpool. Bridgnorth 1870-71 W
John. Ampthill 1701 C
John. Shrewsbury 1686-88 W
John. Wantage 1883
John. Yoxford, Suffolk 1844
Thomas. b c1841 Bilston (son of George of Bridgnorth) Bridgnorth 1870-71- W
Thomas. Liverpool 1781-96 C
Walter. Ironbridge 1868
William. Newbury (believed from Milton, Berks) 1795-1812
GREENFIELD, William. London 1796 W
GREENWOOD, A. Dewsbury c1770 C
Charles. Grantham 1790 sundial
William. Leeds late 18c C
GREGORY, James. Place not known (repairer) 1844-58
Richard. b c1812 Frome, Bridgnorth 1861 W

GREGORY —— Continued
W.T. Gloucester mid 19c W
GRENAT, David. Utrecht early 18c C
GREY, John. Chester repaired T C 1547-5
Richard. Chester repaired T C 1551-52
GRICE, Job. Lancaster c1740 (B 1797-1830), two such?
GRIFFIN, George. London 1824 W
GRIFFITHS, W. Port Talbot early 19c C
Thomas B. Derby 1871- 1912
GRIMADELL, S. Snr & Jnr. 1760-1857
GRIMADALE, Peter. Stamford 1757- 59
GRIMES, E. Bideford c1880 C
C. Ipswich 18c? C
Goodman. Northampton mid-17c (same as William Grives qv)
GRIMWADE, William. Beccles 1839
GROMELS & SCHWAR. Greenwich 1874
GROOM, John. Calton, Edinburgh 1703
GROSVENOR, John. Market Drayton 1822-29
John. b c1783 Whitchurch, Shropshire, Cotton, Wem 1841-61 C
Robert Edwin. b c1830 son of Robert of Market Drayton. Market Drayton until c1856 then Ellesmere -1900
GRUBB, George. Derby 1884- 1908
GRUNERT, R. Beverley 1892
GUANRNERIO, ——. St Ives c1840 prob same as at Huntingdon C
GUEST, George. Place not known early 18c C
GUIHO, ——. Roche Bernard, France c1850 C
GUNNELL, Edmund. Newbury 1832
GWYTHER, Stephen. Bristol 1802-06 Tower Hill C
GLYDE, John. Lambeth, London 1749-51 in 1760s son, Charles, went to Birmingham but d Southwark 1819

H

HADDY, W. Plymouth no date W
HADEN, Henry. Dublin 1809 also jeweller W
Thomas. Place not known c1750 C
HADERER, George. b c1826 Germany. Shrewsbury 1841 C
HADFIELD, Joseph & Son. Chapel-en-le-Frith 1828
HADLEY, Francis. Warwick b c1832 working 1851 C
HAGON, Thomas. d 1796 Kesgrave, Suffolk 'a poor travelling watchmaker' believed from Norwich
HAINES, N. A. Cardiff 19c C?
HALL, Charles. Derby 1854
D. Nottingham 1877
Devance. Alfreton 1852
Henry. Derby 1850
John Jnr. Grimsby (B 1795). Imprisoned 1807 for murder after a man died in a pub brawl
Leo. Southwark, London c1720 C
R. Llangollen c1825-1835. May be the Oswestry man
Samuel. Hocklington, Lincs c1790 C
William. Derby 1828

HALL —— Continued
William. Louth c1740 C
William. Nottingham free 1806, bankrupt 1832 succ by R. Bosworth C
HALLAGAN, John. Dublin 1819 W also goldsmith and jeweller
HALLAM, James. Nottingham 1832
John. Nottingham marr 1791 -1814, later Thomas qv
John & Thomas. Nottingham 1814, later Thomas qv
Thomas. Derby 1853
& STEVENSON. Nottingham 1818- 20
HALLIDAY, James. Dumfries c1850 C
HALLIWELL, ——. Wakefield c1850 C
John. Blackburn c1790 C
HALTON, Jeremy. Place not known N England? 1751 C
HAMBLYN, William. Hartlepool c1840 C
HAM(E)Y, William. Dublin 1802-19 also jeweller W

HAMILTHON, Hieronymus. Scotland 1595 W
HAMILTON, James. Reading 1780
HAMLIN, Richard. London 1677 W
HAMMONDS, John. Caynham, Shropshire 1835 C
HAMNET, Samuel. Hull 1872
HAMPSCHER, William. Bungay repaired T C 1586
HAMPSON, Joseph. Wrexham c1760-72 C
Robert. Dublin 1827-30 W
HANCOCK, Walter. Newbury 1899-1903
HANDLEY, James. Upton, Notts 1828 -d 1831 C
HANLON, William. Dublin 1809-30 W
HANNIWELL, L. London early 18c C
HANSOM, George Jnr. Middlesborough 1872
HARBERT, Morgan. London 18c C
HARDAKER, ——. Salem, N England, prob Yorks
c1800-1820 prob same man as at Bramley, Leeds C
HARDWICK, Richard. Ashwick, Somerset 1696-1770
HARDY, Richard. Nottingham 1834-44
Thomas. Nottingham 1853
Thomas & Son. Newark 1830-50 W
William. Nottingham 1799 W
HARE, William. Place not known c1760 C
HARFORD, John. Bristol c1720 C
HARGRAVES, ——. Colne c1785 (prob Thomas
Hargreaves of Burnley qv) C
James. Derby 1852
William. Settle, Yorks c1750-80 C
HARLAND, R. Bury St Edmunds b c1772 marr
1794, 1819 at Bungay (free at Yarmouth 1796
but prob did not work there) W
Samuel Boulton. Ashbourne b 1751 son of Joseph
Harlow. Working until 1813 (also at
Birmingham?) d 1815
Samuel. Shrewsbury son of William Harley
clockmaker 1766-99 C
William. Shrewsbury b 1766 son of Samuel
Harley clockmaker d 1843 goldsmith & W
HARLOW, Amelia. Ashbourne widow of Robert Harlow
Snr whose business she cont'd after death
of son, Benjamin Wyatt Harlow in 1845 -d 1853
George. Ashbourne c1740
Joseph. Ashbourne (from Birmingham) b c1705
marr 1750
Robert Snr. Ashbourne b 1799 son of Samuel
Boulton Harlow d 1828
Robert Jnr. Ashbourne son of Robert Harlow
Snr. To Stockport by 1842 brassfounder & C
HARMER, Samuel. Beccles 1799 jeweller
HARMAN, Samuel. Tenterden 1789
HARPER, James. Nottingham 1815-34
Mary. Nottingham 1842
Richard. Nottingham 1832-50
Richard. Shrewsbury son of Thomas Harper,
watchmaker, to whom app 1748 -d 1791 W
Thomas. Shrewsbury marr 1723 -1774 W
HARRINGTON, James. Ipswich b c1696 marr
1728 at Lavenham d 1729 founder & C
HARRIS, ——. Newbury repaired T C 1619
Abraham. Hull 1855 W
Christopher. Maidstone 1793'd 1795 W
Edwin. Derby 1895
J. Manchester 1874 W
James. Shoreditch, London 1832,
finisher W
John. Bury St Edmunds d 1742 W
Joseph. Mayfair, London c1770 C
Richard. Wellington, Shropshire b 1738
marr 1765 C
Stephen. Tunbridge (Wells) early 18c C
Thomas. Lidgate, Suffolk c1740-80 C
William. Chippenham c1770-82 C & W

HARRIS —— Continued
William. Mildenhall 1823
William. Shrewsbury app 1723 to Thomas
Harper marr 1736 C
HARRISON, Boehm. Stamford no date W
Frederick. Turnditch, Derbyshire 1876-1900
George. Newark 1840-60
Joseph. Bilston c1790 C
Samuel. Kenilworth b 1820 working Warwick by
1851-54. Later & Sons qv
William. Derby 1849
HARRYS, William. Rye d 1559 C
HARSANT, Walter. Canterbury 1878
HART, C. H. Hull 1846
E. Hull 1846
Samuel. Stonham, Suffolk b 1694 marr 1729
d 1775 (some clocks signed without place-
name, some signed at Stonham Py (= Magpie
Inn, his home) C
William George. Norwich 1854
HARTLEY, John. Blackburn c1800 C
HARVEY, G. W. Wellington late 19c C
HASKAND, Samuel. Derby 1891-1922
HASLAM, William. Sutton in Ashfield, Notts 1740,
clockcase maker
HASLEWOOD, Ann. Stamford b 1729 widow of
Samuel 1744 -d 1811 W
Samuel. Stamford d 1744 W
HASTEY, J. Pudsey, Yorks c1840 C
HATCHISS, ——. Derby 1771
HAUK, ——. Edinburgh c1835 C
HAWKE, Charles. Southwold 1844
James. Bury St Edmunds late 18c C
HAWKES, John Forrester. Albrighton marr 1802
(from Walsall) d 1917 C & W
HAWKINS, James. Bury St Edmunds pre-1810 W prob
error for Hawke qv
Luke. Thatcham, Berks 1715
William. Bury St Edmunds son of Mark Snr
b 1703 d 1775 W & C
William. Newbury c1710
HAWKESWEL, Thomas Snr. Brompton, Yorks
d 1801 C
HAWKSWORTH, John. Wakefield sundial dated 1799
HAWLEY, N. A. Regent Street, London mid 19c C
HAWNEY, W. Sandhurst, Kent sundial dated 1720
HAY, Thomas. b c1756 marr 1787 Bishops Castle,
to Shrewsbury c1801 d 1829 W
HAYES, George. Barlow, Derbys 1710
HAYLES, ——. Ipswich 1629 C
HAYNES, Alfred. Heanor 1888- 95
John. Cork c1800 C
& Son. Cork c1830
HAYOT, ——. Argentan, France c1730 C
HAYWARD, F. Dover 1865
George. Reading 1899
Thomas. b c1788 Birmingham, to Bridg-
north by 1841-61 C & W
HAYWOOD, John. Northwich c1820 C
HAZELDINE, William. b c1646 Rowton, High Ercall,
Shropshire by 1672 to Cold Hatton by 1693 d 1726
T C ——. Berrington, Shropshire 1753 prob
a son of William
HAZLEHURST, James. Mansfield marr 1772 C
HEAFIELD, John. Ashby, Leics? 1773 -c1820 C
HEARSUM, David. East Bergholt, Suffolk 1855
HEATER, John. Wantage 1885
HEATH, Robert Snr. Warwick 1752 -d 1775 C
Robert Jnr. Warwick c1789 -d 1807 C
HEATHCOTE, R. & J. Derby 1895-1932
& Son. Woodville, Derbys 1891

HEDLEY, Amos. Stamford from London 1746 free 1747 (B 1751) W
HEITZMAN, Matthew. Derby 1852
Anthony Joseph. Canterbury 1878-82
HEL(L)IWELL, Richard. Buxton 1876-84
HEMMEN, Edward. London no date W
HENDERSON, E. Middleton (which one?) c1830 C
James. Dublin 1794-1806 C & W
James II. Dublin 1807-26 W also gold-smith & jeweller
James. Edinburgh 1867 W
HENDRIE, J. Place not known 1755 C
HENSON, Thomas. Yaxley, Suffolk c1780 C
HERBERT, Thomas. Ludlow b c1815 -51 W
HERDEN, Harry Jnr. Dartford 1874
HESELTON, James Cape. Beverley son of George b 1831 d 1919 succ to J. C. Fox. Also at Bridlington in summer only
HESSLEWOOD, David. Hull 1747-54
HESTER, ——. Dublin c1710 (believed Wiliam) d 1722 C
HETTICH, J. & Son. Cardiff late 19c C
HEWETT, George. Same man as Hewitt, George qv
HEWISON, John. London no date W
HEXT, Giles. Reading 1864
HIBBINS, J. Stamford 1880-82
HICCOX, John. London c1720 C
HICK, Matthew. York b 1791 d 1834
HICKMAN, Henry. London 1680 not in CC
HICKS__, J. Stamford 1862-68
HIGGIN, John. Peel, Isle of Man c1840 C
HIGGINS, Thomas. Chesterfield 1895-1922
HIGGONS, Joseph. Bridgnorth 1680 T C
HIGGS, John. Wallingford 1780
Robert. London in 1718 took John Cotton-belt as app
Thomas. Ludlow kept T C 1565-1606 when d aged 100
HIGHFIELD, William. Liverpool 1761 then Wrexham, then Oswestry 1778 -d 1782 W
HILL, Albert. Reading 1864
Dennis. Kirkby Moorside 1872
George. Lambourne, Berks 1740- 70
John. Bakewell 1846-76
John. Cromford, Derbys 1835
John. Lewes c1710 C
Thomas. Lambourne, Berks 1721-71
Thomas. Lambourne, Berks 1849
William. Rochester 1847
HILLIER, Benjamin. Place not known (repairer) 1836
HILLS, Edward. b 1840 son of Benjamin of Sudbury. London, the post 1859 Sudbury again as Hill Bros -d 1928
HILTON, Evan. Shrewsbury 1667 W
HIND, John. Derby 1839
HINDERSON, James. Skelton, Cleveland 1872
HINE, James & Co. Exeter c1775 brassfounder, known to have made C
HINES, Edward. Needham Market marr 1748-74, then partner with William Alderman c1776 d 1780 C
John. Prob son of Edward, from Harwich to Ipswich where succ W. Swaine, the to Needham Market by 1783, ret'd 1809 when succ by J. Bethell C
HISCOCK, James. Newbury c1832-68
HITCHCOCK, ——. Hermitage, Berks 1849
E. Bingham, Notts, trade label in clock case, late 19c or early 20c

HOADLEY, William I. Rotherfield, Sussex b 1685, son of Thomas Hoadley d 1756 C
William II. Rotherfield, Sussex marr 1758 to Sarah Dodswell d 1763 clocksmith
HOBART, William. Bury St Edmunds 1823-39
HOBSON, Henry. Goostrey, Cheshire repaired T C 1701
HOCKER, Joseph. Basingstoke 1730, then Reading, 1730-40 C
John. Reading 1682-1729
HODGES, Fred. Dublin 1796-1830 C & W
HODGHON, ——. Preston c1780 C
HODGSON, B. Auchenblae, Kincardine 1879 repairer
J. Middlesborough 1872
see also Hagon
HOLBIN, George. South Cave, Yorks 1851 W also guns and gamekeeper
HOLBROOKE, John. Dublin 1828-30 C & W
HOLDEN, Benjamin. App to D. Read at Ipswich, free 1832, Woodbridge then Ipswich 1841
James. Derby 1849
HOLLIDAY, John Thomas. Horsea, E Yorks 1879, known as 'Clocky Holliday'
HOLLIER, Samuel. London London 1720 (B ante 1777 W)
HOLLINGWORTH, Walter. Sutton on Trent, Notts, 1864 W
HOLLISON, E. H. London no date
HOLLIWELL, Francis. Derby 1843
John. Derby no date W
HOLLOWAY, John. Newbury app to William Hayward c1675-95 C & T C
John. Newbury c1823
HOLLY, John. Selby c1715 C
HOLMES, Christopher. Dublin 1805-09 W & C
James. Derby 1819-57
John. Wolverhampton b 1792 there 1812 -d 1869 C at 9 Cock St (B has Joseph prob error)
Robert. Dublin 1761 -d 1787
?Hutchinson, John Birstall, Yorks c1790 C & Smithard. Derby 1835- 46
Thomas. Cheadle, Staffs no date
HOLT, John. Newark 1820 C
John Henry. Retford 1853
Louis. Hull 1872
Richard Snr. Newark 1810-45 W & C
Richard Jnr. Newark 1840-50
& Unwin. Newark 1805-10, Richard Holt Snr.
Valentine. Newark 1818-21
HOLWILL, Robert. Sidmouth c1805 C
HOME, William. London no date W
HOOD, John. Stroud c1775 C$P
HOPE, Joseph Jackson. Matlock 1884
William. London 1821 W
HOPEWELL, R. London late 18c W
HOPKIN, William. Nottingham 1832
HOPKINS, ——. Dublin 1835 W
Maria. Dublin 1858
Henry. Deptford (B pre-1780) 1783 W
HORLICK, James. Wantage 1899
HORN, William. Glossop 1860
HORNER, William (sometimes Horney). Hutton-le-Hole, N Yorks 1851-56 C
HORNEY, William. See Horner, William
HORTON, J. B. Canterbury 1882-89
HOSE, Charles. From Sneinton, Notts, free at Leicester 1826
HOSKIN, Thomas J. Hull 1872
HOSMER, ——. Tonbridge (B 1790 CC) c1800-10 C
HOTCHKISS, William. Derby 1771

HOUGH, Edwin. Nottingham 1877
HOUGHTON, Richard. Prees, Shropshire repaired
T C1670-71
HOULAHAN, John. Cashel, Eire 1835 W, prob
related to Edward H. qv
HOULDEN, John. Kendal c1750 C
HOULGATT, William. Ipswich marr 1617 d 1637-44 W
HOULSTON, John James. Derby 1888-1922
HOUSTON, Shean. Dublin app 1745 free 1766 -1802
HOW, James III. Bromley son of Thomas 1870
d1910 John. Bromley son of James I 1843-69
Thomas. Bromley son of James I 1843-70
HOWARD, William. Dublin 1767-94 W
HOWELLS, A. Tredegar c1825 C
HOWES, Henry. Ixworth, Suffolk C, prob the
Ballington man
HOWIE, Thomas. Ceres, Fife c1810-15 C
HOY, William. Hingham, Norfolk c1847 repairer
HUBBARD, John. Bury St Edmunds 1823
HUBER, Lawrence. b c1816 Germany, to Shews-
bury by 1841 d 1842 C
HUCKNEY, Peter. Buxton 1891-1932 (& Son after 1922)
HUDSON, ——. Windsor c1815 C
Anthony. Preston c1760 C
William. Thirsk 1872
HUGHES, Christopher. Dublin 1830 watch case maker
HULL, Charles. Dublin 1774-1783 W
F. Dublin 1808
John Charles. Hull 1872
N. F. Dublin 1804 W
HUMLLASON, ——. Gosport c1760-70
HUMMEL, John. Reading 1899
HUMPHREYS, H. Beeston, Notts 1864
Robert. London 1759 W
HUNSDON. See also Hundson

HUNT, J. R. Leicester c1770 C
Robert. b c1802 Horncastle, Ellesmere 1861- 63
partner of Evans & Hunt W
HUNTER, William. Filey 1851
HURLSTON, A. London 1883 W
HURST, Michael. London no date W
HURT, George. Bakewell 1884- 1912
Jesse. Chesterfield 1891, Bolsover 1908
HUTCHINSON, Catherine. Retford 1822
John. Nottingham 1755 W
Thomas. Leeds c1780 C
Thomas. Worksop 1710-64 C
William. Retford 1776-98 C
Benjamin. London app 1682 to Edward
Hutchinson (father?) in Poulter's Co. Lost watch
recorded 1716
HUTTON, ——. Shrewsbury repaired T C 1757
HYAM, Simon. Ipswich 1798-1801 W
HYDE & Sons. Sleaford (Louisa, widow of Thomas,
followed by son, Tom Lelie Hyde, then his widow
Mabel, then his son Tom) 1913-56 C
succ by Hoppers of Boston
& Sons. Sleaford (Anne, widow of John I, together
with sons John II, Thomas & William) 1853-1873,
when John II ret'd C
Thomas. Sleaford 1873 -d 1913
HYAM, Simon. Ipswich 1798-1801 W
HYDE & Sons. Sleaford (Louisa, widow of Thomas,
followed by son, Tom Lelie Hyde, then his widow
Mabel, then his son Tom) 1913-56 C
succ by Hoppers of Boston
& Sons. Sleaford (Anne, widow of John I, together
with sons John II, Thomas & William) 1853-1873,
when John II ret'd C
Thomas. Sleaford 1873 -d 1913 C

I

IMMISCH, Bernhardt Theodore. Hull 1872
INGEMORE, John. Derby 1786
INGHAM, Samuel. Thirsk, Yorks prob same man as
at Ripon, 1822-26
INNOCENT, William. Sheffield 1860

IRWIN, Ormesby. Dublin 1802-04 W also jeweller
William. Dublin 1802 W also jeweller
IVORY, James. London c1760 (perhaps later at Dundee
& see B) C
INGRAM, John. Spalding c1740-50 (B ante 1753) C

J

JACKMAN, David. Ipswich 1868
JACKSON, ——. Newington, Surrey 1677 W
Aaron. Hornsea 1876-92
Emanuel. Derby 1880-95
Isaac. New Mills, Derbys 1891-1932
John. Boston, Lincs c1790 C
John. Dublin 1797-1819 W also jeweller
Jonathan. Repton, Notts 1822
W. Workington c1820 C
William. Derby marr 1780
William. Shrewsbury 1827 W

JACOB, Edmund. Great Yarmouth 1854
JACOBS, Aaron. Hull 1806
Ephraim. Sheffield 1860
H. Nottingham 1818
Joe. Hornsea, E Yorks 1891-1901 W
JACQUES, Abraham. Paris mid-18c C may be same
man as at Eamont Bridge
JAGGER,John. Canterbury 1784-91
JAMES, A. Canterbury 1888
David. b c1841 Brecon, Market Drayton 1861 W
J. Saxmundham 1864

JARED, George. c 1758 at Selattyn, bro of William. Went to Oswestry where d 1788 C
William. b 1744 at Selattyn, Shropshire marr 1770 Ellesmere c 1804 C
JARMIN, John. Long Melford 1683
JEANNERET, Louis. Chaux de Fonds 1890 W
JEFFERIS, G. Market Weighton 1872
JEFFERSON, John. Hull 1872
JEFFERY, Edward. Chatham 1792 (same as Jefferies qv)
JEFFREY, Jac. Carlisle sundial dated 1786
JEFFRIES, Edward. Much Wenlock marr 1786
JENKINS, F. E. Wallingford 1899
JENKINSON, Edward. Nottingham 1714, to South Wingfield, Derbys 1729
Morley. Nottingham 1740, marr 1743, later in Grantham
JENKYNSON, Nicholas. London? 1571
JEPSON, Forrester. Shrewsbury & Bishops-castle 1868 W
JERARD, Henry. Hindon (Hendon?) c1750 C
John. Hindon, Wilts 1741 C
JERRAM, Frederick. Derby 1895-1900
JERVIS, Henry. Newport, Shropshire 1730 C
JOHNSON, David. Dublin 1764 W
F. T. Dublin 1827-30 W
John. Liverpool early 19c W
Joseph. London 1791 W
Mary Anne. Dublin 1827
Samuel. Southwark c1760 C
Samuel. Nottingham 1823 W
Thomas. Dublin 1809-26 W
JOHNSTON, Thomas. Dublin 1741-85 W
JOHNSTONE, Thomas. Dublin 1804-06 W
JOLLEY, C. A. E. Mansfield 1894 succ W. H. Jolly (letter E added, sic)
JOLLIFFE, Robert. & Sons. Derby 1846-52
Simon. Derby c1829-31
JONES, David. Derby 1846-60
Francis. Wellingborough 1723-4 C
H. Chalford, Glos c1760 C
Harry. Nottingham 1864

JONES —— Continued

Henry Charles. Shrewsbury b c1808, 1830 -d 1838 W
James. Shrewsbury b c1841, 1861-65 W
John. Abingdon 1823
Joseph. Nottingham 1850
Thomas. Dublin app 1738 C
Thomas. Oswestry d 1832 W
Wiliam. Derby 1858
William. Beccles c1790-1810 C
William & FOSTER, James. Eccleston, Lancs 1833 W
JOOKES, Philip. Ludlow repaired T C 1542
JOSEPHS, ——. Swansea 1835 W
JOSEPHSON, ——. London 1788 W (may be John, B pre 1751-81?)
JOSLIN, Edward. Reading 1728 (may be same as Josseline, London)
JOYCE, Caleb. East Ilsley, Berks 1846
Caleb. Hampstead Norris, Berks 1840-84, also at Newbury, then Speenhamland after 1884
JOYCE, George. Bristol 1698 W
James. Whitchurch b 1752 Ellesmere son of John Joyce clockmaker, moved from Cockshutt to Whitchurch pre 1782 d 1817 C
John. b 1813 Boxford, Speenhamland, Newbury 1833-66 when succ by A. Stredling, for whom he was foreman d 1896
John. Ellesmere b 1718 son of William Joyce, clockmaker, marr 1737 d 1787 at Cockshutt C
John Barnett. b 1826 Whitchurch, Yorks, son of Thomas Joyce, clockmaker, supposed worked Bradford then back to Whitchurch 1871 T C
John. Tanderagee, Armagh c1790 C
William. b 1811 Boxford, Newbury & Hampstead Norris, where d 1895, bro of John Joyce of Newbury
William. Cockshutt, Ellesmere b 1691, to Wrexham 1718, back to Cockshutt 1723 d 1771 C
JUDD, Henry. Torrington, Connemara 1831- 40 C
JUDGE, Thomas. Frome 18c C
JULLION, John & Son. New Brentford 1762- 78
JURISTHOFF, ——. Bath early 18c C

K

KAISER, Andrew. Richmond 1872
KALTERBACK, George. Hull 1806
KAMMERER, Ignatius. Hull 1872
KAUFMAN, Joseph. Newcastle on Tyne 1866 W
KAYE, Joseph. B c1849 Durham, Market Drayton 1871
KEELING, John. Roddington, Shropshire T Cs 1692-d 1734. Another of this name also a clockmaker d there 1740
KEIR, Peter. Falkirk c1815-1820 C
KELHAM. See also Kilham
——. Epworth c1819-1820 C
prob the Holbeach man
KELLET, Thomas. Thirsk 1850-71
KELLY, Charles Aylmer. Dublin 1784-1815 W
KELSEY, ——. Nottingham 1722-96

KELVEY, James. Shrewsbury marr 1849 d 1850 W
KEMP & BROWN. Yoxford, Suffolk partnership pre 1771 of John Kemp & William Brown C
E.C. & Co. Leicester late 19c C
John. Yoxford, Suffolk 1774 ret'd 1790 d 1801
KENDLE, Richard. London W no date
KENNEDY, Roger & Patrick. Dublin 1794- 6
Roger. Dublin 1785-1814 W (-1825?)
KENT, William. Lynton (Linton, Cambs). 1741-6 perhaps later at Saffron Walden C
KERBER, Joseph. Carlsbad late 17c W
KERBY, Philip. London no date W
KERNETT, Wilfred. Reading 1899
KESSLER, S. Tunbridge Wells 1865
KEYS, Davis. London 1895 W

KIDDY, Joseph. Kirk Ireton, Derbys 1855-60 C

KILBURN, Richard. Liverpool 1826 W

KILHAM, ——. Epworth c 1820 C (see also Kelham)

KIMBER (sometime Kember), Edward. Shaw, nr Newbury c1779 C

KING, Edward. Warwick (b 1776) 1794 -d 1827 W & C
   F. Wokingham, Berks 1899
   George. Norwich 1854
   George. Woodbridge 1802-08 C
   John. Salterforth, W. Yorks later 18c C
   John. Warwick. 1820-1834 then Stratford c1835-50 believed son of Edward
   Robert. Pickering then Scarborough then Hunmanby 1808-d 1813. One time of Washington DC, son of Robert King Jnr of Washington W

KIRK & ABLITT. Stowmarket See Kirk, William
   Ann. Nottingham daughter of Joseph Kirk, clockmaker 1736-60 W

Joseph. Har(d)stoft, Derbys early 18c C

KIRKPATRICK, Martin. Dublin app 1712 to John Burton free 1721 -d 1769 W

KLEISER, Joseph. York 1866-85 C

KLO(C)K, Johannes. Amsterdam c1705-1710 C (same as Jan)

KNAPP, John. Reading c1690-1730

KNEALE, John. London mid 18c C

KNEBEL, ——. Amsterdam c1790-1810 W

KNIGHT, John. Riverhead, Kent c1760 C
   G. Ipswich water clock dated 1639
   Stephen. Ludlow repaired T C 1569 d 1579

KNIGHTS, James. Ipswich 1739-40 Whitesmith and C

KNOWLE, ——. Place not known lantern cl 1679

KREUTZ & FURDERER, Nottingham. 1848-53, later John Furderer

KUNER, Isidor. Canterbury 1878

KYNUIN (or KYNUYN), James. Nr St. Pauls, London 1584, sundial dated 1593

L

LADOUX, Jacques. Amiens 18c C

LAING, T.R. Glasgow 1874 W

LAMB, John. Whitchurch (b c1783) 1803 W

LAMBERT, Frederick. London son of George d 1869 watch case maker
   George, London. b c1830 escapement maker (son of Samuel)
   John. Hadleigh 1855
   Samuel. London son of Thomas of Islington Escapement maker c1820
   Thomas b 1768 Rochford. Islington, London c1800 Pendulum maker

LAMPARD, John. Zeals (parish of Mere, Wilts) 1859 C

LAMY, David. Hoorne, Holland late 17c C

LAND, Thomas. Place not stated. Early 18c C prob Tiverton?

LANE, P. Liverpool. 1795 W
   William. Calne, Wilts c1710 C

LARA, Abraham. London c1775 C

LARGE, S. Leek ?late 18c C

LARRARD, Frederick. Hull 1872

LASSELL, ——. Park. Same as William Lassell qv

LASSETER, Henry. Minsterley, Shropshire 1856

LAVINGTON, Bartholomew. London 1805 W

LAW, Alexander. Errol, Scotland c1820-1835 C

LAWLEY, Thomas. Rugeley c1720-1770 C

LAWLOR, Thomas. London 1804 W

LAWRENCE, Richard. Wellington, Shropshire 1851 W
   Robinson. Halifax c1800 C (see L. Robinson?)

LAWSON, W. Gateshead 1873 W

LAYCOCK, H. Settle, Yorks 1894 W

LEACH, Benjamin. Andover c1770 prob later at Winchester (see B)
   Edwin. b c1837 Saddleworth, Yorks to Dudley by 1866, to Ludlow by 1868- 71 W
   Robert. Stradishall, Suffolk marr 1712 C

LEADBEATTER, Charles. Wigan 1774 (at Hindley) W

LEAH, Samuel. London 1777 W

LEAKE, Daniel. From Ashbourne, Derbys, app London 1703 to Joseph Howes

LEBON, Samuel. London app in CC to Richard George, an orphan in care of parish of St. Mary le Bow

LECOMBER, John. Liverpool 1880 W, same man as at London

LECQUINT, Thomas. Bristol 1806 W at Lawrence Hill

LEDBETTER, Joseph. Brigg, Norfolk 1760 W

LEDERER, Joseph. Norwich 1845 importer of German clocks

LEE, Alfred. Sheffield 1860
   George & Son. Chesterfield 1884-95
   John C. Ashbourne 1895-1932
   Joseph. Newington Butts, London 1764

LEEDHAM, Robert. South Cave Yorks 1891

LEES, David. London b 1811-1851 C
   Philip Henry. Nottingham app 1828 to Charles Lees, free 1835
   William J. Nottingham 1877

LEGGETT, Thomas. Eccles marr 1757-70

LE GROS, Joseph. Chelsey, Berks 1808 prob from London where apr 1784 C

LEIGH, ——. Dunham, Cheshire c1780-c1785 C

LEMAISTRE, Charles. Dublin free 1736-43 W
   Henry. Dublin. 1788-1809 W
   Nicholas. Dublin. a 1739 to Charles Lemaistre free 1751-55 C W
   Peter. Dublin free 1708 -1719 W

LENDON, W. Taunton early 19c

LENSHAM, Anthony. London b1816 -1851 C

LENTHALL, John. Washington, DC 1813 C, son-in-law of Robery King qv

LEONARD, Joseph. Haverhill 1819 W

LEPLASTRIER, ——. Deal. No date W
John. Cockhill, Shadwell, Essex
No date W
LEROLLE,Fréres. Paris c1800 C
LE STRANGE, Anthony. Dublin 1794-1830 C & W
LEVER, Nathaniel. London app 1679 to John Wright
senior
LEVI, Abraham. Wellington, Shropshire 1840- 46
Lion. Plymouth 1810 W
Solomon. Nottingham 1844-53
LEVICK, William. Retford 1830-40
LEVY, George. Ipswich 1804 succ to S. Dammant,
marr 1808
Hyam. Ipswich son of Lazurus d 1822
Isaac. Hull 1806
Lazurus. Ipswich b 1746 1796 -d 1832
LEWIS, Thomas. Gloucester b 1656 -1682 W
William. Shoreditch, London 1832 at
City Road C
LEWTHWAITE (George?). Ulverston (B c1770 -95)
c1790 C
LEY, Alexander. Kincardine O'Neil 1876 W
LIART, Wilhelm. London c1815 W
LICKIS, James. Hull 1804 W
LIECHTI, Ulrich & Andreas. Wintertur,
Switzerland 1596 C
LINDLEY. Thomas. Leicester c1775 C & W
W.G. Melbourne, Derbys. 1888-1932
LINDSEY, John. Nayland, Suffolk b 1677-1733 C
LINGFORD, John. Nottingham 1792-1800 W
LINGWOOD, Samuel. Halesworth marr 1725 d 1741 C
LINTON, Robert. Richmond, Yorks d 1847 C
LISTER, Benjamin. Rotterdam c1670 C
& BROMLEY. Halifax. Partnership c1790-1801
between Thomas Lister Jnr and William
Bromley
LITCHFIELD, Matthew. Sutton in Ashfield 1853
LITHERLAND, John. Badsworth, Yorks. 1682 W
(must be the Liverpool man away from home)
LITHGOW, William. Portobello, Scotland c1825 C
LITTLE, Henry. Lancaster 1890 W
LITTLEMORE. ——. Frodsham c1720-1730 C
(may be same as Jacob Littlemore of Ruabon,
bankrupt 1728)
LIVENS, J.B. London no date W
Lloyd, Morris. Basingstoke c1750 C
Richard. Bridgnorth 1789-1809 W
LOADER, John. Basingstoke c1760 C

LODGE, John. Hull 1872
Thomas. Farnham c1710 C
LOGAN, John. Sheffield d 1809 W
LOHDER, Edward. London early 19c C
LOMAS, Joel. New Mills, Derbys 1891-1912
LONDON, John. Bristol marr 1678 C
LONG, John. b c1789 Hodnet, Shropshire, 1825-
d 1832 W
Leonard John. Dublin 1786-1800 C & W
& SON, Buxton. 1884- 95
LORMIE. Isaac. London 1704 at Vine Court W
LORMIER, ——. Shoreditch, London 1818 W
LOSELEY, Edward. Shifnal, Shropshire 1790 -1823,
(sometimes Loseby)
LOTON, John. Prees, Shropshire repaired T C 1767-70
LOUGHTON, Alfred. Southwell, Notts 1877
LOVERIDGE, Giles. Micheldean c1790 C
LOWRY, Miles. See Lowly
LOVE, John. Northampton 1756 C
W.H. York c1820 C
LOVIDGE, Thomas. Newbury 1764-82
LOWE, Richard. Mansfield 1864
Charles. Arbroath c1808 C
James. Arbroath c1840 C
LOWLY/LOWRY, Miles. Aberford, Yorks b c1723
- 1750 C
LOWREY, ——. Whitehaven c1755-1760 C
LOWTON, George. Winterbourne c1820-1840 C
LUCAS, John. App 1725 to John Seymour of
Wantage, Newbury c173-1750 C & W
LUKE, William. Oxford late 18c C
LUMLEY, George. Bury St Edmonds b 1723 son of
George Lumley wmr of London, app 1737 to
James Lloyd of London. Free in CC 1745. To
Bury by 1754 d 1784 W
Mary. Bury St Edmunds 1784 then c1785
partner with son-in-law John Gudgeon
as Lumley & Gudgeon until she died 1800 C & W
LUNDY, J. Pocklington 1891 W
& SON. Pocklington 1875 W
LUNNON, Frederick. Newbury 1903
LYDDEN, ——. Bristol c1840 (may be Lyddon) C
LYNAS, George. East Row, nr Whitby c1790 C
LYNCH, John. Newbury c1763-96 C
Robert. Newbury and Speenhamland c1760
LYON, Charles. York 1841
LYONS, Aron. Oswestry 1856-58 W, succ by
T. Clay

M

McAVOY, John. Liverpool 1864 W
McCANN, John. Place not known but believed
Scotland. Late 18c C
McLENNAN, A.J. Greenwich 1866-74
McCREADY, Andrew. Mercury Bay, New Zealand,
son of Thomas of Stranraer, emigrated 1853 W
John. Stranraer son of Thomas. Worked with
father, mid 18c W
Thomas. Stranraer marr Margaret Forest,
1836-60

McCREADY —— Continued
Thomas Wallace. Auckland, New Zealand Son
of Thomas of Stranraer. Emigrated 1853, d 1875 W
McCUTCHEON, Hugh. Beaumaris, Wales no date W
McDONALD, Patrick. Forres/Johnshaven, Scotland
1776 W
McGOUAN, David. Strathaven c1810-1820 C
McINNES, James. Glasgow 1844 W
McKENZIE, John. Sheffield 1860
MacKNIGHT, Henry. Alfreton 1884-95

MacMINN, John. London c1770 C
McQUINN, John. Market Drayton 1841 barometers
McTERLAN, John. Oundle 1834 W
MABBOTT, Edwin. Nottingham 1832-40 C
  Samuel. Nottingham 1713 W see
  also Mabit
MABEN, John. Plymouth c1770 also Maiben C
MABIT, William (perhaps Mabbott). Nottingham
  1738-43 W
MACAULEY, James. Scarborough. 1872
MACHER, George. Peterhead c1830 C
MACK, James. Aylsham, Norfolk 1845
MACKIE, ——. London. pre 1810 succ by T. Rust qv
MACKLIN, Peter. Shrewsbury b c1725 d 1800,
  There 50 years C
MACKRILL & LAIT. Hull 1872
MADDEFORD, William. Theale, Berks 1883
MADELAINY, ——. Paris c1680 W
MADOX, Robert. 1801 Name engraved on clock by
  Thomas Benbow of Northwood
MAIBEN. See Maben
MAILING, Robert. Aberdeen, maintained T C 1630
MAKEHAM, J. Reading 1899
MAKIN, ——. Ellesmere 1828 W
MALT, James. Brandon, Norfolk 1844
MANDER, Charles Edward. Windsor 1883
MANLEY, Daniel Snr. App 1650 to Ed. Gilpin of
  London. Free in CC 1660. Then to Yarmouth where
  d 1701 W & C
  Daniel Jnr. Yarmouth son of Daniel
  snr marr 1691 d c1729 C & W
  John. Bury St. Edmunds son of Daniel snr marr
  1686 d 1721 W
MANN, Edward. Coventry 1784. From 1787 partner
  with Wall. W
  Richard. Reading 1800 (may be confusion with
  Robert qv)
MANNING, ——. Enmore, Somerset c1740 C
MANNSFIELD, Thomas. Shaftsbury early 19c C
MANWARING, ——. Burwash. See Mainwaring
MAPSON, William. Hungerford 1899
MARCH, Jonathan. Place not known, walnut C
  (18c?)
MARIS, John. Place not known (Hertford?),
  Serviced T C 1732
MARR, Jonathan. Retford 1770-1814
MARSH, Anthony. Mansfield 1772
  G.T. Ellesmere 1869 W
  Richard. Ipswich b 1636 d 1707 W
MARSHALL, George William. Eckington, Derbys 1895
  Herbert. Beverley 1872
  J. & SON. Ripley, Derbys 1891-1932
  John. Glossop 1860
  John. Nottingham 1818
  Joseph. Nottingham 1844 (free at
  Leicester 1826)
  R. Halifax c1860 C
  Richard. Place not stated. The
  Wolsingham maker wkg 1740 -d 1796
  Samuel. Alfreton 1884
  Sydney. Canterbury 1882-89
  Thomas. Lincoln (B 1791-) d 1821 C
  William. Dunse, Scotland c1825 C
MARATON, James. Tower,London, mid 18c C
MARTIN, Jeremy. Tottenham c1780 C
  John. London (B app 1767) c1780-90 C
  Jonathan. Dublin pre-1677 -post 1685 W
  Thomas. Dublin 1775-95 W
  William. Bristol marr 1689 (B 1703-39) W & C
  Young. Dublin 1774-90 W

MASON, ——. Ashover, Derbys 1784-1812
——. Chesterfield 1743 C
C. & SON. Canterbury 1882-88
Francis. Sheffield. Took a house in
  Pudding Lane 1714, but 'ran away' before
  lease was signed C
Henry. Bisley 1719 C
John. Nailsworth c1750 C
John. Southwell 1822
Robert. Ipswich 1857- 64
Robert. Kelso c1790 C
S. Weaverham. Cabinet maker c1780, made
  clock cases for Thomas Richardson of Weaver-
  ham, Cheshire, some cases signed
& SON. Canterbury 1889
Thomas. Bawtry, Notts 1770-93 C
William. Middlesborough 1872
William W. Canterbury 1878 (& Son from
  1882)
William. Ipswich 1855-57
MASSEY, Charles. B c1837 Liverpool. Ludlow
  1868-71 W
John. Shrewsbury (b c1765) 1816
  -d 1823 C
MASTALGIO, ——. Newcastle upon Tyne 1861 W
MATHESON, ——. Inverness c1840 W
MATHEWS, Thomas. Woodbridge No date (the
  Hadleigh man? see B)
William. Carnarvon c1790 C
MATTHEWS, Thomas. Bishopscastle b 1837 Son
  of John Matthews, watchmaker
MAUD, John. Reading 1864-00
MAULE, Andrew. Hull 1872
MAW, B. Beverley 1876 W, jeweller and
  optician
MAWKES, Thomas. Belper c1800 -d 1854 C
  William. Derby 1791-1832
MAY, Fred. & Son. Dublin (B 1770-) -1774
  (& Son 1800-1813) W
John. Dublin 1814-19 C W
William. Norwich 1845
MAYER, M. Reading 1864
MAYFIELD, Samuel. Dublin 1819 C & W
MAYFOR, C. Cricklade c1820 C
MAYHEW, Francis. Parham, Suffolk son of Henry
  d 1730 C
Henry, Parham and Hacheston, Suffolk
  1686 d 1720 C
John. Saxmundham mid 18c C
Thomas. Place not stated (prob Parham)
  c1710 C
William. b c1725 app 1741 to John Calver
  of Woodbridge. To Parham by 1750, then succ
  Calver at Woodbridge 1751 -d 1791 C
William. Leeds. Date note known C
MAYNARD, George, Boston. Long Melford b 1725
  (stepson to Thomas Moore of Ipswich). To
  Lavenham c1779 d 1789 C
John Jnr. Long Melford son of John
  Snr. b 1670 marr 1692 d 1729 C
John III. Long Melford son of John II
  b 1695 -?
MAYNER, G. London c1760 (B pre 1771) W
MAYS, ——. Foulsham (same as William Mays, qv)
  William. Foulsham, Norfolk c1790 C
MEAKINS, Thomas. Dublin app 1681 to W. Bingham
  free 1699 d 1709 W & C
MEARS, ——. Hempnall, Norfolk, see Meers
MEDLEY, John. Belper 1888-1922
MEE, William E. Church Gresley, Derbys 1884-1900

MEGRETT, David. London 1679 jeweller prob French

MEIGH, Moses. London 1716 had Thomas Adkins as app C

MELLOR, Charles. Derby 1738 W

MELVILL, Robert. Aberdeen 1645 -d 1651. Kept town clocks

MENIALL, James. London 1682 French. Not in CC

MERCER, Edward. Thrapston, Northants. Early 18c prob the London man C
William. Hastings b 1730 marr 1755 W

METCALFE, James. Scarborough 1872

MEW, Joseph. Blandford b c1725 d 1793 C

MICOY, ——. Paris c1859 C

MIDDLETON & CO. (& Son). Sheringham c1842 W
John. Shrewsbury 1656 W

MIDGLEY, Eli. Halifax d 1740 C
William. Sheffield c1740 C

MILLER, David. Bathgate c1790 C
John. Norwich mid 18c C
Robson. Newcastle on Tyne 1835 W
Thomas. Calverton, Notts 1848 W
Thomas. Gazeley, Suffolk c1690-1701 C

MILLIGAN, Thomas. Shrewsbury b c1764 1796 -d 1797 W

MILLINGTON, Robert. London 1814 W
Thomas. From Birmingham, free at Shrewsbury 1796 W
Thomas. Wirksworth 1895-1928

MILLS, James. Long Melford d 1732 C

MILSOM(E), J.P. Newbury c1800-1836

MILTON, ——. Guisborough, Yorks c1790 C
Charles. London no date W

MITCHEL(L) & SON. Glasgow c1850 (prob Alexander & Son) C

MITTEN, Fra.(ncis). Chichester c1710 C

MOHN, John. Buxton 1888-95

MOLINARI, Domenico. Halesworth 1844 C

MOLZ, Godlove (Gotlieb). Derby 1871
John. Derby 1846-52

MONGIN, ——. France c1860

MONRO, George. Cannongate, this is the Edinburgh maker - see Vol 1.

MOODY, John. Oswestry app 1725 to Thomas Nash d 1742 C

MOON, Thomas. b 1826 Tunbridge Wells, to Newbury mid 19c

MOORCROFT, William. Derby 1871

MOORE, Ambrose. Dublin 1780-96 W & goldsmith
Edward. Ipswich son of Thomas b 1731 d 1788 C
Hatley. Ipswich son of Thomas b 1732 d 1796 W
James. Derby 1884- 1928
Rob. Dublin 1784-94 W
Roger Snr. Ipswich marr 1687 d 1727 C
Roger Jnr. Ipswich b 1714 -pre 1762 C
Thomas. Long Melford, marr 1733 d 1754 C
Thomas. Hindon, Wilts app 1741 to John Jerard -c1755 C
William. Brantford, Suffolk 1680 C
William. London. In 1716 took as app John Butterfield C
William. Woodbridge, Suffolk b c1699 -d pre- 1762
James. Derby 1884-1928

MORECOCK, Daniel. Shrewsbury 1775-78 W

MORGAN, Thomas. Oakengates, Shropshire 1868 W
W. Dublin 1796 -1827 W & jeweller

MORISON, Theordore. London 1774 W

MORLAND ——. Masham, Yorks prob Richard M. of Kirkby Malzeard c1800 C
Richard. Masham, N. Yorks 1818-23 kept T C, same man as at Kirkby Malzeard qv

MORLEY, Andrew. Pickering 1872

MORRILL, William. Easingwold c1780 C

MORRIS Bourns & Co. Dublin. 1830 W also jeweller, silversmith, etc.
John. Derby 1862
John. Ludlow (b c1800) 1841 W
Patt. Dublin early 18c C
Richard & John. Shrewsbury, repaired T C 1651 blacksmiths
Robert. Shrewsbury c1790-1821, when succ by nephew, William Baker W
William b c1830 Weston Jones, Staffs, to Newport, Shropshire by 1867-71

MORSE, John. Southampton no date W

MORTIMER, William. Plymouth early 19c C

MORTLOCK, Samuel. Clapham, Surrey b 1758 believed d 1797 W. Left property at Windlesham to son Richard
Bros. Clare, Suffolk 1883
Thomas. Stradishall, Suffolk 1744 C & W

MORTON, Charles. Nottingham 1778 W
John W. Chapel-en-le-Frith 1884- 88

MOSCHE, Jacob. Swansea early 19c C (but of Jacob Moseley?)

MOSELEY, Mrs. Mary E. Derby 1871

MOSES, G. Lincoln c1812-15 C

MOSLEY, John. Shipley c1750 C

MOSS, ——. Basford, Notts 1814 C
T. Nottingham 1790-1814 W & C
Thomas. Frodsham, Cheshire b 1733 d 1785 C

MOTHERSOLE, Thomas. Stowmarket 1892

MOUGIN, A. D. (Paris?) c1880 C

MOULANG, Daniel. Dublin silversmith and watch case maker

MOUNFORD, Charles. London no date W

MUDDLE, Edward. Chatham b 1709 son of Thomas of Rotherfield. At Chatham by 1779 W & C
Nicholas. Lindfield, Sussex c1760 C
Nicholas. Rotherfield, Sussex b 1716 son of Thomas of Rotherfield and bro of Edward. Working 1756 C
Thomas. Rotherfield b 1709 son of Thomas of Rotherfield and bro of Edward d 1785 clockmaker

MUIR, J.G. Matlock 1884-1922
William. Falkirk. Englishman, journeyman of John Russell. Ran away 1792 with 8 watches (then aged between 30 and 40)

MUIRHEAD & ARTHUR. Glasgow c1840 C

MULGRAVE, George. London 1797 W

MULLENEUX. See also Mollineux

MULLER, Hermann. Reading 1864

MUNDEN & SON. Dover 1874

MUNK, James. Tenterden, Kent 1805 C

MURRAY, Joseph. B c1834 Liverpool, to Whitchurch 1870-71 C

MUSSON, ——. Leeds c1800 C
William Martin. Hull 1872

MYERS, E. Cheltenham. Watch, no date
Israel. Newark 1853
Joseph H. New Mills, Derbys 1884-88

N

NACHBAR, Karolus. Pettau 1784 C
NAISBY, Thomas. Marton, E. Yorks c1840 C
(?also shoemaker 1823-51)
NASH, Richard. Shrewsbury repaired T C 1633
Thomas. Shrewsbury (b 1691) 1725 -d 1747 C
NAUNTON, George. Woodbridge 1850
William Jnr. Woodbridge 1855
NAYLOR, John. Lowestoft 1864
NEEDHAM, George. Cuckney, Notts early 19c C
NEEVE, Benjamin. Saxmundham succ. T. Farrer 1798
d 1851 C
NEILSEN, Nicole & Co. London 1885 W
NELLSON, R. Place not known. Sundial dated 1738
NELSON, Robert. Cromford, Derbys 1876
Thomas. Steeple Aston, Oxon. son of
William, working 1854
Thomas. Nottingham 1758-85, then Southwell
1786 C
William. Steeple Aston, Oxon b c1777,
working there from 1801 till d 1844 C
NENNIS, T.W. Tunbridge Wells 1865
NEVINSON, Corbet/Corburt. Liverpool 1759-64 W
NEWALL, William. (& Newhall). Cleobury Mortimer,
Shropshire (b c1762) 1789 -d 1812
Henry. Cleobury Mortimer, Shropshire b 1794
son of William Newall, clockmaker -1846
NEWBERRY, Henry. Somercotes, Derbys 1888
NEWEREN, D.D. London 1760 W
NEWMAN, George. Crediton late 18c C
& Sons. Shields no date W
T. Dublin c1700 C
NEWNES, Samuel. Whitchurch, Shropshire. 1789-
1823 C W
NEWNHAM, Samuel. Shrewsbury (b c1804) 1832-34,
when succ by James Hanny W
NEWSON, George. Wokingham, Berks 1899
NEWTON, Isaac. Bridgnorth, Shropshire. (b c1805)
1841 W

NEWTON —— Continued
Joseph. b c1809 Warks, then Shrewsbury 1851 W.
NICHOLAS, John Samuel. Daventry 1889
W. Birmingham 1775 W
NICHOLDS, Thomas. Shifnal, Shropshire 1840-46,
Albrighton 1851-75 W
NICHOLENI, John. Hull 1872
NICHOLSON, William. Yarm, N. Yorks 1872
NICKISSON, William. Newcastle under Lyme
c1780 (B 1804)
NICOLSON, Thomas. Dunfermline 1867-71 C
NIGHTINGALE, J.T. Shrewsbury 1859-65, when
succ by J. Kent
NOBLE, Nicholas. East Grinstead c1714 C
NODES, John. Nottingham 1750 goldsmith & W
NOEL, Fourquet. Tours c1800 C
NOKE, James. Ludlow app 1718 to Thomas Vernon
NOLLOTH, William. Bungay, Suffolk c1790 C
NORLAND, ——. London c1930 C
NORMAN, Thomas. Eckington, Derbys 1888-1908
NORMINGTON, D. Stokesley, N.Yorks 1872
NORRINGTON, Robert. Finsbury, London, 'In
White's Yard' 1816 W
NORTH, James. Newark 1820-40 clock case maker
NORTON, Edward. London maker of repute - see
Vol 1 but also b 1728 Lincs d 1792
Edward. Berrington, Shropshire. 1680 repaired
T C Perhaps the same man as Warwick
Edward. Warwick c1640 -1700
John. Leek Wootton, Warks made T C
with William 1792
William. Leek Wootton, Warks Made T C
with John 1792
NOYES, Thomas William. Bracknell 1899
NUGENT, Walter Henry. Dublin 1784 W
NUNN, Simon. Lowestoft 1821 W (may be error
for S. Norman qv)
NUSSIE, William. Nottingham repaired T C 1640

O

ODBER, J. Manchester? 1862 dialmaker or
engraver
ODELL, Peter. Repaired clocks at Quickswood
nr Baldock, Herts 1778
OGDEN, Francis. Matlock Bath 1884
OGG, Hendrie. Dunfermline c1809-51 C

OGLE, Cuthbert. London 1774 (B 1805-08) C
OLDINGSAW, Henry. Castle Cary c1800 C
OLDKNOW, John. Derby 1835-75
OLIVER, Isaac. Northill, Beds, sundial dated 1664
OLLIVANT & Bolsford. Manchester 1858 W
OLLIVE, Thomas. Tonbridge c1785 C

O'NEILL, Arthur. Dublin 1822-30 W (may be same man as below)
    Arthur. Dublin 1770 -1814
    Arthur. Dublin son of above app 1788 free 1808-20 W
    John. Dublin free 1776 W
ORAMS, J.E. Newbury 1884
O'RILEY, George. Dublin 1797 W
ORME, Isaac. Clown, Derbys 1891-95
    Michael. Newport, Shropshire 1764 W
ORPWOOD, William. b 1764 at Woodbridge, to Ipswich 1789 -d 1803 C
    Esther. Ipswich widow of William, whom she succ 1803. Partner with John Ablitt until 1808. Later a school owner

ORRE, James. Warminster late 18c C
ORTON, John. Newark 1790
OSBORNE, Richard William & Molyneux, T. Dublin 1819 W & goldsmith
    William. Ixworth, Suffolk 1864-74
OSMAN, William. Bradford ?late 19c C
OVERALL, Thomas. Wellingborough 1808 W
OWEN, Edward. Gravesend 1874
    John. London? sundials dated 1683 & 1697
    William & Thomas. Oswestry sons of William Owen b 1841 and 1845 succ father there 1868-75
OXBROW, Alfred. Canterbury 1878-88
OXLEY, Thomas. Maidstone marr 1760 W

P

PACKER, Thomas. Reading 1780-96
PACQUET, ——. Paris c1780 carriage clocks
PADDOCK, Thomas. Ilkeston 1900
PAGE, Henry. Upper Broughton, Notts 18c ?C
    G. Coventry early 19c W
    John. Ipswich 1752 d 1772 succ by N. Cavell C
PAGET(T), John. Bridgnorth, Shropshire c1720-30 C
PALLISER, John. Pickering, N. Yorks 1780 C
PALMER, Elias Bailey. Madeley, Shropshire 1868-71
    Henry. New Mills 1891
    J. Thetford 1854 T C
    Joseph b c1831 Kidderminster then Ludlow 1851-56 W
    Thomas. b c1846 Coventry to Ludlow c1871-1900
    Walter. Earl Soham, Suffolk 1844
    William. Cleobury Mortimer, Shropshire 1822
    William. London 1837 W
    William. Ludlow 1667 repaired T C
    William Edgar. Tonbridge, Kent b 1868 (Worcester) 1902 electric C
PANTON, ——. Alyth, Scotland c1790 C
PARIS, Albert. Neuchatel 1880 W
    Nicholas, Snr. Warwick, working c1669 -1715 d 1716 C
PARISH, Joseph. Brigg early 19c C
    Joseph. Brigg c1790 C
PARKER, George. Dublin 1701-66 W
    Reuben. Nottingham 1853
    Robert. Derby 1710-48 W
    Stephen. Ipswich 1683-4
    Thomas. Dublin free 1694 d 1751 W C
PARKERSON, A.D. Canterbury c1840 C
PARR, B. Grantham c1830 C
PARTRIDGE, William H. Reading 1899
PASLEY, John. Meltham, Huddersfield ?1782 C
PASSMORE, H. Cricklade c1830-40 C
    William. Town not known c1770 C
PATON, G. Galston. c1830 C
PATTISON, Joseph. Whitchurch (county not known), date not known
    Matthew. Langholm, Scotland 1797 W

PAUL, Philip. Bungay 1844 (see also Poll)
PAYNE, George. Ludlow app 1737 -d 1809 C
    Joseph. Bromsgrove c1830-40 C
    Robert. Newbury, Speenhamland c1820-30
    William. Oswestry 1753-57 W
PEACH, ——. Axminster c1830 C
PEACOCK, Robert. Cornhill, Lincoln no date W
PEAKE, William Henry. Codnor 1884-1922
PEARCE, James. Hull 1872
    John. Stratford early 19c C
    & Son. London mid 19c C
    Thomas. Chard, Somerset early 19c C
    William. Newbury 1892-1903
PEARSON, George. Liverpool c1760 W
    William. Lightcliffe, Yorks no date W
PEAT, Thomas. Crieff, Scotland (B 1784) -c1815 C
PEDEVAL, Henry. Paris mid 18c
PEDLEY, Samuel. Shrewsbury 1802 C
    Thomas. Derby 1818-31
PEDRONI, J.B. Shrewsbury 1840 barometers
PEILE, Joseph. Place not known. Late 17c (prob Southern England) C
PEMBROKE, S.J. San Francisco, USA 1887-1901 clock restorer
PENISTON, John. Horncastle c1830 C
PENLINGTON, John. Ince, Lancs d 1764 (sometimes Pennington) C
    & Hutton. Liverpool 19c C
PENMAN, Margaret. Dunfermline c1835 C (widow of Robert? - see Vol 1)
PEPLOW, Samuel Kirk. Swadlincote, Derbys 1888
PEPPER, ——. Baldock c1790 C
    Charles. Newbury c1764
    John. Beauchamps Rothing (Roding), Essex mid 18c C
    John. Derby 1884-88
PERCE, Frederick. Chatham 1874
PERCIVAL, William. Ludlow free 1722 C
PERCY, Thomas. London no date W
PERINOT, Abraham. London no date C
PERKINS, F. Wakefield 1813 W
PERRY, William. Hockworthy, Devon b c1775 -d 1835 W & Cooper

PERRY —— Continued
William (junior?) Hockworth, Devon b c1805 -57 W & C & mechanic
PERRYMAN, John (?II). b 1813 Barnstaple. To USA 1834-53, then Melbourne, Australia 1853 then Adelaide. d 1871 W & jeweller
PETER & Mockler. Dublin 1813-30 C & W
PETRIE, William. Kirkcaldy, Fife c1850 C
PETT, ——. Tetbury, Glos early 19c C
PETTINGER, Thomas. Hull 1872
PETTIT, Paul. Wellingborough 1735 clocksmith
PEXTON, John Henry. Beverley 1891
PHELPS, ——. Yeovil c1730 C
PHILIPS, Edmund. Derby 1858
Richard Henry. Chesterfield 1884-1928
PHILIPSON, A. Winster, Derbys 1682. (Very doubtful entry, prob confusion of Philipsons of Winster, Westmoreland qv)
PHILLIPS, ——. Loscoe, Derbys early 18c C
Edwin. Derby 1862-71 (?see also Edwin Philips)
Henry. Swansea 1845 W
Joseph. Offton, Suffolk marr 1767, later to Somersham, ret'd 1785 d 1815
William. Ludlow b c1821 -51 W (son of Thomas Phillips of Ludlow, watchmaker)
William. Ludlow 1822-28
PHILO, Ernest. Hungerford and Lambourne 1899
PICARD, James. Geneva. 19c W
PICKERING, David. Hull 1872
PICKLES, John. Patrington, E. Yorks c1830 C
PICKTHALL, Robert. Waberthwaite, Cumbria c1760-90 C
PILCHER, ——. Workington, Cumbria c1770 C
PIMBLETT, T.A. Earlstown, Scotland no date W
PINK, George. London 1757 (B pre 1775) W
PINKNEY, Robert. Guisborough 1872
PIPER, John. Windsor 1824
PITT, Caleb. Frome c1790 C
Charles. Chelsea (prob same as London man qv) W
PIVEN, Alex R. N. Limervady, Ireland early 19c C
PLACE, T. Chorley ?late 18c C
PLANCHION, ——. Paris c1860 C
PLANT, Thomas. Newport, Shropshire marr 1793 -99 C
William. Ilkeston 1895-1900
PLATTS, John. Mansfield 1822-42 C
PLAYER, Mrs. M. Wokingham 1883
PLEASANT, Henry. Colchester 1686, Sudbury by 1694, d 1708 C & bellfounder
PLIMMER, Nathaniel. b 1750 Elerdine, to Wellington by 1783 W
PLUMMER, D. London no date W
Thomas. Newbury believed from London c1705
POCK, Thomas. Derby 1860
POLL, Robert I. Harleston, Nofolk 1741-71 C
Robert II. Metfield, Suffolk d 1815
Robert III. Bungay 1839
POLLARD, William. Canterbury 1888
POLLETT, ——. London, French 1702
POLWORTH, William. Newbury 1823
POND, William. Abingdon 1823

PONTIN, Ernest. Reading 1864
POOL, Mrs. Ann. Derby 1871
POOLE, A.G. Barton on Humber 1880 succ to R. Chapman
James. Dublin 1798-1806 W
James & Adams. Anthony. Dublin 1798 W
POPKINS, John. Dublin free 1681 -d c1707
PORTER, G. Basingstoke late 19c C
PORTLAND, ——. Liverpool 1813 W
POTTS, J.M. Mauchline, Ayrshire c1830-40 C
Robert. Patrington c1770 C
POUND, Stephen. Shrivenham, Berks 1890
POURVIES, James London? Scotland? 1567 C
POWELL, John. Prees, Shropshire 1791
Thomas. Sheffield 1810 W
POWIS, L. Hull 1841
POWLEY, Thomas. Appleby c1750 C
POYNTER, T. Bracknell 1864
POYNTON, George. Willingham c1800
POZZI, Peter. Oswestry 1844 d 1850 barometers
PRECY, Edward. Reading 1864
PREDDY, ——. Martock, Somerset c1800 C (may be same man as William of Taunton)
PREMSELLA & Hamburger. Amsterdam mid 18c
PRESTIDGE, Walter. Towcester 1777 -post 1815 C
PRESTWIDGE, T.P. Southwold, Suffolk 1864
PRICE, Charles. Eckington, Derbys 1888-1900
Edmund. London c1720 W
H.C. Burton on Trent 1885 W
John. Shirebrook, Derbys 1900
R. Wiveliscombe, Somerset c1800-10 C
PRIDHAM, John. Long Eaton, Derbys/ Notts 1884
PRIEST, John. Beywedley Gunners, Shropshire, repaired Oswestry T C 1591
PRIMAVESI, George. Reading 1899
PRIME, William. Riddings, Derbys 1888-1908) (?same person as below)
William. Ripley, Derbys 1884
W. South Normanton, Derbys 1891
PRINCE, Albert. Wantage 1899-1903
Albert. Grove, Berks 1899-1903
& Cattle. York late 18c C
Isaac. Chieveley, Newbury c1864-83
John. London 1810 W
PRIOR, G. Otley, Yorks prob the Ilkley man see Vol 1 C
PRITCHARD, G. Birmingham c1810 C
PROCTOR, George Edwin. South Normanton, 1895-1922
J. Cockermouth 1862-67 cleaned clocks
W. Heanor 1895-1922
PROUDFOOT, Robert. London b 1801-51 W (from Norwich)
PRYOR, R. Hythe 1866
PUGH, Benjamin. From Birmingham free Shrewsbury 1796 watch key maker
PURDUE, James. Newbury 1773
PURMAN, Marcus. Munich sundial dated 1590 (B 1601)
PUTNAM, John. Amersham marr 1714 C
PYBUS, John. Keelby 1830 W
PYNE, William. Reading 1899

Q

QUICK, Paul. Zennor, Cornwall sundial dated 1737

QUINTON, Richard. Downton, Hants 1750-60 C

R

RACTER, ——. London no date W
RAIFE, Maurice. Leeds ?late 19c W
RAILTON, P. Barnard Castle c1840 C
RAINE, Henry. Durham early 18c (c1720) C
RAMMELL, Eliza. Derby 1884
RAMPTON, H.K. Newbury c1903-12
RAMSAY, H. London 1884 W
RAMSBOTTOM, John (?Jnr) b c1770 Wakefield
  d 1850 Scarborough
  Thomas. Barnsley late 18c C
RAND, Joseph. Newbury 1790-1836
  Joseph Jnr. Newbury 1836-64
RANDALL, James. Newbury 1862
  William. App 1754 to Edward Boys, West-
  minster. Reading c1765-90, then Newbury
  c1790-1830 C
RANDELL, Edward. Norton St. Phillips, Wilts
  c1730 C
RANGER, Michael. Newbury c1813 (cf B London
  1774-1820) C
RANKIN, J. Glasgow 1886 W
  Robert. Kilmarnock 19c C
RAPHAEL & Nathan. Hull 1841
RASHER, Joseph. London app transferred 1701 to
  George Etherington
RATHBONE, John. Shrewsbury b 1717 -64 C
RAWLINS, John. Stone c1760 C
RAWSON, Thomas. Penrith b 1734 marr 3 times
  d Middlegate 1811
  William. Kendal b 1731 marr 1753
RAY, Daniel (II). Sudbury b 1701 d 1772 son of
  Daniel I
  Daniel (III) son of William b 1763 London
  then Sudbury from 1788
  Samuel. Wem, Shropshire marr 1747-1770 C
  William. b c1825 at Battle. Leamington
  Spa 1841-61 W
  William Redmore. b 1762 app in London then
  Sudbury from 1783 -d 1841
  William. Sudbury son of Daniel II
  b 1734 d 1808
RAYMENT, Giffin. Bury St Edmunds b 1722
  son of Richard d 1769
  T. Bury St Edmunds pre 1763 W
RAYMOND, ——. Hungerford 1810
  George. marr 1753 Woodbridge then
  Manningtree 1755-75
  Isaac. Dublin 1819 W
  William. Dublin app 1749 free 1762-1802
  (prob -1818) W
REA, Thomas. Walton (on Trent) c1750-60 C
READ, Augustus. Ipswich b c1833 son of Daniel
  -1874
  Daniel (II). Ipswich b c1816-74

READ —— Continued
  Joseph. Town not known c1740-50 C
  Thomas. Beccles 1864
  Thomas. Ispwich b 1733 -d 1817 C
  Thomas. Tarporley, Cheshire c1775 C
READER, John. Cranbrook, Kent marr 1807 W
  (associated with T. Ollive qv)
  John. Hull 1768
  William. Hull 1723-68 C & W
REASON, John. London (B app 1702) -c1770 C
RECKLESS, Joseph. Nottingham free 1710-14
  (prob from London) W
REDSHAW, ——. Newcastle (under Lyme?) 1765 C
REED, Andrew. London 1792 (B c1750) W
REEVE, William. Stonham, Suffolk b 1642
  d 1714 W
REID, ——. Bristol 1884 W
REIDL, A. Vienna, 1853 C
REILLY, John. Dublin (B 1753-80) - 1786 W
RELLAMES, ——. London mid 18c C
RENOIS, ——. Paris c1800 C
RENTCH, Robert. London c1790 C
RENWICK, John. Wigton, Cumbria c1750 -d 1773 C
REVETT, John. Eye, Shropshire marr 1765 d 1798
  (sometimes Rivett)W & C
REVIS, Charles. Cambridge son of Thomas b 1817
  -1840.Later an Excise Officer W
  Thomas. Cambridge b 1782 d 1860 in London
REYNOLDS, David. Broseley, Shropshire 1842 W & C
  John. Barnstaple lantern cl (?17c-
  early 18c)
  Thomas. Chesterfield 1891-1908
  Thomas. Warwick 1751 -d 1783 C
  William. Warwick b 1776 (son of Thomas?)
  -c1795
RICE, Robert. Bristol c1730 C
RICHARDSON, John (?II) Bubwith, Yorks 1817 C
  Richard. Aston.(Cheshire?) d 1756 C
  William. Nottingham 1818-38
RICHIE, James. Hull 1792 (see also Ritchie)
RICHMOND, John. London (B apr 1711) 1799 W
RICHTER, J.E. Gorlitz c1790 C
RIDER, John Tryall. Preston, Lancs b 1720
  marr 1749 C (cf Tryall Rider)
RIDGWAY, Josiah. b c1839 Malpas. Whitchuch where
  app and worked for the Joyces -1885 C
RIGG, John. Guiseley (Guisborough?) 1765 C
RIGMAIDEN, Robert. Dublin free 1686 - d 1723 W
RILEY, A. Eastwood, (?Notts) c1840 C (?same man
  as below)
  J. Eastwood (?Notts) c1848 C
RIMMER, Robert. Childwall, same as Liverpool
  man qv
RINGHAM, Mark. Hull 1872

RINGLAND, John. Dublin 1745 W
RISBRIDGES, William. Dorking took W. Batten as
app 1724 C
RISEAM, Joseph. Cottingham, E. Yorks 1872
RIPLEY, Frank. Eckington, Derbys 1884
RIVA, F. Reading 1800
RIVETT. See Revett
ROBB, William. Montrose b 1727 marr 1757 (B -1776)
ROBBINS, Joseph. Bitten, Glos early 18c C
T. Dover 1865
ROBERTS, A. Canterbury 1888
Charles. Derby c1835-40 C
Edward. Ludlow 1846
James. Ashford 1778 W
John. Bury St Edmunds 1777-91
John. Wrexham joiner & casemaker c1760
Thomas. Derby 1806
William. Amlwch, Anglesey c1780-85 C
William. Derby 1806-35 C
William. b c1826 Derby, Oswestry 1861-91
(somtimes erroniously as Roberto?) W
William. Dunbar c1790-c1800 C
William. Pwllheli c1800 C
ROBERTSON, A. Bridge End, Perth c1790 C
Alexander. Ipswich 1855
Charles. Blairgowrie c1840 C
ROBINSON, Edward. Derby, repaired T C 1663-69
Henry. Clitheroe c1820 C
Lawrence. Halifax c1800 C (of
Lawrence?)
Moss. Derby 1895
Samuel. Hull 1872
Thomas. Chesterfield 1876-1916
& Wells. Shrewsbury 1891-95 later Henry Wells, qv
William. Felliscliffe, Nr. Harrogate
c1820-40 C
ROBSON, Robert. Thirsk 1872
RODWELL, ——. London 1839 W
ROE, John Thomas. Whittington Moor, Derbys
1895-1928

ROGERS, C. Gillingham c1785 C
John. Bridgnorth, Shropshire 1789-1803 W
John. Hadleigh, Essex d 1819 W
Joseph. Oswesty kept T C 1676
Samuel. Marshfield c1810 C
ROHRER, August. Leamington (b c1824 1851
ROLLON, John. Hanworth c1700 C
ROPER, Charles. Chesterfield 1884-1900
Nicholas Snr. Oakhill, Somerset, 1766-1805
Nicholas Jnr. Ashwick, nr Oakhill,
Somerset 1792-1863
William. From London to Lavenham then
Hadleigh 1766-69, also branch at Ipswich
RORKE, Walter. Dublin 1815-18 W & jeweller
ROSE, F. Dorchester early 19c C
Jacob. London b 1816 Poland, -1851
William. Ascot 1899
William. Bracknell, Berks 1883
William. Dublin 1784 W
ROSENBERG, J. Newcastle on Tyne 1874 W
ROSKELL, Eliza. Derby 1860-62
William. Derby 1858-60 son of Robert
Roskell of Liverpool, watchmaker
ROSS, Alexander. Dublin 1821-30 W & C
John. Epworth, Lincs c1800 C
ROSSI, Joseph. Shrewsbury 1860-62 barometers
and alarm clocks
ROUELL, ——. Bayeux mid 17c C
ROURKE, Thomas. Dublin 1781-1830 C & W
ROWELL, ——. Oxford c1830 C
ROWLEY, Henry Jnr. Shrewsbury b 1802 d 1822
son of Henry Snr W
William. Shrewsbury marr 1829-30 W
ROWNING, George. Newmarket b 1751 son of John,
-d 1785
John. Newmarket marr 1745 believed
d 1757 C
ROWTON, John. Cambridge 1800 W
ROYSE, Isaac. Castleton, Derbys 1846

S

SALMON, Colin. Macclesfield c1770 C
SALOMAN, S. Guisborough c1850 C
SALTER, Joseph. b c1725 Oswestry c1760
-d 1800 W & printer
Robert. Oswestry 1789 -marr 1810 W
SALTZ, Samuel. New Brampton, Derbys 1900
SAMPLE & Co. Bedlington, Northumberland 1867 W
SAMUEL, Abraham. Sunderland c1770 C
Lazurus. Sunderland c1800-c1820 C
Moses. Bungay and Ipswich c1790-1808 W
SANDERSON, T. Dublin c1730 C
SANDES, George Henry. Derby 1884-1908
SANDON, Felix. London c1790 (but see also
Vol 1)
SANKEY, John. Coalbrookdale c1790 C
SANSOM, L. London 1806 W
SAUNDERS, ——. Dunbarton c1830 C

SAVAGE, Edward. Hull 1872
Jonathan. Shrewsbury c1820 C
Joseph. Ormskirk c1740-50 C
Richard. Shrewsbury 1698 -d 1728 C
Thomas. Shrewsbury son of Richard Savage,
to whom app 1706 C
William. Shrewsbury son of Richard Savage,
to whom app 1700 -d 1707 C
SAVILE, William. Dublin app 1720 to W. Bingham
free 1748 -d 1761 W
SAWDON, W. Whitby c1875-80 W & Jeweller
SAXBY, ——. Margate c1890 W
SAXTON, Richard H. Buxton 1891-95
SCALON, ——. Wingham, Kent 1464 T C
SCARGILL. See Skargel
SCARLIFF, John. Carlton-le-Moorland, Lincs
c1860-1900

SCHERHARDT, Karl. Blackheath, Kent 1874
SCHMEID, Conrad. Wetterhausen, sundial dated 1750
SCHMIDT, Georg Fidel. Graz, Austria c1770 C
SCHNABEL, Bernard. b Baden, Germany c1818, Leamington Spa1851
SCHOLFIELD, Joseph. Place not known (Lancs?) c1760 C
SCHORTSINGER, ——. Belfast 1878 W
SCHEDULIN, Luis. Vienna 18c W
SCOBELL, H. Place not known c1630 C
SCOTT, Alexander. Leslie, Scotland 1830 C
    Edward. Patrington, E. Yorks 1872-76
    James (& Son). Dublin 1798 -1815(-27?) W
    James. Hull 1872
    Julius. Heanor, Derbys 1855-58
    Morris. Derby 1852-58
    Peter. Lesmahagow, Scotland c1815 C
    Robert. Dublin 1828-30 W
    Thomas. Guisborough 1792 W
SCOVELL, Richard. Newbury 1800-20
    Sarah. Newbury c1820-24
SEABROOK, Samuel. Hornchurch (167 High St) 1871
SEAMAN, Edward. Dawley, Shropshire 1868
SEARE, Albert J. Abingdon 1883
SEARSON, Stephen. Ipswich marr 1730 -d 1776 W
SEASON, Thomas. Ludlow repaired church clock 1549-68
SEDDON, Henry. Whitby 1872
SEDGWICK, Thomas. Derby 1852-72
SELBY, ——. Grantham c1830-40 C
SELLER, Alexander. Reading 1864
SELLWOOD, Percy. Newbury 1899-1961
SELWOOD, William. Carmarthen c1740 -d 18 June 1760 C
SEYMOUR, C.W. Windsor 1883
    William. Hungerford 1725
SHAFTO, ——. Newcastle c1850 C
SHAKESPEARE, Thomas. Swadlincote, Derbys 1846
SHARDLOW, John. Derby 1844
SHARP, John. Bawtry, Notts 1842-53
SHARPE, Anne. London c1777-85 W
    George. Sheffield 1860
SHARRATT, John. Worcester late 18c C
SHAW, David. Derby 1888
    Edmund. Lingards 1810 W
    George. Botesdale, Suffolk marr 1790
    Henry. Heanor, Derbys 1877-1900
    Hugh. Wirksworth, Derbys 1822
    John. Nottingham 1855-64
    Samuel. Radford, Notts 1834
    William. Botesdale 1759-60 also Diss, Norfolk
    William. Gislingham, Suffolk c1720 C
SHAWTER, Michael. Hull c1750 C
SHELTON, Samuel. Ironville, Derbys 1891
    S & H. Ilkeston, Derby 1908-12
    William. Langley Mill, Derbys 1884-1900
    William. Langley Mill 1877
SHEMELD, Joseph. Sheffield d 1714 W
SHENTON, Richard. Matlock Bath 1891-1928
SHEPHERD, ——. Plymouth c1810 C
    Joseph. Sheffield 1711-29 C
    Robert. Eye. Shropshire b c1700 d 1779 C
SHEPLEY, ——. Glossop c1785 C
SHEPPERLEY, Philip. Nottingham 1864
SHERLOCK, Joseph. Dublin 1828-30 W & C
SIMKISS, Samuel. Stourbridge (Bretell Lane) 1825-28 C
SIMPKIN, J. Sundials at Great Sankey 1781
    Burtonwood, Lancs 1791

SIMPKIN —— Continued

    John. See Simpson
    Joseph. Derby 1792
SIMPKINS, William John. Mansfield 1877-85
SIMPSON, ——. Tadcaster c1760 C
    Charles. b c1796 Newport, Shropshire 1841 C
    Christopher. Shrewsbury 1827 W
    James. Askrigg, N. Yorks casemaker of at least some of Wilson's longcases.
    John. (or Simpkin). Rillington, Yorks b 1777 d 1833 C & W
SINCLAIR, ——. Falmouth c1780 C
    D. Edinburgh japanner of dials c1840
SINDERBY, Daniel. See Sunderby
SINGLETON, ——. Preston c1800 C
SINKINSON, John. Doncaster 1817 W
SITHERWOOD, William. Rich Hill, Armagh early 19c C
SKARGEL, Richard. Sheffield 1726 W
SKINNER, John Jnr. Exeter son of John Snr, whom he succ 1818 -d 1846 C
    Nicholas. Newbury kept T C 1602-5
    Thomas. Shoreditch 1832 C
SLEIGHTHOLME, ——. Spalding c1750 C (cf J.S. London)
SLY, Richard. Nottingham b 1706 -30 C
SMALLWOOD, John. Maxfield (=Macclesfield) c1700 C
SMEDLEY, Robert. Matlock Bath 1888
SMITH, Ambrose. Stamford 1682 Invented name found on London watches by Jasper Harmer
    Andrew. Fauldhouse 1827 W
    Andrew. North Berwick 1802 W
    Charles. Dublin free 1733 -d 1743 W
    Charles. Nottingham 1826
    Daniel. Place not known 18c C
    Edmund. Bury St Edmunds (from London) 1753 -d 1779 W
    Edward. Derby 1884
    Edward. Newark b 1736 marr 1760, again 1768 -c1790 C
    Edward. Nottingham 1825
    Francis. Bury St Edmunds 1844
    Henry Alfred. New Mills, Derbys 1857-60
    J. South Stockton b 1807 d 1895 made sundials 1838
    James. Edinburgh 1789 C
    James. Kings Lynn 1760 C
    Jasper. London mid 18c C
    John. Dublin 1784-1819
    John. Southwell, Notts 1848-53
    John James. Kimberley, Notts 1877
    John. Folkestone 1798 W
    John. Halesowen, Cheshire c1750 C
    John. Halifax 1800 W
    John. London 1703 W. A poor man whom CC gave five shillings towards his journey to Carolina.
    John. Reading 1794-1824 also Newbury -1840. Later with sons, Henry and Edward 1853
    John. Shrewsbury 1724-31 C
    Joseph. Bournemouth mid 19c C
    Robert. Nottingham 1789-1804
    William. Shrewsbury 1826 C
    William. App Robert. Oswestry, repaired church clock 1601
    William. Walton, nr Burton-on-Trent 1828-52
    William. Derby 1849-76

SMOULT, T. South Shields c1790 C (prob Thomas Smoult of Newcastle who ret'd 1791)

SMYTHE, Robert. Oswestry repaired church clock 1591

SNAPE, J. Sutton Coldfield sundial dated 1761

SNATT, John. Ashford, Kent 1744 W

SNELLING, James. Newton Abbot b 1802, son of Edward d 1876 C

SNOW, Edward. Woodbridge 1850
R.R. Harewood 1842 W
William. Marlborough 1650 W

SOAR, John. Derby c1843

SOLDINI, G. Wincanton c1830-40 C
Gossu. Wincanton supposed French prisoner working 1805-30 C

SOLLA, Joseph. Hull 1872

SOLLARY, Mary. Nottingham 1844

SOMERVILLE, Francis. Newcastle on Tyne 1797 W

SOMMERS, George. Ilfracombe c1850 C

SONLEY, Richard. Dublin 1753 W

SOPPET, George Robinson. Nothallerton, N. Yorks 1782

SORET, Abraham. Dublin 1685 -d 1715 W
Adam. Dublin free 1675 -1723 W

SPARKE, J. Nairn late 18c C

SPARKES, Robert. Cockfield, Suffolk d 1648 C

SPAULDING & Co. Paris 19c C

SPEAR, George. Old Market, Bristol 1803 W

SPENCE, John. Market Harborough c1760 -d 1783 later succ by widow. C
John. Leicester 1688 C

SPENCER, Joseph. Kirkby-in-Ashfield 1877

SPENDLOVE, John. Brandon, Suffolk no date -?late 18c W

SPENDLOW, James. Brandon, Suffolk 1784-1817

SPENSER, John. Barrington, Shropshire repaired church clock 1719

SPERANTIUS, Sebastian. Nuremberg sundial dated 1503

SPICER, Charles. Haverhill b c1748 marr 1776 W

SPINNER, John. b Stickland, Dorset 1707 marr 1752, later Blandford until c1770 C
John Jnr. Blandford succ father John Snr c1770 -d 1780 C

SPRACKEL, Jan Benjamin. Goor, Holland c1750 C

SPRADLING & Son. Newbury late 19c C

SPRINGFIELD, Jonas. Norwich app 1693 to Samuel Fromanteel and believed C

SPRINGALL, Isaac. Norwich c1850 C (same man as below?)

SPRINGHALL, Isaac. Buxton, Norfolk? 1836

SQUIRREL, William. Bildeston 1810

STABIUS, Johannes. Nuremberg sundial dated 1503

STACEY, Thomas. Farnsfield, Notts marr 1774, to Southwell 1792-1820 C

STAFFORD, ——. Ironbridge ?late 18c
Mary Ann. Dublin 1784-1815 (widow of William?)

STAMFORD, R. London 1870 W

STANCLIFFE, Noah. From Norland to Elland, Yorks where marr 1761 C

STANILAND, John. Malton, N. Yorks 1839 W

STANLEY, Christopher. Itinerant, working in Stradishall, Suffolk in 1744 C

STANTON, George. Derby 1852-58

STAUFFER & Sandoz. Chaux de Fond, Lausanne 1858

STAVELEY, James. Skirlaugh, E. Yorks 1892-97

STEAD, Edward. Ludlow app 1717 free 1722

STEADDY, George. Canterbury 1818 W

STEDMAN, Henry. Dublin 1792 watch casemaker

STEEDS, William. Oakhill, Somerset 1782-1843

STEEL, William Francis. Saxdmundham, Suffolk 1830 C & W

STEER, Mrs. Elma. Derby 1888

STEIGHT, John. Pershore c1730 C

STEINBUCK, Hans. Denmark 1629 -47

STEPHENS, Richard. Bridgnorth 1733 -d 1780 C W

STEPHENSON, George. Warminster, Wilts early 19c C
R. Skelton, N. Yorks Early 19c? C
Thomas. Staindrop, Durham c1830 C

STERLAND, John (sometimes Storland) b c1743 marr Nottingham 1768-88 C

STERLING, William. Dublin 1786-1808 W

STERRY, Robert. Dawley, Shropshire 1840 W

STEVENS, Edward. Saltash, Cornwall 1727 sundial
James. Derby 1858
William. Godalming. See Stephens

STEVENSON, ——. Sheffield c1830 C

STEWARD, ——. Ipswich c1700 C

STEWART, Charles & Sons. Dublin 1830 and later
James. New Langholm, Scotland 1797 C

STICKLAND, Timothy. Kendal c1730-35 C

STIER & Alcher. Bath early 19c C
M. Bath mid 19c C

STILLMAN, James. Newbury 1864-78

STOCKWELL, Thomas. Church Stretton, Derbys 1850

STOKES, Charles. Derby 1843
Isaac. Nottingham 1844

STONE, John. Bakewell 1888-1900

STOKES, John. Bewdley, Worcs c1725 C (B pre 1751)
W. Place not known, may be London, early 18c C

STONE, William. Windsor 1800

STONER, Benjamin. Welling, Kent 1810 W

STOPPINGER, ——. Nuremberg 1565

STORER, Thomas. Derby 1720-43 C
William. Nottingham 1818

STRACHAN, Andrew (or Abraham). London, from Scotland b c1650-60, there 1691 working illicitly
Archibald. Tanfield, nr Gateshead c1760 C (some numbered), prob same man as at Newcastle

STRADLING & Sons. Newbury 1891-1903

STRAFFORD, William. Dublin 1765-84 W and toy seller

STRAHAN, John. Hull 1806-18

STRAY, G.E. Newark 1830 C

STRICKLAND & Richardson. London c1800 C

STRINGER, Josiah. Stockport c1745 C
Thomas. Draton (Drayton?) c1725-30 C

STROMIER, George. Glasgow 1880 W

STROWBRIDGE, Henry. Chudleigh, Devon 1823-30 then Dawlish 1830-57. Succ by William R. Cornelius qv

STRUTT, ——. Blackwell c1750 C

STUBBINGTON, ——. Gosport c1800 C

STUDLEY, Thomas. Ellesmere free at Shrewsbury 1721 C

STURGYS, Edmund. Dublin free 1715-51 W

SUGGATE, George Jnr. Halesworth son of George Snr b 1751 d 1844 C & W

SUGGETT, George. Buttercrambe, Yorks no date W
John. Little Walsingham, Norfolk 1836

SULLEY, Joseph. Nottingham 1850-55
Richard. b c1797 from Huddersfield, marr 1824 Leicester W

SUMMERHAYES, Richard. Taunton c1850 C
  Robert Jnr. b 1809 Buckland St.
  Mary, Somerset. To Wells c1838, Butleigh c1844
  Banwell c1850, Bishop's Lydeard c1852,
  Milverton c1863, Huntspill c1868, where
  d 1869 C W
  William. Bro of Robert Jnr b 1807
  Buckland St. Mary. To Taunton until d 1863 C
SUMMERS, T. London 1810 W
SUNDERBY (or Sinderby), Daniel. Place unknown
  c1720-30 C
SUTCLIFFE, L. Millwood c1830 C
  S. Oxenhope, Yorks c1760 (prob
  Stephen S. qv)
SWAINE, William. Place not known c1770 C
  (prob the Ipswich man qv)

SWAINE —— Continued
  William. b 1749 Woodbridge, to Ipswich 1771
  from London then Woodbridge again by 1783
  -d 1811 C & W
SWANN, George. Glossop 1888-1928
  Hurrel. Foxton, nr Cambridge 1860 W
  William. Derbys 1843
SWIFT, Alfred C. Long Eaton, Derbys 1884
SWINNERTON, Joseph. Place not known c1730-40 C
  Thomas. Place not known, lantern cl
  ?late 17c
SWINTON, James. London 1759 W
SYMON, ——. Wellington, Shropshire repaired
  T C 1619
SYMOND, Thomas. Great Yarmouth 1845
SYMPSON, Charles. London c1735 C

T

TAILOR, John. Bridgnorth b c1810-51 C
TALBOT, C. Tuxford, Notts 1670 C
  N. Canterbury 1882
TANDY, William. Warwick b c1816-50 C
TANSLEY, John. Nottingham 1877
TANTUM, Francis. Loscoe, Derbys early
  18c C
TARRANT, John Frederick. Reading 1864-99
TASKER, Joseph. Place not known c1720 C
TATE, ——. Winterton, Lincs c1815 C
  T. Wolsingham, Durham c1850
TAYLOR, Benjamin. Warwick 1723 repaired T C
  Charles William. Swadlincote, Derbys
  1888-91
  & Davy. Yarmouth late 19c C
  Ebenezer. Derby 1884-1912
  Edward. Albrighton, Shropshire d pre 1855
  G. Place not known 1839 W
  George. Middlesborough 1872
  George. Tunbridge Wells 1865
  Henry. Windsor 1899
  Joseph. Albrighton, Shropshire, same man as at
  Wolverhampton qv
  & Marshall. Nottingham 1822
  R. Ilkeston 1895-1900
  Reuben. Ilkeston 1877
  Reuben. Sandiacre, Derbys 1891-95
  Samuuel (I). To Framlingham, Suffolk 1785
  - c1830
  Samuel (II). Framlingham b c1795, at
  Halesworth, Suffolk 1830-44, then
  Framlingham again 1846
  & Son. Bristol mid 19c C
  Thomas. Anstruther, Fife c1865 C
  Thomas. Belper, Derbys 1852 C
  William. Whitehaven, Cumbria c1775
  -d 1801 C & W
TEAL, J.B. ?London c1725 watch case maker
TEMPLE, John. Whitehaven, Cumbria marr 1751
  - d 1785 C

TEPPER, J. South Molton, Devon c1790 C
TEROLD, Henry. Ipswich c1640 W
TERROUX, Jaques. Place not known c1700 W
TERRY, Thomas. Nottingham 1818
  William. Mildenhall 1823
THACKER, Richmond. Alfreton, Derbys 1888-1922
THACKWELL, ——. Bristol early 19c C
THATCHER, James. Bury St Edmunds 1737 -d 1748 C
  William. Wantage 1770
THAW, Daniel. Derby 1884-95
THERKETTLE, John. London ?early 18c C
THING, Elder. Wrentham 1800
THOMAS, ——. Bury c1785 C
  ——. East Retford, Notts d 1744
  John. Crewkerne 1719-41 C & W
  Richard. Church Streton, Derbys 1822 W
  Richard. Shrewsbury 1824-35 W
  Walter. Chester c1780
THOMPSON, James. Batley, Yorks 1873 W
  John. Acle, Norfolk c1800-36 C
  John Robert. Chesterfield 1884-88
  Matthew. London 1859 W
  Richard. Standish, nr Wigan c1760 C
  William. Newton Stewart, County Tyrone
  1786
THOMSON, W. Leith c1900 C
THORNDIKE, Samuel. Ipswich b 1757 marr 1781
  d 1819 C & W
THORNTON, John. Sudbury Early 18c C
  J. North Shields c1755-60 C
  James. Kirkham, Lancs c1750 C
THORP, Richard. Hull 1814-18
THORPE, James. Dublin app 1772 to Alex Gordon,
  free 1802-32 C
THRELKELD, Deodatus. Newcastle, see Thirkeld
  (Vol 1)
  Ralph. London c1720 C
THURNAM, William Henry. b 1833 Birmingham,
  Warks 1851
THYMEN, P.R. Amsterdam mid 18c C

TIESE, John. London c1610 W
TILBROOK, John. Bury St Edmunds b c1750 d 1810
W & jeweller
Robert. Bury St Edmunds son of John
1810-22
TILL, Thomas Edward. Derby 1871
TILLEY, Edward. Daybrook, Notts 1839-51
TILVEY, George. Nottingham 1864
TIMMINS, Thomas. Derby marr 1791 C
William. Alfreton, Derbys 1791
TINHAM, ?Samuel. New Sarum, Salisbury c1685 C
TINKINGH, Jan. Oozaandam, Holland mid 18c C
TINKLER, John. Lenton, Notts 1853
S. Nottingham 1853
TIPTON, Benjamin. Ludlow app 1726 to Thomas
Nash d 1796 C
TOBINS, Julius. Derby 1858
TOCHER & Mitchell. Aberdeen c1810 C
TODD, James. Bradford 1754 -d 1788 W & C
William. From Barmston, E. Yorks marr
Hornsea, E. Yorks 1704 C
TOLEMAN, William. Castle Lane, Chester 1753 W
TOLLADY, Dallor. Matlock Bath 1884
TOLPUTT, John. Portsmouth Common ?same as
Portsea Common man qv
TOMKINS, ——. Repaired T C at Barford,
Warks 1805-15
George. Buttermarket, Lyme Regis c1750-60 C
TOMLEY, John. Oswestry repaired church clock 1680,
gunsmith
TOMLINSON, Elias. Bakewell 1876-1916
TOMPION, Edward. London 1812 W
Richard. London 1814 W
Thomas. Dublin 1721
TOPHAM, Edward. Dublin 1824-30 W & jeweller
TORKINGTON, Jeffrey (Jesse?). Whitchurch,
Shropshire 1789-99 C ?same man as below
Jesse (Jeffery?). Newcastle under
Lyme c1780-90 (B 1769) C
TORLEY, ——. Hull c1790 C
TORTOREE, James. London c1740 C
TOVEY, George Henry. Derby 1871
TOWLSON, Joseph. Chilwell, Notts 1763-80
TOWNLY, William. Bourton c1750 C
TOWNSEND, Grey. Dublin 1740-58 W & C
Henry. Warwick repaired T C 1706-11
Richard. Warwick (b 1759) 1809 -d 1832
C & W
William. Warwick 1826 W
TRANTER, Thomas. Shrewsbury app 1729 to George
Birchall, d 1784 C (sometimes Traunter)
TRAVERS, G. London 1811 W
TRAVIS, R.W. Kirton in Lindsey, Lincs 1864 W
Samuel. Sheffield 1699 -d 1723 W

TRENDELL, James. Reading 1819-34
TRENTHAM, Thomas. Shrewsbury 1774 W
TRESOLDI, ——. Trowbridge, Wilts c1820-30 C
TRETHWELL, John. Alias of John Thomas app 1701
to S. Sadleir
TRIDE, Daniel. Portsmouth 1773 W
(poss slip for Tribe qv)
TRIMNELL, W.C. Canterbury 1878-89
TRIPPETT, William. Bristol marr 1660
TRISTRAM, Henry. Tamworth (b 1698)
1741 -d 1768 C ?same man as Tristrim qv
TRISTRIM, H. Tamworth c 1750 C
TRITSCHLER, Andrew. Scarborough 1872
TRIVETT, Zenas. North Walsham, Norfolk 1782
-d 1826
TROKE, John. Salisbury c1800 C
TROUP, Alexander. Place not known but believed
Aberdeenshire c1810 C (believed went to
Nova Scotia by 1838 - see other entry)
TRUGARD, James. Plymouth Dock c1720 C
TRUMBALL, P. London c1700 C
Peter. York 1730 W
TRUSCOTT, Thomas. Glossop, Derbys 1884
TRYER, Thomas. ?London 1782
TULEY, John. Bradford c1780-85 C
TULL, Jethro. Newbury 1746-53 repaired T C
TUNNEL, W. Sunderland c1860 C See Tunwell
TUNNELL, W. See Tunwell
TURFORD, James Benjamin. Ludlow 1891-1900 W
TURNBULL, Thomas (II). Whitby b 1793 son
of Thomas
I marr 1818 d 1857 C & shipowner
William (I). Darlington b c1718, New-
castle d 1766. Had a branch at Newcastle
too. C & W
William (II). Whitby b 1789 son of
Thomas T. d 1870 C Bridge St
TURNER, C. Woolwich 1874 (late Townsend)
George. Buxton 1884-88
Henry C. Oswestry (b 1830) 1861 W
James. Buxton 1895
John. Dublin app 1677 to John Martin,
free 1685 -d 1704 W
John. Whaley Bridge, Derbys 1884
R. Tonbridge c1880 C
Richard Henry. Ilkeston 1891-1932
Samuel. Place not known c1650 C
T. Rochford 1826 W
TURTON, ——. Killmarsh (=?Killamarsh) c1800 C
TUTING, William. Newmarket marr 1771-1780 C
TYLER, James. Rotherham 1860
TYRELL, Richard. Ipswich d 1785 W
TYSON, Nicholas. Gosforth, Cumberland
c1850-60

U

UNCLE, Thomas. Hatfield, ?Herts 1733,
repaired T C
UNDERHILL, Thomas. Albrighton, Shropshire
marr 1781 d 1830
USHAW, William. Hull 1872

USMAR, G. West Malling, Kent 1866
UTTING, Thomas. Bungay pre 1731, then succ
uncle Daniel Manley at Yarmouth c1730,
then Beccles from 1743 C

V

VALE, John. London c1730 C
  Thomas. Great Glemham, Suffolk 1742-62
  repaired T C
VALIANT, Samuel. Hoxne, Suffolk 1846,
  Fressingfield, Suffolk 1874 prob
  the Dickleburgh man
VALOGNE, ——. Paris c1850 C prob the
  London man
VAN STRYP, Pierre. Rome early 18c C
VAULOVE, James. London c1760 W
  Matthew. London French CC 1692 W
VEIJNS, ——. France? c1740 W
VERNON, Thomas. Ludlow free 1711, marr 1711
  -d 1740 C
VERRIER, John. Irvine 1882 W
VIALL, James. Walsingham c1760 C & W

VICK & Son. Stroud c1830 C
VICKERS, George. Mansfield 1877
  William. Nottingham marr 1643 W
VICKERY, Henry. Wantage 1888-99
VINCE, John. Hingham, Norfolk d pre 1854
  succ by widow
VINCENT, ——. Weymouth c1830 C
VINER, Joseph. Newbury 1836
VISMARA, John Baptist. Bury St Edmunds 1846
  chronometers
VITU, Peter. London c1740-50 C
VIZER, Barnaby Snr. Dublin 1784-1819 W
VO(C)KINS, John. Newbury c1780 C
VOKINS, George. Newbury marr 1775 C
VOYCE, J. Dean c1730 C

W

WADE, Charles Frederick. Coltishall, Norfolk 1854
  R. Staythorpe, Notts 1844-64 C & cases
WADHAM, ——. Bath mid 19c C
WADMAN, W. Town not known c1860 C
WAGELSWORTH, James. Retford, Notts 1822
WAGSON, Solomon. Bristol. Took Bernard Gernau app
  c1660 (this must be a slip for Solomon
  Wasson qv)
WAIN, William. Derby 1814
  William. Alfreton 1822-30
WAINMAN, William. Howden, E. Yorks late 18c C
WAINWRIGHT, Humphrey. Bunny, Notts marr
  1747-57 C & W
  John. Wellingborough, Northants, prob
  connected with T. Power qv c1710-51,
  then at Northampton C
  Samuel b 1730 son of John of Wellingborough
  marr 1757, then at Northampton after 1763-95 W
  Thomas. b 1876 Saddleworth. Ashbourne
  d 1940
  William. b 1724 son of John of
  Wellingborough, Northants, moved to Northampton
  where d 1768 W & C
  William. Leiston, Suffolk early 19c
  (?c1820) C

WALDON, Samuel. Ascot 1899
WALDRE, ——. Arundel, see Walder
WALER, William. Much Wenlock, Derbys 1828
WALKER, ——. Lancaster late 18c ?C
  ——. Nantwich, Cheshire c1830 C
  F. Kircudbright, Scotland c1790 C
  George. Dublin 1774-1809 (?-1827) W
  George. Nottingham 1828-44
  J. Nottingham 1864
  W.J. Woolwich 1866
  William. Derby 1891-1900
  William G. Derby 1884-1932
  William. Hull 1814 W
  William. Nottingham 1844
  Wiliam. Shrewsbury 1827-29 W (not
  same as above)
WALLER, Robert. Wickham Market, Suffolk 1792
  William. Preston c1730 C
WALLIS, Jacob. London late 17c-early 18c C
WALSH, Ralph. Dublin 1829 W
  Robert. Dublin 1661-79 W
WALTER, Richard. Nayland, Suffolk repaired
  T C 1565
WALTERS, J. Kingston, Hereford c1820 C

WALTHALL, Jonathan. Bishop Hatfield. Late
17c C ?Same man as below
John. Bishop Hatfield c1695 C
WALTON, William Jnr. Chapel-en-le-Frith
1891-95
WANHAGAN, Patrick. Aberdeen kept T C 1651 with
William Cook
WARD, John M. Chesterfield 1884-1932.
Richard. Dublin 1778-1809 W (B -from Bath?)
Robert. Acle, Norfolk 1836
WARDELL, Richard. Appleby early 18c C,
(but see Wastell)
WARMAN, Thomas. b Ilminster, Warks 1797, Warwick
1851 W
WARNER, George. Dublin 1792 -d 1820 W
John. Draycott b 1655 -d 1727 C & T C
John (II) (Believed grandson of John Snr
of Draycott). Draycott, b 1733, to Chipping
Campden, Glos 1751 C
John Jnr. Evesham, Bridge Street
b c1795 -d c1840, when widow, Catherine
succ him -1842 C
Rebecca. Shrewsbury widow of John
1844-49, when marr James Kelvey qv
WARREN, Eleanor. Dublin 1778-1802 W
James. Windsor 1800
John. Ipswich 1855-64
Robert. See Warner, Robert
William. Nottingham 1844
WARRINER, James. Ripley, Derbys 1895
WARTEMBURG, S. Paris early 19c C
WASBRUGH, ——. London c1775 C
WASHBURN, Daniel. Dublin Free 1713 W
WASHFORD, Isaac. Maidstone c1821 W
WASHINGTON, Richard. Kendal c1694-early 18c C
WASTELL, Richard. Appleby marr 1717 -c1740 C
WATERFALL, James. Covenrry b 1786, app to bros
William and John 1801 -d 1840 W
John. Coventry b 1770-1814 & later
bro of William and James W
William. Coventry b 1776 d 1860 W
WATKINS, Ernest. Bracknell, Berks 1899
John. Ripley, Surrey sundial dated 1695
Thomas. Nottingham 1834-40
WATSON, ——. Blackburn Prob Henry Watson qv
Browning. Wangford & Wrentham, Suffolk
1789-93 C
Francis. Roos, E. Yorks 1851
Frederick. Beverley 1872
Thomas. Haslingden, Lancs c1780 C
W. Beverley 1872
Thomas. London (B a1776-94) 1818 W
Walter. Altrincham, Cheshire c1800 C
WATTS, Frances. Lavenham, Suffolk widow of Thomas
Jnr. Partner with G. Maynard 1777-79
James. Taunton Dean, Somerset marr 1697
Thomas (I). Lavenham b 1694 d 1741 C
Thomas (II). Lavenham b 1729 son of Thomas
Snr d 1777
W.E. Derby 1895-1922
WAUGH, John. Langholm, Scotland 1797 C
WEAVER, George. Droitwich early 19c C
William. Newark 1835-53 C
WEB(B) John. Stourton, Shropshire c1770 C
Marmaduke. Dublin app 1736 to Thomas
Crampton, free 1745 d 1757 W
& Son. Newbury, Berks 1800-1850
William. London 1891 W
William. Wellington b 1726, marr 1750
d 1786 C

WEBBER, Daniel, Lavenham, Suffolk d 1814
WEBSTER, James. Shrewsbury 1740, marr 1745,
d 1799 C
James. Shrewsbury b 1806 d 1871 C
John. Ironbridge free at Shrewsbury 1796 C
John Baddeley. Shrewsbury b 1779 son
of Robert Webster, clockmaker. Succ father
1817 marr 1804 d 1829 C
Robert. Shrewsbury b 1755 son of James
Webster clockmaker, marr 1795 d 1832 C
WEHRLE, D. Dundee c1850 C
WEIR, W. Place not known late 17c C
WELCH, Henry. Shrewsbury 1867 W
WELLS. ——. Solihull early 19c C
Henry. Framlingham 1798-1822, furniture
dealer & C
James. Solihull c1780-90 C
Joseph. Reading 1883
William. London 1861 W
WENHAM, Edward. Shipdham & Wymondham,
Norfolk
1839-51
Henry. Bury St Edmunds d 1816
WERNI, G & Son. Reading 1899
WESENCROFT, Joseph. Dublin app 1675 to Adam
Scot, free 1685, d c1688 W
WEST Mrs Emma. Twyford, Berks 1883
Emma & Son. Twyford, Berks 1899
John. Buxton 1884
Matthew. London late 18c C
WESTON, John. Newark, Notts 1820-40
L. Arbroath. 1876 clock restorer
Robert. London 1818 W
Robert John. Bury St Edmunds 1855
Thomas. Bury St Edmunds 1844
WHEATCROFT, Samuel. Ashover 1815-50
William. Ashover 1810-12
WHEATLEY, John. Newbury 1795 C
WHEATLY, Alexander. Dublin 1796-1806 W
WHEELER, Thomas. Dublin Free 1724 -left in
1729 W
WHEELHOUSE, John. Halifax marr 1751 W
WHIELDON, Roger. Derby Repaired chuch
clock 1664
WHISLEY, H. Gloucester c1830 C
WHITE, Andrew. Forres, Moray 1832
E. London c1875 (20 Cockspur St) C
Edward. Nottingham 1844
F. & Sons. Appleton, Berks 1899
John. Ripley, Derbys 1891-1932
Jonathan. Kirton c1775 C
Stephen. Alton, Hants c1750 C
T. Whitney mid 19c C
William. Chesterfield 1822
WHITEHURST, George. Derby 1888-1900
George. Repton marr 1759 bro of
John Whitehurst FRS C
James. Derby marr 1780 C
John II. Derby. Nephew of John
Whitehurst FRS marr 1785-1834 C
John (III) Derby marr 1828 -d 1855 C
Squire. Derby 1831-43
WHITELAW, James. Edinburgh (b 1776) 1820 -d 1846,
but company still trading 1860
WHITERNE, Johnson. Abingdon 1791-1823
WHITHAM, Jonathan. Sheffield (B 1770-90)
-d 1808 W
WHITHORNE, James. Dublin app 1717 free 1725-79
W & goldsmith
WHITT, Thomas. London c1820 W

WHITTORNE, Gordon. Dublin app 1746 to James
?Whitthorne,
free 1772 d 1819 W
William. Dublin app to W. ?Whitthorne,
free 1772 d 1826
WHITWORTH, James. Lussle (sometimes Luzley etc).
Lancs c1770-85 C
WHYTE, Duncan. Oban 1837-60
WHYTOCK, Peter. Dundee 1844
WICKSTEED, Charles. Oswestry 1728-48 W &
goldsmith
WIELAND, Elizabeth. London c1790 W
WIGELSWORTH. See Wiglesworth
WIGNALL, John. Halifax marr 1771 W
W.J. Bromley c1850 or later C
WILBERFORCE, Charles. London 1802 W
WILD, Abraham. Cropwell Bishop, Notts,
installed church clock 1695-98
WILDERS, J. London 1786 (See also Vol 1)
WILKINS, John. Glapthorne, Northants
b 1673 marr 1701 C
Will. Devizes c1720 C ?same as below
William. Devizes c1740 C
WILKINSON, ——. Lancaster c1775-80 C
Abel. Heptonstall, Yorks marr 1754 C
Benjamin. Wantage 1899
John. Driffield 1872
Joseph. Sheffield 1862-67
Samuel. Spalding, Lincs 1781 W
WILLAND, Charles. Norwich 1854
WILLET, James. Dublin 1830
WILLIAMS, ——. Brackley early 19c C
Griffith. Merthyr c1810-15 C
J. Torrington c1780 C
John. Hull clock case maker c1830
Robert. Bristol. At Lawrence Hill 1802 C
Thomas. Axbridge c1790-1800 C
Thomas. Chewstoke c1760 C
W. Dawlish c1800-10 C
William. Bury 1890 W
William T. Sheffield 1860
WILLIAMSON, Ralph. London CC 1709 (B
bankrupt 1723)
WILLMOTT, Thomas. Hucknall Torkard, Notts 1832
WILLOUGHBY, H.A. Derby 1895
Richard. Oswestry repaired T C 1702
WILLOX, Alexander. Aberdeen 1632-45 Kept
town clocks
WILSHIRE & Link. Bristol Mid 18c C See also
under Link
WILSON, Henry. B c1826 London. Wellington,
Shropshire 1851 W
J. Ilkeston, Derbys c1840 C
James. Birmingham b 1755 marr 1776 W
James. Loop of Towie, Scotland, c1810 C
Jonathan. Place not known but signed
'from London' c1740 C
John. Dublin free 1775 -d 1809 W
Max. Kelso, Scotland 1776 C
Samuel. Chesterfield 1876-1908
Thomas. Cambridge 1809
William. Belper 1823
William. Dawley, Shropshire b c1854
-1900 W
WINCH, William Henry. Reading 1883-99
WINDER, William. Wrekenton, Gateshead c1850-
1860 C
WINDLEY, Samuel. Nottingham 1844

WINSTANLEY, ——. Worksworth c1770 C
E. Ulverstone c1820 C
John. Beaumaris, Anglesey late 18c C
Thomas. Ashbourne 1768
WINTER, Richard. Newbury, Berks 1883-1903
WISE, Luke Jnr. Son of Luke Snr app in CC
1698 to father, d at Reading 1704
WISSETT, Henry. Lavenham, Suffolk marr 1768-81 C
WITHAM, Nathaniel. London sundial 1716
WITHERSTON, John. Hereford c1700 C
WITTON. See Whitton
WITTON, Francis. Same as Whitton
WOLLEY, ——. Tong, Shropshire c1790 C prob one
of the Woolleys qv
WOOD & Cooke. Birmingham japanners and
dialmakers c1810-20
George. Shrewsbury d 1741 C
Isaac. Shrewsbury b 1735 son of Richard Wood
of Shrewsbury watchmaker, d 1801 W
J. Brading, Isle of Wight sundial dated 1815
Joseph. Whitby c1760 C
Richard. b c1705 Knutsford, to
Shrewsbury c1737 d 1752.W
T. Cheetham, Bradbury, Lancs 18c C
& Son. Ilkeston, Derby 1884
William. Chapplehall, Scotland c1840 C
WOODCOCK. John F. Ironville, Derbys 1895
John Jnr. Riddings, Derbys 1888-95
WOODLEY, ——. Kingston c1760-70 C
William. Kineton, Warks c1770 C
William. Newbury, Berks 1871-1903
WOODRUFFE, Thomas. Shrewsbury (b c1736) 1767
-d 1801 W C
WOODWARD, Edmund John (b c1800) Bridgnorth
Shropshire 1856-d 1860 W
Edward. London c1720 C
J.V. Derby 1895-1928
J.W. Killamarsh, Derbys 1888-95
John. Warwick 1807 W
Maria. Bridgnorth 1861-68 (?widow
of Edmund John)
WOOLLEY, William Gamble. Tenterden marr 1789 W
William. Tong, Shropshire b 1782 marr 1806
d 1856 C
WOOLRICH, Thomas. Cheswardine, Shropshire,
repaired church clock 1624-26
WOOLTERTON, John. Yarmouth c1790-1800 C
WORDLEY, John. London ?c1880 C
WORLEY, Simon. Starton c1700 C
WORREL, John. Stoneleigh, Warks 1811 W
WORTLEY, Grennans or Crennans, Wootton
under Edge, Glos c1810 C
Thomas. Ripponden, Yorks c1840 C
WRATHEN, Charles. See Wratten
WREGHIT, James. Patrington, Cumbria 1800-26 C
Robert. Patrington 1800-26 C
Thomas. Patrington 1800-26 C
WRENCH, John. Chester (B 1690-1717 W) d 1739 C
WRIGHT, James. Howden, E. Yorks 1872
John. Ipswich. 1813 W
John. Langholm, Scotland 1797 C
John. Linton, Cambs c1780-90 W & C
John. Mansfield, Notts d 1709 C
Joseph Snr. Warwick marr 1731 d 1783
Joseph Jnr. Warwick (son of Joseph
Snr) b 1735 d 1795
Walter Michael. Nottingham b 1857 marr
1886

Y

YATES, E.G. & Co. Liverpool 1859 W
  John. Warrington marr 1753 W
YEATES, Henry. London 1750 W
YEOMANS, William H. Hull 1872
YOUNG, ——. Cleobury Mortimer, Shropshire 1828
  John. Bunbury, Cheshire c1775 C
  John & Co. Ltd. Exeter 1920-30 W & C and
  jewellers & silversmiths

YOUNG   continued
  Samuel. Bunbury, Cheshire free at Chester
  1722, made Bunbury sundial 1749
  John. Dublin 1784-1806, wholesale watches
  William. Nottingham c1795-98 partner
  with Charles Homer. Then alone until c1825
  W & C
YOUNGSON, C. Scampston, Norfolk 1837